Rhizomicrobiome Dynamics in Bioremediation

T0204053

Editor
Vivek Kumar
Sr. Associate Professor
Himalayan School of Biosciences
Swami Rama Himalayan University
Dehradun, Uttarakhand
India

CRC Press
Taylor & Francis Group
Boca Raton London New York

CRC Press is an imprint of the
Taylor & Francis Group, an **informa** business
A SCIENCE PUBLISHERS BOOK

First edition published 2021
by CRC Press
6000 Broken Sound Parkway NW, Suite 300, Boca Raton, FL 33487-2742

and by CRC Press
2 Park Square, Milton Park, Abingdon, Oxon, OX14 4RN

© 2021 Taylor & Francis Group, LLC

CRC Press is an imprint of Taylor & Francis Group, LLC

Library of Congress Cataloging-in-Publication Data
Names: Kumar, Vivek (Sr. Associate Professor), editor.
Title: Rhizomicrobiome dynamics in bioremediation / editor, Vivek Kumar.
Description: First edition. | Boca Raton : CRC Press, Taylor & Francis
    Group, 2021. | Includes bibliographical references and index.
Identifiers: LCCN 2020041201 | ISBN 9780367419660 (hardcover)
Subjects: LCSH: Bioremediation.
Classification: LCC TD192.5 .R48 2021 | DDC 628.1/683--dc23
LC record available at https://lccn.loc.gov/2020041201

ISBN: 978-0-367-41966-0 (hbk)
ISBN: 978-0-367-70348-6 (pbk)
ISBN: 978-0-367-82159-3 (ebk)

DOI: 10.1201/9780367821593

Typeset in Times New Roman
by Shubham Creation

Preface

The necessary development and diversified anthropogenic activities has resulted in the disturbance of the environmental natural biological diversity. Unwarranted human activities and their intrusion has put a huge stress on nature, affecting the balanced dynamics of the ecosystem, by producing toxic contaminants. The presence of these noxious compounds has harmed our biome, which destresses our natural ecosystem. Consequently, there is a worldwide need to diminish environmental pollution produced all over the globe. There should also be a wide cognizance and awareness regarding the detrimental consequences of these noxious contaminants and their remediation approaches. Handling and dumping practices of these toxic contaminants are the prime concern but newly established management approaches are not cost-effective, moreover, result in the generation of lethal by-products which might negatively disturb the ecosystem.

In the field of applied or environmental microbiology, the term bioremediation is not a novel notion. Bioremediation is the practice of using plants, potential microbes, and enzymes to degrade or detoxify toxic contaminants. For many years, microorganisms have been employed in removing organic substances and noxious chemicals from municipal and industrial waste effluents. Over the past many years, based on numerous research findings and field-based research, the bioremediation process has been accepted and also emerged as an industry. This technique has been proved as an operative, cost-effective, and is being applied in the field for cleaning a wide range of toxic contaminants in soils, sediments, and water bodies. Biological remediation approach is now becoming the technology of choice for remediating several polluted environments, predominantly those sites, which are also contaminated with petro-hydrocarbons.

The bioremediation technique has become a demanding research and development field, partially owing to new regulations necessitating firm environmental protection and cleaning of polluted sites. During the past many years, the grants for basic and applied research on bioremediation by private and governmental agencies have increased. Consequently, progress has been made in exploiting efficient and cost-effective microbial bioremediation methods. Later on, this approach has also become a part of the teaching and research disciplines of environmental microbiology/biotechnology/biochemistry/chemistry/metagenomics/engineering.

This book aims to particularly illustrate the significance of synergism between plant and efficient microbes, the 'Dynamics' between microbiome and plant roots leads to the efficient process of rhizoremediation. Besides updated conventional techniques, novel tactics based on approaches of omics, genetic engineering, nanotechnology, and molecular biology are discussed, as well as the realizations of plant science.

The present volume, 'Rhizomicrobiome Dynamics in Bioremediation', comprises chapters dealing with several bioremediation approaches with meticulous pictorial representations. I hope to inculcate the present status, pragmatism, and insinuations of rhizo-microbiome bioremediation aspects to students, academicians, teachers, researchers, ecologists, agrarians, and professionals, together with other passionate people devoted to the safeguarding of nature. I thank all the worthy contributors who have expertise in this field of research for their valuable, timely chapters, and their help in making this a successful venture.

Vivek Kumar

Contents

Fungal Influence on Hydrophobic Organic Pollutants Dynamics within the Soil Matrices

Claire Baranger[1], Isabelle Pezron[1], Anne Le Goff[2] and Antoine Fayeulle*[1]

[1]Université de technologie de Compiègne, ESCOM, TIMR
(Integrated Transformations of Renewable Matter),
Centre de recherche Royallieu–CS 60 319 - 60 203 Compiègne Cedex, France
[2]Université de technologie de Compiègne, CNRS, Biomechanics and Bioengineering,
Centre de recherche Royallieu–CS 60 319 - 60 203 Compiègne Cedex, France.

1. INTRODUCTION

1.1 Hydrophobic Pollutants in Soils

Organic pollutants found in soils include linear, branched and cyclic alkanes, aromatic hydrocarbons including the BTEX family (benzene, toluene, ethylbenzene, xylene) and polycyclic aromatic hydrocarbons, halogenated hydrocarbons, and other synthetic molecules including pesticides, dyes, drugs, etc. Many display low solubility in water and higher affinity for organic phases: their hydrophobicity depends on the length of carbon chains, molecular weight and type of functional groups. Long-term exposure to environmental contaminants, even when present in low amounts, is preoccupying due to carcinogenic, mutagenic, reprotoxic (CMR), or endocrine disrupting effects. Some of these hydrophobic organic compounds (HOC) are considered as persistent organic pollutants (POP), which are a class of environmental contaminants defined

---
*Corresponding author: antoine.fayeulle@utc.fr

by their hazardous nature and their low degradability, causing a high bioaccumulation potential and possible transport over long distances (Ritter et al. 1995). Many persistent contaminants contain one or more aromatic rings, which impart them high chemical stability.

Anthropogenic sources of pollution with organic compounds are diverse. Hydrocarbons occur naturally in crude oil, natural gas and coal tar, which are mainly composed of alkanes of various chain lengths, and contain aromatic molecules such as biphenyls, BTEX and PAH in varying proportions. As a result, they are found in sites contaminated by oil spills or any petroleum derivatives. PAHs are also unintentionally synthesized and released during the combustion of organic matter or fossil fuels at high temperatures, in industrial furnaces, car engines, or domestic waste incinerators. In this case, they are found as particles disseminated through the air and water, rather than associated to other hydrocarbons in a non-aqueous phase liquid (NAPL). Agriculture is an important source of contamination with xenobiotics as well, especially halogenated compounds: pesticides are intentionally spread in crop fields on a large scale and can persist in agricultural soils for decades, leaching into surface and groundwater, while drugs for animal farming are found as contaminants in wastewater. Synthetic dyes and pigments from the textile industry, and chlorinated aromatics released as by-products of pulp and paper bleaching are other examples of significant organic pollutants.

1.2   Bioavailability of Pollutants

Soil contaminants are subjected to a combination of chemical, physical and biological processes affecting their natural attenuation, persistence or transfer between different compartments. Volatile and relatively water-soluble compounds tend to have lower retention times in the soil since they either volatilize or leach into ground waters, potentially leading to an expansion of the contaminated area (Ortega–Calvo et al. 2013). Chemical or biological degradation of organic contaminants lowers the contamination levels by altering their chemical structure. This can lead to the release of metabolites with varying degrees of toxicity and bioavailability, or even to complete mineralization. As natural attenuation occurs, non-volatile hydrophobic pollutants with a high chemical stability tend to persist in soils in higher ratios than other compounds more easily removed through chemical or biological processes. As a result, the remaining fraction in aged contaminated soils is the least bioavailable (Ortega–Calvo et al. 2013).

The bioavailable fraction of a compound in the soil can be defined as the fraction immediately available for assimilation by living organisms (Semple et al. 2004). In environmental science, this concept is used to refer to nutrient uptake by plants, and is extended to interactions between environmental contaminants and organisms. The term is used in toxicology to describe the fraction of a drug able to reach systemic circulation in animals. A given compound can be divided between several soil compartments: in solution, adsorbed or complexed with minerals or organic matter, as part of the minerals themselves. Bioavailability depends on both soil properties and the ability of a given organism to mobilize the molecule of interest. Bioaccessibility is a concept closely related to bioavailability; Semple et al. (2004) propose in the same work a definition of the bioaccessible fraction broader than the bioavailable fraction, as it includes what is potentially bioavailable but may only become so on a longer time scale (from days to years) or after physical or chemical alteration occurs. HOC is typically associated with solid mineral particles, especially clays that display hydrophobic surfaces, or dead organic matter. Association with soil components is driven by adsorption - immobilization by reversible binding onto a surface - and absorption - immobilization by diffusion into a solid phase.

The aim of remediation is to lower the risk associated to a polluted site as much as possible, and restore a sufficient functionality for its intended use (agriculture, housing, natural area, commercial buildings, etc.). This can be achieved by removal of the contaminant, transformation into harmless compounds, and/or stabilization into a non-bioavailable form. Chemical (oxidation, reduction, solvent or tension-active assisted extraction) and physical (venting, sparging, combustion, thermal or electric desorption) treatments are the most widely used to this date. Bioremediation is usually considered a more environmentally friendly approach to pollutant removal, although the longer time needed for soil rehabilitation is a drawback. When they can be implemented *in situ* without excavation, bioremediation techniques preserve the functional and structural integrity of the soil. In contrast, physicochemical treatments can be too expensive or poorly adapted to some cases (Chen et al. 2015). In addition to the technical and financial advantages of bioremediation, these techniques receive a relatively good acceptance from the public. Khalid et al. (2017) reported a high acceptance of phytoremediation techniques for the removal of heavy metals, including microbe-assisted phytoextraction. However, opinions in regard to the open-field use of genetically modified microorganisms appear to be mixed, as showed by a survey conducted in New Zealand about bacterial remediation of the organochloride DDT (Hunt et al. 2003). The release of genetically engineered organisms in the wild raises safety concerns, and is subject to prior authorization in some parts of the world, including the European Union (Directive 2001/18/EC of the European Parliament). As result, most research nowadays focuses on the use of naturally occurring strains, often isolated from contaminated areas.

1.3   Fungi and Environment

Research on the potential of fungi for environmental remediation dates back to 1950s with a study about wastewater cleanup by aquatic fungi (Harvey 1952). Less than two decades later, as concern started to emerge about environmental consequences of the intensification of agricultural practices, the first studies on atrazine degradation by soil ascomycetous fungi were conducted (Kaufman and Blake 1970). Ligninolytic fungi or "white rots" have been extensively studied for the biodegradation of aromatic pollutants, as they express extracellular enzymatic machinery able to non-specifically degrade a wide range of compounds with a similar structure to lignin. Non-ligninolytic soil fungi have drawn interest more recently, especially native strains isolated from contaminated soil and sediment. Fungi are abundant in the soil, making up for 50 to 1000 µg/g dry weight of soil, and developing networks in the order of 10 km hyphae per gram (Osono et al. 2003, Ritz and Young 2004). They play a key role in terrestrial ecosystems, both as a structural component of the soil and through their interactions with other organisms (Ritz and Young 2004). Filamentous fungi are non-motile organisms which colonize their environment through hyphal growth, thereby forming a mycelial network in three dimensions. The filamentous form allows them to grow in confined environments and colonize porous solid matrices, either in the soil or in plant tissues. This morphology is adapted to their heterotrophic mode of nutrition, enabling the fungus to efficiently scout large volumes of soil for heterogeneously distributed nutrients, and develop a high exchange surface.

Fungi affect soil structure and chemistry by breaking down organic matter, modulating surface properties and promoting particle aggregation. As a consequence, the fungal flora could greatly affect pollutant retention and alteration in soils, both directly (through biosorption and biotransformation) and indirectly (by modifying the soil matrix properties). Bioavailability of the pollutants is both a limitation to bioremediation efficiency, and a potential source of environmental risk, as mobile pollutants are more susceptible to diffusing to nearby areas and/or leaching into ground waters (Ortega-Calvo 2013). In this regard, understanding the mechanisms underlying pollutant dynamics in the soil is the key to a better assessment of environmental risk and improvement of remediation strategies. This chapter will discuss how fungi – either native strains or added for bioaugmentation purposes – affect pollutant stability and mobility in soils, as summarized in Fig. 1.

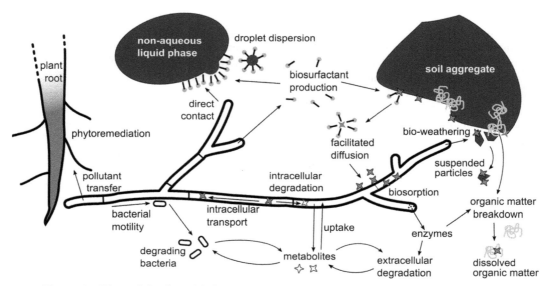

Figure 1   Ways of the fungal influence on hydrophobic organic pollutants dynamics within rhizospheres and soils.

2. FUNGAL INFLUENCE ON SOIL HYDROPHOBICITY AND POLLUTANT RETENTION

Fungi produce several types of surface-active molecules, which play essential functions in their lifestyle and morphogenesis. These compounds modulate both their own cell surface properties and surface properties of their immediate environment, enabling the fungus to cross air/water interfaces, adhere to solid surfaces, and interact with organic substrates such as lignin and oils.

2.1 Extracellular Fungal Proteins Modulating Soil Hydrophobicity

Filamentous fungi are known to produce extracellular surface-active proteins called hydrophobins that regulate fungal interactions with surfaces and air/water interfaces. Hydrophobins are a family of small cystein-rich proteins (<15 kDa) displaying both a hydrophilic and a hydrophobic domains, confering them their surfactant properties. By self-assembling into amphiphilic layers at hydrophilic/hydrophobic interfaces, they lower interfacial tension and allow fungal hyphae to break through air/water interfaces as well as attach to hydrophobic substrates (Wösten et al. 1999). Hydrophobins are divided in two classes with differing structures and displaying various biological functions (Table 1). Class I hydrophobins, defined after the CS3 hydrophobin from *Schizophyllum commune,* form insoluble and extremely stable rodlets. Class II hydrophobins like HFBI and HFBII from *Trichoderma reesei* self-assemble at interfaces into monolayers (Szilvay et al. 2007, Linder 2009). Hydrophobins are the most potent surfactant proteins known to this day (Wösten 2001, Berger and Sallada 2019): class II hydrophobins from *T. reesei* are able to lower the surface tension in water phases to up to 25 mN·m^{-1}, and are active at concentrations in the order of 0.1 μmol.L^{-1} (Cox et al. 2007). They have also been shown to stabilize oil-in-water emulsions and foams (Lumsdon et al. 2005, Tchuenbou-Magaia et al. 2009, Lohrasbi-Nejad et al. 2016).

These modulations of surface properties in the microenvironment of fungi are closely linked to the fungal lifestyle: they are adapted to their heterogeneous habitat, solid porous substrates

Table 1 Diversity of hydrophobins in filamentous fungi and their identified properties

Function/property	Protein name/class	Species	Reference
	Hyd1, Hyd2 (class I)	*Beauveria bassiana*	Zhang et al. 2011
	HCf-1, HCf-2, HCf-3, HCf-4 (class I) HCf-5 and HCf-6 (class II)	*Cladosporium fulvum*	Lacroix and Spanu 2009
Surface hydrophobicity of conidia	Ccg2	*Neurospora crassa*	Bell-Pedersen et al. 1992
	Hyd1, Hyd2, Hyd3 (class I) Hyd4, Hyd5 (Class II)	*Fusarium verticillioides*	Fuchs et al. 2004
	RodA, DewA-E (class I)	*Aspergillus nidulans*	Grünbacher et al. 2014
		Mucor mucedo *Phanerochaete chrysosporium* *Penicillium chrysogenum* *Coprinus cinereus* *Schizophyllum commune*	de Vries et al. 1993
Cell wall of vegetative aerial hyphae and fruiting body	POH1 (fruiting body) POH2, POH3 (vegetative)	*Pleurotus ostreatus*	Ásgeirsdóttir et al. 1998, de Vries et al. 1993
	Class I: ABH1 (fruiting body), ABH3 (vegetative)	*Agaricus bisporus*	Lugones et al. 1996, 1998
Interaction with plant host-pathogenesis	cerato-ulmin (class II) MPG1 (class I)	*Ceratocystis ulmi* *Magnaporthe grisea*	Zhang et al. 2018 Talbot et al. 1996
Interaction with plant host-mycorrhizae	Hyd1 (class I)	*Tricholoma terreum*	Mankel et al. 2002
Stabilization of oil-in-water emulsions and foams	HFBI, HFBII (class II)	*Trichoderma reesei*	Cox et al. 2007, Lohrasbi-Nejad et al. 2016
	Class I and II	*Acremonium sclerotigenum* *Penicillium roseopurpureum* and other marine fungi	Cicatiello et al. 2016
Solubilization of hydrophobic compounds	SC3 (class I) unknown	*Schizophyllum commune* *Aspergillus brasiliensis*	Haas Jimoh Akanbi et al. 2010 Sánchez-Vázquez et al. 2018

partially filled with aqueous phases. Hydrophobins help to retain moisture in cell walls and impart water-repellant properties to the mycelium by forming a coat at the surface of hyphae, allowing it to grow into the air and form aerial reproductive structures (Talbot 1997). Indeed, the formation of aerial mycelium is impaired in hydrophobin-deficient strains of the Basidiomycete *S. commune* (Wösten et al. 1999). Hydrophobins are also involved in the formation of mycorrhizae by mediating hyphal attachment to the plant host cells (Mankel et al. 2002), and form a protective layer on fungal spores, playing a role in spore dispersal and attachment (Zhang et al. 2011).

By coating the surface of soil particles, thereby reversing their hydrophobic or hydrophilic properties, hydrophobins could significantly impact the water flow through pores and the overall wettability of soil matrices (Rillig 2005, Rillig et al. 2007). Hydrophobins self-assemble into membranes at oil/water interfaces as well (Zhang et al. 2018). They are likely to behave in a similar way at liquid/liquid interfaces between soil water and non-aqueous phase liquids (NAPLs) containing hydrocarbons, possibly enabling hyphae to penetrate NAPLs and promoting the access to hydrophobic contaminants.

Glomalin is a glycoprotein found abundantly in soils, produced by arbuscular mycorrhizal fungi of the Glomerales family. It was first identified as a soil protein fraction referred to as glomalin-related soil protein (GRSP), which likely includes a mix of extracellular fungal proteins (Rillig 2004, Gadkar and Rillig 2006). Glomalin itself is thought to be a protein component of the cell wall in arbuscular mycorrhizal fungi that persists in the soil after hyphal death and degradation (Driver et al. 2005). GRSP is known to promote the aggregation of soil particles, affecting soil hydrophobicity and texture. In relation to its role as a structural soil component displaying binding properties, GRSP has been linked to the retention of heavy metals in soils (González-Chávez et al. 2004). Chen et al. (2018) showed that added GRSP in a soil enhances the accumulation of PAHs in ryegrass plants. The authors attributed the increased sorption to plant roots in part to a higher PAH sorbed fraction at the root surface, suggesting that glomalin may act as a "glue" between the pollutant and the root cell wall. Another study demonstrated that the addition of GRSP in a phenanthrene-spiked soil could increase the extractible fraction of phenanthrene in n-butanol (Gao et al. 2017). These results indicate the ability of dissolved GRSP to mobilize PAH and facilitate its transfer to other organisms. In a similar way, cell-wall bound glomalin could mediate the sorption of HOC to the surface of living hyphae.

2.2 Fungal Surfactants and Emulsifiers Enhance Pollutant Partition into Aqueous Phases

Hydrophobins and similar fungal proteins can affect the bioavailability of hydrophobic contaminants in the soil by promoting their dispersion in the aqueous phase. Indeed, a partial solubilization of hydrophobic drugs and increase of their bioavailability have been achieved using the SC3 hydrophobin from the Basidiomycete *Schizophyllum commune* (Haas Jimoh Akanbi et al. 2010), and possible hydrophobin-like proteins able to solubilize PAH have been found in *Aspergillus brasiliensis* (Sánchez-Vázquez et al. 2018). Due to these versatile surface-active properties, hydrophobins have been proposed as a tool for oil recovery (Blesic et al. 2018).

In addition to hydrophobins, fungi produce a wide range of biosurfactants and emulsifiers with diverse chemical natures (Table 2). Although both categories overlap to some extent, surfactants are molecules able to lower interfacial tensions between water and air or organic phases, while emulsifiers are defined by the ability to stabilize oil-in-water emulsions. Most of these molecules or molecular complexes can be classified in three broad categories: glycoproteins/glycopeptides, lipoproteins/lipopeptides, and glycolipids. Biosurfactants from fungi, and especially yeast and yeast-like strains, have been known for decades, yet until recently little information was available about biosurfactants from filamentous fungi (Bhardwaj 2013). However, as the demand for bio-sourced and biodegradable compounds in general increased over the last 10 years, microbial surfactants became a growing field of research.

Based on the overview presented in Table 2, glycolipids appear to be the most widespread among fungal surfactants and emulsifiers. The high molecular weight glycoprotein liposan from *Candida lipolytica* (Cirigliano and Carman 1985), sophorolipids from *Starmerella (Candida) bombicola* (Cooper and Paddock 1984), and ustilagic acid, a cellobiose lipid from the Zygomycete *Ustilago maydis* (Frautz et al. 1986), were the first main biosurfactants characterized in fungi. Mannosylerythritol lipids (MEL) produced by several smut fungi (Arutchelvi et al. 2008) and polyol lipids found in *Aureobasidium pullulans* and oleaginous yeasts of the *Rhodotorula* genus (Garay et al. 2018) are two other more recently identified classes of fungal surfactants. Several functions have been proposed for these extracellular surfactants: in a similar way to hydrophobins, they are likely involved in regulating the interaction of hyphae with surfaces. Polyol lipids and sophorolipids in particular may favor fungal growth on plant tissues by assisting the breakdown of cuticular waxes. In addition, other functions have been identified: the mobilization of hydrophobic substrates, and defense against competing microorganisms through antibacterial or antifungal properties (Puchkov et al. 2002, Arutchelvi et al. 2008, Garay et al. 2018). Some of these molecules appear to be produced constitutively, while others are induced by cultivation in presence of oil, and may be derivatives of lipid catabolism.

The potential industrial applications of biosurfactants was the main driver for research rather than the understanding of fungal physiology and ecology, hence the limited knowledge of their biological function in nature. Interestingly, many surfactant-producing fungal strains were identified as such while investigating their potential for the remediation of hydrocarbons and other hydrophobic soil contaminants (Batista et al. 2010, de Luna et al. 2015, Azin et al. 2018, Zadeh et al. 2018, do Amaral Marques et al. 2019, Pele et al. 2019), including PAH (Deziel et al. 1996, Nikiforova et al. 2009, Deng et al. 2010, Veignie et al. 2012). Rafin et al. showed an increase in Benzo[a]pyrene (BaP) concentration over time in fungal culture filtrates, suggesting the production of extracellular mobilizing agents able to partially stabilize BaP in the water phase (Rafin et al. 2013, Fayeulle et al. 2019). There is also evidence of emulsification of aliphatic and aromatic hydrocarbons by *Penicillium citrinum* (Camargo-de-Morais et al. 2003). Complexation agents are also produced by some strains known for biodegradation of hydrocarbons such as *Fusarium solani* (Veignie et al. 2012).

Remediation enhancement in presence of surfactant is likely driven by two main mechanisms involving i) solubilization and transport of organic compounds into micelles, and ii) displacement of entrapped NAPLs due to interfacial tension reduction (Paria 2008). Indeed, when in presence of surfactant, the fraction of HOC found in micellar form is directly available for degrading bacteria, in a similar way to the dissolved fraction of HOC (Brown 2007). Biosurfactant-enhanced desorption of phenanthrene from suspended clays and humic acid was observed by Garcia-Junco (Garcia-Junco et al. 2003). Similar mechanisms may be involved for fungal surfactants, increasing the bioavailability of HOC in soils to both the fungus itself and other microorganisms.

2.3 Alteration of the Soil Matrix and Effect on the Retention of Hydrophobic Pollutants

Soil structure is determined by the aggregation of its components, including minerals (clays and other silicates, calcium carbonate, metallic oxides) and organic matter. The relative content in particles of various size ranges defines soil texture, regardless of their chemical nature. Fungi greatly contribute to shaping the soil matrix by simultaneously promoting aggregate formation, and altering minerals and organic matter (Gadd et al. 2012, Ritz and Young 2004).

Table 2 Fungal strains identified as sources of extracellular biosurfactants or bioemulsifiers, and chemical nature thereof (when characterized)

		Species	Biosurfactants/bioemulsifiers	References
Zygomycetes	Mucorales	Cunninghamella echinulata	glycolipoprotein	Andrade Silva et al. 2014
		Mucor circinelloides	lipopeptid glycolipid	do Amaral Marques et al. 2019, Zadeh et al. 2018
		Mucor indicus	glycolipid	Oje et al. 2016
		Rhizopus arrhizus	glycoprotein	Pele et al. 2019
		Cunninghamella echinulata	glycolipoprotein	Andrade Silva et al. 2014
Ascomycetes	Saccharomycetales "budding yeasts"	Candida glabrata	unknown	de Luna et al. 2009
		Candida tropicalis	unknown	Batista et al. 2010
		Dipodascus (Candida) ingens	(fatty acids)	Amézcua-Vega et al. 2007
		Kluyveromyces lactis (Candida sphaerica)	unknown	de Luna et al. 2015
		Kluyveromyces marxianus	mannoprotein	Lukondeh et al. 2003
		Lachancea thermotolerans	sophorolipid	Mousavi et al. 2015
		Nakazawaea (Candida) ishiwadae	monoglycerides	Thanomsub et al. 2004
		Saccharomyces cerevisiae	parietal mannoprotein	Cameron et al. 1988
		Starmerella (Candida) bombicola	sophorolipid	Cooper and Paddock 1984
		Wickerhamomyces anomalus	glycolipid	Texeira Souza et al. 2017
		Wickerhamiella domercqiae	sophorolipid	Ma et al. 2014
		Yarrowia (Candida) lipolytica	high molecular weight glycoprotein "Liposan"	Cirigliano and Carman 1985
	Filamentous Ascomycetes	Fusarium fujikuroi	(trehalolipid)	Loureiro dos Reis et al. 2018
		Fusarium neocosmosporiellum	unknown	Azin et al. 2018
		Fusarium proliferatum	fatty amide	Bhardwaj et al. 2015
		Fusarium solani	carbohydrate	Veignie et al. 2012
		Fusarium sp.	unknown	Sena et al. 2018
		Phialemonium sp.	unknown	Guimarães Martins et al. 2006
		Penicillium chrysogenum	lipopeptide	Gautam et al. 2014
		Penicillium citrinum	lipopolysaccharide	Camargo-de-Morais et al. 2003
		Penicillium sp.	Lipids, carbohydrates, protein	Luna-Velasco et al. 2007
		Penicillium sp.	unknown	Sena et al. 2018
		Aspergillus flavus	phenyl glycoside	Ishaq et al. 2015
		Aspergillus fumigatus	unknown	Guimarães Martins et al. 2006
		Aspergillus ustus	glycolipoprotein	Kiran et al. 2009
		Aspergillus niger	unknown	Costa Sperb et al. 2018
		Aureobasidium pullulans	polyol lipids "liamocins"	Price et al. 2013
		Aureobasidium thailandense	fatty acid ester	Meneses et al. 2017
		Exophiala dermatitidis	monoglyceride	Chiewpattanakul et al. 2010
		Trichoderma sp.	unknown	Sena et al. 2018

Table 2 Contd...

Table 2 (contd...) Fungal strains identified as sources of extracellular biosurfactants or bioemulsifiers, and chemical nature thereof (when characterized)

		Species	*Biosurfactants/bioemulsifiers*	*References*
Basidiomycetes	"white rots"	*Ceriporia lacerata*	mannosylerythritol lipids	Niu et al. 2017
		Pleurotus djamor	glycolipoprotein	Velioglu and Urek 2016
		Pleurotus ostreatus	unknown	Nikiforova et al. 2009
		Trametes versicolor	lipopeptid	Lourenço et al. 2018
	Ustilaginales	*Pseudozyma* spp.	mannosylerythritol lipids cellobiose lipids	Kitamoto et al. 1993, Morita et al. 2007, Morita et al. 2013
		Sporisorium sp.	mannosylerythritol lipids	Alimadadi et al. 2018
		Ustilago maydis	cellobiose lipid «ustilagic acid»	Frautz et al. 1986
		Ustilago spp.	mannosylerythritol lipids	Spoeckner et al. 1999, Morita et al. 2008, 2009
	Oleaginous yeasts	*Cutaneotrichosporon curvatum (Cryptococcus curvatus)*	sophorolipid	Ma et al. 2014
		Rhodotorula babjevae	sophorolipid	Sen et al. 2017
		Rhodotorula spp.	polyol lipids	Garay et al. 2018
		Vanrija (Cryptococcus) humicola	cellobiose lipid "mycocin"	Puchkov et al. 2002

In combination with organic matter content, porosity is one of the main parameters affecting the retention of contaminants. Indeed, small pores and high total pore volume mean a high surface available for adsorption, and inaccessibility to degrading organisms and solutions (Ren et al. 2018, Yu et al. 2018). HOC are known to have a greater affinity for fine particles, and tend to sorb onto clay minerals, which exhibit hydrophobic surfaces in a layered structure (Jaynes and Boyd 1991, Yu et al. 2018). As a result, pollutant retention is usually higher in soils with a finer texture (Amellal et al. 2001).

Filamentous fungi develop extensive hyphal networks in the soil and can penetrate solid materials while exploring the substrate. Growing hyphal tips are subjected to swelling/shrinking cycles due to changes in intracellular osmotic pressure, exerting a mechanical force on the surfaces encountered, and cause biomechanical weathering of minerals through penetration into cracks and micropores (Fomina et al. 2006). Fungal weathering is also biochemical: fungi acidify their immediate surroundings by excreting organic acids (acetic, citric, oxalic, formic acid) present in hyphal exsudates, as well as dissolved respiratory CO_2, thus promoting mineral dissolution. Zhang et al. (2016) suggest that organic acids produced through the metabolization of alkanes could in turn enhance the porosity of the solid matrix by dissolving carbonated minerals, thus increasing the accessibility of oil deposits. Metal chelators including siderophores and phenols are also involved in the chemical alteration of soil minerals, destabilizing their chemical structure by solubilizing the metallic elements (Gadd et al. 2012). Such chemical and physical alteration of minerals can enhance the release of inaccessible contaminant droplets or particles trapped in the soil solid matrix. Erosion phenomena lead to the fragmentation of the solid substrate into smaller particles or colloids susceptible to be carried through soil pores by water flows. Colloid-facilitated transport may contribute to the dispersion of sorbed contaminants in soils (de Jonge et al. 2004). Indeed, some studies show that transport in a particle-bound form account for a significant proportion of the leaching of hydrophobic pesticides through macropores (Villholth et al. 2000, Kjær et al. 2011).

One of the major functions of saprotrophic fungi affecting soil structure and chemical cycles is the biodegradation of dead organic matter. Fungi secrete a wide range of extracellular lytic enzymes including cellulases, xylanases, lipases, proteases, and lignin-degrading enzymes in the case of white rots. Through the enzymatic hydrolysis and dissolution of complex macromolecules, fungi disrupt organic matter aggregates, rendering bound HOC more accessible due to two complementary effects: the removal of physical hindrances preventing microorganisms from accessing the source of pollutant, and the release of mobile organic matter fragments promoting the dispersal of sorbed HOC. Soluble humic acids act as carriers increasing the mobility and bioacessibility of HOC despite very low aqueous concentrations and making them directly available for uptake even without desorption and solubilization in the water phase (Smith et al. 2009). Hydrophobic contaminants tend to sorb onto suspended humic acid-clay complexes (Garcia-Junco et al. 2003). Moreover, the biodegradation of organic matter by fungi also increases the dissolved organic carbon content, which has been linked to facilitated diffusion of HOC from NAPLs to the water phase (Smith et al. 2011).

3. MOBILIZATION OF HYDROPHOBIC POLLUTANTS BY FUNGI

Access to the pollutant depends on the ability of microorganisms to explore the substrate. The mechanisms involved differ among organisms: bacteria can rely on motility, while fungi explore the substrate through mycelial growth. Although various utilizing microorganisms exhibit chemotaxy for moderately soluble organic compounds, such mechanisms are not known for the most hydrophobic HOC such as Benzo[a]pyrene. The ability to detect minute concentrations of the pollutant and to grow towards the source could be crucial for its degradation (Ortega-Calvo et al. 2013).

3.1 Surface Properties of the Mycelium and Biosorption of Organic Pollutants

Biosorption refers to the removal of a compound from the solution by a microorganism, through the combination of adsorption onto the cell surface and absorption into the cell. Many studies quantify total biosorption and do not differentiate the respective contribution of both mechanisms. However, adsorption by mycelia could account for a significant part of hydrophobic compound retention in the soil, given the extremely high surface/volume ratio developed by mycelial networks.

Water-insoluble, microfibrillar glucans and chitin are the main components of the fungal cell wall. Parietal α-1,3-glucans appear to play a decisive role in hyphal adhesion to the substrate (He et al. 2018). Ma et al. investigated the adsorption of PAHs to isolated fungal cell wall components, and although the highest affinity was for lipid components, a significant effect of the carbohydrate fractions (chitin, chitosan, glucan and mannan) was found (Ma et al. 2016). Considering their relative abundance, parietal carbohydrates could account for a substantial proportion of PAH adsorption by the whole cell wall. Specific mechanisms may control cell surface properties in presence of HOC, enabling direct contact. First evidence of a modified cell wall structure associated with enhanced hydrocarbon adsorption was reported in *Candida tropicalis* when grown in presence of hexadecane (Käppeli et al. 1984). Similarly, the addition of naphtalene in culture media induces changes in surface hydrophobicity and emulsifying properties in several yeast strains (Deng et al. 2010). In filamentous fungi, hydrophobins modulate the surface hydrophobicity of the cell wall and affect cell adhesion to hydrophobic surfaces (see section 2.1), as well as to hydrocarbons (Zhang et al. 2011).

Adsorption of hydrophobic molecules onto fungal hyphae could be the first step promoting subsequent absorption; however, the mechanisms by which hydrocarbons cross the cell wall and enter fungal cells are still unclear. A mechanism of alkane incorporation involving adsorption onto the mycelium followed by active absorption was first proposed in *Cladosporium resinae* (Lindley and Heydeman 1986). Adsorption of PAHs and crude oil onto the mycelium of a hydrocarbon-degrading *Aspergillus* strain was reported by Al-Hawash et al. (2019), as well as intracellular accumulation: absorption rates of up to 89% were found for pyrene in submerged culture. Incubation with the water-insoluble PAH benzo[*a*]pyrene induces morphological differences at the surface of hyphae in another *Aspergillus* strain, which displayed a roughness attributed to BaP deposits (Wu et al. 2009). Fayeulle (2013) demonstrated that preventing direct contact between the mycelium of *F. solani* and solid BaP flakes strongly decreased biosorption, while the same was not true for the more water-soluble phenanthrene. These results suggest that removal of highly hydrophobic molecules such as BaP from the medium is dependent on their prior adsorption onto the mycelium, while moderately soluble compounds such as phenanthrene could be directly recovered from the aqueous phase.

3.2   Uptake and Intrahyphal Transport of Hydrophobic Molecules

Intracellular uptake of hydrocarbons by fungi has been studied since the 1980s, when the first active hydrocarbon transport mechanism was demonstrated for dodecane in the "kerosene fungus" *C. resinae* (Lindley and Heydeman 1986). Al-Hawash et al. (2019) showed that selective adsorption and absorption of crude oil and low molecular weight PAHs occurred in the mycelium of an *Aspergillus* sp. strain. Uptake and intracellular storage of PAH in lipid bodies has been observed in several other fungal species, including anthracene in the arbuscular mycorrhizal fungus *Glomus intraradices* associated with chicory roots (Verdin et al. 2006) and BaP in the Ascomycete *Fusarium solani* (Verdin et al. 2005, Fayeulle et al. 2014). More precisely, BaP uptake in *F. solani* cells was shown to involve an active transport requiring a functional cytoskeleton network (Fayeulle et al. 2014). The cytochrome P450 oxidation and the β-oxidation of fatty acids are intracellular pathways involved in the biodegradation of alkanes and aromatic hydrocarbons, which means the uptake of organic contaminants is a crucial step preceding their oxidation through these pathways.

Intrahyphal translocation of phenanthrene through storage and transport in lipid vesicles was demonstrated in the Oomycete *Phythium ultimum* (Furuno et al. 2012a). The same species was found to increase the accessibility of fluorene to the PAH-degrading bacteria *Burkholderia sartisoli* after transport by the mycelium (Schamfuß et al. 2013). Although *P. ultimum* is a water mold, it shares a similar filamentous morphology and saprotrophic lifestyle with true fungi. This suggests that similar mechanisms of PAH redistribution could be at work in the mycelium of soil fungi.

Indeed, filamentous fungi are able to redistribute nutrients in the mycelium and propagate chemical signals through intrahyphal cytoplasmic flow. Simultaneous localized response to pathogen attack and glucose translocation was shown in *Coprinopsis cinerea* (Schmieder et al. 2019). This mechanism allows for accommodation of the heterogeneous environment, since a mycelium can cover large volumes and be locally exposed to different conditions in different areas. This is especially important in nutrient-deprived or contaminated lands. The incorporation of PAH into lipid bodies in *F. solani* is dependent on actin, which is consistent with the hypothesis of an intracellular hydrocarbon transport through cytoskeleton-dependent vesicular traffic (Fayeulle et al. 2014). Intracellular vesicle traffic is especially active at the growing tips of hyphae, which are rich in secretory vesicles bringing "building materials" such as membrane lipids and cell wall precursors. Hydrophobic pollutants incorporated into vesicles could thus be released outside of hyphae and thereby become accessible for degradation by other microorganisms.

3.3   Enhancement of the Bioaccessibility for other Organisms

Water films along aerial hyphae of the Oomycete *Phytium ultimum* enable the migration of motile bacteria following chemotactic gradients (Furuno et al. 2010). Here the mycelial network functions as a "highway" promoting the dispersal of PAH-degrading bacteria living at the surface of hyphae. Through the incorporation of fat-soluble pollutants into cell membranes and diffusion thereof, hyphae are likely to act as "conductor wires" transmitting HOC gradients, thus guiding utilizing bacteria towards the source (Furuno et al. 2010, 2012b). This mechanism highlights one of the ways fungi interact with other microorganisms as structural components of the soil, and not solely through trophic exchanges. In combination with intrahyphal translocation of pollutants, it could contribute to enhance the access of unicellular microorganisms to heterogeneously distributed hydrophobic pollutants in the soil (Banitz et al. 2013).

Many fungal species are able to form mutualistic interactions with plants though mycorrhizae. Indeed, 90% of plant species are currently thought to display such symbiotic associations (Bonfante and Genre 2010). Filamentous fungi form mycorrhizes by developing hyphae in plant root tissue and exchanging nutrients with the cells. They provide the plant partner with inorganic phosphorus and nitrogen, while taking up sugars from the plant. In this regard, the mycelium of mycorrhizal fungi functions as an extension of the root network enabling a greater exchange surface for nutrient uptake.

Mycorrhizae may help with bioaccumulation of pollutants in plants, though incorporation, hyphal transport and transfer to plant cells among various exchanged molecules. Wu et al. (2008) found a better removal of phenanthrene from spiked soil with alfalfa plants colonized by the arbuscular mycorrhizal fungus *Glomus etunicatum* compared to plants without mycorrhizae. Enhanced accumulation and degradation of PAH in spiked soils was also shown for the legume *Sesbania cannabina* in triple symbiosis with *Glomus mosseae* and rhizobia (Ren et al. 2017). A similar result was obtained for the accumulation of uranium by another *Sesbania* species, in triple symbiosis with *G. etunicatum* and rhizobia (Ren et al. 2019). In arbuscular mycorrhizal fungi, parietal glomalin or freely dissolved GRSP may be involved in the biosorption of contaminants by mycorrhizal fungi, and its transfer to plant roots (see section 2.2). POP degrading abilities have also been found in various ectomycorrhizal Basidiomycetes, covering a range of aromatic and/or halogenated hydrocarbons, pesticides and pharmaceuticals (Meharg and Cairney 2000). Ectomycorrhizae in poplar tree for instance have been shown to enhance the removal of diesel oil from the soil (Gunderson et al. 2007). Even when they are not known themselves for biodegradation properties, other mycorrhizal fungi could thus contribute to remediation by mediating the transfer of pollutants to plant roots and/or rhizosphere bacteria.

4. CHEMICAL ALTERATION AND FATE OF THE DEGRADATION PRODUCTS

Soil-dwelling fungi are saprotrophic organisms, which contribute to carbon cycling in soils by breaking down decaying organic matter. Their enzymatic machinery is adapted to the biodegradation of a wide range of organic molecules, and some strains have demonstrated biodegradation abilities towards hydrocarbons, synthetic organic compounds and even polymers. Fungi, in cooperation with other microorganisms and plants, thus participate in the natural attenuation of HOC through chemical alteration.

4.1   Biotransformation of Pollutants by Fungi

Hydrocarbon metabolism in fungi is in large part connected to lipid catabolism. Both intracellular lipolytic activity and extracellular enzymatic complexes contribute to the biodegradation of hydrocarbons (Peter et al. 2012, Al-Hawash et al. 2019). Alkanes are first oxidized into fatty acids by mono-oxygenases and NAD-dehydrogenases before entering the intramitochondrial β-oxidation pathway (Singh 2006a), leading either to total mineralization or incorporation into the biomass. Biotransformation of alkanes by fungal enzymes leads to the release of gases and water-soluble organic acids such as oxalate and propionate (Zhang et al. 2016). The use of abundant lipids (or alkanes) as a carbon source could also enhance the biodegradation of more recalcitrant compounds through the release of reactive oxygen species (ROS), as shown for the co-metabolization of vegetable oil and PAH in *F. solani* (Delsarte et al. 2018). This suggests that the fungal degradation of aromatics may be more efficient when they are associated with NAPLs, which contain more easily assimilated linear hydrocarbons. Extracellular lipolytic enzymes also play a significant role in the breakdown of linear and cyclic hydrocarbons as well as aromatics; particularly, heme peroxidases presenting a functional similarity to intracellular P450 oxidases have been found to catalyze the breakdown of a wide range of hydrocarbons and organic compounds (Harms et al. 2011, Peter et al. 2012).

Fungal ligninolytic enzymes have been extensively studied for their potential in the biodegradation of organic contaminants containing aromatic rings, and especially PAHs and dyes, due to their structural similarity with lignin. Many white rot fungi have been identified as depolluting fungi, including *Phanerochaete chrysosporium*, *Pleurotus ostreatus*, *Trametes versicolor* (Lladó et al. 2013), and *Anthracophyllum discolor* (Acevedo et al. 2011). The ligninolytic pathway involves extracellular oxidoreductases– laccases, lignin peroxidases, manganese peroxidases, and tyrosinases– which catalyze the production of oxidizing agents and free radicals leading to the non-specific oxidation of substrates mainly into quinones. The reaction of aromatics with free radicals can also lead to their covalent binding with soil organic matter, as demonstrated during the degradation of trinitrotoluene through fungal laccases purified from *Trametes modesta* (Nyanhongo et al. 2006). The formation of such bound adducts can result in a lower acute toxicity of the degradation products, but also render them less available for further degradation.

Intracellular degradation through the cytochrome P450 mono-oxygenase pathway appears to be the dominant metabolic pathway for aromatics degradation in non-ligninolytic fungi, and yields substituted derivatives that are more reactive and more soluble than the original pollutants: phenol, phthalate, hydroxyl-, carboxy- and dihydrodiol derivates that may, in the second step, be conjugated with sugar moieties (Boll et al. 2015). The fungal metabolization of biphenyls can yield water-soluble conjugates, such as glucuronide and sulfate conjugates as found in *Cunninghamella elegans* (Singh 2006b). PCBs are typically oxidized into hydroxy- and dihydroxybiphenyls, more polar than the original compound (Singh 2006b). Boll et al. showed that PAH metabolites produced by *C. elegans* had radically different behaviors compared to the parent PAHs, and tended to partition into the water phase at much higher rates (Boll et al. 2015). On one hand, the release of such polar metabolites can increase environmental risk as they are more prone to leaching and can spread to surface and groundwater. On the other hand, higher mobility and bioavailability could enhance their further biodegradation by other organisms.

4.2   Biotransformation by Microbial Consortia

In the soil, fungi and bacteria co-exist in close vicinity, as the mycelial network constitutes a favorable microhabitat for soil bacteria. Furuno et al. (2012b) showed that several bacterial strains known for their degradation abilities towards HOC (*Pseudomonas, Xanthomonas, Rhodococcus, Arthrobacter*) could be selectively isolated from the surface of hyphae, highlighting the close link between physical and metabolic interactions regarding the biodegradation of pollutants. Indeed, fungi excrete proteins, polysaccharides and organic acids that can be used as carbon and nitrogen sources by bacteria (de Boer et al. 2005). Dashti et al. showed that bacterial and fungal organisms cooperate in the degradation of alkanes, and that fungi are able to take up and metabolize products released after primary oxidation by alkane-degrading bacteria, such as fatty acids, alkanols and alkanals (Dashti et al. 2008). Co-cultures of fungi and bacteria were also shown to work in synergy for the degradation of aromatic pesticides, by further metabolizing one another's degradation products, thus avoiding the accumulation of toxic intermediates responsible for an inhibition of degradation in pure cultures (Zhao et al. 2016).

Fungi are usually regarded as more efficient than bacteria for the transformation of HOC that are particularly recalcitrant due to their poor mass transfer rates (high molecular weight hydrocarbons including PAH) and/or low nutritional interest (highly oxidized compounds such as nitrated and chlorinated hydrocarbons) (Harms et al. 2011). This higher efficiency is attributed to enzymatic diversity and co-metabolic degradation mechanisms absent from bacteria. However, many bacterial species have been identified as good degraders of lower molecular weight aromatic compounds, mostly members of the genera *Burkholderia, Pseudomonas, Acinetobacter, Rhodococcus,* and *Streptomyces.* Bacterial metabolism of polycyclic aromatics differs from the fungal metabolism by the initiation of ring attack, through a dioxygenase activity as opposed to mono-oxygenase in fungi (Cerniglia 1992). However, more recently dioxygenase activities involved in the biodegradation of lignin and BTEX have also been described in fungi (Gunsch et al. 2005). A new degradation pathway of phenolic compounds that was only known in bacteria thus far has been identified in *Aspergillus* sp. (Yang et al. 2018). Redundancy of metabolic functions in bacteria and fungi may thus be more frequent that was previously thought.

Bacteria and fungi can cooperate for the biodegradation of lignin: biphenyl derivatives from lignin can be further metabolized by *Pseudomonas, Streptomyces* (Vicuña et al. 1993), and *Sphingomonas* strains (Peng et al. 2002) after the first depolymerization step by white rots. Due to the structural similarity between lignin monomers and aromatic pollutants, similar metabolic cooperation mechanisms are likely to occur during biodegradation processes of these compounds. Indeed, metabolic synergies between fungal and bacterial microorganisms can enhance the biodegradation of HOC compared to the yields obtained with individual strains, especially in the case of recalcitrant PAH (Han et al. 2008, Machín-Ramírez et al. 2010), azo-dyes (Lade et al. 2012, Zhou et al. 2014) or pesticides (Hai et al. 2012, Zhao et al. 2016).

Co-metabolization of HOC by several taxa could be of critical importance to avoid the accumulation of toxic intermediates. In soil microcosms, the white-rot fungi *Phanerochaete chrysosporium* and *Trametes versicolor* were able to mineralize 3 and 4 cycles PAH completely, while quinones produced by *Pleurotus ssp* accumulated as a dead-end metabolite (Andersson and Henrysson 1996). This accumulation was attributed to an inhibiting effect of the fungus towards the autochthonous microflora, which appeared to play an essential role in the further degradation of fungal metabolites. Antagonistic effects can also hinder the degradation efficiency, due to the production of anti-microbial compounds and/or competition for resources (Thion et al. 2012).

5. EXPERIMENTAL STRATEGIES TO INVESTIGATE POLLUTANT MOBILIZATION

Bioremediation can be studied at several scales from batch biodegradation tests in the lab to field experiments. Biodegradation is usually assessed by quantifying the remaining pollutants and/or metabolites after treatment. Studying the availability and mobilization of contaminants is a complex problematic as it involves multi-scale interactions between living organisms, dead organic matter, inorganic particles, gas phase and aqueous phase, and contaminants themselves.

Investigating individual mechanisms often requires using simplified models for a specific subsystem, with varying degrees of complexity.

5.1 Characterizing Contaminant Bioavailability in Soils

The bioavailable fraction of a contaminant is by definition the immediately available fraction to microorganisms, and is assumed to be the most soluble and easily extractible in mild conditions. Chemical techniques to measure the bioavailable fraction in contaminated soils include partial extraction with polar organic solvents such as ethanol or butanol, subcritical water extraction, or supercritical CO_2 extraction. Co-solvent extractions using water in combination with a miscible solvent may be used as well. However, solvent-based approaches are subject to discussion regarding their relevance to estimate the fraction actually available for microorganisms. Indeed, their recovery yields can be relatively close to those of exhaustive extractions techniques used to quantify total contamination, suggesting they may be too harsh and lead to an overestimation of the bioavailable fraction (Cachada et al. 2014).

It is important to note that bioavailability is dependent on a number of physical, chemical or biological dynamic processes. That is why some approaches focus on assessing pollutant retention times and desorption kinetics rather than quantifying the available fraction at a single point in time. Other techniques aim at mimicking the effect of pollutant consumption by living organisms through biosorption and degradation, and allow to quantify desorption rates. Desorption behavior can be evaluated directly by measuring the dissolved fraction in water phase over time (Heister et al. 2013) or through resin-aided solid phase extraction (Kan et al. 2000). Solid-phase extraction is an aqueous extraction-based approach where, instead of measuring pollutant concentration at equilibrium in the water phase, a hydrophobic resin such as Tenax is used as a trap to deplete the water phase from HOC. This promotes the desorption of compounds reversibly bound to the soil matrix (Breedveld and Karlsen 2000). The freely dissolved HOC fraction in water or soil slurries can be measured *in situ* using passive samplers for environmental monitoring; however, this type of device is poorly suited for use in unsaturated soils (Cui et al. 2013).

All these methods may give very different results depending on whether they are used on contaminated environmental samples or artificially spiked soil; indeed, the age of a contamination greatly influences sorption to the matrix. The aim can be different as well. In the case of soil samples from a polluted site, the availability of HOC can be evaluated as part of risk assessment and preliminary evaluation before treatment. When using spiked samples, the aim is to evaluate the sorption properties of a given soil, by measuring the differential partitioning of a pollutant between the sample and solvent or water.

5.2 Fungal-mediated Desorption and Solubilization Assays

As developed in part 3, fungi are able to mobilize hydrophobic compounds through several direct and indirect effects, by facilitating their transport in aqueous phase, and affecting their retention in soil. Experimental approaches to characterize fungal mechanisms of emulsification or facilitated transport can focus on the properties of specific molecules (particularly extracellular biosurfactants), or the whole organism.

Surface tension measurements are a simple way to assess the presence of surface-active molecules in microbial extracts, and are commonly used in first approach to identify surfactant producing fungal strains (Kitamoto et al. 1993, Méndez-Castillo et al. 2017, Alimadadi et al. 2018, Al-Hawash et al. 2019). However, surface tension at the air/water interface does not yield direct information about the behavior of biosurfactants in presence of hydrophobic compounds. Alternatively, interfacial tension measurements in contact with an immiscible organic phase can be performed (Kitamoto et al. 1993). Oil displacement assays are a cheap and quick method to compare the surfactant activity of fungal extracts without the need of costly equipment; in a Petri dish, a known volume of surfactant solution is dropped onto an oil film and the area cleared of oil is measured. This test has been used to rapidly detect surfactant activities in cell-free culture supernatants of several fungal species (Gautam et al. 2014, Meneses et al. 2017, Al-Hawash et al. 2019, Pele et al. 2019).

The ability of biosurfactants to solubilize hydrophobic compounds can be estimated by partitioning assays with a principle similar to octanol-water partitioning assay. The octanol-water partitioning coefficient (K_{ow}) is a classical measurement tool used to estimate the relative affinity of a given compound for aqueous or organic phases. It is determined by dissolving the tested compound in a mix of octanol and water, and expressed as the ratio of concentrations at equilibrium in both phases. The same method can be adapted so that culture supernatants or a solution of biosurfactants are used as the hydrophilic phase, and the enhanced partitioning of hydrocarbons into the aqueous phase containing microbial extracts is compared to the partitioning into pure water. In the same way, octanol may be substituted with any other NAPL of interest (Garcia-Junco et al. 2003).

Estimating the emulsifying activity of fungal extracts is a relevant approach to investigate their behavior towards NAPLs. Sánchez-Vázquez et al. (2018) determined the emulsifying activity of a bioemulsifier towards dodecane, diesel oil and PAH dissolved in organic solvents by measuring the optical density of the oil-in-water emulsion or average oil droplet size. A similar protocol was used to estimate the emulsification index towards waste oil (Batista et al. 2010). Emulsification abilities of hydrophobins towards a decane and toluene mix in synthetic sea water were shown by measuring the relative height of the emulsified layer after mixing and resting (Blesic et al. 2018).

The auto-fluorescence properties of some aromatics, and more specifically PAHs, allow for their detection in spectro-fluorometry or fluorescence microscopy. Some desorption assays thus rely on fluorescence measurements in aqueous samples after incubation in direct contact with a solid deposit of PAH (Rafin et al. 2013, Fayeulle et al. 2019). A similar protocol followed by liquid–liquid extraction of PAH from the aqueous phase was used to demonstrate the ability of the bacterial emulsifier alasan to solubilize several PAH (Barkay et al. 1999).

Spatial compartmentalization of the pollutant source and microorganism can be used to demonstrate pollutant translocation. Solid culture of filamentous fungi on agar blocks separated by gap, only bridged by hyphal filaments, showed fungal-mediated transport of PAH from the spiked block to the block inoculated with fungus, without PAH (Furuno et al. 2012a, Schamfuß et al. 2013). Compartmentalization can be achieved with a dialysis membrane to prevent direct contact between the fungus and the solid pollutant deposit, allowing access to the solubilized fraction in aqueous phase only (Fayeulle 2013). The use of a microfluidic device for fungal cultivation has also been proposed to spatially control the contact with a pollutant source (Baranger et al. 2018).

There is a need for new tools to simulate complex environments such as the soil. In this context, the use of micro-fluidic devices to study the interactions between microorganisms is a recent field of research (Stanley et al. 2016). The importance of fungal-bacterial interactions for the degradation of organic pollutants is pointed out by the authors. These devices enable the modeling of heterogeneous, compartmentalized microbial habitats, with or without water flow, and allow for a direct coupling between microbial culture and multi-scale observations from the single cell level to a several millimeter colony (Baranger et al. 2020).

5.3   Surface Properties of Fungi and Biosorption

Adsorption onto the cell surface and absorption into fungal cells both contribute to the removal of HOC from the soil in a non-destructive way, and can be a preliminary step leading to biodegradation. The sum of both mechanisms is quantified as total biosorption; however, some studies attempt to quantify their respective contribution, differentiating between the reversible adsorption of HOC and their incorporation into the biomass. Several techniques have been used to selectively recover the adsorbed fraction and quantify it, the main issue being to find extraction conditions sufficient to recover HOC bound to the extracellular matrix without permeating the cell membrane. Assessing passive biosorption of pesticides and PAH with autoclaved mycelium is one of the used approaches (Hai et al. 2012, Thion et al. 2012). Al-Hawash et al. (2019) used sonic bath extraction to quantify the fraction adsorbed to mycelium pellets. The surface-adsorbed fraction can also be selectively extracted by cold washing the fresh or autoclaved mycelium with organic solvents (Al-Hawash et al. 2019). Chau et al. (2010) demonstrated a correlation between surface hydrophobicity of mycelial mats of several fungal taxa, including *Fusarium*, *Cladosporium*, *Mortierella* and *Penicillium* ssp, and their capacity to absorb ethanol solutions of increasing concentrations: this result suggests a relation between surface hydrophobicity and affinity for organic solvents. The affinity of whole fungal cell walls and insoluble cell wall components for PAH can be tested measuring their partitioning rates in a water-cosolvent mixture (Ma et al. 2016).

Microbial adhesion to hydrocarbons (MATH) tests estimate the surface hydrophobicity of a given microorganism's cell wall, by measuring its affinity for hydrocarbon/water interfaces. The microbial cell suspension is brought into contact with a hydrophobic phase, and the equilibrium concentration of cells that remain suspended in the aqueous phase is quantified by optical density measurements. The decrease of suspended cell concentration, due to increased cell adhesion at the hydrocarbon/water interface, therefore reflects the cell surface hydrophobicity (Hazen 1990, Rosenberg 2006). This test was primarily designed for bacteria, and then extended to yeasts and fungal spores, but is limited to unicellular organisms. Although MATH has been used to assess the hydrophobicity of young hyphal germ tubes of *Candida albicans* (Rodrigues et al. 1999), it is impossible to implement for filamentous fungi at more advanced growth stages, since they form mycelial pellets in submerged culture rather than homogeneous cell suspensions. Hydrophobic interaction chromatography (HIC) has been adapted to assess the affinity of unicellular microorganisms for a hydrophobic matrix: the microbial suspension is filtrated through the column and a retention index is measured. Similarly to MATH assays, HIC can be used for spores and yeast cells (Hazen 1990, Holder et al. 2007)

Surface hydrophobicity of the mycelium can be assessed by measuring the contact angle of water droplets on whole colonies. Several fungal culture methods have been developed to allow for imaging of contact angles in controlled conditions, including growing mycelial mats on cellulose acetate filters (Smits et al. 2003), glass microscope slides that can be mounted directly on a sample stage (Chau et al. 2009), or producing fungal biofilms in submerged culture wells (Siqueira and Lima 2012). The main advantage of contact angle measurements is that they are done on the filamentous form of fungi rather than a homogeneous suspension of germinating spores or broken mycelium fragments, more accurately representing the properties of whole mycelia. The apparent surface hydrophobicity of a mycelial mat does not solely depend on cell wall composition and the coating of hyphae with hydrophobins, but also on geometric parameters: hyphal diameter and tightness of the mycelial network affect the surface roughness of the mycelium and thus contact angles with water. Studies comparing the fungal surface hydrophobicity in different growth conditions highlight differentiated cell wall properties and/ or morphology depending on water saturation and circadian growth cycles (Chau et al. 2009, Siqueira and Lima 2012). However, contact angle measurements do not account for local changes in cell surface properties at the microscale, and whole colony hydrophobicity does not necessarily reflect cell surface properties of individual hyphae. Soil-dwelling mycelia develop in mostly water-unsaturated environments and may thus present frequent variations in the local cell surface properties due to the heterogeneous nature of their microenvironment. Microsphere attachment assays can be implemented with whole hyphae or even small mycelial pellets, and may more accurately reflect local surface hydrophobicity. This method could be a useful tool to identify preferential sites of adsorption and uptake of hydrophobic molecules, and help better understand incorporation mechanisms.

6. CONCLUSION

Fungi are major actors of soil ecology and structure, and affect the retention and transfer of organic pollutants in contaminated soils. They impact pollutant dynamics on three levels: shaping of the soil structure and physico-chemical properties, mobilization of contaminants by facilitated transport and intracellular accumulation, and chemical alteration.

Hydrophobins and glomalin are extracellular fungal proteins able to modulate soil surface properties, possibly affecting pollutant adsorption and transport through water flows. Although the function of GRSP is quite well established from a soil perspective, hydrophobins have been mainly studied for their physiological role and industrial applications, and experimental evidence of their effect on soil properties is still scarce. Fungi produce a wide range of non-protein extracellular surfactants able to emulsify or solubilize HOC; however, it is not known whether these are part of specific transport and incorporation mechanisms. How aromatic hydrocarbons cross the cell wall and membrane before accumulating in intracellular bodies is still unclear as well.

The understanding of microbial communities in the soil and their interactions with plants in the rhizosphere is crucial to get a global picture of pollutant dynamics. Indeed, inter-organismal cooperation can lead to very different outcomes compared to the single organisms in terms of pollutant mobilization, transport, and degradation. Although some communities work in synergy, competition may hinder efficient remediation. Metabolomic and metagenomic approaches could thus help to better understand these interactions and predict the accumulation of toxic metabolites. "Soil-on-chip" microfluidic models have been recently proposed as novel approach to investigate interactions between soil organisms at the microscale in a heterogeneous environment, and certainly have potential to study pollutant mobilization and incorporation.

Finally, soil fungi are essential to the vast majority of terrestrial plants, which form symbiotic associations within their roots in the form of mycorrhizae, and the accumulation properties of some depolluting plants are likely due, in part, to their fungal partners. Interestingly, most fungi described as good degraders of organic contaminants are not known to form mycorrhizae, but research regarding the role of mycorrhizal fungi could be hindered by the difficulty to isolate and cultivate these obligatory symbionts in axenic cultures. Mobilization and transport of pollutants in "fungal pipelines" may not always be an indicator of degradation properties, and hence the cooperation between species is important.

References

Acevedo, F., L. Pizzul, M. del P. Castillo, R. Cuevas and M.C. Diez. 2011. Degradation of polycyclic aromatic hydrocarbons by the Chilean white-rot fungus *Anthracophyllum discolor*. J. Hazard. Mater. 185: 212–219.

Al-Hawash, A.B., X. Zhang and F. Ma. 2019. Removal and biodegradation of different petroleum hydrocarbons using the filamentous fungus *Aspergillus* sp. RFC-1. MicrobiologyOpen 8: e00619.

Alimadadi, N., M.R. Soudi and Z. Talebpour. 2018. Efficient production of tri-acetylated mono-acylated mannosylerythritol lipids by *Sporisorium* sp. aff. sorghi SAM20. J. Appl. Microbiol. 124: 457–468.

do Amaral Marques, N.S.A., T. Alves Lima e Silva, R.F. da Silva Andrade, J.F. Branco Júnior, K. Okada and G.M. de Campos-Takaki. 2019. Lipopeptide biosurfactant produced by *Mucor circinelloides* UCP/WFCC 0001 applied in the removal of crude oil and engine oil from soil. Acta Sci. – Technol. 41(1): e38986..

Amellal, N., J.-M. Portal and J. Berthelin. 2001. Effect of soil structure on the bioavailability of polycyclic aromatic hydrocarbons within aggregates of a contaminated soil. Appl. Geochem. 16: 1611–1619.

Amézcua-Vega, C., H.M. Poggi-Varaldo, F. Esparza-García, E. Ríos-Leal and R. Rodríguez-Vázquez. 2007. Effect of culture conditions on fatty acids composition of a biosurfactant produced by Candida ingens and changes of surface tension of culture media. Bioresour. Technol. 98: 237–240.

Andersson, B.E. and T. Henrysson. 1996. Accumulation and degradation of dead-end metabolites during treatment of soil contaminated with polycyclic aromatic hydrocarbons with five strains of white-rot fungi. Appl. Microbiol. Biotechnol. 46: 647–652.

Andrade Silva, N.R., M.A.C. Luna, A.L.C.M.A. Santiago, L.O. Franco, G.K.B. Silva, P.M. de Souza, et al. 2014. Biosurfactant and bioemulsifier produced by a promising *Cunninghamella echinulata* isolated from caatinga soil in the Northeast of Brazil. Int. J. Mol. Sci. 15: 15377–15395.

Arutchelvi, J.I., S. Bhaduri, P.V. Uppara and M. Doble. 2008. Mannosylerythritol lipids: A review. J. Ind. Microbiol. Biotechnol. 35: 1559–1570.

Ásgeirsdóttir, S.A., O.M.H de Vries and J.G.H Wessels. 1998. Identification of three differentially expressed hydrophobins in *Pleurotus ostreatus* (oyster mushroom). Microbiol. 144: 2961–2969.

Azin, E., H. Moghimi and R. Heidarytabar. 2018. Petroleum degradation, biosurfactant and laccase production by *Fusarium neocosmosporiellum* RH-10: A microcosm study. Soil Sediment Contam. 27: 329–342.

Banitz, T., K. Johst, L.Y. Wick, S. Schamfuß, H. Harms and K. Frank. 2013. Highways versus pipelines: Contributions of two fungal transport mechanisms to efficient bioremediation. Environ. Microbiol. Rep. 5: 211–218.

Baranger, C., L. Creusot, X. Sun, I. Pezron, A. Le Goff and A. Fayeulle. 2018. Microfluidic approaches for improved bioremediation: Monitoring pollutant uptake in a soil fungus. NewTech'18: icepr18.184.

Baranger, C., A. Fayeulle and A. Le Goff. 2020. Microfluidic monitoring of the growth of individual hyphae in confined environments. R. Soc. Open Sci. 7(8): 191535.

Barkay, T., S. Navon-Venezia, E.Z. Ron and E. Rosenberg. 1999. Enhancement of solubilization and biodegradation of polyaromatic hydrocarbons by the bioemulsifier alasan. Appl. Environ. Microbiol. 65: 2697–2702.

Batista, R.M., R.D. Rufino, J.M. de Luna, J.E.G de Souza and L.A. Sarubbo. 2010. Effect of medium components on the production of a biosurfactant from *Candida tropicalis* applied to the removal of hydrophobic contaminants in soil. Water Environ. Res. 82: 418–425.

Bell-Pedersen, D., J.C. Dunlap and J.J Loros. 1992. The *Neurospora* circadian clock-controlled gene, ccg-2, is allelic to eas and encodes a fungal hydrophobin required for formation of the conidial rodlet layer. Genes Dev. 6: 2382–2394.

Berger, B.W. and N.D. Sallada. 2019. Hydrophobins: Multifunctional biosurfactants for interface engineering. J. Biol. Eng. 13: 10.

Bhardwaj, G. 2013. Biosurfactants from fungi: A review. J. Pet. Environ. Biotechnol. 4: 6.

Bhardwaj, G., S.S. Cameotra and H.K. Chopra. 2015. Isolation and purification of a new enamide biosurfactant from *Fusarium proliferatum* using rice-bran. RSC Adv. 5: 54783–54792.

Blesic, M., V. Dichiarante, R. Milani, M. Linder and P. Metrangolo. 2018. Evaluating the potential of natural surfactants in the petroleum industry: The case of hydrophobins. Pure Appl. Chem. 90(2): 305–314.

de Boer, W., L.B. Folman, R.C. Summerbell and L. Boddy. 2005. Living in a fungal world: Impact of fungi on soil bacterial niche development. FEMS Microbiol. Rev. 29: 795–811.

Boll, E.S., A.R. Johnsen and J.H. Christensen. 2015. Polar metabolites of polycyclic aromatic compounds from fungi are potential soil and groundwater contaminants. Chemosphere 119: 250–257.

Bonfante, P. and A. Genre. 2010. Mechanisms underlying beneficial plant–fungus interactions in mycorrhizal symbiosis. Nat. Commun. 1: 1–11.

Breedveld, G.D. and D.A. Karlsen. 2000. Estimating the availability of polycyclic aromatic hydrocarbons for bioremediation of creosote contaminated soils. Appl. Microbiol. Biotechnol. 54: 255–261.

Brown, D.G. 2007. Relationship between micellar and hemi-micellar processes and the bioavailability of surfactant-solubilized hydrophobic organic compounds. Environ. Sci. Technol. 41: 1194–1199.

Cachada, A., R. Pereira, E.F. da Silva and A.C. Duarte. 2014. The prediction of PAHs bioavailability in soils using chemical methods: State of the art and future challenges. Sci. Total Environ. 472: 463–480.

Camargo-de-Morais, M.M., S.A.F. Ramos and M.C.B Pimentel. 2003. Production of an extracellular polysaccharide with emulsifier properties by *Penicillium citrinum*. World J. Microbiol. Biotechnol. 19: 191–194.

Cameron, D.R., D.G. Cooper and R.J. Neufeld. 1988. The mannoprotein of *Saccharomyces cerevisiae* is an effective bioemulsifier. Appl. Environ. Microbiol. 54: 1420–1425.

Cerniglia, C.E. 1992. Biodegradation of polycyclic aromatic hydrocarbons. Biodegradation 3: 351–368.

Chau, H.W., B.C. Si, Y.K. Goh and V. Vujanovic. 2009. A novel method for identifying hydrophobicity on fungal surfaces. Mycol. Res. 113: 1046–1052.

Chau, H.W., Y.K Goh, B.C. Si and V. Vujanovic. 2010. Assessment of alcohol percentage test for fungal surface hydrophobicity measurement. Lett. Appl. Microbiol. 50: 295–300.

Chen, M., P. Xu, G. Zeng, C. Yang, D. Huang and J. Zhang. 2015. Bioremediation of soils contaminated with polycyclic aromatic hydrocarbons, petroleum, pesticides, chlorophenols and heavy metals by composting: Applications, microbes and future research needs. Biotechnol. Adv. 33: 745–755.

Chen, S., J. Wang, M.G. Waigi and Y. Gao. 2018. Glomalin-related soil protein influences the accumulation of polycyclic aromatic hydrocarbons by plant roots. Sci. Total Environ. 644: 465–473.

Chiewpattanakul, P., S. Phonnok, A. Durand, E. Marie and B.W. Thanomsub. 2010. Bioproduction and anticancer activity of biosurfactant produced by the dematiaceous fungus *Exophiala dermatitidis* SK80. J. Microbiol. Biotechnol. 20: 1664–1671.

Cicatiello, P., A.M. Gravagnuolo, G. Gnavi, G. Varese and P. Giardina. 2016. Marine fungi as source of new hydrophobins. Int. J. Biol. Macromol. 92: 1229–1233.

Cirigliano, M.C. and G.M. Carman. 1985. Purification and characterization of Liposan, a bioemulsifier from *Candida lipolytica*. Appl. Environ. Microbiol. 50: 846–850.

Cooper, D.G. and D.A. Paddock. 1984. Production of a biosurfactant from *Torulopsis bombicola*. Appl. Environ. Microbiol. 47: 173–176.

Costa Sperb, J.G., T.M. Costa, S.L. Bertoli and L. Benathar Ballod Tavares. 2018. Simultaneous production of biosurfactants and lipases from *Aspergillus niger* and optimization by response surface methodology and desirability functions. Braz. J. Chem. Eng. 35: 857–868.

Cox, A.R., F. Cagnol, A.B. Russell and M.J. Izzard. 2007. Surface properties of class II hydrophobins from *Trichoderma reesei* and influence on bubble stability. Langmuir 23: 7995–8002.

Cui, X., P. Mayer and J. Gan. 2013. Methods to assess bioavailability of hydrophobic organic contaminants: Principles, operations, and limitations. Environ. Pollut. 172: 223–234.

Dashti, N., H. Al-Awadhi, M. Khanafer, S. Abdelghany and S. Radwan. 2008. Potential of hexadecane-utilizing soil microorganisms for growth on hexadecanol, hexadecanal and hexadecanoic acid as sole sources of carbon and energy. Chemosphere 70: 475–479.

Delsarte, I., C. Rafin, F. Mrad and E. Veignie. 2018. Lipid metabolism and benzo[*a*]pyrene degradation by *Fusarium solani*: An unexplored potential. Environ. Sci. Pollut. Res. 25(12): 12177–12182.

Deng, Y., Y. Zhang, A.-L. Hesham, R. Liu and M. Yang. 2010. Cell surface properties of five polycyclic aromatic compound-degrading yeast strains. Appl. Microbiol. Biotechnol. 86: 1933–1939.

Déziel, E., G. Paquette, R. Villemur, F. Lépine and J.-G. Bisaillon. 1996. Biosurfactant production by a soil *Pseudomonas* strain growing on polycyclic aromatic hydrocarbons. Appl. Environ. Microbiol. 62(6): 1908–1912.

Driver, J.D., W.E. Holben and M.C. Rillig. 2005. Characterization of glomalin as a hyphal wall component of arbuscular mycorrhizal fungi. Soil Biol. Biochem. 37: 101–106.

Fayeulle, A. 2013. Etude des mécanismes intervenant dans la biodégradation des hydrocarbures aromatiques polycycliques par les champignons saprotrophes telluriques en vue d'applications en bioremédiation fongique de sols pollués. Université du Littoral Côte d'Opale, Dunkerque, France.

Fayeulle, A., E. Veignie, C. Slomianny, E. Dewailly, J.-C. Munch and C. Rafin. 2014. Energy-dependent uptake of benzo[*a*]pyrene and its cytoskeleton-dependent intracellular transport by the telluric fungus *Fusarium solani*. Environ. Sci. Pollut. Res. 21: 3515–3523.

Fayeulle, A., E. Veignie, R. Schroll, J.-C. Munch and C. Rafin. 2019. PAH biodegradation by telluric saprotrophic fungi isolated from aged PAH-contaminated soils in mineral medium and historically contaminated soil microcosms. J. Soils Sediments 19(7): 3056–3067.

Fomina, M., E.P. Burford and G.M. Gadd. 2006. Fungal dissolution and transformation of minerals: Significance for nutrient and metal mobility. *In*: G.M. Gadd (ed.), Fungi in Biogeochemical Cycles. Cambridge University Press, Cambridge, UK, pp. 236–266.

Frautz, B., S. Lang and F. Wagner. 1986. Formation of cellobiose lipids by growing and resting cells of *Ustilago maydis*. Biotechnol. Lett. 8: 757–762.

Fuchs, U., K.J. Czymmek and J.A. Sweigard. 2004. Five hydrophobin genes in *Fusarium verticillioides* include two required for microconidial chain formation. Fungal Genet. Biol. 41: 852–864.

Furuno, S., K. Päzolt, C. Rabe, T.R. Neu, H. Harms and L.Y. Wick. 2010. Fungal mycelia allow chemotactic dispersal of polycyclic aromatic hydrocarbon-degrading bacteria in water-unsaturated systems. Environ. Microbiol. 12: 1391–1398.

Furuno, S., S. Foss, E. Wild, K.C. Jones, K.T. Semple, H. Harms, et al. 2012a. Mycelia promote active transport and spatial dispersion of polycyclic aromatic hydrocarbons. Environ. Sci. Technol. 46: 5463–5470.

Furuno, S., R. Remer, A. Chatzinotas, H. Harms and L.Y. Wick. 2012b. Use of mycelia as paths for the isolation of contaminant-degrading bacteria from soil. Microb. Biotechnol. 5: 142–148.

Gadd, G.M., Y.J. Rhee, K. Stephenson and Z. Wei. 2012. Geomycology: metals, actinides and biominerals. Environ. Microbiol. Rep. 4: 270–296.

Gadkar, V. and M.C. Rillig. 2006. The arbuscular mycorrhizal fungal protein glomalin is a putative homolog of heat shock protein 60. FEMS Microbiol. Lett. 263: 93–101.

Gao, Y., Z. Zhou, W. Ling, X. Hu and S. Chen. 2017. Glomalin-related soil protein enhances the availability of polycyclic aromatic hydrocarbons in soil. Soil Biol. Biochem. 107: 129–132.

Garay, L.A., I.R. Sitepu, T. Cajka, J. Xu, H.E. Teh, J.B. German, et al. 2018. Extracellular fungal polyol lipids: A new class of potential high value lipids. Biotechnol. Adv. 36: 397–414.

Garcia-Junco, M., C. Gomez-Lahoz, J.-L. Niqui-Arroyo and J.J. Ortega-Calvo. 2003. Biosurfactant- and biodegradation-enhanced partitioning of polycyclic aromatic hydrocarbons from nonaqueous-phase liquids. Environ. Sci. Technol. 37: 2988–2996.

Gautam, G., V. Mishra, P. Verma, A.K. Panday and S. Negi. 2014. A cost effective strategy for production of bio-surfactant from locally isolated *Penicillium chrysogenum* SNP5 and its applications. J. Bioproces. Biotechniq. 4: 6.

González-Chávez, M.C., R. Carrillo-González, S.F. Wright and K.A. Nichols. 2004. The role of glomalin, a protein produced by arbuscular mycorrhizal fungi, in sequestering potentially toxic elements. Environ. Pollut. 130: 317–323.

Grünbacher, A., T. Throm, C. Seidel, B. Gutt, J. Röhrig, T. Strunk, et al. 2014. Six hydrophobins are involved in hydrophobin rodlet formation in *Aspergillus nidulans* and contribute to hydrophobicity of the spore surface. PLoS One 9: e94546.

Guimarães Martins, V., S.J. Kalil, T.E. Elit and J.A. Vieira Costa. 2006. Solid state biosurfactant production in a fixed-bed column bioreactor. Z. Naturforsch. C 61: 721–726.

Gunderson, J.J., J.D. Knight and K.C.J Van Rees. 2007. Impact of ectomycorrhizal colonization of hybrid poplar on the remediation of diesel-contaminated soil. J. Environ. Qual. 36: 927–934.

Gunsch, C.K., Q. Cheng, K.A. Kinney, P.J. Szaniszlo and C.P. Whitman. 2005. Identification of a homogentisate-1,2-dioxygenase gene in the fungus *Exophiala lecanii-corni*: Analysis and implications. Appl. Microbiol. Biotechnol. 68: 405–411.

Haas Jimoh Akanbi, M., E. Post, A. Meter-Arkema, R. Rink, G.T. Robillard, X. Wang, et al. 2010. Use of hydrophobins in formulation of water insoluble drugs for oral administration. Colloids Surf B Biointerfaces 75: 526–531.

Hai, F.I., O. Modin, K. Yamamoto, K. Fukushi, F. Nakajima and L.D. Nghiem. 2012. Pesticide removal by a mixed culture of bacteria and white-rot fungi. J. Taiwan Inst. Chem. Eng. 43: 459–462.

Han, H.-L., J. Tang, H. Jiang, M.-L. Zhang and Z. Liu. 2008. Synergy between fungi and bacteria in fungi-bacteria augmented remediation of petroleum-contaminated soil. Huan Jing Ke Xue 29: 189–195.

Harms, H., D. Schlosser and L.Y. Wick. 2011. Untapped potential: Exploiting fungi in bioremediation of hazardous chemicals. Nat. Rev. Microbiol. 9: 177–192.

Harvey, J.V. 1952. Relationship of aquatic fungi to water pollution. Sewage Ind. Waste 24: 1159–1164.

Hazen, K.C. 1990. Cell surface hydrophobicity of medically important fungi, especially *Candida* species. *In*: R.J. Doyle and M. Rosenberg (eds), Microbial Cell Surface Hydrophobicity. American Society for Microbiology, Washington DC, USA, pp. 249–295.

He, X., S. Li and S.G.W. Kaminskyj. 2018. Overexpression of *Aspergillus nidulans* α-1,3-glucan synthase increases cellular adhesion and causes cell wall defects. Med. Mycol. 56: 645–648.

Heister, K., S. Pols, J.P.G. Loch and T.N.P. Bosma. 2013. Desorption behaviour of polycyclic aromatic hydrocarbons after long-term storage of two harbour sludges from the port of Rotterdam, The Netherlands. J. Soils Sediments 13: 1113–1122.

Holder, D.J., B.H. Kirkland, M.W. Lewis and N.O. Keyhani. 2007. Surface characteristics of the entomopathogenic fungus *Beauveria* (*Cordyceps*) *bassiana*. Microbiol. 153: 3448–3457.

Hunt, L.M., J.R. Fairweather and F.J. Coyle. 2003. Public understandings of biotechnology in New Zealand: Factors affecting acceptability rankings of five selected biotechnologies. Research report No. 266, Agribusiness and Economics Research Unit, Lincoln University, Canterbury, N.Z.

Ishaq, U., M.S. Akram, Z. Iqbal, M. Rafiq, A. Akrem, M. Nadeem, et al. 2015. Production and characterization of novel self-assembling biosurfactants from *Aspergillus flavus*. J. Appl. Microbiol. 119: 1035–1045.

Jaynes, W.F. and S.A. Boyd. 1991. Hydrophobicity of siloxane surfaces in smectites as revealed by aromatic hydrocarbon adsorption from water. Clays Clay Miner. 39: 428–436.

de Jonge, L.W., C. Kjaergaard and P. Moldrup. 2004. Colloids and colloid-facilitated transport of contaminants in soils: An introduction. Vadose Zone J. 3: 321–325.

Kan, A.T., W. Chen and M.B. Tomson. 2000. Desorption kinetics of neutral hydrophobic organic compounds from field-contaminated sediment. Environ. Pollut. 108: 81–89.

Käppeli, O., P. Walther, M. Mueller and A. Fiechter. 1984. Structure of the cell surface of the yeast *Candida tropicalis* and its relation to hydrocarbon transport. Arch. Microbiol. 138: 279–282.

Kaufman, D.D. and J. Blake. 1970. Degradation of atrazine by soil fungi. Soil Biol. Biochem. 2: 73–80.

Khalid, S., M. Shahid, N.K. Niazi, B. Murtaza, I. Bibi and C. Dumat. 2017. A comparison of technologies for remediation of heavy metal contaminated soils. J. Geochemi. Explor. 182: 247–268.

Kiran, G.S., T.A. Hema, R. Gandhimathi, J. Selvin, T.A. Thomas, T. Rajeetha Ravji and K. Natarajaseenivasan. 2009. Optimization and production of a biosurfactant from the sponge-associated marine fungus *Aspergillus ustus* MSF3. Colloids Surf B Biointerfaces 73: 250–256.

Kitamoto, D., H. Yanagishita, T. Shinbo, T. Nakane, C. Kamisawa and T. Nakahara. 1993. Surface active properties and antimicrobial activities of mannosylerythritol lipids as biosurfactants produced by *Candida antarctica*. J. Biotechnol. 29: 91–96.

Kjær, J., V. Ernstsen, O.H. Jacobsen, N. Hansen, L.W. de Jonge and P. Olsen. 2011. Transport modes and pathways of the strongly sorbing pesticides glyphosate and pendimethalin through structured drained soils. Chemosphere 84: 471–479.

Lacroix, H. and P.D. Spanu. 2009. Silencing of six hydrophobins in *Cladosporium fulvum*: Complexities of simultaneously targeting multiple genes. Appl. Environ. Microbiol. 75: 542–546.

Lade, H.S., T.R. Waghmode, A.A. Kadam and S.P. Govindwar. 2012. Enhanced biodegradation and detoxification of disperse azo dye Rubine GFL and textile industry effluent by defined fungal-bacterial consortium. Int. Biodeterior. Biodegradation 72: 94–107.

Linder, M.B. 2009. Hydrophobins: Proteins that self-assemble at interfaces. Curr. Opin. Colloid Interface Sci. 14: 356–363.

Lindley, N.D. and M.T. Heydeman. 1986. Mechanism of dodecane uptake by whole cells of *Cladosporium resinae*. Microbiol. 132: 751–756.

Lladó, S., S. Covino, A.M. Solanas, M. Viñas, M. Petruccioli and A. d'Annibale. 2013. Comparative assessment of bioremediation approaches to highly recalcitrant PAH degradation in a real industrial polluted soil. J. Hazard. Mater. 248–249: 407–414.

Lohrasbi-Nejad, A., M. Torkzadeh-Mahani and S. Hosseinkhani. 2016. Heterologous expression of a hydrophobin HFB1 and evaluation of its contribution to producing stable foam. Protein Expr. Purif. 118: 25–30.

Lourenço, L.A., M.D. Alberton Magina, L. Benathar Ballod Tavares, S.M.A. Guelli U. de Souza, M. García Román and D. Altmajer-Vaz. 2018. Biosurfactant production by *Trametes versicolor* grown on two-phase olive mill waste in solid-state fermentation. Environ. Technol. 39: 3066–3076.

Lugones, L.G., J.S. Bosscher, K. Scholtmeyer, O.M.H. de Vries and J.G.H. Wessels. 1996. An abundant hydrophobin (ABH1) forms hydrophobic rodlet layers in *Agaricus bisporus* fruiting bodies. Microbiol. 142: 1321–1329.

Lugones, L.G., H.A.B. Wös and J.G.H. Wessels. 1998. A hydrophobin (ABH3) specifically secreted by vegetatively growing hyphae of *Agaricus bisporus* (common white button mushroom). Microbiol. 144: 2345–2353.

Lukondeh, T., N.J. Ashbolt and P.L. Rogers. 2003. Evaluation of *Kluyveromyces marxianus* FII 510700 grown on a lactose-based medium as a source of a natural bioemulsifier. J. Ind. Microbiol. Biotechnol. 30: 715–720.

Lumsdon, S.O., J. Green and B. Stieglitz. 2005. Adsorption of hydrophobin proteins at hydrophobic and hydrophilic interfaces. Colloids Surf. B: Biointerfaces 44: 172–178.

de Luna, J.M., R.D. Rufino, A.M.A.T. Jara, P.P.F. Brasileiro and L.A. Sarubbo. 2015. Environmental applications of the biosurfactant produced by *Candida sphaerica* cultivated in low-cost substrates. Colloids Surf. A: Physicochem. Eng. Asp. 480: 413–418.

de Luna, J.M., L.A. Sarubbo and G.M. de Campos-Takaki. 2009. A new biosurfactant produced by *Candida glabrata* UCP 1002: Characteristics of stability and application in oil recovery. Braz. Arch. Biol. Technol. 52: 785–793.

Luna-Velasco, M.A., F. Esparza-García, R.O. Cañizares-Villanueva and R. Rodríguez-Vázquez. 2007. Production and properties of a bioemulsifier synthesized by phenanthrene-degrading *Penicillium* sp. Process Biochem. 42: 310–314.

Ma, B., X. Lv, Y. He and J. Xu. 2016. Assessing adsorption of polycyclic aromatic hydrocarbons on *Rhizopus oryzae* cell wall components with water-methanol cosolvent model. Ecotoxicol. Environ. Saf. 125: 55–60.

Ma, X.-J., H. Li, D.-X. Wang and X. Song. 2014. Sophorolipid production from delignined corncob residue by *Wickerhamiella domercqiae* var. sophorolipid CGMCC 1576 and *Cryptococcus curvatus* ATCC 96219. Appl. Microbiol. Biotechnol. 98: 475–483.

Machín-Ramírez, C., D. Morales, F. Martínez-Morales, A.I. Okoh and M.R. Trejo-Hernández. 2010. Benzo[*a*]pyrene removal by axenic- and co-cultures of some bacterial and fungal strains. Int. Biodeterior. Biodegradation 64: 538–544.

Mankel, A., K. Krause and E. Kothe. 2002. Identification of a hydrophobin gene that is developmentally regulated in the ectomycorrhizal fungus *Tricholoma terreum*. Appl. Environ. Microbiol. 68: 1408–1413.

Meharg, A.A. and J.W.G. Cairney. 2000. Ectomycorrhizas - extending the capabilities of rhizosphere remediation? Soil Biol. and Biochem. 32: 1475–1484.

Méndez-Castillo, L., E. Prieto-Correa and C. Jiménez-Junca. 2017. Identification of fungi isolated from banana rachis and characterization of their surface activity. Lett. in Appl. Microbiol. 64: 246–251.

Meneses, D.P., E.J. Gudiña, F. Fernandes, L.R.B. Gonçalves, L.R. Rodrigues and S. Rodrigues. 2017. The yeast-like fungus *Aureobasidium thailandense* LB01 produces a new biosurfactant using olive oil mill wastewater as an inducer. Microbiol. Res. 204: 40–47.

Morita, T., M. Konishi, T. Fukuoka, T. Imura, H.K. Kitamoto and D. Kitamoto. 2007. Characterization of the genus *Pseudozyma* by the formation of glycolipid biosurfactants, mannosylerythritol lipids. FEMS Yeast Res. 7: 286–292.

Morita, T., M. Konishi, T. Fukuoka, T. Imura and D. Kitamoto. 2008. Identification of *Ustilago cynodontis* as a new producer of glycolipid biosurfactants, mannosylerythritol lipids, based on ribosomal DNA sequences. J. Oleo Sci. 57: 549–556.

Morita, T., Y. Ishibashi, T. Fukuoka, T. Imura, H. Sakai, M. Abe and D. Kitamoto. 2009. Production of glycolipid biosurfactants, mannosylerythritol lipids, by a smut fungus, *Ustilago scitaminea* NBRC 32730. Biosci. Biotechnol. Biochem. 73: 788–792.

Morita, T., T. Fukuoka, T. Imura and D. Kitamoto. 2013. Accumulation of cellobiose lipids under nitrogen-limiting conditions by two ustilaginomycetous yeasts, *Pseudozyma aphidis* and *Pseudozyma hubeiensis*. FEMS Yeast Res. 13: 44–49.

Mousavi, F., K. Beheshti-Maal and A. Massah. 2015. Production of sophorolipid from an identified current yeast, *Lachancea thermotolerans* BBMCZ7FA20, isolated from honey bee. Curr. Microbiol. 71: 303–310.

Nikiforova, S.V., N.N. Pozdnyakova and O.V. Turkovskaya. 2009. Emulsifying agent production during PAHs degradation by the white rot fungus *Pleurotus ostreatus* D1. Curr. Microbiol. 58: 554–558.

Niu, Y., L. Fan, D. Gu, J. Wu and Q. Chen. 2017. Characterization, enhancement and modelling of mannosylerythritol lipid production by fungal endophyte *Ceriporia lacerate* CHZJU. Food Chem. 228: 610–617.

Nyanhongo, G.S., S.R. Couto and G.M. Guebitz. 2006. Coupling of 2,4,6-trinitrotoluene (TNT) metabolites onto humic monomers by a new laccase from *Trametes modesta*. Chemosphere 64: 359–370.

Oje, O.A., V.E. Okpashi, J.C. Uzor, U.O. Uma, A.O. Irogbolu and I.N.E. Onwurah. 2016. Effect of acid and alkaline pretreatment on the production of biosurfactant from rice husk using *Mucor indicus*. Res. J. Environ. Toxicol. 10: 60–67.

Ortega-Calvo, J.J., M.C. Tejeda-Agredano, C. Jimenez-Sanchez, E. Congiu, R. Sungthong, J.L. Niqui-Arroyo, et al. 2013. Is it possible to increase bioavailability but not environmental risk of PAHs in bioremediation? J. Hazard. Mater. 261: 733–745.

Osono, T., Y. Ono and H. Takeda. 2003. Fungal ingrowth on forest floor and decomposing needle litter of *Chamaecyparis obtusa* in relation to resource availability and moisture condition. Soil Biol. Biochem. 35: 1423–1431.

Paria, S. 2008. Surfactant-enhanced remediation of organic contaminated soil and water. Adv. Colloid Interface Sci. 138: 24–58.

Pele, M.A., D.R. Ribeaux, E.R. Vieira, A.F. Souza, M.A.C. Luna, D.M. Rodríguez, et al. 2019. Conversion of renewable substrates for biosurfactant production by *Rhizopus arrhizus* UCP 1607 and enhancing the removal of diesel oil from marine soil. Electron. J. Biotechnol. 38: 40–48.

Peng, X., E. Masai, H. Kitayama, K. Harada, Y. Katayama and M. Fukuda. 2002. Characterization of the 5-carboxyvanillate decarboxylase gene and its role in lignin-related biphenyl catabolism in *Sphingomonas paucimobilis* SYK-6. Appl. Environ. Microbiol. 68: 4407–4415.

Peter, S., M. Kinne, X. Wang, R. Ullrich, G. Kayser, J.T. Groves, et al. 2012. Selective hydroxylation of alkanes by an extracellular fungal peroxygenase. FEBS J. 278: 3667–3675.

Price, N.P.J., P. Manitchotpisit, K.E. Vermillion, M.J. Bowman and T.D Leathers. 2013. Structural characterization of novel extracellular liamocins (mannitol oils) produced by *Aureobasidium pullulans* strain NRRL 50380. Carbohydr. Res. 370: 24–32.

Puchkov, E.O., U. Zähringer, B. Lindner, T.V. Kulakovskaya, U. Seydel and A. Wiese. 2002. The mycocidal, membrane-active complex of *Cryptococcus humicola* is a new type of cellobiose lipid with detergent features. Biochim. Biophys. Acta Biomembr. 1558: 161–170.

Rafin, C., B. de Foucault and E. Veignie. 2013. Exploring micromycetes biodiversity for screening benzo[a]pyrene degrading potential. Environ. Sci. Pollut. Res. 20: 3280–3289.

Loureiro dos Reis, C.B., L.M.B. Morandini, C.B. Bevilacqua, F. Bublitz, G. Ugalde, M.A. Mazutti, et al. 2018. First report of the production of a potent biosurfactant with α,β-trehalose by *Fusarium fujikuroi* under optimized conditions of submerged fermentation. Braz. J. Microbiol. 49: 185–192.

Ren, C.-G., C.-C. Kong, B. Bian, W. Liu, Y. Li, Y.-M. Luo, et al. 2017. Enhanced phytoremediation of soils contaminated with PAHs by arbuscular mycorrhiza and rhizobium. Int. J. Phytoremediat. 19: 789–797.

Ren, C.-G., C.-C. Kong, S.-X. Wang and Z.-H. Xie. 2019. Enhanced phytoremediation of uranium-contaminated soils by arbuscular mycorrhiza and rhizobium. Chemosphere 217: 773–779.

Ren, X., G. Zeng, L. Tang, J. Wang, J. Wan, Y. Liu, et al. 2018. Sorption, transport and biodegradation – An insight into bioavailability of persistent organic pollutants in soil. Sci. Total Environ. 610–611: 1154–1163.

Rillig, M.C. 2004. Arbuscular mycorrhizae, glomalin, and soil aggregation. Can. J. Soil. Sci. 84: 355–363.

Rillig, M.C. 2005. A connection between fungal hydrophobins and soil water repellency? Pedobiologia 49: 395–399.

Rillig, M.C., B.A. Caldwell, H.A.B. Wösten and P. Sollins. 2007. Role of proteins in soil carbon and nitrogen storage: Controls on persistence. Biogeochemistry 85: 25–44.

Ritter, L., K.R. Solomon, J. Forget, M. Stemeroff and C. O'Leary. 1995. Persistent organic pollutants: An assessment aeport on DDT, aldrin, dieldrin, endrin, clordane, heptachlor-, hexachlorobenzene, mirex, toxaphene, polychlorinated biphenyls, dioxins, and furans. The International Programme on Chemical Safety 95.39. WHO.

Ritz, K. and I.M. Young. 2004. Interactions between soil structure and fungi. Mycologist 18: 52–59.

Rodrigues, A.G., P.-A. Mårdh, C. Pina-Vaz, J. Martinez-de-Oliveira and A.F. Fonseca. 1999. Germ tube formation changes surface hydrophobicity of *Candida* cells. Infect. Dis. Obstet. Gynecol. 7: 222–226.

Rosenberg, M. 2006. Microbial adhesion to hydrocarbons: Twenty-five years of doing MATH. FEMS Microbiol. Lett. 262: 129–134.

Sánchez-Vázquez, V., K. Shirai, I. González and M. Gutiérrez-Rojas. 2018. Polycyclic aromatic hydrocarbon-emulsifier protein produced by *Aspergillus brasiliensis* (*niger*) in an airlift bioreactor following an electrochemical pretreatment. Bioresour. Technol. 256: 408–413.

Schamfuß, S., T.R. Neu, J.R. van der Meer, R. Tecon, H. Harms and L.Y. Wick. 2013. Impact of mycelia on the accessibility of fluorene to PAH-degrading bacteria. Environ. Sci. Technol. 47(13): 6908-6915.

Schmieder, S.S., C.E. Stanley, A. Rzepiela, D. van Swaay, J. Sabotič, S.F. Nørrelykke, et al. 2019. Bidirectional propagation of signals and nutrients in fungal networks via specialized hyphae. Curr. Biol. 29: 217–228.

Semple, K.T., K.J. Doick, K.C. Jones, P. Burauel, A. Craven and H. Harms. 2004. Defining bioavailability and bioaccessibility of contaminated soil and sediment is complicated. Environ. Sci. Technol. 38: 228–231.

Sen, S., S.N. Borah, A. Bora and S. Deka. 2017. Production, characterization, and antifungal activity of a biosurfactant produced by *Rhodotorula babjevae* YS3. Microb. Cell Fact. 16: 95

Sena, H.H., M. A. Sanches, D.F. Silva Rocha, W.O. P. Filho Segundo, É.S. de Souza and J.V. B. de Souza. 2018. Production of biosurfactants by soil fungi isolated from the Amazon forest. Int. J. Microbiol. 2018: 5684261.

Singh, H. 2006a. Fungal metabolism of petroleum hydrocarbons. *In*: H. Singh (ed.), Mycoremediation: Fungal Bioremediation. John Wiley & Sons, Hoboken NJ, USA, pp. 115–148.

Singh, H. 2006b. Fungal degradation of polychlorinated biphenyls and dioxins. *In*: H. Singh (ed.), Mycoremediation: Fungal Bioremediation. John Wiley & Sons, Hoboken NJ, USA, pp. 149–180.

Siqueira, V. and N. Lima. 2012. Surface hydrophobicity of culture and water biofilm of *Penicillium* spp. Curr. Microbiol. 64: 93–99.

Smith, K.E.C., M. Thullner, L.Y. Wick and H. Harms. 2009. Sorption to humic acids enhances polycyclic aromatic hydrocarbon biodegradation. Environ. Sci. Technol. 43: 7205–7211.

Smith, K.E.C., M. Thullner, L.Y. Wick and H. Harms. 2011. Dissolved organic carbon enhances the mass transfer of hydrophobic organic compounds from nonaqueous phase liquids (NAPLs) into the aqueous phase. Environ. Sci. Technol. 45: 8741–8747.

Smits, T.H.M., L.Y. Wick, H. Harms and C. Keel. 2003. Characterization of the surface hydrophobicity of filamentous fungi. Environ. Microbiol. 5: 85–91.

Spoeckner, S., Wray, V., M. Nimtz and S. Lang. 1999. Glycolipids of the smut fungus *Ustilago maydis* from cultivation on renewable resources. Appl. Microbiol. Biotechnol. 51: 33–39.

Stanley, C.E., G. Grossmann, X. Casadevall i Solvas and A.J. deMello. 2016. Soil-on-a-Chip: Microfluidic platforms for environmental organismal studies. Lab Chip 16: 228–241.

Szilvay, G.R., A. Paananen, K. Laurikainen, E. Vuorimaa, H. Lemmetyinen, J. Peltonen, et al. 2007. Self-assembled hydrophobin protein films at the air–water interface: Structural analysis and molecular angineering. Biochem. 46: 2345–2354.

Talbot, N.J., M.J. Kershaw, G.E. Wakley, O.M.H. de Vries, J.G.H. Wessels and J.E. Hamer. 1996. MPG1 encodes a fungal hydrophobin involved in surface interactions during infection-related development of *Magnaporthe grisea*. Plant Cell 8: 985–999.

Talbot, N.J. 1997. Fungal biology: Growing into the air. Curr. Biol. 7: R78–R81.

Tchuenbou-Magaia, F.L., I.T. Norton and P.W. Cox. 2009. Hydrophobins stabilised air-filled emulsions for the food industry. Food Hydrocoll. 23: 1877–1885.

Texeira Souza, K.S., E.J. Gudiña, Z. Azevedo, V. de Freitas, R.F. Schwan, L.R. Rodrigues, et al. 2017. New glycolipid biosurfactants produced by the yeast strain *Wickerhamomyces anomalus* CCMA 0358. Colloids Surf B Biointerfaces 154: 373–382.

Thanomsub, B., T. Watcharachaipong, K. Chotelersak, P. Arunrattiyakorn, T. Nitoda and H. Kanzaki. 2004. Monoacylglycerols: Glycolipid biosurfactants produced by a thermotolerant yeast, *Candida ishiwadae*. J. Appl. Microbiol. 96: 588–592.

Thion, C., A. Cébron, T. Beguiristain and C. Leyval. 2012. PAH biotransformation and sorption by *Fusarium solani* and *Arthrobacter oxydans* isolated from a polluted soil in axenic cultures and mixed co-cultures. Int. Biodeterior. Biodegradation 68: 28–35.

Veignie, E., E. Vinogradov, I. Sadovskaya, C. Coulon and C. Rafin. 2012. Preliminary characterizations of a carbohydrate from the concentrated culture filtrate from *Fusarium solani* and its role in benzo[*a*] pyrene solubilization. Adv. Microbiol. 02: 375–381.

Velioglu, Z. and R.O. Urek. 2016. Physicochemical and structural characterization of biosurfactant produced by *Pleurotus djamor* in solid-state fermentation. Biotechnol. Bioprocess Eng. 21: 430–438.

Verdin, A., A. Lounès-Hadj Sahraoui, R. Newsam, G. Robinson and R. Durand. 2005. Polycyclic aromatic hydrocarbons storage by *Fusarium solani* in intracellular lipid vesicles. Environ. Pollut. 133: 283–291.

Verdin, A., A. Lounès-Hadj Sahraoui, J. Fontaine, A. Grandmougin-Ferjani and R. Durand. 2006. Effects of anthracene on development of an arbuscular mycorrhizal fungus and contribution of the symbiotic association to pollutant dissipation. Mycorrhiza 16: 397–405.

Vicuña, R., B. González, D. Seelenfreund, C. Rüttimann and L. Salas. 1993. Ability of natural bacterial isolates to metabolize high and low molecular weight lignin-derived molecules. J. Biotechnol. 30: 9–13.

Villholth, K.G., N.J. Jarvis, O.H. Jacobsen and H. de Jonge. 2000. Field investigations and modeling of particle-facilitated pesticide transport in macroporous soil. J. Environ. Qual. 29: 1298–1309.

de Vries, O.M.H., M.P. Fekkes, H.A.B. Wösten and J.G.H. Wessels. 1993. Insoluble hydrophobin complexes in the walls of *Schizophyllum commune* and other filamentous fungi. Arch. Microbiol. 159: 330–335.

Wösten, H.A.B., M.-A. van Wetter, L.G. Lugones, H.C. van der Mei, H.J. Busscher and J.G.H. Wessels. 1999. How a fungus escapes the water to grow into the air. Curr. Biol. 9: 85–88.

Wösten, H.A.B. 2001. Hydrophobins: Multipurpose proteins. Ann. Rev. Microbiol. 55: 625–646.

Wu, N., S. Zhang, H. Huang and P. Christie. 2008. Enhanced dissipation of phenanthrene in spiked soil by arbuscular mycorrhizal alfalfa combined with a non-ionic surfactant amendment. Sci. Total Environ. 394: 230–236.

Wu, Y.-R., T.-T. He, J.-S. Lun, K. Maskaoui, T.-W. Huang and Z. Hu. 2009. Removal of benzo[*a*]pyrene by a fungus *Aspergillus* sp. BAP14. World J. Microbiol. Biotechnol. 25: 1395–1401.

Yang, Z., Y. Shi, Y. Zhang, Q. Cheng, X. Li, C. Zhao, et al. 2018. Different pathways for 4-n-nonylphenol biodegradation by two *Aspergillus* strains derived from estuary sediment: Evidence from metabolites determination and key-gene identification. J. Hazard. Mater. 359: 203–212.

Yu, L., L. Duan, R. Naidu and K.T. Semple. 2018. Abiotic factors controlling bioavailability and bioaccessibility of polycyclic aromatic hydrocarbons in soil: Putting together a bigger picture. Sci. Total Environ. 613–614: 1140–1153.

Zadeh, P.H., H. Moghimi and J. Hamedi. 2018. Biosurfactant production by *Mucor circinelloides*: Environmental applications and surface-active properties. Eng. Life Sci. 18: 317–325.

Zhang, J.-H., Q.-H. Xue, H. Gao, X. Ma and P. Wang. 2016. Degradation of crude oil by fungal enzyme preparations from *Aspergillus spp*. for potential use in enhanced oil recovery. J. Chem. Technol. Biotechnol. 91: 865–875.

Zhang, S., Y.X. Xia, B. Kim and N.O. Keyhani. 2011. Two hydrophobins are involved in fungal spore coat rodlet layer assembly and each play distinct roles in surface interactions, development and pathogenesis in the entomopathogenic fungus, *Beauveria bassiana*. Mol. Microbiol. 80: 811–826.

Zhang, X., S.M. Kirby, Y. Chen, S.L. Anna, L.M. Walker, F.R. Hung, et al. 2018. Formation and elasticity of membranes of the class II hydrophobin cerato-ulmin at oil-water interfaces. Colloids Surf B Biointerfaces 164: 98–106.

Zhao, J., Y. Chi, Y. Xu, D. Jia and K. Yao. 2016. Co-metabolic degradation of β-cypermethrin and 3-phenoxybenzoic acid by co-culture of *Bacillus licheniformis* B-1 and *Aspergillus oryzae* M-4. PLoS One 11: e0166796.

Zhou, D., X. Zhang, Y. Du, S. Dong, Z. Xu and L. Yan. 2014. Insights into the synergistic effect of fungi and bacteria for reactive red decolorization. J. Spectrosc (Hindawi) 2014: 237346.

Directive 2001/18/EC of the European Parliament and of the Council on the deliberate release into the environment of genetically modified organisms. OJ L106, 17.04.2001: 1–39.

CHAPTER 2

Bioindication and Bioremediation of Mining Degraded Soil

Danica Fazekašová* and Juraj Fazekaš,

Department of Environmental Management, Faculty of Management, University of Prešov,
Konštantínova 16, 080 01 Prešov, Slovakia.

1. INTRODUCTION

Contamination of all environmental spheres with chemicals is steadily increasing and currently being addressed with increased attention. The quantitative incidence of such contamination, along with the qualitative diversity, is on the rise. New compounds are introduced into the environment which exhibit considerable chemical and biological stability. Current analytical methods make it possible to accurately characterize the biological effects of individual pollutants, and they possibly quantify the risks associated with their occurrences (Piatrik et al. 2007). Contaminants are organic as well as inorganic compounds that are not naturally present in the environment, known as xenobiotics (foreign environmental substances, especially organic "manmade" compounds), or are present in unnaturally high concentrations in the individual components of the environment (e.g. heavy metals). Experts and laymen focus primarily on hazardous substances which do not easily degrade, and are persistent and toxic. The soil system is a very specific component and to some extent can naturally eliminate various foreign substances (Fazekašová et al. 2014).

Risk elements are most often associated with metallic elements, especially heavy metals. Heavy metals are those with a density greater than 5 g/m^3. These are considered one of the most important sources of environmental pollution since their significant impact on the ecological quality of the environment has been identified (Sastre et al. 2002). Increasing concentrations of heavy metals, especially mobile forms, can cause serious environmental problems related to soil and water biota contamination (Chopin and Alloway 2007). Heavy metal contamination is a

*Corresponding author: danica.fazekasova@unipo.sk
Juraj Fazekaš email: juraj.fazekas@unipo.sk

global concern due to the toxicity of metals and their potential threat to human health (Kashem et al. 2007). Heavy metals are associated with soil components in various ways—these linkages indicate their mobility in soils as well as their bioavailability (Ahumada and Mendoza 2001). The environmental impact of heavy metals is highlighted by their non-degradability. Heavy metals are undergoing a global ecological cycle in which soil and water play a major role. Soil does not act as a passive heavy metal acceptor, and contaminated soil becomes a source of pollution of other components of the environment as for example water, air, and biota ultimately enter the food chain (Fazekašová et al. 2014).

The natural levels of heavy metals in soil are related to the content of the elements in the parent rock. High concentrations of hazardous heavy metals are present in soils of ore deposits. The highest concentrations of heavy metals contain igneous rocks. In the group of sedimentary rocks, the highest levels of heavy metals are found in clays and shales (Ďurža 2003). In addition to natural geochemical resources, the accumulation of heavy metals in soils also results from industrial and agricultural activities (metallurgy, combustion of fossil fuels, automotive, organic and mineral fertilizers, liming, pesticides, sewage sludge, household and industrial waste). Atmospheric deposition is one of the sources of soil contamination with heavy metals (Kafka and Punčochářová 2002). Regular and long-term monitoring of heavy metals in soils and various international projects points to a significant increase in the soil concentration in urban and industrial areas. Excessive levels of risk elements in the food chain can cause various health problems and illnesses, ranging from less severe than allergic reactions to severe respiratory, cardiovascular, nervous system, and tumour-related diseases (Kabata–Pendias and Pendias 2001). The effect of such risk elements is mainly due to their low concentrations and long-term exposure, which causes many of the aforementioned diseases. Heavy metals include essential, vital elements for the nutrition of organisms such as Cu, Fe, Zn, Co and Se, as well as non-essential metals that are potentially toxic: Hg, Pb and Cd (Beneš 1994).

Risk elements characterized by ecotoxicity, non-degradability, and cumulation capability are subject to global ecological cycles in which the main role is attributed to soil. At the expense of the soil, both air and water are purified (through the buffering and filtering properties of the soils). Soil is a source of contamination of other components of the environment—it is a start of risk elements for food. Risk elements cause negative changes in soil properties, called soil metallization. This is manifested by a decrease in the production potential of soils, especially by a threat to food safety (Noskovič et al. 2012). These risk elements have become one of the most serious stress factors for plant and animal organisms. Evidence of carcinogenicity of some of the risk elements in humans is based on clinical studies (Kafka and Punčochářová 2002). Heavy metals in plants and their transfer to the food chain will affect their bioavailability, which depends on soil properties (pH, cation exchange capacity, redox potential, organic carbon content, and soil species). Bioavailability of Zn, Pb, Cd decreases as soil pH rises. Individual hazardous chemical elements are more active in acidic soil reaction. The mobility of individual chemical elements is high at pH: Cd – 6.5; Mn – 6.5; Ni – 5.5; Co – 5.5; Cu – 4.5; As – 4.5; CrIII – 4.5; Pb – 4.0; Hg – 4.0 FeIII – 3.5; Zn – 6.0.

Soil is an irreplaceable natural resource that requires protection from degradation; it should be maintained and preserved for future generations. Soil quality and function, especially its internal biodiversity, has been the subject of increasing attention at both scientific and political levels. Organisms and pollutants interact closely with other ecosystem components. The life activity of the organism (bioindicator) is influenced by abiotic and biotic factors and can often be exposed to number of pollutants (Havlicek 2012).

Soils are biologically active; they are habitats for living organisms. In fact, soils are formed by these organisms, and soil development is limited without their presence. Soil health, biodiversity and soil resilience are critically limited in extreme environments and respond very sensitively to anthropogenic impacts (Doran and Zeiss 2000). Soil health depends on its biotic component, and the state of biota is also a direct indicator of the soil ecosystem quality. The composition of the flora, just like other parts of the bio-component, reflects the current state of soil conditions. Various sensitive plant species by their occurrence draw attention to changes in soil chemistry (Jurko 1990). The importance of edaphone as one of the basic components of soil biota in ecosystems is not limited to the decomposition of dead organic matter, but it is understood in a broader context. In addition to degradation processes, its importance in the paedogenesis and development of soil properties, as well as in their stabilization, including the maintenance of soil fertility, is emphasized. In natural as well as anthropogenic ecosystems (e.g. arable land), the individual components of edaphone are sensitive to a wide range of toxic substances.

The decline in soil biodiversity is one of the eight identified threats to European soil. Despite the Rio Conference in 1992 and the popularization of the concept of biodiversity, there are no laws or regulations for specifically addressing soil biodiversity either at international or domestic level. In a broad sense, biodiversity refers to 'diversity of life', which includes 'diversity within species, between species and ecosystems' (Convention on Biological Diversity 1992, Havlicek 2012).

Soil quality indicators currently used are mainly based on chemical and physical parameters. Numerous critical 'threshold' levels are accepted nationally for chemical factors. In general, chemical and physical indicators require long periods if the effects on humans or management practices are to be identified. In contrast, soil biota reacts in a sensitive way to changes and therefore biological indicators are suitable for early diagnosis of degradation processes (Abdu et al. 2017, Havlicek 2012). There are two concepts: bioindicators and biomonitoring. Bioindicators are organisms or communities of organisms that provide information about the quality of the environment. Biomonitoring is a regular, systematic use of sensitive organisms to monitor the quality of the environment; it stores quantitative information on the quality of the environment (Havlicek 2012).

Knowledge of how organisms react to direct anthropogenic effects (contamination, erosion) or indirect effects (increase in atmospheric carbon dioxide, etc.) is only partial. Available data shows that soil organisms respond not only to the organic matter content in the soil, but also to chemical inputs like heavy metals or organic contaminants as well as to any change in the physical properties (e.g. compaction). As part of long-term monitoring, it is necessary to consider soil biodiversity at the level of species diversity as well as functional biodiversity. The structure and activity of soil microbial communities is difficult to elucidate through a single monitoring approach. It is, therefore, necessary to use different approaches to better understand and fully depict the soil microbial situation (Khan et al. 2010). More than 80 methods related to species diversity or diversity related to biological functions are currently relevant. The selection of precise and appropriate methods is subject to several criteria. The parameter should be simple, practical, and comprehensible. The parameters considered should also be readily available and must conform to standardized methods; they must be based on scientific knowledge and ensure comparability of data between sites and studies; the indicators must be accurate and easy to interpret; cost-effectiveness and time-limited consumption are also important aspects (Havlicek 2012).

Most terrestrial ecosystem functions and services depend heavily on soil and soil biota. In fact, biological elements are the key to ecosystem functions—the identification of bioindicators is critical to meeting soil protection objectives (Wang et al. 2010, Havlicek 2012). Nonetheless, soil, soil biodiversity, and the link between ground and above–ground organisms have remained poorly understood despite the interest of scientists and experts.

Decontamination of areas contaminated with toxic metals is one of the important research subjects of contemporary science. Cleaning such sites using conventional physicochemical methods is costly and often non–ecological. In recent years, technologies with biological systems are being developed. The application of biotechnological methods represents a new potential way of solving the problems related to heavy metals. Bioremediation is defined as the use of living organisms as well as their components (e.g. their enzymes) to eliminate or reduce the environmental hazard accumulation of toxic xenobiotics (Dercová et al. 2004).

Remediation of the contaminated environment has been a major concern in recent years. This is because soil and water contaminated with toxic substances pose a serious threat to public health. The remediation of toxic metals from the soil by means of sorption to organo-mineral complexes comprising natural organic and inorganic components (zeolite or colinite with bound humic acids) is studied experimentally (Barančíková and Makovníková 2003). Humified soil organic matter is one of the main factors controlling the physical, chemical, and biological properties of soil and plays a significant role in soil hygiene. Humic acids affect the mobility, bioavailability, degradation and phytotoxicity of various organic and inorganic contaminants (Fazekašová et al. 2014). The ability to bind metals is one of the most important properties of humic substances. In natural systems, these substances can bind polluting metals and significantly affect transport phenomena, toxicity, regeneration, and purification processes. They are used in plant and animal production to increase productivity. They are natural substances that do not pollute the environment (Skybová 2006). Activated carbon is one of the most widely used sorbents for removing contaminants due to its high sorption capacity (Pipíška and Remenárová 2014).

Three types of remediation techniques are used in heavy metal remediation: pollutant removal, pollutant stabilization, and natural pollutant weakening. Inorganic pollutants are mostly chemical elements and are non-degradable; their remediation does not make use of the biodegradation method. Phytoremediation techniques using plants have been developed for inorganic pollutants. Phytoremediation is an emerging group of technologies that use green plants to clean the environment of pollutants; it is a cost-effective and non-invasive alternative to conventional remediation methods (Mahmood 2010). There are different versions of phytoremediation techniques – e.g. phytoextraction and phytostabilization (Hegedűsová et al. 2008) – which are most often used for the remediation of soils contaminated with toxic heavy metals. In the phytoextraction process, plants extract pollutants from the soil through the root system and store them mainly in green biomass. The entire process can be periodically repeated until the desired level of pollution reduction. The obtained biomass is then processed by microbial composting, thermal incineration or incineration, or chemically by extraction. Phytostabilization exploits the ability of plants to chemically fix or stabilize pollutants in the soil. Phytostabilization is particularly suitable for soil extractants that are difficult to extract, e.g. lead. These methods can be combined with each other. Phytoremediation has several major advantages. In addition to its low cost and environmental friendliness, it is suitable for various types of contaminants, has low energy requirements, and is well received by the public. The disadvantages include a long processing time, limiting the possibilities of soil decontamination by the depth of contamination, and the potential entry of contaminants into the food chain of plant-fed animals. The various aspects of phytoextraction and the scope of this technology for the remediation of contaminated soil by heavy metals are a challenge as well as an opportunity to realize phytoextraction as an economically viable remediation method (Mahmood 2010).

Some plants are referred as the so-called hyperaccumulators and bind metals in large quantities. About 400 species are described which can be classified as heavy metal hyperaccumulators (Baker et al. 1994, 2000). Among the plants, *Thlaspi caerulescens* subsp. *Caerulescens* can accumulate 30 g/kg of nickel, 43 g/kg of zinc, and 2 g/kg of cadmium, and *Thlaspi caerulescens* subsp. *Tatrense* accumulates 20 g of zinc per kg of the dry matter (Dercová et al. 2004). The *Agrostis stolonifera* can drain 300 times more arsenic than other plants growing in the same habitat. The genus Brassica is represented in the accumulation of heavy metals by a large number of species and accumulate mainly Pb, Cd, Cr, Zn, and Cu. Other plant species may be mentioned – e.g. *Thlaspi orbifolia*, *Chlorocyperus esculentus*, *Avena sativa*, and

Glycine soy, all of which accumulate Ni, while Cd is especially accumulated in *Thlaspi arvenze*, *Artisis campris*, and *Sorghum dochna* war. *saccharatum* (Schwarczová et al. 2011). Phytoremediation is best applied on sites with surface location. Improvement of the effect of the decontamination of inorganic pollutants by biotechnological pathways can also be achieved in conjunction with physical and chemical methods (Dercová et al. 2004).

This chapter evaluates the chemical parameters of soils, biological activity, functional diversity of microorganisms, and the transfer of heavy metals in the soil-plant system of mining degraded soil. It also presents the results of experimental studies dealing with the possibilities of remediation of toxic metals by using natural substances based on humic acids.

2. TOXICITY, BIOLOGICAL ACTIVITY AND BIODIVERSITY

2.1 The Methodological Approach

The study area. The research was carried out in the metallically, acid-burdened, and the alkaline-burdened region of north–eastern and central Slovakia, which, according to the environmental regionalization of the Slovak Republic, represents those regions with a slightly disturbed environment (Klinda et al. 2016).

As a result of long-term and intensive mining and treatment activities, the area of Middle Spiš (MS) was contaminated by heavy metals, and the landscape was deformed by extensive anthropogenic forms. The mining of mineral resources together with the utility component also relocated large volumes of residual material, which was subsequently stored in the form of mining deposits, heaps, and tailings ponds—all these significantly disturbed the landscape structure and contaminated individual components of the environment.

The Jelšava–Lubeník (JL) area, with specific alkaline emissions, is one of the most devastated areas of Slovakia and has an alarming degree of environmental damage. The processing of the magnesite raw material and the production of magnesite clinkers was accompanied by enormous emission of MgO dust particles into the atmosphere. Magnesium fumes with a significant proportion of reactive caustic MgO in contact with the soil under wet conditions form aggressive alkaline solutions that negatively affect the value of the soil reaction. The result is secondary feverish salinity, chemical intoxication, and soil devastation (Hronec et al. 2012, Fazekaš et al. 2018).

The main emission source in Nižná Slaná (NS) was an iron ore mining and processing plant focused on siderite mining. The main useful component of the deposit is siderite (Grecula et al. 1995). Nizhnyan siderite is highly ferrous, with an increased level of the Mn content. Undesirable impurities in the bearing include As, S, Pb, and Zn. The most important undesirable heavy metal admixture is arsenic (Hančuľák et al. 2007). The acidification and metallization of soil present the risks of chemical soil degradation in the emission area of iron ore mines. The acidification effect is accelerated by an anthropogenic load by the acidic emissions of sulphur and nitrogen oxides.

2.2 Sampling Procedure

Soil sampling was carried out in 2014–2019 once a year in summer from the A horizon (0.15–0.20 m) to determine the total heavy metals content and the selected biological and chemical indicators. The sampling sites were situated on permanent grasslands (PG) located in the polluted fields of pile and tailings ponds in the investigated areas of Middle Spiš (MS), Jelšava and Lubeník (JL), and Nižná Slana (NS) (Fig. 1).

Figure 1   Localization of the investigated areas – Middle Spiš (MS), Jelšava and Lubeník (JL), and Nižná Slana (NS) (Slovakia).

2.3   Data Collection and Analysis

The soil samples were homogenized and dried at room temperature. After drying the aggregate state, they were sieved through a sieve (Ø 2 mm) and stored in plastic bags. The total heavy metal content (Hg, Cd, Cr, Ni, Pb, Cu, Zn, As, Mg, Mn, Fe) was determined by AAS (atomic absorption spectrometry) and RFS (X-ray fluorescence spectrometry) (Fiala et al. 1999). The soil samples were extracted in 68% nitric acid. The total mercury content was measured directly from a 0.2 g dry soil sample with a Milestone DMA–80 Direct Mercury Analyser. The measured values of heavy metals were compared with the limit values set by the Act of the National Council of the Slovak Republic No. 220/2004 Coll. for sand-clay soils of Slovakia.

Aboveground organs of the plant species *Phragmites australis*, *Elytrigia repens*, and *Agrostis stolonifera* were collected for the determination of heavy metals in plants. The collected vegetation samples were dried at 40°C, homogenized to a fraction <0.09 mm, burned in an oven at 550°C, and quenched with a mixture of HCl and HNO_3. Heavy metals Cd, Pb, Zn, Cr, Mn, and Mg were determined in the samples. The measured values of the heavy metals were compared with the limit values set by the Act of the National Council of the Slovak Republic No. 220/2004 Coll. for plants in Slovakia. The samples were analysed by AAS (atomic absorption spectrometry) and RFS (X-ray fluorescence spectrometry) according to the method of Fiala et al. (1999).

The activity of selected soil enzymes was determined from biological properties. Urease activity (URE) was measured by method described by Chazijev (1976). Alkaline phosphatase (ALP) and acid phosphatase (ACP) were measured by the method described by Chazijev, and modified method by Grejtovský (1991).

The metabolic profiles of the microbial communities were evaluated in fresh soil samples using soil biology (Garland 1996, Hofman et al. 2004, Gömöryová 2008) for fresh soil samples. Data were normalized for the average well colour development (AWCD) parameter and calculated in accordance with Garland (1996). The functional group diversity of microorganisms was evaluated using the Shannon diversity (H') index.

From the soil chemical properties, soil reaction (pH/KCl) and the redox potential (Eh) were determined using the Mettler–Toledo and the organic carbon content (Cox) according to Ťurin (Fiala et al. 1999).

The transfer of heavy metals from the soil to the plant was calculated according to the methodology of Garg et al. (2014), as a heavy metal transfer factor (MTF):

$$\text{MTF} = \frac{C_\text{vegetation}}{C_\text{soil}} \tag{1}$$

where $C_\text{vegetation}$ and C_soil represent the concentration of heavy metals in the plant and soil extracts.

As the heavy metal sorbent, 100% natural HUMAC Enviro was made from a pure source of oxihumolite (brown coal) at 1% and 2% concentrations. The active ingredients of the formulation are humic acids, which have a large absorption capacity that binds them to toxic substances.

The results were logarithmically transformed and evaluated in STATISTICA 12, PAST 3, and SPSS Statistics.

2.4   Potential Toxic Elements in Soils and Plants

One of today's most significant environmental problems, which affects all environmental compartments, is global environmental contamination. Scientists focus primarily on hazardous substances (heavy metals) that are difficult to degrade in nature, with high persistence and often exhibiting toxic effects on the environment.

High concentrations of Hg and Cu were reported in the soils in the Middle Spiš region (MS) because of the long-term extraction and processing of mercury and copper in the investigated area. At the highest mercury concentration, the limit value was exceeded by 69.6 times, and in the case of copper it is – 14.95 times. High values were recorded for Zn, Cd, Pb and Cr. The highest observed concentrations exceeded the Zn – 6.9 times, Cd – 6.4 times, Pb – 4.6 times, and Cr – 1.1 times (Table 1).

The processing of magnesite raw material and the production of magnesite clinkers in Jelšava and Lubeník (JL) were accompanied by an enormous emission of MgO dust particles into the atmosphere and the escape of gaseous compounds, especially SO_2 and NO_x. Heavy metals, especially Cd, Pb, Zn, Mn, and Cr, are also present in the dust particles. The chromium content in the soil in the investigated territory of Jelšava and Lubeník was within the range of 140.00 ± 279.49 mg/kg (median ± standard deviation) (Table 1). The mean level of chromium in Slovak Republic soils is 85 mg/kg in Horizon A (Šefčík et al. 2008). Hexavalent chromium (Cr^{6+}) is classified as one of the most important environmental contaminants (Kafka and Punčochářová 2002). Magnesium is considered the fifth major plant nutrient. The values of the available magnesium in the upper layer of the agricultural land in Slovakia are within the range of 200–400 mg/kg Mg, which represents a high content of this element in the soil (Kobza et al. 2009). In the study area, we found significant soil contamination with magnesium—it is, on average, 18 to 493 times higher than the threshold and comparable to the results derived by Wang et al. (2015). The measured manganese contents show a similar course. The average manganese content in Slovak Republic soils ranges from 0.85 to 112.90 mg/kg, which indicates a significant spatial heterogeneity of the elements, but the mean stock of this element in the soil predominates (Fazekaš et al. 2018). Kabata–Pendias (2011) reports 1500 mg kg^{-1}, which shows symptoms of manganese toxicity. Based on the results obtained, it can be concluded that the Cr, Mn and Mg contents are above the toxicity level. Their significant exceedance indicates contamination, which can result in harmfulness and toxicity.

The measured values of heavy metals on selected sites in Nižná Slaná (Table 1) point at the increased contents of As, Hg, and Fe. The highest measured concentrations exceeded the limit of As – 37.0 times, Hg – 4.0 times, Cd – 2.7 times, Pb – 1.9 times, and Ni – 1.4 times. The amount of Fe in the soil is 3.5% on average and is likely to increase in heavy clay soils and some organic soils (Kabata–Pendias 2011). Background value of the Fe content in Slovak Republic soils is 2.64%. Our research showed that the highest values exceeded the average Fe content up to six times. Iron has many biochemical functions, but excess intake of iron may be toxic (Čurlík and Šefčík 1999).

Table 1 Measured values of the heavy metals (mg/kg) of the mining-degraded soil expressed by descriptive statistics

Area	Parameter	Mean	Min	Max	Median	SD	Limit*
Middle Spiš	Hg	12.17	1.10	34.80	6.54	12.61	0.50
	Cd	1.55	0.50	4.50	0.60	1.67	0.70
	Pb	112.75	26.00	324.00	54.50	115.50	70
	Cr	108.25	57.00	169.00	89.50	42.82	150
	Zn	365.25	66.00	1036.00	187.00	371.71	150
	Cu	325.38	53.00	897.00	171.50	350.51	60
	As	86.25	24.00	181.00	71.00	55.92	25
	Mn	1900.00	600.00	3400.00	1700.00	1017.00	
Jelšava and Lubeník	Hg	0.08	0.04	0.10	0.08	0.03	0.50
	Cd	0.50	0.50	0.50	0.50	0.00	0.70
	Pb	32.42	17.00	45.00	32.00	8.16	70.00
	Cr	231.08	83.00	1055.00	140.00	279.49	150.00
	Zn	88.33	48.00	108.00	88.50	15.17	150.00
	Cu	23.08	11.00	44.00	20.00	9.40	60.00
	Mn	1410.00	220.00	2300.00	1550.00	607.62	
	Mg	49841.67	7000.00	197000.00	26150.00	59039.25	
Nižná Slaná	Hg	0.94	0.43	2.00	0.71	0.49	0.50
	Cd	0.83	0.50	1.90	0.50	0.46	0.70
	Pb	47.36	23.00	136.00	33.00	33.20	70.00
	Ni	43.45	27.00	69.00	50.00	15.43	50.00
	As	165.19	23.00	924.00	40.00	270.66	25.00
	Fe	6.21	3.19	16.00	4.55	4.10	–
	Mn	0.35	0.12	1.08	0.20	0.35	–

*Act No. 220/2004, SD – Standard deviation, Fe (%)

For the evaluation of vegetation contamination, some species were selected on the basis of their importance in the studied sites. Pb, Zn, Cr, Mn, and Mg were analysed in above-ground plant organs. The highest contamination compared to the limit values was found for *Elytrigia repens*, where the zinc content exceeded the limit 47 times and the lead content 33 times. Zinc limits were also found in the dry matter of *Phragmites australis* (Table 2, Fig. 2).

Table 2 Values of heavy metals (mg/kg) in plant species (*Phragmites australis, Elytrigia repens, Agrostis stolonifera*) expressed by descriptive statistics

Parameter	Mean	Min	Max	Median	SD	Limit*
Cd	0.09	0.09	0.10	0.10	0.01	0.10
Pb	1.77	1.00	3.30	3.30	1.33	0.10
Zn	39.63	7.80	95.00	85.00	48.13	2.00
Cr	2.13	1.00	4.30	4.30	1.88	–
Mn	206.67	78.00	400.00	78.00	170.46	–
Mg	11162.33	5419.00	21208.00	21.21	8729.59	–

*Act No. 220/2004, SD – Standard deviation

The magnesium concentration in the test plants showed a very high content (Fig. 2). The highest magnesium content was found in *Elytrigia repens* (21,208 mg/kg) > *Phragmites australis* (6,860 mg/kg) > *Agrostis stolonifera* (5,419 mg/kg). High concentrations of manganese also occurred in *Elytrigia repens* (400 mg/kg) > *Agrostis stolonifera* (142 mg/kg) > *Phragmites australis* (78 mg/kg). Most plants are affected by Mn contents above 400 mg/kg. Any accumulation above 1000 mg/kg has been found for several other resistant species. Hyperaccumulative plants (*Phytoacca americana* L.) absorbed Mn from contaminated soil up to 13,400 mg/kg in leaves (Kabata–Pendias 2011). Excessive amounts of magnesium cause signs of toxicity. Plants that received 10,000 mg Mg^{2+}/kg died on Day 20 after the toxic dose and plants that received 5,000 mg Mg^{2+}/kg died on Day 45 (Venkatesan and Jayaganesh 2010).

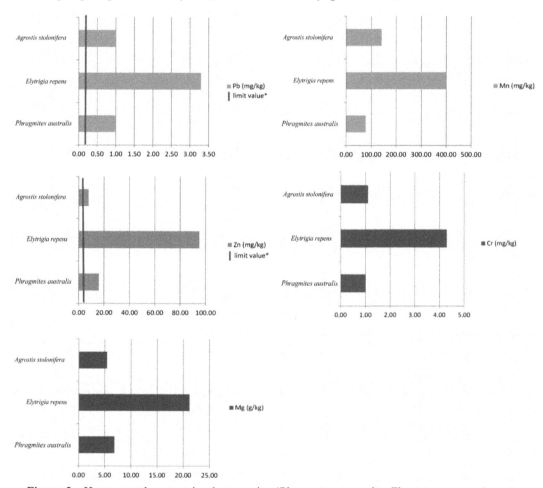

Figure 2 Heavy metal content in plant species (*Phragmites australis, Elytrigia repens, Agrostis stolonifera*) in the research areas Jelšava and Lubeník (Slovakia).

Phragmites australis can tolerate many habitat types and grow in areas of broad ecological amplitude, including wetlands and soil. Many studies have reported *Phragmites australis* as one of the best plant organisms for the detection and adsorption of heavy metals (Wang et al. 2015). It is often used to reduce metal concentrations in soil, sediment, and water (Bragato et al. 2006). *Phragmites australis* is also used to identify the presence of Cd, Cu, Pb, and Zn in the mouth of rivers, suggesting that it can be used as a bioindicator (Cicero–Fernandes et al. 2017, Bragato et al. 2006).

Long-term observations and investigations have shown that *Phragmites australis*, originally a hygrophilous plant, has appeared in dry contaminated areas of Jelšava and Lubeník (JL) where groundwater is at a depth of several metres. The strong vitality of the common reed has been documented in several locations where the pH value reached 8–9, which corresponds to strong alkalinisation of soil (Fazekaš et al. 2019). It is a dominant, invasive, resistant, and technically available species that provides an alternative solution for the sanitation and fertilization of alkaline soils. It can be considered a renewable natural resource because its energy–calorific value is comparable to lignite. Biomass from this plant is considered a suitable material for matting, cellulose, paper, as well as an additive to cattle nutrition (Hronec et al. 2012).

Table 3 shows the results of the heavy metal transfer factor (MTF) from soil to plant using the examples of the plant species *Phragmites australis*, *Elytrigia repens*, and *Agrostis stolonifera*. The MTF is considered an important parameter to assess the bioavailability of metals on a particular soil substrate (Garg et al. 2014). The highest MTF values for the metals evaluated (Cd, Pb, Zn, Cr, Mn and Mg) were found in *Elytrigia repens* compared to *Phragmites australis* and *Agrostis stolonifera*.

Table 3 The heavy metal transfer factor in soil/plant in Jelšava and Lubeník

Cd	*Pb*	*Zn*	*Cr*	*Mn*	*Mg*
<0.20	0.03–0.06	0.17–0.34	0.01	0.03–0.05	0.03–0.11
		Phragmites australis			
<0.20	0.09–0.19	1.02–1.98	0.03–0.05	0.17–0.25	0.11–0.60
		Elytrigia repens			
<0.20	0.03–0.06	0.08–0.16	0.01	0.06–0.09	0.03–0.15
		Agrostis stolonifera			

Agrostis stolonifera and *Elytrigia repens* can produce tolerant ecotypes and are characterized by high ecological valency and resistance to heavy metals (Banásová 2004, Fazekašová et al. 2016). *Elytrigia repens* is most resistant to Pb, Cr, Zn, Mn, Ni, and Cu contamination and is characterized by a low content of the monitored elements compared to other plant species (Minkina et al. 2017). Ranieri et al. (2013) reported that *Phragmites australis* has high potential for Cr adsorption from contaminated soil. Charlesworth et al. (2016) reported that *Agrostis stolonifera* has high storage potential for heavy metals. These findings will help to select the best species for tackling environmental pollution by pollutants.

2.5 Biochemical Parameters of Mining Degraded Soil

Soil reaction is considered one of the main chemical properties since it affects all biochemical reactions in the soil environment (Hohl and Varma 2010). The range of soil reaction values in the metallically contaminated area of Middle Spiš (MS) defines the soil as extremely acidic to weakly acidic and thereby an environment for easy passage of all monitored risk elements and subsequent accumulation of contaminants in plants. The main component of environmental pollution in Jelšava–Lubeník (JL) is magnesite powder. Mg drifts with a significant proportion of reactive caustic magnesite are aggressive in the natural environment. Even when in small amounts in contact with soil or plant moisture, they form saturated solutions with a highly alkaline pH value (Baluchová et al. 2011). Dust particles strongly influence the dynamic properties of soils, especially pH (Fazekasova et al. 2016). Continuous magnesite crust covers part of the soil and has influenced ecologically important soil functions, according to Wang et al. (2015). The range of soil reaction values indicates that soil is slightly acidic to strongly alkaline. Risks of the chemical degradation of soil in the emission area of iron ore mines in Nižná Slaná (NS) are represented by soil acidification and metallization. The acidification effect is accelerated by anthropogenic load by acidic emission of sulphur and nitrogen oxides. The range of soil reaction values indicates that the soil is extremely acidic to strongly alkaline (Fazekaš et al. 2018).

The soil oxidation–reduction potential (Eh – redox potential) is an important indicator of the soil environment, as the bioavailability and toxicity rate of chemicals often change during oxidation–reduction reactions. If the Eh value falls below 200 mV, reduction processes begin to develop in the soil (Husson et al. 2016). The measured values dropped below the given value mostly in the middle Spiš region (MS)—this indicates that reduction processes predominate in the monitored area.

Humified soil organic matter (Cox) is one of the main factors controlling the physical, chemical, and biological properties of soil; its quantity and composition affect the fertility functions of soil and also play an important role in soil hygiene (Fazekašová et al. 2014). The soils in the Spiš region (MS) have a very good supply of humus. In Jelšava and Lubeník (JL) and Nižná Slana (NS), the supply of humus in soil is moderate to very good. The monitored soil pH, Eh, and Cox parameters influence the release dynamics and the possible mobility of heavy metals, as pointed out by El-Naggar et al. (2018). The results of the soil's chemical parameters monitored, expressed by descriptive statistics, are shown in Table 4.

We used a BIOLOG® Eco plate to analyse changes in microbial communities. This method is known for its sensitivity and speed; it has been used inter alia for the ecotoxicological evaluation of contaminated soils (Tischer et al. 2008). To calculate the Shannon diversity index (Shannon 1948), we used the absorbance at a given AWCD (Average well colour development) in samples examined after 168 hours. AWCD is an important and sensitive indicator reflecting the metabolic profiles of the soil microbial community, especially in the presence of toxic heavy metals in the soil. The lowest average values of the metabolic activities of microorganisms (AWCD) were found in Jelšava and Lubeník (0.50 ± 0.13) and the highest in Nižná Slaná (0.98 ± 0.38).

Based on the results of the Shannon index H', we can conclude that the diversity at the surveyed sites in Middle Spiš (MS) and in the Jelšava and Lubeník areas (JL) is low to medium (2.3–3.1), while in the region of Nižná Slaná (NS) it is very low (0.9) to moderate (3.4) (Table 4). Greater diversity stabilizes the ecosystem's functional properties, which are more stable, productive, and more resistant to stress factors and disorders (Torsvik and Øvreas 2002). Our findings were consistent with the results of Xie et al. (2011), who found that a clear inhibitory effect on the functional activity of soil microorganisms occur with an increasing level of the metal content. On this basis, we can conclude that the soil ecosystem in a deteriorated environment is unstable.

Soil enzymatic activity is considered a microbial indicator of soil because its activity is closely related to important soil characteristics (Šarapatka and Kršková 1997). Enzymes catalyse biochemical reactions and form an integral part of the nutrient cycle in the soil ecosystem (Yang et al. 2016). Soil biochemical properties, including enzymatic activities, respond rapidly to anthropogenic environmental disorders (Paz–Ferreiro and Fu 2016). Many scientific studies have pointed out that enzymatic activity is an important biological indicator of soil quality and soil contamination rates as well as an indicator of soil contamination by heavy metals (Nannipieri et al. 2012, Trasar–Cepeda et al. 2008, Gao et al. 2010). Heavy metals are toxic on soil microorganisms; they inhibit microbial activities and change the diversity of microbial communities (Hu et al. 2014).

Until now, no quantitative standard has been set to assess the level of heavy metal contamination using the numerical values of soil enzyme activities (Yang et al. 2016). High enzyme activity represents good soil quality, while low activity may be related to the toxicity of pollutants on biological processes (Tang et al. 2019). The sudden exposure of microorganisms to heavy metals causes reduction in enzymatic activity. Later, the microorganisms adapt to the polluted environment—this usually results in enzymatic activity recovery. Currently, the enzyme catalase, dehydrogenase, urease, phosphatase, etc. are commonly used as bioindicators (Tang et al. 2019).

Soil enzymes have catalytic activities responsible for soil biochemical processes and are susceptible to heavy metal contamination. These are considered an excellent biological indicator of soil quality and pollution assessment. Soil enzyme activity is used as a reliable biological indicator to assess soil contamination (Wang et al. 2010).

One of the most widespread enzymes is urease. It is loosely present in the soil solution in the soil and more often is firmly bound to soil organic matter or clay particles. It plays an important role in cycle N (Tang et al. 2019). Soil ecosystem stability is influenced by soil depth, soil type, heavy metal content, pH, etc. (Fazekašová 2012). The mean value of soil urease activity was 0.08–0.10 mg $NH_4^+ - N \cdot g^{-1} \cdot 24$ h^{-1} in the studied areas.

Soil phosphatases (acid and alkaline phosphatases) refer to the key enzyme that hydrolyzes organic phosphate to an inorganic form, thereby increasing the soil phosphorus supply. Therefore, it plays a major role in soil phosphorus cycling (Wang et al. 2010, Nannipieri et al. 2011, Dick and Tabatabai 1983). Mean acid phosphatase (ACF) values ranged within 76.23–230.48 µg P g^{-1} 3 h^{-1}. The lowest ACF values were found in the territory of Jelšava and Lubeník (JL), which is extremely contaminated with Cr, Mg, and Mn. Here soils are mostly alkaline. Similar results were found for alkaline phosphatase (ALP), with mean values ranging within 108.88–199.97 µg P g^{-1} 3 h^{-1}. Our studies have shown that soil enzyme activity decreases when the heavy metal content reaches critical levels.

Table 4 Measured values of selected biochemical parameters of the mining–degraded soil expressed by descriptive statistics

Area	Parameter	Mean	Min	Max	Median	SD
Middle Spiš	pH/KCl	4.91	3.83	6.09	4.89	0.73
	Eh (mV)	176.75	17.00	363.00	155.00	127.46
	Cox (%)	5.59	3.99	8.78	5.20	1.66
	AWCD	0.57	0.14	1.03	0.57	0.29
	H′	2.9	2.4	3.1	3.0	0.3
	URE (mg $NH_4^+ - N \cdot g^{-1}$ 24 h^{-1})	0.10	0.01	0.24	0.09	0.07
	ALP (µg P g^{-1} 3 h^{-1})	199.97	22.10	336.65	222.71	90.42
	ACP (µg P g^{-1} 3 h^{-1})	230.48	62.57	321.15	258.16	80.56
Jelšava and Lubeník	pH/KCl	7.99	6.50	9.15	7.82	0.98
	Eh (mV)	162.50	83.00	233.00	175.50	52.90
	Cox (%)	3.69	1.27	6.30	3.33	1.66
	AWCD	0.55	0.41	0.95	0.50	0.16
	H′	2.7	2.3	3.1	2.8	0.2
	URE (mg $NH_4^+ - N \cdot g^{-1}$ 24 h^{-1})	0.08	0.03	0.15	0.06	0.03
	ALP (µg P g^{-1} 3 h^{-1})	108.88	14.64	251.13	96.43	55.57
	ACP (µg P g^{-1} 3 h^{-1})	76.23	11.19	210.08	74.05	39.69
Nižná Slaná	pH/KCl	5.99	4.25	8.75	5.77	1.34
	Eh (mV)	228.36	65.00	334.00	213.00	72.06
	Cox (%)	3.18	1.40	9.80	2.28	2.36
	AWCD	0.96	0.01	1.37	0.98	0.38
	H′	2.98	0.9	3.4	3.2	0.7
	URE (mg $NH_4^+ - N \cdot g^{-1}$ 24 h^{-1})	0.10	0.06	0.20	0.08	0.04
	ALP (µg P g^{-1} 3 h^{-1})	197.92	67.73	373.96	178.23	86.71
	ACP (µg P g^{-1} 3 h^{-1})	224.79	98.73	312.26	240.94	73.85

Notes: Cox – organic carbon, Eh – redox potential, AWCD – Average Well Colour Development, *H′* – Shannom index, SD – Standard deviation, URE – Urease, ALP – Alkaline phosphatase, ACP – Acid phosphatase

Many studies (Hu et al. 2014, Papa et al. 2010, Kızılkaya et al. 2004) confirmed there is a strong correlation between soil enzyme activity (URE, ALP, ACP) and the heavy metal content, suggesting that heavy metals have toxic effects on microbial processes. A negative correlation between phosphatase activity and the Cd and As contents was confirmed in our study. Positive correlations were found between urease activity and Hg, Ni, and Mn contents (Table 5). Soil enzyme responses to heavy metal stress are complicated. The synergistic and antagonistic effects depend on the type and content of heavy metals as well as the physico-chemical properties of the soils, and therefore, they require greater attention by scientists (Tang et al. 2019).

Table 5 Spearman correlations between the concentrations of heavy metals and biochemical soil parameters

Parameter	Hg	Cd	Pb	Cr	Zn	Cu	Ni	As	Mn	Mg
pH/KCl	−0.70**	−0.27	0.18	0.31	−0.15	−0.45*	−0.37	−0.42*	−0.106	0.52**
Cox	0.31	0.18	0.30	0.10	0.39*	0.38*	0.13	0.28	0.21	0.36
Eh	0.08	0.34	0.26	−0.40*	0.09	0.01	−0.09	0.10	−0.42*	−0.20
AWCD	0.26	−0.06	−0.08	−0.13	−0.33	0.13	0.40*	0.29	0.56**	−0.15
H′	0.33	0.01	0.04	−0.29	−0.21	0.15	0.34	0.26	0.48**	−0.32
URE	0.39*	−0.27	−0.10	0.10	−0.04	0.20	0.38*	0.08	0.62**	−0.10
ALP	−0.06	−0.51**	−0.08	0.56**	0.18	−0.21	−0.15	−0.42*	0.04	0.01
ACP	0.01	−0.47**	−0.14	0.52**	0.07	−0.18	−0.07	−0.39*	0.10	−0.01

Notes: Cox – organic carbon, Eh – redox potential, AWCD – Average Well Colour Development, H′ – Shannom index, URE – Urease, ALP – Alkaline phosphatase, ACP – Acid phosphatase, $P < 0.01$**, $P < 0.05$*

Using eight biochemical parameters and eight heavy metals as variables, cluster analysis was used to describe the difference in soil–monitoring points. The results indicated that the points were categorized into four groups (Fig. 3). Clusters 1 and 2 include metallic contaminated points (NS10, SS1, SS3, SS5–SS8), Cluster 3 includes acidification contaminated points (SS2, SS4, NS1–NS9), and Cluster 4 includes alkaline contaminated points (JL1–JL12). The cluster analysis showed the diversity of the areas examined.

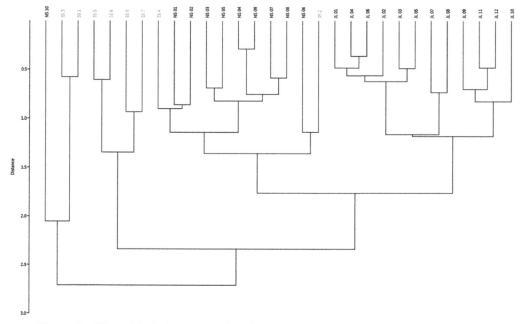

Figure 3 Hierarchical cluster analysis using biochemical parameters and heavy metals.

2.6 Environmental Options for Reducing Soil Toxicity

The ability of metal binding is one of the most important properties of humic substances. In natural systems, these substances can bind metal pollutants and significantly affect transport phenomena, toxicity, regeneration, and cleaning processes. They have application in crop and animal production for increasing the productivity. They are natural substances and environment-friendly (Skybová 2006).

In recent years, the remediation of the contaminated environment is at the centre of attention because soil and water contaminated with toxic substances pose a serious public health hazard. Activated carbon is one of the most widely used sorbents to remove contaminants due to a high sorption capacity (Pipíška and Remenárová 2014, Fazekašová et al. 2016).

As a sorbent, we used 100% natural substance HUMAC Enviro made from net source oxihumolite (brown coal) in concentrations of 2%. The active substance is a humic acid that has a high absorption capacity of binding to each different toxic substance. The content of humic acids in the dry matter of the used preparation was 62% (Fazekašová et al. 2016).

Extraction and processing of raw materials for the production of copper production, the electrotechnical industry, and the production of alloys are among the most important sources of copper in the environment. The natural content in soils is 20–30 mg/kg. The recommended limit for decontamination is 1500 mg/kg for soil and 500 mg/l for groundwater. In contaminated soil, it occurs in the form of Cu^{2+} ions, often in complexes with organic ligands (Pipíška and Remenárová 2014). Copper is bound inaccessibly, mainly due to the formation of complexes with humic acids, which are characterized by a very slow dissolution rate and penetrate to those sites in the humic acid structure from where they are very difficult to release (Makovníková et al. 2006).

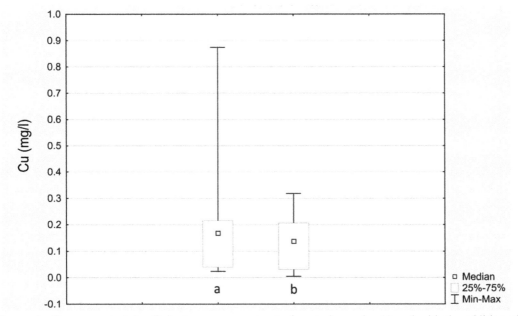

Figure 4 Copper content (mg/l) in the aqueous extract of control samples (a) and with the addition of 2% HUMAC Enviro after five days (b).

In the studied region of Middle Spiš (MS), copper has been mined and processed for a long time, and therefore, it reaches the highest value nationwide. The highest copper contamination was recorded in permanent grassland (1271 mg/kg). The results of experimental measurements showed that in the water extract, which was made from samples containing above-limit Cu contents, there is an above-limit Cu content in all samples (0.02–0.87 mg/l, limit 0.02 Government Regulation No. 269/2010 on surface water), which is 13 times the limit values. Subsequent application of 2% HUMAC Enviro in the aqueous extract resulted in 0.00–0.32 mg/l Cu after five days. The average value was 0.14 mg/l Cu and exceeded seven times. Compared with the controls, we found a 35% reduction in the Cu content (Fig. 4), confirming our assumption that an increasing level of the organic matter in the soil significantly increased metal sorption, especially sorption of metals with a high affinity for the organic matter. According to our findings, many studies have shown that natural humic acid substances are considered as highly potent sorbents of various contaminants (Cattani et al. 2009, Olu–Owolabi et al. 2010, Xu and Zhao 2013, Xu et al. 2013, Sounthararajah et al. 2015).

The structure and function of soil microbial communities are used as indicators of the soil quality, soil health, and soil fertility (Zhu et al. 2017). The effects of the natural substance HUMAC Enviro on the metabolic activity of microorganisms were tested under laboratory conditions.

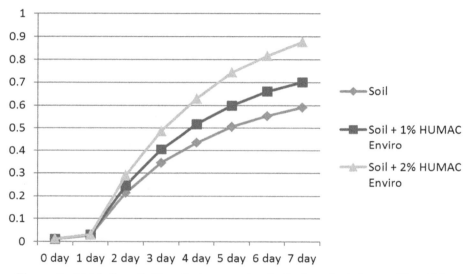

Figure 5   Metabolic activities of microorganism in the investigated areas (Slovakia).

By applying a 1% and 2% concentration of the HUMAC Enviro natural substance to fresh soil samples from the contaminated sites, we found positive effects on the increase in the metabolic activity of microorganisms (Fig. 5). The highest metabolic activity was found in the territory of Nižná Slaná after the addition of the 2% concentration. At the recommended dose of 400 kg/ha of HUMAC Enviro, the cost is €304/ha for regular application (every four to five years). Increased application (1,000 kg/ha) of this conditioner will result in a significantly higher metabolic activity of the soil microbial community as well as a simultaneous increase in metal sorption. It is important for farms operating in a contaminated area to improve the soil quality by increasing the biological activity by applying HUMAC Enviro. This helps them achieve an increasing level of heavy metal sorption, plant nutrient intake, and cultivated production.

3.   CONCLUSION

Decontamination of areas contaminated by toxic metals is one of the important research subjects of contemporary science. Cleaning these locations' established physicochemical methods is costly and too often non-ecological. Therefore, of late, some approaches have been formed to develop technologies by utilizing biological systems. Bioremediation is defined as the use of living organisms and their parts (e.g. enzymes) to eliminate or reduce the environmental hazards of toxic elements. Bioindicators are living organisms (plants, planktons, animals, and microbes) that provide information about the quality of the environment. The identification of bioindicators is critical to meeting soil protection objectives. Bioindication uses the properties of living organisms to indicate the state of the environment. The soil biota, biodiversity and its activity play a key role in soil ecosystem. Soil enzymatic activity, and the structure and function of microbial communities are important biological indicators of soil quality, soil contamination rates as well as of soil contamination by heavy metals. The results of this study on the environmental possibilities of soil degradation have shown that humic acid-based natural substances are effective sorbents of contaminants. The *Phragmites australis* is an appropriate environmental tool for restoring degraded soil.

Acknowledgments

The study was supported by VEGA 1/0313/19 Ecosystem approach as a parameter of the modern environmental research of contaminated areas and KEGA 011PU–4/2019 Implementation of environmental education and research into the teaching of management courses in the study programme management.

References

Abdu, N., A.A. Abdullahi and A. Abdulkadir. 2017. Heavy metals and soil microbes. Environ. Chem. Lett. 15: 65–84.

Act of the National Council of the Slovak Republic No. 220/2004 Coll.

Ahumada, I. and J. Mendoza. 2001. Effect of acetate, citrate and lactate in corporation on distribution of Cd and Cu chemical forms in soils. Commun. Soil Sci. Plant Anal. 32: 771–785.

Baker, A.J.M., S.P. McGrath, C.M.D. Sidoli and R.D. Reeves. 1994. The possibility of in situ heavy metal decontamination of polluted soils using crops of metal–accumulating plants. Resour. Conserv. Recy. 11: 41–49.

Baker, A.J.M., S.P. McGrath, R.D. Reeves and J.A.C. Smith. 2000. Metal hyperaccumulator plants: A review of the ecology and physiology of a biological resource for phytoremediation of metal-polluted soils. *In:* N. Terry and G. Bañuelos (eds), Phytoremediation of Contaminated Soil and Water. CRC Press, Boca Raton, Florida, USA, pp. 85–108.

Baluchová, B., P. Bačík, P. Fejdi and M. Čaplovičová. 2011. Mineralogický výskum prašných spadov z rokov 2006–2008 v oblasti Jelšavy/Mineralogical research of the mineral dust fallout from the years 2006–2008 in the area of Jelšava (Slovak Republic). Mineral. Slov. 43: 327–334.

Banásová, V. 2004. The unique vegetation on old mining dumps/Unikátna vegetácia na starých banských haldách. Protected areas in Slovakia – Plant species protection/Chránené územia Slovenska – druhová ochrana rastlín. 62: 42–43.

Barančíková, G. and J. Makovníková. 2003. The influence of humic acid quality on the sorption and mobility of heavy metals. Plant Soil Environ. 49: 565–571.

Beneš, S. 1994. Obsahy a bilance prvku ve sférach ŽP/Contents and balance of the element in the spheres of the environment. Ministerstvo zemědelství ČR, Praha.

Bragato, C., H. Brix and M. Malagoli. 2006. Accumulation of nutrients and heavy metals in *Phragmites australis* (Cav.) Trin. ex Steudel and *Bolboschoenus maritimus* (L.) Palla in a constructed wetland of the Venice lagoon watershed. Environ. Pollut. 144: 967–975.

Cattani, I.H., B.M. Beone, A.A. Del Re, R. Boccelli and M.J. Trevisan. 2009. The role of natural purified humic acids in modifying mercury accessibility in water and soil. J. Environ. Qual. 38: 493–501.

Cicero–Fernandes, D., M. Peña–Fernandez, J.A. Expósito–Camargo and B. Antizar–Ladislao. 2017. Long term (two annual cycles) phytoremediation of heavy metal–contaminated estuarine sediments by *Phragmites australis*. New Biotechnol. 38: 56–64.

Čurlík, J. and P. Šefčík. 1999. Geochemický atlas Slovenskej republiky/Geochemical atlas of the Slovak Republic. SSCRI, Bratislava.

Dercová, K., M. Čertík, A. Maľová and Z. Sejáková. 2004. Effect of chlorophenols on the membrane lipids of bacterial cells. Int. Biodeter. Biodegr. 54: 251–254.

Dick, W.A. and M.A. Tabatabai. 1983. Activation of soil pyrophosphatase by metal ions. Soil Biol. Biochem. 15: 359–363.

Dohovor o biologickej diverzite/Convention on Biological Diversity. 1992. CBD, Rio de Janeiro, Brazil.

Doran, J.W. and M.R. Zeiss. 2000. Soil health and sustainability: Managing the biotic component of soil quality. Appl. Soil Ecol. 15: 3–11.

Ďurža, O. 2003. Utilization of the soil magnetometry for study of heavy metals soil contamination. Acta geol. univ. comen. 58: 29–55.

El-Naggar, A., S.M. Shaheen, Y.S. Ok and J. Rinkleb. 2018. Biochar affects the dissolved and colloidal concentrations of Cd, Cu, Ni,and Zn and their phytoavailability and potential mobility in a mining soilunder dynamic redox–conditions. Sci. Total Environ. 624: 1059–1071.

Fazekaš, J., D. Fazekašová, P. Adamišin, P. Huličová and E. Benková. 2019. Functional diversity of microorganisms in metal– and alkali–contaminated soils of Central and North–eastern Slovakia. Soil Water Res. 14: 32–39.

Fazekašová, D. 2012. Evaluation of soil quality parameters development in terms of sustainable land use. *In*: S. Curkovic (ed.), Sustainable Development—Authoritative and Leading Edge Content for Environmental Management. InTech, Rijeka, Croatia, pp. 435–458..

Fazekašová, D., G. Barančíková, S. Torma, M. Ivanová and P. Manko. 2014. Chemické a environmentálne aspekty zložiek životného prostredia a krajiny/Chemical and environmental aspects of the environment and landscape. University of Prešov, Prešov.

Fazekašová, D., Z. Boguská, J. Fazekaš, J. Škvareninová and J. Chovancová. 2016. Contamination of vegetation growing on soil and substrates in the unhygienic region of central Spiš (Slovakia) polluted by heavy metals. J. Environ. Biol. 37: 1335–1340.

Fazekašová, D., J. Fazekaš, P. Huličová, E. Benková, M. Karahuta. 2016. The effect of natural substance HUMAC Enviro on reducing the concentration of cooper and mercury in contaminated soils. Int. J. Metall. Mater. Eng. 2: 4.

Fiala, K., G. Barančíková, V. Brečková, V. Burik, B. Houšková, A. Chomaničová, et al. 1999. Záväzné metódy rozborov pôd/Binding methods of soil analysis. SSCRI, Bratislava.

Gao, Y., P. Zhou, L. Mao, Y. Zhi and W. Shi. 2010. Assessment of effects of heavy metals combined pollution on soil enzyme activities and microbial community structure: Modified ecological dose–response model and PCR–RAPD. Environ. Earth Sci. 60: 603–612.

Garg, V.K., P. Yadav, S. Mor, B. Singh and V. Pulhani. 2014. Heavy metals bioconcentration from soil to vegetables and assessment of health risk caused by their ingestion. Biol. Trace Elem. Res. 157: 256–265.

Garland, J.L. 1996. Analysis and interpretation of community–level physiological profiles in microbial ecology. FEMS Microbiol. Ecol. 24: 289–300.

Gömöryová, E. 2008. Diversity of soil microorganisms. *In:* B. Pálka (ed.), Piate Pôdoznalecké Dni: Pôda – Národné Bohatstvo. SSCRI, Bratislava, Slovakia, pp. 155–160.

Grecula, P., A. Abonyi, M. Abonyiová, J. Antáš, B. Bartalský, J. Bartalský, et al. 1995. Ložiská nerastných surovín Slovenského Rudohoria/Mineral deposits of Slovak Ore Mountains. Miner. Slov. Monogr., Bratislava.

Grejtovský, A. 1991. Effect of soil improve processes to enzymatic activity of heavy alluvial soil. Rostlinná Výroba. 37: 289–295.

Hančuľák, J., M. Bobro, E. Fedorová, O. Šestinová, J. Brehuv, T. Špaldon, et al. 2007. Monitoring depozície ťažkých kovov z prašného spadu v oblasti pôsobenia železorudného banského závodu v Nižnej Slanej. *In:* K. Střelcová, J. Škvarenina and M. Blaženec (eds), Bioclimatology and Natural Hazards. International Scientific Conference, Poľana nad Detvou, Slovakia, pp. 1–7. Online.

Havlicek, E. 2012. Soil biodiversity and bioindication: From complex thinking to simple acting. Eur. J. Soil Biol. 49: 80–84.

Hegedüsová, A., J. Švikruhová, P. Boleček and O. Hegedüs. 2008. Dekontaminácia pôd využitím techniky fytoextrakcie/Soil decontamination using phytoextraction technique. *In:* T. Bubeníková (ed.), Earth in a Trap? 2008: Analysis of Environmental Components. International Scientific Conference, Technical University in Zvolen, pp. 193–197.

Hofman, J., J. Švihálek and I. Holoubek. 2004. Evaluation of functional diversity of soil microbial communities – a case study. Plant Soil Environ. 50: 141–148.

Hohl, H. and A. Varma. 2010. Soil: The living matrix. *In:* I. Sherameti and A. Varma (eds), Soil Heavy Metals, Soil Biology. Springer–Verlag Berlin, Heidelberg, Germany, pp. 1–18.

Hronec, O., J. Vilček, P. Adamišin, P. Andrejovský and E. Huttmanová. 2012. Use of *Phragmites australis* (Cav.) trin and its reproduction in the revitalization of contaminated soils. J. Prod. Eng. 15: 107–111.

Husson, O., B. Husson, A. Brunet, D. Babre, K. Alary, J.P. Sarthou, et al. 2016. Practical improvements in soil redox potential (Eh) measurement forcharacterisation of soil properties. Application for comparison of conventional and conservation agriculture cropping systems. Anal. Chim. Acta. 906: 98–109.

Hu, X.F., Y. Jiang, Y. Shu, X. Hu, L. Liu and F. Luo. 2014. Effects of mining wastewater discharges on heavy metal pollution and soil enzyme activity of the paddy fields. J. Geochem. Explor. 147: 139–150.

Charlesworth, S.M., J. Bennett and A. Waite. 2016. An evaluation of the use of individual grass species in retaining polluted soil and dust particulates in vegetated sustainable drainage devices. Environ. Geochem. Health. 38: 973–985.

Chazijev, F.Ch. 1976. Fermentativnaja aktivnost' počv. Metodičeskoje Posobje. Moskva.

Chopin, E.I.B. and B.J. Alloway. 2007. Distribution and mobility of trace elements in soils and vegetation around the mining and smelting areas of Tharsis, Riotinto and Huelva, Iberian Pyrite Belt, SW Spain. Water Air Soil Poll. 182: 245–261.

Jurko, A. 1990. Ekologické a socioekonomické hodnotenie vegetácie / Ecological and socioeconomical assessment of vegetation. Príroda, Bratislava.

Kabata–Pendias, A. and H. Pendias. 2001. Trace Elements in Soil and Plants. CRC Press, London.

Kabata–Pendias, A. 2011. Trace Elements in Soils and Plants, 4th Ed. Taylor & Francis Group, Boca Raton, London New York.

Kafka, Z. and J. Punčochářová. 2002. Toxicity of Heavy Metals in Nature. Chemické Listy. 96: 611–617.

Kashem, A., B.R. Singh and S. Kawai. 2007. Mobility and distribution of cadmium, nickel and zinc in contaminated profiles from Bangladesh. Nutr. Cycl. Agroecosyst. 77: 187–198.

Khan, S., E.A. Hesham, Q. Min, R. Shafiqur and J. He. 2010. Effects of Cd and Pb on soil microbial community structure and activities. Environ. Sci. Pollut. Res. Int. 17: 288–96.

Kızılkaya, R., T. Aşkin, B. Bayraklı and M. Sağlam. 2004. Microbiological characteristics of soils contaminated with heavy metals. Eur. J. Soil Biol. 40: 95–102.

Klinda, J., T. Mičík, M. Némethová and M. Slámková. 2016. Environmentálna regionalizácia Slovenskej republiky/Environmental regionalization of the Slovak Republic. Ministerstvo životného prostredia, Bratislava.

Kobza, J., G. Barančíková, L. Čumová, R. Dodok, K. Hrivňaková, J. Makovníková, et al. 2009. Soil Monitoring of Slovak Republic. SSCRI, Bratislava.

Mahmood, T. 2010. Review Phytoextraction of heavy metals – the process and scope for remediation of contaminated soils. Soil Environ. 29: 91–109.

Makovníková, J., G. Barančíková, P. Dlapa and K. Dercová. 2006. Inorganic contaminants in soil ecosystems. Chemické listy. 100: 424–432.

Minkina, T.M., S.S. Mandzhieva, V.A. Chaplygin, T.V. Bauer, M.V. Burachevskaya, D.G. Nevidomskaya, et al. 2017. Content and distribution of heavy metals in herbaceous plants under the effect of industrial aerosol emissions. J. Geochem. Explor. 174: 113–120.

Morel, J.L. 1997. Bioavailability of trace elements to terrestrial plants. *In:* J. Tarradellas, G. Bitton and D. Rossel (eds), Soil Ecotoxicology. CRC Press, Boca Raton, Florida, USA, pp. 141–176.

Nannipieri, P., L. Giagnoni, L. Landi, and G. Renella. 2011. Role of phosphatase enzymes in soil. *In:* E.K. Bunemann et al. (eds), Phosphorus in Action, Soil Biology 26. Springer–Verlag Berlin, Heidelberg, Germany, pp. 215–243.

Nannipieri, P., L. Giagnoni, G. Renella, E. Puglisi, B. Ceccanti, G. Masciandaro, et al. 2012. Soil enzymology: Classical and molecular approaches. Biol. Fert. Soils 48: 743–762.

Noskovič, J., J. Chlpík, L. Jedlovská, M. Lacko–Bartošová, L. Nozdrovický, P. Ondrišík, et al. 2012. Ochrana a tvorba životného prostredia/Protection and Creating of the Environment. Slovak University of Agriculture in Nitra, Nitra.

Olu–Owolabi, B.I., D.B. Popoola and E.I. Unuabonah. 2010. Removal of Cu^{2+} and $Cd^{2+}$from aqueous solution by bentonite clay modified with binary of goethite and humic acid. Water Air Soil Pollut. 211: 459–474.

Papa, S., G. Bartoli, A. Pellegrino and A. Fioretto. 2010. Microbial activities and trace element contents in an urban soil. Environ. Monit. Assess. 165: 193–203.

Paz–Ferreiro, J. and S. Fu. 2016. Biological Indices for Soil Quality Evaluation: Perspectives and Limitations. Land Degrad. Develop. 27: 14–25.

Piatrik, M., V. Kollár, J. Marenčáková and Z. Juríčková. 2007. Základy Environmentalistiky/Basics of Environmental Studies. University of Central Europe, Skalica.

Pipíška, M. and L. Remenárová. 2014. Environmentálne biotechnológie. Biosorpcia toxických látok/ Environmental biotechnology. Biosorption of toxic substances. University of Ss. Cyril and Methodius, Faculty of Natural Sciences, Trnava.

Ranieri, E., U. Fratino, D. Petruzzelli and A.C. Borges. 2013. A comparison between *Phragmites australis* and *Helianthus annuus* in chromium phytoextraction. Water Air Soil Poll. 224: 1465–1474.

Sastre, J., A. Sahuquillo, M. Vydal and G. Rauert. 2002. Determination of Cd, Cu, Pb and Zn in environmental samples: Microwave–assisted total digestion versus aqua regia and nitric acid extraction. Anal. Chim. Acta. 462: 59–72.

Shannon, C.E. 1948. A mathematical theory of communication. Bell Sys. Tech. J. 27: 379–423.

Schwarzová, H., O. Hronec, P. Adamišin, L. Bekeová, E. Huttmanová and E. Zabavníková. 2011. Monitoring a sanácia v životnom prostredí/Monitoring and remediation in the environment. University of Central Europe, Skalica.

Skybová, M. 2006. Humic acids – the contribution to environmental research. Acta Montan. Slovaca. 11: 362–366.

Sounthararajah, D.P., P. Loganathan, J. Kandasamy and S. Vigneswaran. 2015. Effects of humic acid and suspended solids on the removal of heavy metals from water by adsorption onto granular activated carbon. Int. J. Environ. Res. Public Health. 12: 10475–10489.

Šarapatka, B. and M. Kršková. 1997. Interactions between phosphatase activity and soil characteristic from some locations in the Czech Republic. Rostlinná výroba. 43: 415–419.

Šefčík, P., S. Pramuka and A. Gluch. 2008. Assesment of soil contamination in Slovakia according index of geoaccumulation. Agriculture (Poľnohospodárstvo). 54: 119–130.

Tang, J., J. Zhang, L. Ren, Y. Zhou, J. Gao, L. Luo, et al. 2019. Diagnosis of soil contamination using microbiological indices: A review on heavy metal pollution. J. Environ. Manage. 242: 121–130.

Tischer, S., H. Tannaberg and G. Guggenberger. 2008. Microbial parameters of soils contaminated with heavy metals: Assesment for ecotoxicological monitoring. Pol. J. Ecol. 56: 471–479.

Torsvik, V. and L. Øvreås. 2002. Microbial diversity and function in soil: From genes to ecosystems. Curr. Opi. Microbiol. 5: 240–245.

Trasar-Cepeda, C., M.C. Leirós and F. Gil-Sotres. 2008. Hydrolytic enzyme activities in agricultural and forest soils. Some implications for their use as indicators of soil quality. Soil Biol. Biochem. 40: 2146–2155.

Venkatesan, S. and S. Jayaganesh. 2010. Characterisation of magnesium toxicity, its influence on amino acid synthesis pathway and biochemical parameters of tea. Res. J. Phytochem. 4: 67–77.

Wang Q., J. Dai, Y. Yu, Y. Zhang, T. Shen, J. Liu, et al. 2010. Efficiencies of different microbial parameters as indicator to assess slight metal pollutions in a farm field near a gold mining area. Environ. Monit. Assess. 161: 495–508.

Wang, L., P. Tai, Ch. Jia, X. Li and X. Xiong. 2015. Magnesium contamination in soil at a magnesite mining region of liaoning province, China. Bull. Environ. Contam. Toxicol. 95: 90–96.

Xie, X., M. Liao, A. Ma and H. Zhang. 2011. Effect of contamination of single and combined cadmium and mercury on the soil microbial community structural diversity and functional diversity. Chin. J. Geochem. 30: 366–374.

Xu, R. and A. Zhao. 2013. Effect of biochars on adsorption of Cu(II), Pb(II) and Cd(II) by three variable charge soils from southern China. Environ. Sci. Pollut. Res. 20: 8491–8501.

Xu, X., X. Cao, L. Zhao, H. Wang, H. Yu and B. Gao. 2013. Removal of Cu, Zn and Cd from aqueous solutions by the dairy manure–derived biochar. Environ. Sci. Pollut. Res. 20: 358–368.

Yang, J.S., F.L. Yang, Y. Yang, G.L. Xing, Ch.P. Deng, Y.T. Shen, et al. 2016. A proposal of "core enzyme" bioindicator in long–term Pb–Zn ore pollution areas based on topsoil property analysis. Environ. Pollut. 213: 760–769.

Zhu, L.X., Q. Xiao, Y.F. Shen and S.Q. Li. 2017. Microbial functional diversity responses to 2 years since biochar application in silt-loam soils on the Loess Plateau. Ecotox. Environ. Safe. 144: 578–584.

CHAPTER 3

Nanobioremediation of Contaminated Agro-ecosystems: Applications, Challenges and Prospect

Busiswa Ndaba[1], Maryam Bello-Akinosho[1*],
Ashira Roopnarain[1], Emomotimi Bamuza-Pemu[1],
Rosina Nkuna[2], Haripriya Rama[1] and Rasheed Adeleke[3]

[1]Microbiology and Environmental Biotechnology Research Group, Agricultural Research Council – Soil, Climate and Water (ARC-SCW), Arcadia, Pretoria 0001, South Africa.

[2]Institute for the Development of Energy for African Sustainability (IDEAS), University of South Africa's College of Science, Engineering and Technology Florida, 1710, South Africa.

[3]Unit for Environmental Science and Management, North-West University (Potchefstroom Campus), Potchefstroom 2520, South Africa.

1. INTRODUCTION

The industrial revolution with its associated scientific, technological, social and economic advances has transformed the course of humanity. However, the multitude of technological innovations that were implemented since the 19th century did not take into account environmental implications. Resources, such as coal and oil, were excessively extracted, processed and disseminated with minimal waste management strategies in place. This has resulted in toxic waste contamination of water and soil reserves as well as the atmosphere (Tejada et al. 2011, Cecchin et al. 2017, Tripathi et al. 2018). Apart from contamination of the environment by waste, certain ecosystems, such as agro-ecosystems, are predisposed to contamination by the amendments that are used to stimulate plant growth and protect crops, for example excess chemical fertilizers, pesticides and herbicides (Tejada et al. 2011). With the escalating human population and dwindling arable land, it is imperative that existing agro-ecosystems be protected to ensure that sufficient food will be produced to support the growing population. The escalating environmental, social and

*Corresponding author: belloakinoshoM@arc.agric.za

economic implications associated with improper waste disposal and excessive use of chemical agricultural inputs have motivated the search for environmentally friendly methods to remediate agro-ecosystems and protect the natural environment (Tripathi et al. 2018).

Several remediation methods have been developed for both *in situ* and *ex situ* applications. Irrespective of the location of remediation, sustainability of the remediation process has received much attention in recent times. Sustainable remediation involves the reduction in the contaminant/ pollutant to concentrations that do not pose any risk, in addition to minimising any associated detrimental environmental impacts, such as waste generation, greenhouse gas emissions and the consumption of natural resources (Reddy and Adams 2010). Furthermore, for any remediation process to be sustainable, it is imperative that the process is economically viable and socially acceptable (Wei et al. 2009). Remediation methods that are presently in use include physical, chemical and biological approaches. Each method of remediation has associated advantages and disadvantages (Ingle et al. 2014). Physical and chemical remediation methods result in rapid decontamination; however, these processes are generally cost intensive and may result in secondary contamination as well as disturbance of the environment that is being treated (Xia et al. 2019). Due to the disadvantages associated with physical and chemical remediation, focus has been drawn to more sustainable remediation methods such as bioremediation.

Bioremediation refers to the application of living organisms, such as plants and microorganisms, for the degradation/transformation of pollutants from contaminated environments into non-toxic or less toxic substances (Ingle et al. 2014, Verma and Sharma 2017). Whilst bioremediation has numerous associated advantages such as cost effectiveness, environmental preservation and sustainability (Cecchin et al. 2017, Davis et al. 2017, Jagdale et al. 2018), there are some limitations to the process. For effective bioremediation, extensive treatment duration may be necessary and the process may not be feasible on sites where the contaminant concentrations are at levels that are toxic to bioremediating agents (Cecchin et al. 2017, Davis et al. 2017). Furthermore, it has been established that oftentimes, the use of a single remediation method may not result in complete remediation of the contaminated site (Rao et al. 2001). The integration of remediation methods would result in overcoming the disadvantages/limitations associated with each method while maximising the benefits (Singh et al. 2020). One such integrated method that has received much attention of late is nanobioremediation.

While the use of organisms in environmental clean-up has long been established, the use of nanoparticles (NPs) in remediation is still relatively new (Kumari and Singh 2016). Nanoparticles are particles that range in size from 1 to 100 nm and portray properties that may differ from those of bulk samples of the same material that they are derived from (Auffan et al. 2009, Kumari and Singh 2016). Nanoparticles are either inorganic metal-based, organic polymeric-based or carbon-based (Guerra et al. 2018). Their minute sizes, unique properties, large surface area to volume ratios, elevated chemical reactivity and wide availability have warranted their assessment as remediating agents (Araújo et al. 2015, Kumari and Singh 2016). When NPs are applied for the detoxification and transformation of environmental contaminants, the process is known as nanoremediation (Karn et al. 2009). However, a hybrid technology that encompasses biological synthesis of NPs (biosynthesis) and remediation processes is known as nanobioremediation (Davis et al. 2017) (Fig. 1a). Biosynthesis of NPs is the preferred nanoparticle preparation method because physico-chemical preparation methods have associated disadvantages such as elevated cost implications, use of hazardous chemicals and generation of toxic by-products (Konishi 2006). Nanobioremediation has also been defined as the integration of two remediation technologies, i.e. nanoremediation and bioremediation (Cecchin et al. 2017, Singh et al. 2020). The second definition involves a two stage process, where NPs are introduced in phase one, to reduce contaminant levels to concentrations that are conducive to biodegradation processes. Biodegradation then commences, in phase two, to further reduce contaminant levels or the by-products of the first phase to concentrations that do not pose any environmental risk (Fig. 1b) (Cecchin et al. 2017). Both definitions are elaborated on from section 2 to 4 of the present chapter.

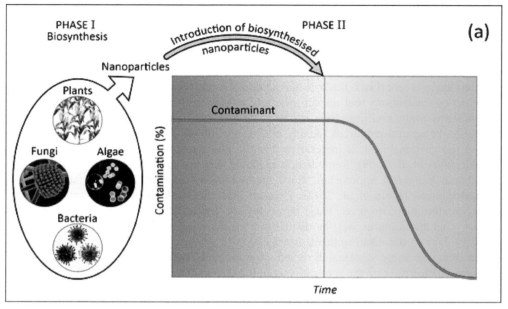

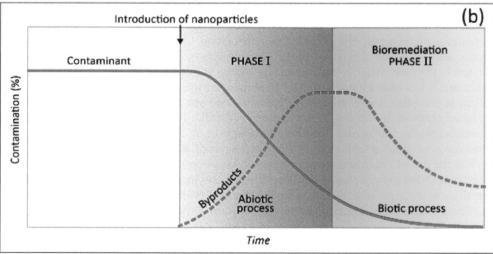

Figure 1(a, b)   Overview of nanobioremediation process
(Adapted and modified from Cecchin et al. 2017)

2.   THE COMBINED UTILIZATION OF NANOPARTICLES AND MICROORGANISMS FOR REMEDIATION

For some time, bioremediation has been used as an alternative to traditional remediation methods such as landfilling or incineration (Tabak et al. 2005), mainly because this method involves the use of suitable microorganisms in the correct amounts and combinations (Tabak et al. 2005). Bioremediation works by either transforming or degrading contaminants to non- or less hazardous chemicals (Murphy 2010, Singh et al. 2014). In biotransformation, the microorganisms are involved in the alteration of the molecular or atomic structure of a compound, whereas during biodegradation, the microorganisms breakdown the organic compounds into smaller organic or inorganic components (Tabak et al. 2005). Although bioremediation is environmentally friendly and cost-effective, the activity of microorganisms is usually limited by the toxicity of the contaminants treated (Gerhardt et al. 2009). In addition, long retention times are required for bioremediation processes to be effective (Cecchin et al. 2017). Therefore, the search for ways to improve the process is still on-going. Recently, the development of novel materials such as NPs for application in remediation strategies has also gained attention (Gong et al. 2018). These nanoparticles have been observed to have advantages over microbial remediation because they can tolerate higher contaminant concentrations and have the potential to be applied alongside microorganisms for increased efficiency (Fang et al. 2012).

Currently, the use of NPs to treat different kinds of environmental contaminants such as pesticides, organic solvents, fertilizers and heavy metals has been proven to be feasible. This has broadened the field of remediation for removing contaminants in the agro-ecosystems (Araújo et al. 2015, Dong et al. 2016, Choi et al. 2017, Jiang et al. 2018). Examples of elemental NPs include iron, nickel and palladium (Cecchin et al. 2017). Iron NPs are the most commonly used NPs in remediation due to their ability to perform as strong reducing agents during oxidation-reduction (redox) reactions (Galdames et al. 2017). When in contact with the contaminant, iron particles are rapidly oxidized, while reducing the contaminant by donating their electrons (Eqs. 1 and 2) (Babuponnusami and Muthukumar 2012, Chen et al. 2017). This process ensures that the contaminants are stabilized and less mobile with reduced toxicity. An example of iron nanoparticle used as remediating agent is found in a study by Tratnyek and Johnson (2006), where successful remediation of chlorinated hydrocarbons, metals and metalloids was reported using zero-valent irons (ZVI).

$$O_2 + Fe^0 + 2H^+ \rightarrow Fe^{2+} + H_2O_2 \tag{1}$$

$$Fe^{2+} + H_2O_2 \rightarrow Fe^{3+} + OH^- + \cdot OH \tag{2}$$

Apart from interacting with contaminants, the unique properties of NPs also allow them to interact with microbial communities in the environment (Galdames et al. 2017, Gong et al. 2018). The positive influence of NPs on the microbial community is related to their ability to bio-stimulate microbes such as methanogens and sulphate reducers (Weathers et al. 1997, Liu and Lowry 2006). For example, when nZVI are used for degradation purposes, hydrogen ions (H^+) are formed, implying that certain bacteria and methanogens can use nZVI as an electron donor (Weathers et al. 1997, Dinh et al. 2004). The negative influence of certain NPs is based on their toxicity towards microbes and this was evidenced by the inhibition of microbial growth that was observed during toxicity studies (Morones et al. 2005, Nel et al. 2006). The microbial cell inhibition in the presence of NPs occurs as a result of impaired permeability of the cell membrane (Nel et al. 2006, Cecchin et al. 2017). This is detrimental to the microbes as the cell membrane is responsible for transfer of substances in and out of the cell. If the cell membrane is impaired, cell starvation and subsequent death can occur due to blocked membrane preventing nutrients access. Furthermore, inability of the cell membrane to control what enters the cell would result in uncontrolled entry of toxic compounds, which would subsequently destroy the cell (Nel et al. 2006, Cecchin et al. 2017). However, it is worthy to note that the toxicity or bio-stimulation ability of the NPs is dependent on the nanoparticle type and concentration, in addition to type of microbe, soil, environmental condition and contaminants to be treated (Pawlett et al. 2013).

The ultimate goal in combining NPs and microorganisms for remediation is to have an efficient approach to treat contaminants (Table 1) (Fang et al. 2012).

For example, NPs are effective in reducing contaminant concentration through various mechanisms including adsorption, transformation and/or catalysis to a level suitable for biodegradation by microorganisms (Zou et al. 2016, Galdames et al. 2017). The efficiency of the combined remediation strategies has been reported by Tungittiplakorn et al. (2005), where NPs made from polyethylene glycol modified urethane acrylate (PMUA) precursor chain were

Table 1 Types of NPs used in combination with different microorganisms for improved removal of contaminants

Contaminants	NPs	Microorganisms	Efficiency	Mechanism of action	Reference
Lead	Iron oxide magnetic nanoparticles and calcium alginate	*Phanerochaete chrysosporium*	Removed 185.25 mg/g of lead	Adsorption	Xu et al. (2012)
2,3,7,8-tetrachlorodibenzo-p dioxin (2,3,7,8-TeCDD)	Pd/Fe NPs	*Sphingomonas wittichii*	100 % dechlorination of 2,3,7,8-TeCDD and 100% degradation of the end product dibenzo-p-dioxin after 10 hours	Dechlorination and degradation	Bokare et al. (2012)
Phenol	Nano zero-valent iron and Ni/Fe NPs	*Bacillus fusiformis*	Degradation of 94.3% at pH 8 with nZVI and 97.6% at pH 6 with Ni/Fe NPs after 7 hours	NPs serves as an electron donor to improve enzymatic degradation of phenol by bacteria	Kuang et al. (2013)
Crude oil	Iron oxide NPs	*Halomonas xianhensis* strain A-1 (I_1) and *Halomonas salifodinae* strain BC7 (I_4) as a consortium with biosurfactant	90% degradation of crude oil after 7 days	Biosurfactant increased solubility of crude oil, while NPs improved reaction rate for enzymatic degradation and utilization of oil by bacteria	El-Sheshtawy et al. (2017)
Crude oil	Polyvinylpyrrolidone-coated iron oxide NPs	*Halomonas* sp., *Vibrio gazogenes* and *Marinobacter hydrocarbonoclasticus* SP17, separately	100% degradation of crude oil	Addition of NPs increased reaction rate for enzymatic degradation and utilization of oil by bacteria	Alabresm et al. (2018)
Low-density polyethylene (LDPE)	Silicon dioxide NPs	*Bacillus* sp. strain V8 and *Pseudomonas* sp. strain C 2 5, separately	Increased efficiency reported as shifts in FT-IR λ-max	NPs improved bacterial growth and accelerated the hydrolysis of the polymer by bacteria	Pathak and Kumar (2017)
1,2,3,4,5,6-hexachloro-cyclohexane (γ-HCH)	Pd/Fe[0] bimetallic NPs (CMC-Pd/nFe[0])	*Sphingomonas* sp. strain NM05	Approximately 1.7–2.1 times greater efficiency compared to using bacteria alone	NPs provided an electron donor to bacteria, thus promoting its ability to dechlorinate γ-HCH	Singh et al. (2013)

found to enhance phenanthrene bioavailability for degradation by *Comamonas testosteroni* GZ39. This was also evidenced in a study conducted by Fang et al. (2012), where soil pollution was simulated with 2,4-dichlorophenoxyacetic acid (2,4-D) and the effect of iron oxide (Fe_3O_4) NPs combined with a group of soil indigenous microorganisms was investigated. Their results showed higher degradation efficiency as compared to the individual treatments of 2, 4-D with Fe_3O_4 NPs or indigenous microbes. As highlighted by the previous examples, a combination of NPs and microorganisms for remediation of environmental contaminants has been effectively employed; however, there is a dearth of information in this area. More research focus has been directed to the use of NPs in combination with plants for remediation purposes.

3. THE COMBINED UTILIZATION OF NANOPARTICLES AND PLANTS FOR REMEDIATION

Several studies have reviewed phytoremediation as a strategy for stabilization, extraction, volatilization, degradation and/or removal of contaminants, particularly from soils, through application of plants (Alkorta and Garbisu 2001, Rascio and Navari-Izzo 2011, Sarwar et al. 2017). These phytoremediation processes are described in Table 2.

Table 2   The description of key phytoremediation processes

Phytoremediation processes	Description	Reference
Phytoextraction	The plant roots take up contaminants where they accumulate and are sequestered in the plant tissues	Sarwar et al. (2017)
Phytostabilization	The roots of the plants decrease the mobility and bioavailability of contaminants within the rhizosphere	Bolan et al. (2011)
Phytovolatilization	Plants convert contaminants to a less toxic and volatile state which is released into the environment	Sarwar et al. (2017)
Phytodegradation	Plants take up contaminants and metabolize them to a less toxic form	Cunningham et al. (1996)

Although phytoremediation may be used for remediating contaminated sites, the process may take several years to completely remediate the site. The integration of nanotechnology with phytoremediation is a promising approach for treatment of soil contaminants as it has the potential to enhance phytoremediation efficiency due to the unique properties of the nano-sized materials (Siddiqui et al. 2015, Nwadinigwe and Ugwu 2018). Nanophytoremediation is an emerging technology being explored for the remediation of contaminants in the environment using the combination of NPs and plants (Srivastav et al. 2018). Nanoparticles may directly degrade or remove contaminants from the environment to make contaminants bioavailable and facilitate phytoremediation processes (Song et al. 2019). Nanoparticles may also penetrate plant tissues to enhance phytoremediation processes (Nwadinigwe and Ugwu 2018). This is possibly due to the high surface area to volume ratios of NPs, which may consequently improve the reactivity between the plant and NPs as well as their interaction with the contaminant (Guerra et al. 2018). The uptake of NPs by plants and their combined effectiveness in remediation are also determined by the morphology, chemical composition, concentration applied and the physico-chemical properties of the NPs (Srivastav et al. 2018). External environmental factors that may influence the efficiency of the nanophytoremediation process include temperature, pH, soil moisture content and soil microbial biodiversity (Srivastav et al. 2018). The characteristics of the plant such as plant type, root system and ability to accumulate contaminants as well as the properties of the contaminant such as solubility, mobility, reactivity and molecular weight also affect the nanophytoremediation process (Srivastav et al. 2018).

A recent study by Kumari and Khan (2018) reported that the fluoride accumulation efficiencies by honey mesquite (*Prosopis juliflora*) from contaminated soil was 34.13 mg/kg, whereas using the combination of the plant and iron oxide (Fe_3O_4) NPs improved accumulation efficiency of the plant to 63.07 mg/kg. Other nanophytoremediation studies are described in Table 3.

Table 3 Types of NPs used in combination with different plant species for improved removal of contaminants

Contaminants	NPs	Plant species	Efficiency	Mechanism of action	Reference
Hydrocarbons					
Endosulfan	nZVI NPs	Galangal (*Alpinia calcarata*)	≈100% removed after 28 days	Hydrogenolysis and dehalogenation	Pillai and Kottekottil (2016)
		Holy basil (*Ocimum sanctum*)	76.28% removed after 28 days		
		Lemon grass (*Cymbopogon citratus*)	86.16% removed after 28 days		
Fipronil	*Brassica* synthesized-Silver NPs		68.8% degradation efficiency	Dissipation through transformation and uptake	Romeh (2018)
	Plantago synthesized-Silver NPs	Broadleaf plantain (*Plantago major*)	54.64% degradation efficiency		
	Ipomoea synthesized-Silver NPs		43.75% degradation efficiency		
	Camellia synthesized-Silver NPs		30.99% degradation efficiency		
Trichloro-ethylene (TCE)	Fullerene NPs	Eastern cottonwood (*Populus deltoides*)	82% of TCE was taken up by the plant	Adsorption and uptake	Ma and Wang (2010)
Explosives					
Trinitro-toluene (TNT)	nZVI NPs	Guinea grass (*Panicum maximum*)	86.07-100% degradation of TNT	Transformation	Jiamjitrpanich et al. (2012)
Heavy metals					
Lead (Pb)	Nano-Hydroxyapatite (NHAP)	Ryegrass (*Lolium*)	2.86 – 21.1% reduction in roots; 13.19 – 20.3% reduction in shoots	Transformation	Ding et al. (2017)
Cadmium (Cd)	Titanium dioxide NPs	Soybeans (*Glycine max*)	507.6 µg/plant Cd was removed	Transformation	Singh and Lee (2016)
Arsenic (As)	Salicylic acid nanoparticles SANPs	*Isatis cappadocica*	705 mg/kg accumulation in the roots, 1188 mg/kg accumulation in the shoots	Transformation	Souri et al. (2017)

In addition to their remediation abilities, NPs play a role in improving plant growth and stress tolerance to indirectly enhance phytoremediation efficiency (Siddiqui et al. 2015). For example, NPs could possibly trigger the induction of nitric oxide synthesis, which subsequently activates stress tolerance genes and the plant defence system (Khan et al. 2017). Plant hormones involved in stress tolerance may also be augmented by NPs to activate signalling pathways involved in retaining plant growth plasticity (Bücker–Neto et al. 2017). Such beneficial effects of NPs improve the uptake of contaminants during phytoremediation. Khan and Bano (2016) reported the augmentation of abscisic acid (ABA) and gibberellic acid (GA) as well as bioaccumulation of Nickel (Ni) and Lead (Pb) in maize (*Zea mays* L.) by Silver (Ag) NPs. The high presence of ABA enabled the plant to tolerate heavy metal (Ni and Pb) contaminated environments. Gibberellic acid production led to the increased uptake of water and nutrients by the maize plant. High concentrations of Ni and Pb were shown to accumulate in the maize shoots through the application of AgNPs. Another study performed by Rizwan et al. (2019) found that application of zinc oxide and iron NPs reduced cadmium toxicity and thereby enhanced growth and development of cadmium-stressed wheat (*Triticum aestivum*). In a separate study, the energy crops, reed canary grass (*Phalaris arundinacea* L.) and rapeseed (*Brassica napus* L.) were used in combination with compost and nano silicon dioxide (SiO_2) fertilizer for the degradation of polycyclic aromatic hydrocarbons (PAHs) in soils (Włóka et al. 2019). The results from the study indicated that the addition of nano Silicon dioxide (SiO_2) improved the PAH degradation efficiency as well as plant biomass production. These studies indicate the benefits of incorporating NPs with plants for contaminant remediation as well as plant growth and stress tolerance.

On the contrary, it was also reported that NPs could have negative impacts on plant growth and phytoremediation abilities (Zhu et al. 2019). For instance, Song and Lee (2016) found that Zinc oxide (ZnO) NPs induced significant phytotoxicity in water thyme (*Hydrilla verticillata*) and common reed (*Phragmites Australis*). Varying concentrations of ZnO NPs induced different toxicity levels observed as reduced growth and influenced the absorption of nutrients and water by the plant structures (Song and Lee 2016). In another study, Fernandes et al. (2017) reported the negative impact of AgNPs on plant growth promoting microorganisms (PGPM) and phytoremediation processes in the presence of rhizosediment (sediments associated with plant roots). Aggregation of AgNPs with rhizosediment prevented efficient metal phytoremediation. Integration of metal-based NPs with phytoremediation also showed potential reduction of contaminant uptake from the environment and also prevented PGPM from indirectly promoting phytoremediation processes within the rhizosphere (Fernandes et al. 2017).

Despite negative effects associated with NPs on plants, nanophytoremediation as a treatment strategy shows great potential for the removal of contaminants from the environment. Research investigating the development of eco-friendly NPs that have minimal negative impacts on plant health, soil microbial diversity and/or groundwater quality is essential. Recent studies are exploring an environmentally friendly approach for the production of NPs through biosynthesis for remediation purposes.

4. NANOBIOREMEDIATION USING BIOSYNTHESIZED NANOPARTICLES

Biological synthesis of NPs for environmental remediation is considered eco-friendly because it is not associated with toxic chemicals which are conventionally used during chemical synthesis of NPs (Yadav et al. 2017). These biologically synthesized NPs can be produced from microorganisms and/or plants using various mechanisms (Rafique et al. 2017). In microbial synthesis, metallic ions attach onto the surface or are captured into the microbial cells where they are reduced by microbial enzymes into metal NPs (Singh et al. 2018). The two main

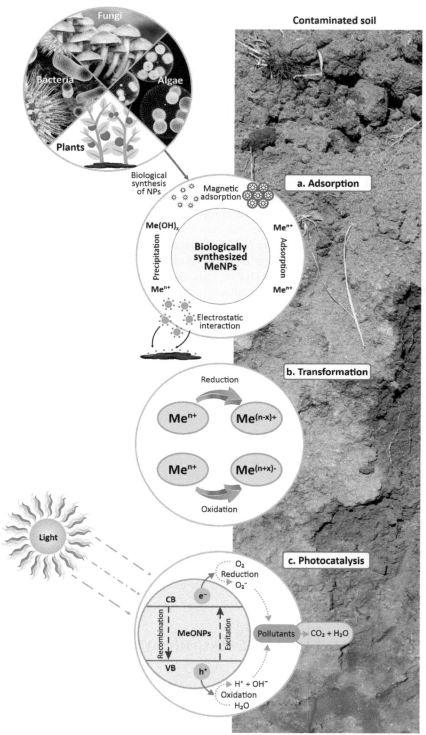

Figure 2   Mechanisms of contaminant removal by biologically synthesized nanomaterials
(Adapted with modifications from Das et al. (2018))

Legend: Metals (Me), Metal oxide nanoparticles (MeONPs) CB and VB represent conduction band and valence band, respectively.

features in the process are NADH (nicotinamide adenine dinucleotide) and NADH-dependent nitrate reductase (Ali et al. 2019). During plant extract-mediated nanoparticle synthesis, the plant extract is mixed with a metal precursor solution to reduce the metal ions in solution to metal NPs. Phytochemicals such as flavonoids, contained in the extract, have a potential to reduce metal ions in a shorter period in comparison to fungi and bacteria, which require longer incubation times (Prathna et al. 2010). Nanoparticles synthesized using different microorganisms or plants possess various sizes and morphologies, which can be exploited for diverse applications in pharmaceutical industries, wastewater treatments and agro-ecosystems (Wang et al. 2014, Rafique et al. 2017, Silva et al. 2017). However, this section only focuses on application of biologically synthesized NPs for remediation of different contaminants in the agro-ecosystems.

Nanobioremediation of contaminants in the environment using biologically synthesized NPs includes three major mechanisms, which are adsorption, transformation and catalysis (Fig. 2) (Das et al. 2017, 2018). During adsorption, contaminants bind onto the surface of NPs. This is termed an exothermic process because it is facilitated by ionic and surface interactions. During transformation, oxidation or reduction takes place leading to metal speciation and reduced metal toxicity in the environment (such as in soil). Electron-transfer between contaminants and NPs leads to a changed state of toxicity in the environment. The catalysis mechanism involves modification of the remediation rate by a catalyst for overall acceleration of the reaction. The catalyst interacts with the pollutant, and sometimes remains unmodified or can dissolve during the process, which is mainly dependent on the homogeneous or heterogeneous nature of the catalyst. The process is also termed photocatalysis when photocatalysts are involved (Huang et al. 2012, Das et al. 2018).

4.1 Nanobioremediation Using Microbially Synthesized Nanoparticles

Microbial synthesis of NPs for nanobioremediation is a promising application (Pandey 2018). Microorganisms including bacteria, fungi and algae have been utilized to synthesize NPs for environmental remediation of different contaminant categories (Table 4) (Das et al. 2018). These microorganisms are involved in metal ion reduction, which leads to synthesis of nanoscale materials. In addition, NPs are formed because microorganisms secrete extracellular enzymes, which are important as reducing agents in the process. Out of all microorganisms, bacteria have been proven to have high binding affinities for metals, thus making them useful precursors for remediation (Pantidos and Horsfall 2014).

Synthesising NPs using various bacterial species is a common concept due to their abundance in the environment and the ability of some to adapt in extreme conditions (Nair et al. 2002, He et al. 2007, Saravanan et al. 2018). Bacteria have been used for synthesis of NPs such as gold, silver, platinum, iron and palladium for different applications (Pantidos and Horsfall 2014). In a study by Schlüter et al. (2014), synthesis of palladium NPs was conducted using *Pseudomonas* species isolated from Alpine sites contaminated with heavy metals. In their findings, they concluded that palladium NPs synthesized using *Pseudomonas* degraded polychlorinated dioxin.

Fungal species have also received interest as synthesising agents for metallic NPs. They secrete higher quantities of proteins such as hydrophobins, which are considered to play a major role in fungal hyphae aggregation to form multicellular tissues such as fruit bodies, thereby amplifying the productivity of nanoparticle synthesis (Mukherjee et al. 2001). For example, in a study by Zhao et al. (2015), silver NPs produced by *Trichoderma hamatum* showed photocatalytic degradation of approximately 96% against carcinogenic Rhodamine B dye.

Iron oxide NPs have also been synthesized using extracts from algal species *Padina pavonica* and *Sargassum acinarium*. The extracts were utilized as reductants for iron chloride. It was found that NPs prepared using *P. pavonica* could remove approximately 91% of lead, while NPs synthesized using *S. acinarium* had a 78% removal capacity in wastewaters after 75 min of treatment (El-Kassas et al. 2016). Mandal et al. (2016) used Spirulina (*Arthrospira platensis*), a blue-green alga, to biosynthesize cadmium sulphide (CdS) NPs by converting toxic Cd(II) ions to less toxic CdS NPs. The synthesized CdS NPs also functioned to remediate malachite green dyes.

Table 4 Types of microorganisms used to synthesize NPs for removal of contaminants

Contaminants	Microorganisms used for synthesis	NPs	Mechanism of action	Efficiency	Reference
Organic contaminants					
Carcinogenic Rhodamine B dye	*Trichoderma hamatum*	Silver NPs	Photocatalysis	Degradation properties of approximately 96%	Zhao et al. (2015)
Azo dyes	*Klebsiella oxytoca*	Palladium NPs	Transformation	Reduction of azo dyes by approximately 96%	Wang et al. (2018)
Organophosphorus pesticides	*Rhizopus oryzae*	Gold NPs	Adsorption	85–99% adsorption of organophosphorus	Das et al. (2009)
Hydrocarbons					
Anthracene, phenanthrene, fluorine, chrysene and benzo[a]pyrene	*Sapindus-mukorossi*	Iron NPs	Photocatalysis and transformation	Anthracene and phenanthrene degradation reached 80 to 90% in both water and soils, while degradations of fluorine, chrysene and benzo[a]pyrene were approximately 70–80%.	Shanker et al. (2017)
Polychlorinated dioxin	*Pseudomonas* species	Palladium NPs	Transformation	Degraded polychlorinated dioxin	Schlüter et al. (2014)
Heavy metals					
Lead	*Padina pavonica*	Iron oxide NPs	Adsorption	Removal of Lead by approximately 91%	El-Kassas et al. (2016)
Lead	*Sargassum acinarium*	Iron oxide NPs	Adsorption	78% lead removal capacity	El-Kassas et al. (2016)
Cadmium	*Spirulina (Arthrospira) platensis*	Cadmium sulphide NPs	Transformation	Cd (II) was transformed to a less toxic form of CdS NPs	Mandal et al. (2016)

4.2   Nanobioremediation Using Phytosynthesized Nanoparticles

It has also been established that green synthesis of metal NPs using various plant parts including leaves, stems, seeds and roots is simple, cost effective, reproducible and result in uniform

Table 5 Phytosynthesized nanoparticles for nanobioremediation

Contaminants	Plants used for synthesis	NPs	Mechanism of action	Efficiency	Reference
Organic contaminants					
4-Nitrophenol	Stem extract of Breyna (*Breynia rhamnoides*)	Silver and Gold	Transformation	Efficient reduction of 4-nitrophenol (4-NP) to 4-aminophenol (4-AP)	Gangula et al. (2011)
Ibufren	Black tea (*Camellia sinensis*) leaf, grape marc and vine leaf extracts	nZVI NPs	Transformation	Ibuprofen degradation by 54–66% from initial concentration	Machado et al. (2013)
Safranine O, Methyl red, Methyl orange and Methylene blue	Winged prickly ash (*Zanthoxylum armatum*) leaf extract	Silver NPs	Transformation	Reduction of dyes was confirmed by a decrease in dye absorbance values	Jyoti and Singh (2016)
Hydrocarbons					
Phenanthrene, anthracene and pyrene	Garlic (*Allium Sativum*) extract	Silver NPs	Adsorption	Removal efficiency of 80% on average	Abbasi et al. (2014)
Total petroleum hydrocarbons	Mortino berry (*Vaccinium floribundum*) extract	Iron NPs	Transformation	88.34% TPH removal efficiency reached in contaminated water, while 81.90% TPH removal efficiency was achieved in contaminated soil	Murgueitio et al. (2018)
Anthracene	Coriander (*Corriandrum sativum*) leaf extract	Zinc oxide NPs	Photocatalysis	96% photocatalytic degradation	Hassan et al. (2015)
Heavy metals					
Chromium	Water hyacinth (*Eichhornia crassipes*) extracts	Iron NPs	Transformation	89.9% of chromium was removed through adsorption and reduction	Wei et al. (2017)
Malachite	Green tea (*Camellia sinensis*) extracts	Iron based NPs	Adsorption	96% removal of malachite	Weng et al. (2013)
Lead	Emblic (*Emblica officinalis*) leaf extract	nZVI NPs	Adsorption and transformation	99% reduction of lead in 24 hours	Kumar et al. (2015)

dispersibility (Saif et al. 2016). The use of plant parts for nanoparticle synthesis has proven to be more feasible for large-scale synthesis in comparison to microorganisms due to easy cultivation in high quantities. However, the activity of those NPs lie in their properties as well as natural composition of various reducing agents (Iravani 2011).

Examples of plants used for biosynthesis include saltwort (*Noaea mucronata),* which is a common plant species used for accumulation of heavy metals such as cadmium, copper, lead and nickel. The NPs synthesized using this plant can also be used for remediation of heavy metal contaminants (Rizwan et al. 2014). Whilst biosynthesis of NPs using several plant extracts has been discussed (Khalil et al. 2017, Shaik et al. 2018, Hassan et al. 2019), there are still limited studies on the effect of such NPs on contaminant degradation efficiency in agro-ecosystems. Examples of such limited studies for removal of different contaminants are presented in Table 5.

Ibuprofen drug release into the agro-ecosystems is considered to have negative effects on the growth of plants. Machado et al. (2013) synthesized nZVIs from black tea (*Camellia sinensis*) leaf, grape marc and vine leaf extracts, which were found to degrade ibuprofen by 54–66%. In another study, iron NPs were synthesized using water hyacinth (*Eichhornia crassipes*) antioxidants extract. During synthesis, the extracts were used as both reducing agents and as capping agents for coating the surface of iron NPs for improved stability. The findings suggested that iron NPs were able to remove chromium compounds that are frequently used in numerous industries such as textile dying, tanneries and metallurgy (Wei et al. 2017). Weng et al. (2013) synthesized iron based NPs using green tea (*Camellia sinensis*) extracts for successful 96% degradation of malachite. In another study, the NPs synthesized from green tea extract were used as a catalyst for the successful degradation of methylene blue and methyl orange dyes (Shahwan et al. 2011). Silver NPs synthesized using *Z. armatum* leaf extract have been shown to enhance photocatalytic degradation of various dyes (Jyoti and Singh 2016).

5. RECENT ADVANCES AND HURDLES IN NANOBIOREMEDIATION

Undoubtedly, there is an emerging interest in nanobioremediation for treatment of contaminated agro-ecosystems due to the advantage of enhanced remediation efficiency (Karn et al. 2009, Ghasemzadeh et al. 2014, Yadav et al. 2017, Das et al. 2018). Within the last two decades, many advances have been made in the field of nanobioremediation. The realization that NPs can be synthesized biologically has gained immense popularity and acceptance as a viable and eco-friendly alternative to conventionally synthesized NPs in nanobioremediation. For instance, biogenic uraninite NPs have been used in bioremediation of subsurface uranium(VI) contamination (Bargar et al. 2008). Furthermore, nZVI NPs are currently applied extensively and effectively in nanobioremediation (Stefaniuk et al. 2016, Chen et al. 2017). They are able to reduce both organic and inorganic contaminants because metallic iron acts as an electron donor (Chen et al. 2017). For example, Xi and co-workers (2010) biosynthesized nZVI and applied it in nanobioremediation of lead ions in aqueous phase. The biosynthesized nZVI, as compared to commercial zero valent iron, had higher reactivity towards lead ions which afforded removal in a very short time.

In addition, polymeric NPs are applied to contaminants prior to bioremediation in order to improve bioavailability of contaminants. This is particularly important in the nanobioremediation of hydrophobic environmental contaminants such as polycyclic aromatic hydrocarbons (PAHs) with high sorption to soil particles and low solubility in aqueous phase (Rizwan and Ahmed 2018). An example of this application was demonstrated by Tungittiplakorn et al. (2005), where NPs prepared from poly(ethylene) glycol modified urethane acrylate (PMUA) precursor chain were developed for bioremediation of phenanthrene, a PAH, in soil. However, it is surprising that no recorded recent use of the PMUA NPs is available for this compilation, which is a challenge to be overcome in the emerging field of nanobioremediation. Polymers are often employed to overcome limitations of NPs, enhance their mechanical and thermal stability as well as impart properties such as durability and recyclability to them (Guerra et al. 2018). They are particularly important for conjugation of NPs. Biopolymers such as poly lactic acid (PLA), polyhroxyalkanoates (PHA), carboxymethyl cellulose (CMC) and chitosan are used in the synthesis of biopolymer-based NPs which are biodegradable, biocompatible and non-toxic (Das et al. 2018). A group of polymeric NPs called dendrimers have also been utilized in remediation of Cu^{2+} – contaminated wastewater (Diallo et al. 2005). Dendrimers are particularly of importance to researchers in the medical field; nonetheless, they have found relevance in nanobioremediation for water treatment (Diallo 2009). For example, poly (amidoamine) dendrimers, with their characteristic structure, are used as efficient and innocuous water treatment agents (Rizwan et al. 2014).

Recently, single enzyme NPs (SENPs) are being deployed to advance nanobioremediation (Duran 2008, Mukherjee et al. 2013, Sharma et al. 2018). Microbial enzymes are the machinery that facilitate microbial degradation of contaminants (Karigar and Rao 2011, Sharma et al. 2018). However, once isolated, the short catalytic lifespan of enzymes and their instability limit their use. To resolve this, enzymes are capped/encapsulated with NPs that possess high magnetic properties in order to enhance their stability, efficiency, reusability and lifespan (Yang et al. 2008, Rizwan et al. 2014). The process is more economical because the enzymes can then be separated from reactants and products by applying a magnetic field (Rizwan et al. 2014), thus facilitating their reuse. For instance, laccase immobilized on magnetic core-shell (Fe_3O_4–SiO_2) NPs was used for detection and determination of catechol, a hazardous phenolic environmental contaminant, as a nanobioremediation mitigation strategy (Tang et al. 2008). The same immobilized enzyme was earlier used, as a nanobioremediation strategy, to biosense hydroquinone in compost extracts (Zhang et al. 2007).

In spite of the potentials of NPs in bioremediation, several hurdles are encountered in their use. A prominent hurdle borders around safety of NPs in the environment as it is suggested that their uncontrolled release into the environment could lead to deterioration of the quality of abiotic as well as impede biotic components of the environment (Varadhi et al. 2005, Handy et al. 2008, Kumari and Singh 2016). Nanoparticles such as zinc oxide, silver oxide, titanium dioxide and copper oxide are known for their cytotoxicity (Karlsson et al. 2008); hence, incorporating such NPs during nanobioremediation could also result in toxicity to soil microbes. Apart from toxicity of NPs in soil, unmodified NPs also display a blocking effect in soil by clogging the pores and restricting fluids from passing through (Pandey 2018). Although, nZVI NPs have been demonstrated to have limited toxicity compared to other NPs (Reijnders 2006, Mukherjee et al. 2016), their modification, for instance, with organic polymers such as polyasparaginate, further reduces their toxicity because direct contact with microbial cells is limited (Chen et al. 2017, Gong et al. 2018).

Generally, strategies to overcome such challenges are necessary to broaden the application of NPs in nanobioremediation. However, considering the small size of NPs, their fate after participation in agro-ecosystem nanobioremediation is still a concern, as high concentrations of the NPs could potentially accumulate in the food chain. Extensive downstream research is therefore required to evaluate the extent of their environmental fate and trophic transfer.

6. FUTURE PROSPECTS AND CONCLUSION

Nanobioremediation technology offers an efficient alternative approach for treating contaminants in agro-ecosystems (Zhang et al. 2019, Cecchin et al. 2017). Biological synthesis of NPs is advocated as an eco-friendly option for obtaining NPs for agro-ecosystem remediation, especially as the fate of NPs in soil matrices is not yet fully understood (Cecchin et al. 2017, Panpatte et al. 2016).

In spite of the reported successful applications of nanobioremediation, the potential of this technology for remediation of contaminated agro-ecosystems is yet to be fully explored. Future studies for the use of NPs in agro-ecosystems could explore their applicability in pest and weed control in place of chemical pesticides and herbicides (Bhattacharyya et al. 2016). These applications will act as preventative strategies for chemical contamination ensuring sustainable management of agro-ecosystems. In addition, future studies should focus on unravelling the interactions of NPs with contaminants. Such studies would provide guidance for future optimization processes (Srivastav et al. 2018). In addition, adoption of engineered NPs for targeted applications as well as investigation of potential effects of adding other biological agents to the process are important (El-Sheshtawy et al. 2014, Zhang et al. 2019, Hedge 2016). Overall, nanobioremediation is a promising remediation strategy; however, more research is still required for full beneficial application of NPs in the agro-ecosystems.

References

Abbasi, M., F. Saeed and U. Rafique. 2014. Preparation of silver nanoparticles from synthetic and natural sources: Remediation model for PAHs. IOP Conf. Ser. Mater. Sci. Eng. 60: 012061.

Alabresm, A., Y.P. Chen, A.W. Decho and J. Lead. 2018. A novel method for the synergistic remediation of oil-water mixtures using nanoparticles and oil-degrading bacteria. Sci. Total Environ. 630: 1292–1297

Ali, J., N. Ali, L. Wang, H. Waseem and G. Pan. 2019. Revisiting the mechanistic pathways for bacterial mediated synthesis of noble metal nanoparticles. J. Microbiol. Methods. 159: 18–25.

Alkorta, I. and C. Garbisu .2001. Phytoremediation of organic contaminants in soils. Bioresour. Technol. 79: 273–276.

Araújo, R., A.C.M. Castro and A. Fiúza. 2015. The use of nanoparticles in soil and water remediation processes. Mater. Today. 2: 315–320.

Auffan, M., J. Rose, J.Y. Bottero, G.V. Lowry, J.P. Jolivet and M.R. Wiesner. 2009. Towards a definition of inorganic nanoparticles from an environmental, health and safety perspective. Nat. Nanotechnol. 4: 634.

Babuponnusami, A. and K. Muthukumar. 2012. Removal of phenol by heterogenous photo electro Fenton-like process using nano-zero valent iron. Sep. Purif. Technol. 98: 130–135.

Bargar, J.R., R. Bernier-Latmani, D.E. Giammar, and B.M. Tebo. 2008. Biogenic uraninite nanoparticles and their importance for uranium remediation. Element. 4: 407–412.

Bhattacharyya, A., Duraisamy, P., Govindarajan, M., Buhroo, A.A. and R. Prasad. 2016. Nano-biofungicides: Emerging trend in insect pest control. *In:* R. Prasad (ed.), Advances and Applications Through Fungal Nanobiotechnology. Springer, Cham, pp. 307–319.

Bokare, V., K. Murugesan, J.H. Kim, E.J. Kim and Y.S. Chang. 2012. Integrated hybrid treatment for the remediation of 2,3,7,8-tetrachlorodibenzo-p-dioxin. Sci. Total Environ. 435–436: 563–566

Bolan, N.S., J.H. Park, B. Robinson, R. Naidu and K.Y. Huh. 2011. Phytostabilization: A green approach to contaminant containment. *In:* D.L. Sparks (eds), Advances in agronomy. Academic Press, USA, pp. 145–204.

Bücker-Neto, L., A.L.S. Paiva, R.D. Machado, R.A. Arenhart and M. Margis-Pinheiro. 2017. Interactions between plant hormones and heavy metals responses. Genet. Mol. Biol. 40: 373–386.

Cecchin, I., K.R. Reddy, A. Thomé, E.F. Tessaro and F. Schnaid. 2017. Nanobioremediation: Integration of nanoparticles and bioremediation for sustainable remediation of chlorinated organic contaminants in soils. Int. Biodeter. Biodegr. 119: 419–428.

Chen, X., D. Ji, X. Wang and L. Zang. 2017. Review on nano zerovalent iron (nZVI): From modification to environmental applications. IOP Conf. Ser. Earth Environ. Sci. 51: 012004.

Choi, M.H., S.W. Jeong, H.E. Shim, S.J. Yun, S. Mushtaq, D.S. Choi, et al. 2017. Efficient bioremediation of radioactive iodine using biogenic gold nanomaterial-containing radiation-resistant bacterium, Deinococcus radiodurans R1. Chem. Commun. 53: 3937–3940.

Cunningham, S.D., T.A. Anderson, A.P. Schwab and F.C. Hsu. 1996. Phytoremediation of soils contaminated with organic pollutants. Adv. Agron. 56: 55–114.

Das, S.K., A.R. Das and A.K. Guha. 2009. Gold nanoparticles: Microbial synthesis and application in water hygiene management. Langmuir. 25: 8192–8199.

Das, R.K., V.L. Pachapur, L. Lonappan, M. Naghdi, R. Pulicharla, S. Maiti, et al. 2017. Biological synthesis of metallic nanoparticles: Plants, animals and microbial aspects. Nanotech. Environ. Eng. 2: 18.

Das, S., J. Chakraborty, S. Chatterjee and H. Kumar. 2018. Prospects of biosynthesized nanomaterials for the remediation of organic and inorganic environmental contaminants. Environ. Sci. Nano. 5: 2784–2808.

Davis, A.S., P. Prakash and K. Thamaraiselvi. 2017. Nanobioremediation technologies for sustainable environment. *In:* M. Prashanthi, R. Sundaram, A. Jeyaseelan and T. Kaliannan (eds), Bioremediation and Sustainable Technologies for Cleaner Environment. Springer, Cham, pp. 13–33..

Diallo, M. S., S., Christie, P., Swaminathan, J. H., Johnson and W. A. Goddard. 2005. Dendrimer enhanced ultrafiltration. 1. Recovery of Cu (II) from aqueous solutions using PAMAM dendrimers with ethylene diamine core and terminal NH_2 groups. Environ. Sci. Technol. 39: 1366–1377.

Diallo, M.S. 2009. Water treatment by dendrimer-enhanced filtration: Principles and applications. *In:* A. Street, R. Sustich, J. Duncan, N. Savage (eds), Nanotechnology Applications for Clean Water. William Andrew Publishing, pp. 143–155.

Ding, L., J. Li, W. Liu, Q. Zuo and S.X. Liang. 2017. Influence of nano-hydroxyapatite on the metal bioavailability, plant metal accumulation and root exudates of ryegrass for phytoremediation in lead-polluted soil. Int. J. Environ. Res. Public Health. 14: 532.

Dinh, H.T., J. Kuever, M. Muszmann, A.W. Hassel, M. Stratmann, F. Widdel. 2004. Iron corrosion by novel anaerobic microorganisms. Nature. 427: 829–832.

Dong, H., K. Ahmad, G. Zeng, Z. Li, G. Chen, Q. He, et al. 2016. Influence of fulvic acid on the colloidal stability and reactivity of nanoscale zero-valent iron. Environ. Pollut. 211: 363–369.

Duran, N. 2008. Use of nanoparticles in soil-water bioremediation processes. Revista de la ciencia del suelo y nutrición vegetal, 8: 33–38.

El-Kassas, H.Y., M.A. Aly-Eldeen and S.M. Gharib. 2016. Green synthesis of iron oxide (Fe_3O_4) nanoparticles using two selected brown seaweeds: Characterization and application for lead bioremediation. Acta Oceanol. Sin. 35: 89–98.

El-Sheshtawy, H.S., N.M. Khalil, W. Ahmed and R.I. Abdallah. 2014. Monitoring of oil pollution at Gemsa Bay and bioremediation capacity of bacterial isolates with biosurfactants and nanoparticles. Mar. Pollut, 87(1–2): 191–200.

El-Sheshtawy, H., N. Khalil, W., Ahmed and N. Amin. 2017. Enhancement the bioremediation of crude oil by nanoparticle and biosurfactants. Egypt. J. Chem. 60: 835–848.

Fang, G., Y. Si, C. Tian, G. Zhang and D. Zhou. 2012. Degradation of 2,4-D in soils by Fe_3O_4 nanoparticles combined with stimulating indigenous microbes. Environ. Sci. Pollut. R. 19: 784–793.

Fernandes, J.P., A.P. Mucha, T. Francisco, C.R. Gomes and C.M.R. Almeida. (2017). Silver nanoparticles uptake by salt marsh plants-Implications for phytoremediation processes and effects in microbial community dynamics. Mar. Pollut. 119: 176–183.

Galdames, A., A. Mendoza, M. I.S. Orueta, de Soto García, M. Sánchez, I. Virto, et al. 2017. Development of new remediation technologies for contaminated soils based on the application of zero-valent iron nanoparticles and bioremediation with compost. Resour.-Effic. Technol. 3: 166–176.

Gangula, A., R. Podila, L. Karanam, C. Janardhana and A.M. Rao. 2011. Catalytic reduction of 4-nitrophenol using biogenic gold and silver nanoparticles derived from *Breynia rhamnoides*. Langmuir. 27: 15268–15274.

Gerhardt, K.E., X.D. Huang, B.R. Glick and B.M. Greenberg. 2009. Phytoremediation and rhizoremediation of organic soil contaminants: Potential and challenges. Plant Sci. 176(1): 20–30.

Ghasemzadeh, G., M. Momenpour, F. Omidi, M.R. Hosseini, M. Ahani and A. Barzegari. 2014. Applications of nanomaterials in water treatment and environmental remediation. Front. Env. Sci Eng. 8: 47–482.

Gong, X., D. Huang, Y. Liu, Z. Peng, G. Zeng, P. Xu, et al. 2018. Remediation of contaminated soils by biotechnology with nanomaterials: Bio-behavior, applications, and perspectives. Crit. Rev. Biotechnol. 38: 455–468.

Guerra, F., M. Attia, D. Whitehead and F. Alexis. 2018. Nanotechnology for environmental remediation: Materials and applications. Molecules. 23: 1760.

Handy, R.D., R. Owen and E. Valsami-Jones. 2008. The ecotoxicology of nanoparticles and nanomaterials: Current status, knowledge gaps, challenges, and future needs. Ecotoxicol. 17: 315–325.

Hassan, S.S., W.I. El Azab, H.R. Ali and M.S. Mansour. 2015. Green synthesis and characterization of ZnO nanoparticles for photocatalytic degradation of anthracene. Adv. Nat. Sci.: Nanosci. Nanotech. 6: 045012.

Hassan, D., A.T. Khalil, A.R. Solangi, A. El-Mallul, Z.K. Shinwari and M. Maaza. 2019. Physiochemical properties and novel biological applications of *Callistemon viminalis* -mediated α-Cr_2O_3 nanoparticles. Appl. Organomet. Chem. 33: e5041.

He, S., Z. Guo, Y. Zhang, S. Zhang, J. Wang and N. Gu. 2007. Biosynthesis of gold nanoparticles using the bacteria *Rhodopseudomonas capsulata*. Mater. 61: 3984–3987.

Hegde, K., Brar, S.K., Verma, M. and R.Y. Surampalli. 2016. Current understandings of toxicity, risks and regulations of engineered nanoparticles with respect to environmental microorganisms. Nanotech. Environ. Eng. 1(1): 5.

Huang, J., Y. Cao, Z. Liu, Z. Deng, F. Tang and W. Wang. 2012. Efficient removal of heavy metal ions from water system by titanate nanoflowers. Chem. Eng. J. 180: 75–80.

Ingle, A.P., A.B. Seabra, N. Duran and M. Rai. 2014. Nanoremediation: A new and emerging technology for the removal of toxic contaminant from environment. *In:* S. Das (ed.), Microbial Biodegradation and Bioremediation. Elsevier, London, pp. 233–250.

Iravani, S. 2011. Green synthesis of metal nanoparticles using plants. Green Chem. 13: 2638–2650.

Jagdale, S., A. Hable and A. Chabukswar. 2018. Nanobiotechnology for bioremediation: Recent trends. *In:* A.K. Rathoure (ed.), Biostimulation Remediation Technologies for Groundwater Contaminants. IGI Global, PA, USA, pp. 259–284..

Jiamjitrpanich, W., P. Parkpian, C. Polprasert and R. Kosanlavit. 2012. Enhanced phytoremediation efficiency of TNT-contaminated soil by nanoscale zero valent iron. Int. Proc. Chem. Biol. Environ. Eng. 35: 82–86.

Jiang, D., G. Zeng, D. Huang, M. Chen, C. Zhang, C. Huang, et al. 2018. Remediation of contaminated soils by enhanced nanoscale zero valent iron. Environ. Res. J. 163: 217–227.

Jyoti K. and A. Singh. 2016. Green synthesis of nanostructured silver particles and their catalytic application in dye degradation. J. Genet. Eng. Biotechnol. 14: 311–7.

Karigar, C.S. and S.S. Rao. 2011. Role of microbial enzymes in the bioremediation of pollutants: A review. Enzyme Res. 2011: 1–11.

Karlsson, H.L., P. Cronholm, J. Gustafsson and L. Moller. 2008. Copper oxide nanoparticles are highly toxic: A comparison between metal oxide nanoparticles and carbon nanotubes. Chem. Res. Toxicol. 21(9): 1726–1732.

Karn, B., T. Kuiken and M. Otto. 2009. Nanotechnology and in situ remediation: A review of the benefits and potential risks. Environ. Health Perspect. 117: 1813–1831.

Khalil, A.T., M. Ovais, I. Ullah, M. Ali, Z.K. Shinwari and M. Maaza. 2017. Biosynthesis of iron oxide (Fe_2O_3) nanoparticles via aqueous extracts of *Sageretia thea* (Osbeck.) and their pharmacognostic properties. Green Chem. Lett. Rev. 10: 186–201.

Khan, N. and A. Bano. 2016. Modulation of phytoremediation and plant growth by the treatment with PGPR, Ag nanoparticle and untreated municipal wastewater. Int. J. Phytorem. 18: 1258–1269.

Khan, M.N., M. Mobin, Z.K. Abbas, K.A. AlMutairi and Z.H. Siddiqui. 2017. Role of nanomaterials in plants under challenging environments. Plant Physiol. Biochem. 110: 194–209.

Konishi, Y. 2006. Microbial synthesis of noble metal nanoparticles using metal-reducing bacteria. J. Soc. Powder Technol. 43: 515–521.

Kuang, Y., Y. Zhou, Z. Chen, M. Megharaj and R. Naidu. 2013. Impact of Fe and Ni/Fe nanoparticles on biodegradation of phenol by the strain *Bacillus fusiformis* (BFN) at various pH values. Bioresour. Technol. 136: 588–594.

Kumar, R., N. Singh and S.N. Pandey. 2015. Potential of green synthesized zero-valent iron nanoparticles for remediation of lead-contaminated water. Int. J. Sci. Environ. Technol. 12: 3943–3950.

Kumari, B. and D.P. Singh. 2016. A review on multifaceted application of nanoparticles in the field of bioremediation of petroleum hydrocarbons. Ecol. Eng. 97: 98–105.

Kumari, S. and S. Khan. 2018. Effect of Fe_3O_4 NPs application on fluoride (F) accumulation efficiency of *Prosopis juliflora*. Ecotoxicol. Environ. Saf. 166: 419–426.

Liu, Y. and G.V. Lowry. 2006. Effect of particle age (Fe0 content) and solution pH on NZVI reactivity: H_2 evolution and TCE dechlorination. Environ. Sci. Technol. 40: 6085–6090.

Ma, X. and C. Wang. 2010. Fullerene nanoparticles affect the fate and uptake of trichloroethylene in phytoremediation systems. Environ. Eng. Sci. 27: 989–992.

Machado, S., W. Stawiński, P. Slonina, A.R. Pinto, J.P. Grosso, H.P.A. Nouws, et al. 2013. Application of green zero-valent iron nanoparticles to the remediation of soils contaminated with ibuprofen. Sci. Total Environ. 461: 323–329.

Mandal, R.P., S. Sekh, N.S. Sarkar, D. Chattopadhyay and S. De. 2016. Algae mediated synthesis of cadmium sulphide nanoparticles and their application in bioremediation. Mater. Res. Express. 3: 055007.

Morones, J.R., J.L. Elechiguerra, A. Camacho, K. Holt, J.B. Kouri, J.T. Ramírez, et al. 2005. The bactericidal effect of silver nanoparticles. Nanotechnology. 16: 2346–2353.

Mukherjee, P., A. Ahmad, D. Mandal, S. Senapati, S.R. Sainkar, M.I. Khan, et. al. 2001. Fungus-mediated synthesis of silver nanoparticles and their immobilization in the mycelial matrix: A novel biological approach to nanoparticle synthesis. Nano Lett. 1: 515–519.

Mukherjee, S., B. Basak, B. Bhunia, A. Dey and B. Mondal. 2013. Potential use of polyphenol oxidases (PPO) in the bioremediation of phenolic contaminants containing industrial wastewater. Rev. Environ. Sci. Biotechnol. 12: 61–73.

Mukherjee, R., R. Kumar, A. Sinha, Y. Lama and A.K. Saha. 2016. A review on synthesis, characterization, and applications of nano zero valent iron (nZVI) for environmental remediation. Crit. Rev. Env. Sci. Tec. 46: 443–466.

Murgueitio, E., L. Cumbal, M. Abril, A. Izquierdo, A. Debut and O. Tinoco. 2018. Green synthesis of iron nanoparticles: Application on the removal of petroleum oil from contaminated water and soils. J. Nanotechnol. 2018: 1–8.

Murphy, C.D. 2010. Biodegradation and biotransformation of organofluorine compounds. Biotechnol. Lett. 32(3): 351–359.

Nair, B. and T. Pradeep. 2002. Coalescence of nanoclusters and formation of submicron crystallites assisted by *Lactobacillus* strains. [Cryst. Growth Des. 2: 293–298.

Nel, A., T. Xia, L. Mädler and N. Li. 2006. Toxic potential of materials at the nanolevel. Sci. 311: 622–627.

Nwadinigwe, A.O. and E.C. Ugwu. 2018. Overview of nano-phytoremediation applications. *In:* A.A. Ansari, S.S. Gill, R. Gill, G.R. Lanza and L. Newman (eds). Phytoremediation. Springer, Cham, pp. 377–382.

Pandey, G. 2018. Prospects of Nanobioremediation in environmental cleanup. Orient. J. Chem. 34(6): 2838–2850.

Panpatte, D.G., Y.K. Jhala, H.N. Shelat and R.V. Vyas. 2016. Nanoparticles: The next generation technology for sustainable agriculture. *In:* D.P. Singh, H.B. Singh and R. Prabha (eds), Microbial Inoculants in Sustainable Agricultural Productivity, Vol. 2: Functional Applications. Springer, New Delhi, pp. 289–300.

Pantidos, N. and L.E. Horsfall. 2014. Biological synthesis of metallic nanoparticles by bacteria, fungi and plants. J. Nanomed. 5: 1.

Pathak, V.M. and N. Kumar. 2017. Implications of SiO_2 nanoparticles for in vitro biodegradation of low-density polyethylene with potential isolates of *Bacillus, Pseudomonas*, and their synergistic effect on *Vigna mungo* growth. Energy Ecol. Environ. 2: 418–427.

Pawlett, M., K. Ritz, R.A. Dorey, S. Rocks, J. Ramsden and J.A. Harris. 2013. The impact of zero-valent iron nanoparticles upon soil microbial communities is context dependent. Environ. Sci. Pollut. R. 20: 1041–1049.

Pillai, H.P. and J. Kottekottil. 2016. Nano-phytotechnological remediation of endosulfan using zero valent iron nanoparticles. J. Environ. 7: 734.

Prathna, T.C., L. Mathew, N. Chandrasekaran, A.M. Raichur and A. Mukherjee. 2010. Biomimetic synthesis of nanoparticles: Science, technology and applicability. *In:* A. Mukherjee (ed.), Biomimetics Learning from Nature. Intechopen, Croatia, pp. 1–20.

Rafique, M., I. Sadaf, M.S. Rafique and M.B. Tahir. 2017. A review on green synthesis of silver nanoparticles and their applications. Artif. Cells Nanomed. Biotechnol. 45: 1272–1291.

Rao, P.S.C., J.W. Jawitz, C.G. Enfield, R.W. Falta Jr, M.D. Annable and A.L. Wood. 2001. Technology integration for contaminated site remediation: Clean-up goals and performance criteria. *In:* S.F. Thornton and S.E. Oswald (eds), Groundwater Quality: Natural and Enhanced Restoration of Groundwater Pollution. IAHS Publication no. 275, Wallingford, Oxforshire, UK, pp. 571–578.

Rascio, N. and F. Navari-Izzo. 2011. Heavy metal hyperaccumulating plants: How and why do they do it? And what makes them so interesting?. Plant Sci. 180: 169–181.

Reddy, K.R. and J.A. Adams. 2010. Sustainable Remediation of Contaminated Sites. Momentum Press, New York.

Reijnders, L. 2006. Cleaner nanotechnology and hazard reduction of manufactured nanoparticles. J. Clean. Prod. 14: 124–133.

Rizwan, M., M. Singh, C.K. Mitra and R.K. Morve. 2014. Ecofriendly application of nanomaterials: Nanobioremediation. J. Nanomater. 2014: 1–7.

Rizwan M. and M.U. Ahmed. 2018 Nanobioremediation: Ecofriendly application of nanomaterials. *In:* L.M.T. Martínez, O.V. Kharissova and B.I. Kharisov (eds), Handbook of Ecomaterials. Springer, Cham, pp. 3523–3535.

Rizwan, M., S. Ali, B. Ali, M. Adrees, M. Arshad, A. Hussain, et al. 2019. Zinc and iron oxide nanoparticles improved the plant growth and reduced the oxidative stress and cadmium concentration in wheat. Chemosphere 214: 269–277.

Romeh A.A.A. 2018. Green silver nanoparticles for enhancing the phytoremediation of soil and water contaminated by fipronil and degradation products. Water Air Soil Pollut. 229: 1–13.

Saif, S., A. Tahir and Y. Chen. 2016. Green synthesis of iron nanoparticles and their environmental applications and implications. Nanomaterials 6: 209.

Saravanan, M., S.K. Barik, D. MubarakAli, P. Prakash and A. Pugazhendhi. 2018. Synthesis of silver nanoparticles from *Bacillus brevis* (NCIM 2533) and their antibacterial activity against pathogenic bacteria. Microb. Pathog. 116: 221–226.

Sarwar, N., M. Imran, M.R. Shaheen, W. Ishaque, M.A. Kamran, A. Matloob, et.al. 2017. Phytoremediation strategies for soils contaminated with heavy metals: Modifications and future perspectives. Chemosphere 171: 710–721.

Schlüter, M., T. Hentzel, C. Suarez, M. Koch, W.G. Lorenz, L. Böhm, et.al. 2014. Synthesis of novel palladium (0) nanocatalysts by microorganisms from heavy-metal-influenced high-alpine sites for dehalogenation of polychlorinated dioxins. Chemosphere 117: 462–470.

Shahwan, T., S.A. Sirriah, M. Nairat, E. Boyacı, A.E. Eroğlu, T.B. Scott, et al. 2011. Green synthesis of iron nanoparticles and their application as a Fenton-like catalyst for the degradation of aqueous cationic and anionic dyes. Chem. Eng. Trans. 172: 258–266.

Shaik, M., M. Khan, M. Kuniyil, A. Al-Warthan, H. Alkhathlan, M. Siddiqui, et.al. 2018. Plant-extract-assisted green synthesis of silver nanoparticles using *Origanum vulgare* L. extract and their microbicidal activities. Sustainability 10: 913.

Shanker, U., V. Jassal and M. Rani. 2017. Green synthesis of iron hexacyanoferrate nanoparticles: Potential candidate for the degradation of toxic PAHs. J. Environ. Chem. Eng. 5: 4108–4120.

Sharma, B., A.K. Dangi and P. Shukla. 2018. Contemporary enzyme based technologies for bioremediation: A review. J. Environ. Manage. 210: 10–22.

Siddiqui, M.H., M.H. Al-Whaibi, M. Firoz and M.Y. Al-Khaishany. 2015. Role of nanoparticles in plants. *In:* M.H. Siddiqui, M.H. Al-Whaibi and M. Firoz (eds), Nanotechnology and Plant Sciences. Springer, Cham, pp. 19–35.

Silva, L.P., A.P. Silveira, C.C. Bonatto, I.G. Reis and P.V. Milreu. 2017. Silver nanoparticles as antimicrobial agents: Past, present, and future. *In:* A. Ficai and A.M. Grumezescu (eds), Nanostructures for Antimicrobial Therapy. Elsevier, Amsterdam, Netherlands, pp. 577–596.

Singh, R., N. Manickam, M.K.R. Mudiam, R.C. Murthy and V. Misra. 2013. An integrated (nano-bio) technique for degradation of γ-HCH contaminated soil. J. Hazard. Mater. 258: 35–41.

Singh, R., P. Sing and R. Sharma. 2014. Microorganism as a tool of bioremediation technology for cleaning environment: A review. Proc. Int. Acad. Ecol. Environ. Sci. 4(1): 1–6.

Singh, J. and B.K. Lee. 2016. Influence of nano-TiO_2 particles on the bioaccumulation of Cd in soybean plants (*Glycine max*): A possible mechanism for the removal of Cd from the contaminated soil. J. Environ. Manage. 170: 88–96.

Singh, J., T. Dutta, K.H. Kim, M. Rawat, P. Samddar and P. Kumar. 2018. 'Green'synthesis of metals and their oxide nanoparticles: Applications for environmental remediation. J. Nanobiotechnol. 16: 84.

Singh, R., M. Behera and S. Kumar. 2020. Nano-bioremediation: An innovative remediation technology for treatment and management of contaminated sites. *In:* G. Saxena and R.N. Bharagava (eds), Bioremediation of Industrial Waste for Environmental Safety. Springer, Singapore, pp. 165–182.

Song, U. and S. Lee. 2016. Phytotoxicity and accumulation of zinc oxide nanoparticles on the aquatic plants *Hydrilla verticillata* and *Phragmites Australis*: Leaf-type-dependent responses. Environ. Sci. Pollut. R. 23: 8539–8545.

Song, B., P. Xu, M. Chen, W. Tang, G. Zeng, J. Gong, et.al. 2019. Using nanomaterials to facilitate the phytoremediation of contaminated soil. Crit. Rev. Env. Sci. Tec. 49: 791–824.

Souri, Z., N. Karimi, M. Sarmadi and E. Rostami. 2017. Salicylic acid nanoparticles (SANPs) improve growth and phytoremediation efficiency of *Isatis cappadocica* Desv., under As stress. IET Nanobiotechnol. 11: 650–655.

Srivastav, A., K.K. Yadav, S. Yadav, N. Gupta, J.K. Singh, R. Katiyar, et.al. 2018. Nano-phytoremediation of pollutants from contaminated soil environment: Current scenario and future prospects. *In:* A.A. Ansari, S.S. Gill, R. Gill, G.R. Lanza and L. Newman (eds), Phytoremediation. Springer, Cham, pp. 383–401.

Stefaniuk, M., P. Oleszczuk and Y.S. Ok. 2016. Review on nano zerovalent iron (nZVI): From synthesis to environmental applications. Chem. Eng. J. 287: 618–632.

Tabak, H.H., L. Lens, E.D van Hullebusch and W. Dejonghe. 2005. Developments in bioremediation of soils and sediments polluted with metals and radionuclides–1. Microbial processes and mechanisms affecting bioremediation of metal contamination and influencing metal toxicity and transport. Rev Environ Sci Bio, 4(3): 115–156.

Tang, L., G. Zeng, J. Liu, X. Xu, Y. Zhang, M Liu, et al. 2008. Catechol determination in compost bioremediation using a laccase sensor and artificial neural networks. Anal. Bioanal. Chem. 391: 679–685.

Tejada, M., I. Gómez and M. del Toro. 2011. Use of organic amendments as a bioremediation strategy to reduce the bioavailability of chlorpyrifos insecticide in soils. Effects on soil biology. Ecotoxicol. Environ. Saf. 74: 2075–2081.

Tratnyek, P.G. and R.L. Johnson. 2006. Nanotechnologies for environmental cleanup. Nano Today. 1: 44–48.

Tripathi, S., R. Sanjeevi, J. Anuradha, D.S. Chauhan and A.K. Rathoure. 2018. Nano-bioremediation: nanotechnology and bioremediation. *In*: A.K. Rathoure (eds), Biostimulation Remediation Technologies for Groundwater Contaminants. IGI Global, pp. 202-219.

Tungittiplakorn, W., C. Cohen and L.W. Lion. 2005. Engineered polymeric nanoparticles for bioremediation of hydrophobic contaminants. Environ. Sci. Technol. 39: 1354–1358.

Varadhi, S.N., H., Gill, L.J., Apoldo, K., Liao, R.A., Blackman and W.K. Wittman. 2005. Full-scale nanoiron injection for treatment of groundwater contaminated with chlorinated hydrocarbons. Natural Gas Technologies Conference, Feb. 2005, Orlando, FL.

Verma, N. and R. Sharma. 2017. Bioremediation of toxic heavy metals: A patent review. Recent Pat. Biotechnol. 11: 171–187.

Wang, T., X. Jin, Z. Chen, M. Megharaj and R. Naidu. 2014. Green synthesis of Fe nanoparticles using eucalyptus leaf extracts for treatment of eutrophic wastewater. Sci. Total Environ. 466: 210–213.

Wang, P.T., Y.H. Song, H.C. Fan and L. Yu. 2018. Bioreduction of azo dyes was enhanced by in-situ biogenic palladium nanoparticles. Bioresour. Technol. 266: 176–180.

Weathers, L.J., G.F. Parkin and P.J. Alvarez. 1997. Utilization of cathodic hydrogen as electron donor for chloroform cometabolism by a mixed, methanogenic culture. Environ. Sci. Technol. 31: 880–885.

Wei, Y., B. Davidson, D. Chen and R. White. 2009. Balancing the economic, social and environmental dimensions of agro-ecosystems: An integrated modeling approach. Agr. Ecosyst. Environ. 131: 263–273.

Wei, Y., Z. Fang, L. Zheng and E.P. Tsang. 2017. Biosynthesized iron nanoparticles in aqueous extracts of *Eichhornia crassipes* and its mechanism in the hexavalent chromium removal. Appl. Surf. Sci. 399: 322–329.

Weng, X., L. Huang, Z. Chen, M. Megharaj and R. Naidu. 2013. Synthesis of iron-based nanoparticles by green tea extract and their degradation of malachite. Ind. Crops Prod. 51: 342–347.

Włóka, D., A. Placek, M. Smol, A. Rorat, D. Hutchison and M. Kacprzak. 2019. The efficiency and economic aspects of phytoremediation technology using *Phalaris arundinacea* L. and *Brassica napus* L. combined with compost and nano SiO$_2$ fertilization for the removal of PAHs from soil. J. Environ. Manage. 234: 311–319.

Xi, Y., M. Mallavarapu and R. Naidu. 2010. Reduction and adsorption of Pb^{2+} in aqueous solution by nano-zero-valent iron-a SEM, TEM and XPS study. Mater. Res. Bull. 45: 1361–1367.

Xia, S., Z. Song, P. Jeyakumar, S.M. Shaheen, J. Rinklebe, Y.S. Ok, et.al. 2019. A critical review on bioremediation technologies for Cr(VI)-contaminated soils and wastewater. Crit. Rev. Env. Sci. Tec. 49: 1027–1078.

Xu, P., G.M. Zeng, D.L. Huang, C. Lai, M.H. Zhao, Z. Wei, et al. 2012. Adsorption of Pb(II) by iron oxide nanoparticles immobilized *Phanerochaete chrysosporium*: Equilibrium, kinetic, thermodynamic and mechanisms analysis. Chem. Eng. J. 203: 423–431.

Yadav, K.K., J.K. Singh, N. Gupta and V. Kumar. 2017. A review of nanobioremediation technologies for environmental cleanup: A novel biological approach. J. Mater. Environ. Sci. 8: 740–757.

Yang Z., S. Si and C. Zhang. 2008 Magnetic single enzyme nanoparticles with high activity and stability. Biochem. Biophys. Res. Commun. 367: 169–175.

Zhang, Y., G.M. Zeng, L. Tang, H.Y. Yu and J.B. Li. 2007. Catechol biosensor based on immobilizing laccase to modified core-shell magnetic nanoparticles supported on carbon paste electrode. Huan Jing ke Xue 28: 2320–2325.

Zhang, T., G.V. Lowry, N.L. Capiro, J. Chen, W. Chen, Y. Chen, et al. 2019. *In-situ* remediation of subsurface contamination: Opportunities and challenges for nanotechnology and advanced materials. Environ. Sci.: Nano 6: 1283–1302.

Zhao, X., J. Zhang, B. Wang, A. Zada and M. Humayun. 2015. Biochemical synthesis of Ag/AgCl nanoparticles for visible-light-driven photocatalytic removal of colored dyes. Materials 8: 2043–2053.

Zhu, Y., F. Xu, Q. Liu, M. Chen, X. Liu, Y. Wang, et al. 2019. Nanomaterials and plants: Positive effects, toxicity and the remediation of metal and metalloid pollution in soil. Sci. Total Environ. 662: 414–421.

Zou, Y., X. Wang, A. Khan, P. Wang, Y. Liu, A. Alsaedi, et al. 2016. Environmental remediation and application of nanoscale zero-valent iron and its composites for the removal of heavy metal ions: A review. Environ. Sci. Technol. 50(14): 7290–7304.

CHAPTER 4

Challenges of Multi-omics in Improving Microbial-assisted Phytoremediation

Camilla Fagorzi

Lab of Molecular and Microbial Evolution, Dep. of. Biology, University of Florence, Via Madonna del Piano, 6 – 50019 Sesto Fiorentino (Fi), Italy.

1. INTRODUCTION

Soil has a fundamental role on the Earth, sustaining human life through agriculture, controlling ecological balance and maintaining biodiversity (Banwart 2011, Basu et al. 2018). The constant and sometimes uncontrolled use of pesticides and the dispersion of inorganic and organic pollutants in the environment imbalance the ecosystem by damaging the soil (Basu et al. 2018). By 2050, the world's population is expected to reach 9.1 billion and nearly all this population increase will occur in developing countries and will be associated with a huge urbanization process. This larger, more urban and richer population will be needed to be fed and this means that food production should increase by around 70% (FAO 2009, "How to Feed a Hungry World" 2010). In this context, it becomes crucial to develop innovative agrobiotechnological approaches to optimize the use of lands and arable soils in agriculture. On the other side, soluble soil pollutants also represent a source of contamination for hydraulic systems (i.e. nitrogen and phosphorous contamination that bring eutrophication of coastal areas and rivers), leading to a growing desire to safeguard water resources (Le Moal et al. 2019). Strategies to treat contaminations that nowadays arouse great interest include the use of biological agents (Abhilash et al. 2009). Bioremediation and phytoremediation represents an eco-friendly and low- cost microbe- or plant-based approach to remove contaminants from soil, water and sludges using physiological processes of living organisms like bacteria or plants (Anderson et al. 1993, Basu et al. 2018). The harnessing of plant-microbe interactions is a promising solution to remove

*Corresponding author: camilla.fagorzi@unifi.it

toxic elements from the soil. The points in favor of the phytoremediation are evident, especially from an economic and ecological point of view; nevertheless, the uncertainty of a full and rapid remediation prevents large investments in these technologies. To optimize this potential, in addition to a classical methodology, an integrative systems biology approach could be pivotal to analyse omics datasets, understanding gene functions and unravelling candidate genes involved in the signalling network between plants and microbes in the polluted context. These studies could bring to the deciphering of complex plant-microbiome-soil interactions and allow to predict the effect of the microbiome and rationally optimize bacterial-assisted phytoremediation technologies.

2. BACKGROUND

To improve phytoremediation, during the last decades, researchers worked on the idea to develop hybrid or transgenic plants with increased tolerance to pollutants or enhanced contaminants uptake abilities and to harness microbial contribution with strains capable of interacting with the plant by promoting its growth or by increasing pollutant degradation and/or transformation (Abhilash et al. 2012, Bell et al. 2014a, Doty 2008, Fagorzi et al. 2018). One of the first and most relevant examples was the use of recombinant bacterial strains (namely, a *Burkholderia cepacia*) able to degrade toluene, which was used to inoculate lupin plants. The resulting association was able to degrade toluene, providing a strong proof-of-principle for such an approach (Barac et al. 2004). Transgenic plants have also been used. For instance, poplars overexpressing a mammalian cytochrome P450, a family of enzymes commonly involved in the metabolism of toxic compounds, showed enhanced performance in the removal of a range of toxic volatile organic pollutants, including vinyl chloride, carbon tetrachloride, chloroform and benzene (Van Aken 2008). In another example, regarding the treatment of diesel polluted areas, the soil inoculation with plant-growth promoting bacteria promoted plant growth (Afzal et al. 2012).

3. MICROBIAL OMICS

3.1 The Bugs of Classical Methodology for the Study of Plant-microbes Interactions

One of the main limitation in the study of microbial communities remains the frequently cited bias that only 1–2% of total taxa can be studied by culture-dependent methods (Pham and Kim 2012, Stefani et al. 2015). Despite this limit being sometimes questioned, omics technologies permit the analysis of virtually total microbial community composition and activity, while preserving the fingerprint of biotic and abiotic factors present *in situ* (Martiny 2019). Indeed, the inoculation of plants, with microorganisms positively influencing their growth or the remediation ability, has often been unsuccessful, probably because of the exclusion by adapted indigenous communities (Thompson et al. 2005). Furthermore, it has been demonstrated that even in case of transient and unsuccessful colonization, a strain can impact microbial community of a certain environment, modifying both the diversity and functioning of resident communities and influencing future invasion attempts (Mallon et al. 2018).

3.2 The Advent of Omics Approaches

In 2005, next-generation sequencing technologies appeared in the market. Large dataset were created and our understanding of plants, microbes and their interaction increased (Morozova and Marra 2008) and now, we need to translate these data into usable technologies. The introduction of high-throughput technologies combined with mass spectrometry had an important contribution in understanding cellular processes and signalling networks involved in the response to environmental stress (Rabara et al. 2017).

3.3   Single Organism Omics

Culture-dependent screening of remediation efficient bacteria permits the isolation of interesting strains, with the ability to transform or degrade pollutants from an environment. An example is represented by the isolation and characterization of rhizospheric bacteria to enhance growth and metal accumulation by the chromium hyperaccumulator, *Vetiveria zizanoides*. The inoculation with *Bacillus cereus* T1B3 strain improved the phytoremediation efficiency of heavy metals (Nayak et al. 2018). *Pseudomonas* sp. Ps29C and *Bacillus megaterium* Bm4C, isolated from serpentine soil, showed their ability to protect plants against the growth-inhibitory effects of nickel; thus, their inoculation in nickel contaminated soil is described as a novel method to improve phytoremediation of this element (Rajkumar and Freitas 2008). The sequencing of the entire genome of cultivated isolates can reveal unknown genes or transformation pathways (Bell et al. 2014b). Taghavi et al. provided a genome survey of endophytic bacteria improving the biomass of *Populus* and identified putative mechanisms affecting growth and development of host plants (Taghavi et al. 2009). Similarly, the genome of *Pantoea ananatis* GB1, isolated from the roots of poplars in a diesel contaminated soil, provided information about the ability of the bacteria to enhance plant growth such as the presence of genes encoding proteins for inorganic phosphorus solubilization, nitrogen fixation, and siderophore production, as well as genes for alkanes utilization (Gkorezis et al. 2016).

Transcriptomic or proteomic analysis can be assessed to identify differentially expressed genes in presence of contaminants (Han et al. 2016). Lu et al., through RNA-seq technology, proved the role of three operons (MCO, YedYZ and CopG) in the tolerance to copper, zinc, lead and cadmium of the *Sinorhizobium meliloti* CCNWSX00200. The symbiosis of this strain with *Medicago lupulina* has been considered an efficient tool for bioremediation of heavy metal-polluted soils (Lu et al. 2017). Comparative transcriptomic studies were performed to gain insights on the expression levels of bacterial functions involved in the life-style of the plant-growth promoting endophytic bacteria *Enterobacter* sp. 638. The growth in sucrose as sole carbon source resulted as the best strategy to mimic the endophytic life-style of the bacteria, resulting in a decrease of motility and increase of the transcription of genes involved in colonization, adhesion and biofilm formation (Taghavi et al. 2015). More innovative technologies, as the droplet-based microfluidic devices now permit to directly isolate not-cultivated cells from the environment. It is therefore possible to proceed with omics approaches on the selected bacteria or to identify targets for isolation or manipulation in the field (Lasken 2012). With this approach, it is thus possible to screen effective pollutant degraders and selective-isolate species from a contaminated environment. Chen et al. screened and isolated imidazoline degraders from a polluted area, identifying members of genus *Ochrobactrum* as important contributors to the degradation of this class of herbicides (Chen et al. 2019).

3.4   Multiple Organism Omics

Meta-omics studies of DNA, RNA and proteins are often used to have a picture of the structure and the activity of a mixed population of microorganisms in a plant environment. The main advantages of these methodologies are that cultivation of microorganisms is not necessary and the result obtained is a snapshot of the bacterial activity *in situ*. A possible strategy to test and apply new discoveries derived from meta-omics studies (i.e. new candidate genes) is the construction of plasmids with large genomic fragment that can be transferred in *E. coli* and tested. Clones can undergo single organism omics analysis to be functionally characterized (Mirete et al. 2016, Uchiyama and Miyazaki 2009). The combined use of meta-omics approaches can produce a multi-layers information on microbe-mediated stress response of plants (Mukhtar et al. 2019).

Metagenomics

Targeted or untargeted metagenomics can be used to analyse bacterial composition of an environment at different levels. Data obtained from a plethora of studies showed, for example, that microbial communities of bulk and rhizospheric soil, as well as that of different plant organs, are different (Bulgarelli et al. 2012, 2013, Maggini et al. 2018). Horizontal gene transfer is a key agent in the spread of resistance gene, and it is mainly driven by processes such as conjugation (by bacterial plasmids and conjugative transposons), transformation (by acquisition of free DNA from the environment), and transduction (by bacteriophages). In addition to the identification at phyla/genus/strain level of the microorganisms present in a certain environment, metagenomic sequencing can be harnessed to study the mobilome of a community (Lekunberri et al. 2018). Through metagenomic, it is also possible to assign functional properties to microorganisms, with the possibility to appreciate whether and how functional pathways involved in the presence of soil contaminants change with microbial structure variation (Feng et al. 2018, Fuhrman 2009). In recent years, to test ecological theories, co-occurrence among members in a community has been studied through microbial network analysis and keystone taxa have been statistically identified (Barberán et al. 2012, Fuhrman et al. 2015, Berry and Widder 2014). According to Banerjee et al. (2018), microbial keystone taxa are highly connected taxa that exert (individually or in association) considerable influence on microbiome structure and functioning, without taking into account their abundance across space and time (Banerjee et al. 2018). In the plant-microbe dual system for example, nitrogen fixing bacteria have been proposed as keystone taxa as their abundance improves plant productivity and community evenness (Van Der Heijden et al. 2006). Rhizobiales, *Nitrospira*, Pseudomonadales and Actinobacteria are described as keystone taxa in ecosystem characterized by contaminations (Chao et al. 2016, Jiao et al. 2016). Nevertheless, following the identification of keystone taxa by network analysis, it is important to link such taxa to ecosystem processes by isolating the organisms through culture-dependent methods and characterization with metatrascriptomics and/or metaproteomics (Banerjee et al. 2018).

Metatranscriptomics

The ability of plant species used in phytoremediation to tolerate and even prosper in contaminated soils depend on a highly complex and mostly cryptic interacting community of microbial organisms (Gonzalez et al. 2018). While the metagenomic analysis gives indications on the composition and functional potentiality of the groups present in the community, it is with the transcriptomics that we understand what are the contributions of the single genes to the transformation or reduction of contaminants. Transcriptomic analysis can shed light on the alterations of the expression of microbial genes in a certain polluted environment after the introduction of plant. Yergeau et al. analysed, at metagenomic and metatranscriptomic levels, bacterial communities before and after the introduction of willow in a contaminated field, finding that the plant introduction only marginally impacted the composition of the bacterial community, but influenced the gene expression (Yergeau et al. 2014). It is therefore possible to state that there is no necessary correlation between the abundance of a genus and its contribution in the remediation or in the plant growth. The combination of metagenomic and metatranscriptomic analysis can also clarify if the shifting of a microbial community is compulsory to obtain an overexpression of functions bioremediation related. Particularly interesting is the fact that among the results that can be obtained from a transcriptome sequencing, there is the possibility of observing variations in gene expression over time. The rapid mRNA turnover was observed by Klatt et al. in a study about daily variations of Chloroflexi in a microbial mat in a geothermal spring (Klatt et al. 2013). Furtado et al. analysed the transcriptome of *Salicornia europaea*, an extremely salt-tolerant succulent obligatory halophyte, representing an ideal candidate for phytoremediation. Genes involved in salt tolerance mechanisms can be used for the improvement of crops. Seasonal variation in gene expression was observed and related to the role of season-dependent acclimation of the plant to salinity (Furtado et al. 2019). In a bioremediation context, transcriptomic shifts can be studied to understand the gene expression changed over different weather conditions, plant health status or introduction of exogenous compounds in the environment.

Metaproteomics
Metaproteomics has been used to describe plant-microbe and microbe-microbe interactions in different environments. When an environment presents a low diversity, metaproteomics can provide information on the presence of proteins produced by or in presence of specific microorganisms (Delmotte et al. 2009). If combined with stable probing (e.g. metaproteomic coupled to stable isotope probing, SIP), this approach can indicate which microorganisms act in the metabolism of contaminants (Jehmlich et al. 2016). Proteomic studies can also provide information about the response of a crop to abiotic stresses, allowing the possibility to use protein markers in breeding processes (Kosová et al. 2015).

Metabolomics
The analysis of the metabolome of a target microorganism acting in the bioremediation process as well as the study of the entire meta-metabolome of a microbial community associated with different plant organs can unravel the mechanisms involved in the increased heavy metal resistance in a crop (Berni et al. 2019). However, the study of metabolome presents some critical points. Information obtained from the metabolic pattern of a cell can be hardly correlated with particular genes, since a metabolite can be participate and regulate different pathways and is subjected to a fast turnover. Furthermore, in the metabolome of an organism, a plethora of different molecules are present (i.e. proteins, fatty acids, lipids, carbohydrates, inorganic molecules) (Silas et al. 2005). In the field of bioremediation, metabolomic analysis finds application associated with the isotope distribution analysis of metabolites and the molecular connectivity analysis. In the first case, substrates are labelled with an isotope suitable molecule and metabolites are then followed in the cells to study their distribution and characterize the metabolic changes that they undergo (Breitling et al. 2006, Del Carratore et al. 2019, Creek et al. 2012). In the second case, metabolites are followed through the fine measurement of their mass through mass spectrometry (Silas Granato Villas-Bôas and Bruheim 2007).

4. PLANT OMICS

The role of microbial communities in phytoremediation has been described as central in different studies; nevertheless, some plants can naturally contribute to this process though contaminants accumulation, degradation, stabilization or volatilization (Bellabarba et al. 2019). Several mechanisms behind these processes have been elucidated (and many other are still to be discovered and understood), thanks to the new omics approaches. Genomics approaches are often used to identify new candidate genes involved in phytoremediation and plant-tolerance. Going further with the investigations, activity focused analysis regarding transcriptomes, proteomes and metabolomes allow the examination of the functions of candidate plants under different environmental stressed conditions (Bell et al. 2014b). For example, plant-cadmium interaction has been well described through omics approaches, revealing the advantages of their use to compare species and stresses at the whole plant level (Villiers et al. 2012). Finally, another advantage of the omics approaches in the study of phytoremediation from a plant-centric view is the fact that plants are often characterized by extreme polyploidy and characters responsible for bioremediation are frequently polygenic (Zivy et al. 2015).

5. THE PLANT-MICROBE METAORGANISM

Rather than separating microbial and plant biological material, metaorganism omics can be the solution to obtain a full picture of the integrated responses of the interdependent organisms (Thijs et al. 2016). The metaorganism/holobiont is considered from different areas of biology, from medicine, to zoology and botany, as an integrated entity (Sieber et al. 2019, Hassani et al. 2019, Esser et al. 2019, Rausch et al. 2019). The different subcomponents of the metaorganisms – microbiome, interactome and plantome – can help targeting new aspects of the plant-microbe association. The microbiome can provide information about the intermicrobial interaction and relationship between gene content, expression and activity. On the other side, the plantome supply answers as the physiological response of the plant across different environments and contaminant gradients and variation in gene expressions, content and activity between tissues and conditions. It is with the interactome that it is possible to integrate information from the other two sub-components to understand mechanisms like the microbial suppression or promotion by the plant and, conversely, of the plant growth promotion or suppression exerted by microbes. Plant and microbes can cooperate for the removal/ transformation of contaminants through translocation of molecules and their relationship can be so strong that an inter-kingdom gene transfer can be observed (Bell et al. 2014a). How plant shapes the beneficial microbial community and microorganisms respond to plant cues in a contaminated environment? To answer this question, experiments have been performed by assessing defined contaminated gradients, to understand in which condition the remediation effect of the metaorganism can be maximised (Shen et al. 2018). A proteomic approach has been chosen to verify the response of *Pennisetum* (fountain grass) to the inoculation of the manganese tolerant endophytic bacterium *Bacillus megaterium* 1Y31. *B. megaterium* 1Y31 increased the growth and manganese uptake of the plant by increasing the efficiency of photosynthesis and energy metabolism (Zhang et al. 2015). Some symbiotic interrelationship, characterized by a close and well-defined connection between microorganism and plant, has been deeply studied. One example is represented by the legume-rhizobium partnership: the molecular basis of genome x genome interactions in the legume-rhizobium mutualistic model were studied by Heath et al., by using gene expression microarrays. The transcriptomic variations detected implicates regulatory changes in both species as drivers of symbiotic gene expression variation (Heath et al. 2012). Recently, the *in silico* reconstruction of the integrated metabolic network of a *Medicago truncatula* plant nodulated by the rhizobium *Sinorhizobium meliloti* has been reported. Here, the relevance of integrated models to understand the functioning of legume nodules has been highlighted, as also its potential for hypothesis generation for applicative studies and engineering of symbiotic nitrogen fixation (diCenzo et al. 2019).

6. A DESIGN–BUILD–TEST–LEARN (DBTL) CYCLE

The -omic approach for the study of plant-microbe interactions in phytoremediation finds its position in a potentially recursively used pipeline, a DBTL (Design–Build–Test–Learn) cycle. Here, the design is represented by the inoculation of the plant with microorganisms to observe their phenotypic changes and to collect dataset generated with -omics approaches (build) that can be integrated, thanks to system biology, identifying candidate genes. These functions can be introduced in a system, genes can undergo gene editing (test and learn) and results are thus evaluated to consider further improvement (Islam and Saha 2018, Petzold et al. 2015) (Fig. 1). One of the most recent frontiers in genome editing in phytoremediation is represented by the CRISPR technology, systems focused on a guide RNA coupled with an endonuclease, Cas9 or Cfp1 (Arora and Narula 2017, Kashtwari et al. 2019). CRISPR-based genome editing of plants is still a poorly explored field for the production of phytoremediators with augmented efficiency; nonetheless, increasing the synthesis of metal ligand, metal transporters, hormones and root exudates can represent a promising perspective to improve the yield of a plant (Basharat et al. 2018). Furthermore, phytoremediation can also take advantage of the employment of the CRISPR technology to create more efficient plant growth promoting rhizobacteria, with enhanced synthesis of phytohormones, siderophores and nitrogenase (Basharat et al. 2018). The genetic manipulation of the "hologenome" of the plant-microbe metaorganism can exert significant environmental effects in terms of optimization of the remediation process (Fasani et al. 2018, Macdonald and Singh 2013, Mueller and Sachs 2015, Thijs et al. 2016).

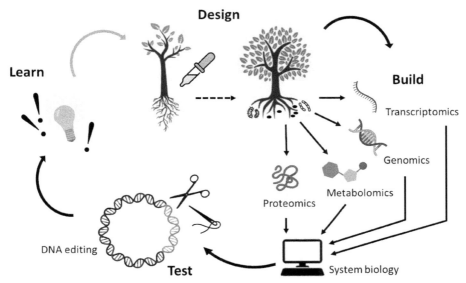

Figure 1 The potentially recursive pipeline of a DBTL cycle.

7. CONCLUSION

Phytoremediation is a green and low-cost technology, a precious resource to reduce the impact of pollutants. In the ideal situation, phytoremediators should be characterized by fast growth rate and high biomass production, wide-expanded root system and tolerance to heavy metals and toxic compounds, together with the capacity to immobilize and/or uptake them (Padmavathiamma and Li 2007). Considering phytoremediation as a process in which microorganisms and plants cooperate is crucial to develop new technologies that maximize results. To harness the potentiality of this methodology, a system-wide approach needs to be developed and applied: the outcome of genomic and metagenomic sequencing, transcriptomic, metabolomics and proteomics can shed light on unknown processes at the base of phytoremediation, thus leading to its wider and more significant application.

References

Abhilash, P.C., S. Jamil and N. Singh. 2009. Transgenic plants for enhanced biodegradation and phytoremediation of organic xenobiotics. Biotechnol. Adv. 27(4): 474–488. doi:10.1016/j.biotechadv. 2009.04.002.

Abhilash, P.C., J.R. Powell., H.B. Singh and B.K. Singh. 2012. Plant-microbe interactions: Novel applications for exploitation in multipurpose remediation technologies. Trends Biotechnol. 27(14): R713–R715. doi:10.1016/j.tibtech.2012.04.004.

Afzal, M., S. Yousaf., T.G. Reichenauer and A. Sessitsch. 2012. The inoculation method affects colonization and performance of bacterial inoculant strains in the phytoremediation of soil contaminated with diesel oil. Int. J. Phytoremed. 14(1): 35–47. doi:10.1080/15226514.2011.552928.

Anderson, T.A., E.A. Guthrie and B.T. Walton. 1993. Bioremediation in the rhizosphere. Env. Sci. Technol. 27: 2630–2636. doi:10.1021/es00049a001.

Arora, L. and A. Narula. 2017. Gene editing and crop improvement using CRISPR-Cas9 system. Frontiers Plant Sci. 7(1): 181–193. doi:10.3389/fpls.2017.01932.

Banerjee, S., K. Schlaeppi, G.A. Marcel and M.G.A. van der Heijden. 2018. Keystone taxa as drivers of microbiome structure and functioning. Nat. Rev. Microbiol. 16: 567–576. doi:10.1038/s41579-018-0024-1.

Banwart, S. 2011. Save our soils. Nature. 474: 151–152. doi:10.1038/474151a.

Barac, T., S. Taghavi, B. Borremans, A. Provoost, L. Oeyen, J.V. Colpaert, et al. 2004. Engineered endophytic bacteria improve phytoremediation of water-soluble, volatile, organic pollutants. Nature Biotechnol. 22(5): 583–588. doi:10.1038/nbt960.

Barberán, A., S.T. Bates, E.O. Casamayor and N. Fierer. 2012. Using network analysis to explore co-occurrence patterns in soil microbial communities. ISME J. 6(2): 343–351. doi:10.1038/ismej.2011.119.

Basharat, Z., L.A.B. Novo and A. Yasmin. 2018. Genome editing weds CRISPR: What is in it for phytoremediation? Plants. 7(3): 51. doi:10.3390/plants7030051.

Basu, S., R.C. Rabara. S. Negi and P. Shukla. 2018. Engineering PGPMOs through gene editing and systems biology: A solution for phytoremediation? Trends Biotechnol. 36(5): 499–510. doi:10.1016/j.tibtech.2018.01.011.

Bell, T.H., S. El-Din Hassan, A.L.Moreau, F. Al-Otaibi, M. Hijri, E. Yergeau, et al. 2014a. Linkage between bacterial and fungal rhizosphere communities in hydrocarbon-contaminated soils is related to plant phylogeny. The ISME Journal. 8: 331–343. doi:10.1038/ismej.2013.149.

Bell, T.H., S. Joly, F.E. Pitre and E. Yergeau. 2014b. Increasing phytoremediation efficiency and reliability using novel omics approaches. Trends Biotechnol. 32(5): 271–280 doi:10.1016/j.tibtech.2014.02.008.

Bellabarba, A., C. Fagorzi, G.C. diCenzo, F. Pini., C. Viti and A. Checcucci. 2019. Deciphering the symbiotic plant microbiome: Translating the most recent discoveries on rhizobia for the improvement of agricultural practices in metal-contaminated and high saline lands. Agron. 9(9): 529–539. doi:10.3390/agronomy9090529.

Berni, R., G. Guerriero and G. Cai. 2019. One for all and all for one increased plant heavy metal tolerance by growth-promoting microbes: A metabolomics standpoint. In: G. Sablok (ed.), Plant Metallomics Functional Omics. Springer Nature, Switzerland, pp. 39–51. doi:10.1007/978-3-030-19103-0_3.

Berry, D. and S. Widder. 2014. Deciphering microbial interactions and detecting keystone species with co-occurrence networks. Front. Microbiol. 5: 219. doi:10.3389/fmicb.2014.00219.

Breitling, R., A.R. Pitt and M.P. Barrett. 2006. Precision mapping of the metabolome. Trends Biotechnol. 12: 543–548. doi:10.1016/j.tibtech.2006.10.006.

Bulgarelli, D., M. Rott, K. Schlaepp, E.V.L. van Themaat, N. Ahmadinejad, F. Assenza, et al. 2012. Revealing structure and assembly cues for *Arabidopsis* root-inhabiting bacterial microbiota. Nature. 4(10): 4332–4339. doi:10.1038/nature11336.

Bulgarelli, D., K. Schlaeppi, S. Spaepen, E.V.L. van Themaat and P.S. Lefert. 2013. Structure and functions of the bacterial microbiota of plants. Ann. Rev. Plant Biol. 169(6): 2769–2773. doi:10.1146/annurev-arplant-050312-120106.

Chao, Y., W. Liu, Y. Chen, W. Chen, L. Zhao, Q. Ding, et al. 2016. Structure, variation, and co-occurrence of soil microbial communities in abandoned sites of a rare earth elements mine. Env. Sci. Technol. 50(21): 11481–11490. doi:10.1021/acs.est.6b02284.

Chen, D., S.J. Liu and W. Du. 2019. Chemotactic screening of imidazolinone-degrading bacteria by microfluidic slipchip. J. Hazard. Mat. 366: 512–519. doi:10.1016/j.jhazmat.2018.12.029.

Creek, D.J., A. Chokkathukalam, A. Jankevics, K.E.V. Burgess, R. Breitling and M.P. Barrett. 2012. Stable isotope-assisted metabolomics for network-wide metabolic pathway elucidation. Analytical Chem. 84: 8442–8447. doi:10.1021/ac3018795.

Carratore, F.D., K Schmidt, M. Vinaixa, K.A. Hollywood, C.G. Bews, E. Takano, et al. 2019. Integrated probabilistic annotation (IPA): A bayesian-based annotation method for metabolomic profiles integrating biochemical connections, isotope patterns and adduct relationships. Analytical Chem. 91(20): 12799–12807. doi:10.1021/acs.analchem.9b02354.

Delmotte, N., C. Knief, S. Chaffron, G. Innerebner, B. Roschitzki, R. Schlapbach, et al. 2009. Community proteogenomics reveals insights into the physiology of phyllosphere bacteria. Proceedings Nat. Acad. Sci. USA. 106(38): 16428–16433. doi:10.1073/pnas.0905240106.

diCenzo, G.C., M. Tesi, T. Pfau, A. Mengoni and M. Fondi. 2019. A virtual nodule environment (vine) for modelling the inter-kingdom metabolic integration during symbiotic nitrogen fixation. bioRxiv. 765271. doi:10.1101/765271.

Doty, S.L. 2008. Enhancing phytoremediation through the use of transgenics and endophytes. New Phytol. 179(2): 318–333. doi:10.1111/j.1469-8137.2008.02446.x.

Esser, D., J. Lange, G. Marinos, M. Sieber, L. Best, D. Prasse, et al. 2019. Functions of the microbiota for the physiology of animal metaorganisms. J. Innate Immunity. 11(5): 393–404. doi:10.1159/000495115.

Fagorzi, C., A. Checcucci, G.C. Dicenzo, K. Debiec-Andrzejewska, L. Dziewit, F. Pini, et al. 2018. Harnessing rhizobia to improve heavy-metal phytoremediation by legumes. Genes. 9(11): 542. doi:10.3390/genes9110542.

FAO. 2009. How to feed the world in 2050. Insights from an expert meeting at FAO. 1–35. doi:10.1111/j.1728-4457.2009.00312.x.

Fasani, E., A. Manara, F. Martini, A. Furini and G. DalCorso. 2018. The potential of genetic engineering of plants for the remediation of soils contaminated with heavy metals. Plant Cell Environ. 41(5): 1201–1232. doi:10.1111/pce.12963.

Feng, G., T. Xie, X. Wang, J. Bai, L. Tang, H. Zhao, et al. 2018. Metagenomic analysis of microbial community and function involved in cd-contaminated soil. BMC Microbiol. 18(1): 11. doi:10.1186/s12866-018-1152-5.

Fuhrman, J.A. 2009. Microbial community structure and its functional implications. Nature. 459(7244): 193–199. doi:10.1038/nature08058.

Fuhrman, J.A., J.A. Cram and D.M. Needham. 2015. Marine microbial community dynamics and their ecological interpretation. Nat. Rev. Microbiol. 13(3): 133–146. doi:10.1038/nrmicro3417.

Furtado, B.U., I. Nagy, T. Asp, M. Skorupa, M. Go, P. Hulisz, et al. 2019. Transcriptome profiling and environmental linkage to salinity across *Salicornia europaea* vegetation. BMC Plant Biol. 19: 1–14. doi: 10.1186/s12870-019-2032-3

Gkorezis, P., J.D. Van Hamme, E.M. Bottos, S. Thijs, M. Balseiro-Romero, C. Monterroso, et al. 2016. Draft genome sequence of *Pantoea ananatis* gb1, a plant-growth-promoting hydrocarbonoclastic root endophyte, isolated at a diesel fuel phytoremediation site planted with *Populus*. Genome Announc. 4(1): e00028–16. doi:10.1128/genomeA.00028-16.

Gonzalez, E., F.E. Pitre, A.P. Pagé, J. Marleau, W. Guidi Nissim, M. St-Arnaud, et al. 2018. Trees, fungi and bacteria: Tripartite metatranscriptomics of a root microbiome responding to soil contamination. Microbiome. 6(1): 53. doi:10.1186/s40168-018-0432-5.

Han, X., H. Yin, X. Song, Y. Zhang, M. Liu, J. Sang, et al. 2016. Integration of small RNAs, degradome and transcriptome sequencing in hyperaccumulator *Sedum alfredii* uncovers a complex regulatory network and provides insights into cadmium phytoremediation. Plant Biotechnol. J. 14(6): 1470–1483. doi:10.1111/pbi.12512.

Hassani, M.A., E. Özkurt, H. Seybold, T. Dagan and E.H. Stukenbrock. 2019. Interactions and coadaptation in plant metaorganisms. Ann. Rev. Phytopathol. 18(78): 110. doi:10.1146/annurev-phyto-082718-100008.

Heath, K.D., P.V. Burke and J.R. Stinchcombe. 2012. Coevolutionary genetic variation in the legume-*Rhizobium* transcriptome. Mol. Ecol. 21(19): 4735–4747. doi:10.1111/j.1365-294X.2012.05629.x.

Islam, M.M. and R. Saha 2018. Computational approaches on stoichiometric and kinetic modeling for efficient strain design. *In*: M.K. Jensen and J.D. Keasling (eds), Synthetic Metabolic Pathways. Methods in Molecular Biology, vol 1671. Humana Press, New York, NY, pp. 63–82. https://doi.org/10.1007/978-1-4939-7295-1_5.

Jehmlich, N., C. Vogt., V. Lünsmann, H.H. Richnow and M. von Bergen. 2016. Protein-SIP in environmental studies. Curr. Opin. Biotechnol. 41: 26–33. doi:10.1016/j.copbio.2016.04.010.

Jiao, S., Z. Liu., Y. Lin, J. Yang, W. Chen and G. Wei. 2016. Bacterial communities in oil contaminated soils: Biogeography and co-occurrence patterns. Soil Biol. Biochem. 98: 64–73. doi:10.1016/j. soilbio.2016.04.005.

Kashtwari, M., A.A. Wani and R.N. Rather. 2019. TILLING: An alternative path for crop improvement. J. Crop Improve. 33(1): 83–109. doi:10.1080/15427528.2018.1544954.

Klatt, C.G., Z. Liu., M. Ludwig., M. Kühl, S.I. Jensen., D.A. Bryant, et al. 2013. Temporal metatranscriptomic patterning in phototrophic chloroflexi inhabiting a microbial mat in a geothermal spring. ISME J. 7(9): 1775–1789. doi:10.1038/ismej.2013.52.

Kosová, K., P. Vítámvás., M.O. Urban., M. Klíma., A. Roy and I.T. Prášil. 2015. Biological networks underlying abiotic stress tolerance in temperate crops-a proteomic perspective. Int. J. Mol. Sci. 16(9): 20913–20942. doi:10.3390/ijms160920913.

Lasken, R.S. 2012. Genomic sequencing of uncultured microorganisms from single cells. Nature Rev. Microbiol. 10(9): 631–640. doi:10.1038/nrmicro2857.

Le Moal, M., C. Gascuel-Odoux, A. Ménesguen, Y. Souchon, C. Étrillard, A. Levain, et al. 2019. Eutrophication: A new wine in an old bottle? Sci Total Env. 651: 1–11. doi:10.1016/j.scitotenv.2018.09.139.

Lekunberri, I., J.L. Balcázar and C.M. Borrego. 2018. Metagenomic exploration reveals a marked change in the river resistome and mobilome after treated wastewater discharges. Environ. Poll. 27(14): R713–R715. doi:10.1016/j.envpol.2017.12.001.

Lu, M., S. Jiao, E. Gao, X. Song, Z. Li, X. Hao, et al. 2017. Transcriptome response to heavy metals in *Sinorhizobium meliloti* CCNWSX0020 reveals new metal resistance determinants that also promote bioremediation by *Medicago lupulina* in metal contaminated soil. Appl. Environ. Microbiol. 83(20): e01244-17. doi:10.1128/AEM.01244-17.

Macdonald, C. and B. Singh. 2013. Harnessing plant-microbe interactions for enhancing farm productivity. Bioengineered. 5(1): 5–9. doi:10.4161/bioe.25320.

Maggini, V., E. Miceli, C. Fagorzi, I. Maida, M. Fondi, E. Perrin, et al. 2018. Antagonism and antibiotic resistance drive a species-specific plant microbiota differentiation in *Echinacea* spp. FEMS Microbiol. Ecol. 94(8): fiy118. doi:10.1093/femsec/fiy118.

Mallon, C.A., X. Le Roux, G.S. Van Doorn, F. Dini-Andreote, F. Poly and J.F. Salles. 2018. The impact of failure: Unsuccessful bacterial invasions steer the soil microbial community away from the invader's niche. ISME J. 12(3): 728–741. doi:10.1038/s41396-017-0003-y.

Martiny, A.C. 2019. High proportions of bacteria are culturable across major biomes. ISME J. 13(8): 2125–2128. doi:10.1038/s41396-019-0410-3.

Mirete, S., V. Morgante and J.E. González-Pastor. 2016. Functional metagenomics of extreme environments. Curr. Opinion Biotechnol. 38: 143–149. doi:10.1016/j.copbio.2016.01.017.

Morozova, O. and M.A. Marra. 2008. Applications of next-generation sequencing technologies in functional genomics. Genomics. 92(5): 255–264. doi:10.1016/j.ygeno.2008.07.001.

Mueller, U.G. and J.L. Sachs. 2015. Engineering microbiomes to improve plant and animal health. Trends Microbiol. 23(10): 606–617. doi:10.1016/j.tim.2015.07.009.

Mukhtar, S., S. Mehnaz and K.A. Malik. 2019. Microbial diversity in the rhizosphere of plants growing under extreme environments and its impact on crop improvement. Environ. Sustain. 2: 329–338. doi:10.1007/s42398-019-00061-5.

Nayak, A.K., S.S. Panda, A. Basu and N.K. Dhal. 2018. Enhancement of toxic Cr (VI), Fe, and other heavy metals phytoremediation by the synergistic combination of native *Bacillus cereus* strain and *Vetiveria zizanioides* L. Int. J. Phytoremed. 20(7): 682–691. doi:10.1080/15226514.2017.1413332.

Padmavathiamma, P.K. and L.Y. Li. 2007. Phytoremediation technology: Hyper-accumulation metals in plants. Water Air Soil Pollut. 184: 105–126. doi:10.1007/s11270-007-9401-5.

Petzold, C.J., L.J.G. Chan, M. Nhan and P.D. Adams. 2015. Analytics for metabolic engineering. Front. Bioeng. Biotechnol. 3: 135. doi:10.3389/fbioe.2015.00135.

Pham, Van H.T. and J. Kim. 2012. Cultivation of unculturable soil bacteria. Trends Biotechnol. 30(9): 475–484. doi:10.1016/j.tibtech.2012.05.007.

Rabara, R.C., P. Tripathi and P.J. Rushton. 2017. comparative metabolome profile between tobacco and soybean grown under water-stressed conditions. BioMed. Res. Int. Article ID 3065251. p. 12. doi:10.1155/2017/3065251.

Rajkumar, M. and H. Freitas. 2008. Effects of inoculation of plant-growth promoting bacteria on ni uptake by indian mustard. Bioresource Technol. 99: 3491–3498. doi:10.1016/j.biortech.2007.07.046.

Rausch, P., M. Rühlemann, B. Hermes, S. Doms, T. Dagan, K. Dierking, et al. 2019. Comparative analysis of amplicon and metagenomic sequencing methods reveals key features in the evolution of animal metaorganisms. Bio. Rxiv. 7(1): 1–19. doi:10.1101/604314.

Shen, Y., Y. Ji, C. Li, P. Luo, W. Wang, Y. Zhang, et al. 2018. Effects of phytoremediation treatment on bacterial community structure and diversity in different petroleum-contaminated soils. Int. J Env. Res. Public Health 15(10): 2168. doi:10.3390/ijerph15102168.

Sieber, M., L. Pita, N. Weiland-Bräuer, P. Dirksen, J. Wang, B. Mortzfeld, et al. 2019. Neutrality in the metaorganism. PLoS Biol. 17(6): e3000298. doi:10.1371/journal.pbio.3000298.

Stefani, F.O.P., T.H. Bell, C. Marchand, I.E. De La Providencia, A. El Yassimi, M. St-Arnaud, et al. 2015. Culture-dependent and -independent methods capture different microbial community fractions in hydrocarbon-contaminated soils. PLoS ONE. 10(4): e0124260. doi:10.1371/journal.pone.0128272.

Taghavi, S., C. Garafola, S. Monchy, L. Newman, A. Hoffman, N. Weyens, et al. 2009. Genome survey and characterization of endophytic bacteria exhibiting a beneficial effect on growth and development of poplar trees. Appl. Env. Microbiol. 75(3): 748–757. doi:10.1128/AEM.02239-08.

Taghavi, S., X. Wu, L. Ouyang, Y.B. Zhang, A. Stadler, S. McCorkle, et al. 2015. transcriptional responses to sucrose mimic the plant-associated life style of the plant growth promoting endophyte *Enterobacter* sp. 638. PLoS ONE. 9(12): e113571. doi:10.1371/journal.pone.0115455.

Thijs, S., W. Sillen, F. Rineau, N. Weyens and J. Vangronsveld. 2016. Towards an enhanced understanding of plant-microbiome interactions to improve phytoremediation: Engineering the metaorganism. Front. Microbiol. 7: 341. doi:10.3389/fmicb.2016.00341.

Thompson, I.P., C.J. van Der Gast, L. Ciric and A.C. Singer. 2005. Bioaugmentation for bioremediation: The challenge of strain selection. Environ. Microbiol. 4: 27–64. doi:10.1111/j.1462-2920.2005.00804.x.

Uchiyama, T. and K. Miyazaki. 2009. Functional metagenomics for enzyme discovery: Challenges to efficient screening. Curr. Opi. Biotechnol. 20(6): 616–622. doi:10.1016/j.copbio.2009.09.010.

Van Aken, B. 2008. Transgenic plants for phytoremediation: Helping nature to clean up environmental pollution. Trends Biotechnol. 26(5): 225–227. doi:10.1016/j.tibtech.2008.02.001.

Van der Heijden, M.G.A., R. Bakker, J. Verwaal, T.R. Scheublin, M. Rutten, R.S.P. van Logtestijn, et al. 2006. Symbiotic bacteria as a determinant of plant community structure and plant productivity in dune grassland. FEMS Microbiol. Ecol. 33(3): 761–787. doi:10.1111/j.1574-6941.2006.00086.x.

Villas-Bôas, S.G., S. Mas, M. Åkesson, J. Smedsgaard and J. Nielsen. 2005. Mass spectrometry in metabolome analysis. Mass Spectro. Rev. 24(5): 613–646. doi:10.1002/mas.20032.

Villas-Bôas, S.G. and P. Bruheim. 2007. The potential of metabolomics tools in bioremediation studies. OMICS A Journal Integrative Biol. 11(3): 305–313.. doi:10.1089/omi.2007.0005.

Villiers, F., V. Hugouvieux, N. Leonhardt, A. Vavasseur, C. Junot, Y. Vandenbrouck, et al. 2012. Exploring the plant response to cadmium exposure by transcriptomic, proteomic and metabolomic approaches: Potentiality of high-throughput methods, promises of integrative biology. *In*: D.K. Gupta and L.M. Sandalio (eds), Metal Toxicity in Plants: Perception, Signaling and Remediation. Springer-Verlag, Berlin, Heildelberg, pp. 119–132. doi:10.1007/978-3-642-22081-4_6.

Yergeau, E., S. Sanschagrin, C. Maynard, M. St-Arnaud and C.W. Greer. 2014. Microbial expression profiles in the rhizosphere of willows depend on soil contamination. ISME J. 8(2): 344–358. doi:10.1038/ismej.2013.163.

Zhang, W.H., L.Y. He, Q. Wang and X.F. Sheng. 2015. Inoculation with endophytic *Bacillus megaterium* 1Y31 increases Mn accumulation and induces the growth and energy metabolism-related differentially-expressed proteome in Mn hyperaccumulator hybrid *Pennisetum*. J. Hazard. Mat. 300: 513–521. doi:10.1016/j.jhazmat.2015.07.049.

Zivy, M., S. Wienkoop, J. Renaut, C. Pinheiro, E. Goulas and S. Carpentier. 2015. The quest for tolerant varieties: The importance of integrating 'omics' techniques to phenotyping. Front. Plant Sci. 6: 448. doi:10.3389/fpls.2015.00448.

CHAPTER 5

Microbiomes and Metallic Nanoparticles in Remediation of Contaminated Environments

**Ana Maria Queijeiro López[1]*, Amanda Lys dos Santos Silva[1],
Elane Cristina Lourenço dos Santos[1]
and Jean Phellipe Marques do Nascimento[2]**

[1]Federal University of Alagoas, Institute of Chemistry and Biotechnology (IQB),
Laboratory of Biochemistry of Parasitism and Environmental Microbiology (LBPMA),
57072-900, Maceió-AL, Brazil.
[2]Federal University of Alagoas, Institute of Biological and Health Sciences, Laboratory of
Genetic and Applied Microbiology, 57072-900, Maceió-AL, Brazil.

1. INTRODUCTION

About 3.5 billion years ago, microorganisms appeared, multiplied, differentiated, occupied all territories, dominated and shaped the evolution of the biosphere on our planet, including interfering in the hydrosphere, lithosphere, and atmosphere. So, the health of the environment and of the living beings is dependent on their microbial partners. Together, they purify water from rivers, seas, streams, lakes, reservoirs, and aquifers, regulating the flow of nutrients derived from carbon, nitrogen, phosphorus, sulphur and oxygen, which regulate the stability of ecosystems and complex food chains. In turn, when the balance between populations of microorganisms (and between their genes) in nature (microbiomes) changes, due to natural or human interference, damage to the entire ecosphere can occur (Blaser et al. 2016). With a current world population of about 7.8 billion human beings (many of whom are living in urban centers), the planet is experiencing a difficult phase for all living beings, including all microbiomes. Several challenges have arisen related to the sustainable production of food, energy, and chemicals to favor the life of the human population, with an urgent need to understand, predict and combat the Earth's

*Corresponding author: amql@qui.ufal.br

environmental changes, preventing and reversing the degradation of ecosystems and better controlling diseases (plant, human and other animals) of microbial etiology.

One of the main tenets of sustainable development is that man should not lead the planet beyond the environmental limits of its living forms, including that of humans themselves. With each passing day, this premise becomes a more urgent cause, especially when the world's human population is expected to increase to about ten billion by 2050 (Barea 2015). This will require not only more renewable forms of energy than those derived from finite fossil fuels, and the burning of which generates pollutants toxic to the health of all life forms, but mainly drinking water for their consumption, water for irrigation, and unsalted fertile soil to produce all the food required for its survival. Consequently, many challenges arise, such as the incentive to create processes that reduce the generation of pollutants, the treatment of water to be reused and sustainably intensive agricultural production, with minimal use of recalcitrant agrochemicals, while preserving the biosphere as much as possible. In this sense, several practices are currently implemented on a global scale and different strategies are being addressed to meet this requirement. Commonly, in the remediation of contaminated areas, technologies such as chemical oxidation, thermal desorption, confinement in geotechnical cells, incineration and bioremediation are used. So, different microbial strains, with high genetic diversity, residing for instance in soil or contaminated effluents, provide a variety of metabolic capacities that are useful in many aspects for ecological issues, degrading, removing, altering, immobilizing, or detoxifying several types of "wastes" from the environment through diverse ways (Abatenh et al. 2017). They can produce bioactive compounds (Russo et al. 2012, Tanvir et al. 2016), biodegradable plastics (Wei and Zimmermann 2017), promote plant growth and, as mentioned above, degrade pollutants, and this is the process called bioremediation. The biological communities used in this strategy comprise a very promising and fascinating area, but several environmental factors – such as levels of humidity, temperature, pH, salinity, and presence of organic matter, along with the structure and concentration of contaminants, may affect this and other strategies of remediation.

Compared to chemical remediation, bioremediation has been prioritized in many circumstances (Zhou et al. 2018, Li et al. 2015, 2018), especially when the BOD/COD ratio is > 0.6, in addition to its low cost and no secondary pollution (Liu et al. 2017, Chen et al. 2018). Since the sixth decade of the last century, many researchers have worked on isolating microorganisms from the soil capable of degrading efficiently, for example, the herbicide atrazine, such as species of the bacteria *Klebsiella* (Yang et al. 2010), *Pseudomonas* (Klein et al. 2012), *Rhodococcus* (Meyer et al. 2014) and *Acinetobacter* (Yang et al. 2017), or of the actinomycetes *Nocardia* (Smith et al. 2005) and *Arthrobacter* (Getenga et al. 2009), as well as strains of the fungi *Penicillium* (which have great potential for remediation of recalcitrant organic matter due to their enzymatic system via cytochrome P450, that performs dealkylation and dehalogenation reactions (Yang et al. 2008, Yu et al. 2018)). However, bio-based techniques are generally slower and sensitive to high concentrations of pollutants, which can become toxic to those involved in this remediation.

In general, bioremediation is based on *in situ* stimulation of the indigenous microorganisms (biostimulation) or additions of specific degrading ones (bioaugmentation) (Alabresm et al. 2018), but the synergism between nanotechnology and microbiome could improve this process, which means, for the success of this process, it is crucial to know the microbiota to understand its degradation mechanisms, the benefits of its presence and whether the addition of nanoparticles (<100 nm) provides it with a healthy substrate for its activity, speeding up the cleaning of the environment ecologically and economically (Davis et al. 2017). Species such as *Acinetobacter calcoaceticus*, *Bacillus amyloliquefaciens*, *B. licheniformis*, *Escherichia coli* and *Pseudomonas aeruginosa*, for instance, are efficient in the production of silver nanoparticles (Siddiqi et al. 2018).

Therefore, the nanoscale size of metal nanoparticles (MtNPs) produced by microorganisms allowed them to have new or better properties than the larger particles from which they originated, with particular physicochemical and biological characteristics according to their size, distribution, morphology, phase, composition, among others. Nanobioremediation was used by Alabresm et al. (2018) to degrade oil spills, and this can be used in contaminated marine and non-marine systems. On the other hand, microbes highly resistant to the toxicity of gold ions were isolated from mines of this metal and could be used efficiently in the synthesis of nanoparticles of the same (AuNPs). In face of its ability to generate bioelectricity and reduce metals, the strain of *Shewanella oneidensis* MR-1 is broadly used in the recovery of precious metals, bioremediation, and AuNPs fabrication. Huang et al. (2019) reported, for the first time in this strain, the mechanism for the formation of AuNPs through the accumulation of photons based on the intensity and wavelength of light. Srinath et al. (2018) used *B. subtilis* isolated from Hatti Gold Mine (India) for the synthesis of AuNPs, which were then used as catalysts for the degradation of methylene blue dye, with a view to later use in the decomposition of other dyes toxic to the environment.

According to the report titled "Metal & Metal Oxide Nanoparticles Market: Global Industry Analysis and Opportunity Assessment, 2016-2026", the world production of MtNPs in 2017 was valued at US$ 13.7 billion, and their widespread use contributed significantly to robust macroeconomic development. By the end of 2026, it is projected that the global market of metal oxide nanoparticles will be worth ten times less than that of MtNPs when this will reach 50 billion US dollars. The report reveals that the demand for metal (and metal oxide) nanoparticles will remain considerably high in three regions of the world – North America, Western Europe and the Asia-Pacific (excluding Japan) towards the end of 2026, when North America will dominate the global market of MtNPs, with 30% share, while Europe will be at the forefront of global metal oxide nanoparticles market (FMI 2017).

To determine the physicochemical properties of nanoparticles in the environmental remediation field, particles as self-organized monolayers on mesoporous supports, dendrimers and carbon nanotubes, as well as their path, distribution, persistence and their toxicological effects on biological systems have been studied. Additionally, as microorganisms are considered potential "biofactory" for the biosynthesis of eco-friendly, cost-effective and stable nanoparticles, and most of bacteria and archaea have special metal-binding abilities and two-dimensional arrays of proteinaceous subunits forming the surface layers (S-layers) of their envelope, these organisms are useful for applications in bioremediation and nanotechnology (Yadav et al. 2017). Therefore, the present chapter presents a review of nanoparticles and microbes used in bioremediation.

2. CONTAMINATED ENVIRONMENTS AND MICROBIOMES

Human activities are creating more and more obstacles to the planet's health: antibiotics and household chemicals are pollutants that are exponentially entering sensitive ecosystems, causing imbalances in them. In 2019, the interagency coordination group on the antimicrobial resistance of the United Nations (IACGAMR) declared that "AMR" is a worldwide epidemic since antibiotic-resistant bacteria already cause about 700,000 deaths worldwide each year, including 230,000 deaths from multidrug-resistant tuberculosis, predicting that by 2050 about 2.4 million people can die because of them in rich countries, and if no action is taken, that number could increase to 10 million deaths a year worldwide (WHO-IACGAMR 2019).

Research co-led in 2019 by Boxall and Wilkinson, from the University of York (England), showed that 65% of rivers in 711 sites sampled in 72 countries were contaminated with antibiotics, and their presence exceeded the safe level in 111 of those sites. Poor countries were the most impacted, due to their high consumption of antibiotics associated with the lack of adequate wastewater treatment technologies and sanitation facilities (Boxall and Wilkinson 2019, Wilkinson and Boxall 2019). In the same way, the application of pesticides to crops, as they penetrate ecosystems and contaminate our soils, rivers, and wetlands, also negatively impacted native microorganisms as well as insect and wildlife populations.

Due to the nutritional versatility of microorganisms, they can often produce different compounds (i.e. gases, enzymes, acids, solvents, biosurfactants, and biopolymers, etc.) that improve the transformation of contaminants into less toxic or non-toxic elemental forms (Abatenh et al. 2017, Russo et al. 2012, Varjani and Upasani 2016). Biosurfactants, for instance, play a key role in increasing the bioavailability of polluting hydrocarbons for degrading enzymes (Santos et al. 2019, Souza et al. 2014). Bacterial strains of the genera *Aeromicrobium, Bacillus, Brevibacterium, Burkholderia, Corynebacterium, Desulfotomaculum, Desulfovibrio, Dietzia, Escherichia, Gordonia, Methanosaeta, Micrococcus, Moraxella, Mycobacterium, Pandoraea, Pelatomaculum, Pseudomonas, Rhodococcus, Staphylococcus, Stenotrophomonas* and *Sphingobium* (Chowdhury et al. 2008, Mikeskov'a et al. 2012, De Roy et al. 2014, Chakraborty and Das 2016), as well fungal strains of the genera *Amorphoteca, Graphium, Irpex, Neosartorya* and *Talaromyces,* for instance, were found able to degrade persistant organic products (POPs) (Gupta et al. 2016a and b, Lenoir et al. 2016).

Studies *in vivo*, however, indicated that individual strains can be affected or cause more marked changes in the native microbial community than the consortium, which can compromise the functionality of the ecosystem to be corrected (Teng and Chen 2019). Exogenous microorganisms are difficult to survive and grow in some ecosystems, such as soil environments, not only because of persistent abiotic stressors, but also the autochthonous microorganisms that can produce or resist antibiotic have a higher growth rate and compete for space and nutrients, etc. (Perez-Garcia et al. 2016, Varjani and Upasani 2017). The advantage of microbial consortia concerning single strains is that their high diversity can favor exogenous functional microorganisms to survive in new environments (Großkopf and Soyer 2014). Sathishkumar et al. (2008) studied the biodegradation of crude oil and found that the degradation rate by the individual strains of four bacteria (*Pseudomonas* sp. BPS1-8, *Bacillus* sp. IOS1-7, *Pseudomonas* sp. HPS2-5 and *Corynebacterium* sp. BPS2-6), were, respectively, 69%, 64%, 45%, and 41%, but the consortium removed 77% of such pollutant. Kang et al. (2016) also analyzed the synergistic effect of a bacterial consortium formed by *Enterobacter cloacae* KJ-46, *Enterobacter cloacae* KJ-47, *Sporosarcina soli* B-22 and *Viridibacillus arenosi* B-21 on the remediation of the combination of cadmium, copper, and lead in soils, and evidenced that the bacterial consortium was more resistant to toxic heavy metal than the single strains.

However, the knowledge that carefully balanced communities of microorganisms from a particular environment (microbiota) plus their genes, which is called microbiome, are crucial to the health and prosperity of different ecosystems (different parts of humans and other animals, oceans, soils, and rivers) is not so widespread. Such populations form intricate interactive cell-to-cell networks through such quorum sensing communication (Kylilis et al. 2018) and may cooperate to form biofilms or compete, making the microbiome efficient and stable. This increases the resistance and resilience of microorganism processes in such environments, and they can proliferate in toxic organic compounds, such as polycyclic aromatic hydrocarbons (PAHs), agrochemicals, polychlorinated biphenyls (PCBs), plastics, etc. The microbial gene mutation, rearrangement, and differential regulation in the microbiome populations may help their survival in many of these unfavorable contaminated environments (Teng et al. 2015).

The health of aquatic or soil microbiomes should be taken into account if crop and wildlife growth is to be expected to feed and sustain a wide variety of species of different trophic levels. Therefore, balancing the regulatory mechanisms of microbiome communities (Fig. 1) is extremely important for their existence and ecosystem remediation. This is expected to occur, for example, in oceans containing more than eight billion tons of petroleum-based plastics that have been produced worldwide in the last 70 years. Biodegradation of plastics, however, is restricted by the low surface-to-volume ratio of their debris, and research on their micronization, increasing surface area accessibility to microbial enzymes (such as polyester hydrolases), may improve their subsequent degradation (Wei and Zimmermann 2017).

Microbial polyethylene terephthalate hydrolase (PETase) and mono(2-hydroxyethyl) terephthalic acid (TPA) hydrolase, for instance, biodegrade PET to TPA, and this is the goal of the TPA-dioxygenase. The resulting product is converted by the protocatechuic acid dioxygenase in 3-carboxy-muconic acid (Yoshida et al. 2016). Apart from these, several other hydrolases (for example, lipases, cellulases, and proteases), which cleave the main chemical bonds in recalcitrant molecules, reducing their toxicity or making them more accessible to other different enzymes, also play a role in biodegradation (Hiraishi 2016). The degradation of many kinds of halogenated organic compounds (i.e. plasticizers, halogenated methanes, ethanes, ethylenes, insecticides, herbicides, fungicides, etc.) can be obtained by specific oxidoreductases (Karigar and Rao 2011).

Figure 1 Interferences for the formation, operation, and integration of different microbiomes and the impact of this knowledge on their responsible use in technology (Adapted from Blaser et al. 2016).

Mono-oxygenases, such as those from the P450 system, are types of heme-containing oxygenase present in eukaryotic and prokaryotic organisms, and incorporate one atom of the molecular oxygen into the substrate (Arora et al. 2010), while dioxygenases use two oxygen atoms to oxidize organic molecules as aromatic compounds, resulting in the release of aliphatic molecules (Chakraborty and Das 2016). Laccases, in another way, can oxidize decarboxylate and demethylate substrates such as polyaromatic hydrocarbons, polyphenols, ortho and para-diphenols (Zeng et al. 2016), while peroxidases are enzymes present in practically all organisms, that lead to the oxidation of lignin and other phenolic compounds at the expense of H_2O_2 in the presence of a mediator (Karigar and Rao 2011).

The microbiome metabolism could also change the environment pH, release siderophores and organic acids, including volatiles, being then able to enhance the complexation or removal of metals, their mobility, and bioavailability. Such microbial metal "resistance" may be due to individual sequestration of toxic metals by cell wall components or intracellular metal-binding proteins/ peptides of some strains of the group, alteration of pathways to stop metal uptake, immobilization and/or change of their redox state to decrease the toxicity to enzymes, and reduction of intracellular concentrations of these metals using efflux systems (Teng and Chen 2019, Teng et al. 2015). Moreover, microorganisms that promote plant growth in the soil, for example, employ different strategies for their action, including nitrogen fixation, phosphorus solubilization, synthesis and release of siderophores, indoleacetic acid, and other phytohormones, 1-aminocyclopropane-acid deaminase 1-carboxylic and volatile compounds (Hao et al. 2014), which make them valuable tools to increase the crops' productivity and assist phytoremediation (Ojuederie and Babalola 2017, Teng and Chen 2019).

Therefore, more research is needed to improve the microbiome efficiency for better water quality and soil fertility, while making bioremediation not only a widely accepted but economically efficient technique (Table 1).

Table 1 Advantages and disadvantages of bioremediation using microorganisms.

Advantages	*Disadvantages*
• A natural and cheap process that takes a little time; • The residues for the treatment are usually harmless products; • It requires much less effort and assists in the complete removal of pollutants (many of the dangerous compounds are converted into products without toxicity); • It does not use any dangerous chemicals; • Eco-friendly and sustainable; • The relatively easy implementation.	• Not all compounds are susceptible to rapid and complete degradation; • The success of bioremediation relies on the existence of metabolically active microbial populations in the appropriate environmental conditions for their growth, including concentrations of nutrients and contaminants; • It is difficult to take bench studies to the pilot scale and field-scale operations; • It is generally more time consuming than other environmental quality correction strategies, such as excavation, removal or incineration.

In this context, studies using nanotechnology through microbiomes, i.e. the removal of heavy metals, organic and inorganic pollutants from contaminated sites with the help of nanoparticles/nanomaterial produced by microbiomes, or their effect in microbiomes, has gained great prominence. For example, the growing use of antimicrobial silver nanoparticles (AgNPs) in industrial and agricultural household products is also rising in ecosystems (Grün et al. 2018). Sun et al. (2014) estimated annual European production of 32.4 tons of AgNPs and an annual increase of these in soils and sediments in the range of 1.2 ng/(kg year) to 2.3 ng/(kg year). The significant negative effect of these AgNPs on biomass and bacterial ammonia oxidants after 1 year of exposure to 0.01 mg AgNP/kg was found. In the long run, the tested AgNP concentrations of 0.01–1.00 mg AgNP/kg significantly decreased the activity of leucine aminopeptidase, as well as the population of nitrogen-fixing bacteria (Grün et al. 2018). Hänsch and Emmerling (2010) had already found a decrease in the microbial biomass of sandy soil after 4 months of increasing the concentrations of AgNP from 0.0032 to 0.320 mg

AgNP/kg. These studies demonstrate the effect of AgNPs on soil functions, such as the transformation of organic matter and the cycle of energy and nutrients.

For all these reasons, the important impact of the interaction microbiomes-nanotechnology in biodegradation of pollutants in soil, water or air, has gained increased recognition over the past few decades (Yadav et al. 2017).

3. NANOPARTICLES AND BIOREMEDIATION

Through evolutionary mechanisms, the microbiota of environments containing high levels of metals adapts to deal with them, and this may involve changing the chemical nature of the substrate, ceasing or reducing toxicity, with the formation of nanoparticles of the metal in question (Pantidos and Horsfall 2014). Among all the metals commonly detected in waste, gold is a noble metal used in jewelry, electronics and medical applications (Wu and Ng 2017). Silver, on the other hand, is not as expensive as gold but is also widely used in electronics, chemical and jewelry industries; so, its presence on industrial wastes is frequent.

On the other hand, the concept of nanotechnology was first mentioned by Richard Feynman in 1959, and its current use is often referred to as an innovative tool, or the "new industrial revolution", and is characterized by the use of atomic or molecular aggregates called nanoparticles or ultrafine particles (1 to 100 nm in size – about 1000 times smaller than the width of a human hair). Such particles can be produced from a variety of crude materials, and their performance also differs according to their chemical composition, size or shape (Yadav et al. 2017). Then, although nanomaterials may have the same chemical composition as their parent material (micrometer or millimeter equivalent), they may be significantly more reactive than it (Mueller and Nowack 2010). For example, aluminum cans and unrefined gold are not chemically hazardous or catalytically active materials, but their nanoscale equivalents are, respectively, highly explosive or have catalytic activity (Suchomel et al. 2018). Another important example is the zero-valent iron (ZVI), a strong reducing agent able to abiotically dehalogenate several common chlorinated solvents. Reductive technologies based on ZVI and nanoscale zerovalent iron (nZVI, nanoFe0) have become widely used due to their ability to degrade a wide spectrum of contaminants (Jang et al. 2014).

Nanomaterials also show quantum effect and surface plasmonium, i.e. a larger amount of them can contact the surrounding products, requiring less activation energy to enable chemical reactions, so that they may be used to detect the presence of toxic compounds (Rizwan et al. 2014). Therefore, the high reactivity and large surface area to mass ratio of metallic nanoparticles (MtNPs) give them different properties (electronic, magnetic, optical, antimicrobial, anti-cancer, and catalyzing) and make them the tools of choice in many fields, including the remediation of environmental pollutants (Davis et al. 2017). Several nano-applications for ecoefficient remediation of environments are quickly growing from the pilot to full scale, treating, for instance, chlorinated sites that are environmentally demanding. Preliminarily, such MtNPs were synthesized through different physical (lithography, pyrolysis, vapor pressure, etc.) and chemical (micro-emulsion, irradiation, and electrochemical reduction) methods. Those methods are orthodox but have high energy consumption, high cost and involve toxic chemical compounds (Ovais et al. 2018). Otherwise, the use of many bacteria, fungi and plants able to synthesize MtNPs, i.e. the biological route or "natural" biogenic MtNPs synthesis (Pantidos and Horsfall 2014), has shown to be an advantageous approach – fast, economical, biocompatible, non-toxic and eco-friendly (Ovais et al. 2018), despite their yield still being lower than that of chemical synthesis.

AgNPs are essential and not replaceable components for different processes in several industries (automobile appliances, cosmetics, electronics, medical devices, imaging techniques, textiles, etc.) (Güzel and Erdal 2018), and they can be biosynthesized by reduction (Hulkoti and Taranath 2014), exhibiting also bactericidal activity higher than conventional antibiotics, even against multiresistant bacteria through synergistic effects (Siddiqi et al. 2018).

In 2014, Makky et al. described, for the first time, the production of AgNPs from disposed X-ray film, which was used as the sole carbon for the growth of *Bacillus* strains, which were successfully able to reduce silver nitrates to nitrites. In another report, Kulkarni et al. (2015) took into account the fact that *Deinococcus radiodurans* can withstand high conditions of radiation and desiccation, suggesting that it could be used in bioremediation, and elucidated the extracellular biosynthesis of AgNPs by such bacteria, reducing silver chloride solution. Regarding α-Fe_2O_3 NPs, Khatoon and Rai (2018) showed that their combination with *Bacillus badius* ABP6 and *B. encimensis* ABP8 cultures was more efficient to degrade atrazine [2-chloro-4-(ethylamine)-6-(isopropyl amino)-1,3,5-triazine] as carbon and nitrogen source than the free cells of these strains. It is important to emphasize that in bioremediation, free bacterial cells can have limited degradation. However, such difficulty is minimized when they are immobilized.

Nanoscale TiO_2, carbon nanotubes (CNTs), dendrimers, swellable organically modified silica (SOMS) and metallo-porphyrinogens are potential nanoproducts for pollutants' remediation, *ex situ* or *in situ* (He et al. 2015, Joshi et al. 2019, Wang et al. 2017). TiO_2 nanoparticles can remove a range of chemical agrochemicals through photo-catalysis as well as for *ex situ* management of infected ground-water resources. Biological production of NPs of copper, iron, and titanium in combination with a metal-catalyst (such as Au, Ni, Pd, and Pt) improves the redox-reaction rate (He et al. 2015). The biosynthesized iron NPs and Fe-Pd NPs have shown wider application in the treatment of agrochemicals, dyes, hydrocarbons, 2,3,7,8-tetrachlorodibenzo-p dioxin, trichloroethene (TCE), polychlorinated biphenyl (PCB), etc., using bacterial metabolism (Asmel et al. 2017, Pandi et al. 2017, Zhang et al. 2018).

PdNPs catalyze the reduction of trichloroethene in ethane without intermediates such as vinyl chloride. A parallel metal-glass fusion material, Pd-Osorb has been successfully tested for *ex situ* remediation of chlorinated volatile organic compounds (VOCs) (Song et al. 2017, Su et al. 2016). Silica NPs remove lead, while zinc NPs remove CS_2 from the air, and nanocrystalline hydroxyl-apatite removes lead and cadmium, while zerovalent nano-iron, CNTs, fullerenes, TiO_2 and ZnO NPs and bimetallic nano-metals are used for removal of dichlorodiphenyltrichloroethane (DDT), carbamates, and heavy metals (like chromium, lead, arsenic and cadmium) from soil (Campbell et al. 2015, Pandi et al. 2017).

4. MICROBIAL SYNTHESIS OF METALLIC-NANOPARTICLES

Different microbes can generate MtNPs (Table 2) and produce biomolecules that can help their capture, stabilization, and determine their properties (such as shape, size, etc.) and functionalities, making them more effective in relation to the NPs that are chemically synthesized (Singh et al. 2013). These microbial MtNPs derived from cadmium sulfide, copper, gold, magnetite, platinum, palladium, silver, tellurium, titanium, titanium dioxide, zinc, etc., can occur via both intracellular and/or extracellular strategies (Hulkoti and Tanarath 2014) and have received considerable attention in the field of material synthesis (Güzel and Erdal 2018, Ovais et al. 2018, Yadav et al. 2017).

So, the microbial synthesis of MtNPs may be influenced by many factors, such as the concentration of the previous substrate, the pH, the oxygenation, the temperature, the interval of incubation and reactions as well as the NPs solubility, biosorption, metal complexation, extracellular precipitation, toxicity via oxidation-reduction, the absence of specific transporters and efflux pumps (Mukherjee and Patra 2017, Pantidos and Horsfall 2014, Patra et al. 2014). In the intracellular strategy, the microbial cell has a sophisticated ion transport system, and due to electrostatic interaction, the negatively charged cell wall of the bacteria attracts positively charged metal ions, in addition to containing enzymes that reduce metal ions to their respective NPs. On the other hand, in the extracellular strategy, the fungal mycelium or bacterial cell secretes mainly reductases and other compounds used in the bioreduction of metal ions into the corresponding MtNPs (Hulkoti and Tanarath 2014).

Table 2   List of examples (from 2011) of bacteria and yeast/fungi involved in the synthesis of mainly metallic nanoparticles (MtNPs) and their possible use in bioremediation

Nanoparticles (NPs)	*Bacteria Species*	*References*
Cadmium	*Pseudomonas* spp.	Plaza et al. 2016
	Psychrobacter spp.	Plaza et al. 2016
	Shewanella spp.	Plaza et al. 2016
Gold	*Shewanella oneidensis*	Suresh et al. 2011, Kane et al. 2016,
	Shewanella loihica	Ahmed et al. 2018
Paladium	*Desulfovibrio desulfuricans*	Macaskie et al. 2012, Mikheenko
	Escherichia coli	et al. 2019, Macaskie et al. 2012,
	Shewanella loihica	Ahmed et al. 2018
Platinum	*Shewanella loihica*	Ahmed et al. 2018
Selenium	*Bacillus cereus*	Kora 2018
Silver	*Bacillus licheniformis*	Farias et al. 2014
	Bacillus cereus	Das et al. 2014
	Corynebacterium	Gowramma et al. 2015
	Escherichia coli	Farias et al. 2014, Müller et al. 2016
	Pseudomonas aeruginosa	Farias et al. 2014
Nanoparticles (NPs)	*Yeast and filamentous Fungi*	
Cadmium	*Trametes versicolor*	Manna et al. 2018
Gold	*Phaenerochaete chrysosporium*	Sanghi et al. 2011
	Aspegillus sp.	Shen et al. 2017
	Fusarium solani	Gopinath and Arumugam 2014
	Neurospora crassa	Longoria et al. 2011
Paladium	*Saccharomyces cerevisiae*	Sriramulu and Sumati 2018
Platinum	*Fusarium oxysporum*	Gupta and Chundawat 2019
	Neurospora crassa	Longoria et al. 2012
Selenium	*Trichoderma atroviride*	Joshi et al. 2019
Silver	*Aspergillus flavus*	Vala et al. 2014
	Fusarium oxysporum	Gholami-Shabani et al. 2014
	Rhodotorula glutinis	Cunha et al. 2018
	Rhodotorula mucilaginosa	Cunha et al. 2018

Srivastava et al. (2012) found that *P. aeruginosa* produces several nanoparticles intracellularly, such as those of Pd, Ag, Rh, Ni, Fe, Co, Pt, and Li. In another previous work, when exposed to a concentrated solution of $AgNO_3$, a strain of *P. stutzeri* AG259 resistant to this metal, reduced Ag^+ ions in its periplasmic space, through reduced Nicotinamide Adenine Dinucleotide (NADH)-dependent reductase, and formed NAD^+ and AgNPs in the size range from a few nm to 200 nm (Klaus-Joerger et al. 2001). Likewise, the exposure of a biomass of the acidophilic fungus *Verticillium* sp., to an aqueous solution of Ag^+ ions, resulted in the in situ reduction (trapping of the Ag^+ ions on the surface of the fungal cells, possibly via electrostatic interaction involving negatively charged carboxylate groups in enzymes present in the cell wall of the mycelia), leading to the formation of silver nuclei, which subsequently grow by further reduction of Ag+ ions and accumulation on these nuclei (Mukherjee et al. 2001). In technological terms (Fig. 2), however, extracellular synthesis of MtNPs is preferred over intracellular synthesis, since it does not require such complex further separation processing (Singh et al. 2013).

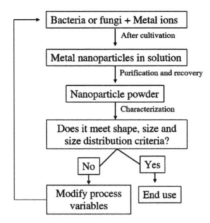

Figure 2 Flowchart outlining the biosynthesis of nanoparticles.

In 2014, Karthik et al. incubated cultures of the actinomycete *Streptomyces* sp. LK3 with silver ions, at room temperature, and found that nitrate reductase was responsible for the reduction of nitrate to nitrite, and of nitrite to nitrogenous gases, producing stable AgNPs that rest for months, without using any capping agents. According to these authors, no particle aggregation was observed in the mixture which further strengthens the stability of AgNPs. This study provided an easy, cost-effective, and eco-friendly approach for the biological synthesis of small size (5 nm) AgNPs. Saravanan et al. (2018), on the other hand, used *B. brevis* for the production of spherical AgNPs (size range of 41–62 nm). In another study, Divya et al. (2019) biosynthesized monodispersed AgNPs using a strain of *Alcaligenes faecalis*, and showed that reducing agents, such as NADH and NADH-dependent reductases secreted in the supernatant, were the main ones responsible for mediating the production of AgNPs, which presented antimicrobial activity against urinary tract infection caused by clinical isolates such as *Bacillus* sp., *C. Albicans*, *E. coli*, *K. pneumoniae*, and *S. aureus*.

Numerous reports emphasize the identification of two distinct ATP-producing mechanisms, i.e. oxidative phosphorylation and substrate-level phosphorylation (Hunt et al. 2010, Kane et al. 2016, Pinchuk et al. 2011). Oxidative phosphorylation is typically associated with respiration, in which the reduction of terminal electron acceptors is coupled to proton motive force (PMF) generation, which subsequently contributes to ATP-synthesis via ATP synthase. Substrate level phosphorylation is associated with the production of ATP through direct transfer of a phosphoryl group to ADP, using enzymes such as phosphotransacetylase and acetate kinase. In the context of gold-NPs, bacteria able to reduce the Au^{3+} ions into $Au°$ in the nanoscale are important, such as *Shewanella oneidensis* MR-1 and other *Shewanella* species that utilize a broad range of electron acceptors, making this genus a model for research of anaerobic respiration and metabolic energy conservation. The substrate-level phosphorylation of *S. oneidensis* strain MR-1, for instance, is the primary source of ATP during anaerobic growth, while ATPase has either minor contributions to ATP production or acts as an ATP-driven proton pump that generates PMF (Hunt et al. 2010). This was surprising, given that *Shewanella* bacteria are obligated to utilize terminal electron acceptors when growing under anaerobic conditions.

A poorly understood aspect refers to how the production of ATP, the generation of PMF and the redox reactions interact, and together contribute to the metabolic pathways and energy conservation in *Shewanella* strains. Shi et al. (2009, 2012) analyzed the possibility of the strain *S. oneidensis* MR-1 using ferric oxide minerals as electron terminal receptors in its anaerobic respiration, as well as the role of type c-cytochromes in this pathway, and concluded that this and other related *Shewanella* strains have evolutionarily developed a metal reducing machinery (via Mtr) to affect this electron transfer across cell membranes to the surface. Such researchers proposed a model for this mechanism involving type c-cytochromes in the transfer of electrons from quinol in the inner membrane (IM) to the periplasmic space (PS) and through the outer membrane (OM) to the surface of the metal oxide (oxide de Fe(III)), identifying the protein components CymA, MtrA, MtrB, MtrC, and OmcA (Figure 3), so that a similar mechanism can occur for bacterial synthesis of MtNPs outside the bacterial cell surface, i.e. with CymA being the tetraheme cytochrome (c-Cyt) of the NapC/NrfH family of quinol dehydrogenases, oxidizing the quinol in quinone in IM and transferring the electrons to MtrA, directly or indirectly, through other periplasmic proteins. MtrA, a c-Cyt decaheme, is incorporated into the OM trans and MtrB protein-type porin. Together with MtrB, MtrA transfers electrons from OM to MtrC and OmcA, located on the outermost surface. OmcA and MtrC are two decaheme c-Cyt of OM that are translocated to the cell surface by the type II bacterial secretion pathway. MtrC and OmcA function as terminal reducing agents and can bond to the surface of Fe(III) oxides and transfer electrons directly to these oxides through the exposed heme part. MtrC and OmcA can also use flavins secreted by *S. oneidensis* MR-1 cells as diffusible factors for reducing Fe(III) oxides at faster rates, and MtrC and OmcA can also serve as terminal reducing agents for soluble forms of Fe(III) because of its wide redox potentials and extracellular location (Shi et al. 2009, 2012).

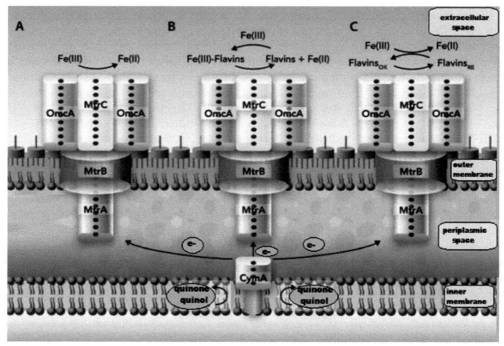

Figure 3. The functions of MtrC and OmcA in the extracellular reduction of Fe(III) oxides mediated by *Shewanella oneidensis* MR-1. The proteins directly involved in the reduction are: (a) c-Cyt CymA (a homolog of the NapC/NirT family of quinol dehydrogenases) of the inner membrane (IM); (b) the periplasmic c-Cyt MtrA; (c) the protein MtrB of the outer membrane (OM); (d) the c-Cyts MtrC and OmcA of the OM. These proteins transfer electrons from the quinone/quinol pool in the IM to the periplasm space (PS) and then to the OM, where MtrC and OmcA can transfer electrons directly to the surface of the solid Fe(III) oxides (A). MtrC and OmcA might also reduce Fe(III) oxides indirectly, by transferring electrons to either flavin-chelated Fe(III) (B) or to oxidized flavins (C). (Adapted from Shi et al. 2009).

A mutant strain of *S. oneidensis* without cytochrome genes (MtrC and OmcA) was used by Ng et al. (2013) to explain the role of type c cytochromes in the extracellular synthesis of AgNPs, verifying that in these mutants there was less production of these NPs and in smaller size than in the wild strain of *S. oneidensis*, suggesting that c-type cytochromes of this species also aid in the transfer of electrons to extracellular metal ions. Liu et al. (2014, 2015) found the direct involvement of OM c-type cytochrome protein complexes (ombB, omaB and omcB) in extracellular reduction of Fe(III) citrate and ferrihydrite using the metal-reducing bacterium *Geobacter sulfurreducens* PCA. Vasylevskyi et al. (2019) verified that electrons from the respiration of *G. sulfurreducens* were transferred by Fe^{2+}/hemes of c-cytochromes via PS, to Fe^{3+}/hemes of the OM cytochromes (Omc), and from these to the surface attached Ag^+ ions, synthesizing AgNPs while oxidizing the cytochromes again to Fe^{3+}/hemes.

Buchman et al. (2019) investigated the effects of iron oxide NPs (IONPs) and mesoporous silica-coated iron oxide NPs (msIONPs) on *S. oneidensis* under aerobic conditions (such as the environments where these NPs will be after use in chemical products or therapeutic applications). After exposure of *S. oneidensis* to the NPs, it was found that IONPs promoted bacterial survival, while msIONPs did not affect it, and there was a correlation between the NPs and the bacterial membrane, as revealed by transmission electron microscopy (TEM) and inductively coupled plasma-mass spectrometry (ICP–MS) studies, and upregulation of membrane-associated genes. However, similar survival was observed when the bacteria was exposed to equivalent concentrations of released ions from each tested NP, indicating that aqueous NPs transformations were responsible for the observed changes in their viability, and a simple mesoporous silica coating regulated the dissolution of the IONP core by drastically reducing the concentration of released iron ions, so that these msIONPs represent a very sustainable option to remediate ecosystems upon release of nanoparticles into the environment.

Furthermore, bacterial exopolysaccharides (EPS), such as pullulan, xanthan gum or dextran, whose structures are mainly composed by monomers that confer to them an anionic nature, such as carboxyl, phosphate, sulfate, and pyruvate substituents, are explored as agents for the production of several MtNPs. The compounds with this function are d-glucose, d-galactose, d-mannose, d-galacturonic acid, d-glucuronic acid, d-mannuronic acid, l-guluronic acid, l-fucose, l-rhamnose, *N*-acetyl-d-glucosamine, and *N*-acetyl-d-galactosamine, as well as noncarbohydrate components. This is possible since their aldehyde and hemiacetal groups are able to reduce metal ions to synthesize and stabilize NPs, acting as capping agents. Choudhury et al. (2014) developed an approach for the production of AuNPs using pullulan as reducing agent, and in-depth analysis of thermodynamic parameters further revealed that this synthesis followed first order kinetics, with small size NPs formed at high temperature (100°C). Fourier-transform infrared spectroscopy (FTIR) revealed that the main structure of pullulan is not affected during production of AuNPs (α-1,4 and α-1,6 linkages are intact after the reaction), and it was hence concluded that the biomineralization of Au^{3+} ions results from the oxidation of side chain aliphatic alcoholic groups of the monossacharides (free CH_2OH groups) which were oxidized to COO^- carboxyl group, while simultaneous reduction of Au^{3+} to Au^0 produces AuNPs. The same was seen by Gahlawat et al. (2016) for production and capping of AgNPs by EPS of *Ochrobactrum rhizosphaerae*, where the groups hydroxyl, carboxyl, phosphoric, hemiacetal, and the amino end groups were able to reduce metal ions from the precursor salts to obtain the respective NPs.

Regarding the extracellular approach for the synthesis of NPs, the bacterium *Rhodopseudomonas capsulata*, for instance, produced AuNPs through the secretion of NADH and NADH-reliant enzymes, so that gold ions accept electrons and get reduced (Au^{3+} to Au^0), leading to the formation of AuNPs (He et al. 2007). Beside these enzymes, different compounds, including naphthoquinones, anthraquinones and hydroquinones, are involved in their production (Patra et al. 2014).

Also, different fungal extracellular enzymes are used in the bioreduction of metals (Senapati et al. 2005, Ingle et al. 2008, Kumar et al. 2007a, b), but NADH and NADH-dependent nitrate reductase enzymes are the most relevant in the biosynthesis of metallic NPs (Zomorodian et al. 2016, Baymiller et al. 2017).

For many years, for instance, the extracellular reductases of the fungus *Fusarium oxysporum* have been used *in vitro* to reduce Au^{3+} and Ag^+ in the Au–Ag alloys of NPs (Senapati et al. 2005, Kumar et al. 2007a, Karbasian et al. 2008). The fungal biomass produced in media for extracellular synthesis of AgNPs is then washed in distilled/deionized water for release of the compounds that act in this process, and after filtration, silver nitrate is added to the filtrate (AbdelRahim et al. 2017, Costa Silva et al. 2017, Guilger et al. 2017, Mekkawy et al. 2017, Ottoni et al. 2017). Different bacteria that could also have a high tolerance to metals and fungal cultures form an immense quantity of biomass able to secrete large amounts of extracellular proteins, which contributes to the stability and catalysis of the overall process, so that additional steps to the supernatant filtration are not required. Then, fungi are more easy to manipulate and more attractive agents for biological production of NPs (Du et al. 2015, Netala et al. 2016). In an *in vitro* study, in the presence of oxygen, NADPH as coenzyme, a stabilizer protein (phytochelatin) and an electron carrier (4-hydroxyquinoline), this fungus exhibited suitable extracellular production of AgNPs, being considered an extraordinary candidate for extracellular synthesis of other MtNPs (Kumar et al. 2007a). The mycelia of *F. oxysporum*, harvested after 96 h of incubation (at 25 ± 1°C, 180 rpm) in a medium containing malt and yeast extracts plus glucose, was used to convert silver nitrate solution into nano-silver. The bioconversion was optimized through response surface methodology-central composite design, and the factors which affected the process were concentration of $AgNO_3$ solution mycelia biomass that positively affected the AgNPs production, whereas negative factors were pH, temperature, rate of agitation, time of incubation and the interaction between biomass and temperature (Karbasian et al. 2008).

Enzymes from other fungal species of *Fusarium*, such as *F. semitectum* and *F. solani*, were also studied for the extracellular production of AgNPs, and the researchers concluded that specific proteins might be responsible for the reduction of Ag^+ forming AgNPs (Basavaraja et al. 2008, Ingle et al. 2009). Some species like *F. moniliforme* have already been studied and failed to generate AgNPs, even upon the release of the reductase, indicating the Ag^+ reduction via the redox reactions of electron-carriers, as NADP-reliant nitrate reductase (Duran et al. 2005). *F. oxysporum* was also used for the extracellular synthesis of semiconductor CdS-NPs, and extremely luminescent CdSe-NPs were synthesized using its reductase enzyme (Kumar et al. 2007b, Ahmad et al. 2002). By the same way, Hamedi et al. (2017) used *F. oxysporum* to produce highly monodispersed AgNPs, altering the process conditions such as incubation time, temperature, metal salt concentration and C : N ratio. The authors found that an increase in C : N ratio resulted in production of small size AgNPs with high monodispersity and productivity. In another report, El Domany et al. (2018), using extracellular filtrate of *Pleurotus ostreatus* culture, synthesized stable AuNps (10–30 nm) with moderate dispersity, and studied the effect of different variables in this process, which are the concentration of tetrachloro auric acid (HAuCl4), pH, temperature, the ratio of Extracellular filtrate (ECF) and $HAuCl_4$ by volume, agitation and incubation time on the quantity of biosynthesized AuNPs. Its production rate increased proportionally with the concentration of $HAuCl_4$ salt, time of incubation and agitation, whereas the ratio and pH showed negative relation with the NPs formation (lowering both the pH and ratio increase the biosynthesis).

The extracellular synthesis of AgNPs by *Cladosporium cladosporioides* and *Coriolus versicolor* involved organic acids, polysaccharides and proteins, which affect the shape and dimension of the nanocrystals (Balaji et al. 2009). When the fungus *Aspergillus niger* was incubated in a solution of $AgNO_3$, the AgNPs extracellularly produced were also stabilized by fungal proteins (Gade et al. 2008). Likewise, *A. fumigatus* extracellularly produced AgNPs more quickly (10 min) than other physical and chemical techniques (Bhainsa et al. 2006), being it another ideal candidate for large scale production of NPs. The same time was required for the marine fungus *Penicillium fellutanum* to reduce Ag^+ ions through the action of its extracellular nitrate-reductase (Kathiresan et al. 2009). It has also been reported that *P. brevicompactum* can reduce Ag^+ ions by releasing NADH-dependent nitrate reductases (Shaligram et al. 2009). AbdelRahim et al. (2017) evidenced that protein molecules within aqueous mycelial extract of *Rhizopus stolonifer* make the production and the coat cover of AgNPs (capping proteins) easy in solutions of $AgNO_3$. Parameter optimization showed the smallest size of AgNPs (2.86 ± 0.3 nm) obtained with 10^{-2} M $AgNO_3$ at 40°C.

The main problem in synthesizing the NPs is the control of their physical properties, i.e. uniform particle size distribution, identical shape, morphology, nanoparticle coating or stabilizing agent, chemical composition or type and crystal structure. Farias et al. (2014), for example, successfully synthesized spherical AgNPs with uniform distribution in water-oil microemulsion using a strain of the bacteria *P. aeruginosa*, which also produced a biosurfactant in a low-cost medium of agro-industrial substrates able to stabilize the AgNPs. In another report, Huang et al. (2019) reported for the first time photo-induction as a trigger to stimulate gold nanoparticles (AuNPs) formation by *S. oneidensis* MR-1. Suchomel et al. (2018) carried out a procedure for one-step preparation of AuNPs, providing a straightforward regulation of the dimension of the produced NPs by changing, in the reaction mixture, the concentration of the nonionic surfactant Tween 80 (polyethylene glycol sorbitan monooleate).

Therefore, apart from optimizing the above-mentioned parameters, another strategy that provides a favorable environment for the efficient and easily estimated large-scale biosynthesis of MtNPs, especially in aquatic systems, maybe the use of biofilms, since they exhibit several characteristics relevant to the process, such as high biomass and surface area, catalytic activity, very reducing matrix formed by natural reducers (proteins, peptides, and heterocyclic compounds) and, finally, protective nature with limited diffusion of external compounds, keeping the process free from contamination. In general, this makes the use of biofilms a promising approach for the biosynthesis of MtNPs in aqueous systems. However, the use of this technique has not been well explored yet, and little is known about its stabilization mechanism. Therefore, studies on the molecular mechanism of the synthesis of nanoparticles in biofilms will allow the genetic modification of bacterial and fungal strains, aiming at the rapid and optimized biosynthesis of MtNPs, with defined shapes and sizes and high yields (Gahlawat et al. 2016, Khan et al. 2013).

5. CONCLUSION

Over the past 50 years, there have been spectacular advances in understanding energy-producing microbial reactions, including their ability to use insoluble metal oxides as final acceptor of electrons and hydrocarbons produced by anthropogenic activities as preliminary electron donors or acceptors. Most of these processes are carried out by joint efforts of microbial populations that do not act as individuals, but rather in communities, and their populations dynamically change in number and diversity, with all cells genomically interacting and communicating with each other. Thus, microbiomes involve not only populations but genes that cooperate and possibly alter each other's biochemical phenotypes. Soil microbiomes, for example, contribute to the degradation and conversion of soil contaminants and are the engines for the cycles of carbon, oxygen, nitrogen, sulfur, phosphorus, and heavy or potentially toxic metals. So, microbiome engineering, i.e. the design and built of simple or complex artificial microbial systems to execute tasks that are difficult or impossible for individual microbes, allows the development of new outfits and methodologies in focused operations, such as bioremediation or even nanobioremediation, if the same relates to nanotechnology. To apply the soil microbiome knowledge in bioremediation projects, for instance, it is necessary to characterize the composition of the communities of microorganisms and their cellular and molecular activities, as well as the normal behavior of soil microbiomes which were changed by the presence of toxic contaminants. In this review, we presented the most recent discoveries in the field of environmental microbiomes and contributed to elucidating their potential for environmental preservation, remediation, as well as its use in the production of nanoparticles that, among other possibilities, can minimize the large-scale use of different toxic metals. In this context, the combination of microbiology and nanotechnology should generate new eco-efficient products, which will act as excellent adsorbents for bioremediation of solid waste, polluted groundwater and effluents contaminated by hydrocarbons and other heavy metals. For now, it is certainly possible to conclude that there is a very promising future for the microbial synthesis of MtNPs and their broad spectrum of applications in comparison to chemically generated NPs, due to their low cost, low toxicity of residues generated in their synthesis, high degradability, and use of wastes as substrates to be remedied or recycled and used in various other activities, such as biomedical applications. It is also encouraging that several researchers focus on decoding the detailed aspects of the biosynthesis of MtNPs and their action mechanisms, in addition to the different microbiological origins, such as site of production (intracellular or extracellular), and growth conditions (pH, temperature, oxygenation and incubation time) which may interfere in their size, shape, synthesis-yield, extraction facilities, concentration of synthesized material versus concentration of material removed from sample, etc. This will eventually lead to better understanding and wiser industrial applicability in the nearby future.

References

Abatenh, E., B. Gizaw, Z. Tsegaye and M. Wassie. 2017. Application of microorganisms in bioremediation-review. J. Environ. Microbiol. 1: 2–9.

Abdel Rahim, K., S.Y. Mahmoud, A.M. Ali, K.S Almaary, A.E. Mustafa and S.M. Husseiny. 2017. Extracellular biosynthesis of silver nanoparticles using *Rhizopus stolonifer*. Saudi J. Biol. Sci. 24: 208–216.

Ahmad, A., P. Mukherjee, D. Mandal, S. Senapati, M.I. Khan, R. Kumar, et al. 2002. Enzyme mediated extracellular synthesis of CdS nanoparticles by the fungus, *Fusarium oxysporum*. J. American Chem. Soc. 124: 12108–12109.

Ahmed, E., S. Kalathil, L. Shi, O. Alharbi and P. Wang. 2018. Synthesis of ultra-small platinum, palladium and gold nanoparticles by *Shewanella loihica* PV-4 electrochemically active biofilms and their enhanced catalytic activities. J. Saudi Chem. Soc. 22: 919–929.

Alabresm, A., Y.P. Chen, A.W. Decho and J. Lead. 2018. A novel method for the synergistic remediation of oil-water mixtures using nanoparticles and oil-degrading bacteria. Sci. Total Environ. 630: 1292–1297.

Arora, P.K., A. Srivastava and V.P. Singh. 2010. Application of monooxygenases in dehalogenation, desulphurization, denitrification and hydroxylation of aromatic compounds. J. Bioremed. Biodegrad. 1: 112–119.

Asmel, N.K., A.R.M. Yusoff, L.S. Krishna, Z.A. Majid and S. Salmiati. 2017. High concentration arsenic removal from aqueous solution using nano-iron ion enrich material (NIIEM) super adsorbent. Chem. Eng. J. 317: 343–355.

Balaji, D., S. Basavaraja, R. Deshpande, D.B. Mahesh, B.K. Prabhakar and A. Venkataraman. 2009. Extracellular biosynthesis of functionalized silver nanoparticles by strains of *Cladosporium cladosporioides* fungus. Colloids Surf. B Biointerfaces 68: 88–92.

Barea, J.M. 2015. Future challenges and perspectives for applying microbial biotechnology in sustainable agriculture based on a better understanding of plant-microbiome interactions. J. Soil Sci. Plant Nutr. 15: 261–282.

Basavaraja, S., S. Balaji, A. Lagashetty, A. Rajasab and A. Venkataraman. 2008. Extracellular biosynthesis of silver nanoparticles using the fungus *Fusarium semitectum*. Mater. Res. Bull. 43: 1164–1170.

Baymiller, M., F. Huang and S. Rogelj. 2017. Rapid one-step synthesis of gold nanoparticles using the ubiquitous coenzyme NADH. Matters. 2017: 1–4.

Bhainsa, K.C. and S. D'souza. 2006. Extracellular biosynthesis of silver nanoparticles using the fungus *Aspergillus fumigatus*. Colloids Surf. B Biointerfaces 47: 160–164.

Blaser, M.J., Z.G. Cardon, M.K. Cho, J.L. Dang, T.J. Donohue, J.L. Green, et al. 2016. Toward a predictive understanding of Earth's microbiomes to address 21st century challenges. mBio 7: 1–16.

Boxall, A. and J. Wilkinson. 2019. Identifying hotspots of resistance selection from antibiotic exposure in urban environments around the world. https: //helsinki.setac.org/wp-content/uploads/2019/05/SETAC-Helsinki-Abstract-Book-2019.pdf (accessed November 12, 2019).

Buchman, J.T., T. Pho, R.S. Rodriguez, V.Z. Feng and C.L. Haynes. 2019. Coating iron oxide nanoparticles with mesoporous silica reduces their interaction and impact on *S. oneidensis* MR-1. Chemosph. 237: 124511.

Campbell, K.M., T.J. Gallegos and E.R. Landa. 2015. Biogeochemical aspects of uranium mineralization, mining, milling, and remediation. Appl. Geochem. 57: 206–235.

Chakraborty, J. and S. Das. 2016. Molecular perspectives and recent advances in microbial remediation of persistent organic pollutants. Environ. Sci. Pollut. Res. 23: 16883–16903.

Chen, Y., L. Feng, H. Li, Y. Wang, G. Chen and Q. Zhang. 2018. Biodegradation and detoxification of direct black G textile dye by a newly isolated thermophilic microflora. Bioresour. Technol. 250: 650–657.

Choudhury, A.R., A. Malhotra, P. Bhattacharjee and G. Prasad. 2014. Facile and rapid thermo-regulated biomineralization of gold by pullulan and study of its thermodynamic parameters. Carbohydr. Polym. 106: 154–159.

Chowdhury, A., S. Pradhan, M. Saha and N. Sanyal. 2008. Impact of pesticides on soil microbiological parameters and possible bioremediation strategies. Indian J. Microbiol. 48: 114–127.

Costa-Silva, L.P., J.P. Oliveira, W.J. Keijok, A.R. da Silva, A.R. Aguiar, M.C.C. Guimarães, et al. 2017. Extracellular biosynthesis of silver nanoparticles using the cell-free filtrate of nematophagus fungus *Duddingtonia flagans*. Int. J. Nanomed. 12: 6373–6381.

Cunha, F.A., M.D.C.S.O. Cunha, S.M. Frota, E.J.J. Mallmann, T.M. Freire, L.S. Costa, et al. 2018. Biogenic synthesis of multifunctional silver nanoparticles from *Rhodotorula glutinis* and *Rhodotorula mucilaginosa*: Antifungal, catalytic and cytotoxicity activities. World J. Microbiol. Biotechnol. 34: 127–142.

Das, V.L., R. Thomas, R.T. Varghese, E. Soniya, J. Mathew and E. Radhakrishnan. 2014. Extracellular synthesis of silver nanoparticles by the *Bacillus* strain CS 11 isolated from industrialized area. 3 Biotech. 4: 121–126.

Davis, A.S., P. Prakash and K. Thamaraiselvi. 2017. Nanobioremediation technologies for sustainable environment. *In*: M. Prashanthi, R. Sundaram, A. Jeyaseelan and T. Kaliannan (eds), Bioremediation and Sustainable Technologies for Cleaner Environment, Environmental Science and Engineering. Springer International Publishing AG, Cham, pp. 13–33.

De Roy, K., M. Marzorati, P. Van den Abbeele, T. Van de Wiele and N. Boon. 2014. Synthetic microbial ecosystems: An exciting tool to understand and apply microbial communities. Environ. Microbiol. 16: 1472–1481.

El-Domany, E.B., T.M. Essam, A.E. Ahmed and A.A. Farghali. 2018. Biosynthesis physico-chemical optimization of gold nanoparticles as anti-cancer and synergetic antimicrobial activity using *Pleurotus ostreatus* fungus. J. Appl. Pharm. Sci. 8: 119–128.

Du, L., Q. Xu, M. Huang, L. Xian and J.X. Feng. 2015. Synthesis of small silver nanoparticles under light radiation by fungus *Penicillium oxalicum* and its application for the catalytic reduction of methylene blue. Mater. Chem. Phys. 160: 40–47.

Durán, N., P.D. Marcato, O.L. Alves, G.I. De Souza and E. Esposito. 2005. Mechanistic aspects of biosynthesis of silver nanoparticles by several *Fusarium oxysporum* strains. J. Nanobiotechnol. 3: 8–15.

Divya, M., S.K. George, H. Saqib and S. Joseph. 2019. Biogenic synthesis and effect of silver nanoparticles (AgNPs) to combat catheter-related urinary tract infections. Biocatal. Agric. Biotechnol. 18: 101037.

Farias, C.B.B., A.F. Silva, R.D. Rufino, J.M. Luna, J.E.G. Souza and L.A. Sarubbo. 2014. Synthesis of silver nanoparticles using a biosurfactant produced in low-cost medium as stabilizing agent. Electron. J. Biotechnol. 17: 122–125.

FMI (Future Market Insights). 2017. Global Market for Metal & Metal Oxide Nanoparticles to Surge at More Than 10% CAGR. http://markets.businessinsider.com/news/stocks/Global-Market-forMetal-Metal-Oxide-Nanoparticles-to-Surge-at-More-Than-10CAGR-1001862836 (accessed September 20, 2019).

Gade, A., P. Bonde, A. Ingle, P. Marcato, N. Duran and M. Rai. 2008. Exploitation of *Aspergillus niger* for synthesis of silver nanoparticles. J. Biobased. Mater. Biol. 2: 243–247.

Gahlawat, G., S. Shikha, B.S. Chaddha, S.R. Chaudhuri, S. Mayilraj and A.R. Choudhury. 2016. Microbial glycolipoprotein capped silver nanoparticles as emerging antibacterial agents against cholera. Microb. Cell Fact. 15: 25.

Getenga, Z., U. Dörfler, A. Iwobi, M. Schmid and R. Schroll. 2009. Atrazine and terbuthylazine mineralization by an *Arthrobacter* sp. isolated from a sugarcane-cultivated soil in Kenya. Chemosphere 77: 534–539.

Gholami-Shabani, M., A. Akbarzadeh, D. Norouzian, A. Amini, Z. Gholami-Shabani, A. Imani, et al. 2014. Antimicrobial activity and physical characterization of silver nanoparticles green synthesized using nitrate reductase from *Fusarium oxysporum*. Appl. Biochem. Biotechnol. 172: 4084–4098.

Gopinath, K. and A. Arumugam. 2014. Extracellular mycosynthesis of gold nanoparticles using *Fusarium solani*. Appl. Nanosci. 4: 657–662.

Gowramma, B., U. Keerthi, M. Rafi and D. Muralidhara Rao. 2015. Biogenic silver nanoparticles production and characterization from native stain of *Corynebacterium* species and its antimicrobial activity. 3 Biotech. 5: 195–201.

Großkopf, T. and O.S. Soyer. 2014. Synthetic microbial communities. Curr. Opin. Microbiol. 18: 72–77.

Grün, A-L., S. Straskraba, S. Schulz, M. Schloter and C. Emmerling. 2018. Long-term effects of environmentally relevant concentrations of silver nanoparticles on microbial biomass, enzyme activity, and functional genes involved in the nitrogen cycle of loamy soil. J. Environ. Sci. 69: 12–22.

Guilger, M., T. Pasquoto-Stigliani, N. Bilesky-Jose, R. Grillo, P.C. Abhilash, L.F. Fraceto, et al. 2017. Biogenic silver nanoparticles based on *Trichoderma harzianum*: Synthesis, characterization, toxicity evaluation and biological activity. Sci. Rep. 7: 1–13.

Gupta, A., J. Joia, A. Sood, R. Sood, C. Sidhu and G. Kaur. 2016a. Microbes as potential tool for remediation of heavy metals: A review. J. Microb. Biochem. Technol. 8: 364–372.

Gupta, G., V. Kumar and A.K. Pal. 2016b. Biodegradation of polycyclic aromatic hydrocarbons by microbial consortium: A distinctive approach for decontamination of soil. Soil Sediment Contam. J. 25: 597–623.

Gupta, K. and T.S. Chundawat. 2019. Bio-inspired synthesis of platinum nanoparticles from fungus *Fusarium oxysporum*: Its characteristics, potential antimicrobial, antioxidant and photocatalytic activities. Mater. Res. Express. 6: 1–12.

Güzel, R. and G. Erdal. 2018. Synthesis of silver nanoparticles. *In*: M. Khan (ed.), Silver Nanoparticles: Fabrication, Characterization and Applications. IntechOpen, United Kingdom, pp. 3–20.

Hamedi, S., M. Ghaseminezhad, S. Shokrollahzadeh and S.A. Shojaosadati. 2017. Controlled biosynthesis of silver nanoparticles using nitrate reductase enzyme induction of filamentous fungus and their antibacterial evaluation. Artif. Cells Nanomed. Biotechnol. 45: 1588–1596.

Hänsch, M. and C. Emmerling. 2010. Effects of silver nanoparticles on the microbiota and enzyme activity in soil. J. Plant Nutr. Soil Sci. 173: 554–558.

Hao, X., S. Taghavi, P. Xie, M.J. Orbach, H.A. Alwathnani, C. Rensing, et al. 2014. Phytoremediation of heavy and transition metals aided by legume-rhizobia symbiosis. Int. J. Phytoremed. 16: 179–202.

He, L., M. Wang, G. Zhang, G. Qiu and X. Zhang. 2015. Remediation of Cr (VI) contaminated soil using long duration sodium thiosulfate supported by micro-nano networks. J. Hazard. Mater. 294: 64–68.

He, S., Z. Guo, Y. Zhang, S. Zhang, J. Wang and N. Gu. 2007. Biosynthesis of gold nanoparticles using the bacteria *Rhodopseudomonas capsulata*. Mater. Lett. 61: 3984–3987.

Hiraishi, T. 2016. Poly (aspartic acid) (PAA) hydrolases and PAA biodegradation: Current knowledge and impact on applications. Appl. Microbiol. Biotechnol. 100: 1623–1630.

Huang, B.C., Y.C. Yi, J.S. Chang and I.S. Ng. 2019. Mechanism study of photo-induced gold nanoparticles formation by *Shewanella oneidensis* MR-1. Sci. Rep. 9: 1–11.

Hulkoti, N.I. and T.C. Taranath. 2014. Biosynthesis of nanoparticles using microbes – a review. Colloids Surf. B Biointerfaces. 121: 474–483.

Hunt, K.A., J.M. Flynn, B. Naranjo, I.D. Shikhare and J.A. Gralnick. 2010. Substrate-level phosphorylation is the primary source of energy conservation during anaerobic respiration of *Shewanella oneidensis* strain MR-1. J. Bacteriol. 192: 3345–3351.

Ingle, A., M. Rai, A. Gade and M. Bawaskar. 2009. *Fusarium solani*: A novel biological agent for the extracellular synthesis of silver nanoparticles. J. Nanopart. Res. 11: 2079–2085.

Ingle, A., A. Gade, S. Pierrat, C. Sonnichsen and M. Rai. 2008. Mycosynthesis of silver nanoparticles using the fungus *Fusarium acuminatum* and its activity against some human pathogenic bacteria. Curr. Nanosci. 4: 141–144.

Jang, M.H., M. Lim and Y.S. Hwang. 2014. Potential environmental implications of nanoscale zero-valent iron particles for environmental remediation. Environ. Health Toxicol. 29: 1–9.

Joshi, S.M., S. De Britto, S. Jogaiah and S.I. Ito. 2019. Mycogenic selenium nanoparticles as potential new generation broad spectrum antifungal molecules. Biomolecules 9: 419–435.

Kane, A.L., E.D. Brutinel, H. Joo, R. Maysonet, C.M. VanDrisse, N.J. Kotloski, et al. 2016. Formate metabolism in *Shewanella oneidensis* generates proton motive force and prevents growth without an electron acceptor. J. Bacteriol. 198: 1337–1346.

Kang, C.H., Y.J. Kwon and J.S. So. 2016. Bioremediation of heavy metals by using bacterial mixtures. Ecol. Eng. 89: 64–69.

Karbasian, M., S. Atyabi, S. Siadat, S. Momen and D. Norouzian. 2008. Optimizing nano-silver formation by *Fusarium oxysporum* PTCC 5115 employing response surface methodology. American J. Agric. Biol. Sci. 3: 433–437.

Karigar, C.S. and S.S. Rao. 2011. Role of microbial enzymes in the bioremediation of pollutants: A review. Enzyme Res. 2011: 1–11.

Karthik, L., G. Kumar, A.V. Kirthi, A.A. Rahuman and K.V. Bhaskara Rao. 2014. *Streptomyces* sp. LK3 mediated synthesis of silver nanoparticles and its biomedical application. Bioprocess Biosyst. Eng. 37: 261–266.

Kathiresan, K., S. Manivannan, M. Nabeel and B. Dhivya. 2009. Studies on silver nanoparticles synthesized by a marine fungus, *Penicillium fellutanum* isolated from coastal mangrove sediment. Colloids Surf. B Biointerfaces 71: 133–137.

Khan, M.M., S. Kalathil, T.H. Han, J. Lee and M.H. Cho. 2013. Positively charged gold nanoparticles synthesized by electrochemically active biofilm – a biogenic approach. Nanosc. Nanotechnol. 13: 6079–6085.

Khatoon, H. and J.P.N. Rai. 2018. Augmentation of Atrazine biodegradation by two *Bacilli* immobilized on α-Fe_2O_3 magnetic nanoparticles. Sci. Rep. 8: 1–12.

Klaus-Joerger, T., R. Joerger, E. Olsson and C.G. Granqvist. 2001. Bacteria as workers in the living factory: Metal accumulating bacteria and their potential for materials science. Trends Biotechnol. 19: 15–20.

Klein, S., R. Avrahami, E. Zussman, M. Beliavski, S. Tarre and M. Green. 2012. Encapsulation of *Pseudomonas* sp. ADP cells in electrospun microtubes for atrazine bioremediation. J. Ind. Microbiol. Biotechnol. 39: 1605.

Kora, A.J. 2018. *Bacillus cereus*, selenite-reducing bacterium from contaminated lake of an industrial area: A renewable nanofactory for the synthesis of selenium nanoparticles. Bioresour. Bioprocess. 5: 1–12.

Kulkarni, R.R., N.S. Shaiwale, D.N. Deobagkar and D.D. Deobagkar. 2015. Synthesis and extracellular accumulation of silver nanoparticles by employing radiation-resistant *Deinococcus radiodurans*, their characterization and determination of bioactivity. Int. J. Nanomed. 10: 963–974.

Kumar, S.A., M.K. Abyaneh, S.W. Gosavi, S.K. Kulkarni, R. Pasricha, A. Ahmad, et al. 2007a. Nitrate reductase-mediated synthesis of silver nanoparticles from $AgNO_3$. Biotechnol. Lett. 29: 439–445.

Kumar, S.A., A.A. Ansary, A. Ahmad and M. Khan. 2007b. Extracellular biosynthesis of CdSe quantum dots by the fungus, *Fusarium oxysporum*. J. Biomed. Nanotechnol. 3: 190–194.

Kylilis, N., Z.A. Tuza, G.B. Stan and K.M. Polizzi. 2018. Tools for engineering coordinated system behaviour in synthetic microbial consortia. Nat. Commun. 9: 1–9.

Lenoir, I., A. Lounes-Hadj Sahraoui and J. Fontaine. 2016. Arbuscular mycorrhizal fungal-assisted phytoremediation of soil contaminated with persistent organic pollutants: A review. Eur. J. Soil Sci. 67: 624–640.

Li, X.N., W.T. Jiao, R.B. Xiao, W.P. Chen and A.C. Chang. 2015. Soil pollution and site remediation policies in China: A review. Environ. Rev. 23: 263–274.

Li, X., F. Fan, B. Zhang, K. Zhang and B. Chen. 2018. Biosurfactant enhanced soil bioremediation of petroleum hydrocarbons:Design of experiments (DOE) based system optimization and phospholipid fatty acid (PLFA) based microbial community analysis. Int. Biodeter. Biodegrad. 132: 216–225.

Liu, S.H., G.M. Zeng, Q.Y. Niu, Y. Liu, L. Zhou, L.H. Jiang, et al. 2017. Bioremediation mechanisms of combined pollution of PAHs and heavy metals by bacteria and fungi: A mini review. Bioresor. Technol. 224: 25–33.

Liu, Y., Z. Wang, J. Liu, C. Levar, M.J. Edwards, J.T. Babauta, et al. 2014. A trans-outer membrane porin-cytochrome protein complex for extracellular electron transfer by *Geobacter sulfurreducens* PCA. Environ. Microbiol. Rep. 6: 776 —785.

Liu, Y., J.K. Fredrickson, J.M. Zachara and L. Shi. 2015. Direct involvement of *ombB*, *omaB*, and *omcB* genes in extracellular reduction of Fe(III) by *Geobacter sulfurreducens* PCA. Front. Microbiol. 6: 1075.

Longoria, E.C., A.R.V. Nestor and M. Avalos-Borja. 2011. Biosynthesis of silver, gold and bimetallic nanoparticles using the filamentous fungus *Neurospora crassa*. Colloids Surf B Biointerfaces 83: 42–48.

Longoria, E.C., S.D.M. Velázquez, A.R.V. Nestor, E.A. Berumen and M.A. Borja. 2012. Production of platinum nanoparticles and nanoaggregates using *Neurospora crassa*. J. Microbiol. Biotechnol. 22: 1000–1004.

Macaskie, L.E., A.C. Humphries, I.P. Mikheenko, V.S. Baxter-Plant, K. Deplanche, M.D. Redwood, et al. 2012. Use of *Desulfovibrio* and *Escherichia coli* Pd-nanocatalysts in reduction of Cr(VI) and hydrogenolytic dehalogenation of polychlorinated biphenyls and used transformer oil. J. Chem. Technol. Biotechnol. 87: 1430–1435.

Makky, E.A., S.H.M. Rasdi, J.B. Al-Dabbagh, G.F. Najmuldeen and A.R.M. Hasbi. 2014. Bioremediation of hazardous waste for silver nanoparticles production. Int. J. Curr. Microbiol. App. Sci. 3: 364–371.

Manna, A., E.J. Sundaram, C. Amutha and E.S. Vasantha. 2018. Efficient removal of cadmium using edible fungus and its quantitative fluorimetric estimation using (Z)-2-(4H-1,2,4-triazol-4-yl) iminomethylphenol. ACS Omega 3: 6243–6250

Mekkawy, A.I., M.A. El-Mokhtar, N.A. Nafady, N. Yousef, M.A. Hamad, S.M. El-Shanawany, et al. 2017. *In vitro* and *in vivo* evaluation of biologically synthesized silver nanoparticles for topical applications: Effect of surface coating and loading into hydrogels. Int. J. Nanomed. 12: 759–777.

Meyer, D.D., S.A. Beker and F. Bücker. 2014. Bioremediation strategies for diesel and biodiesel in oxisol from southern Brazil. Int. Biodeter. Biodegr. 95: 356–363.

Mikeskova, H., C. Novotny´ and K. Svobodova. 2012. Interspecific interactions in mixed microbial cultures in a biodegradation perspective. Appl. Microbiol. Biotechnol. 95: 861–870.

Mikheenko, I.P., J. Gomez-Bolivar, M.L. Merroun, L.E. Macaskie, S. Sharma, M. Walker, et al. 2019. Upconversion of cellulosic waste into a potential "drop in fuel" via novel catalyst generated using *Desulfovibrio desulfuricans* and a consortium of acidophilic sulfidogens. Front. Microbiol. 10: 1–20.

Mukherjee, S. and C.R. Patra. 2017. Biologically synthesized metal nanoparticles: Recent advancement and future perspectives in cancer theranostics. Future Sci. 26: 1–4.

Mukherjee, P., A. Ahmad., D. Mandal, S. Senapati, S.R. Sainkar, M.I. Khan, et al. 2001. Fungus-mediated synthesis of silver nanoparticles and their immobilization in the mycelial matrix: A novel biological approach to nanoparticle synthesis. Nano. Lett. 1: 515–519.

Müller, A., D. Behsnilian, E. Walz, V. Gräf, L. Hogekamp and R. Greiner. 2016. Effect of culture medium on the extracellular synthesis of silver nanoparticles using *Klebsiella pneumoniae, Escherichia coli* and *Pseudomonas jessinii*. Biocat. Agric. Biotech. 6: 107–115.

Mueller, N.C. and B. Nowack. 2010. Nanoparticles for remediation: Solving big problems with little particles. Elements 6: 395–400.

Netala, V.R., M.S. Bethu, B. Pushpalatah, V.B. Baki, S. Aishwarya, J.V. Rao, et al. 2016. Biogenesis of silver nanoparticles using endophytic fungus *Pestalotiopsis microspora* and evaluation of their antioxidant and anticancer activities. Int. J. Nanomed. 11: 5683–5696.

Ng, C.K., K. Sivakumar, X. Liu, M. Madhaiyan, L. Ji, L. Yang, et al. 2013. Influence of outer membrane c-type cytochromes on particle size and activity of extracellular nanoparticles produced by *Shewanella oneidensis*. Biotechnol. Bioeng. 110: 1831–1837.

Ojuederie, O.B. and O.O. Babalola. 2017. Microbial and plant-assisted bioremediation of heavy metal polluted environments: A review. Int. J. Environ. Res. Public Health. 14: 1–26.

Ottoni, C.A., M.F. Simões, S. Fernandes, J.G. Dos Santos, E.S. da Silva, R.F.B. de Souza, et al. 2017. Screening of filamentous fungi for antimicrobial silver nanoparticles synthesis. AMB Express 7: 31–41.

Ovais, M., A. Khalil, M. Ayaz, I. Ahmad, S. Nethi and S. Mukherjee. 2018. Biosynthesis of metal nanoparticles via microbial enzymes: A mechanistic approach. Int. J. Mol. Sci. 19: 1–20.

Pandi, K., S. Periyasamy and N. Viswanathan. 2017. Remediation of fluoride from drinking water using magnetic iron oxide coated hydrotalcite/chitosan composite. Int. J Biol. Macromol. 104B: 1569–1577.

Pantidos, N. and L.E. Horsfall. 2014. Biological synthesis of metallic nanoparticles by bacteria, fungi and plants. J. Nanomed. Nanotech. 5: 1–11.

Patra, C.R., S. Mukherjee and R. Kotcherlakota. 2014. Biosynthesized silver nanoparticles: A step forward for cancer theranostics? Nanomed. 9: 1445–1448.

Perez-Garcia, O., G. Lear and N. Singhal. 2016. Metabolic network modeling of microbial interactions in natural and engineered environmental systems. Front. Microbiol. 7: 1–30.

Pinchuk, G.E., O.V. Geydebrekht and E.A. Hill. 2011. Pyruvate and lactate metabolism by *Shewanella oneidensis* MR-1 under fermentation, oxygen limitation and fumarate respiration conditions. Appl. Environ. Microbiol. 77: 8234–8240.

Plaza, D.O., C. Gallardo, Y.D. Straub, D. Bravo and J.M. Pérez-Donoso. 2016. Biological synthesis of fluorescent nanoparticles by cadmium and tellurite resistant Antarctic bacteria: Exploring novel natural nanofactories. Microb. Cell Fact. 15: 76–85.

Rizwan, M., M. Singh, C.K. Mitra and R.K. Morve. 2014. Ecofriendly application of nanomaterials: Nanobioremediation. J. Nanop. 2014: 1–7.

Russo, P., D. Acierno, M. Palomba, G. Carotenuto, R. Rosa, A. Rizzuti, et al. 2012. Ultrafine magnetite nanopowder: Synthesis, characterization, and preliminary use as filler of polymethylmethacrylate nanocomposites. J. Nanotechnol. 2012: 1–8.

Sanghi, R., Verma, P. and S. Puri. 2011. Enzymatic formation of gold nanoparticles using *Phanerochaete chrysosporium*. Adv. Chem. Eng. Sci. 1: 154–162.

Santos, E.C.L., D.A.R. Miranda, A.L.S. Silva and A.M.Q. Lopez. 2019. Biosurfactant Production by *Bacillus* strains isolated from sugar cane mill wastewaters. Braz. Arch. Biol. Technol. 62: 1–12.

Saravanan, M., S.K. Barik, D. MubarakAli, P. Prakash and A. Pugazhendhi. 2018. Synthesis of silver nanoparticles from *Bacillus brevis* (NCIM 2533) and their antibacterial activity against pathogenic bacteria. Microb. Patho. 116: 221–226.

Sathishkumar, M., A.R. Binupriya, S.H. Baik and S.E. Yun. 2008. Biodegradation of crude oil by individual bacterial strains and a mixed bacterial consortium isolated from hydrocarbon contaminated areas. Clean-Soil air Water. 36: 92–96

Senapati, S., A. Ahmad, M.I. Khan, M. Sastry and R. Kumar. 2005. Extracellular biosynthesis of bimetallic Au–Ag alloy nanoparticles. Small. 1: 517–520.

Shaligram, N.S., M. Bule, R. Bhambure, R.S. Singhal, S.K. Singh, G. Szakacs, et al. 2009. Biosynthesis of silver nanoparticles using aqueous extract from the compactin producing fungal strain. Process Biochem. 44: 939–943.

Shen, W., Y. Qu, X. Pei, S. Li, S. You, J. Wang, et al. 2017. Catalytic reduction of 4-nitrophenol using gold nanoparticles biosynthesized by cell-free extracts of *Aspergillus* sp. WL-Au. J. Hazard. Mater. 321: 299–306.

Shi, L., D.J. Richardson, Z. Wang, S. Kerisit, K. Rosso, J.M. Zachara, et al. 2009. The roles of outer membrane cytochromes of *Shewanella* and *Geobacter* in extracellular electron transfer. Environm. Microb. Rep. 1(4): 220–227.

Shi, L., K.M. Rosso, T.A. Clarke, D.J. Richardson, J.M. Zachara and J.K. Fredrickson. 2012. Molecular underpinnings of Fe(III) oxide reduction by *Shewanella oneidensis* MR-1. Front. Microbiol. 3: 50–54.

Siddiqi, K.S., A. Husen and R.A.K. Rao. 2018. A review on biosynthesis of silver nanoparticles and their biocidal properties. J. Nanobiotechnol. 16(1): 14–42.

Singh, R., P. Wagh, S. Wadhwani, S. Gaidhani, A. Kumbhar, J. Bellare, et al. 2013. Synthesis, optimization, and characterization of silver nanoparticles from *Acinetobacter calcoaceticus* and their enhanced antibacterial activity when combined with antibiotics. Int. J. Nanomed. 8: 4277–4290.

Smith, D., S. Alvey and D.E. Crowley 2005. Cooperative catabolic pathways within an atrazine-degrading enrichment culture isolated from soil. FEMS Microbiol. Ecol. 53: 265–273.

Song, B., G. Zeng, J. Gong, J. Liang and X. Ren. 2017. Evaluation methods for assessing effectiveness of in situ remediation of soil and sediment contaminated with organic pollutants and heavy metals. Environm. Int. 105: 43–55

Souza, E.C., T.C. Vessoni-Penna and R.P. de Souza-Oliveira. 2014. Biosurfactant-enhanced hydrocarbon bioremediation: An overview. Int. Biodeterior. Biodegrad. 89: 88–94.

Sriramulu, M. and S. Sumathi. 2018. Biosynthesis of palladium nanoparticles using *Saccharomyces cerevisiae* extract and its photocatalytic degradation behaviour. Adv. Nat. Sci.: Nanosci. Nanotechnol. 9(2): 1–6.

Srinath, B.S., K. Namratha and K. Byrappa. 2018. Eco-friendly synthesis of gold nanoparticles by *Bacillus subtilis* and their environmental application. Adv. Sci. Lett. 24: 5942–5946.

Srivastava, S.K. and M. Constanti. 2012. Room temperature biogenic synthesis of multiple nanoparticles (Ag, Pd, Fe, Rh, Ni, Ru, Pt, Co, and Li) by *Pseudomonas aeruginosa* SM1. J. Nanopart. Res. 14: 831.

Su, H., Z. Fang, P.E. Tsang, J. Fang and D. Zhao. 2016. Stabilization of nanoscale zero-valent iron with biochar for enhanced transport and in-situ remediation of hexavalent chromium in soil. Environ. Pollu. 214: 94–100.

Suchomel, P., L. Kvitek, R. Prucek, A. Panacek, A. Halder and S. Vajda. 2018. Simple size-controlled synthesis of Au nanoparticles and their size-dependent catalytic activity. Sci. Rep. 8: 4589.

Sun, T.Y., F. Gottschalk, K. Hungerbühler and B. Nowack. 2014. Comprehensive probabilistic modelling of environmental emissions of engineered nanomaterials. Environ. Pollut. 185: 69–76.

Suresh, A.K., D.A. Pelletier, W. Wang, M.L. Broich, J.W. Moon, B. Gu, et al. 2011. Biofabrication of discrete spherical gold nanoparticles using the metal-reducing bacterium *Shewanella oneidensis*. Acta Biomater. 7: 2148–2152.

Tanvir, R., I. Sajid, S. Hasnain, A. Kulik and S. Grond. 2016. Rare actinomycetes *Nocardia caishijiensis* and *Pseudonocardia carboxydivorans* as endophytes, their bioactivity and metabolites evaluation. Microb. Res. 185: 22–35.

Teng, Y. and W. Chen. 2019. Soil microbiomes, a promising strategy for contaminated soil remediation: A review. Pedosphere. 29: 283–297.

Teng, Y., X.M. Wang, L.N. Li, Z.G. Li and Y.M. Luo. 2015. Rhizobia and their bio-partners as novel drivers for functional remediation in contaminated soils. Front. Plant Sci. 6: 1–11.

Vala, A.K., S. Shah and R. Patel. 2014. Biogenesis of silver nanoparticles by marine derived fungus *Aspergillus flavus* from Bhavnagar coast, Gulf of Khambhat, India. J. Mar. Biol. Oceanogr. 3: 1–3.

Varjani, S.J. and V.N. Upasani. 2016. Core flood study for enhanced oil recovery through ex-situ bioaugmentation with thermo- and halo-tolerant rhamnolipid produced by *Pseudomonas aeruginosa* NCIM 5514. Bioresour. Technol. 220: 175–182.

Varjani, S.J. and V.N. Upasani. 2017. A new look on factors affecting microbial degradation of petroleum hydrocarbon pollutants. Int. Biodeterior. Biodegrad. 120: 71–83.

Vasylevskyi, S.I., S. Kracht, P. Corcosa, K.M. Fromm, B. Giese and M. Füeg. 2017. Formation of silver nanoparticles by electron transfer in peptides and c-cytochromes. Angew. Chem. Int. Ed. 56: 5926–5930.

Wang, X., D. Zhang, X. Pan, D.J. Lee and G.M. Gadd. 2017. Aerobic and anaerobic biosynthesis of nano-selenium for remediation of mercury contaminated soil. Chemosph. 170: 266

Wei, R. and W. Zimmermann. 2017. Microbial enzymes for the recycling of recalcitrant petroleum-based plastics: How far are we? Microb. Biotechnol. 10: 1308–1322.

WHO-IACG-AMR (World Health Organization-Interagency Coordination Group on the Antimicrobial Resistance of the United Nations). 2019. No Time to Wait: Securing the future from drug-resistant infections. Report to the secretary-general of the united nations. https://www.who.int/antimicrobial-resistance/interagency-coordination-group/final-report/en/ (accessed July 20, 2019).

Wilkinson, J. and A. Boxall. 2019. The first global study of pharmaceutical contamination in riverine environments. https://helsinki.setac.org/wp-content/uploads/2019/05/SETAC-Helsinki-Abstract-Book-2019.pdf (accessed December 19, 2019).

Wu, J.W. and I.S. Ng. 2017. Biofabrication of gold nanoparticles by *Shewanella* species. Bioresour. Bioprocess. 4: 50–59.

Yadav, K.K., J.K. Singh, N. Gupta and V. Kumar. 2017. A review of nanobioremediation technologies for environmental cleanup: A novel biological approach. J. Mat. Environ. Sci. 8: 740–757.

Yang, C., Y. Li, K. Zhang, X. Wang, C. Ma and H. Tang. 2010. Atrazine degradation by a simple consortium of *Klebsiella* sp. A1 and *Comamonas* sp. A2 in nitrogen enriched medium. Biodegrad. 21: 97–105.

Yang, F., Q. Jiang, M. Zhu, L. Zhao and Y. Zhang. 2017. Effects of biochars and MWNTs on biodegradation behavior of atrazine by *Acinetobacter lwoffii* DNS32. Sci. Total Environ. 577: 54–60.

Yang, J., M. Liao and M. Shou. 2008. Cytochrome p450 turnover: Regulation of synthesis and degradation, methods for determining rates, and implications for the prediction of drug interactions. Curr. Drug Metab. 9: 384–94.

Yoshida, S., K. Hiraga and T. Takehana. 2016. A bacterium that degrades and assimilates poly (ethylene terephthalate). Sci. 351: 1196–1199.

Yu, J., H. He, W.L. Yang, C. Yang, G. Zeng and X. Wu. 2018 Magnetic bionanoparticles of *Penicillium* sp. yz11-22N2 doped with Fe_3O_4 and encapsulated within PVA-SA gel beads for atrazine removal. Biores. Technol. 260: 196–203.

Zeng, J., Q.H. Zhu, Y.C. Wu and X.G. Lin. 2016. Oxidation of polycyclic aromatic hydrocarbons using *Bacillus subtilis* CotA with high laccase activity and copper independence. Chemosphere 148: 1–7.

Zhang, W., I.M.C. Lo, L.Hu, C.P. Voon, B.L. Lim and W.K. Versaw. 2018. Environmental risks of nano zerovalent iron for arsenate remediation: Impacts on cytosolic levels of inorganic phosphate and MgATP2– in *Arabidopsis thaliana.* Environ. Sci. Tech. 52: 4385–4392.

Zhou, Y., S. Wu, H. Zhou, H. Huang, J. Zhao, Y. Deng, et al. 2018. Chiral pharmaceuticals: Environment sources, potential human health impacts, remediation technologies and future perspective. Environ. Int. 121: 523–537.

Zomorodian, K., S. Pourshahid, A. Sadatsharifi, P. Mehryar, K. Pakshir, M.J. Rahimi, et al. 2016. Biosynthesis and characterization of silver nanoparticles by *Aspergillus* species. Biomed. Res. Int. 1–6. Article ID: 5435397. doi:10.1155/2016/5435397

CHAPTER 6

Roles of Nanoparticles in Bioremediation Rate Enhancement

Wael A. Aboutaleb[1*] and Huda Saleh El-Sheshtawy[2]

[1]Catalysis Lab, Refining Department, Egyptian Petroleum Research Institute, Nasr City, Cairo, Egypt, PO 11727.
[2]Biotechnology Lab, Processes Development Department, Egyptian Petroleum Research Institute, Nasr City, Cairo, Egypt, PO 11727.

1. INTRODUCTION

As long as the bioremediation process employs the microbial diversity having hydrocarbon metabolizing capability, can be utilized to biodegrade hydrocarbons contamination. The process has high environmental potential due to many advantages such as eco-friendliness, economic viability and degradation of wastes and contaminants to non-toxic or volatile by-products or even to H_2O and CO_2. The process has been extensively studied and applied to eliminate the plastic waste, heavy metals, petroleum hydrocarbons contaminants and particularly the polyaromatic compounds.

Despite the high efficiency and advantages of bioremediation, it still has many obstacles such as selecting right microbe and also challenges related to the process. The bioremediation challenges can be attributed to the factors concerning the process conditions, bacterium strains and/or the physical and chemical features of the contaminants. Regarding the process condition, the process is multi-variable dependent and so, it would be time consuming to optimize all these parameters following the classical approaches (Mohajeri et al. 2010). Moreover, the synergistic effect of bioremediation controlling parameters may cause biased results.

To meet the bioremediation challenge, application of potential microbes or already acclimatized microbes could be a better approach (Harayama et al. 2004), represents the core of biodegradation challenges. Furthermore, the bacterium metabolization of a specific and/or narrow range of petroleum hydrocarbons represents bioremediation's second dilemma.

*Corresponding author: waelaboutaleb@yahoo.com; Second author: hudaelsheshtawy@yahoo.com

The procedure to cultivate the adapted bioremediation bacteria and hence enhance the process rate is also an important point to be considered. For example, in marine environments, most of the bacteria (90–99%) are uncultivable (Rozsak, and Colwell 1987, Amann et al. 1995) and so, the analysis of microbial communities applied to direct hydrocarbon bioremediation activities is considered a challenge to microbiologists (Rollins and Colwell 1986, Sadiq and McCain 1993). Although numerous efforts have been made to solve those problems by applying a bacterium consortium and enrich the bacterium cultivation with additional nitrogen and phosphoric chemicals, the process still has the drawback of very low rate. Additionally, when applying the experimental condition in marine field environment, the bioremediation effectiveness is limited due to the surrounding factors such as the oil's physical properties, salinity, pH, dissolved oxygen (DO), temperature and competition with other microbial communities.

Chemically, the pollutants persistence in contaminated sites depends on many factors such as type and concentration of pollutants at particular site and their interaction with the naturally occurring chemicals at that site and physical features. Indeed, the aliphatic fraction of petroleum hydrocarbons is much more easily degraded than the aromatic compounds. Furthermore, the pollutant hydrophobicity is the main bio-remediation challenge that normally hinders the pollutant in water solubility and impedes hydrocarbons-bacterium cell interaction. However, the majority of applied hydrocarbon-degrading micro-organisms can produce emulsifying agents (biosurfactants) that facilitate the utilization of oil-water interface breakdown and hence faster oil decomposition by micro-organisms. Biosurfactants offer an attractive solution to enhance the hydrocarbons-bacteria cell wall interaction and hence influence the bioremediation process. Correspondingly, efforts have been made to enrich the *in situ* biosurfactant production from bacterial communities. For the last few decades, biosurfactants have received great attention due to their advantages of being an environmentally friendly alternative to conventional synthetic surfactants, their biodegradability, low toxicity, production from renewable resources, and functionality under extreme conditions (Desai and Banat 1997). On the contrary, their extensive pilot scale application is limited due to its relatively high production costs. Furthermore, biosurfactants exhibited a low biodegradation efficiency toward the high concentrated pollutant and/or large polycyclic aromatic hydrocarbons (PAHs) molecules. The PAHs recalcitrance occurred due to its high hydrophobicity, which assists their strong interaction to soil, long precipitation on the sea-floor and accumulation in the marine environment. Unfortunately, the PAHs normally exist in high concentration in petroleum pollutants in both soil and water environments. PAHs are hazardous contaminants because of their mean potential risk to human health since they cause different types of cancer (De´ziel et al. 1996).

The utilization of alternative substrates, such as agro-industrial wastes like carbon source, represents another possible economic strategy. Finding a waste with the suitable nutrients balance that influences the cell growth and product accumulation poses another challenge (Makkar and Cameotra 1999).

Talking about the influence of reaction rate directly brings to mind the concept of catalysis. The catalysis concept deals with introducing a foreign material to the reaction, which reduces the reaction activation energy and hence facilitates the reaction rate. Considering the bioremediation again, it could be simply described as hydrocarbon reduction reaction, whereas, chemically, the reduction process normally occurs in the presence of solid catalysts that are able to validate hydrogen to the reaction environment through adsorption technique. Additionally, the catalyst with high surface area would be another important aspect to accelerate the bioremediation reaction rate. Furthermore, as is well known, the type of catalyst applied potentially affects the reaction products and its formation rate. So, the bioremediation assisted solid catalysts will offer a faster rate with higher efficiency by pollutant adsorption on the catalyst surface.

Nanomaterials have been known as efficient catalysts in many reactions. Furthermore, nanotechnology is still a hot topic because of its expected impact on areas such as energy, medicine, electronics, and space industries. The use of nanoparticles in bioremediation process could be fruitful because the nanoparticles will be able to treat a larger contaminated area, hence better and faster bioremediation process. Nanoparticles, with unique properties in chemistry, optics, electronics, and magnetics, have led to an increasing interest in their synthesis and widening its application. Very recently, the integration of nanotechnology with biotechnology, termed as bio-nanotechnology, described the application of nanomaterials in combination with biological agents were employed in bioremediation process. Such technologies combination offered answers to handle complicated environmentally related problems. In this context, all the above arguments provide a reliable solution to influence the bioremediation rate.

2. ADVANTAGES OF NANOMATERIALS APPLICATION IN BIOREMEDIATION

Nanotechnology is a term that describes the synthesis and characterization of small particles of matter in the range of 1–100 nm. Normally, the nanoparticles (NPs) are characterized by new features and/or enhanced properties such as higher surface area, higher conductivity, powerful magnetism and enhanced catalytic efficiency, which is more than the bulk materials (Gupta et al. 2011). The nanomaterials could play several direct roles in bioremediation, where it may be utilized to cover large surface area and influences the soil conditions. Nanomaterials may also exhibit indirect actions in bioremediation process through the hydrocarbon adsorbent and/or oxidant, so that the features influenced by nanomaterials can efficiently frost the bioremediation process through several routes. (Bhattacharya et al. 2013). Hence, NPs act as catalyst for microbe's growth and metabolism, and decrease hydrocarbon adsorption on microbe's cell wall, hence enhancing the biomass growth. Furthermore, the NPs' powerful magnetism facilitates the removal of immobilized enzymes or proteins from the reaction media by applying a magnetic field (Khoshnevisan et al. 2011, Ranjbakhsh et al. 2012.). Additionally, nanomaterials could be used to immobilize the microbe's cells and/or the produced enzymes, whereas the NPs exhibit enzymes bio-compatibility and microbe's inert media behavior that does not hinder the enzymes' original properties and so keeps their biological activity (Ansari and Husain 2012).

Not only the nanomaterials' properties affect the bioremediation rate but also the type of nanomaterials applied. Several types of nanomaterials such as zeolites, carbon nanotubes, bimetallic nanoparticles and metal oxides were employed to mediate bioremediation of polyaromatic hydrocarbons (PAHs) and polychlorinated biphenyls pollutants in contaminated site (Mehndiratta et al. 2013, Shao et al. 2010, Gotovac et al. 2006). However, a careful selection of nanomaterials for remediation process is a crucial step as they may be toxic to the micro-organisms involved in the remediation process (Rizwan et al. 2014).

In the next sections, a detailed description of the reversible relation between nanoparticles and bioremediation process would be illustrated. The discussion will include, on one side, the bioremediation rate enhancement using the nanoparticles, the mechanism, methods of nanoparticles application, pilot scale studies and the indirect roles of nanoparticles in influencing the bioremediation. On the other hand, the biosynthesis of nanoparticles using the commonly known bioremediation capability micro-organisms will be illustrated. Such points would represent complete cycle of self-assembly process, resulting in hydrocarbon elimination and eco-friendly technique for nanoparticles production, using the metal nitrates as microbial nutrients. Finally, the limitations and prospective work considering application of nanoparticles to bioremediation will be discussed.

3. DIRECT ROLES OF NANOPARTICLES IN BIOREMEDIATION PROCESS

3.1 Enhancement of Hydrocarbon Solubility and Mobility

As mentioned above, the bioremediation process is normally hindered by the hydrocarbons' hydrophobicity feature, which limits its solubility in aquatic environment and/or increases its sorption in soil micelles (Cameotra and Makkar 2010). Meanwhile, the produced biosurfactant can alter the hydrocarbons water solubility by decreasing the surface tension and increasing the hydrocarbons' bioavailability. However, the biosurfactant production process is usually limited due to the micro-organism's metabolic pathways (Desai and Banat 1997). Furthermore, the biosurfactant micelles usually contact the degrading micro-organisms liposomes and change the cell membrane properties, which hinder the bioremediation process (Kaczorek and Olszanowski 2011). Also, the surfactants' application in the presence of PAHs results in increasing its toxicity to the microbe's cells which interrupt the microbial cell membrane contact, which reduces its biosurfactant productivity (Kumari, et al. 2014).

The advances and use of nanotechnology offers an attractive approach to biosurfactant production limitations and/or biosurfactant alternative and therefore can be effectively utilized in bioremediation process. Very recently, literature stated that use of some specific metals can control the microbe's metabolic pathways to enrich the biosurfactant production (Kiran et al. 2011). Especially, Fe has commonly been considered as a crucial nutritional element for biosurfactant synthesis using several micro-organisms (Haferburg and Kothe 2007). The Fe high activity can attribute to the Fe compounds' low toxicity (Liu et al. 2013) and its inveterate biocompatibility (Perez et al. 2002). Kiran et al. reported that the addition of Fe nanoparticles increased the biosurfactant production by 80% from the Nocardiopsis MSA13A species with no deformation in fungi filamentous structure morphology, which proves Fe nontoxicity (Kiran et al. 2014). El-Sheshtawy et al. (2014) and El-Sheshtawy (2017) have investigated the effect of Fe_2O_3 nanoparticles' addition on *Pseudomonas xanthomarina* KMM 1447, *Pseudomonas stutzeri* ATCC 17588 species to produce the biosurfactant and bioremediate the crude oil contaminated soil and sew water. The results showed that, because of nanoparticles' co-existence, the optimum biodegradation of crude oil was demonstrated at 60% of microcosms containing biosurfactant and nanoparticles after only 7 days. The bacterial strain is highly potential to consume the total paraffins (iso- and *n*-paraffins) in crude oil samples. Additionally, the consumption of specific member rings of polyaromatic hydrogen compounds depends on the type and the concentration of nanoparticles. In one study, the workers also studied the bioremediation of crude oil contaminated soil using *Bacillus licheniformis* species in the presence of NiO nanoparticles (El-Sheshtawy and Ahmed 2017). The results indicated up to 80% of oil content reduction after 7 days due to the bioremediation process in the presence of the nanoparticles. Additionally, a maximum bioremediation affinity of 90% was obtained in the presence of bacterium consortium with biosurfactant and NiO after 7 days.

On the other hand, the advances in nanoparticles' synthesis methods allow to produce nanoparticles as surfactant alternative. The idea concerns the synthesis of inorganic materials based on polymers that contain polar groups such as OH or CO–NH. The produced nanoparticles have the ability to increase the PAHs mobility and solubility through the nanoparticle-PHAs-suspension stat formation to solute the hydrophobic materials. Tungittiplakorn et al. reported a high solubility and easy desorption and mobility to phenanthrene in the contaminated soil using the polyurethane acrylate anionmer or polyethylene glycol-based nanoparticles (Tungittiplakorn et al. 2005). The NPs' superiority over the surfactant application arises from the simultaneous enhancement of NPs affinity to contact contaminants through alteration of polymeric chain and increasing the contaminants' mobility via controlling the charge density of modified nanoparticles. The polymer-based nanoparticles can hence handle the problem of surfactant to microbe's liposomes interaction, leading to a decline in the contaminants' bioavailability.

3.2 Soil Conditions Improvement

The application of nanoparticles in bio-reactions can serve in microbe's nutrients' reservation and hence significantly enhance the total microbial growth. Several nanomaterials were studied to enhance the micro-organisms' nutrition through different routes. The materials scale is the first factor affecting the improvement of soil conditions. Rangaraj et al. reported that application of nano silica with particles size of 20–40 nm has highly influenced the total microbial biomass in soil (Rangaraj et al. 2013). The authors claimed that the nano-silica highly promotes the microbial growth and validates more carbon and nitrogen to the soil than other silica sources, such as sodium silicate and micro-silica. Nano-silica coated Fe or carbon and some metal traces such as Ca, Mg, Na and Al were also reported to influence the phosphorus elimination from ultisol (Rick and Arai 2011).

Nanomaterials can also effectively increase the nutrient-to-soil supply duration. The zeolites and clays' naturally occurring materials are examples of such materials that are used to deliver nutrients to micro-organisms and so increase their growth. The zeolites' efficiency is related to their porous structured crystals that can be loaded with important nutrients such as nitrogen, potassium, phosphors and calcium with advantage of declining its leaching from soil (Veronica et al. 2015. DeRosa et al. 2010). Improving the solubility and dispersion of insoluble nutrients in soil is another mechanism of nanoparticles to improve the soil conditions. This behavior would play an important role in enriching soil-nutrients and hence increase the bioavailability and the micro-organisms' growth.

3.3 Metabolism and Growth Catalysts

Nanoparticles can be an added value to the bioremediation process through catalyzing the micro-organisms' metabolism and microbe's growth rate. Different nanoparticles such as fullerene-60 and super-magnetic iron oxide have been reported to control the growth phases, whereas the existence of such nanoparticles has been able to decrease the lag-phase duration while supporting and elongating the exponential and stationary growth phases (Bhatia et al. 2013). Also, it is reported that *E. coli* can reach its growth stationary phase after only 4 hours without adaptation time in presence of ZnO nanoparticles (Merced et al. 2005). Additionally, the nanoparticles can maintain the microbe's growth acquiring energy from oxidation-reduction transfection or reacting with some microbe's surrounding pollutants and hence catalyzing its growth. For example, the presence of tiny amount of iron oxide NPs (10 mg/L) influenced the *Nocardiopsis* MSA13A micro-organism's growth due to the microbe's capability to gain the growth and energy via the Fe^{2+} to Fe^{3+} oxidation (Kiran et al. 2014). Moreover, the *in vivo* and *in vitro* high microbial reduction of soluble Pd^{2+} ions to nano-sized metallic Pd utilized 90% polychlorinated biphenyl dehalogenation in 5 hours at ambient temperature (De Windt et al. 2005). Also, 5g/L of Fe^0 nanoparticles were sufficient to nitrates ions removal in only 3 days, while just 20% removal in 7 days was recorded for non-assisted Fe^0 sample blank (Shin and Cha 2008). The nanoparticles can assist the microbes to increase their biological reaction rate, whereas the existence of tiny amounts of nanoparticles can duplicate or even triplicate the biodegradation kinetics of PAHs by catalyzing its metabolism (Masy et al. 2012). Also, it is reported that the anatase phase of TiO_2 nanoparticles declined the biodegradation of pyrene compound duration from 45.9 to 31.36 h, which indicates the increased reaction rate (Dong et al. 2010). Jorfi et al. also reported about 7% biological pyrene removal in 6 hours in acidic media with existence of 30 mM Fe_2O_3, H_2O_2 and sodium pyrophosphate as chelating agent (Jorfi et al. 2013).

3.4 Microbial Cells and Enzymes Immobilizer

The bioremediation process validity depends extensively on the cell survival in the contaminated sites (J'ez'equel and Lebeau 2008). Cell survival is facing many problems that hinder the process efficiency when considering the process environment. First of all, the growth rate of efficiently degrading bacteria in soil or aquatic media is normally a long-term step. (Roselien et al. 2007). All these features cause a decline in the bioremediation rate and represent challenges for application to degrade the pollutants in both the soil and aquatic environment. In this framework, several studies claimed that immobilization of either the micro-organisms or their enzymes on nanoparticles provide efficient bioremediation (Hu and Qi 2015) and prevents the microbial cell loss (Wen et al. 2010), whereas the immobilization enhances the microbial cell efficiency as compared to the soil natural microflora and protect the cells from the soil pH and toxicity (Zhang et al. 2008).

3.4.1 Micro-organisms' Immobilization

As mentioned earlier, the presence of nanoparticles in bioremediation work as catalyst and enrich the growth media, which enhances the microbial growth. Another possible important role of nanoparticles is to immobilize the bacterial cells. Several researches have reported that magnetite immobilized bacterial species such as *Rhodococcus erythropolis* IGST8 (Ansari et al. 2009) and *Pseudomonas delafieldii* (Shan et al. 2005), which were 56% more efficient than free cells on desulphurization of Dibenzo-thiophene. Also, the complete degradation of carbazole was reported using the immobilized *Sphingomonas* sp. strain XLDN2–5 cells on magnetite in 16–18 hours, which was too fast than the free cells (Wang et al. 2007). The apparent synergism of magnetite in immobilization of bacterial cells has been attributed not only to magnetite effect as catalyst but also to the ease of cells' separation by applying a magnate. Non-magnetic nanoparticles such as rice husk derived silica were also used to immobilize several *Pseudomonas* strains for oil contaminated soil (Zhubanova and Mansurov 2008). The results showed that 80–90% of bacterial cells were immobilized on silica with high oil contamination utilization.

On the other hand, the bacterial cells' immobilization achievements have utilized the bioremediation efficiency but not the effective process duration. So, another technique was developed by using the biostimulation of microbes and then immobilize the produced enzymes on nanoparticles. The suggested technique solves the microbial cells-contaminants direct interaction problem, and hence prevents the cell toxicity (Wenyu et al. 2013) and accelerates the bioremediation duration.

3.4.2 Immobilization of Enzymes

The enzymes are key factor in bioremediation process. It represents the main function in microbe-pollutant contact and hence the pollutants bioavailability enhancement. Instead of microbial growth problems, freshly prepared microbial enzymes can be used directly to eliminate the pollution in contaminated sites. Meanwhile, the direct application of enzymes also faces many limitations such as high production cost, low stability, difficult to reuse, low stability and high sensitivity to reaction conditions. All those problems can be resolved through immobilization tool (Hernandez and Fernandez-Lafuente 2011). Enzymes immobilization technique has been known since 1916, when Nelson and Griffin reported the sucrose hydrolysis by invertase absorbed on charcoal (Nelson and Griffin 1916). The enzymes' immobilization was initially processed through biological methods but after a while, and due to the high developing rate in nanotechnology, the advantageous combination of the two technologies resulted in superior enzymes activity (Kim et al. 2008).

The enzymes' immobilization plays a dual function in bioremediation efficiency due to the enzymes' freshly obtained properties and also the role of nanoparticles existence. The immobilized enzymes can enhance the bioremediation through many roles. The first role is that immobilized enzymes can overcome the kinetic limitations enforced by free enzymes considering their purity, activity, stability, selectivity and durable efficiency (Garcia-Galan et al. 2011). As a practical example, immobilization utilized the lipase enzyme separation from the reaction mixture and has kept 70% activity of enzyme after 21 reaction cycles (Ren et al. 2011). Also, the lipase immobilized on sol gel/calcium nanoparticles has a reactive and stable performance for 100 continuous days (Jeganathan et al. 2007). The second advantage for immobilization is that it influences its application along wide range of pH values and temperature if compared with the native enzymes (Ansari and Husain 2012). Additionally, the immobilization on nanoparticles guarantee the enzyme-substrate interaction in solution because nanoparticles prevent the potential enzyme aggregation (Dutta et al. 2013). Furthermore, nanoparticles in immobilization technique performs as molecular chaperons that influence the native protein structure maintenance (Singha et al. 2010). Enhancement in enzyme thermal stability during the bioremediation is another important role of enzyme on nanoparticles immobilization process (Chronopoulou et al. 2011). The most important role of enzyme immobilization, especially on nanoparticles matrix, is that it obviously increases activation energy, whereas Acevedo et al. reported that the immobilization of manganese peroxidase on nano-clay possesses more efficient bioremediation with activation energy of 51.9 kJ mol^{-1}, where the free enzyme has 34.6 kJ mol^{-1}. Furthermore, the immobilization on nan-clay assisted the enzyme to do the enzymatic conversion of polyaromatic hydrocarbons efficiently, even under non-optimum bioremediation conditions (Acevedo et al. 2010). Such immobilization process speeds up the electron migration from the enzyme active sites to the substrate.

A huge diversity of nanoparticles such as metal oxides (SnO_2, TiO_2, ZrO_2, ZnO) graphene, and carbon nanotubes (CNT) were investigated as enzymes immobilization. Nanoparticles offer a large specific surface area and low diffusion limitations which increases the mass transfer required for the micro-organism-pollutant interaction. The immobilization on nanoparticles technique to increase the enzymes activity towards petroleum hydrocarbons degradation was successfully applied for several enzymes. Table 1 shows examples of some enzymes loaded with different nanoparticles matrices. Among the large nanoparticle's community, the magnetic nanoparticles are the most recommended particles for application in enzyme immobilization due to its high specific surface area, which allows more enzyme loading, and higher mass transfer ability that allow more cell binding, biocompatible nature, nontoxic feature and ease of recovery by applying external magnate (Wang et al. 2012).

The catalytic activity of the Fe_2O_3 immobilized lipase enzyme was found to be duplicate as the free enzyme. The immobilization steps of lipase enzyme on magnetite nanoparticles as an immobilization example is illustrated in Fig. 1. The preparation procedure involves the functionalization of magnetite surface with multifunctional group's molecule such as polydopamine at alkaline media in the first step, whereas the dopamine polymer performs as a substrate that ensures the loading of enzyme on the magnetite surface and prevents the enzyme dissolution during the degradation process. The procedure is then extended as a second step to load the enzyme at the surface due to the charge difference between the enzyme and dopamine polymeric chain end. At the end of preparation procedure, the enzyme found to form a sick shell totally covers the polymer functionalized magnetic nanoparticles' core.

Not only nanoparticles but natural polymers, such as chitosan loaded nanoparticles, were also used to immobilize the enzymes. The chitosan/nanoparticles matrix is a highly efficient enzyme immobilizer, the efficiency of it can be attributed to many reasons such as its biocompatibility, film forming ability, high mechanical strength, economic and nontoxicity (Leitgeb et al. 2014). Additionally, the chitosan/nanoparticles matrix's high conductivity increases the bacterial activity. Also, González-Campos et al. (2013) have investigated the relation between the nanoparticle/chitosan conductivity and the bacterial activity. The author proved that the nanocomposite had a considerable bacteriocidic effect on both gram-positive and gram-negative bacterial species. The authors explained that the nanoparticles covered with chitosan prevents the nanoparticles aggregation, which in turn increases the nanoparticles availability to cell wall interaction or 2) the chitosan gain positive charges from the nanoparticles which increases its binding efficiency to the negative charge on th product (i.e. a pure product is acquired e cell wall). Very recently, Mady et al. investigated the chitosan loaded Ag, CdS, Fe and CuO nanoparticles with electrical conductivity (Mady et al. 2018). The authors found that the Ag and CdS nanoparticles had a much higher electrical conductivity than the pure chitosan and nanoparticles.

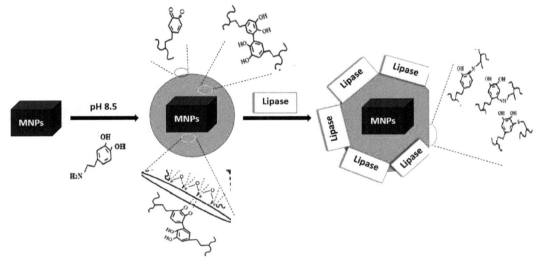

Figure 1 Two step synthesis of lipase immobilized polydopamine film on magnetic nanoparticles

Table 1 Immobilized enzymes involved in the degradation of petroleum hydrocarbons

Immobilized enzyme	Nanoparticle	References
Mn Peroxidase	Nano clay	Acevedo et al. 2010
Alcohol dehydrogenase	Magnetic nanoparticles Oxidized diamond nanoparticles Citrate stabilized silver (Ag-citr) nanoparticles, functionalized with carboxyl groups Citrate stabilized gold (Au-citr) nanoparticle stabilized by cetyltrimethylammonium bromide (CTAB), functionalized with amine groups Au-COOH (CTAB) and (Au-citr) functionalized with amine group stabilized with CTAB (Au–NH$_2$–(CTAB) nanoparticles	Ren et al. 2011 Li et al. 2008 Petkova et al. 2012
Horseradish peroxidaze Laccase	Magnetic nanoparticles Chitosan coated magnitite Platinum nanoparticles Titanium dioxide Large-pore magnetic mesoporous silica	Garcia et al. 2011 Kalkan et al. 2012 Li et al. 2011 Hou et al. 2019 Wang et al. 2010
Lipase	PolyDopamine coated magnetite nanoparticles Alkyl silane modified magnetic nanoparticles Poly acetic acid nanoparticles	Acevedo et al. 2010 Wang et al. 2012 Coronopoulou et al. 2011
Catalase	Gold nanoparticles	Chirra et al. 2011
Glucose oxidase	Silica-encapsulated magnetite nanoparticles	Ashtari et al. 2012
Pectate lyase	Hydroxy apatite nanoparticles	Dutta et al. 2013

The enzymes immobilization on nanoparticles can be achieved by several techniques. Berna and Batista have classified the enzymes immobilization to irreversible and reversible methods as illustrated in Fig. 2. The irreversible immobilization techniques are covalent binding and entrapment, while the reversible methods include adsorption, chemical interaction, affinity binding and metal binding (Berna and Batista 2006), where the stability of immobilized enzymes depends mainly on the number of bonds formed between nanoparticles and the enzyme.

In covalent binding method, a direct contact of enzyme and nanoparticles through covalent bond is achieved (Kim et al. 2008). The covalent binding produces a strong and stable binding with inorganic materials such as porous glass, aerosols and silica or polymers such as polyacrylamide (Wong et al. 2008). This method is normally used when enzyme is not favorable to the end product (i.e. a pure product is acquired). The covalent binding technique for enzymes' immobilization is normally achieved in two steps. The first step includes the addition of reactive materials which generates an electrophilic group to activate the support surface. The activation step is then followed by modification of polymer matrix to interact with the enzyme via the nucleophilic groups in proteins. In other words, the covalent binding is a chemical interaction between the support's (polymer or nanoparticles) generated functional group and the enzyme's amino acids terminals such as sulfhydryl or hydroxyl groups in cysteine or threonine, respectively (Quirk et al. 2001). The exact attachment of enzymes and support can occur via a direct link formation or space arm mechanism. The space arm mechanism is more efficient since it is characterized by high enzyme mobile affinity, which results in powerful enzyme activity.

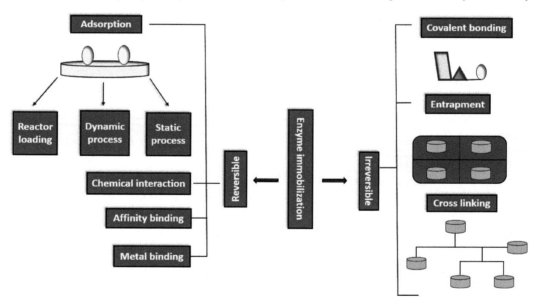

Figure 2 Methods of enzyme immobilization on substrate matrices (Sneha et al. 2019).

Entrapment is another irreversible enzyme's immobilization technique in which the enzyme is trapped inside the polymers' network or pores. In this technique, selective membranes used as support allow the substrate and the products to migrate into/out of it and prevent the enzyme mobility. The entrapment technique can be applied by using gel, fibers entrapping or microencapsulation (Bernfeld and Wan 1963). The enzyme immobilization by entrapment technique is advantageous because it is a fast and cheap process operated at mild conditions. On the contrary, the process is practically hindered because of its mass transfer limitations. In microencapsulation method, the enzyme is trapped in spherical shaped and selective semipermeable membrane (Rosevear et al. 1987). Although this method allows large quantity of enzyme immobilization due to the membrane high surface area, it suffers the enzyme deactivation through the immobilization procedure.

Adsorption is another simple methodology used to immobilize the enzymes. The process has a reversible pathway between the carrier and surroundings (Brady et al. 2009). Several cheap materials like activated charcoal, alumina, and ion exchange resins have been used as adsorbent for enzymes. The procedure is simple and cheap to be applied but has the drawback of enzyme to substrate weak binding. The carrier-enzyme interaction may occur via weak connection generated from van der Waals force, salt formation, hydrophobic bonding or stronger attraction due to hydrogen or ionic bonding formation. The generation of a strong or weak bonding through adsorption process depends mainly on the charge's rearrangement on both carrier matrix and the protein from enzyme, whereas a strong bonding would be formed when many opposite charges are found on their surface (Tosa et al. 1966). The enzyme adsorption technique is easy and cheap since it requires less activation and chemicals.

Immobilization through ionic binding is another reversible enzyme loading technique, where the bonding forms according to the salt linkage between the enzyme and the substrate. The bonding action can easily reverse if factors such as temperature, polarity and/or the ionic strength have changed during the reaction. The enzyme adsorption due to the polarity concept is like protein separation by gas chromatography (Gusian et al. 1997). Affinity binding is one of the reversible methods for enzyme immobilization methodology. The affinity technique can be achieved through one of the two pathways. The first procedure is to activate the substrate (the polymer or nanomaterial matrix) by adding functional groups coupled with enzyme, followed by the addition of enzyme to such groups. This methodology provides the main feature to prevent the enzyme exposure to harmful chemicals. On the other hand, the second procedure includes the enzyme modification with another molecule which has the ability to contact the substrate surface (Porath 1992). The enzymes can be also loaded to substrate matrices by metal linked enzyme technique. In this technique, a metal salt, which has affinity to bind the nucleophilic groups in matrix, is precipitated on the carrier surface. The precipitation of metal salt can be obtained by simple procedure of heating. The metal linked technique is a simple procedure that enhances the enzyme activity by (30–80%), while enzyme and carrier can be further separated by altering the pH value, which facilitates the enzyme separation and/or recombination (Yücel 2011).

4. NANOMATERIALS' INDIRECT ROLES IN BIOREMEDIATION

As the nanoparticles are normally used as highly efficient adsorbent, catalysts or photocatalysts for oxidation reactions due their high surface area and unique electronic structure, they can also be nominated to perform this action in bioremediation process. The next titles will illustrate the role of nanoparticles in oxidation and adsorption of hydrocarbons during the bioremediation process.

4.1 Adsorption of PAHs

The high adsorption ability of nanoparticles enables it to powerfully decrease the soil and water pollutant toxicity (Nowack and Bucheli 2007). For example, hematite, magnetite and ZnO nanoparticles were widely investigated as PAHs and non-biodegradable asphaltene adsorbent. The gold modified with fused silica fibers is another inorganic nanoparticle used to eliminate some pure PAHs such as naphthalene, phenanthrene, fluoranthene and pyrene (Masooleh et al. 2010).

But on the other hand, the nanoparticles cannot resist the flow of water stream if applied in fresh water environment and also may not highly contact the hydrophobic pollutants. To increase the nanoparticles' activity to adsorb much more petroleum hydrocarbons, it's better to be modified with organic molecules. The hybrid organic-inorganic nanocomposites then combine both the advantages of nanoparticles, that is, their large surface area and good electronic structure properties, together with the hydrophobicity of organic molecule to contact the nonpolar petroleum hydrocarbons. Several researches have claimed that the organic-inorganic hybrid materials exhibited a powerful adsorption of hydrocarbons. For example, organic modified clay can remove hydrocarbons more than 5 times its weight (Karimi et al. 2013). The clay modification with either alkyl or aryl constituent surfactant molecules increases the hydrocarbons' adsorption because of the inorganic cations exchange to organic centers, which results in better pollutants contact (Safaei et al. 2008). As the next step in nanoparticles' modification to enhance the petroleum hydrocarbons absorptivity, hydrophilic-hydrophobic co-polymers were suggested. Several dual functional co-polymers were investigated as nanoparticles modifiers to absorb PAHs. Magnetite modification with blocked dual hydrophilic polyacrylic acid and hydrophobic polystyrene co-polymers were investigated to eliminate the hydrocarbons' pollutants (Pavía-Sanders et al. 2013), whereas the prepared nanocomposite eliminates crude oil about 10 times their applied weight, which duplicates the effect of one polymer-nanoparticles composite. Also, polyacrylic acid-oleic acid modified hematite nanoparticles efficiently adsorbed crude oil from aquatic environment.

Carbon nanotube (CNT) is another considerable organic powerful adsorptive material that can remove many pollutants such as PAHs, like chlorophenol, phthalate esters, etc. (Nowack and Bucheli 2007, Gotovac et al. 2006). Comprisable CNT-graphene hybrid material is another investigated organic adsorbent that has been used to remove the crude oil from water stream (Hu et al. 2014). The hybrid showed activity of around 95% removal of such large aromatic hydrocarbons. Hence, it is generally acceptable to say that application of such nanocomposites emphasizes the bioremediation process through the nanoparticle's adsorption ability mechanism.

4.2 Chemical Hydrocarbons' Oxidation and Reduction

It is well known that several nanoparticles, like F_2O_3, TiO_2, CeO_2 and Al_2O_3, are active catalysts for chemical oxidation of hydrocarbons such as CH_4, C_2H_5OH and CH_3OH, to produce many useful materials such as H_2, acetaldehyde CNT and petrochemicals. This known property can also assist the bioremediation process by oxidizing the hydrocarbons' pollutants in contaminated sites, thus decreasing the contaminants concentration, toxicity to microbes and influencing the microbial growth during the bio-remediation.

Several nanoparticles were investigated as bioremediation process improver. The nano metallic iron particles have linearly assisted the soil contaminated hydrocarbons' bio-remediation by increasing the nanoparticles concentration (Jameia et al. 2013). Nanoparticles such as Co and Mn followed different oxidation mechanisms to enhance the PAHs bioremediation of 1-substituted naphthalene derivative (Nador et al. 2010). The Co nanoparticles assisted the reduction pathway of naphthalene to the corresponding tetraline compound, while Mn nanoparticles derived the reaction to produce hydrogenated 5,8-derivatives. The Co nanoparticles were found more efficient than Mn nanoparticles in naphthalene and phenanthrene conversion to 1,2,3,4-tetra-hydronaphthalene at room temperature and 9,10-dihydrophenanthrene within 3 hours. Tiny amount of TiO_2 nano-particles (100 mg/L) was reported to remove 78% of hydrocarbon effluents of petroleum refinery wastewater at 45°C operating temperature for around 90 min in presence of UV irradiation (Saien and Shahrezaei 2012).

Besides the nanoparticles catalytic hydrocarbons oxidation activity, some nanoparticles like CaO_2 can enhance the bioremediation process by releasing oxygen to the aquatic system to form O_2 and H_2O_2 molecules (Wang et al. 2016). The CaO_2 dissolute in presence of water to generate hydrogen peroxide and calcium hydroxide. Also, the addition of Fe^{3+} to the solution enhances the OH radical formation and increases the Fenton reaction according to the equations:

$$CaO_2 + 2H_2O \rightarrow H_2O_2 + Ca(OH)_2 \tag{1}$$

$$2H_2O_2 \rightarrow 2H_2O + O_2 \tag{2}$$

$$H_2O_2 + e^- \rightarrow OH^{\cdot} + OH^- \tag{3}$$

$$H_2O_2 + Fe_2^+ \rightarrow OH^{\cdot} + OH^- + Fe^{3+} \tag{4}$$

The liberated oxygen then affects the contaminated site by hydrocarbon contaminants oxidation and also emphasizes the biodegraded micro-organisms aeration (Mosmeri et al. 2017). Other compounds such as pure H_2O_2 are also known to liberate the OH radicals in aqueous solutions. The CaO_2 nanoparticles differ from pure H_2O_2 basically in the quantity of generated oxygen. Researchers found that CaO_2 is much more effective in liberating oxygen molecules to fresh water (Ndjou'ou and Cassidy 2006). Moreover, CaO_2 particles are more effective oxidant than H_2O_2 at all pH values because it powerfully produces H_2O_2 only when dissolute in water, which conserves the O_2 deficiency carried out in case of H_2O_2, which decomposes in air (Northup and Cassidy 2008).

CaO_2 nanoparticles have been investigated as oxygen releasing compound with simultaneous bio-remediation of single pollutant like benzene, xylene and ethylbenzene (BTEX) as well as mixture of those pollutants. It is reported that the CaO_2 efficiency in individual BTEX bioremediation follows the order of ethyl-benzene >toluene >pxylene >benzene (Xue et al. 2018). In addition, it is also demonstrated that CaO_2 based Fenton reaction efficiency for simultaneous remediation of BTEX increased up to 98% removal in fresh water and 88% in groundwater by increasing the $CaO2:Fe^{3+}$: BTEX ratio up to 40:40:1, respectively (Yeh et al. 2010). For more effective bioremediation of pollutants like benzene, toluene, ethyl-benzene and xylene, CaO_2 nanoparticles' combination with organic molecules like citrate, polyvinyl alcohol, and bamboo biochar with one specific ratio was investigated. The biochar modified CaO_2 beds showed around 99% removal of toluene in the contaminated site (Wu et al. 2015). Effective organic contaminates degradation of 18% in sediment was attributed to the O_2 liberation from addition of CaO_2 while no-degradation was associated with no addition of CaO_2 (Nykanen et al. 2012).

Although great attention has recently been paid to the simultaneous application of oxygen releasing compounds and bioremediation for petroleum hydrocarbons, there are still dark areas in this topic. For example, the impact of CaO_2 application on the microbial community structure of groundwater contaminated sites to enhance has seldom been investigated. However, very recently the next generation sequencing technique has been used to investigate the effect of CaO_2 on the microbial diversity during the bioremediation of groundwater hydrocarbons' contaminants (Patel et al. 2012).

Despite the monitoring of nanoparticles in environment during the bioremediation by electron microscopy, the CaO_2 precipitation and sedimentation properties is very rare as these factors dramatically affect the oxygen releasing property when studied (Hsiung et al. 2016). Additionally, the application of CaO_2 for bioremediation of long hydrocarbon molecules and/or hydrocarbon mixture such as diesel petroleum fraction is quite limited.

Considering all the above mentioned activities and roles of nanoparticles in developing the bioremediation process, it is important to discuss the methods used to synthesize nanoparticles. Practically, the nanoparticles have been synthesized by various physical and chemical processes; however, some chemical methods cannot avoid the use of toxic chemicals in the synthesis process (Azim et al. 2009). Other methods such as physical vapor deposition (PVD) and chemical vapor synthesis (CVs) need high temperature and/or ultra-vacuum approaches. Thus, much attention has been paid to develop green synthesis procedures of nanoparticle. Bio-synthesis methods of nanoparticle using either micro-organisms or plant extracts have provided a reliable, environmental alternative to chemical and physical methods. Furthermore, some researches approved the ability of micro-organisms to use nitrates as nutrients and production of small size nanoparticles inside and on the cell wall during micro-organism growth for hydrocarbon bioremediation (De Windt et al. 2005). Also, bioremediation is a well-known eco-friendly technique normally used for removal of heavy metals from soil and fresh water. All these aspects recommend the biological method to synthesize different types of nanoparticles. The next subtitle would briefly discuss the nanoparticle biosynthesis process considering the technique steps, type of micro-organisms, efficiency and process disadvantages.

5. BIOSYNTHESIS OF NANOPARTICLES

Nanoparticles' biosynthesis is the process where the micro-organism converts, either inside or outside the cell, its surrounded metal ions to nanoscale metallic or metal oxide form through reductant enzymes produced by the cell activities. In contrast to chemical or physical method to make nanoparticles, green biosynthesis technique possesses many advantages like being ecofriendly, economic, uses no harmful chemicals and is achieved mostly at ambient temperature and pH values. Also, the biosynthesized nanoparticles commonly have higher surface area, are more catalytically active and influence the pollutants and enzymes, as mentioned previously. The nanoparticles biosynthesis can occurred inside or outside the microbe's cell, hence it is classified into intra- and extracellular synthesis (Li et al. 2011). In intracellular method, the micro-organism introduces the metal's ions into its cell and reduces the ions to metal or metal oxide nanoparticles using the cell enzyme (Simkiss and Wilbur 1989), whereas adsorption and reduction of metal ions to the corresponding metals, metal oxides or sulfides on the microbial cell's surface is called the extracellular synthesis technique (Mann 1996). Recently, scientists have given great attention to associate the micro-organisms with inorganic materials, therefore many workers biologically synthesized. Several micro-organism species such as fungi (Riddin et al. 2006), actinomycete (Ahmad et al. 2003), yeast (Kowshik et al. 2002), viruses (Shenton et al. 1999) and bacterial species (Lengke et al. 2006b) have been exploited to make nanoparticles. The microbial ability to synthesize nanoparticles can be attributed to microbe's immunity to most of the toxic heavy metals. This immunity is obtained either by their chemical toxification or due to ionic flux through the cell wall protein, which acts as ATPase or osmotic cation or proton transfer inhibitor. Microbes can also synthesize the nanoparticles by changing the solubility, where the microbes can convert the soluble toxic metal ions to non-soluble metals through reduction or precipitation (Beveridge et al. 1997, Bruins et al. 2000). The biosynthesis of nanoparticles can occur by extracellular demineralization, biosorption, precipitation, complexation and accumulation inside the microbial cell. Extracellular synthesis is a commercially superior technique because of its poly-dispersibility, whereas intracellular technique normally produces particles with specific size and less poly-dispersibility (Mann 1992).

Because the bioremediation process normally occurs using bacteria, in the following subtitles the intra and extracellular synthesis of nanoparticles using bacterial strains will be briefly considered.

5.1 Bacterial Synthesis of Nanoparticles

In the last few decades, bacterial activity was extensively involved in ores deposition. The superiority of bacterial synthesis of nanoparticles can be partially attributed to high bacterial growth rate. Among many species, the prokaryotic bacteria have been widely exploited for metals biological synthesis. Several bacterium strains were reported to reduce the metal ions to metallic spherical shaped nanoparticles. Slawson and his coworkers reported the ability of *Pseudomonas* stutzeri AG259, isolated from silver mines, to intracellularly synthesize Ag^0 with 35–46 nm particle size (Slawson et al. 1992). Klaus and his group also reported the biosynthesis of metallic Ag nanoparticles using the *Pseudomonas* stutzeri AG259 with well-defined size and distinct topography up to 200 nm particle size (Klaus et al. 1999). The authors have proved the synthesis of Ag^0 nanoparticles by the TEM image, where the particles can be seen as dark spherical and trigonal points in the bacterial periplasmic space. Recently, an airborne *Bacillus* sp. isolated from the atmosphere was also found to reduce Ag^+ ions to Ag^0. This bacterium accumulated metallic silver of 5–15 nm in size in the periplasmic space of the cell (Pugazhenthiran et al. 2009).

Parikh et al. also reported the extracellular synthesis of metallic Ag with nano-size of around 20 nm from the $Ag(NO_3)$ precursor (Parikh et al. 2008).

Nair and Pradeep (2002) synthesized Ag^0, Au^0 and Ag–Au nanoparticles with well-defined morphology from concentrated solutions using the *Lactobacillus* strains located in buttermilk. The authors also reported the retainment of bacterial strain higher count even in high ions concentration and after the crystal growth. *Bacillus subtilis* 168 also reported the intracellular reduction of aqueous Au ions to octahedral 5–20 nm Au^0 particles (Southam and Beveridge 1994). Sulfate-reducing bacterial strain, isolated from the gold mines, was applied to intracellular precipitation of >10 nm Au^0 from its thiosulfate complex and released H_2S gas as by-product (Lengke and Southam 2006). *Shewanella algae* has also been reported to reduce Au^{3+} under anaerobic conditions to 10–20 nm Au^0 particles at 25°C and pH value of 7, while particles with 50–200 nm were obtained at 2.8 pH value (Konishi et al. 2007). Lengke and coworkers investigated the cyanobacterium, *Plectonema boryanum* UTEX485, to synthesize the Au^0 nanoparticles in abiotic and cyano-bacterial systems in temperature range of 25°C and 200°C by using aqueous gold thiosulphate and chloride solution (Lengke et al. 2006a). In case of thiosulphate solution, cubic particles of 10-25 nm size Au^0 were obtained in periplasmic cell and >10 nm particles and 10–25 nm Au_2S were precipitated in solution. Applying the $AuCl_4^-$ precursor resulted in precipitating >10 nm octahedral Au^0 inside the cells with 1–10 μm platelets morphology at the solution.

Metal oxides and magnetic oxides such as Fe_3O_4, γ-Fe_2O_3 and FeS are also important materials that can be biologically synthesized and applied in biological reaction due to their ease of separation with applied magnetic field. Iron oxides and/or sulfides can be intracellularly produced using the Magnetotactic bacteria (BacMPs). The magnetic particles, aligned normally in chain within the bacterium, have been hypothesized to support the bacterium migration along the oxygen constituents in aquatic environment under the geomagnetic force. Additionally, the BacMPs are normally enveloped by phospholipids and proteins organic membranes, which facilitate its aquatic dispersion. Furthermore, an individual BacMP contains a single matter, such as magnetite, that yields superior magnetic properties (Thornhill et al. 1995). Since 1975, several bacterial morphologies such as *cocci, spirilla, vibrios, ovoid, rod-shaped* and *multicellular* bacteria were identified and colonized in the marine environment (Spring and Schleifer 1995). The only cultivated Magnetotactic cocci MC-1,are normally found to populate the aquatic sediments surface, which prove its aerobic nature. Also, three anaerobic cultivated vibrio bacteria, namely MV-1, 2 and -4, have been isolated from salt marshes.

These bacteria, which are identified as α-Proteobacteria members and possibly belong to the Rhodospirillaceae family, can synthesize the hexa-octahedral shaped BacMPs. On the other hand, several members of the Magnetospirillaceae bacterial family such as *Magnetospirillum magnetotacticum* MS-1, *M. gryphiswaldense* strain MSR-1 were isolated from fresh water sediments and cultivated with the use of growth medium and magnetic isolation techniques, to produce magnetic nanoparticles (Arakaki et al. 2008). Moreover, *Magnetospirillum magneticum* AMB-1 species isolated by Arakaki et al. (2008) were anaerobically facultative magnetotactic spirilla (Watson et al. 1999). Several new magnetotactic bacteria have been identified since the beginning of the 21st century. Meanwhile, most of those bacteria are mesophilic and have inhibited growth over 30°C, whereas thermophilic magnetotactic bacteria such as HSMV-1 have been rarely reported by Lefèvre et al. (2010). The author reported the growth of the HSMV-1 species, isolated from the hot springs with a temperature range of 32–63°C. Additionally,

Watson et al. (1999) demonstrated that sulfate-reducing bacteria synthesize strongly magnetic iron sulfide (FeS) nanoparticles on their surfaces. The magnetic nanoparticles (about 20 nm in size) were separated from the solution by a high gradient field of 1 Tesla.

Recently, the concept of two technologies' combination or self-assembly of bioremediation and biosynthesis of nanoparticles has attracted much attention. Vitor et al. (2015) have synthesized the ZnS nanoparticles using the excess sulfur produced through the two steps bioremediation by applying the sulfate reducing bacteria. The produced ZnS were in nanoscale range of 30–50 nm. Recently, Eltarahony et al. (2018) represented the ability of producing the Ag^0 nanoparticles by using the immobilized denitrifying bacteria and its enzyme periplasmic nitrate reductase enzyme (NAP). The bacteria can completely denitrify the $AgNo_3$ in around 192 h to produce Ag metal with particle size ranged from 26–58 nm.

5.2   Mechanisms of Nanoparticles' Biosynthesis

The mechanism of how the nanoparticles are bio-synthesized normally differs according to the micro-organism used for synthesis method. However, nanoparticles are normally produced in the same sequence, which includes the transfer of metal ions' precursor from the solution into the microbe cell. The collected ions were then reduced to metallic or metal oxide forms by action of secreted enzymes. The strategy of how the microbe affects the precursor involves the altering of the solution to become super saturated solution and hence precipitates the metals. The other possible pathway includes the microbial cells' production of polymer(s) molecule, which affects the nanoparticles' nucleation by enhancing or preventing the early formed particle stabilization.

Considering the silver and gold nanoparticles' biosynthesis, several synthesis mechanisms for different micro-organisms have been reported. For example, the mechanism for metallic silver synthesis by *B. licheniformis* depends mainly on the species secretion of NADH and NADH-dependent enzyme nitrate reductase, which is responsible of extracellular reduction of silver nitrates to Ag^0. The ions reduction normally goes through the electron shuttle enzymatic metal process (Kalishwaralal et al. 2008).

The biological heavy metals nanoparticles formation such as Ag, Cu, Cd and Ni can be explained via the metallophilic micro-organisms mutation, which generates new genes and proteomic responses to toxic environment (Reith et al. 2007). The generated genes allow the micro-organisms to detoxify the rich metal surrounding different mechanisms like complexation, efflux, or reductive precipitation. The metallophilic bacteria can be applied to prepare heavy metals' nanoparticles in high metallic concentrated places such as mining waste rocks, and effluents from metal processing plants (Mergeay et al. 2003).

A multi-step mechanism for magnetic nanoparticles' (BacMP) biosynthesis has been postulated by Arakaki et al. (2008). In the first step, the cytoplasmic membrane is invaginated to form vesicle, which acts as BacMP precursor's membrane. Up till now, the hypothesized mechanism for vesicle formation is that a specific GTPase facilitate the start of invagination process. The formed vesicle then forms linear chain along the filament's skeleton. The second step in BacMP preparation is to accumulate the external iron precursor into the vesicles using the protein transports and siderophores. Meanwhile, the internal iron ions can be controlled through the redox processes. Finally, the BacMP bonded proteins form the magnetite crystal nuclei and/or shape the particles.

Shewanella oneidensis synthesizes the magnetic nanoparticles through another mechanism, which was suggested by Perez-Gonzalez et al. (2010). As the first step, the bacterial utilization of ferrihydrites as terminal electron acceptor and Fe^{2+} ions formed, while the pH value increased due to the amino acids metabolization. The accumulated Fe^{2+} and Fe^{3+} ions at the negative cell wall then passively lead to supersaturation, which then causes the magnetite precipitation on the cell wall.

6. METHODS FOR APPLYING THE NANOPARTICLES IN BIOREMEDIATION

The bioremediation is a well-known process applied to eliminate the hydrocarbon pollutants, especially in aquatic environment, but with a very slow rate (Chen et al. 2010). For faster processing, the bioremediation can be carried out according to biostimulation technique by providing nutrients and oxygen, the most important factor in microbial growth, to the native micro-organisms in the contaminated site. Additionally, the application of nanotechnology as process modifier has played many roles (see sec. 1.3) to enhance the bioremediation rate. Among many applied nanoparticles, the oxygen releasing compounds (ORCs) such as CaO_2 and MgO_2 can slowly form the H_2O_2, which converts (see sec. 1.4.2) to water, liberates oxygen and forms OH radicals.

To apply the ORCs materials in bioremediation, several application techniques were used. The ORCs' nanomaterials can be injected directly into ground water through the soil pores or encapsulated in other substrates' permeable reactive barrier (PRB). The PRB is a highly porous material that has pore sizes larger than the substrate and can remediate the ground water pollutants through different procedures (Erto et al. 2014).

Yeh et al. (2018) have prepared the CaO_2 nanoparticles as ORC material by wet and dry chemical methods and investigated the effect of preparation method on the material activity towards diesel bio-remediation enhancement. The prepared nanoparticles' structures and morphologies were proved using the TEM, XRD and SEM analysis tools. The results indicated that the CaO_2 prepared by the wet method is the most active material as bioremediation improver. The high activity of wet-method prepared CaO_2 is attributed to its high dissolved oxygen ability in process media, which hasten the microbial growth.

Mosmeri et al. (2017) have studied the bioremediation of benzene and toluene contaminated ground water in presence of CaO_2 nanomaterial. The authors controlled the CaO_2 release to the water by encapsulation in sodium alginate permeable reactive barrier (PRB). The results reflected the encapsulation of CaO_2 advantages since it enhances the nanoparticles' durability in the ground water and inhibits the harmful effect of H_2O_2 and OH radicals on the micro-organism.

Although numerous researches have investigated the bioremediation in the presence of ORCs, the impact of such materials on the micro-organism diversity and structure is rare (Jiang et al. 2018). However, the investigation of impact of solid peroxides on the microbes provides data which make the bioremediation more effectively understood. To study the impact of ROCs on microbial diversity, the next generation sequencing (NGS) technique was developed. The technique consists of serial and continuous isolation of the microbes from the media and studying the microbial count with time, which allows a huge data in sequence to follow the bioremediation (Patel and Jain 2012).

Very recently, a new trend of using continuous flow bioreactor to investigate the CaO_2 activity in bioremediation and its impact on the biodiversity through the NGS methodology was suggested by Mosmeri et al. (2019). The authors studied the remediation of (50ml/L) benzene on a sand backed reactor. The results showed that the benzene remediation reached 93% in around 30 days, while the OH radicals' generation rate was decreased when the pH value was increased. Also, the addition of CaO_2 affected the bacterial growth by around two folds than the CaO_2 non-supplied reactor, which reflects the role of oxygen supply by the nanomaterial. Considering the effect of CaO_2 on the microbial biodiversity, using the NGS technique, the authors reported a high biofilm growth on the sand surface and its continuous attachment to the surface even after the process. These observations confirmed the complete positive impact of CaO_2 on biodiversity.

Mosmeri et al. (2019) also investigated the bioremediation mechanism and the impact of PRB encapsulated MgO_2 addition on the microbial diversity during the toluene and naphthalene bioremediation in ground water (Gholami et al. 2019). The authors stimulated the ground water pollutants' bioremediation by using a sand packed continuous flow reactor. The results implemented that naphthalene removal was faster than the toluene by around 66% by the micro-organism in aqueous phase. Also, the NGS analysis of the pollutants revealed the attachment of the *P. putida and P. mendocina* degrading bacteria on the substrate surface, which proves the role of ORCs' material on the microbial diversity

7. BIOREMEDIATION ASSISTED NANOPARTICLES' PROSPECTIVE AND PROBLEMS

The main problem of applying the nanoparticles as bioremediation improver is the uncontrolled release of those materials to environment. Such hazardous release would inhibit the process upscaling and affect its quality in abiotic or even biotic conditions (Handy et al. 2008). Moreover, despite the nanoparticles' beneficial surface area in microbial growth and pollutants adsorption affinity, the nanoparticles are normally reactive and have the same impact as the bulk materials on the environment. The application of nanomaterials in biochemical processes negatively affects the environment, which reverses the acquired target of bioremediation process. Furthermore, some of the effective nanoparticles in bioremediation such as ZnO, CuO, AgO and Fe_2O_3 are known for their high antimicrobial activity and toxicity, if they exist in high concentration.

8. CONCLUSION

To conclude this chapter, the combination of nano- and bio- technologies offers an eco-friendly technique to fast removal of organic pollutants (specially the PAHs) from soils and aquatic systems. The unique properties of nanoparticles such as high surface area and charged particles play several important direct roles in enhancing the bioremediation. Nanoparticles can be used for the enhancement of hydrocarbon solubility and mobility, soil condition's improvement, as metabolism and growth catalysts, and for the immobilization of bacteria and enzymes. Also, it acts as adsorbent or oxidant for hydrocarbons' pollutants, which facilitate the bioremediation indirectly.

Additionally, the bioremediation process can simultaneously be applied to bio-synthesize the nanoparticles by introducing the metal oxide precursor in the growth media of micro-organisms. The pilot scale application of nanoparticles in bioremediation is a facile procedure which can be effectively carried out using with potential microbes or with enzymes.

Acknowledgments

The Authors would like to sincerely express their deep acknowledgement and gratitude to Dr. Ramadan Soliman, researcher at Biotechnology Lab., Processes Development Department in Egyptian Petroleum Research Institute, for taking out time to read this chapter. His hints have helped the work to reach such acceptable final form.

References

Acevedo, F., L. Pizzul, M.D. Castillo, M.E. Gonzáleza, M. Cea, L. Gianfreda, et al. 2010. Degradation of polycyclic aromatic hydrocarbons by free and nano-clay immobilized manganese peroxidase from *Anthracophyllum discolor*. Chemosph. 80(3): 271–278.

Ahmad, A., S. Senapati, M.I. Khan, R. Kumar and M. Sastry. 2003. Extracellular biosynthesis of mono-dispersed gold nanoparticles by a novel extremophilic actinomycete, *Thermomono spora*, sp. Langmuir. 19(8): 3550–3553.

Amann, R.I., W. Ludwig and K.H. Schleifer. 1995. Phylogenetic identification and *in situ* detection of individual microbial cells without cultivation. Microbiol. Rev. 59(1): 143–169.

Ansari, F., P. Grigoriev, S. Libor, I.E. Tothill and J.J. Ramsden. 2009. DBT degradation enhancement by decorating *Rhodococcus erythropolis* IGST8 with magnetic Fe3O4 nano- particles. J. Biotechnol. Bioeng. 102(5): 1505–1512.

Ansari, S.A. and Q. Husain. 2012. Potential applications of enzymes immobilized on/in nano materials: A review. J. Biotechnol. Adv. 30(3): 512–523.

Arakaki, A., H. Nakazawa, M. Nemoto, T. Mori and T. Matsunaga. 2008. Formation of magnetite by bacteria and its application. J. Roy. Soc. Inter. 5(26): 977–999.

Ashtari, K., K. Khajeh, J. Fasihi, P. Ashtari, A. Ramazani and H. Vali. 2012. Silica-encapsulated magnetic nanoparticles: Enzyme immobilization and cytotoxic study. Int. J. Biol. Macromol. 50(4): 1063–1069.

Azim, A., Z. Davood, F. Ali, R.M. Mohammad, N. Dariush, T. Shahram, et al. 2009. Synthesis and characterization of gold nanoparticles by tryptophane. Am. J. Appl. Sci. 6(4): 691–695.

Berna B.M. and F. Batista. 2006. Enzyme immobilization literature survey methods in biotechnology: Immobilization of enzymes and cells, 2nd Ed. pp. 16–30.

Bernfeld, P. and J. Wan. 1963. Antigens and enzymes made insoluble by entrapping them into the lattice of synthetic polymers. Science. 142(3593): 678–679.

Beveridge, T.J., M.N. Hughes, H. Lee, K.T. Leung, R.K. Poole, I. Savvaidis, S. Silver and J.T. Trevors. 1997. Metal-microbe interactions: Contemporary approaches. Adv. Microb. Physiol. 38: 177–243.

Bhatia, M., A. Girdhar, B. Chandrakar and A. Tiwari. 2013. Implicating nanoparticles as potential biodegradation enhancers: A review. J. Nanomed. Nanotechol. 4: 175–181.

Bhattacharya, S., I. Saha, A. Mukhopadhyay, D. Chattopadhyay, U.C. Ghosh and D. Chatterjee. 2013. Role of nanotechnology in water treatment and purification: Potential applications and implications. Int. J. Chem. Sci. Technol. 3(3): 59–64.

Brady, D. and A. Jordan. 2009. Advances in enzyme immobilization. Biotechnol. Lett. 31(11): 1639–1650.

Bruins R.M., S. Kapil and S.W. Oehme. 2000. Microbial resistance to metals in the environment. Ecotoxicol Environ Saf. 45(3): 198–207.

Cameotra, S.S. and R.S. Makkar. 2010. Biosurfactant enhanced bioremediation of hydrophobic pollutants. Pure Appl. Chem. 82(1): 97–116.

Chen, K.F., C.M. Kao, C.W. Chen, R.Y. Surampalli and M.S. Lee. 2010. Control of petroleum hydrocarbon contaminated groundwater by intrinsic and enhanced bioremed-iation. J. Environ. Sci. (China) 22(6): 864–871.

Chirra H.D., T. Sexton, D. Biswal, L.B. Hersh and J.Z. Hilt. 2011. Catalase coupled gold nanoparticles: Comparison between carbodiimide and biotin-streptavidin methods. Acta Biomater. 7(7): 2865–2872.

Chronopoulou, L., G. Kamel, C. Sparago, F. Bordi, S. Lupi, M. Diociaiuti, et al. 2011. Structure-activity relationships of Candida rugosa lipase immobilized on polylactic acid nanoparticles. Sof. Mat. 7(6): 2653–2662.

Desai, J.D. and I.M. Banat. 1997. Microbial production of surfactants and their commercial potential. Microbio. Molec. Biol. Rev. 61(1): 47–64.

De Rosa, M.R., C. Monreal, M. Schnitzer, R. Walsh and Y. Sultan. 2010. Nanotechnology in fertilizers. Nat. Nanotechnol. J. 5(2): 91–96.

De Windt, W., A. Peter and V. Willy. 2005. Bioreductive deposition of palladium (0) nanoparticles on *Shewanella oneidensis* with catalytic activity towards reductive dechlorination of polychlorinated biphenyls. Environ. Microbiol. 7(3): 314–325.

De'ziel, E., G. Paquette, R. Villemur, F. Le'pine and J.G. Bisaillon. 1996. Biosurfactant production by a soil *Pseudomonas* strain growing on polycyclic aromatic hydrocarbons. Appl. Environ. Microbiol. 62(6): 1908–1912.

Dong, D., P. Li, X. Li, Q. Zhao, Y. Zhang, C. Jia, et al. 2010. Investigation on the photocatalytic degradation of pyrene on soil surfaces using nanometer anatase TiO_2 under UV irradiation. J. Hazard. Mater. 174: 859–863.

Dutta, N., A. Mukhopadhyay, A.K. Dasgupta and K. Chakrabarti. 2013. Nanotechnology enabled enhancement of enzyme activity and thermostability: Study on impaired pectate lyase from attenuated *Macrophomina phaseolina* in presence of hydroxyapatite nanoparticle. PLoS One 8(5): 63567.

El-Sheshtawy, H.S., N.M. Khalil, W. Ahmed and R.I. Abdallah. 2014. Monitoring of oil pollution at Gemsa Bay and bioremediation capacity of bacterial isolates with biosurfactants and nanoparticles Mar. Poll. Bull. 87: 191–200.

El-Sheshtawy, H.S., N.M. Khalil, W. Ahmed and A.A. Nabila. 2017. Enhancement of bioremediation of crude oil by nanoparticles and biosurfactant. Egypt. J. chem. 5(60): 835–848.

El-Sheshtawy, H.S. and W. Ahmed. 2017. Bioremediation of crude oil by *Bacillus licheniformis* in the presence of different concentration nanoparticles and produced biosurfactant Int. J. Environ. Sci. Technol. 4(8): 1603–1614.

Eltarahony, M., S. Zaki, Z. Kheiralla and D. Abd-El-Haleema. 2018. NAP enzyme recruitment in simultaneous bioremediation and nanoparticles synthesis. Biotechnol. Rep. (Amst). 18: e00257. doi:10.1016/j.btre.2018.e00257

Erto, A., I. Bortone, A. Di Nardo, M. Di Natale and D. Musmarra. 2014. Permeable adsorptive barrier (PAB) for the remediation of groundwater simultaneously contaminated by some chlorinated organic compounds. J. Environ. Manag. 140: 111–119.

Garcia-Galan, C., A., Berenguer-Murcia, R., Fernandez-Lafuente and R.C. Rodrigues. 2011. Potential of different enzyme immobilization strategies to improve enzyme performance. Adv. Synth. Catal. 353(16): 2885–2904.

Gholami, F., H. Mosmeri, M. Shavandi, S.M. Mehdi and M.A. Dastgheib. 2019. Application of encapsulated magnesium peroxide (MgO_2) nanoparticles in permeable reactive barrier (PRB) for naphthalene and toluene bioremediation from groundwater. Sci. Total Environ. 655: 633–640.

González-Campos, J.B., J.D. Mota-Morales, S. Kumar, D. Zárate-Trivino, M. Hernández-Iturriaga, Y. Prokhorov, et al. 2013. New insights into the bactericidal activity of chitosan-Ag bionanocomposite: The role of the electrical conductivity Coll. Surf. B: Biointerfaces 111: 741–746

Gotovac, S., Y. Hattori, D. Noguchi, J. Miyamoto, M. Kanamaru, S. Utsumi, et al. 2006. Phenanthrene adsorption from solution on single wall carbon nanotubes J. Phys. Chem. B. 110: 16219–16224.

Gupta, K., S. Bhattacharya, D.J. Chattopadhyay, A. Mukhopadhyay, H. Biswas, J. Dutta, et al. 2011. Ceria associated manganese oxide nanoparticles: Synthesis, characterization and arsenic (V) sorption behavior. Chem. Eng. J. 172: 219–229.

Gusian, J.M., G. Penzol, P. Armisen, A. Bastida, R.M. Blanco, R. Fernandez-Lafuente, et al. 1997. Immobilization of enzyme acting on macromolecular substrate, immobilization of enzymes and cells. Humana Press, Totowa, NJ. pp. 261–275.

Haferburg, G. and E. Kothe. 2007. Microbes and metals: Interactions in the environment. J. Basic Microbiol. 47: 453–467.

Harayama, S., Y. Kasai and A. Hara. 2004. Microbial communities in oil contaminated sea water. Curr. Opin. Biotech. 15(3): 205–214.

Hernandez, K. and R. Fernandez-Lafuente. 2011. Control of protein immobilization: Coupling immobilization and site-directed mutagenesis to improve biocatalyst or biosensor performance. Enzyme Microbial. Technol. 48(2): 107–122.

Hou, J., L. Wang, C. Wang, S. Zhang, H. Liu, S. Li, et al. 2019. Toxicity and mechanisms of action of titanium dioxide nanoparticles in living organisms. J. Environ. Sci. (China). 75: 40–53. doi:10.1016/j.jes.2018.06.010

Hu, H., Z. Zhao, Y. Gogotsi and J. Qiu. 2014. Compressible carbon nanotube graphene hybrid aerogels with super hydrophobicity and super oleophilicity for oil sorption. Environ. Sci. Technol. Lett. 1(3): 214–220.

Hu, J. and Y. Qi. 2015. Microbial degradation of di-n-butyl phthalate by *Micrococcus* sp. immobilized with polyvinyl alcohol. Desalin. Water Treat. 56(9): 2457–2463.

Hsiung, C.E., H.L. Lien, A.E. Galliano, C.S. Yeh and Y.H. Shih. 2016. Effects of water chemistry on the destabilization and sedimentation of commercial TiO$_2$ nanoparticles: Role of double-layer compression and charge neutralization. Chemosph. 151: 145–151.

Handy, R.D., R. Owen and E. Valsami-Jones. 2008. The ecotoxicology of nanoparticles and nanomaterials: Current status, knowledge gaps, challenges and future needs. Ecotoxicol. 17(5): 315–325.

Jameia, M.R., M.R. Khosravib and B. Anvaripoura. 2013. Degradation of oil from soil using nano zero valent iron. Sci. Int. (Lahore) 25(4): 863–867.

Jeganathan, J., G. Nakhla and A. Bassi. 2007. Hydrolytic pre-treatment of oily wastewater by immobilized lipase. J. Hazard. Mater. 145(1–2): 127–135.

Jézéquel, K. and T. Lebeau. 2008. Soil bioaugmentation by free and immobilized bacteria to reduce potentially phytoavailable cadmium. Bioresour. Technol. 99: 690–698.

Jiang, W., P. Tang, S. Lu, Y. Xue, X. Zhang, Z. Qiu, et al. 2018. Comparative studies of H2O2/Fe (II)/formic acid, sodium percarbonate/Fe (II)/formic acid and calcium peroxide/Fe (II)/formic acid processes for degradation performance of carbon tetrachloride. Chem. Eng. J. 344: 453–461.

Jorfi, S., A. Rezaee, G. Mohebali and N.A. Jaafarzadeh. 2013. Pyrene removal from contaminated soils by modified Fenton oxidation using iron nano particles. J. Environ. Health Sci. Eng. 11: 17.

Kaczorek, E. and A. Olszanowski. 2011. Uptake of hydrocarbon by *Pseudomonas fluorescens* (P1) and *Pseudomonas putida* (K1) strains in the presence of surfactants: A cell surface modification. Water Air Soil Pollut. 214(1–4): 451–459.

Kalishwaralal, K., V. Deepak, S. Ramkumarpandian, H. Nellaiah and G. Sangiliyandi. 2008. Extracellular biosynthesis of silver nanoparticles by the culture supernatant of *Bacillus licheniformis*. Mater. Lett. 62(29): 4411–4413.

Kalkan, N.A., S. Aksoy, E.A. Aksoy and N. Hasirci. 2012. Preparation of chitosan-coatedmagnetite nanoparticles and application for immobilization of laccase. J. Appl. Polym. Sci. 123(2): 707–716.

Karimi, M., F. Aboufazeli, H. Zhad, O. Sadeghi and E. Najafi. 2013. Determination of polycyclic aromatic hydrocarbons in Persian Gulf and Caspian Sea: Gold nanoparticles fiber for a head space solid phase micro extraction. Bull. Environ. Contam. Toxicol. 90(3): 291–295.

Kim, J., J.W. Grate and P. Wang. 2008. Nanocatalysis and its potential applications. Trends Biotechnol. 26(11), 639-646.

Kiran, G.S., J. Selvin, A. Manilal and S. Sujith. 2011. Biosurfactants as green stabilizer for the biological synthesis of nanoparticles. Crit. Rev. Biotechnol. 31: 354–364.

Kiran, G.S., L.A. Nishanth, S. Priyadharshini, K. Anitha and J. Selvin. 2014. Effect of Fe nanoparticle on growth and glycolipid biosurfactant production under solid state culture by marine *Nocardiopsis* sp. MSA13A. BMC Biotechnol. 14: 48–53.

Khoshnevisan, K., A.K. Bordbar, D. Zare, D. Davoodi, M. Noruzi, M. Barkhi, et al. 2011. Immobilization of cellulase enzyme on super paramagnetic nanoparticles and determination of its activity and stability. Chem. Eng. J. 171(2): 669–673.

Klaus T.R., J.E. Olsson and C.G. Granqvist. 1999. Silver based crystalline nanoparticles, microbially fabricated. Proc. Natl. Acad. Sci. USA. 96: 13611–13614.

Konishi Y., T. Tsukiyama, T. Tachimi, N. Saitoh, T. Nomura and S. Nagamine. 2007. Microbial deposition of gold nanoparticles by the metal reducing bacterium *Shewanella* algae. Electrochim. Acta. 53(1): 186–192.

Kowshik M., N. Deshmukh, W. Vogel, J. Urban, S.K. Kulkarni and K.M. Paknikar. 2002. Microbial synthesis of semiconductor CdS nanoparticles, their characterization, and their use in the fabrication of an ideal diode. Biotechnol. Bioeng. 78(5): 583–588.

Kumari, B., S. Rajput, P. Gaur, S.N. Singh and D.P. Singh. 2014. Biodegradation of pyrene and phenanthrene by bacterial consortium and evaluation of role of surfactant. Cell. Mol. Biol. 60(5): 21–26.

Lefèvre, C.T., F. Abreu, M.L. Schmidt, U. Lins, R.B. Frankel, B.P. Hedlund and D.A. Bazylinski et al. 2010. Moderately thermophilic magnetotactic bacteria from hot springs in Nevada. Appl. Environ. Microbiol. 76(11): 3740–3743.

Leitgeb, M., K. Heržič, G.H. Podrepšek, A. Hojski, A. Crnjac and Z. Knez. 2014. Toxicity of magnetic chitosan micro and nanoparticles as carriers for biologically active substances. Acta Chim. Slov. 61(1): 145–152.

Lengke M. and G. Southam. 2006. Bioaccumulation of gold by sulfate-reducing bacteria cultured in the presence of gold(I)-thiosulfate complex. Geochim. Cosmochim. Acta. 70(14): 3646–3661.

Lengke M., M.E. Fleet and G. Southam. 2006a. Synthesis of platinum nanoparticles by reaction of filamentous cyanobacteria with platinum(IV)-Chloride complex. Langmuir. 22(17): 7318–7323.

Lengke M., B. Ravel, M.E. Fleet, G. Wanger, R.A. Gordon and G. Southam. 2006b. Mechanisms of gold bioaccumulation by filamentous cyanobacteria from gold(III)-chloride complex. Environ. Sci. Technol. 40(20): 6304–6309.

Li GY., K.L. Huang., Y.R. Jiang., D.L. Yang and P. Ding. 2008. Preparation and characterization of Saccharomyces cerevisiae alcohol dehydrogenase immobilized on magnetic nanoparticles. Int. J. Biol. Macromol. 42(5): 405–412.

Li, X., H. Xu, Z. Chen and G. Chen. 2011. Biosynthesis of nanoparticles by microorganisms and their applications. J. Nanomaterials. 1–15. doi:10.1155/2011/270974

Liu, J., C. Vipulanandan, T.F. Cooper and G. Vipulanandan. 2013. Effects of Fe nanoparticles on bacterial growth and biosurfactant production. J. Nanopart. Res. 15: 1405–1414.

Makkar, R.S. and S.S. Cameotra. 1999. Biosurfactant production by microorganisms on uncon-ventional carbon sources – a review. J. Surf. Det. 2: 237–241.

Mady, M.F., H.H. El-Shiekh, M.M. Elaasser, H.S. El-Sheshtawy, W.A. Aboutaleb, H.H.H. Hefni. 2018. Biosynthesis of Ag, CdS, Fe and CuO nanoparticles by *Salmunella typhimurim* as chitosan conductivity improvers. African J. Mycol. Biotech. 23(1): 21–39.

Mann, S. 1992. Bacteria and the Midas touch. Nature 357: 358–360.

Mann, S. 1996. Biomimetic Materials Chemistry. Wiley-VCH, New York. Vol. 119(4):

Masooleh, M.S., S. Bazgir, M. Tamizifar and A. Nemati. 2010. Adsorption of petroleum hydrocarbons on organoclay. J. Appl. Chem. Res. 4(14): 19–23.

Masy, T., W. Wannoussa, S. Lambert, B. Heinrichs, J.P. Pirard, H. Serge, et al. 2012. Investigation of nanoparticles as potential activators for optimization of PAH biodegradation. Environ. Eng. Manage. J. 11(3) Supplement, S9.

Merced, T., S. Santos, O. Rivera, N. Villalba, Y. Baez, J. Gaudier, et al. 2005. Effect of zinc oxide nanocrystals in media containing *E. coli* and *C. xerosis* bacteria. In: C.J. Zhong, N.A. Kotov, W. Daniell, F.P. Zamborini (eds), Material Research Society Symposium Proceedings 2005, 0900-O03-08.

Mergeay, M., S. Monchy, T. Vallaeys, V. Auquier, A. Benotmane, P. Bertin, S. Taghavi, J. Dunn, D. van der Lelie and R. Wattiez. 2003. A bacterium specifically adapted to toxic metals: Towards a catalogue of metal-responsive genes, FEMS Microbiol. Rev. 27(2–3): 385–410.

Mehndiratta, P., A. Jain, S. Srivastava and N. Gupta. 2013. Environmental pollution and nano-technology. Environ. Pollut. 2(2): 49–58.

Mohajeri, S., H.A. Aziz, M.H. Isa, M.A. Zahed and M.N. Adlan. 2010. Statistical optimization of process parameters for landfill leachate treatment using Electro-Fenton technique, J. Hazard. Mater. 176(1-3): 749–758.

Mosmeri, H., E. Alaie, M. Shavandi, S.M.M. Dastgheib and S. Tasharrofi. 2017. Benzene contaminated groundwater remediation using calcium peroxide nanoparticles: Synthesis and process optimization. Environ. Monit. Assess. 189(9): 452. doi: 10.1007/s10661-017-6157-2.

Mosmeri, H., F. Gholamib, M. Shavandi, S.M.M. Dastghei and E. Alaie. 2019. Bioremediation of benzene contaminated groundwater by calcium peroxide (CaO_2) nanoparticles: Continuous flow and biodiversity studies. J. Hazard. Mate. 371: 183–190.

Nelson, J.M. and E.G. Griffin. 1916. Adsorption of Invertase. J. Am Chem. Soc. 38: 1109–1115.

Nowack, B. and T.D. Bucheli. 2007. Occurrence, behavior and effects of nanoparticles in the environment. Environ. Pollut. 150: 5–22.

Nador, F., Y. Moglie, C. Vitale, M. Yus, F. Alonso and G. Radivoy. 2010. Reduction of polycyclic aromatic hydrocarbons promoted by cobalt or manganese nanoparticles. Tetrahedron 66(24): 4318–4325.

Ndjou'ou, A.C. and D. Cassidy. 2006. Surfactant production accompanying the modified Fenton oxidation of hydrocarbons in soil. Chemosph. 65: 1610–1615.

Northup, A. and D. Cassidy. 2008. Calcium peroxide (CaO$_2$) for use in modified Fenton chemistry. J. Hazard. Mater. 152: 1164–1170.

Nykanen, A., H. Kontio, O. Klutas, O.P. Penttinen, S. Kostia, J. Mikola, et al. 2012. Increasing lake water and sediment oxygen levels using slow release peroxide. Sci. Total Environ. 429: 317–324.

Nair, B. and T. Pradeep. 2002. Coalescence of nanoclusters and formation of submicron crystallites assisted by *Lactobacillus* strains. Cryst. Growth Des. 4(2): 293–298.

Parikh R.P., S. Singh, B.L.V. Prasad, M.S. Patole, M. Sastry and Y.S. Shouche. 2008. Extracellular synthesis of crystalline silver nanoparticles and molecular evidence of silver resistance from *Morganella* sp.: Towards understanding biochemical synthesis mechanism. Chem. Biochem. 9(9): 1415–1422.

Patel, R.K. and M. Jain. 2012. NGS QC toolkit: A toolkit for quality control of next generation sequencing data. PloS One 7(2): 30619.

Pavía-Sanders, A., S. Zhang, J.A. Flores, J.E. Sanders, E. Raymond and K.L.J. Wooley. 2013. Robust magnetic/polymer hybrid nanoparticles designed for crude oil entrapment and recovery in aqueous environments. ACS Nano. 7(9): 7552–7561.

Perez, J.M., T. O'Loughin, F.J. Simeone, R. Weissleder and L. Josephson. 2002. DNA-based magnetic nanoparticle assembly acts as a magnetic relaxation nano switch allowing screening of DNA cleaving agents. J. Am. Chem. Soc. 124: 2856–2857.

Perez-Gonzalez, T., C. Jimenez-Lopez, A.L. Neal, F. Rull-Perez, A. Rodriguez-Navarro, A. Fernandez-Vivas, et al. 2010. Magnetite biomineralization induced by *Shewanella oneidensis*. Geochim. Cosmoch. Acta. 74(3): 967–979.

Petkova G.A., K. Záruba, P. Žvátora and V. Králcorresponding. 2012. Gold and silver nanoparticles for biomolecule immobilization and enzymatic catalysis. Nanoscale Res. Lett. 7(1): 287. doi: 10.1186/1556-276X-7-287

Porath, J. 1992. Immobilized metal ion affinity chromatography. Protein Expr. Purif. 3(4): 263–281.

Pugazhenthiran N., S. Anandan, G. Kathiravan, N.K.U, Prakash, S. Crawford and A.M. Kumar. 2009. Microbial synthesis of silver nanoparticles by *Bacillus* sp. J. Nanopart. Res. 11: 1811.

Quirk, R.A., W.C. Chan, M.C., Davies, S.J.B. Tendler and K.M. Shakeshef. 2001. Poly (l-lysine)– GRGDS as a biomimetic surface modifier for poly(lactic acid). Biomat. 22: 865–872.

Rangaraj, S., K. Gopalu, Y. Rathinam, P. Periasamy and R. Venkatachalam. 2013. Effect of silica nanoparticles on microbial biomass and silica availability in maizerhizo sphere. Biotechnol. Appl. Biochem. 61(6): 668–675.

Ranjbakhsh, E., A.K. Bordbar, M. Abbasi, A.R. Khosropour and E. Shams. 2012. Enhancement of stability and catalytic activity of immobilized lipase on silica-coated modified magnetite nanoparticles. Chem. Eng. J. 179: 272–276.

Ren, Y., J.G.L. Rivera He, H. Kulkarni, D.K. Lee and P.B. Messersmith. 2011. Facile high efficiency immobilization of lipase enzyme on magnetic iron oxide nanoparticles via a bio-mimetic coating. BMC Biotechnol. 11: 63.

Reith, F., M.F. Lengke, D. Falconer, D. Craw and G. Southam. 2007. The geomicrobiology of gold. ISME J. 1(7): 567–584.

Rick, A.R. and Y. Arai. 2011. Role of natural nanoparticles in phosphorus transport processes in ultisols. Soil Sci. Soc. Am. J. 75(2): 335–347.

Riddin T.L., M. Gericke and C.G. Whiteley. 2006. Analysis of the inter-and extracellular formation of platinum nanoparticles by *Fusarium oxysporum* f. sp. *lycopersici* using response surface methodology. Nanotechnol. 17(14): 3482–3489.

Rizwan, Md., M. Singh, C.K. Mitra and R.K. Morve. 2014. Review: Ecofriendly application of nano-materials: Nano bioremediation. J. Nanopart. 7: Article ID 431787.

Rollins, D.M. and R.R. Colwell. 1986. Viable but non culturable stage of *Campylobacter jejuni* and its role in survival in the natural aquatic environment. Appl. Environ. Microbio. 52(3): 531–538.

Roselien, C., A. Yoram, D. Tom, P. Bossier and W. Verstraete. 2007. Nitrogen removal techniques in aquaculture for a sustainable production. Aquacul. 270(1–4): 1–14.

Rosevear, A., J.F. Kennedy and J.M.S. Chabral. 1987. Immobilized Enzymes and Cells, Adam Hilger, Bristol and Philadelphia.

Rozsak, D.B. and R.R. Colwell. 1987. Survival strategies of bacteria in the natural environment. Microbiol. Rev. 51(3): 365–379.

Sadiq, M. and J. McCain. 1993. An environmental catastrophe. *In*: J. McCain (ed.), The Gulf War Aftermath: An Environmental Tragedy. Kluwer Academic Publishers, Netherlands, pp. 225–232.

Safaei, M., H.R. Aghabozorg and H. Shariatpanahi. 2008. Modification of domestic clays for preporation of polymer nanocomposites. Nashrieh Shimi vaMohandesi Shimi Iran (NSMSI) 27 (1).

Saien, J. and F. Shahrezaei. 2012. Organic pollutants removal from petroleum refinery wastewater with nanotitania photocatalyst and UV Light emission. Int. J. Photoenergy. 5: Article ID 703074.

Shan, G.B., J.M. Xing, H.Y. Zhang and H.Z. Liu. 2005. Bio-desulfurization of dibenzo-thiophene by microbial cells coated with magnetic nanoparticles. Appl. Environ. Microbiol. 71(8): 4497–4502.

Shao, D., G. Sheng, C. Chen, X. Wang and M. Nagatsu. 2010. Removal of polychlorinated biphenyls from aqueous solutions using cyclodextrin grafted multi walled carbon nanotubes. Chemosphere 79: 679–685.

Shenton W., T. Douglas, M. Young, G. Stubbs and S. Mann. 1999. Inorganic-organic nanotube composites from templated mineralization of tobacco mosic virus. Adv. Mater. 11(3): 253–256.

Shin, K.H. and D.K. Cha. 2008. Microbial reduction of nitrate in the presence of nanoscal zero-alent iron. Chemosphere. 72(2): 257–262.

Simkiss K. and K.M. Wilbur. 1989. Biomineralization, Academic Press, New York, NY, USA. 1989.

Singha, S., H. Datta and A.K. Dasgupta. 2010. Size dependent chaperon properties of gold nanoparticles. J. Nanosci. Nanotechnol. 10(2): 826–832.

Slawson, R.M., M.I. Van Dyke, H. Lee and J.T. Trevor. 1992. Germanium and silver resistance, accumulation and toxicity in microorganisms. Plasmid. 27: 73–79.

Sneha, H.P., K.C.B. Pushpa and S. Murthy. 2019. Enzyme immobilization methods and applications in the food industry. *In*: M. Kuddus (ed.), Enzymes in Food Biotechnology: Production, Applications, and Future Prospects. Elsevier, pp. 645–658.

Southam G. and T.J. Beveridge. 1994. The in vitro formation of placer gold by bacteria Geochim Cosmochim. Acta. 58(20): 4527–4530.

Spring S. and K.H. Schleifer. 1995. Diversity of magnetotactic bacteria. System. Appl. Microbiol. 18(2): 147–153.

Thornhill, R.H., J.G. Burgess and T. Matsunaga. 1995. PCR for direct detection of indigenous uncultured magnetic cocci in sediment and phylogenetic analysis of amplified16S ribosomal DNA. Appl. Environ. Microbiol. 61(2): 495–500.

Tosa, T., T. Mori, N. Fuse and I. Chibata. 1966. Studies on continuous enzyme reactions. I. Screening of carriers for preparation of water-insoluble aminoacylase. Enzymologia. 31: 214–224.

Tungittiplakorn, W., C. Cohen and L.W. Lion. 2005. Engineered polymeric nanoparticles for bioremediation of hydrophobic contaminants. Environ. Sci. Technol. 39: 1354–1358.

Veronica, N., T. Guru, R. Thatikunta and S.N. Reddy. 2015. Role of nano fertilizers in agricultural farming. Int. J. Environ. Sci. Technol. 1(1): 1–3.

Vitor, G., T.C. Palma, B. Vieira, J.P. Lourenco, R.J. Barros and M.C. Costa. 2015. Start-up, adjustment and long-term performance of a two-stage bioremediation process, treating real acid mine, coupled with biosynthesis of ZnS nanoparticles and ZnS/TiO$_2$ nanocomposites, Minerals Eng. 75: 85–93.

Wang, F., C. Guo, L.R. Yang, C.Z. Liu. 2010. Magnetic mesoporous silica nanoparticles: Fabrication and their laccase immobilization performance. Bioresour. Technol. 101(23): 8931–8935. doi:10.1016/j. biortech.2010.06.115

Wang, H., Y. Zhao, T. Li, Z. Chen, Y. Wang and C. Qin. 2016. Properties of calcium peroxide for release of hydrogen peroxide and oxygen: A kinetics study. Chem. Eng. J. 303: 450–457.

Wang, J., G. Meng, K. Tao, M. Feng, X. Zhao, Z. Li, et al. 2012. Immobilization of lipases on alkyl silane modified magnetic nanoparticles: Effect of alkyl chain length on enzyme activity. PLoS One. 7(8): 43478.

Wang, X., Z. Gai, B. Yu, J. Feng, C. Xu, Y. Yuan, et al. 2007. Degradation of carbazole by microbial cell immobilized in magnetic gellan gum gel beads. Appl. Environ. Microbiol. 73(20): 6421.

Watson, J.H.P., D.C. Ellwood, A.K. Soper and J. Charnock. 1999. Nanosized strongly magnetic bacterially-produced iron sulfide materials. J. Magn. Magn. Mater. 203: 69–72.

Wen, Q., Z.Q. Chen, Y. Zhao, H. Zhang and Y. Feng. 2010. Bioremediation of polyacrylamide by bacteria isolated from activated sludge and oil contaminated soil. J. Hazard. Mater. 175(3): 955–959.

Wenyu, L., Z. Linlin and S. Luhua. 2013. Biological treatment technology of oily sludge. Int. J. Eng. Res. Dev. 8(3): 52–55.

Wong, L.S., J. Thirlway and J. Micklefield. 2008. Direct site-selective covalent protein immobilization catalyzed by a phosphopantetheinyl transferase. J. Am. Chem. Soc. 130(37): 12456–12464.

Wu, C.H., S.H. Chang and C.W. Lin. 2015. Improvement of oxygen release from calcium peroxide-polyvinyl alcohol beads by adding low-cost bamboo biochar and its application in bioremediation. Clean Soil Air Water. 43: 287–295.

Xue, Y., S. Lu, X. Fu, V.K. Sharma, I. Mendoza-Sanchez, Z. Qiu, et al. 2018. Simultaneous removal of benzene, toluene, ethylbenzene and xylene (BTEX) by CaO_2 based Fenton system: Enhanced degradation by chelating agents. Chem. Eng. J. 331: 255–264.

Yücel, Y. 2011. Biodiesel production from pomace oil by using Lipase immobilized onto olive pomace. Bioresor. Tech. 102: 3977–3980.

Yeh, C.H., C.W. Lin and C.H. Wu. 2010. A permeable reactive barrier for the bioremediation of BTEX-contaminated groundwater: Microbial community distribution and removal efficiencies. J. Hazard. Mater. 178: 74–80.

Yeh, C., R. Wang, W. Chang and Y. Shih. 2018. Synthesis and characterization of stabilized oxygen-releasing CaO_2 nanoparticles for bioremediation. J. Environ. Manag. 212: 17–22.

Zhang, K., Y.Y. Xu, X.F. Hua, H.L. Han, J.N. Wang, J. Wang, et al. 2008. An intensified degradation of phenanthrene with microporous alginate-lignin beads immobilized *Phanerochaete chrysosporium*. Biochem. Eng. J. 41: 251–257.

Zhubanova, A.A. and Z.A. Mansurov. 2008. The creation of new nano bio-preparates on base nano particles of carbonized rice husk and microorganism's cell for bioremediation of oil contamination soils. ID302. *In*: 4th European Bioremediation Conference, held at Chania. Crete 2–6 September.

CHAPTER 7

Biosynthesized Metal Nanoparticles in Bioremediation

Jaison Jeevanandam, Yiik Siang Hii and Yen San Chan*

Department of Chemical Engineering, Faculty of Engineering and Science,
Curtin University, CDT 250, Miri, Sarawak 98009, Malaysia.

1. INTRODUCTION

In the past decade, nanoparticles have gained the consideration of researchers due to their smaller size, enhanced surface to volume ratio and unique properties (Khan et al. 2017). The exclusive properties of nanoparticles have made them to be an interesting research material and utilizable in various applications ranging from biomedical (McNamara and Tofail 2017) and pharmaceutical (Bhatia 2016) to electronics (Stark et al. 2015) and environment (Haider et al. 2017). Among nanoparticles, nanosized metals are extensively utilized in environmental applications due to their stability and enhanced reactivity (Ge et al. 2016). Currently, environmental pollution has emerged as a major global challenge which affects the quality of abiotic factors that are required for living organisms such as air, water and soil (Filippelli and Taylor 2018). Conventional environment remediation methods utilize chemicals, which helps in cleaning of pollutants from abiotic factors and may also be toxic to living organism (Ma et al. 2017). Thus, metal-based nanoparticles are specifically employed to replace conventional environmental remediation materials for the treatment of ecological abiotic entities such as water, air and soil by modifying their conditions and degrading target pollutants (Patil et al. 2016). It is noteworthy that different types of metal nanoparticles, including nanosized stand-alone metals (Kamat and Meisel 2003), metal oxides (Singh et al. 2017), metal core-shell structures (Mondal and Sharma 2016), dopants (Lamba et al. 2015) and composites (Manatunga et al. 2018) are involved in the environmental remediation applications. Moreover, metal nanoparticles possess ability to convert complex pollutants into useful components via a reaction between electron-hole pairs, which makes them a potential environmental remediation material (Fujiwara et al. 2017). However, nanoparticles

---

*Corresponding author: chanyensan@curtin.edu.my

synthesized via chemicals were also proved to be noxious to humans, other organisms and the environment, which paved the way for the advancement of biosynthesis methods for the fabrication of nontoxic, bioactive and biodegradable nanoparticles (Andra et al. 2019).

In general, the utilization of biological organisms including plants and microbes for environmental remediation application is known as bioremediation (Hazen 2018). Natural or genetically engineered microbes possess ability to convert complex pollutants to simple compounds, which eventually reduce environmental contaminations (Liu et al. 2019). However, there is a possibility of mutations in these microbes which may lead to unexpected environmental catastrophes (Paul et al. 2005). Hence, biosynthesized nanoparticles have been introduced in bioremediation applications known as nanobioremediation to overcome the limitations of conventional bioremediation approaches using microbes (Cecchin et al. 2017). Biosynthesized nanoparticles using microbial or plant extracts have the ability to be involved directly in bioremediation (Vasantharaj et al. 2019b) or they can activate enzyme secretion in microbes, which helps in bioremediation applications (Sutherland et al. 2004). In certain cases, microbial or plant enzymes are encapsulated into nanoformulations for enhanced bioremediation applications (Deshpande 2019). In addition to nontoxicity towards humans and the environment, biosynthesized nanoparticles can target specific pollutants and degrade them into beneficial compounds (He et al. 2017). Further, large-scale synthesis of metal nanoparticle is possible via biosynthesis approach, which is significant in terms of huge demand for these pollution degradation materials in the commercial market (Kitching et al. 2015). Thus, the aim of the present chapter is to enlist different metal nanoparticles that are used in bioremediation and the role of metal nanoparticles synthesized via biological agents such as microbes and plants, as an efficient, non-toxic bioremediation agent. In addition, the significance of metal nanoparticles in altering microbes and the use of magnetosomes for efficient bioremediation was also discussed.

2. METAL NANOPARTICLES IN REMEDIATION

Since early '90s, usage of nanoparticles for water, soil and air remediation has been extensively studied. Generally, nanoscale zero-valent iron (nZVI) is widely utilized for the remediation of soil as well as water. Conventionally, nanoscale titanium dioxide has been used for the contaminant degradation in water and air via photocatalytic property. Table 1 is the summary of various nanoparticles that are employed in environmental remediation application.

Table 1   Conventional nanoparticles in environmental remediation (Mueller and Nowack 2010)

Nanomaterials	Target compounds	Process exploited
TiO_2	Organic pollutants NOx and VOCs in air	Photocatalysis
Iron oxides	Metals, arsenic, organic compounds	Adsorption
Nanoscale zero-valent iron (nZVI)	Chlorinated hydrocarbons, metals, nitrate, sulfates, arsenate, oil	Redox reaction

This section will focus on soil and water remediation via nZVI and some notable metal nanoparticles. Application of nanosized oxides of metal particles and nanocomposites in remediation will be further discussed in the consecutive sections.

2.1   Nanoscale Zero-Valent Iron (nZVI)

Application of iron for environmental remediation has been accepted by numerous regulatory agencies as it is cheap, and there are scarce reports on the toxicity of iron towards the environment. Regardless of its particle size, zero-valent iron (ZVI) is normally used in all applications related to remediation; however, nanoscale ZVI (nZVI) has grabbed the attention of researchers. Due to its greater specific surface area, nZVI exhibits superior reactivity and enhanced field deployment abilities in comparison with granular or micro-sized ZVI particles (Zhang and Elliott 2006). ZVI has been known as an excellent reducing reagent as it has a strong tendency to release electrons to its surrounding environment as described in the Eq. (1) (Zhang 2003)

$$Fe^0 \rightarrow Fe^{2+} + 2e^- \tag{1}$$

Besides, ZVI can react with hydrogen ions, dissolved oxygen, nitrate and sulfate, making it a robust remediation catalyst. The reaction between ZVI and dissolved oxygen can be expressed as in Eq. (2).

$$2Fe^0_{(s)} + 4H^+_{(aq)} + O_{2(aq)} \rightarrow 2Fe^{2+}_{(aq)} + 2H_2O_{(l)} \tag{2}$$

In an environment with low dissolved oxygen, such as groundwater, ZVI forms effective redox couple with water which can be described as in Eq. (3).

$$2Fe^0_{(s)} + 2H_2O_{(aq)} \rightarrow 2Fe^{2+}_{(aq)} + H_{2(g)} + 2OH^-_{(aq)} \tag{3}$$

In addition, ZVI also readily reacts with various contaminants such as chlorinated hydrocarbons. Generally, various chlorinated hydrocarbons, which include aliphatic, aliphatic cyclic and aromatic compounds, can be abridged, depending on their stoichiometry as shown in Eq. (4).

$$RCL + H^+ + Fe^0 \rightarrow RH + Fe^{2+} + Cl^- \tag{4}$$

Contaminants such as tetrachloroethene ($C_2H_2Cl_4$) can be reduced to ethene according to Eq. (4). In addition, nZVI can also lower the toxicity such as heavy metal contaminants (Gil-Díaz et al. 2017) and inorganic complexes (Vilardi et al. 2017) or transform it into inert compounds. Presently, nZVI is extensively used in soil, water and air remediation, which will be further discussed in the following sections.

2.1.1 Soil

Soil contamination has become a critical issue due to improper waste management from industries (Wall et al. 2015). Contaminants in soil, including toxic high concentration of heavy metals and organic complexes (Ding et al. 2017, Huang et al. 2017), possess risks and hazards to environment and human beings (Suominen et al. 2014). Therefore, there is a need to reinstate polluted soils. Among all treatment techniques, nano-remediation has gained the attention of researchers for contaminated soil remediation. Remarkably, zero-valent iron (ZVI) nanoparticles are widely considered for their potential in soil and water remediation (Xue et al. 2018).

The usage of insecticide dichlorodiphenyltrichloroethane (DDT) was banned in 1972, which was massively used throughout the world. Even though, DDT was not used for over decades, its components are still detected in soils and poses risk to human health and the environment. Thus, El-Temsah et al. (2016) evaluated the remediation of aged DDT from contaminated soil using nZVI. Two nZVI types were utilized in this experiment. First, nZVI type B (nZVI-B) was prepared by dissolving iron sulfate ($FeSO_4$) and carboxymethyl cellulose (CMC) in sodium borohydride to form nZVI, whereas nZVI type T (nZVI-T) was fabricated using iron oxide reduction method in hydrogen gas phase. In this study, soil contaminated for over 45 years was used, which contained 24.7 mg DDT/kg. Their results showed that nZVI-B managed to degrade DDT by 22.4%, yet only 9.2% degradation can be achieved using nZVI-T after 48 hours.

Polychlorinated biphenyls (PCB) are categorized as persistent organic pollutants, carcinogenic and has strong affinity to soils. Gomes et al. (2015) compared the effectiveness of dechlorination of PCB by nZVI methods, which were the electro-dialytic and conventional electro-kinetic remediation. A soil sample with a primary concentration of 258 PCB mg/kg of soil was treated with 10 ml of nZVI for five days. The results revealed that 83% of PCB were removed by electrodialysis, when compared with electrokinetic remediation. Additionally, electrodialysis is proven to require lesser nZVI consumption for enhanced removal of PCB.

2.1.2 Water

Due to the ability of nZVI to migrate below ground, these materials are employed in contaminated groundwater remediation, which targets organic chlorinated contaminants that includes pesticides and ions of metals (Mueller et al. 2012). Full-scale nZVI application for remediation of groundwater in Horice, Czech Republic, was reported by Mueller et al. (2012). The full-scale application targeted an area that was contaminated with perchloroethylene, trichloroethene, and dichloroethane with concentration of contaminants, which was escalated up to 70 mg/L. In the first stage, 300 kg of nZVI was inoculated and 60–75% reduction of the original contaminant concentration was achieved. Later, 300 kg of nZVI was again introduced and the chlorinated hydrocarbon concentration was reported to be decreasing gradually.

Nitrate is the most widespread groundwater contaminant, due to its enhanced water solubility (Nujić and Habuda-Stanić 2017). Thus, nitrates cause serious ecological disturbances and impose a great risk to drinking water supplies (Liu et al. 2005, Kapoor and Viraraghavan 1997). When the concentration of nitrate is below 50 mg/L, it does not possess any risk; however, a slight increase in the concentration is toxic. Recently, Ghanei Ardekani and Hassani (2018) investigated the removal of nitrates using nZVI from polluted water with nitrate concentration of up to 105 mg/L. Their results showed that 86% of nitrate was removed via remediation after 7th day.

Additionally, phosphorus built up in aquatic environment was commonly caused by municipal waste, industrial wastewaters and run-offs from agriculture. The excess phosphorus in runoff caused eutrophication, which potentially harms the aquatic ecosystem (Penn and Warren 2009). Almeelbi and Bezbaruah (2012) investigated the removal of phosphate polluted water by using nZVI and micro-ZVI. They tested water with distinct phosphate concentrations, such as 1, 5, and 10 mg PO_4^{3-} per liter with 400 mg nZVI/l or micro-ZVI/l. In this study, swift exclusion of ~88 to 95% phosphate was achieved for all concentrations by nZVI in 10 min and around 96 to 100% of phosphate removal in 30 min, while micro-ZVI only removed 23% of phosphate in 30 mins. Thus, the efficiency of nZVI in removing phosphate was demonstrated to be 13.9 times greater than micro-ZVI with identical concentration of surface area that are utilized in the testing process.

2.2 Metal Nanoparticles

Recently, nanosized silver particles (AgNPs) are extensively under research due to their exceptional electronic, catalytic and antibacterial properties. This has led AgNPs to be widely used as biosensors, catalysts, in surface-enhanced spectroscopy and in active energy devices (Tran et al. 2013). However, it is noteworthy that the AgNPs are widely utilized as nanocatalyst in environmental remediation applications. AgNPs possess capability to degrade harmful pollutants such as heavy metals, pesticides and dyes (Isa and Lockman 2019). Dyes are broadly used in textiles and poor management in wastewater treatment cause these effluents to be discharged into water bodies. These effluents generally consist of various toxic compounds, which include heavy metals, acids, and alkalis, that pose great risks to aquatic life and humans. Therefore, it is important to remediate wastewater to minimize the damage to the environment (Carvalho and Carvalho 2017). Herein, Singhal and Gupta (2019) utilized disposed sheet of X-ray extracted silver for AgNPs' fabrication. The ability of these AgNPs in discoloring dyes were tested against five azo dyes that are soluble in water, namely evans and methylene blue, congo red, eriochrome black T and methyl orange. In their study, individual dyes were treated in static condition with AgNPs at room temperature and the experiment was carried for 90 mins. Results obtained showed that AgNPs were able to degrade all dyes by ~96–99%, except eriochrome black T.

3. METAL OXIDE NANOPARTICLE IN REMEDIATION

3.1 Iron Oxides Nanoparticles

In the past decades, iron oxide nanoparticles have garnered interest in environmental remediation owing to their excellent adsorption ability to arsenic (As), which is a very important water contaminant (Dixit and Hering 2003). Various nanosized oxides of iron particles are tested for As-contaminated water. Remarkably, Liang and Zhao (2014) tested the ability of iron oxide nanosized particles for As(V) remediation. As(V) laden soil with arsenic loading of 31.45 mg/kg was treated with 0.1 g of Fe/L. Their results showed that small fraction inclusions of the iron oxide nanoparticles reduced the As concentration by approximately 98%. Rajput et al. (2017) investigated the lead and copper ion remediation from groundwater via superparamagnetic iron oxide nanoparticles. The groundwater was spiked with 5 mg/L of lead and copper ions separately. 0.2 g/L of iron oxide nanoparticles was added to lead ions spiked groundwater and 0.1 g/L to copper ions spiked groundwater. Their results showed that 74% of lead ions and 28% of copper ions were removed from the groundwater. Besides, their findings also showed that there is no effect on the remediation efficiency of nanoparticles by other groundwater ions.

Magnetic iron oxide nanoparticles display great competence in wastewater treatment due to its high precise surface area, as it can remove contaminants from wastewater rapidly. Additionally, magnetic iron oxide can be reused by magnetic separation after treatment. Nassar et al. (2015) studied the application of nanosized magnetic oxides of iron particles to remove the presence of synthetic dyes in wastewater released from textile industry. Their results demonstrated that magnetic nanosized oxides of iron particles effectively adsorbed dye. The equilibrium of adsorption was attained within 50 min and 125 min for the prototype dyes and real textile wastewater, respectively. Zhao et al. (2018) investigated the use of porous iron oxides (Fe_2O_3) microcubes in organic pollutants and ion removal of heavy metals from polluted samples. The adsorption capacity of porous Fe_2O_3 microcubes was tested against humic acid, methyl blue, chromium ions [Cr(VI)] and lead ions [Pb(II)]. The synthesized porous Fe_2O_3 microcubes have large precise surface area of approximately 155 m^2/g and consists of huge amount of Fe_2O_3 nanoparticles. Results obtained showed that porous Fe_2O_3 microcubes exhibited great capacity of adsorbing Cr(VI), Pb(II), humic acid and methyl blue with value of 175.5 mg/g, 97.8 mg/g, 159.4 mg/g and 4 Pb(II) 25.9 mg/g, respectively. Thus, porous-Fe_2O_3 was emphasized to be useful as a potential nanocatalyst in organic contaminant and ions of heavy metal remediation from water.

3.2 Silica (SiO$_2$) Nanoparticles

Mesoporous silica (SiO_2) materials have gained attention in environmental remediation owing to their enhanced surface area, large volume of pores, facile surface modification and adjustable pore size which aid in adsorption (Tsai et al. 2016). Due to their excellent potential as adsorbents, SiO_2 has been extensively studied for the contaminant gas phase remediation. Moreover, various mesoporous SiO_2 with altered surfaces were tested to improve its adsorption capacity (Guerra et al. 2018). Grafting of functional groups is a noteworthy strategy to design new adsorbents in removing contaminants of interest (Huang et al. 2003). For example, silica has been functionalized with carboxylic acid, thiol, amino groups for heavy metals remediation from wastewater. Amine groups are generally incorporated into silica for acid removal from natural gas such as carbon dioxide (CO_2) and hydrogen sulfide (H_2S). In this regard, Huang et al. (2003) synthesized the 3-aminopropyl-functionalized ordered mesoporous silica (MCM-48) in H_2S and CO_2 removal from natural gas. Their results showed that the high efficiency of H_2S and CO_2 removal was credited to high surface amine group availability of 3-aminopropyl-functionalized MCM-48. 80% of CO_2 was removed within 30 min at room temperature. Similarly, 80% of H_2S was removed in 35 min. Wang et al. (2015) studied the amino-functionalized magnetic silica mesoporous structure for the lead (Pb^{2+}) ion remediation. Magnetic silica was functionalized with aminopropyltriethoxysilane and thus, large quantity of amino groups was grafted on the surface. The amino-functionalized magnetic mesoporous composite exhibited outstanding adsorption capacity for Pb^{2+} ions with a value of 243.9 mg/g. Recovery of the adsorbent could be done easily with an external magnetic field followed by acid treatment. Drese et al. (2011) and Choi et al. (2011) also emphasized the ability of amine-modified silica for the capture of CO_2. Their results showed that CO_2 absorbs reversibly to amine-modified silica and it remained stable after adsorption-desorption process of 50 cycles, making it a cost-effective absorbent. The absorption of CO_2 by amine-modified silica was fast and 90% of CO_2 was removed within the first few minutes of treatment. Thus, their results indicated that grafting of amine groups to silica exhibit greater performance and stability and is cheaper, making it a feasible substitute to traditional CO_2 capture (Qi et al. 2011).

Besides, materials with silica are broadly studied for dye remediation from contaminated wastewater. Moreover, functional groups such as carboxylic (–COOH) groups can form hydrogen bonds with various pollutants. In this regard, Tsai et al. (2016) examined the silica mesoporous nanostructures that are functionalized with –COOH groups for methylene blue remediation from water samples. Their results revealed that the methylene blue remediation was highly dependent on the pH of the sample. The maximum methylene blue uptake was achieved at pH 9 indicating that carboxylic acid functionalized silica can be an effective adsorbent at basic pH conditions. As for this reason, these materials might not be able to usable in all real textile wastewater treatment due to the limited pH range.

3.3 Titanium Oxide (TiO_2) Nanoparticles

Titanium oxide (TiO_2) is another frequently studied metal-based remediation material of water and air due to its nontoxicity, cost effectiveness and photocatalytic properties (Li et al. 2008). TiO_2 nanoparticles can be easily activated by UV light making it an effective photocatalyst to remove contaminants in various media. TiO_2 nanoparticles also produce oxidative agents with enhanced reactivity such as radicals of hydroxyl that destroy microorganisms such as fungi, bacteria, viruses, and algae (Li et al. 2014). Alizadeh Fard et al. (2013) investigated the photocatalytic petroleum aromatic hydrocarbon degradation in groundwater via nanofilms fabricated by TiO_2 powder. For their experiment, 1.5 g/L of TiO_2 nanoparticles was coated on glass beads. The initial concentration of benzene, toluene, ethylbenzene and xylene in the groundwater used was 508, 778, 934 and 631 µg/L, respectively. Their results showed that under natural light photocatalytic remediation, the total removal of benzene, toluene, ethylbenzene and xylene achieved was 61.2%, 77.4%, 86.7% and 58.9%, respectively. Similarly, Mammadov et al. (2017) also demonstrated the photocatalytic ethylene degradation in air by nanosized TiO_2 particles coated on beads of glass. The maximum efficiency in degradation was reported to be 70%.

Due to high demand in agricultural products, the application of pesticides has been increasing worldwide. Soil and water contamination by these toxic non-biodegradable pesticides such as 2,4-dichlorophenoxyacetic acid (2,4-D) and 2,4-dichlorophenoxypropionic acid (2,4-DP) have caused serious environmental problems. Abdennouri et al. (2016) synthesized TiO_2 and investigated their photocatalytic degradation ability of 2,4-D and 2,4-DP in water samples. Initial concentration of 20 mg/L of 2,4-D and 2,4-DP was used and after 90 mins of exposing to UV light, approximately 85% of 2,4-D and 75% of 2,4-DP were removed. This showed that these photocatalysts can efficiently degrade 2,4-D and 2,4-DP. Similarly, Abdennouri et al. (2015) also studied the photodegradation of herbicides using TiO_2. Their results showed that almost half from the initial concentration of 2,4-D and 2,4-DP were degraded within 90 min of irradiation time.

3.4   Graphene Oxide

Pollutants in water samples such as heavy metals, pesticides and halogens needed to be specifically treated, due to their toxicity and prevalent occurrence in soil and water (Koushik et al. 2016). Graphene oxide (GO), owing to its plentiful oxygenous functional groups which include carboxylate, hydroxyl and epoxide, made it an excellent candidate in removing heavy metal ions and pesticides from wastewater (Liu et al. 2016). Therefore, numerous researchers are keen to study the application of GO in remediation field. Particularly, Zhao et al. (2011) synthesized adsorbents using GO for cobalt (Co(II)) and cadmium (Cd(II)) ion remediation from water. The elevated capacities of Cd(II) and Co(II) ion sorption over nanosheets of GO at pH 6 was 106.3 mg/g and at room temperature was 68.2 mg/g. Sun et al. (2012) also investigated the application of GO in the europium (Eu(III)) remediation from water samples. Eu(III) is commonly used in nuclear control applications because it is a good neutron absorber and is generally selected as a chemical trivalent actinides (An(III)) and lanthanides (Ln(III)) analogue. The results revealed that the capacity of Eu(III) adsorption on GO nanoparticles was highest at pH 6 with a value of 175.44 mg/g at room temperature. Besides, they also revealed that adsorption capacity of Eu(III) by GO is superior when compared with other adsorbents such as TiO_2, aluminosilicate zeolite and multi-walled carbon nanotube. Thus, it is proved that GO nanoparticles can be utilized in the trivalent actinide and lanthanide remediation of pollution in environment. Additionally, Wang and Chen (2015) examined the adsorption of cadmium (Cd^{2+}), 1-naphthol and naphthalene by graphene oxide. Their results showed that GO showed exclusive capacities of adsorption for a wide range of pH. Numerous functional carboxyl, hydroxyl and carbonyl groups are formed on the surface due to incomplete GO reduction. These functional groups displayed a strong attraction to Cd^{2+} when compared with naphthalene and 1-naphthol. In addition, Liu et al. (2016) studied the oxides of graphene with distinct degrees of oxidation property for Co(II) ion remediation. They synthesized oxidative GOs via treatment with ozone at times of bubbling such as 0, 2, 4 and 6 hours. Adsorption capacity of ozonized GOs towards Co(II) was tested with initial Co(II) concentration of 20mg/L at pH 6.8 and room temperature. Their results showed that 30%, 35%, 40% and 42% of Co(II) were removed by GO samples after 18 hours. It has been revealed that the fabricated GOs with regulated oxidation not only increased the absorption capacity, it also decreased the size of GO sheet which improved their abilities of dispersion in water samples. Therefore, it is deduced that GO are better adsorbents for heavy metal remediation from water. Maliyekkal et al. (2013) have experimented the absorbent property of reduced GO for pesticide removal from water. The results revealed that the reduced GO exhibited extraordinary potential in removing pesticides from water.

4.   METAL NANOCOMPOSITES AND OTHER NOVEL METAL NANOPARTICLES IN REMEDIATION

4.1   Iron-based Nanocomposites

Even though nZVI is widely used for various environmental remediation, its application in wastewater or groundwater remediation is inadequate owing to particles aggregation, low stability, and iron leaching. Besides, due to the high activity of nZVI, its surfaces should be air and oxygen protected to prevent the formation of oxide layer (García et al. 2018). Thus, to address these issues various researchers have functionalized nZVI with various materials such as biochar (BC), activated carbon (AC), sulfide, and organic acids to improve its performance. Recently, Wei et al. (2019) synthesized nZVI on oak sawdust-derived biochar (nZVI/BC) for the removal of nitrobenzene (NB). In their study, they experimented the efficacy of BC, nZVI and nZVI/BC to remove NB. For BC, 28% of NB was removed in 30 min and reached equilibrium, whereas the removal of NB by nZVI was approximately 37% in 360 min, which was slower than BC. As for nZVI/BC, 94% of NB removal was achieved in 360 min. These showed that nZVI and BC nanocomposites had synergistic effect and thus, enhanced the removal efficiency of NB.

Compounds containing sulfur such as pyrite (FeS_2), greigite (Fe_3S_4) and iron sulfide (FeS) are known to have strong affinity for mercury (Hg), owing to its elevated capacity of mercury sorption and lasting steadiness in contaminated soil (Jeong et al. 2007). Nevertheless, FeS nanoparticles tend to agglomerate, making it unsuitable for *in situ* soil remediation. To address this concern, numerous methods such as polymer-stabilized wet-chemical synthesis, reverse micelle and poly(amidoamine) dendrimer-stabilization have been developed to synthesize nanosized FeS particles with precise distribution of morphology and size particle (Liu et al. 2015). Remarkably, Xiong et al. (2009) fabricated a stable nanosized FeS particles with carboxymethyl cellulose (CMC). Their results demonstrated that nanosized FeS/CMC particles were highly operative in removing mercury (Hg^{2+}) from sediment. The results obtained for batch tests showed that 97% of Hg^{2+} concentration was reduced, when 26.5 is the molar ration of FeS/Hg. Then, the column sediment treatment tests with a 0.5 g/L of nanosized FeS particle suspension presented that the concentration of Hg^{2+} was reduced by 67%.

Iron nanoparticles are generally used in remediation of heavy metal contaminated soils. However, their ability to remove lead (Pb^{2+}) ions from contaminated soil is not extensively studied. Recently, Peng et al. (2019) experimented the potential of numerous absorbents for Pb ion removal from contaminated soil. The absorbents studied included nZVI, nZVI supported by BC (nZVI/BC), FeS nanoparticles, BC (FeS/BC) supported FeS, ferrous oxide (Fe_3O_4) nanoparticles and Fe_3O_4 supported by BC (Fe_3O_4/BC). Their findings revealed that BC could advance the configuration of nanoparticles and increase the surface area for adsorption. The efficient removal of Pb^{2+} by these nanosized iron-based particles showed significant difference, where 45.80%, 54.68%, 2.70%, 5.13%, 47.47%, and 30.51% were obtained by nZVI, nZVI/BC, FeS, FeS/BC, Fe_3O_4 and Fe_3O_4/BC, respectively.

Remarkably, Mahmoud et al. (2017) have reported the application of silayting agents as appropriate reagents for surface nZVI functionalization. Therefore, Mahmoud et al. (2019) studied the functionalization of nZVI via encapsulation to produce $nZVI/NH_2$ and nZVI/ED, respectively. The metal ions removal including Co^{2+}, Cu^{2+}, Zn^{2+}, Cd^{2+}, Hg^{2+} and Pb^{2+} from various water samples were studied and assessed using the micro-column technique. The results obtained by $nZVI/NH_2$ nanocomposite were tabulated in Table 2 and nZVI/ ED nanocomposite in Table 3. The data established that $nZVI/NH_2$ possess high selectivity towards Hg^{2+} and Pb^{2+}. Besides, the ability of nanosized $nZVI/NH_2$ and nZVI/ED composites in removing radioactive ^{60}Co and ^{65}Zn from artificial liquid wastewater was tested. $nZVI/NH_2$ was able to extract 94.57% and 97.30%, and nZVI/ED 96.31% and 98.57% of ^{60}Co and ^{65}Zn, respectively. Overall, their studies showed that $nZVI/NH_2$ and nZVI/ED were able to remove various metal ions and radioactive isotope incredibly.

Table 2 Metal ions removal efficiency by $nZVI/NH_2$ nanocomposite from various waters (Mahmoud et al. 2019)

	Co^{2+}	Cu^{2+}	Zn^{2+}	Cd^{2+}	Hg^{2+}	Pb^{2+}
Tap water	92.21	93.32	90.10	96.77	93.10	100.0
Sea water	94.90	91.42	97.20	98.20	92.00	100.0
Wastewater	94.79	98.34	98.00	96.93	99.55	96.25

Table 3   Metal ions removal efficiency by nZVI/ED nanocomposite
from various waters (Mahmoud et al. 2019)

	Co^{2+}	Cu^{2+}	Zn^{2+}	Cd^{2+}	Hg^{2+}	Pb^{2+}
Tap water	93.88	96.38	97.88	96.16	94.11	100.0
Sea water	94.12	98.50	97.96	97.00	95.52	100.0
Wastewater	93.80	91.20	97.15	93.71	93.80	100.0

Nevertheless, persulfate activation method has been widely used for environmental remediation. Persulfate generates sulfate radical ($SO_4\bullet-$), which possesses high reactivity in several pollutant degradation (Hao et al. 2014). Thus, Su et al. (2019) synthesized a novel iron oxyhydroxides (α-FeOOH) functionalized on the surface of graphene oxide carbon nanotubes (GO-CNTs) matrix as an activator of persulfate for decolorization treatment of Orange II O(II). Approximately 99% of OII was decolorized by α-FeOOH/GO-CNTs in comparison with α-FeOOH of 44% and GO-CNTs of 18%.

4.2   Nickel–Cobalt Alloy Nanoparticles

Arsenite with triple valency (As(III)) is highly noxious and is tedious to eliminate from water, compared to arsenate with pentavalency (As(V)). Thus, As(III) to (V) conversion is considered as a feasible method to solve arsenic pollution problems. Guo et al. (2019) studied the electrocatalytic arsenite conversion into arsenate using nickel-cobalt (NiCo) nanosized alloy particles loaded onto carbon nanotubes (NiCoNPs/C). The oxidation potential of NiCoNPs/C was compared with bare carbon electrode in 0.1 M electrolyte of KOH, in the presence or absence of sodium arsenite (0.01 M of $NaAsO_2$). The results obtained showed that no oxidation occurred for the bare carbon electrode, signifying that As(III) was not converted to As(V). However, an oxidation peak at 1.1 V was observed when NiCoNPs/C electrode was tested in KOH electrolyte without $NaAsO_2$. The appearance of oxidation peak was contributed by the bivalent species (M^{2+}) to trivalent species (M^{3+}) (M = Ni, Co) oxidation in KOH. Then, $NaAsO_2$ was added and the potential increased to 1.15 V, which can be attributed to the As(III) to As(V) oxidation. M^{3+} was reduced to M^{2+} in the existence of As(III) and thus, oxidized to As(V). Their findings showed that NiCoNPs/C nanocomposites have great potential to be used in toxic As(III) into As(V) conversion, providing a promising method to tackle arsenic pollution issues.

4.3   Titanium Oxide-based Nanocomposites

As discussed in Section 3.3, titanium oxide (TiO_2) was largely applied for water and air remediation. Herein, Truppi et al. (2019) examined the efficiency of TiO_2/Au nanorods in degrading contaminants in water using UV–vis light. The efficacy of TiO_2/Au nanorods as photocatalysts was tested with methylene blue (MB) and nalidixic acid (NA). Results collected showed that TiO_2/Au nanorods had highest efficiency when they were synthesized under 450°C calcination temperature. Under UV light, TiO_2/Au nanorods successfully catalyzed MB and NA, 2.5 and 3.2 times higher, respectively, when compared with commercially available TiO_2 P25 Evonik. Furthermore, the reaction time of TiO_2/Au nanorods was faster by 13 times in comparison with bare TiO_2-based catalysts for MB and NA. Albay et al. (2016) also synthesized the photocatalystic TiO_2/phthalocyanine- copper(II) nanosized composite for dye removal in aqueous solution. $CuPc/TiO_2$ showed 100% removal of Cr(VI) after 150 mins of irradiation time, yet only 58% of Cr(VI) was removed by TiO_2. Their results showed that $CuPc/TiO_2$ nanocomposites is a feasible and promising catalyst that eliminates Cr(VI) ions up to 10 mg/L via visible light irradiation.

4.4 Carboxymethyl Cellulose and Polyacrylamide Nanocomposites

Godiya et al. (2019) prepared novel nanocomposites of CMC and PAM (polyacrylamide) for heavy metal remediation of wastewater. The CMC/PAM nanosized composites showed strong affinity towards ions of copper (Cu^{2+}), Pb^{2+} and Cd^{2+}. The capacity of adsorption in CMC/PAM is superior with a value of 227.3 mg/g, 312.5 mg/g, 256.4 mg/g for Cu^{2+}, Pb^{2+} and Cd^{2+}, respectively. Then, the adsorbed Cu^{2+} was condensed *in situ* for the formation of nanosized copper particles that are loaded with CMC/PAM (Cu–CMC/PAM) and the catalytic ability was tested with 4-nitrophenol. Their results showed that Cu–CMC/PAM effectively eliminated 4-nitrophenol and 4-aminophenol, thus verifying the dual functionalities of Cu–CMC/PAM as adsorbent and catalyst in remediation.

4.5 Fe–Mn Oxide-based Nanocomposites

Fe–Mn oxide is a binary metal oxide that is present largely in the environment and offers high adsorption capacity towards As and selenium (Se) oxyanions due to their elevated precise surface area, anion exchange and concurrent redox reactions (Liu et al. 2015). An and Zhao (2012) studied a novel nanosized binary Fe–Mn oxide particles stabilized with starch and CMC for the elimination of As(III) and As(V), to enhance the binary metal oxide deliverability for water and soil remediation. The extreme capacity of adsorption for As(III) and As(V) was reported to be 338 and 272 mg/g, respectively, at 5.5 pH. The results were superior when compared with other non-stabilized Fe–Mn nanocomposites reported by other researchers. McCann et al. (2018) also studied the sorption and oxidation of *in situ* arsenic using binary Fe–Mn oxide on arsenic contaminated soil. According to their initial results, it was deduced that As(III) was initially modified to As(V) by the Mn oxide via oxidation, then simultaneous adsorption of As(V) to iron oxide. Adsorption of As(III) and (V) on Fe–Mn oxides was rapid with removal rate of 93% within the first 2 hours and reached equilibrium at 97% after 24 hours. Maximum capacity of Fe–Mn oxide to adsorb As(III) and (V) was achieved with value of 70 and 32 mg/g, respectively. Similarly, Lin et al. (2019) synthesized Fe–Mn oxide supported on BC (FM/BC) in the study of arsenic removal. Various different concentration ratios of $FeSO_4 \cdot 7H_2O$ and $KMnO_4$ solutions were used to synthesize FMBCs and they found that As(III) removal efficiency improved when FMBC dosage increased and weight ratios of BC:$FeSO_4$:$KMnO_4$ of 18:3:1 had the highest As(III) removal efficiency. Their results showed that this nanocomposite possessed great potential in remediation of arsenic contamination.

4.6 Graphene Oxide-based Nanocomposites

Koushik et al. (2016) studied the dehalogenation of pesticides and organics using condensed nanosized silver-oxide of graphene (rGO/Ag) composite. The results indicated that graphene oxide silver nanocomposite was able to degrade pesticides such as chlorpyrifos (CP), endosulfan (ES), and dichlorodiphenyldichloroethylene (DDE) completely in 60 min. Sarno et al. (2017) also emphasized the elimination of persistent pesticide, chlordane, using rGO/Ag nanocomposite. It was found that rGO/Ag completely removed chlordane in 11 min. The fast degradation of chlordance can be attributed to the absorbing characteristic of graphene and outstanding catalytic performance of the Ag. Dong et al. (2018) also synthesized a nanosized functional composite hydrogel, which was made from rGO with Fe_3O_4 nanoparticles and polyacrylamide (PAM). The proposed Fe_3O_4/rGO/PAM hydrogel exhibits high photocatalytic reaction for organic pollutant degradation, great capacity for heavy metal adsorption and outstanding mechanical strength. The ability of Fe_3O_4/rGO/PAM hydrogel to degrade organic dye was tested with Rhodamine B (RhB). Their results showed that 90% of 20 mg/L of RhB was removed under irradiation of visible light within 60 min. Noteworthy is the fact that the degradation efficiency of

Fe$_3$O$_4$/rGO/PAM hydrogel remained at 90% even after 10 cycles. In addition, synchronous removal of RhB and heavy metal ions was experimented. The results obtained showed that Fe$_3$O$_4$/rGO/PAM hydrogel was able to remove 90% of RhB and the heavy metal removal efficiency was in the range of 34.8 to 66.3%.

5 BIOSYNTHESIZED METAL NANOPARTICLES IN BIOREMEDIATION APPLICATION

The fabrication of metal nanoparticles via biosynthesis approaches was introduced due to the limitations in chemical synthesis approaches, including toxic reactions towards humans and the environment as well as their high cost (Singh et al. 2016a). Biological agents such as plants and microbes are widely used as organisms for the biogenic metal nanoparticle fabrication (Jeevanandam et al. 2016) that are beneficial in bioremediation application as summarized in Table 4.

5.1 Metal-based Nanoparticles via Bacteria

In the past decades, metal-based nanoparticles, including standalone metals, metal oxides and metal composites, were synthesized by using bacterial cell and their extracts to yield smaller sized nanoparticles with exclusive environmental properties as depicted in this section.

5.1.1 Metal Nanoparticles via Bacteria

The bacteria have been utilized to fabricate metal nanoparticles for a long time, due to the ease in culturing them, rapid multiplication time and ability to yield smaller nanoparticles. Bacterial strains such as *Pseudomonas aeruginosa*, *P. denitrificans*, *P. fluo-rescens*, *P. veronii*, *Rhodopseudomonas capsulate*, *Shewanella algae*, *Bacillus* species, *Escherichia coli*, *Serratia marcescens*, *Citrobacter freundii*, *Proteus vulgaris*, *C. koseri*, *P. mirabilis*, *Enterobacter* species, *Klebsiella pneumoniae*, *K. oxytoca*, *Lactobacillus amylotrophicious*, *Salmonella enterica*, *Bacillus stearothermophilus*, *Stenotrophomonas maltophilia*, *Geobacillusstearo thermophilus*, *Magnetospirillum gryphiswaldense* MSR-1, *Sporosarcina koreensis* DC4 and *Staphylococcus epidermidis* are used for the fabrication of gold nanoparticles with enhanced properties and less toxicity. It is noteworthy that the gold nanoparticles that are synthesized via these bacterial strains are in the range of 5–50 nm, which is beneficial to exhibit quantum properties (Menon et al. 2017). Further, nanosized metal particles are fabricated via two distinct approaches, such as intracellular and extracellular methods via bacteria (Singh et al. 2016b). The nanoparticle synthesis by reducing the precursor inside a bacterial cell is termed as intracellular bacterial biosynthesis approach (Park et al. 2016). Even though this approach is highly beneficial in yielding monodispersed nanoparticles, extraction of nanoparticle from bacterial cell and purification is tedious, which stands as a major limitation of this method (Rai et al. 2015). Contrarily, the bacterial enzymes are used to fabricate nanoparticles and this is known as extracellular approach (Singh et al. 2016b). This method possesses more advantages than intracellular approach, especially in yielding pure nanoparticles (Das et al. 2014). However, the instability of nanoparticles and limitations in scale-up production are the drawbacks of this method (Durán et al. 2011). These bacterial synthesized standalone gold nanoparticles are not a better agent for bioremediation applications. However, they are beneficial as a sensor to perceive pollutants in air, water and soil (Apte et al. 2016). Further, they are also combined with other semiconductor nanoparticles as composites to enhance their photocatalytic properties for the degradation of noxious dyes and chemicals in the environment (Sowani et al. 2016).

Table 4 Biosynthesis of metal nanoparticles using different living organisms for bioremediation application

Microbes/plants	Nanoparticles	Bioremediation application	Reference
Bacteria			
Pseudomonas aeruginosa, *P. denitrificans,* *P. fluo-rescens,* *P. veronii,* *Rhodopseudo43978monas capsulate,* *Shewanella algae,* *Bacillus* species, *Escherichia coli,* *Citrobacter freundii,* *Citrobacter koseri,* *Proteus vulgaris,* *P. mirabilis,* *Serratia marcescens,* *Enterobacter* species, *Klebsiella pneumoniae,* *K. oxytoca,* *Lactobacillus amylotrophicious,* *Salmonella enterica,* *Bacillus stearothermophilus,* *Stenotrophomonas maltophilia,* *Geobacillusstearo thermophilus,* *Magnetospirillum gryphiswaldense* MSR-1, *Sporosarcina koreensis* DC4 and *Staphylococcus epidermidis*	Gold nanoparticles	Sensor to detect pollutants in air, water and soil	(Menon et al. 2017) (Apte et al. 2016)
Aeromonas species, *E. coli,* *Bordetella* species, *Enterobacter aerogenes,* *Gluconobacter roseus,* *Geobacter sulfurreducens,* *Klebsiella pneumoniae,* *Pseudomonas aeruginosa,* *P. stutzeri* AG259, *Rhodobacter sphaeroides,* *Rhodopseudomonas palustris,* *Salmonella typhimurium,* *Vibrio algionlyticus,* *Xanthomonas oryzae* and *Yersinia enterocolitica*	Silver nanoparticles	Heavy metal removal, photocatalytic organic pollutant degradation, air purification and wastewater treatment	(Singh et al. 2015) (Huang et al. 2018) (Bhakya et al. 2015) (Le et al. 2015) (Tan et al. 2015)
Bacillus licheniformis M09	Silver nanoparticles	Degrade methylene blue	(Momin et al. 2019)
Cocci and *bacillus*	Lead sulfide and zinc sulfide	Reduce organic dyes from environment	(Zhou et al. 2009)
Actinobacter species, *Thermoanaerobacter* species, *Bacillus subtilis* and *Thiobacillus thioparus*	Iron oxide nanoparticles	Heavy metal removal, dye degradation, pollutant degradation, wastewater treatment and antibacterial applications	(Saif et al. 2016)

Table 4 (Contd...)

Table 4 (Contd...)   Biosynthesis of metal nanoparticles using different living organisms for bioremediation application

Microbes/plants	Nanoparticles	Bioremediation application	Reference
Bacillus subtilis	Titanium dioxide nanoparticles	Photocatalytic activity to degrade environmental pollutant	(Dhandapani et al. 2012)
Sachharomyces cerevisae and *Lactobacillus* species	Titanium dioxide nanoparticles	Air purification, wastewater treatment and bioremediation	(Jha et al. 2009) (Luo et al. 2005) (Waghmode et al. 2019)
Aeromonas hydrophila	Zinc oxide nanoparticles	Antibacterial activity against water-borne pathogens	(Jayaseelan et al. 2012)
Gluconacetobacter xylinum	Iron oxide – bacterial cellulose nanocomposites	Removal of heavy metal ions and recyclable after heavy metal elution	(Zhu et al. 2011)
Acetobacter xylinum	Bacterial cellulose fibers-iron oxide-silver nanocomposites	Antibacterial activity to purify water	(Sureshkumar et al. 2010)
Shewanella oneidensis	Graphene-magnetite nanocomposite	Adsorb chromium and toxic dyes	(Ramalingam et al. 2018)
Algae			
Sargassum wightii, *S. incisifolium,* *S. muticum,* *Porphyrra* species, *Laminaria japonica,* *Padina pavonica* and *Rhizoclonium hieroglyphicum*	Gold nanoparticles	Sensor to detect pollutants in air, water and soil	(Khan et al. 2019)
Blue green algae namely *Plectonema boryanum,* *Spirulina platensis,* *Phormidium valderianum,* *Calothrix* species and *Microcoleus chthonoplastes*	Gold nanoparticles	Sensor to detect pollutants in air, water and soil	(Khan et al. 2019)
Chlorella pyrenoidosa	Silver nanoparticles	Photocatalytic efficiency in wastewater treatment	(Aziz et al. 2015)
Ulva lactuca	Silver nanoparticles	Photocatalytically degrades toxic methyl orange dye	(Kumar et al. 2013)
Chlorella	Zinc oxide nanoparticles	97% photocatalytic degradation of dibenzothiophene with recyclability after five runs	(Khalafi et al. 2019)
Sargassum wighitii	MgO nanoparticles	Photocatalytic ability to degrade toxic organic methylene blue dye	(Pugazhendhi et al. 2019)
Cystoseira trinoids	Copper oxide nanoparticles	Degrade toxic methylene blue dye	(Gu et al. 2018)
Sargassum muticum	Iron oxide nanoparticles	Wastewater treatment agent, photocatalytic toxic heavy metal and dye degradation application	(Mahdavi et al. 2013, Fawcett et al. 2017)

Table 4 (Contd...)

Table 4 (Contd...) Biosynthesis of metal nanoparticles using different living organisms for bioremediation application

Microbes/plants	Nanoparticles	Bioremediation application	Reference
Chlorella pyrenoidosa	TiO_2-graphene oxide nanocomposite	Photocatalytic activity to degrade crystal violet dye	(Sharma et al. 2018)
Padina gymnospora	PVP-platinum nanocomposite	Antibacterial activity against bacteria that contaminate air, water and soil	(Ramkumar et al. 2017)
Ulva lactuca	Titanium dioxide – zinc oxide nanocomposite	Antibacterial activity against bacteria that contaminates environment	(Gurusamy et al. 2019)
Brown algae	Silver-alginate nanocomposites	Catalytic ability to reduce organic dye pollutants, namely methylene blue, 4-nitrophenol and reactive red	(Thangaraj et al. 2018)
Green algae *Enteromorpha* biomass	Copper-hydrogel	Copper heavy metal ion removal and methylene blue and p-nitrophenol catalytic reduction ability	(Su et al. 2018)
Algal biochar	Lanthanum, copper and zirconium trimetallic structure	Photocatalytic ability against environmental contaminants	(Sharma et al. 2019)
Fungi			
Aspergillus niger, *A. fumigatus*, *Verticillium* species and *Phoma glomerata*	Silver nanoparticles	Reduce heavy metal contamination	(Salvadori et al. 2018)
Rhizopus oryzae, *Verticillium luteoalbum*, *Aureobasidium pullulans* and *Colletotrichum* species	Gold nanoparticles	Reduce heavy metal contamination	(Salvadori et al. 2018)
Hypocrea lixii and *Trichoderma koningiopsis*	Copper nanoparticles	Reduce copper contamination	(Salvadori et al. 2018)
Fusarium oxysporum	Cadmium nanoparticles	Reduce cadmium contamination	(Salvadori et al. 2018)
Phaenerochaete chrysosporium, *Trametes versicolor*, *Pleurotus sajor-caju*, *Schizophyllum commune*, *P. ostreatus* and *Stereum hirsutum*	Silver, gold, copper, selenium and cadmium sulfide nanoparticles	Bioremediation application	(He et al. 2017)
Flammulina velutipes	Gold nanoparticles	Reduce toxic organic pollutant such as methylene blue dye and 4-nitrophenol	(Narayanan et al. 2015)
Saccharomyces cerevisiae	Silver nanoparticles	Catalytic degradation ability against methylene blue organic dye	(Roy et al. 2015)
Aspergillus foetidus	Silver nanoparticles	Efficient reduction of arsenic in contaminated aqua environment	(Mukherjee et al. 2017)

Table 4 (Contd...)

Table 4 (Contd...)   Biosynthesis of metal nanoparticles using different living organisms for bioremediation application

Microbes/plants	Nanoparticles	Bioremediation application	Reference
Aspergillus aculeatus	Iron containing nanoparticles	Enhanced plant growth from mungbean seed	(Bedi et al. 2018)
Stereum hisutum	Copper and copper oxide nanoparticles	Enhanced bioremediation and photocatalytic degradation ability against organic pollutant	(Cuevas et al. 2015)
Dead fungal biomass of *Hypocrea lixii*	Nickel oxide nanoparticles	Enhanced bioremediation and photocatalytic degradation ability against organic pollutant	(Salvadori et al. 2015)
Fungal chitosan	Fungal derived chitosan-based metal nanocomposites	Remove heavy metal contaminants	(Pattnaik and Busi 2018)
Plant leaf extract			
Aloysia triphylla	Gold nanoparticles	Catalytically degrade organic pollutants such as Congo red and methylene blue	(López-Miranda et al. 2019)
Avicennia marina	Gold nanoparticles	Reduction of 4-nitrophenol organic pollutant	(Nabikhan et al. 2018)
Cressa cretica	Gold nanoparticles	Catalytic activity to reduce 4-nitrophenol	(Balasubramanian et al. 2019)
Myxopyrum serratulum A.W. Hill	Gold nanoparticles	Catalytically reduce organic pollutant dyes such as Congo red, 4-nitrophenol and methylene blue	(Vijayan et al. 2018)
Mussaenda erythrophylla	Silver nanoparticles	Catalytic degradation ability of azo methyl orange dye	(Varadavenkatesan et al. 2016)
Helicteres isora	Silver nanoparticles	Catalytic degradation property against organic dyes such as safranin, methyl orange, methylene blue and methyl violet	(Bhakya et al. 2015)
Passiflora edulis f. flavicarpa	Silver nanoparticles	Photocatalytic degradation property against organic pollutants such as methyl orange and methylene blue	(Thomas et al. 2019)
Justicia adhatoda	Silver nanoparticles	Photocatalytic activity against methylene blue organic dye	(Latha et al. 2019)
Spinacia oleracea	Iron nanoparticles	Wastewater treatment and catalytic pollutant degradation application	(Dauthal and Mukhopadhyay 2013)
Delonix regia	Palladium nanoparticles	Wastewater treatment and catalytic pollutant degradation application	(Turakhia et al. 2018)

Table 4 (Contd...)

Table 4 (Contd...) Biosynthesis of metal nanoparticles using different living organisms for bioremediation application

Microbes/plants	Nanoparticles	Bioremediation application	Reference
Murraya koenigii	Zinc oxide nanoparticles	Remove microbial contaminants in wastewater treatment facilities	(Elumalai et al. 2015)
Corymbia citriodora	Zinc sulfide nanoparticles	Photocatalytic degradation agent of organic pollutant methylene blue dye	(Chen et al. 2016)
Ruellia tuberosa	Iron oxide nanoparticles	Photocatalytically degrade 80% of crystal violet dye	(Vasantharaj et al. 2019a)
Amaranthus spinosus	Iron oxide nanoparticles	Degrade 75% of methyl orange and 69% of methylene blue	(Muthukumar and Matheswaran 2015)
Fraxinus chinensis Roxb	Iron oxide nanoparticles	Degrade toxic crystal violet and eriochrome black T dyes	(Ali et al. 2019)
R. tuberosa and *Azadirachta indica*	Copper oxide nanoparticles	Photocatalytic bioremediation	(Vasantharaj et al. 2019b) (Thirumurugan et al. 2017)
Jatropha curcas	TiO_2 nanoparticles	Photocatalytic bioremediation	(Goutam et al. 2018)
Withania coagulans	Palladium-reduced graphene oxide-iron oxide nanocomposites	Reduce 4-nitrophenol organic dye pollutant from water	(Atarod et al. 2016a)
Euphorbia heterophylla	Silver-titanium dioxide nanoparticles	Reduce organic pollutant dyes such as Congo red, 4-nitrophenol, methylene blue and methyl orange	(Atarod et al. 2016b)
Abutilon hirtum	Silver-reduced graphene oxide nanocomposites	Catalytic reduction property against Rhodomine B, 4-nitrophenol and Congo red	(Maryami et al. 2016)
Euphorbia wallichii	Copper-reduced graphene oxide-iron oxide nanocomposites	Enhanced catalytic reduction agent of organic pollutants in water	(Atarod et al. 2015)
Euphorbia heterophylla	Silver-HZSM-5 nanocomposite	Enhanced catalytic reduction agent of organic pollutants in water	(Tajbakhsh et al. 2016)
Euphorbia peplus Linn	Silver-iron oxide nanocomposite	Enhanced catalytic reduction agent of organic pollutants in water	(Sajjadi et al. 2017)
Other plant extracts			
Banana pith extract	Gold nanoparticles	Catalytic reduction ability against organic malachite green pollutant dye	(Nayak et al. 2018)
Bridelia retusa fruit extract	Silver nanoparticles	Reduce Congo red dye	(Vinayagam et al. 2017)

Table 4 (Contd...)

Table 4 (Contd...)   Biosynthesis of metal nanoparticles using different living organisms for
bioremediation application

Microbes/plants	Nanoparticles	Bioremediation application	Reference
Vaccinium floribundum and *Anthemis pseudocotula* fruit extract	Iron and iron oxide nanoparticles	Removal of total petroleum hydrocarbons from soil and water	(Murgueitio et al. 2018) (Abdullah et al. 2018)
Pittosporum undulatum, *Schinus molle*, *Melia azedarach* and *Syzygium paniculatum* plant extract	Iron oxide nanoparticles	Remove hexavalent chromium contaminants	(Truskewycz et al. 2018)
Grape fruit extract	Reduced graphene oxide-zinc oxide nanocomposite	Degrade Rhodamine B dye	(Ramanathan et al. 2019)
Crataegus pentagyna fruit extract	Iron oxide-silicon dioxide-copper oxide-silver nanocomposite	Degrade methylene blue and rhodamine B dyes	(Ebrahimzadeh et al. 2019)
Cassytha filiformis fruit extract	Copper-magnesium oxide nanocomposite	Methylene blue, Congo red, 4-nitrophenol and 2, 4-dinitrophenylhydrazine degradation	Nasrollahzadeh et al. 2018)
Valeriana officinalis root extract	Silver-zinc oxide nanocomposite	Reusable catalyst for organic dye reduction	(Yeganeh-Faal et al. 2017)
Silybum marianum seed extract	Silver-gold bimetallic nanocomposites	Catalytic pollutant reduction	(Gopalakrishnan et al. 2015)
Salix alba bark extract	Cobalt ferrite silica magnetic nanocomposite	Catalytic reduction of malachite green dye	(Amiri et al. 2017)
Centaurea cyanus flower extract	Silver-iron oxide-zirconium oxide	Catalytic organic dye pollutant reduction	(Rostami-Vartooni et al. 2019)

Apart from gold, nanosized silver particles are widely fabricated via bacteria for bioremediation applications. Bacterial biomass (Mahdieh et al. 2012), culture supernatant (Saifuddin et al. 2009) and cell-free extract (Singh et al. 2013) are the different extracellular approaches that are used for the bacterial nanosized silver particle fabrication. Likewise, intracellular methods (Srivastava et al. 2013) and bacterial-derived components such as biosurfactants, enzymes, exopolysaccharides, spores, actinorhodin pigment, polysaccharide bioflocculant, native and repolymerized flagella were also reported to be beneficial for the bacteria mediated nanosized silver particle fabrication (Singh et al. 2015). Strains of bacteria including *Aeromonas* species, *E. coli*, *Bordetella* species, *Enterobacter aerogenes*, *Gluconobacter roseus*, *Geobacter sulfurreducens*, *Klebsiella pneumoniae*, *Pseudomonas aeruginosa*, *P. stutzeri* AG259, *Rhodobacter sphaeroides*, *Rhodopseudomonas palustris*, *Salmonella typhimurium*, *Vibrio algionlyticus*, *Xanthomonas oryzae* and *Yersinia enterocolitica* are utilized for nanosized silver particle preparation (Singh et al. 2015). These biogenic silver nanoparticles are extensively employed in bioremediation for heavy metal removal (Huang et al. 2018), photocatalytic organic pollutant degradation (Bhakya et al. 2015), air purification (Le et al. 2015) and wastewater treatment (Tan et al. 2015). Recently, 10–30 nm sized nano-silver particles were fabricated by using supernatant of mutant *Bacillus licheniformis* M09 that are free from cells. The photocatalytic results revealed that these nanoparticles possess ability to reduce methylene blue dye within 3 h under sunlight irradiation, and can be beneficial in next generation dye degradation and effluent treatment applications (Momin et al. 2019). In addition, metal chalcogen nanoparticles such as lead sulfide and zinc sulfide are fabricated via bacteria *cocci* and *bacillus* as morph-templates. These nanoparticles are proved to possess photocatalytic properties which will be beneficial in preventing toxic organic dyes from contaminating the environment (Zhou et al. 2009). Moreover, several other metal nanoparticles such as platinum (Konishi et al. 2007) and copper (Saif Hasan et al. 2008) were also prepared by using bacteria-mediated synthesis approach to bioremediation applications. The chemical and other biogenic version of these nanoparticles are employed in environmental applications. The limitations of bacterial synthesis methods including instability of resultant nanoparticles acts as a hurdle, which blocks their utilization in environmental applications.

5.1.2   Metal Oxide Nanoparticles from Bacteria

Nanosized oxides of metal particles are semiconductor in nature that are widely utilized for photocatalytic toxic effluent degradation applications. Iron oxide nanoparticles are the most common nanosized metal oxide that is employed in effective photocatalytic dye degradation and wastewater treatment, as they are naturally magnetic and can be easily removed from the environment after its purpose (Xu et al. 2012, Lassoued et al. 2018). Bacterial strains, namely *Actinobacter* species, were used to fabricate spherical shaped iron oxide nanoparticles, γ-Fe_2O_3 (maghemite) and Fe_3S_4 (greigite) under aerobic conditions. *Thermoanaerobacter* species, *Bacillus subtilis* and *Thiobacillus thioparus* are the other bacterial strains that are beneficial for nanosized iron oxide particle synthesis. Such nanosized particles were explored and proved to possess exclusive properties that are beneficial in heavy metal removal, dye degradation, pollutant degradation, wastewater treatment and antibacterial applications (Saif et al. 2016). In recent times, titanium dioxide is another semiconductor metal oxide nanoparticles that are synthesized by *Bacillus subtilis* with exceptional photocatalytic activity to degrade environmental pollutants and for hydrogen production (Dhandapani et al. 2012). Further, bacterial species such as *Sachharomyces cerevisae* and *Lactobacillus* species are utilized for nanosized dioxides of titanium particle fabrication (Jha et al. 2009), which can be useful for air purification (Luo et al. 2005), wastewater treatment and bioremediation applications (Waghmode et al. 2019). Zinc oxide (ZnO) nanoparticles are another popular nanosized metal oxides that are used as a pollutant degrading photocatalyst (Jang et al. 2006), wastewater treatment (Puay et al. 2015) and bioremediation agent (Bhandari 2018). However, less literature reports are available on the bacterial nanosized ZnO particle fabrication for environmental applications. Recently, Jayaseelan et al. (2012) synthesized 60 nm sized spherical and oval ZnO nanoparticles using *Aeromonas hydrophila* bacteria. The result from this study demonstrated that these nanoparticles are useful in inhibiting growth of harmful bacteria and fungi in water (Jayaseelan et al. 2012). Tin dioxide (Gorai 2018), copper oxide (Saif Hasan et al. 2008) and manganese dioxide (Sinha et al. 2011) are the other nanosized metal oxide particles that are prepared from bacteria-mediated biosynthesis approach, which have the potential to be employed in bioremediation applications.

5.1.3   Metal Nanocomposites Using Bacteria

Metal nanocomposites synthesized via bacteria are unique nanostructures that are formed by combining bacterial biomolecules and nanoparticles. These nanostructures are better than standalone metal or metal oxide nanostructures and bacterial synthesis will further facilitate their reduction in toxicity towards the environment. Zhu et al. (2011) synthesized iron oxide–bacterial cellulose nanocomposites using *Gluconacetobacter xylinum* via agitation fermentation. The pH-controlled biosynthesis approach yielded iron oxide nanoparticles that are encapsulated homogeneously with spherical shaped bacterial cellulose. Further, the study emphasized that these nanocomposites possess enhanced adsorption properties for manganese, lead and chromium heavy metal elimination, and are recyclable after heavy metal elution (Zhu et al. 2011). Likewise, *Acetobacter xylinum* was used to extract bacterial cellulose fibers and these fibers were made into magnetic composite with iron oxide and silver nanoparticles to form nanocomposites. These magnetic nanocomposites are proven to have an enhanced antimicrobial property which can be useful to inhibit microbes that can cause diseases in the environment and can be recycled via magnetic extraction (Sureshkumar et al. 2010). In recent times, Ramalingam et al. (2018) fabricated a novel graphene-magnetite (iron oxide) nanocomposite by functionalizing living cell of *Shewanella oneidensis*. The results revealed that the nanocomposite possesses hydrophilic property, super paramagnetic behavior, excellent adsorption capacity towards dyes as well as chromium and high adsorption efficiency for toxic dyes and heavy metal pollutant removal applications (Ramalingam et al. 2018). However, extensive research in this direction is required to synthesize stable metal nanocomposites using bacteria for efficient environmental applications.

5.2 Algae and Fungal Synthesis of Metal-based Nanoparticles for Bioremediation

Algae and fungi are the alternative microbes, next to bacteria, for nanosized metal-based particle synthesis including metals, oxides of metals and composites. Algae and fungi possess a wide variety of biomolecules which can be extracted in large quantities. Thus, these microbial mediated synthesis approaches are highly beneficial in the up-conversion of nanoparticle production, in addition to reduction in toxicity towards the environment (Zhao et al. 2018).

5.2.1 Algal Synthesis of Metal and Metal Oxide Nanoparticles

Algae, intracellular methods and extracellular extracts are highly beneficial in nanosized metal particles synthesis, similar to bacteria. Gold nanoparticles are synthesized using algae, namely *Sargassum wightii*, *S. incisifolium*, *S. muticum*, *Porphyrra* species, *Laminaria japonica*, *Padina pavonica* and *Rhizoclonium hieroglyphicum*. It is noteworthy that the nanosized gold obtained from these algal species is below 50 nm in size. In addition, blue-green algae such as *Plectonema boryanum*, *Spirulina platensis*, *Phormidium valderianum*, *Calothrix* species and *Microcoleus chthonoplastes* were also utilized for 10–100 nm sized nano-gold particle fabrication (Khan et al. 2019). However, gold nanoparticles are used to detect toxic heavy metals or pollutants and are not used to eliminate them, as mentioned in the earlier section. Nanosized silver particles are the most significant standalone nanosized metallic particles that are synthesized via algal extracts for bioremediation applications. Aziz et al. (2015) fabricated nano-silver using *Chlorella pyrenoidosa* extracts. The result showed that the algal extracts offered an exceptional consistency in nanoparticles' morphology with intrinsic crystallinity and functional moieties for surface stabilization. In addition, the study also showcased the photocatalytic efficiency of algal extract synthesized silver nanoparticles, which will be useful for the treatment of wastewater (Aziz et al. 2015). Likewise, Kumar et al. (2013) fabricated nano-silver using extract of *Ulva lactuca* at room temperature. They also proved that these biogenic, spherical shaped, ~50 nm sized silver nanoparticles possess ability to photocatalytically degrade toxic methyl orange dye in water via silver as nanocatalyst under the illumination of visible light (Kumar et al. 2013).

Metal oxides such as zinc, magnesium, copper and iron are also synthesized via algal extracts for enhanced bioremediation application. In recent times, Khalafi et al. (2019) synthesized highly pure and stable zinc oxide nanoparticles using microalgal extracts of *Chlorella*. The characterization results showed that 20 nm sized, monodispersed ZnO with hexagonal Wurtzite structure was formed. The gas chromatography results revealed that the oxide nanoparticle possess enhanced photocatalytic activity of Dibenzothiophene degradation (97%) as an organosulfur contaminant prototype at neutral pH. Further, the biogenic oxide nanoparticle exhibited exclusive recyclability after five runs and rapid separation of contaminants which proved their durability (Khalafi et al. 2019). Similarly, magnesium oxide (MgO) nanoparticles were synthesized using the algal extract of marine brown algae named *Sargassum wightii*.

The results revealed that the average size of the obtained nanoparticle with structure of cube with face-centers is 68 nm and possess enhanced antibacterial and antifungal properties against human pathogens. In addition, these MgO nanoparticles also possess a photocatalytic ability to degrade toxic organic methylene blue dye after exposing it to sunlight and ultraviolet irradiation (Pugazhendhi et al. 2019). Likewise, Gu et al. (2018) synthesized nanosized oxides of copper particles by using brown alga *Cystoseira trinoids* and ultrasonication. The results demonstrated that spherical shaped, 6–8 nm sized copper oxide nanoparticles were fabricated via algal extract with enhanced antibacterial and 1,1-Diphenyl-2-picrylhydrazyl (DPPH) free radical scavenging property. Furthermore, these nanoparticles also possess ability to degrade toxic methylene blue dye via sunlight and ultraviolet light mediated catalytic property at acidic pH 4.0 (Gu et al. 2018). It is noteworthy that 18–25 nm sized, cubic shaped, nanosized oxides of iron particles can be fabricated using *Sargassum muticum*, which can be utilized as magnetic waste water treatment agent, photocatalytic toxic heavy metal and dye degradation application as mentioned in the earlier section (Mahdavi et al. 2013, Fawcett et al. 2017).

5.2.2 Fungal Synthesis of Metal and Metal Oxide Nanoparticles

Fungal synthesis, specifically dead biomass extracts, is extensively utilized for nanosized metal particle fabrication for bioremediation applications (Rai et al. 2016). Filamentous fungi including *Aspergillus niger, A. fumigatus, Verticillium* species and *Phoma glomerata* are used for silver nanoparticle synthesis, whereas *Rhizopus oryzae, Verticillium luteoalbum, Aureobasidium pullulans* and *Colletotrichum* species are utilized for gold nanoparticles synthesis. In addition, copper nanoparticles are synthesized via *Hypocrea lixii* and *Trichoderma koningiopsis*, whereas cadmium nanoparticles were synthesized using *Fusarium oxysporum* extracts (Salvadori et al. 2018). Fungi also serve as a pollutant or contamination in the environment (Andersen et al. 2017). Thus, the utilization of fungal dead biomass as a source for nanoparticle preparation will serve as a better bioremediation approach in industries to avoid fungal contaminants towards the environment. Similarly, He et al. (2017) reported that white rot fungi group, such as *Phaenerochaete chrysosporium, Trametes versicolor, Pleurotus sajor-caju, Schizophyllum commune, P. ostreatus* and *Stereum hirsutum*, are also extensively used for the fabrication of nano-metal particles such as silver, copper, gold, selenium and cadmium sulfide nanoparticles. Also, these fungal extracts' fabricated nanosized particles were proved to possess enhanced bioremediation properties (He et al. 2017). Moreover, Narayanan et al. (2015) demonstrated a novel fabrication method of nanosized gold particles by intracellular approach using *Flammulina velutipes* mushroom. The result showed that the fungi lead to the development of 20 nm sized nano-gold particles and possess the ability of heterogeneous catalyst to reduce the toxic organic pollutant such as methylene blue dye and 4-nitrophenol, which can serve as a better bioremediation agent (Narayanan et al. 2015). Furthermore, Roy et al. (2014) fabricated nanosized silver particles via *Saccharomyces cerevisiae* yeast extract. The result reported that the yeast extract yielded spherical shaped, quantum sized nano-silver particles with 10 nm of average size. These biogenic nanoparticles were proved to possess enhanced catalytic degradation ability against organic methylene blue dye under irradiation of visible sunlight for a few hours (Roy et al. 2015). Recently, porous silver nanoparticles with 2.35 nm as pore width were synthesized via *Aspergillus foetidus* and were utilized for the efficient reduction of arsenic in contaminated aqua environment (Mukherjee et al. 2017).

Iron ore tailing is a mining waste, which causes hazard to the environment and is required to be remediated from the contaminated sites. The fungal species that are present in these sites, especially *Aspergillus aculeatus,* possess enhanced leaching efficiency and produce iron containing nanoparticles under optimum condition, which reduces contamination in the site. Further, these iron nanoparticles are capped with proteins of fungi, which make them highly biocompatible and serve as a bioavailable iron source for the enhanced growth of plants from the seed of mungbean (Bedi et al. 2018). Likewise, nanosized copper and copper oxide particles are synthesized by the extracellular biomolecules of a white rot fungus from Chilean forests, namely *Stereum hisutum* (Cuevas et al. 2015), whereas both extra and intracellular extracts from the dead fungal biomass of *Hypocrea lixii* were utilized for the synthesis of nanosized oxides of nickel particles (Salvadori et al. 2015). Both these nanoparticles possess enhanced bioremediation and photocatalytic degradation ability against organic pollutants (Bokare et al. 2008, Gong et al. 2011, Devi and Singh 2014).

5.2.3 Metal Nanocomposites Synthesized via Algae and Fungi

In recent times, numerous metal nanocomposites were developed using algae and fungal extracts for bioremediation applications. Recently, titanium dioxide (TiO_2) nanoparticles were synthesized via extracts of green algae named *Chlorella pyrenoidosa* and were deposited over graphene oxide sheets to form nanocomposites. The result revealed that the TiO_2 nanoparticles possess enhanced photocatalytic activity, which elevates the potential of nanocomposite to degrade crystal violet dye in aqueous medium under irradiation of visible light (Sharma et al. 2018). Likewise, platinum nanoparticles were synthesized using the extracts of Indian brown seaweed named *Padina gymnospora* and were embedded with polyvinylpyrrolidone (PVP) for the formation of nanocomposites. The results showed that the seaweed extract yielded 5–50 nm sized nanoparticles with truncated octahedral shape and the nanocomposite possessed enhanced antibacterial activity against seven disease causing bacteria which can contaminate air, water and soil (Ramkumar et al. 2017). Moreover, Gurusamy et al. (2019) utilized chemical synthesized, 12 nm sized titanium dioxide – zinc oxide nanocomposite for biodiesel production from *Ulva lactuca* seaweeds. Further, the leftover biomass of *U. lactuca* was utilized to synthesize 12 nm sized silver nanoparticle with enhanced antibacterial activity against *Proteus vulgaris* (Gurusamy et al. 2019). Likewise, silver nanoparticles were synthesized by encapsulating alginate from the cell wall of brown algae to form the silver-alginate nanocomposites. The resultant nanocomposites were 8 nm in size with spherical shape and a crystal structure of face cube. Further, the results emphasized that these nanocomposites possess catalytic ability to reduce organic dye pollutants such as methylene blue, 4-nitrophenol and reactive red with the existence of sodium borohydride (Thangaraj et al. 2018). In recent times, a novel hydrogel was fabricated from the green algae *Enteromorpha* biomass polymerization with acrylic acid for three different applications. These hydrogels possess enhanced swelling property, which makes them act as an effective adsorbent for copper based heavy metal removal application. The adsorbed metal particles were below 100 nm in size, which makes them metal-hydrogel nanocomposites. These nanocomposites were reported to possess catalytic activity to reduce organic contaminants, including methylene blue and p-nitrophenol with catalytic activity (80–90%) after five times of reuse and storage for 30 days (Su et al. 2018). Similarly, it was recently reported that the fungal derived chitosan-based metal nanocomposites are widely used in the heavy metal biosorption application to eliminate contaminants of heavy metals from the environmental entities. It was emphasized that the biosorption property of individual fungal derived chitosan and metal nanoparticles will increase drastically, when they are synthesized as a nanocomposite (Pattnaik and Busi 2018). Furthermore, trimetallic structure with lanthanum, copper and zirconium was reinforced with algal biochar via microwave method to form nanocomposite. These nanocomposites were employed for the remediation of green malachite dye from the contaminated site, which revealed their exclusive adsorption and photocatalytic ability against environmental contaminants (Sharma et al. 2019). Even though all these studies supported fungal and algal synthesized metal nanoparticles for bioremediation process, the time taken for the nanoparticle production, typically 24–120 hours, and intracellular approach hindering the downstream process still remain as limitations to develop simple and cheap biogenic metal nanoparticles for bioremediation purpose (Jeevanandam et al. 2016).

5.3 Plant Extracts Mediated Synthesis of Metal-based Nanoparticles for Bioremediation

Plant extracts in the form of phytochemicals are the most common and widely utilized reducing and stabilizing agent for the formation of nanosized biogenic metal particles. The major advantage of this approach is the wide availability of plants and the extract can be obtained from several parts of the plants (Dauthal and Mukhopadhyay 2016). It is noteworthy that the phytochemicals can be extracted from the plant parts including leaves, fruits and other parts such as root, stem and bark for less toxic nanometal particle synthesis for bioremediation applications.

5.3.1 Leaf Extracts

Leaves are the most common plant part that was used for the extraction of phytochemicals to serve as a better agent with reducing and stabilizing property for nanosized particle formation (Kumar et al. 2012). Recently, aqueous leaf extracts of *Aloysia triphylla* for nanogold particle fabrication were shown to exhibit antibacterial and catalytic property. The results emphasized that the extracted phytochemicals yielded 40–60 nm sized, spherical shaped nanogold particles with the ability to catalytically reduce organic pollutants within 10 mins (López-Miranda et al. 2019). Likewise, spherical shaped, 4–13 nm sized nanogold particles were prepared via mangrove leaf *Avicennia marina* extract for bioremediation application. These nanoparticles were encapsulated with sodium alginate to form nanobeads and were employed for 4-nitrophenol organic pollutant elimination with their enhanced heterogeneous catalytic property (Nabikhan et al. 2018). Similarly, aqueous leaf extract of *Cressa cretica* was utilized for the fabrication of hexagonal, spherical, pentagonal and rod shaped, 15–22 nm sized gold nanoparticles. These biogenic nanoparticles were proved to possess exclusive catalytic activity to reduce 4-nitrophenol under the influence of sodium borohydride solution (Balasubramanian et al. 2019). Moreover, silver and gold nanoparticles were fabricated using an extract from the leaf of medicinal plant named *Myxopyrum serratulum A.W. Hill*. The resultant nanoparticles were proved to possess bioremediation ability to catalytically reduce organic pollutant dyes such as Congo red, 4-nitrophenol and methylene blue (Vijayan et al. 2018). Nanosized silver particles are the other major metal nanoparticles that are synthesized via leaf extracts. The *Mussaenda erythrophylla* leaf extract was recently utilized to fabricate nanosized silver particles with the exceptional catalytic degradation ability of azo methyl orange dye along with sodium borohydride (Varadavenkatesan et al. 2016). Similarly, nano-silver particles were prepared via leaf extract of a medicinal plant of India, namely *Helicteres isora*. These biosynthesized silver nanoparticles have been proven to possess catalytic degradation property against organic dyes such as safranin, methyl orange, methylene blue and methyl violet (Bhakya et al. 2015). Further, aqueous leaf extract of *Passiflora edulis f. flavicarpa* (*P. edulis*) was used to prepare silver nanoparticles with effective photocatalytic degradation property against organic pollutants such as methyl orange and methylene blue (Thomas et al. 2019). Furthermore, *Justicia adhatoda* extract from leaf was utilized as agent with reducing property for nano-silver particle preparation with size dependent photocatalytic activity against methylene blue organic dye (Latha et al. 2019). Apart from noble metals, iron nanoparticles were also synthesized from the leaf extract of *Spinacia oleracea* and palladium nanoparticles via *Delonix regia* leaf extract to be useful for wastewater treatment and catalytic pollutant degradation application (Dauthal and Mukhopadhyay 2013, Turakhia et al. 2018).

Oxides of zinc, iron, copper and titanium are the most common metal oxide nanoparticles that are synthesized via plant leaf extracts for bioremediation and environmental applications. In recent times, stable, hexagonal and spherical shaped zinc oxide nanoparticles with an average size of 12 nm are fabricated via *Murraya koenigii* leaf extracts. These nanostructures exhibited enhanced antibacterial activity against *Staphylococcus aureus* and *Bacillus subtilis*, which can be beneficial as a latent agent to eliminate microbial contaminants in the treatment of wastewater facilities (Elumalai et al. 2015). Similarly, zinc sulfide nanoparticles were prepared via leaf extract of *Corymbia citriodora* to exhibit photocatalytic property. The result revealed that the nanoparticle size was 45 nm with surface plasmon resonance and quantum confinement property, which makes them an excellent photocatalytic degradation agent of organic pollutant methylene blue dye under UV light irradiation (Chen et al. 2016). Further, hexagonal rod shaped, 52 nm sized nano-iron oxide particles were fabricated via extract of *Ruellia tuberosa* from leaves to exhibit photocatalytic activity. The result demonstrated that these oxide nanoparticles possess ability to photocatalytically degrade 80% of crystal violet dye under the influence of solar irradiation (Vasantharaj et al. 2019b). Moreover, nanosized iron oxide particles are prepared via extract of *Amaranthus spinosus* from leaf to exhibit improved photocatalytic ability. The result emphasized that the biogenic nanosized iron oxide possesses ability to degrade 75% of methyl orange and 69% of methylene blue under solar irradiation (Muthukumar and Matheswaran 2015). Likewise, leaf extract of *Fraxinus chinensis* Roxb synthesized magnetic nanosized iron oxide particles were proved to possess ability to degrade toxic crystal violet and eriochrome black T dyes to save environment from these pollutants (Ali et al. 2019). Leaf extracts of *R. tuberosa* (Vasantharaj et al. 2019b), *Azadirachta indica* (Thirumurugan et al. 2017) for copper oxide nanoparticles synthesis and TiO$_2$ nanoparticles via *Jatropha curcas* (Goutam et al. 2018) are the other oxide nanoparticles that are used as photocatalyst and in bioremediation applications.

Nanocomposites were also extensively synthesized using plant leaf extracts for environmental bioremediation applications. Recently, nanosized palladium-reduced GO-oxide of iron composites were synthesized via extract of *Withania coagulans* from leaf for photocatalytic pollutant degradation. These novel biogenic nanocomposites exhibit enhanced ability as magnetically separable and reusable catalyst to reduce 4-nitrophenol organic dye pollutant from water at room temperature (Atarod et al. 2016b). Similarly, nanosized titanium dioxide-silver particles were synthesized via leaf extract of *Euphorbia heterophylla* for bioremediation application. The result revealed that the nanocomposite possesses excellent catalytic activity to reduce organic pollutant dyes such as Congo red, 4-nitrophenol, methylene blue and methyl orange in water, which will be beneficial to remediate polluted wastewater (Atarod et al. 2016). Likewise, nanosized silver-reduced GO composites were fabricated via leaf extract from *Abutilon hirtum,* which exhibited excellent recoverable catalytic reduction property against Rhodomine B, 4-nitrophenol and Congo red in aqueous medium (Maryami et al. 2016). Moreover, copper-reduced graphene oxide-iron oxide nanocomposites from *Euphorbia wallichii* leaf extract (Atarod et al. 2015), silver-H-zeolite Socony Mobil-5 (HZSM) nanocomposite via *Euphorbia heterophylla* (Tajbakhsh et al. 2016) and silver-iron oxide nanocomposite from *Euphorbia peplus* Linn (Sajjadi et al. 2017) are used as enhanced catalytic reduction agents of organic pollutants in water.

5.3.2 Other Extracts

Fruit, root, bark and flower extracts are the other reducing agents that are used to form nanosized metal particles to use them as bioremediation agent. Gold nanoparticles are fabricated using the Banana pith extract to exhibit efficient catalytic reduction ability against organic malachite green pollutant dye in water (Nayak et al. 2018). Nano-silver particles with enhanced stability and an average size of 69 nm are synthesized via extracts of fruits from *Bridelia retusa* to reduce Congo red dye (Vinayagam et al. 2017). Likewise, nanosized iron and oxides of iron particles from extracts of fruit *Vaccinium floribundum* (Murgueitio et al. 2018) and *Anthemis pseudocotula* (Abdullah et al. 2018), respectively, were used for the elimination of total petroleum hydrocarbons from soil and water. Moreover, iron oxide nanoparticles were synthesized by obtaining extracts from traditional plant species such as *Pittosporum undulatum*, *Schinus molle*, *Melia azedarach* and *Syzygium paniculatum*. The resultant nanoparticles were proved to possess ability to reduce and act as an excellent bioremediation agent to remove hexavalent chromium contaminants from the environment (Truskewycz et al. 2018). Further, nanocomposites such as reduced graphene oxide-zinc oxide from grapefruit extract to degrade Rhodamine B dye (Ramanathan et al. 2019), iron oxide-silicon dioxide-copper oxide-silver from *Crataegus pentagyna* fruit extract to degrade methylene blue and Rhodamine b (Ebrahimzadeh et al. 2019), and copper-magnesium oxide from *Cassytha filiformis* for methylene blue, Congo red, 4-nitrophenol and 2, 4-dinitrophenylhydrazine degradation (Nasrollahzadeh et al. 2018) are fabricated to be a potential bioremediation agent. In addition, root (Yeganeh-Faal et al. 2017), seed (Gopalakrishnan et al. 2015), bark (Amiri et al. 2017) and flower extracts (Rostami-Vartooni et al. 2019) were also used to synthesize nanoparticles for bioremediation applications. However, stability and reusability are the major limitation for utilizing these nanoparticles as commercial and large-scale bioremediation agents.

6. ROLE OF METAL NANOPARTICLES IN MICROBES TO ENHANCE BIOREMEDIATION

There are numerous microbes that are used to eliminate environmental pollutants as bioremediation agents. Microbes possess the ability to convert complex environment pollutants to simple ions that can be utilized either for the microbe or other living organisms, especially plants, as nutrients to enhance their growth (Rajendran and Gunasekaran 2019). Recently, it was reported that the microbes that belong to species such as *Rhodococcus, Alcaligenes, Corynebacterium, Bacillus, Pseudomonas, Arthrobacter, Azotobacter, Mycobacterium, Flavobacterium, Nocardia* and *Methosinus* are used in the bioremediation of soils that are contaminated by the heavy metals, namely zinc, gold, lead, nickel, copper, arsenic, mercury, cadmium and chromium (Girma 2015). Likewise, microbes were also used in the bioremediation of environmental sites that are contaminated by oil spills (Villela et al. 2019), petroleum hydrocarbons (Varjani 2017), radioactive wastes (Roh et al. 2015), polycyclic aromatic compound contaminated soils (Biache et al. 2017) Biache et al. 2017 and wastewater with textile dyes (Kumar et al. 2016). However, the microbes require nutrients for their growth at initial stages to form colonies or biofilms for the conversion of toxic pollutants into nontoxic useful compounds (Kumar et al. 2018). Thus, nanoparticles can serve as nutrients for the initial growth of microbes during the bioremediation process. In addition, biosynthesized nanoparticles with the ability to act as bioremediation will reduce environment pollutants and serve as nutrients or activate certain enzymes in microbes to elevate their pollutant degradation ability.

Apart from nanoparticles activating toxic pollutant degradation ability in microbes, the heavy metals that are degraded and used as nutrients by microbes will be present as intracellular ions, which may help in the formation of intracellular nanoparticles. Thus, degradation of pollutants by microbes can also be useful in the formation of nanoparticles, which can be extracted and used for biomedical as well as other novel applications (Pollmann et al. 2006). Recently, microbial recovery of metallic nanoparticles from industrial waste was proved with mechanisms which is beneficial in environmental cleanup applications. Further, these metal nanoparticles possess ability similar to green synthesized nanoparticles with enhanced biological properties and less toxicity (Pat-Espadas and Cervantes 2018). Magnetosomes are the recent trends in the microbial bioremediation applications that are used to clean up heavily contaminated sites (Vargas et al. 2018). It was reported that certain bacteria possess certain genes with ability to produce magnetic iron oxide nanoparticles called magnetosome crystals, intracellularly, which will be beneficial in heavy metal removal, wastewater treatment and photocatalytic degradation of pollutants (Tajer-Mohammad-Ghazvini et al. 2016). These magnetosome nanoparticles are formed when the growth medium of these bacteria contains a high concentration of iron ions. These magnetic nanoparticles can be extracted from the bacteria as biogenic magnetic nanoparticles that can be employed in bioremediation and biomedical applications (Dieudonné et al. 2019).

Moreover, bacteria with magnetosomes will help as coagulating agents to agglomerate heavy metals present in the contaminated site and remediate them. In addition, these magnetosomes can be removed from the site via magnets which can be reused for further bioremediation (Arakaki et al. 2018). These magnetosomes will be the future of microbial bioremediation applications to transform the contaminated site into a cleaner habitable environment as shown in Fig. 1.

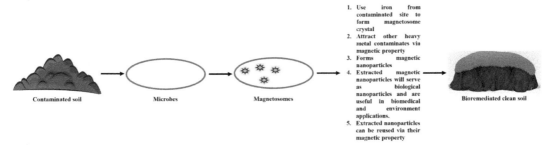

Figure 1   Schematics of magnetosomes and their application in bioremediation of contaminated soil.

7. FUTURE PERSPECTIVE

Currently, nanoparticles are used as standalone bioremediation agents with photocatalytic degradation of pollutants, antimicrobial, wastewater treatment ability and as nutrients and enzyme triggering agents among microbes to act as bioremediation agents. In future nanoformulation consisting of a combination of different types of nanoparticles will replace the standalone nanoparticles in environment. These nanoformulations will be prepared by polymers extracted from microbes or plants or dendrimers with biomolecules. Further, the nanoformulation will contain compartments to hold various nanoparticles with distinct properties such as antimicrobial, photocatalytic degradation, biofertilizer and enzyme-triggering abilities. The application of these biogenic nanoformulation will degrade in the contaminated environment to release the specific nanoparticles. The nanoparticle with antimicrobial property will help in inhibiting toxic microbes

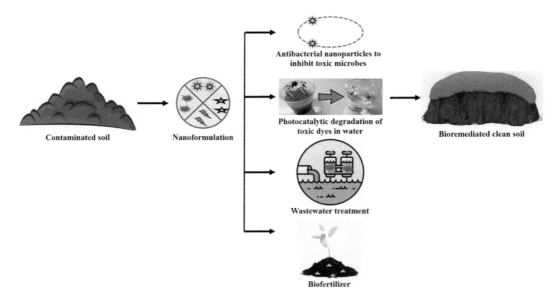

Figure 2   Schematic representation of nanoformulations in the environmental bioremediation

from the environment. Likewise, nanoparticle with photocatalytic degradation ability will reduce toxic organic dyes from the environment using solar irradiation. Further, nanoparticles with wastewater treatment ability will help to coagulate pollutants present in surface or ground water bodies and reduce their effective toxicity by removing them. Furthermore, nanoparticles can also serve as biofertilizers by dissolving in the contaminated site, serving as nutrients for the growth of plants to reduce pollutants via phytoremediation process and helping microbes in bioremediation by triggering specific enzymes. Thus, a single nanoformulated particle will clean the contaminated site and transform them into a habitable environment, as shown in Fig. 2, instead of several nanoparticles. In future, these nanoformulated agents will replace the conventional bioremediation nanoparticles to elevate their efficiency in bioremediation applications. Moreover, synthesizing each encapsulated nanoparticle via biological approach and coating gold nanoparticles to detect contamination to release specific concentration of nanoparticles will enhance the effectiveness of the nanoformulated bioremediation agents in future for environmental applications.

8. CONCLUSION

The present chapter is an overview of different metal nanoparticles that are used in bioremediation and the role of metal nanoparticles synthesized via biological agents such as microbes and plants, as an efficient, non-toxic bioremediation agent. In addition, the significance of biosynthesized metal nanoparticles and its ability in altering microbes and the use of magnetosomes for efficient bioremediation were also discussed. These metal nanoparticles will reduce the limitations of conventional microbial and phytoremediation methods and replace them to effectively reduce pollutants in environments. Further, the future of bioremediation will be the nanoformulated bioremediation agents with potential to encapsulate several nanoparticles with environmental bioremediation property to clean up the contaminated sites.

References

Abdennouri, M., M. Baâlala, A. Galadi, M. El-Makhfouk, M. Bensitel, K. Nohair, et al. 2016. Photocatalytic degradation of pesticides by titanium dioxide and titanium pillared purified clays. Arabian J. Chem. 9: S313–S318.

Abdennouri, M., A. Elhalil, M. Farnane, H. Tounsadi, F.Z. Mahjoubi, R. Elmoubarki, et al. 2015. Photocatalytic degradation of 2,4-D and 2,4-DP herbicides on Pt/TiO$_2$ nanoparticles. J. Saudi Chem. Soc. 19(5): 485–493.

Abdullah, M.M., M.A. Atta, A.H. Allohedan, Z.H. Alkhathlan, M. Khan and O.A. Ezzat. 2018. Green synthesis of hydrophobic magnetite nanoparticles coated with plant extract and their application as petroleum oil spill collectors. Nanomaterials. 8(10): 855. doi: 10.3390/nano8100855.

Albay, C., M. Koç, İ. Altın, R. Bayrak, İ. Değirmencioğlu and M. Sökmen. 2016. New dye sensitized photocatalysts: copper(II)-phthalocyanine/TiO$_2$ nanocomposite for water remediation. J. Photochem. Photobiol. A: Chem. 324: 117–125.

Ali, I., C. Peng, Z.M. Khan, M. Sultan, I. Naz, M. Ali, et al. 2019. Removal of crystal violet and eriochrome black T dyes from aqueous solutions by magnetic nanoparticles biosynthesized from leaf extract of *Fraxinus Chinensis Roxb*. Polish J. Environ. Stud. 28(4): 2027–2040.

Alizadeh Fard, M., B. Aminzadeh and H. Vahidi. 2013. Degradation of petroleum aromatic hydrocarbons using TiO$_2$ nanopowder film. Environ. Technol. (9): 1183–1190.

Almeelbi, T. and A. Bezbaruah. 2012. Aqueous phosphate removal using nanoscale zero-valent iron. J. Nanopart. Res. 14(7): 900–909.

Amiri, M., M. Salavati-Niasari, A. Akbari and T. Gholami .2017. Removal of malachite green (a toxic dye) from water by cobalt ferrite silica magnetic nanocomposite: Herbal and green sol-gel autocombustion synthesis. Int. J. Hydrogen Energy. 42(39): 24846–24860.

An, B. and D. Zhao. 2012. Immobilization of As(III) in soil and groundwater using a new class of polysaccharide stabilized Fe–Mn oxide nanoparticles. J. Hazard. Mat. 211: 332–341.

Andersen, B., I. Dosen, A.M. Lewinska and K.F. Nielsen. 2017. Pre-contamination of new gypsum wallboard with potentially harmful fungal species. Indoor Air. 27(1): 6–12.

Andra, S., S.K. Balu, J. Jeevanandham, M. Muthalagu, M. Vidyavathy, Y.S. Chan, et al. 2019. Phytosynthesized metal oxide nanoparticles for pharmaceutical applications. Naunyn-Schmiedeberg's Arch. Pharmacol. 18: 1–17.

Apte, M., P. Chaudhari, A. Vaidya, A.R. Kumar and S. Zinjarde. 2016. Application of nanoparticles derived from marine Staphylococcus lentus in sensing dichlorvos and mercury ions. Colloid Surface A: Physicochem. Eng. Aspects 501: 1–8.

Arakaki, A., M. Tanaka and T. Matsunaga. 2018. Molecular mechanism of magnetic crystal formation in magnetotactic bacteria. *In*: T. Matsunaga, T. Tanaka and D. Kisailus (eds), Biological Magnetic Materials and Applications. Springer Singapore, pp. 23–51.

Atarod, M., M. Nasrollahzadeh and S.M. Sajadi. 2015. Green synthesis of a Cu/reduced graphene oxide/ Fe_3O_4 nanocomposite using *Euphorbia wallichii* leaf extract and its application as a recyclable and heterogeneous catalyst for the reduction of 4-nitrophenol and rhodamine B. RSC Advances 5(111): 91532–91543.

Atarod, M., M. Nasrollahzadeh and S. Mohammad Sajadi. 2016a. Euphorbia heterophylla leaf extract mediated green synthesis of Ag/TiO_2 nanocomposite and investigation of its excellent catalytic activity for reduction of variety of dyes in water. J. Colloid Interface Sci. 462: 272–279.

Atarod, M., M. Nasrollahzadeh and S. Mohammad Sajadi. 2016b. Green synthesis of $Pd/RGO/Fe_3O_4$ nanocomposite using *Withania coagulans* leaf extract and its application as magnetically separable and reusable catalyst for the reduction of 4-nitrophenol. J. Colloid Interface Sci. 465: 249–258.

Aziz, N., M. Faraz, R. Pandey, M. Shakir, T. Fatma, A. Varma, et al. 2015. Facile algae-derived route to biogenic silver nanoparticles: Synthesis, antibacterial, and photocatalytic properties. Langmuir. 31(42): 11605–11612.

Balasubramanian, S., S.M.J. Kala, T.L. Pushparaj and P. Kumar. 2019. Biofabrication of gold nanoparticles using *Cressa cretica* leaf extract and evaluation of catalytic and antibacterial efficacy. Nano. Biomed. Eng. 11(1): 58–66.

Bedi, A., B.R. Singh, S.K. Deshmukh, A. Adholeya and C.J. Barrow. 2018. An *Aspergillus aculateus* strain was capable of producing agriculturally useful nanoparticles via bioremediation of iron ore tailings. J. Environ. Manag. 100–107.

Bhakya, S., S. Muthukrishnan, M. Sukumaran, M. Muthukumar, S.T. Kumar and M.V. Rao. 2015. Catalytic degradation of organic dyes using synthesized silver nanoparticles: a green approach. J. Biomed. Biodegrd. 6(5): 11–19.

Bhandari, G. 2018. Environmental nanotechnology: Applications of nanoparticles for bioremediation. *In*: R. Prasad, A. Aranda (eds), Approaches in Bioremediation. Springer, Swizerland, pp. 301–315.

Bhatia, S. 2016. Nanoparticles types, classification, characterization, fabrication methods and drug delivery applications. *In*: Natural Polymer Drug Delivery Systems. Springer, Cham, pp. 33–93.

Biache, C., S. Ouali, A. Cébron, C. Lorgeoux, S. Colombano and P. Faure. 2017. Bioremediation of PAH-contamined soils: Consequences on formation and degradation of polar-polycyclic aromatic compounds and microbial community abundance. J. Hazard. Mat. 329: 1–10.

Bokare, A.D., R.C. Chikate, C.V. Rode and K.M. Paknikar. 2008. Iron-nickel bimetallic nanoparticles for reductive degradation of azo dye Orange G in aqueous solution. Appl. Catalysis B: Environ. 79(3): 270–278.

Carvalho, S.S. and N.M. Carvalho. 2017. Dye degradation by green heterogeneous Fenton catalysts prepared in presence of *Camellia sinensis*. J. Environ. Manag. 187: 82–88.

Cecchin, I., K.R. Reddy, A. Thomé, E.F. Tessaro and F. Schnaid. 2017. Nanobioremediation: Integration of nanoparticles and bioremediation for sustainable remediation of chlorinated organic contaminants in soils. Int. Biodeteriorat. Biodegrad. 119: 419–428.

Chen, J., B. Hu and J. Zhi .2016. Optical and photocatalytic properties of Corymbia citriodora leaf extract synthesized ZnS nanoparticles. Physica E: Low-dimensional Sys. Nanostruc. 79: 103–106.

Choi, S., J.H. Drese, P.M. Eisenberger and C.W. Jones. 2011. Application of amine-tethered solid sorbents for direct CO_2 capture from the ambient air. Environ. Sci. Technol. 45(6): 2420–2427.

Cuevas, R., N. Durán, M.C. Diez, G.R. Tortella and O. Rubilar. 2015. Extracellular biosynthesis of copper and copper oxide nanoparticles by *Stereum hirsutum*, a native white-rot fungus from chilean forests. J. Nanomat. 16(1): 57–63.

Das, V.L., R. Thomas, R.T. Varghese, E.V. Soniya, J. Mathew and E.K. Radhakrishnan .2014. Extracellular synthesis of silver nanoparticles by the Bacillus strain CS 11 isolated from industrialized area. 3 Biotech 4(2): 121–126.

Dauthal, P. and M. Mukhopadhyay. 2013. Biosynthesis of palladium nanoparticles using Delonix regia leaf extract and its catalytic activity for nitro-aromatics hydrogenation. Ind. Eng. Chem. Res. 52(51): 18131–18139.

Dauthal, P. and M. Mukhopadhyay. 2016. Noble metal nanoparticles: Plant-mediated synthesis, mechanistic aspects of synthesis, and applications. Ind. Eng. Chem. Res. 55(36): 9557–9577.

Deshpande, M.V. 2019. Nanobiopesticide perspectives for protection and nutrition of plants. *In*: O. Koul (ed.), Nano-Biopesticides Today and Future Perspectives. Elsevier, pp. 47–68.

Devi, H.S. and T.D. Singh. 2014. Synthesis of copper oxide nanoparticles by a novel method and its application in the degradation of methyl orange. Adv. Electron Electr. Eng. 4(1): 83–88.

Dhandapani, P., S. Maruthamuthu and G. Rajagopal. 2012. Bio-mediated synthesis of TiO_2 nanoparticles and its photocatalytic effect on aquatic biofilm. J. Photochem. Photobiol. B: Biology 110: 43–49.

Dieudonné, A., D. Pignol and S. Prévéral. 2019. Magnetosomes: Biogenic iron nanoparticles produced by environmental bacteria. Appl. Microbiol. Biotechnol. 103(9): 3637–3649.

Ding, Q., G. Cheng, Y. Wang and D. Zhuang. 2017. Effects of natural factors on the spatial distribution of heavy metals in soils surrounding mining regions. Sci. Total Environ. 578: 577–585.

Dixit, S. and J.G. Hering. 2003. Comparison of arsenic (V) and arsenic (III) sorption onto iron oxide minerals: Implications for arsenic mobility. Environ. Sci. Technol. 37(18): 4182–4189.

Dong, C., J. Lu, B. Qiu, B. Shen, M. Xing and J. Zhang. 2018. Developing stretchable and graphene-oxide-based hydrogel for the removal of organic pollutants and metal ions. Catalysis B: Environ. 222: 146–156.

Drese, J.H., A.D. Talley and C.W. Jones. 2011. Aminosilica materials as adsorbents for the selective removal of aldehydes and ketones from simulated bio-oil. ChemSusChem. 4(3): 379–385.

Durán, N., P.D. Marcato, M. Durán, A. Yadav, A. Gade and M. Rai. 2011. Mechanistic aspects in the biogenic synthesis of extracellular metal nanoparticles by peptides, bacteria, fungi, and plants. Appl. Microbiol. Biotechnol. 90(5): 1609–1624.

Ebrahimzadeh, M.A., S. Mortazavi-Derazkola and M.A. Zazouli. 2019. Eco-friendly green synthesis and characterization of novel $Fe_3O_4/SiO_2/Cu_2O$–Ag nanocomposites using *Crataegus pentagyna* fruit extract for photocatalytic degradation of organic contaminants. J. Mat Sci. Mat. Electron. 30(12): 10994–11004.

El-Temsah, Y.S., A. Sevcu, K. Bobcikova, M. Cernik and E.J. Joner. 2016. DDT degradation efficiency and ecotoxicological effects of two types of nano-sized zero-valent iron (nZVI) in water and soil. Chemosphere 144: 2221–2228.

Elumalai, K., S. Velmurugan, S. Ravi, V. Kathiravan and S. Ashokkumar. 2015. Bio-fabrication of zinc oxide nanoparticles using leaf extract of curry leaf (*Murraya koenigii*) and its antimicrobial activities. Mater. Sci. Semicond. Process. 34: 365–372.

Fawcett, D., J.J. Verduin, M. Shah, S.B. Sharma and G.E.J. Poinern. 2017. A review of current research into the biogenic synthesis of metal and metal oxide nanoparticles via marine algae and seagrasses. J. Nanoscience. Article ID 8013850. https://doi.org/10.1155/2017/8013850

Filippelli, G.M. and M.P. Taylor. 2018. Addressing pollution-related global environmental health burdens. GeoHealth 2(1): 2–5.

Fujiwara, K., K. Okuyama and S.E. Pratsinis. 2017. Metal–support interactions in catalysts for environmental remediation. Environ. Sci.: Nano. 4(11): 2076–2092.

García, F.E., J. Plaza-Cazón, V.N. Montesinos, E.R. Donati and M.I. Litter. 2018. Combined strategy for removal of Reactive Black 5 by biomass sorption on *Macrocystis pyrifera* and zerovalent iron nanoparticles. J. Environ. Manage. 207: 70–79.

Ge, M., C. Cao, J. Huang, S. Li, Z. Chen, K.Q. Zhang, et al. 2016. A review of one-dimensional TiO_2 nanostructured materials for environmental and energy applications. J. Materials Chem. A 4(18): 6772–6801.

Ghanei Ardekani, J. and Z. Hassani. 2018. Study of the environmental impacts of nitrate pollution and its removal by nanoscale zero-valent iron (nZVI) at the south of Shahre-Kord aquifer (Chaharmahal and Bakhtiari province, Iran). Arabian J. Geosci. 11(22): 708–712.

Gil-Díaz, M., J. Alonso, E. Rodríguez-Valdés, J. Gallego and M.C. Lobo. 2017. Comparing different commercial zero valent iron nanoparticles to immobilize As and Hg in brownfield soil. Sci. Total Environ. 584: 1324–1332.

Girma, G. 2015. Microbial bioremediation of some heavy metals in soils: An updated review. Indian J. Sci. Res. 6(1): 147.

Godiya, C.B., X. Cheng, D. Li, Z. Chen and X. Lu. 2019. Carboxymethyl cellulose/polyacrylamide composite hydrogel for cascaded treatment/reuse of heavy metal ions in wastewater. J. Hazard. Mat. 364: 28–38.

Gomes, H.I., L.M. Ottosen, A.B. Ribeiro and C. Dias-Ferreira. 2015. Treatment of a suspension of PCB contaminated soil using iron nanoparticles and electric current. J. Environ. Manage. 151: 550–555.

Gong, N., K. Shao, W. Feng, Z. Lin, C. Liang and Y. Sun. 2011. Biotoxicity of nickel oxide nanoparticles and bio-remediation by microalgae *Chlorella vulgaris*. Chemosphere 83(4): 510–516.

Gopalakrishnan, R., B. Loganathan and K. Raghu. 2015. Green synthesis of Au–Ag bimetallic nanocomposites using *Silybum marianum* seed extract and their application as a catalyst. RSC Advances. 5(40): 31691–31699.

Gorai, S. 2018. Bio-based synthesis and applications of SnO_2 nanoparticles-an overview. J. Env Mat Sci. 9(10): 2894–2903.

Goutam, S.P., G. Saxena, V. Singh, A.K. Yadav, R.N. Bharagava and K.B. Thapa. 2018. Green synthesis of TiO_2 nanoparticles using leaf extract of *Jatropha curcas* L. for photocatalytic degradation of tannery wastewater. Chem. Eng. J. 336: 386–396.

Gu, H., X. Chen, F. Chen, X. Zhou and Z. Parsaee. 2018. Ultrasound-assisted biosynthesis of CuO-NPs using brown alga Cystoseira trinodis: Characterization, photocatalytic AOP, DPPH scavenging and antibacterial investigations. Ultrasonics Sonochem. 41: 109–119.

Guerra, F., M. Attia, D. Whitehead and F. Alexis. 2018. Nanotechnology for environmental remediation: materials and applications. Molecules. 23(7): 1760–1765.

Guo, J., S. Ci, P. Cai and Z. Wen. 2019. Loading NiCo alloy nanoparticles onto nanocarbon for electrocatalytic conversion of arsenite into arsenate. Electrochem. Commun. 104: 106477.

Gurusamy, S., M.R. Kulanthaisamy, D.G. Hari, A. Veleeswaran, B. Thulasinathan, J.B. Muthuramalingam, et al. 2019. Environmental friendly synthesis of TiO_2–ZnO nanocomposite catalyst and silver nanomaterials for the enhanced production of biodiesel from *Ulva lactuca* seaweed and potential antimicrobial properties against the microbial pathogens. J. Photochem. Photobiol. B: Biology 193: 118–130.

Haider, A.J., R.H. Al-Anbari, G.R. Kadhim and C.T. Salame. 2017. Exploring potential environmental applications of TiO_2 nanoparticles. Energy Procedia. 119: 332–345.

Hao, F., W. Guo, A. Wang, Y. Leng and H. Li. 2014. Intensification of sonochemical degradation of ammonium perfluorooctanoate by persulfate oxidant. Ultrasonics sonochem. 21(2): 554–558.

Hazen, T.C. 2018. Bioremediation. Microbiology of the Terrestrial Deep Subsurface. CRC Press, pp. 247–266.

He, K., G. Chen, G. Zeng, Z. Huang, Z. Guo, T. Huang, et al. 2017. Applications of white rot fungi in bioremediation with nanoparticles and biosynthesis of metallic nanoparticles. Appl. Microbiol. Biotechnol. 101(12): 4853–4862.

Huang, D., C. Hu, G. Zeng, M. Cheng, P. Xu, X. Gong, et al. 2017. Combination of Fenton processes and biotreatment for wastewater treatment and soil remediation. Sci. Total Environ. 574: 1599–1610.

Huang, H.Y., R.T. Yang, D. Chinn and C.L. Munson. 2003. Amine-grafted MCM-48 and silica xerogel as superior sorbents for acidic gas removal from natural gas. Industrial Eng. Chem. Res. 42(12): 2427–2433.

Huang, Z., Z. Zeng, A. Chen, G. Zeng, R. Xiao, P. Xu, et al. 2018. Differential behaviors of silver nanoparticles and silver ions towards cysteine: Bioremediation and toxicity to *Phanerochaete chrysosporium*. Chemosphere 203: 199–208.

Isa, N. and Z. Lockman. 2019. Methylene blue dye removal on silver nanoparticles reduced by Kyllinga brevifolia. Environ. Sci. Poll. Res. 26(11): 11482–11495.

Jang, Y.J., C. Simer and T. Ohm. 2006. Comparison of zinc oxide nanoparticles and its nano-crystalline particles on the photocatalytic degradation of methylene blue. Mater. Res. Bull. 41(1): 67–77.

Jayaseelan, C., A.A. Rahuman, A.V. Kirthi, S. Marimuthu, T. Santhoshkumar, A. Bagavan, et al. 2012. Novel microbial route to synthesize ZnO nanoparticles using Aeromonas hydrophila and their activity against pathogenic bacteria and fungi. Spectrochimica Acta Part A: Mol. Biomol. Spectroscopy 90: 78–84.

Jeevanandam, J., Y.S. Chan and M.K. Danquah. 2016. Biosynthesis of metal and metal oxide nanoparticles. Chem. Biol. Eng. Rev. 3(2): 55–67.

Jeong, H.Y., B. Klaue, J.D. Blum and K.F. Hayes. 2007. Sorption of mercuric ion by synthetic nano-crystalline mackinawite (FeS). Environ. Sci. Technol. 41(22): 7699–7705.

Jha, A.K., K. Prasad and A.R. Kulkarni. 2009. Synthesis of TiO_2 nanoparticles using microorganisms. Colloid. Surface. B: Biointerface. 71(2): 226–229.

Kamat, P.V. and D. Meisel. 2003. Nanoscience opportunities in environmental remediation. C.R. Chim. 6(8–10): 999–1007.

Kapoor, A. and T. Viraraghavan. 1997. Nitrate removal from drinking water: review. J Environ Eng 123(4): 371–380.

Khalafi, T., F. Buazar and K. Ghanemi. 2019. Phycosynthesis and enhanced photocatalytic activity of zinc oxide nanoparticles toward organosulfur pollutants. Scien. Rep. 9(1): 6866.

Khan, A.U., M. Khan, N. Malik, M.H. Cho and M.M. Khan. 2019. Recent progress of algae and blue-green algae-assisted synthesis of gold nanoparticles for various applications. Bioprocess Biosyst. Eng. 42(1): 1–15.

Khan, I., K. Saeed and I. Khan. 2017. Nanoparticles: Properties, applications and toxicities. Arabian J. Chem. 12(7): 908–931.

Kitching, M., M. Ramani and E. Marsili. 2015. Fungal biosynthesis of gold nanoparticles: Mechanism and scale up. Microbial. Biotechnol. 8(6): 904–917.

Konishi, Y., K. Ohno, N. Saitoh, T. Nomura, S. Nagamine, H. Hishida, et al. 2007. Bioreductive deposition of platinum nanoparticles on the bacterium *Shewanella* algae. J. Biotechnol. 128(3): 648–653.

Koushik, D., S. Sen Gupta, S.M. Maliyekkal and T. Pradeep. 2016. Rapid dehalogenation of pesticides and organics at the interface of reduced graphene oxide-silver nanocomposite. J. Hazard. Mat. 308: 192–198.

Kumar, P., S. Senthamil Selvi, A. Lakshmi Prabha, K. Prem Kumar, R.S. Ganeshkumar and M. Govindaraju. 2012. Synthesis of silver nanoparticles from *Sargassum tenerrimum* and screening phytochemicals for its antibacterial activity. Nano. Biomed. Eng. 4(1): 12–16.

Kumar, P., M. Govindaraju, S. Senthamilselvi and K. Premkumar. 2013. Photocatalytic degradation of methyl orange dye using silver (Ag) nanoparticles synthesized from *Ulva lactuca*. Colloid. Surface. B: Biointerface. 103: 658–661.

Kumar, S.S., S. Shantkriti, T. Muruganandham, E. Murugesh, N. Rane and S.P. Govindwar. 2016. Bioinformatics aided microbial approach for bioremediation of wastewater containing textile dyes. Ecol. Informatics. 31: 112–121.

Kumar, A., S. Devi and D. Singh. 2018. Significance and approaches of microbial bioremediation in sustainable development. *In*: J. Singh, D. Sharma, G. Kumar and N. Sharma (eds), Microbial Bioprospecting for Sustainable Development. Springer, Singapore, pp. 93–114.

Lamba, R., A. Umar, S.K. Mehta and S.K. Kansal. 2015. ZnO doped SnO_2 nanoparticles heterojunction photo-catalyst for environmental remediation. J. Alloys Comp. 653: 327–333.

Lassoued, A., M.S. Lassoued, B. Dkhil, S. Ammar and A. Gadri. 2018. Photocatalytic degradation of methylene blue dye by iron oxide (α-Fe$_2$O$_3$) nanoparticles under visible irradiation. J. Mat. Sci. Mat. Electron. 29(10): 8142–8152.

Latha, D., P. Prabu, G. Gnanamoorthy, S. Sampurnam, R. Manikandan, C. Arulvasu, et al. 2019. Facile *Justicia adhatoda* leaf extract derived route to silver nanoparticle: Synthesis, characterization and its application in photocatalytic and anticancer activity. Mater. Res. Exp. 6(4): 045003.

Le, T.S., T.H. Dao, D.C. Nguyen, H.C. Nguyen and I.L. Balikhin. 2015. Air purification equipment combining a filter coated by silver nanoparticles with a nano-TiO$_2$ photocatalyst for use in hospitals. Adv. Nat. Sci. Nanosci. Nanotechnol. 6(1): 015016.

Li, D., F. Cui, Z. Zhao, D. Liu, Y. Xu, H. Li and X. Yang. 2014. The impact of titanium dioxide nanoparticles on biological nitrogen removal from wastewater and bacterial community shifts in activated sludge. Biodegrad. 25(2): 167–177.

Li, Q., S. Mahendra, D.Y. Lyon, L. Brunet, M.V. Liga, D. Li and P.J. Alvarez. 2008. Antimicrobial nanomaterials for water disinfection and microbial control: Potential applications and implications. Water Res. 42(18): 4591–4602.

Liang, Q. and D. Zhao. 2014. Immobilization of arsenate in a sandy loam soil using starch-stabilized magnetite nanoparticles. J. Hazard. Mat. 271: 16–23.

Lin, L., Z. Song, Y. Huang, Z. H. Khan and W. Qiu. 2019. Removal and oxidation of arsenic from aqueous solution by biochar impregnated with Fe–Mn oxides. Water Air Soil Poll. 230(5): 105.

Liu, A., J. Ming R.O. Ankumah. 2005. Nitrate contamination in private wells in rural Alabama, United States. Sci Tot Environ 346(1–3):112–120

Liu, W., S. Tian, X. Zhao, X. Xie, Y. Gong and D. Zhao. 2015. Application of stabilized nanoparticles for in situ remediation of metal-contaminated soil and groundwater: A critical review. Curr. Poll. Reports 1(4): 280–291.

Liu, X., Y. Huang, S. Duan, Y. Wang, J. Li, Y. Chen, et al. 2016. Graphene oxides with different oxidation degrees for Co(II) ion pollution management. Chem. Eng. J. 302: 763–772.

Liu, L., M. Bilal, X. Duan and H.M.N. Iqbal. 2019. Mitigation of environmental pollution by genetically engineered bacteria–current challenges and future perspectives. Sci. Total Environ. 667:444–454.

López-Miranda, J.L., R. Esparza, G. Rosas, R. Pérez and M. Estévez-González. 2019. Catalytic and antibacterial properties of gold nanoparticles synthesized by a green approach for bioremediation applications. 3 Biotech. 9(4): 135–139.

Luo, M.L., J.Q. Zhao, W. Tang and C.S. Pu. 2005. Hydrophilic modification of poly (ether sulfone) ultrafiltration membrane surface by self-assembly of TiO$_2$ nanoparticles. Appl. Surface Sci. 249(1–4): 76–84.

Ma, Q., Y. Yu, M. Sindoro, A.G. Fane, R. Wang and H. Zhang. 2017. Carbon-based functional materials derived from waste for water remediation and energy storage. Adv. Mat. 29(13): 1605361.

Mahdavi, M., F. Namvar, M. Ahmad and R. Mohamad. 2013. Green biosynthesis and characterization of magnetic iron oxide (Fe$_3$O$_4$) nanoparticles using seaweed (*Sargassum muticum*) aqueous extract. Molecules 18(5): 5954–5964.

Mahdieh, M., A. Zolanvari and A.S. Azimee. 2012. Green biosynthesis of silver nanoparticles by *Spirulina platensis*. Scien. Iran. 19(3): 926–929.

Mahmoud, E., A. Saad, M.A. Soliman and M.S. Abdelwahab. 2017. Encapsulation of nano zerovalent iron with ethylenediamine and diethylenetriamine for removing cobalt and zinc and their radionuclides from water. J. Environ. Chem. Eng. 5(5): 5157–5168.

Mahmoud, E., A. Saad, M.A. Soliman and M.S. Abdelwahab. 2019. Environmental water remediation using covalently functionalized zerovalent iron nanocomposites with 2-pyridinecarboxaldehyde via 3-aminopropyltrimethoxysilane and ethylenediamine. Sep. Sci. Technol. 54(7): 1125–1140.

Maliyekkal, S.M., T. Sreeprasad, D. Krishnan, S. Kouser, A.K. Mishra, U.V. Waghmare, et al. 2013. Graphene: A reusable substrate for unprecedented adsorption of pesticides. Small. 9(2): 273–283.

Mammadov, G., M. Ramazanov, A. Kanaev, U. Hasanova and K. Huseynov. 2017. Photocatalytic degradation of organic pollutants in air by application of titanium dioxide nanoparticles. Chem. Eng. Transact. 60: 241–246.

Manatunga, D.C., R.M. de Silva, K.M. Nalin de Silva, N. de Silva and E.V.A. Premalal. 2018. Metal and polymer-mediated synthesis of porous crystalline hydroxyapatite nanocomposites for environmental remediation. Royal Soc. Open Sci. 5(1): 171557.

Maryami, M., M. Nasrollahzadeh, E. Mehdipour and S.M. Sajadi. 2016. Preparation of the Ag/RGO nanocomposite by use of *Abutilon hirtum* leaf extract: A recoverable catalyst for the reduction of organic dyes in aqueous medium at room temperature. Int. J. Hydrog. Energy. 41(46): 21236–21245.

McCann, C.M., C.L. Peacock, K.A. Hudson-Edwards, T. Shrimpton, N.D. Gray and K.L. Johnson. 2018. In situ arsenic oxidation and sorption by a Fe–Mn binary oxide waste in soil. J. Hazard. Mat. 342: 724–731.

McNamara, K. and S.A.M. Tofail. 2017. Nanoparticles in biomedical applications. Adv. Phy. X 2(1): 54–88.

Menon, S., S. Rajeshkumar and S. Venkat Kumar. 2017. A review on biogenic synthesis of gold nanoparticles, characterization, and its applications. Resour. Effic. Technol. 3(4): 516–527.

Momin, B., S. Rahman, N. Jha and U.S. Annapure. 2019. Valorization of mutant *Bacillus licheniformis* M09 supernatant for green synthesis of silver nanoparticles: Photocatalytic dye degradation, antibacterial activity, and cytotoxicity. Bioprocess Biosyst. Eng. 42(4): 541–553.

Mondal, K. and A. Sharma. 2016. Recent advances in the synthesis and application of photocatalytic metal–metal oxide core-shell nanoparticles for environmental remediation and their recycling process. RSC Adv. 6(87): 83589–83612.

Mueller, N.C. and B. Nowack. 2010. Nanoparticles for remediation: Solving big problems with little particles. Elements 6(6): 395–400.

Mueller, N.C., J. Braun, J. Bruns, M. Černík, P. Rissing, D. Rickerby et al. 2012. Application of nanoscale zero valent iron (nZVI) for groundwater remediation in Europe. Environ. Sci. Poll. Res. 19(2): 550–558.

Mukherjee, T., S. Chakraborty, A.A. Biswas and T.K. Das. 2017. Bioremediation potential of arsenic by non-enzymatically biofabricated silver nanoparticles adhered to the mesoporous carbonized fungal cell surface of *Aspergillus foetidus* MTCC8876. J. Environ. Manag. 201: 435–446.

Murgueitio, E., L. Cumbal, M. Abril, A. Izquierdo, A. Debut and O. Tinoco. 2018. Green synthesis of iron nanoparticles: Application on the removal of petroleum oil from contaminated water and soils. J. Nanotechnol. Article ID 4184769 https://doi.org/10.1155/2018/4184769

Muthukumar, H. and M. Matheswaran. 2015. *Amaranthus spinosus* leaf extract mediated FeO nanoparticles: Physicochemical traits, photocatalytic and antioxidant activity. ACS Sustain. Chem. Eng. 3(12): 3149–3156.

Nabikhan, A., S. Rathinam and K. Kandasamy. 2018. Biogenic gold nanoparticles for reduction of 4-nitrophenol to 4-aminophenol: An eco-friendly bioremediation. IET Nanobiotechnol. 12, 479–483.

Narayanan, K.B., H.H. Park and S.S. Han. 2015. Synthesis and characterization of biomatrixed-gold nanoparticles by the mushroom *Flammulina velutipes* and its heterogeneous catalytic potential. Chemosphere. 141: 169–175.

Nasrollahzadeh, M., Z. Issaabadi and S.M. Sajadi. 2018. Green synthesis of a Cu/MgO nanocomposite by *Cassytha filiformis* L. extract and investigation of its catalytic activity in the reduction of methylene blue, congo red and nitro compounds in aqueous media. RSC Adv. 8(7): 3723–3735.

Nassar, N.N., N.N. Marei, G. Vitale and L.A. Arar. 2015. Adsorptive removal of dyes from synthetic and real textile wastewater using magnetic iron oxide nanoparticles: Thermodynamic and mechanistic insights. Can. J. Chem. Eng. 93(11): 1965–1974.

Nayak, S., S.P. Sajankila and C.V. Rao. 2018. Green synthesis of gold nanoparticles from banana pith extract and its evaluation of antibacterial activity and catalytic reduction of malachite green dye. The J. Microbiol. Biotechnol. Food Sci. 7(6): 641–647.

Nujić M. and M. Habuda-Stanić. 2017. Nitrates and nitrites, metabolism and toxicity. Food Health Dis. Sci. Pof. J. Nutri. Diet. 6(2): 48–89.

Park, T.J., K.G. Lee and S.Y. Lee. 2016. Advances in microbial biosynthesis of metal nanoparticles. Appl. Microbiol. Biotechnol. 100(2): 521–534.

Pat-Espadas, A.M. and F.J. Cervantes. 2018. Microbial recovery of metallic nanoparticles from industrial wastes and their environmental applications. J. Chem. Technol. Biotechnol. 93(11): 3091–3112.

Patil, S.S., U.U. Shedbalkar, A. Truskewycz, B.A. Chopade and A.S. Ball. 2016. Nanoparticles for environmental clean-up: A review of potential risks and emerging solutions. Environ. Technol. Innov. 5: 10–21.

Pattnaik, S. and S. Busi. 2018. Fungal-derived chitosan-based nanocomposites: A sustainable approach for heavy metal biosorption and environmental management, *In*: R. Prasad (ed.), Mycoremediation and Environmental Sustainability, Vol. 2. Springer International Publishing, Cham, pp. 325–349.

Paul, D., G. Pandey and R.K. Jain. 2005. Suicidal genetically engineered microorganisms for bioremediation: Need and perspectives. Bioessays 27(5): 563–573.

Peng, D., B. Wu, H. Tan, S. Hou, M. Liu, H. Tang, et al. 2019. Effect of multiple iron-based nanoparticles on availability of lead and iron, and micro-ecology in lead contaminated soil. Chemosphere 228: 44–53.

Penn, C.J. and J.G. Warren. 2009. Investigating phosphorus sorption onto kaolinite using isothermal titration calorimetry. Soil Sci. Soc. Am. J. 73(2): 560–568.

Pollmann, K., J. Raff, M. Merroun, K. Fahmy and S. Selenska-Pobell. 2006. Metal binding by bacteria from uranium mining waste piles and its technological applications. Biotechnol. Adv. 24(1): 58–68.

Puay, N.Q., G. Qiu and Y.P. Ting. 2015. Effect of Zinc oxide nanoparticles on biological wastewater treatment in a sequencing batch reactor. J. Clean. Product. 88: 139–145.

Pugazhendhi, A., R. Prabhu, K. Muruganantham, R. Shanmuganathan and S. Natarajan. 2019. Anticancer, antimicrobial and photocatalytic activities of green synthesized magnesium oxide nanoparticles (MgONPs) using aqueous extract of *Sargassum wightii*. J. Photochem. Photobiol. B: Biology 190: 86–97.

Qi, G., Y. Wang, L. Estevez, X. Duan, N. Anako, A.H.A. Park, et al. 2011. High efficiency nanocomposite sorbents for CO_2 capture based on amine-functionalized mesoporous capsules. Energy Environ. Sci. 4(2): 444–452.

Rai, M., I. Maliszewska, A. Ingle, I. Gupta and A. Yadav. 2015. Diversity of microbes in synthesis of metal nanoparticles: Progress and limitations. *In*: O.V. Singh (ed.), Bio-Nanoparticles: Biosynthesis and Sustainable Biotechnological Implications. Wiley, Chichester, pp. 1–30.

Rai, M., A. Ingle, S. Gaikwad, I. Gupta, A. Yadav, A. Gade, et al. 2016. Fungi: Myconanofactory, mycoremediation and medicine. *In*: S.K. Deshmukh, J.K. Misra, J.P. Tiwari and T. Papp (eds), Fungi: Applications and Management Strategies. CRC Press, USA, pp. 201–219.

Rajendran, P. and P. Gunasekaran. 2019. Microbial Bioremediation, MJP Publisher, Chennai, India.

Rajput, S., L.P. Singh, C.U. Pittman and D. Mohan. 2017. Lead (Pb^{2+}) and copper (Cu^{2+}) remediation from water using superparamagnetic maghemite (γ-Fe_2O_3) nanoparticles synthesized by Flame Spray Pyrolysis (FSP). J. Colloid Interface Sci. 492: 176–190.

Ramalingam, B., T. Parandhaman, P. Choudhary and S.K. Das. 2018. Biomaterial functionalized graphene-magnetite nanocomposite: A novel approach for simultaneous removal of anionic dyes and heavy-metal ions. ACS Sustain. Chem. Eng. 6(5): 6328–6341.

Ramanathan, S., S.P. Selvin, A. Obadiah, A. Durairaj, P. Santhoshkumar, S. Lydia, et al. 2019. Synthesis of reduced graphene oxide/ZnO nanocomposites using grape fruit extract and Eichhornia crassipes leaf extract and a comparative study of their photocatalytic property in degrading Rhodamine B dye. J. Environ. Health Sci. Eng. 17(1): 195–207.

Ramkumar, V.S., A. Pugazhendhi, S. Prakash, N.K. Ahila, G. Vinoj, S. Selvam, et al. 2017. Synthesis of platinum nanoparticles using seaweed *Padina gymnospora* and their catalytic activity as PVP/PtNPs nanocomposite towards biological applications. Biomed. Pharmacotherapy. 92: 479–490.

Roh, C., C. Kang and J.R. Lloyd. 2015. Microbial bioremediation processes for radioactive waste. Korean J. Chem. Eng. 32(9): 1720–1726.

Rostami-Vartooni, A., A. Moradi-Saadatmand, M. Bagherzadeh and M. Mahdavi. 2019. Green synthesis of Ag/Fe_3O_4/ZrO_2 nanocomposite using aqueous *Centaurea cyanus* flower extract and its catalytic application for reduction of organic pollutants. Iranian J. Catalysis. 9(1): 27–35.

Roy, K., C.K. Sarkar and C.K. Ghosh. 2015. Photocatalytic activity of biogenic silver nanoparticles synthesized using yeast (*Saccharomyces cerevisiae*) extract. Appl. Nanosci. 5(8): 953–959.

Saif Hasan, S., S. Singh, R.Y. Parikh, M.S. Dharne, M.S. Patole, B.L.V. Prasad, et al. 2008. Bacterial synthesis of copper/copper oxide nanoparticles. J. Nanosci. Nanotechnol. 8(6): 3191–3196.

Saif, S., A. Tahir and Y. Chen. 2016. Green synthesis of iron nanoparticles and their environmental applications and implications. Nanomaterials. 6(11): 209–214.

Saifuddin, N., C.W. Wong and A.A. Yasumira. 2009. Rapid biosynthesis of silver nanoparticles using culture supernatant of bacteria with microwave irradiation. J. Clean. Prod. 6(1): 61–70.

Sajjadi, M., M. Nasrollahzadeh and S. Mohammad Sajadi. 2017. Green synthesis of Ag/Fe$_3$O$_4$ nanocomposite using *Euphorbia peplus* Linn leaf extract and evaluation of its catalytic activity. J. Colloid Interface Sci. 497: 1–13.

Salvadori, M.R., R.A. Ando, C.A.O. Nascimento and B. Corrêa. 2018. Biosynthesis of metal nanoparticles via fungal dead biomass in industrial bioremediation process. *In*: R. Prasad, V. Kumar, M. Kumar and S. Wang (eds), Fungal Nanobionics: Principles and Applications. Springer Singapore, Singapore, pp. 165–199.

Salvadori, M.R., R.A. Ando, C.A. Oller Nascimento and B. Corrêa. 2015. Extra and intracellular synthesis of nickel oxide nanoparticles mediated by dead fungal biomass. PLOS ONE 10(6): e0129799.

Sarno, M., M. Casa, C. Cirillo and P. Ciambelli. 2017. Complete removal of persistent pesticide using reduced graphene oxide-silver nanocomposite. Chem. Eng. Transac. 60: 151–156.

Sharma, M., K. Behl, S. Nigam and M. Joshi. 2018. TiO$_2$-GO nanocomposite for photocatalysis and environmental applications: A green synthesis approach. Vacuum 156: 434–439.

Sharma, G., S. Bhogal, V.K. Gupta, S. Agarwal, A. Kumar, D. Pathania, et al. 2019. Algal biochar reinforced trimetallic nanocomposite as adsorptional/photocatalyst for remediation of malachite green from aqueous medium. J. Mol. Liq. 275: 499–509.

Singh, R., P. Wagh, S. Wadhwani, S. Gaidhani, A. Kumbhar, J. Bellare, et al. 2013. Synthesis, optimization, and characterization of silver nanoparticles from *Acinetobacter calcoaceticus* and their enhanced antibacterial activity when combined with antibiotics. Int. J. Nanomed. 8: 4277–4281.

Singh, R., U.U. Shedbalkar, S.A. Wadhwani and B.A. Chopade. 2015. Bacteriagenic silver nanoparticles: Synthesis, mechanism, and applications. Appl. Microbiol. Biotechnol. 99(11): 4579–4593.

Singh, P., Y.J. Kim, D. Zhang and D.C. Yang. 2016a. Biological synthesis of nanoparticles from plants and microorganisms. Trends Biotechnol. 34(7): 588–599.

Singh, P., Y.J. Kim and D.C. Yang. 2016b. A strategic approach for rapid synthesis of gold and silver nanoparticles by Panax ginseng leaves. Artificial cells, Nanomed. Biotechnol. 44(8): 1949–1957.

Singh, P., H. Singh, Y.J. Kim, R. Mathiyalagan, C. Wang and D.C. Yang. 2016c. Extracellular synthesis of silver and gold nanoparticles by *Sporosarcina koreensis* DC4 and their biological applications. Enzyme Microbial. Technol. 86: 75–83.

Singh, S., N. Kumar, M. Kumar, A. Agarwal and B. Mizaikoff. 2017. Electrochemical sensing and remediation of 4-nitrophenol using bio-synthesized copper oxide nanoparticles. Chem. Eng. J. 313: 283–292.

Singhal, A. and A. Gupta. 2019. Sustainable synthesis of silver nanoparticles using exposed X-ray sheets and forest-industrial waste biomass: Assessment of kinetic and catalytic properties for degradation of toxic dyes mixture. J. Environ. Manage. 247: 698–711.

Sinha, A., V.N. Singh, B.R. Mehta and S.K. Khare. 2011. Synthesis and characterization of monodispersed orthorhombic manganese oxide nanoparticles produced by *Bacillus* sp. cells simultaneous to its bioremediation. J. Hazard. Mat. 192(2): 620–627.

Sowani, H., P. Mohite, H. Munot, Y. Shouche, T. Bapat, A.R. Kumar, et al. 2016. Green synthesis of gold and silver nanoparticles by an actinomycete *Gordonia amicalis* HS-11: Mechanistic aspects and biological application. Process Biochem. 51(3): 374–383.

Srivastava, P., J. Bragança, S.R. Ramanan and M. Kowshik. 2013. Synthesis of silver nanoparticles using haloarchaeal isolate *Halococcus salifodinae* BK 3. Extremophiles 17(5): 821–831.

Stark, W.J., P.R. Stoessel, W. Wohlleben and A. Hafner. 2015. Industrial applications of nanoparticles. Chem. Soc. Rev. 44(16): 5793–5805.

Su, R., Q. Li, R. Huang, L. Zhao, Q. Yue, B. Gao, et al. 2018. Biomass-based soft hydrogel for triple use: Adsorbent for metal removal, template for metal nanoparticle synthesis, and a reactor for nitrophenol and methylene blue reduction. J. Taiwan Inst. Chem. Eng. 91: 235–242.

Su, S., Y. Liu, W. He, X. Tang, W. Jin and Y. Zhao. 2019. A novel graphene oxide-carbon nanotubes anchored α-FeOOH hybrid activated persulfate system for enhanced degradation of Orange II. J. Environ. Sci. 83: 73–84.

Sun, Y., Q. Wang, C. Chen, X. Tan and X. Wang. 2012. Interaction between Eu (III) and graphene oxide nanosheets investigated by batch and extended X-ray absorption fine structure spectroscopy and by modeling techniques. Environ. Sci. Technol. 46(11): 6020–6027.

Suominen, K., M. Verta and S. Marttinen. 2014. Hazardous organic compounds in biogas plant end products-Soil burden and risk to food safety. Sci. Total Environ. 491: 192–199.

Sureshkumar, M., D.Y. Siswanto and C.K. Lee. 2010. Magnetic antimicrobial nanocomposite based on bacterial cellulose and silver nanoparticles. J. Mat. Chem. 20(33): 6948–6955.

Sutherland, T.D., I. Horne, K.M. Weir, C.W. Coppin, M.R. Williams, M. Selleck, et al. 2004. Enzymatic bioremediation: From enzyme discovery to applications. Clinical Exptl. Pharmacol. Physiol. 31(11): 817–821.

Tajbakhsh, M., H. Alinezhad, M. Nasrollahzadeh and T.A. Kamali. 2016. Green synthesis of the Ag/HZSM-5 nanocomposite by using *Euphorbia heterophylla* leaf extract: A recoverable catalyst for reduction of organic dyes. J. Alloys Comp. 685: 258–265.

Tajer-Mohammad-Ghazvini, P., R. Kasra-Kermanshahi, A. Nozad-Golikand, M. Sadeghizadeh, S. Ghorbanzadeh-Mashkani and R. Dabbagh. 2016. Cobalt separation by *Alphaproteobacterium* MTB-KTN90: Magnetotactic bacteria in bioremediation. Bioprocess Biosyst. Eng. 39(12): 1899–1911.

Tan, J.M., G. Qiu and Y.P. Ting. 2015. Osmotic membrane bioreactor for municipal wastewater treatment and the effects of silver nanoparticles on system performance. J. Cleaner Prod. 88: 146–151.

Thangaraj, V., S. Mahmud, W. Li, F. Yang and H. Liu. 2018. Greenly synthesised silver-alginate nanocomposites for degrading dyes and bacteria. IET Nanobiotechnol. 12: 47–51.

Thirumurugan, A., E. Harshini, B. Deepika Marakathanandhini, S. Rajesh Kannan and P. Muthukumaran. 2017. Catalytic degradation of reactive red 120 by copper oxide nanoparticles synthesized by *Azadirachta indica*. *In*: M. Prashanthi, R. Sundaram, A. Jeyaseelan and T. Kaliannan (eds), Bioremediation and Sustainable Technologies for Cleaner Environment. Springer International Publishing, Cham, pp. 95–102.

Thomas, B., B.S.M. Vithiya, T.A.A. Prasad, S.B. Mohamed, C.M. Magdalane, K. Kaviyarasu, et al. 2019. Antioxidant and photocatalytic activity of aqueous leaf extract mediated green synthesis of silver nanoparticles using *Passiflora edulis* f. flavicarpa. J. Nanosci. Nanotechnol. 19(5): 2640–2648.

Tran Q.H., V.Q. Nguyen and A.T. Le. 2013. Silver nanoparticles: Synthesis, properties, toxicology, applications and perspectives. Adv. Nat. Sci. 4: 20. 10.1088/2043-6262/4/3/033001

Truppi, A., F. Petronella, T. Placido, V. Margiotta, G. Lasorella, L. Giotta, et al. 2019. Gram scale synthesis of UV-Vis light active plasmonic photocatalytic nanocomposite based on TiO_2/Au nanorods for degradation of pollutants in water. Appl. Catalysis B: Environ. 243: 604–613.

Truskewycz, A., R. Shukla and A.S. Ball. 2018. Phytofabrication of iron nanoparticles for hexavalent chromium remediation. ACS Omega. 3(9): 10781–10790.

Tsai, C.H., W.C. Chang, D. Saikia, C.E. Wu and H.M. Kao. 2016. Functionalization of cubic mesoporous silica SBA-16 with carboxylic acid via one-pot synthesis route for effective removal of cationic dyes. J. Hazard. Mat. 309: 236–248.

Turakhia, B., P. Turakhia and S. Shah. 2018. Green synthesis of zero valent iron nanoparticles from *Spinacia oleracea* (spinach) and its application in waste water treatment. J. Adv. Res. Appl. Sci. 5(1): 46–51.

Varadavenkatesan, T., R. Selvaraj and R. Vinayagam. 2016. Phyto-synthesis of silver nanoparticles from *Mussaenda erythrophylla* leaf extract and their application in catalytic degradation of methyl orange dye. J. Mol. Liquids. 221: 1063–1070.

Vargas, G., J. Cypriano, T. Correa, P. Leão, A.D. Bazylinski and F. Abreu. 2018. Applications of magnetotactic bacteria, magnetosomes and magnetosome crystals in biotechnology and nanotechnology: Mini-review. Molecules 23(10): E2438. doi:10.3390/molecules23102438

Varjani, S.J. 2017. Microbial degradation of petroleum hydrocarbons. Biores. Technol. 223: 277–286.

Vasantharaj, S., S. Sathiyavimal, M. Saravanan, P. Senthilkumar, K. Gnanasekaran, M. Shanmugavel, et al. 2019a. Synthesis of ecofriendly copper oxide nanoparticles for fabrication over textile fabrics: Characterization of antibacterial activity and dye degradation potential. J. Photochem. Photobiol. B: Biology. 191: 143–149.

Vasantharaj, S., S. Sathiyavimal, P. Senthilkumar, F. LewisOscar and A. Pugazhendhi. 2019b. Biosynthesis of iron oxide nanoparticles using leaf extract of *Ruellia tuberosa*: Antimicrobial properties and their applications in photocatalytic degradation. J. Photochem. Photobiol. B: Biol. 192: 74–82.

Vijayan, R., S. Joseph and B. Mathew. 2018. Green synthesis, characterization and applications of noble metal nanoparticles using *Myxopyrum serratulum* AW hill leaf extract. Bionanosci. 8(1): 105–117.

Vilardi, G., N. Verdone and L. Di Palma. 2017. The influence of nitrate on the reduction of hexavalent chromium by zero-valent iron nanoparticles in polluted wastewater. Desalin Water Treat. 86: 252–258.

Villela, H.D.M., R.S. Peixoto, A.U. Soriano and F.L. Carmo (2019). Microbial bioremediation of oil contaminated seawater: A survey of patent deposits and the characterization of the top genera applied. Sci. Total Environ. 666: 743–758.

Vinayagam, R., T. Varadavenkatesan and R. Selvaraj. 2017. Evaluation of the anticoagulant and catalytic activities of the *Bridelia retusa* fruit extract-functionalized silver nanoparticles. J. Cluster Sci. 28(5): 2919–2932.

Waghmode, M.S., A.B. Gunjal, J.A. Mulla, N.N. Patil and N.N. Nawani. 2019. Studies on the titanium dioxide nanoparticles: Biosynthesis, applications and remediation. SN Appl. Sci. 1(4): 310.

Wall, D.H., U.N. Nielsen and J. Six. 2015. Soil biodiversity and human health. Nature. 528(7580): 69–73.

Wang, J. and B. Chen. 2015. Adsorption and coadsorption of organic pollutants and a heavy metal by graphene oxide and reduced graphene materials. Chem. Eng. J. 281: 379–388.

Wang, S., K. Wang, C. Dai, H. Shi and J. Li. 2015. Adsorption of Pb^{2+} on amino-functionalized core-shell magnetic mesoporous SBA-15 silica composite. Chem. Eng. J. 262: 897–903.

Wei, G., J. Zhang, J. Luo, H. Xue, D. Huang, Z. Cheng, et al. 2019. Nanoscale zero-valent iron supported on biochar for the highly efficient removal of nitrobenzene. Front. Environ. Sci. Eng. 13(4): 61–66.

Xiong, Z., F. He, D. Zhao and M.O. Barnett. 2009. Immobilization of mercury in sediment using stabilized iron sulfide nanoparticles. Water Res. 43(20): 5171–5179.

Xu, P., G.M. Zeng, D.L. Huang, C.L. Feng, S. Hu, M.H. Zhao, et al. 2012. Use of iron oxide nanomaterials in wastewater treatment: A review. Sci. Total Environ. 424: 1–10.

Xue, W., D. Huang, G. Zeng, J. Wan, M. Cheng, C. Zhang, et al. 2018. Performance and toxicity assessment of nanoscale zero valent iron particles in the remediation of contaminated soil: A review. Chemosphere 210: 1145–1156.

Yeganeh-Faal, A., M. Bordbar, N. Negahdar and M. Nasrollahzadeh. 2017. Green synthesis of the Ag/ZnO nanocomposite using *Valeriana officinalis* L. root extract: Application as a reusable catalyst for the reduction of organic dyes in a very short time. IET Nanobiotechnol. 11, 669–676.

Zhang, W.X. 2003. Nanoscale iron particles for environmental remediation: An overview. J. Nanopart. Res. 5(3-4): 323–332.

Zhang, W.X. and D.W. Elliott. 2006. Applications of iron nanoparticles for groundwater remediation. Rem. J. 16(2): 7–21.

Zhao, G., J. Li, X. Ren, C. Chen and X. Wang. 2011. Few-layered graphene oxide nanosheets as superior sorbents for heavy metal ion pollution management. Environ. Sci. Technol. 45(24): 10454–10462.

Zhao, X., L. Zhou, M.S. Riaz Rajoka, L. Yan, C. Jiang, D. Shao, et al. 2018. Fungal silver nanoparticles: Synthesis, application and challenges. Critical Rev. Biotechnol. 38(6): 817–835.

Zhou, H., T. Fan, T. Han, X. Li, J. Ding, D. Zhang, et al. 2009. Bacteria based controlled assembly of metal chalcogenide hollow nanostructures with enhanced light-harvesting and photocatalytic properties. Nanotechnol. 20(8): 085603.

Zhu, H., S. Jia, T. Wan, Y. Jia, H. Yang, J. Li, et al. 2011. Biosynthesis of spherical Fe_3O_4/bacterial cellulose nanocomposites as adsorbents for heavy metal ions. Carb. Polymer. 86(4): 1558–1564.

CHAPTER 8

Nanoparticles, Biosurfactants and Microbes in Bioremediation

Charles O. Nwuche[1*], Victor C. Igbokwe[1],
Daniel D. Ajagbe[1,2] and Chukwudi O. Onwosi[1]

[1]Department of Microbiology, University of Nigeria, Nsukka, Enugu State, Nigeria.
[2]Department of Microbiology, Obafemi Awolowo University, Ile-Ife, Osun State, Nigeria.

1. INTRODUCTION

Today, many industrial processes and the technological revolution driven by the modern age have brought about huge increase in the volume of pollutants released into the ecology. Although many chemical and physical methods of clean up or removal have been practiced for decades, most are expensive, requires the expertise of well-trained personnel and may be ineffective in handling lowly concentrated residues. Some chemical treatment methods leave behind more toxic intermediates and are generally damaging to the environment. The most sustainable way to eliminating pollutants from the environment is still the subject of many scientific studies and the pertinent approaches are perhaps still locked away in some very astute minds awaiting to be birthed and harnessed. Many trends are presently emerging relating to the adoption of biological methods in the treatment and management of existing environmental challenges. The exploration and exploitation of the metabolic capacity of microorganisms and plants to degrade or immobilize pollutants thus rendering them less toxic and bioavailable in the environment have been widely reported (Sajna et al. 2015, Pandey and Shrivastava 2018) as significant milestones.

The strategy for good and effective cleanup of any environmental pollutant is guided by a string of limiting factors, which may include efficiency of the process, relative cost, complexity of the cleanup exercise, associated hazard(s), resource availability and expected time of completion (Pandey and Shrivastava 2018). Sometimes, the use of single strategy model limits the efficiency of a cleanup process to only the merits of such technology. Singh et al. (2019)

*Corresponding author: charles.nwuche@unn.edu.ng; Tel: +2348033728524 in his report opined that integrated cleanup technologies could overcome the shortcomings of single step applications and provide a better alternative to conventional remediation methods. Although bioremediation has been reported to be a green and sustainable method of treating polluted sites, it requires long treatment time and may not be effective at elevated pollutant concentration. The integration of nanoparticles in bioremediation offers great promise of an effective, efficient and sustainable alternative to chemical and physical methods (Cecchin et al. 2016). According to Singh et al. (2019), nanobioremediation is one of such integrated cleanup technologies that is currently receiving a lot of attention because it offers the combined benefits of reducing the pollutants to the least detectable concentration and alleviating their potentially negative environmental impacts.

2. NANOPARTICLES IN NATURE

Nanoparticles (NPs) are atomic or molecular masses with dimensions greater than 1 but less than 100 nm and are capable of radically altering their elemental characteristics. This definition places nanoparticles in the same size domain as ultrafine particles (airborne particles) and as sub-set of colloidal particles (Scenihr 2005). They are more reactive and more mobile compared to their respective bulk material. They possess high surface area and their enhanced reactivity adds to the unique physical and chemical properties of the nanomaterials. They are used as adsorbents and catalysts, as well as in the detection of pollutants and the removal of infective agents such as bacteria, yeasts, parasites and viruses (Tewari 2019). In addition, metal and metal oxide nanoparticles display interesting characteristics such as sorption, chemical reduction, and ligand sequestration (Tewari 2019). Thus, they can be adapted for use in the bioremediation of contaminants. Nanoparticles can be synthesized from numerous materials and their activity depends largely on their chemical composition, size and shape (Yadav et al. 2017). Nanoparticles are made up of three layers: the surface layer, which may be functionalized with different molecules, metal ion, surfactant and polymers; the shell layers, which are composed of chemically different materials and the core, which is the nucleus of the nanoparticles (Shin et al. 2016).

Amongst numerous nanomaterials in current applications, the nanoscale zero-valent iron, also known as nanoiron (nZVI), has been found to be very active at subsurface spaces of contaminated sites and economical to use owing to its low toxicity and cost of production. Due to these important features, nZVI has been the subject of numerous scientific studies, accounting for over 90% of the research output on nanoparticles (Cecchin et al. 2016). Nanoparticles can be classified into organic (e.g. fullerenes), inorganic (metals, e.g. gold and silver) and semiconductor (e.g. titanium dioxide and zinc oxide) types.

3. SOURCES OF NANOPARTICLES

Three sources of nanoparticles have been identified and are classified based on their origin. They include;

3.1 Incidental Nanomaterials

This class of nanomaterials are produced as a result of combustion activities such as forest fires and other natural processes such as skin and hair shedding by animals and plants. Industrial operations such as welding and fumes from vehicle exhaust and heavy equipment also contribute to the makeup of nanoparticles. The stated conditions significantly increase the concentration of nanoparticles in the environment, thereby bringing about reduction in the health quality of humans, animals and environment, respectively.

3.2 Engineered Nanomaterials

Many anthropogenic activities such as transportation, manufacturing, smelting and ore refining are some of the operations that promote the formation and dissemination of nanoparticles. Combustion from cooking, charcoal burning and the engines of airplanes also contribute to 10% of the atmospheric nanoparticles said to result from human activities. The remaining 90% are attributed to natural courses. Nanoparticles can also be generated from metallic compounds such as titanium oxide (TiO_2) and aluminum oxide (Al_2O_3).

3.3 Naturally Produced Nanomaterials

Nanoparticles occur in living things, which include microorganisms such as bacteria, fungi and viruses (particles), to complex systems such as plants, animals and man. Recently, the use of sensitive equipment to study nanomaterials has helped in the precise descriptions of the structures of naturally formed nanoparticles, thus providing fuller understanding of the nanostructures present in microorganisms. Often, due to the similarities found amongst nanomaterials from diverse sources, clear cut classifications into incidental, engineered or naturally occurring category poses some difficulty because incidental nanomaterials can equally become grouped as naturally produced nanomaterials.

4. BIOSYNTHESIS OF NANOPARTICLES

Nanoparticles can be synthesized by biological or chemo-physical methods. In selecting a preferred means of synthesizing nanoparticles, many researchers agree that the cost-effectiveness and environmental sustainability of the process are important considerations. Flowing from the foregoing, the synthesis of nanoparticles by biological methods has received greater attention than alternative chemical and physical methods (Abdelghany et al. 2017). Biosynthesis offers a simple, compatible, eco-friendly and more productive approach to the production of nanoparticles than the physical or chemical methods, which are reported to be eco-toxic, of high cost and low productivity (Singh et al. 2015).

4.1 Microorganisms

A number of microorganisms, which include bacteria, yeasts, algae and even fungi, have been investigated for their potential application in the synthesis of nanoparticles. It is envisaged that the use of biological materials makes the process sustainable, albeit nanotechnology itself is a new and budding field. In order to fully exploit the biological synthesis of nanomaterials, Punjabi et al. (2015) opined that knowledge of the basic mechanisms of biological development is crucial in order to acquire control of the synthetic process. In their views, the general principle underlying the biosynthesis of nanoparticles is linked to the involvement of proteins like enzymes and co-factors that have redox potential as well as act as electron shuttles in metal reduction. The use of microorganisms offers numerous comparative benefits especially in the area of space and costs. Microbes offer better size control and are easier to optimize process conditions for maximum product formation. Fungi have the ability to stabilize nanoparticles because they contain biological products which can convert metal ions to nanoparticles. Copious quantities of proteins are found in fungi; hence, they are able to rapidly convert metal salts into metal nanoparticles (Anwar 2018). Examples of fungi that have been used to synthesize nanoparticles include *Aspergillus fumigates*, *Phoma* sp., *Aspergillus niger*, and *Cylindrocladium floridanum* (Anwar 2018). Bacteria also possess the ability to reduce metal ions and are potential candidates for the production of nanoparticles. Bacteria such as *Thiobacillus ferrooxidans*, *T. thiooxidans*, and *Sulfolobus acidocaldarius, Delftia acidovorans,* have been used but yeast strains offer more benefits because of their high productivity and ease of process control.

4.2 Plants

Coastal plants, especially mangroves, have equally been identified as potential resources for the synthesis of nanoparticles (Singh et al. 2015). Punjabi et al. (2015) showed that plant extracts have many reducing and stabilizing agents that are essential for the formation of nanoparticles. However, the different nature of plants significantly affects their type of extracts and the nanoparticles synthesized because plant extracts contain different biochemical properties as well as a variety of phytochemical products such as carboxylic acids, aldehydes, flavones, ketones, terpenoids and amides. The marine plants produce different forms of bioactive compounds including alkaloids, polyphenols, tannins and flavonoids.

4.3 Algae

Algae is another machinery that have been evaluated for application in the biosynthesis of nanoparticles. Silver, gold, and silica nanoparticles have been made using different polysaccharides such as starch, chitosan, natural gums and hyaluronan from algae (Anwar 2018). Different species of algae such as seaweed, diatoms, and blue-green algae are able to synthesize nanoparticles because they contain polysaccharides, which can be used in the reduction and stabilization of nanoparticles. Polysaccharides provide a stabilization pattern that is dependent on the presence of multiple binding sites on the molecule. The binding sites promote adhesion to the nanoparticle as well as protection against physical and chemical alteration.

5. BIOSURFACTANT-ASSISTED NANO-BIOREMEDIATION

Nano-bioremediation has gained popularity in recent times due to the synergy between nanotechnology and bioremediation. However, the application of bioremediation technologies in the cleanup of hydrocarbon contaminated sites have hitherto been faced with the challenge of limited bioavailability of the pollutant compound(s). Often, the slow release of nonpolar compounds from the soil matrix into the aqueous phase represents a rate-limiting factor in the whole bioremediation effort (Franzetti et al. 2010) and could diminish both the potency of the bioremediation agent and the ability of the participating microbes to reach the target pollutant. In order to combat this challenge, the incorporation of biosurfactants, i.e. surface-active agents produced by microorganisms, offers a veritable solution.

Biosurfactants are compounds with both interfacial and biological activities which are synthesized by microorganisms. Biosurfactants have gained prominence owing to their vast importance in a variety of environmental applications and biomedical uses (Singh et al. 2017). They have been extensively employed in the biodegradation of hydrocarbon compounds and other oleophilic pollutants. Biosurfactants work by emulsifying hydrocarbons, thereby increasing their water solubility potential and decreasing their surface tension (El-Sheshtawy et al. 2014). Surfactants are made up of two parts, the hydrophobic head and a hydrophilic tail. In chemically synthesized surfactants, the most common hydrophobic components are the alcohols, olefins, paraffins, alkylbenzenes and alkylphenols, while the hydrophilic moiety consists of anionic (sulphate, sulphonate and carboxylate groups), cationic (quaternary ammonium group) and nonionic (polyoxyethylene, sucrose or polypeptide) surfactants (Franzetti et al. 2010).

Despite the documented benefits of biosurfactants in accelerating bioremediation, high doses negatively affect the integrity of the process (Sajna et al. 2015). Biosurfactants display antimicrobial properties which inhibit the activities of microorganisms during remediation. Apart from their effect on the biota, biosurfactants could adversely modify the environment and hence impede pollutant degradation. Although microorganisms may vary in their responses to increased concentrations of biosurfactants, understanding the growth limitation kinetics of biosurfactants is crucial in determining the optimal concentration necessary to promote microbial participation during remediation and to circumventing potentially limiting conditions.

As with biosurfactants, many nanoparticles (e.g. ZnO) show activity against several microorganisms such as *Staphylococcus aureus* and *Escherichia coli* (Ismail et al. 2013). Many reports, however, dispute this antimicrobial propensity of nanoparticles but rather affirms their stimulatory effects as electron donors of microbes (Pandey and Shrivastava 2018). The use of biosurfactants to enhance the bioremediation of pollutants is not restricted to the cleanup of oil-rich pollutants. Due to their polar and non-polar characteristics, surfactants can modify phase dispersal of many pollutants via diverse mechanisms such as emulsification, micellarization, sorption to soil and desorption of contaminants (Franzetti et al. 2010).

6. ROLE OF NANOPARTICLES IN BIOREMEDIATION OF CONTAMINATED SITES

The development of sustainable and environmentally acceptable technologies in the removal of pollutants from contaminated sites is one of the biggest challenges of environmental scientists in the 21st century. With recent advances in the field of nanotechnology, biosynthetic nanoparticles offer promise of potential application in the remediation of soils, water and sediments contaminated with heavy metals as well as organic and inorganic pollutants. This emerging technique is environmentally sensitive and economically feasible alternative to the clean-up of contaminants from the environment.

Nanotechnology is presently at the forefront of restoration efforts covering a wide spectrum of environmental contamination. Specifically, several sites previously polluted with hydrocarbons, chlorinated compounds and even heavy metals have been treated successfully using carbon nanomaterials. The application of nanotechnology in the removal or cleanup of environmental pollutants is termed nanoremediation. It is a new development that engages the use of nanomaterials in the recovery and restoration of impacted environments. Nanoremediation is a rapid, cost effective and efficient replacement to the old, tedious and expensive bioremediation approaches. It equally offers clear strategies for preventing, detecting and monitoring the progress of remediation (Rajan 2011). Nanoremediation involves the use of nanomaterials and nanoscale particles such as carbon nanotubes, zeolites and fibres in the transformation and detoxification of pollutants. Nanomaterials have been used to overcome countless limitations often associated with '*in situ*' remediation when conventional treatment methods such as chemical oxidation and thermal treatment were applied. Presently, nanoscale zero-valent iron (nZVI's) is the most widely used nanoparticle (Garner and Keller 2014). Their large surface area and excellent reactivity make them best suited for the cleanup or removal of contaminated materials. They have low reduction potential and other beneficial attributes that promote mobility across the soil matrix (Tosco et al. 2014).

Nanomaterials display catalytic properties and the quality of chemical reduction. Both processes are important in extenuating the impacts of very many pollutants. The use of nZVI in the clean-up and conversion of many pollutants such as chlorinated and organic compounds, pesticides and heavy metals has been reported (Karn et al. 2009). Also, certain bi-metallic nanoparticles, e.g. Ag/Fe, Ni/Fe and Cu/Fe, are useful in the removal of heavy metals and organochlorines from water bodies (Koutsospyros et al. 2012, Nie et al. 2013).

During remediation, nZVIs get initially converted from ferrous oxide (FeO) to ferrous ions (Fe^{2+}) and subsequently to ferric ions (Fe^{3+}) as determined by the equation below:

$$2FeO_{(s)} + 2H_2O_{(l)} \rightarrow 2Fe^{2+} + H_{2(g)} + 2OH^-$$

The ferrous ion (Fe^{2+}) gets further oxidized to a more stable ferric ion (Fe^{3+}), according to either of the following equations by Crane and Scott (2012):

$$2Fe^{2+}_{(s)} + 2H^+_{(aq)} + \frac{1}{2}O_{2(aq)} \rightarrow 2Fe^{3+} + 2H_2O_{(l)}$$

$$2Fe^{2+}_{(s)} + 2H_2O_{(l)} \rightarrow 2Fe^{3+} + H_{2(g)} + 2OH^-_{(aq)}$$

The ferrous ion (Fe^{2+}) is the most noxious state of iron to all invertebrates and microbiota. However, under neutral to alkaline conditions, they quickly convert to their ferric (Fe^{3+}) states. In oxygen deficient environments like the aquifers, oxidation reactions cause the formation of compounds such as magnetite (Fe_3O_4) or maghemeite (Fe_2O_3, γ-Fe_2O_3) in the presence of iron II oxide (FeO) (Grieger et al. 2010).

7. PRINCIPLES OF ACTION OF NANOCATALYSTS

7.1 Catalysis

Catalysis is the increase in the speed with which a chemical reaction proceeds due to the participation of another compound (the catalyst) which is not used up in the reaction process. Nanocatalysis, therefore, is a type of catalytic process in which nanomaterials and/or nanoparticles are used as catalysts. The nanomaterials are referred to as nanocatalysts. Catalysis is one of the earliest applications of nanoparticles. Indeed, many elements such as iron, aluminum and silica have been used as catalysts in nanoscale measures for many years. Nanoparticles are the most important industrial catalysts. A lot have been used over the years in such fields as manufacturing, energy and storage.

7.2 Nanocatalysis and Nanocatalysts

Nanocatalysis developed as a catalytic system with a domain that links homogeneous catalysis (when the catalyst and reacting compounds are in identical phase) with heterogeneous reactions (when the catalysts and the reacting compounds are in separate phases) (Xiufang 2017). Nanocatalysts have large surfaces of catalytically charged materials, which increase access of particles to active site of reactants enabling reactions to proceed in identical phases. They are made up of tiny particles with dimensions in the range of 1 to 100 nm. When they are insoluble in solvents, the reactants easily separate from reaction mixtures because of different phase systems occurring in the reaction (Xiufang 2017).

Nanomaterials have unique properties which influence their morphology, particle size, electronic structure, geometry and surface area charge. These characteristics make them chemically reactive, increasing their capacity for adsorption, interaction and catalytic activity (Rizwan et al. 2014). The surface catalytic property of nanocatalysts is due to the distinctive features of the wide surface areas, which are shape-dependent and enhance the degradation of contaminants (Prachi and Madathil 2013). The nanoscale sizes, and hence the large surface area, offer greatly increased contact between reactants and catalysts, while the insolubility of the reaction solvents indicates that the catalyst being of different phase dimensions can separate easily out of the reaction mixtures (Tandon and Singh 2014, Yan et al. 2014). The activity and sensitivity of nanocatalysts can be activated when certain changes occur to their elemental characteristics (Xiufang 2017).

8. ROLE OF NANOSORBENTS IN NANOBIOREMEDIATION

Nanosorbents are organic or inorganic nanoscale particles that show high absorption capacity to substances following physical or chemical reactions. Nanomaterials exhibit good adsorbent properties toward modified chemical groups (Zamboulis et al. 2011) due to their wide surface area and affinity. Adsorption is considered a better remediation strategy in the removal of heavy metals than the conventional treatment methods, which include chemical precipitation (Durate-Vazquez et al. 2003), reverse osmosis (Bodalo-Santoyo et al. 2003), electrochemical treatment (Wash and Reade 1994), coagulation (Zhang et al. 2003), ion-exchange (Xing et al. 2007) and irradiation (Batley and Farrar 1978). Some of the benefits of adsorption are their high efficiency, simplicity of operation, and low cost (Zamboulis et al. 2011). Several efforts have led to the development of selective sorbents with high removal capacity for toxic metal ions from groundwaters. According to Kumar et al. (2016), nanomaterials for use as nanosorbents must:

1. Be non-toxic
2. Have high sorption capability and sensitivity to low concentrated pollutants
3. The nanosorbent surface must be capable of easy removal of adsorbed pollutants and
4. Have capacity for repeated recycling.

8.1 Categories of Effective Nanosorbents

8.1.1 Carbon-based Nanosorbents

The high sorption and non-toxic property of carbon-based nanomaterials is important to the removal of heavy metals from drinking water. Carbon-based nanosorbents such as aerogels, coatings, fibres and graphenes have been extensively studied for their application in water treatment (Nnaji et al. 2018). Carbon nanotubes (CNTs) are important in the removal of heavy metals from waste waters during treatment, while graphenes are selected due to their excellent mechanical and thermal characteristics (Nnaji et al. 2018). Activated carbon is also a good adsorbent for organic and inorganic pollutants but its application is limited to most heavy metals, particularly arsenic As(V) (Daus et al. 2002). On the other hand, CNTs exhibit greater efficiency in the absorption of many organic compounds compared to activated carbon (Pan and Xing 2008). This ability is mainly due to the interactions between the pollutant and the large surface areas of CNTs.

8.1.2 Inorganic-based Nanosorbents

Metallic nanoparticles represent another class of nanomaterials that enjoy wide applications in the treatment of waste waters for the removal of heavy metals due to their favourable physical characteristics (Nnaji et al. 2018). Some of the investigated metallic nanoparticles and their oxides include silver (Ag), (Fabrega et al. 2011), manganese (Gupta et al. 2011), ferric oxide (Feng et al. 2012), titanium (Luo et al. 2010), copper (Goswani et al. 2012), magnesium (Gao et al. 2008), and ceric (Cao et al. 2010) oxides. The nanoparticles of some metallic oxides display amazing ability in the removal of arsenate from water (Hristovski et al. 2007). Titanate nanoparticles are equally effective in the removal of cadmium (Cd^{2+}), a potentially toxic metal ion. However, one major limitation in the use of inorganic nanomaterials is the difficulty of phase separation after use in the treatment of wastewaters (Kumar and Gopinath 2016).

8.1.3 Iron-based Nanomaterials

Iron nanoparticles play a notable role in the removal of Chromium (Cr)(VI) and Arsenic (As)(III) (Abdollahi et al. 2015, Mu et al. 2015). The commonest iron-based nanomaterials used for remediation are nanoscale zero-valent iron (nZVI), iron sulphide, bimetallic Fe and nanoscale FeO nanoparticles (Ludwig et al. 2007). Iron-oxide nanomaterials used in water treatment are of two categories: the nano-adsorbent or immobilized carriers, which promotes the removal efficiency of pollutants, and the photocatalysts, which convert hazardous contaminants to more innocuous materials. The iron oxide nanomaterials are considered to be important in the absorption of heavy metals and organic pollutants but one major concern is the potential long-term effect of these toxic materials on human health and the environment.

8.1.4 Polymer-based Nanosorbents

Polymer nanosorbents are efficient sorbents with remarkable capability and rapid rate of activity. Polymers, e.g. dendrimers, are effective in the removal of heavy metals and organic compounds. Polymer-layered silicate nanocomposites fused with pyromellitic acid dianhydride (PMDA) and phenylamino methyl trimethoxy saline (PAMTMS) were very useful in the selective withdrawal of metallic ions such as copper (Cu^{2+}) and lead (Pb^{2+}) during water treatment (Liu et al. 2010). Electrostatic interactions, hydrophobic effects, bond formation between hydrogen atoms, and complexation are examples of mechanisms through which polymers absorb heavy metals and organic compounds (Wang et al. 2015).

9. NANOCATALYSTS AND REDOX ACTIVE NANOPARTICLES

Nanocatalysts are nanomaterials with catalytic characteristics. The nanoscale sizes of nanocatalysts contribute to their high catalytic capability because of the larger surfaces available for interaction with the reactants.

9.1 Zero-valent Iron (ZVIs)

Zero-valent iron (ZVI) is a reactive nanometal with a redox potential of –0.44 V. The suitability and efficiency of ZVIs in the remediation of contaminated environments and treatment of hazardous wastes have long been reported (Garner and Kelly 2014). The mechanism of action involves the transfer of electrons from ZVI to the contaminants, thereby bringing about their conversion from the toxic state to innocuous species. In addition, ZVI can bring about the oxidation of several organic compounds by the transfer of two electrons to oxygen to produce hydrogen peroxide (H_2O_2). Hydrogen peroxide is then reduced to water by a second two-electron transfer from ZVI or combines with ferrous (Fe^{2+}) ion (from waters) to generate hydroxyl radicals (OH) for the oxidation of organic compounds. The reactions as first described by Fu et al. (2014) are reproduced below:

$$FeO + O_2 + 2H^+ \rightarrow Fe^{2+} + H_2O_2$$
$$FeO + H_2O_2 + 2H^+ \rightarrow Fe^{2+} + 2H_2O$$
$$Fe^{2+} + H_2O_2 \rightarrow Fe^{3+} + OH^-$$

ZVI have numerous interesting qualities that make them the nanocatalyst of choice in the removal of many contaminants and in the treatment of ground water. They are nontoxic, abundant, cheap and easy to produce (Joo et al. 2004). The core of ZVI particles consists of zero-valent or metallic iron, while the surface is formed from oxidized metallic iron (Li et al. 2006). ZVIs are electron donors which increase their reactivity, making them excellent materials for use in remediation activities (Garner and Kelly 2014).

9.2   Titanium Dioxide (TiO$_2$)

Titanium dioxide (TiO$_2$) is a popular catalyst due to its high reactivity under UV irradiation. Other major advantages include high stability, low cost and safety to both humans and the environment (Gupta and Tripathi 2011). Photocatalytic reaction occurs when light absorption induces a charge separation. At this state, TiO$_2$ particles become unstable and react with several pollutants or even water to produce highly reactive hydroxyl radicals. TiO$_2$ exists in three basic crystalline forms, which include the most common anatase and rutile types as well as the much less known brookite (Bagheri et al. 2014). TiO$_2$ is most often used in the powdery forms rather than liquid. However, its photoactivity is significantly improved when applied as nanoparticles (Han and Bai 2009). The application of TiO$_2$ nanoparticles is limited on a large scale due to the requirement for filtration after each treatment. This makes the catalyst expensive to use. However, this condition can be overcome by immobilizing the nanoparticles on matrices (Ma et al. 2001, Rao et al. 2003). Titanium dioxide (TiO$_2$) nanoparticles are effective in the removal of organic contaminants but their lack of selectivity is a fundamental shortcoming of their potential wide scale applicability.

10.   USE OF NANOMATERIALS AS IMMOBILIZING AGENTS IN BIOREMEDIATION

Pollution of soil by heavy metals (i.e. Cd, Pb, Zn and Cr) is a significant global challenge because crops grown in contaminated areas retain considerable amounts of the cations, which accumulate and concentrate along the food chain, leading to morbidity and mortality in man and animals. Many techniques aimed at the removal or stabilization of heavy metals (i.e. vitrification, soil capping, etc.) are ineffective on a large scale. Nanotechnology has been successfully applied in the removal of heavy metals from polluted systems by immobilization.

10.1   Nanoscale Zero-valent Iron

Nanoscale zero-valent iron (nZVI) nanoparticles enhance the breakdown or immobilization of many high-profile contaminants due to their high reactivity, mobility and large specific surface area. They have been widely investigated for their roles in the remediation of contaminated soil and water. Stabilized nZVI nanoparticles are used in immobilized forms for the removal of heavy metals by converting more mobile/soluble metal species such as Chromium VI and Uranium VI to their immobile forms (Xu and Zhao 2007).

10.2   Iron Sulphide Nanoparticles

Iron sulphide (FeS) nanoparticles have been successfully used in the removal of mercury from the environment by immobilization. Many immobilization processes such as concurrent surface complexation, structural incorporation and precipitation have been described (Gong et al. 2012).

10.3   Carbon Nanotubes

Carbon nanotubes (CNTs) are used in the production of immobilized screen filters to protect water filtration systems from developing biofilms. Many reports show that the filters were successful in removing bacteria and even viruses (Mostafavi et al. 2009) during water treatment. Thus, CNTs could play very important roles in municipal wastewater treatment.

10.4   Iron Phosphate Nanoparticles

Iron phosphate ($FePO_4$) nanoparticles have been applied in the immobilization of heavy metals in contaminated soils and water (Raicevic et al. 2006). When phosphate nanoparticles are used, they bring about the generation of insoluble metal species which are immobile and stable across a wide temperature and pH conditions (Cundy et al. 2008). Iron phosphate nanoparticles are more effective in heavy metal immobilization than solution phosphates because they offer lower phosphate leaching and longer lasting reactivity. In order words, the nanoparticles act as a sink for the removal of metal cations from polluted systems (Reynolds and Davies 2001).

11.   ENGINEERING POLYMERIC NANOPARTICLES FOR ENHANCED BIOREMEDIATION

11.1   Engineered Nanoparticles in the Remediation of Organic Contaminants

The rate of solubilization, mobility and bioavailability of most organic contaminants in soils is hindered by sorption. Polymer nanoparticles have been engineered to promote the partitioning of polyaromatic hydrocarbons (PAHs) such as phenanthrene into the aqueous phase. The nanoparticles thus enhance their dissolubility and effective removal from contaminated waters. Also, the degradation rate of phenanthrene crystals in water is significantly improved by the addition of precursors made from poly(ethylene)glycol modified urethane acrylate (PMUA). These PMUA nanoparticles increase the effectiveness of remediation, particularly under *in situ* conditions, because bacterial populations are provided with access to the contaminants within the particles. Extracted nanoparticles are stable and could be recycled and reused in many remediation schemes, thus enabling significant cost savings and controlled release of the nanoparticles to the environment (Tungittiplakorn et al. 2004).

11.2   Engineered Nanoparticles in Soil Remediation

It has been stated that the sorption of organic contaminants into soils make their clean-up difficult. Today, the synthesis and application of polyurethane nanoparticles has greatly improved the effectiveness of remediation efforts in soils contaminated with polyaromatic hydrocarbons. The particles increase desorption of the pollutants by reducing their surface tension. The polyurethane nanoparticles can be modified to promote their affinity for any particular pollutant and hence increase their mobility in the soil. Many reports indicate that this could be achieved by altering the polar characteristics of the core and peripheral surfaces of the particles. Thus, by maintaining a non-polar or hydrophobic interior, the propensity for attachment of the polyurethane nanoparticles to organic pollutants is significantly boosted. On the other hand, manipulating the charge density or size of the peripheral hydrophilic spaces characteristically increases the mobility of the polyurethane suspensions in soil. Engineering nanoparticles to achieve custom-made properties is a revolutionary approach to their application because it provides opportunity to develop nanoparticles which are adapted to different soil and pollutant characteristics (Tungittiplakorn et al. 2004).

12. TOXICITY OF NANOMATERIALS

Although nanoparticles occur naturally, they are nowadays produced synthetically because of their numerous potential appeals. Despite the monumental breakthroughs and advancements in the use of nanotechnology, new vistas of applications open each day. This emerging trend justifies the continued production and use of engineered nanoparticles to meet up with the ever-increasing requirements in our daily lives. Lately, concerns have emerged that the basic characteristics of nanomaterials, which are exploited in their wide applications in industry, medicine and remediation, could potentiate unanticipated dangers to health and environment (Khan and Shanker 2015). Engineered nanomaterials persist within organisms. They have easy tissue penetration and a tendency to accumulate along the food chain. They are readily dispersible and have a propensity to initiate complex and irreversible biochemical reactions within biological systems and the environment. This type of characteristics is implicated in toxicity studies involving new chemical species.

12.1 Characteristics of Nanomaterials that Influence their Toxicity

According to Khan et al. (2019), the potential toxicity of nanoparticles can be credited to the following attributes:
- Their large surface area to volume ratio which increases their rate of reaction
- Their diverse chemical composition which increases the reactivity of the particles
- The surface charge of the particles which increases their electrostatic interactions
- The hydrophobic and lipophilic groups which interact with the proteins and membranes within the tissues and
- Inert particles could trigger tissue formation and lead to the formation of scar tissues when they accumulate in the body.

The nanoparticle size and large surface area to volume ratio are crucial properties that make nanomaterials toxic. As the particle sizes diminish, the surface areas increase, hence distributing considerable percentage of its atoms or molecules on the superficial rather than the core areas of the material. This change in the structural features of nanomaterials is responsible for numerous reported biological reactions that could bring about toxicological effects.

12.2 Toxic Effects

Nanomaterials can bring about both biological and environmental toxicity.

12.2.1 Biological Toxicity

Nanoparticles gain entrance into the body through six principal routes, namely, oral, respiratory, dermal, intraperitoneal, subcutaneous and intravenous passages. Although the respiratory tract is the major portal of entry, the gastrointestinal tract is equally a very important gateway. Nanoparticles access the body directly through food, water and drugs or indirectly through mucociliary movement (Meng et al. 2007). When the nanoparticles encounter biological materials such as proteins and cells, they are absorbed and distributed to the various organs where they may be modified, metabolized or excreted. The study of the interactions of nanoparticles with biological systems is termed nanotoxicology. It highlights how the basic characteristics of nanoparticles govern the initiation of toxic biological responses. The following toxic effects have been reported: allergy (Maynard et al. 2006), fibrosis, organ failure, inflammation and cytotoxicity (Nel et al. 2006), tissue and DNA damage (Singh et al. 2009) and generation of reactive oxygen species (Meng et al. 2007).

12.2.2 Environmental Toxicity

The application of nanotechnology, particularly in remediation, leaves behind significant residues, which could result in nanoparticle pollution of the soil and ground water environment (Seetharam and Sridhar 2007, Colvin 2007). Presently, the precise transport mechanism by which nanoparticles are removed during waste water treatment is not clear, nor are their potential effects on plants and microbiota fully understood. This is because the fate of nanomaterials in many spaces is a complex interplay of different interactions involving many biotic and abiotic components in the ecosystem (Khan et al. 2019). Thus, at present, complete knowledge of the toxic effects of nanoparticles in the environment is very limited. Therefore, continuous ecological risk assessment is advocated in order to further understand the environmental risks associated with the use of nanomaterials. Although current toxicity testing protocols are important in identifying the injurious effects of nanoparticles, research into new methods are necessary in order to accommodate the numerous properties of nanomaterials and their potential interaction with other compounds in the environment (Khan and Shanker 2015).

13. INTERACTION OF NANOPARTICLES WITH SOIL AND MICROBES

Nanoparticles get introduced into the soil as a result of a number of anthropogenic activities such as remediation, fertilization and releases via air, water and sewage sludge application on land. Nanoparticles have the ability to adhere to the surface of contaminants or even disperse into air in the form of aerosols; these small molecular particles may enter the surface of contaminated soils easily without any active energy (Van Der et al. 2011). The toxicity of nanoparticles on microorganisms has been proven based on reduction in microbial activity, distribution and diversity (Jiang et al. 2009). Soil microorganisms play vital roles in the maintenance of soil health and the promotion of numerous ecosystem functions. The antimicrobial property of many nanoparticles, e.g. titanium dioxide and fullerene, have been reported (Lovern and Klaper 2006). Some, particularly silver nanoparticles, cause significant damage to bacterial cell wall (Soni and Bondi 2004). In *Escherichia coli* for instance, nanoparticles accumulate in the cell wall, causing increased cell permeability and subsequently cell death. The lethality of zinc and magnesium oxide nanoparticles is a property that is presently exploited in the preservation of food products. Iron and copper nanoparticles react with peroxides in the environment to generate free radicals, which show activity against many microbes (Saliba et al. 2006). Presently, silver nanoparticles (AgNPs) are increasingly applied as bacteriostatic agents in a variety of coatings to control microbial growth. The copper nanoparticles, which come as alternatives to the silver nanoparticles, are considerably more toxic to the nitrogen fixing bacteria and cause disruption of their cell membranes (Karlsson et al. 2008).

Recently, Simon and Richaume (2015) evaluated the effect of engineered nanoparticles on soil microbial activities. Inorganic (metal and their oxides) nanoparticles potentiate higher toxicological effects than organic (e.g. fullerenes) nanoparticles on soil biota. An exception perhaps is iron oxide nanoparticles whose biocidal effects appear negligible even under high dose applications. Other metal (and oxide) species show significant lethality at concentrations lower than 1 mg per kg of soil. Silver nanoparticles negatively affect microbial degradation of soil organic matter, which is key to the biogeochemical cycling of mineral elements by interfering with the activity of critical enzymes. On the other hand, copper and zinc-based nanoparticles are known to be active against bacterial growth and biomass development but the concentration at which these effects manifest is predicted to be higher than the permitted environmental limit of 250 mg per kg. Although nZVI are important in the context of removal and cleanup of soil pollutants, their effects on the microbial transformation of organic matter need to be further verified. Equally, the impacts of ecological factors, i.e. clay, organic matter and pH on the bioavailability of nanoparticles, have not been sufficiently investigated and might contain valuable information for improving the quality and sensitivity of environmental risk assessment (Cornelis et al. 2014).

14. FATE OF NANOPARTICLES IN THE ENVIRONMENT

Without doubt, nanotechnology offers a lot of novel possibilities in medicine, bioremediation and industry, but knowledge about their fate, transport as well as physiochemical and biological transformation, which are possible in the environment, is presently not established. Often, the fate may also change as a result of interactions with certain components of the environment. Thus, determining the real fates of each compound involves a study of their cycles from initiation of production to final application (Klaine et al. 2008). The properties of nanomaterials as well as those of the receiving systems, i.e. water or soil, can also interact based on their unique physical and chemical characteristics to determine the final fate of nanoparticles in the environment (Baalousha et al. 2011).

15. CHALLENGES OF NANO-BIOREMEDIATION AND CONCLUSION

The application of nanotechnology in bioremediation is emerging, sustainable and innovative due to their high efficiency and resource conservation compared to conventional methods (Nnaji 2018). Nevertheless, its widescale application is still currently plagued by numerous challenges. Although the primary objective of remediation is the removal and elimination of potential pollutants, concerns have emerged that nanostructured materials could lead to unanticipated health and environmental dangers (Khan and Shanker 2015). These concerns are often based on the high mobility of nanoparticles. They are dispersed over great distances and their persistence in the environment brings about concerns for worry. Although the chemical nature of nanomaterials indicate they are mostly benign, a primary challenge of the continued application is the protection of the environment, human and animal health from potential dangers (Ejeta et al. 2017).

Apart from the potential toxic effects of nanomaterials on health and ecology, many other factors limit their widescale application. For instance, while the developed nations have enormous funds for the implementation of the new technologies, poor funding and dilapidated infrastructures appear to be the case with the developing economies (Nnaji 2018). Aside infrastructure, there is little knowledge of the technology, as well as their application for human and environmental safety. Nigeria is presently at the kindergarten stage of the nanotechnology revolution (Ejeta et al. 2017, Ezema et al. 2014). As such, little or no information presently exists on the application by public and private organizations, particularly in the downstream oil and gas engagements (Nnaji 2017).

To this end, training programmes are needed to develop the human capital base in nanotechnology and its applications, particularly as it pertains to bioremediation (Ezema et al. 2014). Issues such as public awareness and economic viability of nanotechnology-based approaches are some of the issues that need to be addressed in order to promote public acceptance of the initiatives in developing countries (Qu et al. 2013). Nano-based approaches can only compete with conventional technologies if the costs of application are lower, more effective and viable (Sahu et al. 2019).

References

Abdelghany, T.M., A.M.H. Al-Rajhi, A.M. Al Abboud, M.M. Alawlaqi, A.G. Magdah, E.A.M. Helmy, et al. 2017. Recent advances in green synthesis of silver nanoparticles and their applications: about future directions. A review. BioNanoSci. 8: 5–16.

Abdollahi, M., S. Zeinali, S. Nasirimoghaddam and S. Sabbaghi. 2015. Effective removal of As(III) from drinking water samples by chitosan-coated magnetic nanoparticles. Desal Water Treat. 56: 2092–2104.

Anwar, S.H. 2018. A brief review on nanoparticles: Types of platforms, biological synthesis and applications. Res. Rev. J. Mat. Sci. 6: 109–116.

Baalousha, M., J. Lead and Y. Ju-Nam. 2011. Natural colloids and manufactured nanoparticles in aquatic and terrestrial systems. Treat. Water. Sci. 81: 89–129.

Bagheri, S., J.N. Muhd and A.H.S. Bee. 2014. Titanium dioxide as a catalyst support in heterogeneous catalysis. Sci World J. 727496. doi:10.1155/2014/727496.

Batley, G. and Y. Farrar. 1978 Irradiation techniques for the release of bound heavy metals in natural waters and blood. Anal. Chimica. Acta. 99: 283–292.

Bodalo-Santoyo, A., J. Gómez-Carrasco, E. Gomez-Gomez, F. Maximo-Martin and A. Hidalgo-Montesinos. 2003. Application of reverse osmosis to reduce pollutants present in industrial wastewater. Desalin. 155: 101–108.

Cao, C.Y., Z.M. Cui, C.Q. Chen, Q.G. Song and W. Cai. 2010. Ceria hollow nanospheres produced by a template-free microwave-assisted hydrothermal method for heavy metal ion removal and catalysis. J. Phys. Chem. C 114: 9865–9870.

Cecchin, I., K. Reddy, A. Thomé, E.F. Tessaro and F. Schnaid. 2016. Nanobioremediation: Integration of nanoparticles and bioremediation for sustainable remediation of chlorinated organic contaminants in soils. Int. Biodet. Biodegr. 119: 419–428.

Colvin, V.L. 2003. The potential environmental effect of engineered nanomaterials. Nat. Biotechnol. 21: 1166–1170.

Cornelis, G., K. Hund-Rinke, T. Kuhlbusch, N. Van den Brink and C. Nickel. 2014. Fate and bioavailability of engineered nanoparticles in soils: A review. Crit Rev Environ. Sci. Technol. 44: 2720–2764.

Crane, R.A. and T.B. Scott. 2012. Nanoscale zero-valent iron: Future prospects for an emerging water treatment technology. J. Hazard. Mat. 211: 112–125.

Cundy, A.B., L. Hopkinson and R.L. Whitby. 2008. Use of iron-based technologies in contaminated land and groundwater remediation: A review. Sci. Total Environ. 400: 42–51.

Daus, B., R. Wennrich and H. Weiss. 2004. Sorption materials for arsenic removal from water: A comparative study. Water Res. 38(12): 2948–2954.

Duarte-Vázquez, M.A., M.A. Ortega-Tovar, B.E. García-Almendarez and C. Regalado. 2003. Removal of aqueous phenolic compounds from a model system by oxidative polymerization with turnip (*Brassica napus* L. var purple top white globe) peroxidase. J Chem. Technol. Biotechnol. 78: 42–47.

Ejeta, K.O., G.A. Dolo, G.I. Ndubuka, K.I. Nkuma-Udah, T.O. Azeez and O. Odugu. 2017. Impacts of nanotechnology in Nigeria: a short survey. Int. J. Biosens. Bioelectro. 2: 1–5.

El-Sheshtawy, H.S., N.M. Khalil, W. Ahmed and R.I. Abdallah. 2014. Monitoring of oil pollution at Gemsa Bay and bioremediation capacity of bacterial isolates with biosurfactants and nanoparticles. Marine Pollut. Bull. 87: 191–200.

Ezema, I.C., P.O. Ogbobe and A.D. Omah. 2014. Initiatives and strategies for development of nanotechnology in nations: A lesson for Africa and other least developed countries. Nanoscale Res. Lett. 9: 1–8.

Fabrega, J., S.N. Luoma, C.R. Tyler, T.S. Galloway and J.R. Lead. 2011. Silver nanoparticles: Behaviour and effects in the aquatic environment. Environ. Int. 37: 517–531.

Feng, L., M. Cao, X. Ma, Y. Zhu and C. Hu. 2012. Superparamagnetic high-surface-area Fe_3O_4 nanoparticles as adsorbents for arsenic removal. J. Hazard Mater. 217: 439–446.

Franzetti, A., I. Gandolfi, G. Bestetti and I.M. Banat. 2010. (Bio)surfactant and bioremediation, successes and failures. Trends Biorem. Phytorem. 145–156.

Fu, F., D.D. Dionysiou and H. Liu. 2014. The use of zero-valent iron for groundwater remediation and wastewater treatment: A review. J. Hazard. Mater. 267: 194–205.

Gao, C., W. Zhang, H. Li, L. Lang and Z. Xu. 2008. Controllable fabrication of mesoporous MgO with various morphologies and their absorption performance for toxic pollutants in water. Crystal Growth Design. 8: 3785–3790.

Garner, K.L. and A.A. Keller. 2014. Emerging patterns for engineered nanomaterials in the environment: A review of fate and toxicity studies. J. Nanopart Res. 16: 2503.

Gong, Y., Y. Liu, Z. Xiong, D. Kaback and D. Zhao. 2012. Immobilization of mercury in field soil and sediment using carboxymethyl cellulose stabilized iron sulfide nanoparticles. Nanotechnol. 23: 294007.

Goswami, A., P. Raul and M. Purkait. 2012. Arsenic adsorption using copper (II) oxide nanoparticles. Chem. Eng. Res. Design. 90: 1387–1396.

Grieger, K.D., A. Fjordøge, N.B. Hartmann, E. Eriksson, P.L. Bjerg and A. Baun. 2010. Environmental benefits and risks of zero-valent iron particles (nZVI) for in situ remediation: risk mitigation or trade-off? J. Contam. Hydrol. 118: 165–183.

Gupta, K., S. Bhattacharya, D. Chattopadhyay, A. Mukhopadhyay, H. Biswas, J. Dutta, et al. 2011. Ceria associated manganese oxide nanoparticles: Synthesis, characterization and arsenic(V) sorption behavior. Chem. Eng. J. 172: 219–229.

Gupta, S.M. and M. Tripathi. 2011. A review of TiO_2 nanoparticles. Chinese Sci. Bull. 56: 1639–1657.

Han, H. and R. Bai. 2009. Buoyant photocatalyst with greatly enhanced visible-light activity prepared through a low temperature hydrothermal method. Ind. Eng. Chem. Res. 48: 2891–2898.

Hristovsk, K., A. Baumgardner and P. Westerhoff. 2007. Selecting metal oxide nanomaterials for arsenic removal in fixed bed columns: From nanopowders to aggregated nanoparticle media. J. Hazard. Mater. 147: 265–274.

Ismail, W., N.A. Alhamad, W.S. El-Sayed, A.M. El Nayal, C. Yin-Ru and Y.H. Riyad. 2013. Bacterial degradation of the saturate fraction of Arabian light crude oil: biosurfactant production and the effect of ZnO nanoparticles. J. Petrol. Environ. Biotechnol. 4: 163.

Jiang, W., H. Mashayekhi and B. Xing. 2009. Bacterial toxicity comparison between nano- and micro-scaled oxide particles. Environ. Pollut. 157: 1619–1625.

Joo, S.H., A.J. Feitz and T.D. Waite. 2004. Oxidative degradation of the carbothioate herbicide, molinate, using nanoscale zero-valent iron. Environ. Sci. Technol. 38: 2242–2247.

Karlsson, H.L., P. Cronholm, J. Gustafsson and L. Moeller. 2008. Copper oxide nanoparticles are highly toxic: A comparison between metal oxide nanoparticles and carbon nanotubes. Chem. Res. Toxicol. 21: 1726–1732.

Karn, B., T. Kuiken and M. Otto. 2009. Nanotechnology and in situ remediation: A review of the benefits and potential risks. Environ. Health Persp. 117: 1823–1831.

Khan, H.A. and R. Shanker. 2015. Toxicity of nanomaterial. BioMed. Res. Int. Article ID: 521014 doi: 10.1155/2015/521014

Khan, I., K. Saeed and I. Khan. 2019. Nanoparticles: Properties, applications and toxicities. Arab. J. Chem. 12: 908–931.

Klaine, S.J., P.J.J. Alvarez, G.E. Batley, T.F. Fernandes, R.D. Handy, D.Y. Lion, et al. 2008. Nanomaterials in the environment: Behaviour, fate, bioavailability and effects. Environ. Toxicol. Chem. 27: 1825–1851.

Koutsospyros, A., J. Pavlov, J. Fawcett, D. Strickland, B. Smolinski and W. Braida. 2012. Degradation of high energetic and insensitive munitions compounds by Fe/Cu bi-metal reduction. J. Hazard. Mater. 219–220: 75–81

Kumar, S.R. and P. Gopinath. 2016. Nano-Bioremediation applications of nanotechnology for bioremediation, remediation of heavy materials in the environment, In: J.P. Chen, L.K. Wang, Mu-Hao S. Wang, Y.T. Hung and N.K. Shammas (eds), Remediation of Heavy Metals in the Environment. CRC Press, pp. 27–48.

Li, X., D.W. Elliott and W. Zhang. 2006. Zero-valent iron nanoparticles for abatement of environmental pollutants: materials and engineering aspects. Crit. Rev. Solid State Mater. Sci. 31: 111–122.

Liu, J., Y. Ma, T. Xu and G. Shao. 2010. Preparation of zwitterionic hybrid polymer and its application for the removal of heavy metal ions from water. J. Hazard. Mater. 178: 1021–1029.

Lovern, S.B. and R. Klaper. 2006. Daphnia magna mortality when exposed to titanium dioxide and fullerene (C60) nanoparticles. Environ. Toxicol. Chem. 25: 1132–1137.

Ludwig, R.D., C. Su, T.R. Lee, R.T. Wilkin, S.D. Acree, R.R. Ross, et al. 2007. *In situ* chemical reduction of Cr(VI) in groundwater using a combination of ferrous sulfate and sodium dithionite: A field investigation. Environ. Sci. Technol. 41: 5299–5305.

Luo, T., J. Cui, S. Hu, Y. Huang and C. Jing. 2010. Arsenic removal and recovery from copper smelting wastewater using TiO_2. Environ. Sci. Technol. 44: 9094–9098.

Ma, Y., J. Qiu, and Y. Cao. 2001. Photocatalytic activity of TiO_2 films grown on different substrates. Chemosphere. 44: 1087–1092.

Maynard, A.D., R.J. Aitken, T. Butz, V. Colvin, K. Donaldson, G. Oberdörster, et al. 2006. Safe handling of nanotechnology. Nature. 444: 267–269.

Meng, H., Z. Chen, G. Xing, H. Yuan, C. Chen, F. Zhao, et al. 2007. Ultrahigh reactivity provokes nanotoxicity: Explanation of oral toxicity of nano-copper particles. Toxicol. Lett. 175: 1–3.

Mostafavi, S., M. Mehrnia and A. Rashidi. 2009. Preparation of nanofilter from carbon nanotubes for application in virus removal from water. Desalination. 238: 271–280.

Mu, Y., Z. Ai, L. Zhang and F. Song. 2015. Insight into core-shell dependent anoxic Cr(VI) removal with $Fe@Fe_2O_3$ nanowires: Indispensable role of surface bound Fe(II) ACS Appl. Mater. Interf. 7: 1997–2005.

Nel, A., T. Xia, L. Madler and N. Li. 2006. Toxic potential of materials at the nanolevel. Science 311: 622–627.

Nie, X., J. Liu, X. Zeng and D. Yue. 2013. Rapid degradation of hexachlorobenzene by micron Ag/Fe bimetal particles. J. Environ. Sci. 25: 473–478.

Nnaji, J. 2017. Nanomaterials for bioremediation of petroleum contaminated soil and water. Umudike J. Eng. Technol. (UJET) 3: 23–29.

Nnaji, C.O., J. Jeevanandam, Y.S. Chan, M.I.K. Danquah, S. Pan and A. Barhoum. 2018. Engineered nanomaterials for wastewater treatment: current and future trends. *In*: A. Barhoum and A.S.H. Makhlouf (eds), Fundamentals of Nanoparticles Classifications, Synthesis Methods, Properties and Characterization Micro and Nano Technologies. Elsevier, Amsterdam, Netherlands, pp. 129–168.

Pan, B. and B.S. Xing. 2008. Adsorption mechanisms of organic chemicals on carbon nanotubes. Environ. Sci. Technol. 42: 9005–9013.

Pandey, A.K. and A. Shrivastava. 2018. Bioremediation of lead contaminated soil using bacteria. Res. J. Life Sci. Bioinfor. Chem. Sci. 4: 355–361.

Parachi, P.G., D. Madathil and A.B. Nair. 2013. Nanotechnology in wastewater treatment: A review. Int. J. Chem. Tech. Res. 5: 2303–2308.

Punjabi, K., P. Choudhary, L. Samant, S. Mukherjee, S. Vaidya and A. Chowdhary. 2015. Biosynthesis of nanoparticles: A review. Int. J. Pharm. Sci. Rev. Res. 30: 219–226.

Qu, X., P.J. Alvarez and Q. Li. 2013. Applications of nanotechnology in water and wastewater treatment. Water Res. 47: 3931–3946.

Raicevic, S., J. Wright, V. Veljkovic and J. Conca. 2006. Theoretical stability assessment of uranyl phosphates and apatites: selection of amendments for in situ remediation of uranium. Sci. Total Environ. 355: 13–24.

Rajan, S. 2011. Nanotechnology in ground water remediation. Int. J. Environ. Sci. Dev. 2: 182–187.

Rao, K.V.S., A. Rachel, M. Subrahmanyam and P. Boule. 2003. Immobilization of TiO_2 on pumice stone for the photocatalytic degradation of dyes and dye industry pollutants. Appl. Catal. B Environ. 46: 77–85.

Reynolds, C.S. and P.S. Davies. 2001. Sources and bioavailability of phosphorus fractions in freshwaters: A British perspective. Biol. Rev. 76:27–64.

Rizwan, M.D., M. Singh, C.K. Mitra and R.K. Morve. 2014. Ecofriendly application of nanomaterials: nanobioremediation. J. Nanoparticles. Article ID 431787, http://dx.doi.org/10.1155/2014/431787.

Sahu, J.N., R.R. Karri, H.M. Zabed, S. Shams and X. Qi. 2019. Current perspectives and future prospects of nano-biotechnology in wastewater treatment. Separat Purific. Rev. DOI:10.1080/15422119.2019.16 30430.

Sajna, K.V., R.K. Sukumaran, L.D. Gottumukkala and A. Pandey. 2015. Crude oil biodegradation aided by biosurfactants from *Pseudozyma* sp. NII 08165 or its culture broth. Biores. Technol. 191: 133–139.

Saliba, A.M., R. Nishi, B. Raymond, E.A. Marques, U.G. Lopes, L. Touqui, et al. 2006. Implications of oxidative stress in the cytotoxicity of *Pseudomonas aeruginosa* ExoU. Microb. Infect. 2: 450–459.

Scenihr. 2005. The appropriateness of existing methodologies to assess the potential risks associated with engineered and adventitious products of nanotechnologies. SCENIHR/002/05.

Seetharam, R.J. and K.R. Sridhar. 2007. Nanotoxicity: Threat posed by nanoparticles. Curr. Sci. 93: 769–770.

Shin, W.K., J. Cho, A.G. Kannan, Y.S. Lee and D.W. Kim. 2016. Cross-linked composite gel polymer electrolyte using mesoporous methacrylate functionalized SiO_2 nano particles for lithium-ion polymer batteries. Sci. Rep. 6: 26332.

Simon, M. and A. Richaume. 2015. Impacts of engineered nanoparticles on the activity, abundance, and diversity of soil microbial communities: A review. Environ. Sci. Pollut. Res. 22: 13710–13723.

Singh, N., B. Manshian, G.J.S. Jenkins, S.M. Griffiths, P.M. Williams, T.G.G. Maffeis, C.J. Wright and S.H. Doak. 2009. Nanogenotoxicology: The DNA damaging potential of engineered nanomaterials. Biomat 30: 3891–3914.

Singh, C.R., K. Kathiresan and S. Anandhan. 2015. A review on marine based nanoparticles and their potential applications. Afr. J. Biotechnol. 14: 1525–1532.

Singh, P., S. Ravindran, J.K. Suthar, P. Deshpande, R. Rokade and V. Rale. 2017. Production of biosurfactant stabilized nanoparticles. Int. J Pharm. Bio. Sci. 8: 701–707.

Singh, N., A. Kumar and B. Sharma. 2019. Role of fungal enzymes for bioremediation of hazardous chemicals. *In*: A.N. Yadav, A.N. Singh, S. Mishra, S. Gupta (eds), Rec Advanc White Biotechnol Through Fungi, Vol 3. Springer, Berlin, Germany, pp. 237–256.

Soni, I. and S.B. Bondi. 2004. Silver nanoparticles as antimicrobial agent: A case study on *E. coli* as a model for Gram-negative bacteria. J. Colloid Interf. Sci. 275: 1770–1782.

Tandon, P.K. and S.B. Singh. 2014. Catalytic applications of copper species in organic transformations: A review. J. Catal. Catal. 1: 1–14.

Tewari, B.B. 2019. Critical reviews on engineered nanoparticles in environmental remediation. Curr. J. Appl. Sci. Technol. 36: 1–21.

Tosco, T., M.P. Papini, C.C. Viggi and R. Sethi. 2014. Nano scale zero valent iron particles for ground water remediation: A review. J. Cleaner Prod. 77: 10–21.

Tungittiplakorn, W., L.W. Lion, C. Cohen and J.Y. Kim. 2004. Engineered polymeric nanoparticles for soil remediation. Environ. Sci. Technol. 38: 1605–1610.

Tungittiplakorn, W., C. Cohen and L.W. Lion. 2005. Engineered polymeric nanoparticles for bioremediation of hydrophobic contaminants. Environ. Sci. Technol. 39: 1354–1358.

Van der Zande, M., R.J.B. Peters, A.A. Peijnenburg and H. Bouwmeester. 2011. Biodistribution and toxicity of silver nanoparticles in rats after subchronic oral administration. Toxicol. Lett. 205: S289–S289.

Walsh, F.C. and G.W. Reade. 1994. Electrochemical techniques for the treatment of dilute metal-ion solutions. Stud. Environ. Sci. 59: 3–44.

Wang, D., C. Deraedt, J. Ruiz and D. Astruc. 2015. Magnetic and dendritic catalysts. Accts. Chem. Res. 48: 1871–1880.

Xing, Y., X. Chen and D. Wang. 2007. Electrically regenerated ion exchange for removal and recovery of Cr(VI) from wastewater. Environ. Sci. Technol. 41: 1439–1443.

Xiufang, C. 2017. Nanocatalysis, Encyclopedia of Physical Organic Chemistry. John Wiley & Sons, Inc. ISBN 978-1-118-46858-6.

Xu, Y. and D. Zhao. 2007. Reductive immobilization of chromate in water and soil using stabilized iron nanoparticles. Water Res. 41: 2101–2108.

Yadav, K.K., J.K. Singh, N. Gupta and V. Kumar. 2017. A review of nanobioremediation technologies for environmental cleanup: A novel biological approach. J. Mater. Environ. Sci. 8: 740–757.

Yan, K., G. Wu, C. Jarvis, J. Wen and A. Chen. 2014. Facile synthesis of porous microspheres composed of TiO_2 nanorods with high photocatalytic activity for hydrogen production. Appl. Catal B: Environ. 148: 281–287.

Zamboulis, D., E.N. Peleka, N.K. Lazaridis and K.A. Matis. 2011. Metal ion separation and recovery from environmental sources using various flotation and sorption techniques. J. Chem. Technol. Biotechnol. 86: 335–344.

Zhang, P., H.H. Hahn and E. Hoffmann. 2003. Different behavior of iron(III) and aluminum(III) salts to coagulate silica particle suspension. Acta Hydroch. Hydrobiol. 31: 145–151.

CHAPTER 9

Recent Updates on the Role of Biosurfactants for Remediation of Various Pollutants

Nilanjana Das[1], Sanjeeb Kumar Mandal[2], Devlina Das[3], Jagannathan Madhavan[4] and Adikesavan Selvi[5*]

[1]Bioremediation Laboratory, School of Bio Sciences and Technology, Vellore Institute of Technology, Vellore-632014, Tamil Nadu, India.

[2]Department of Biotechnology, Sri Shakthi Institute of Engineering and Technology, Coimbatore, Tamil Nadu, India

[3]Department of Biotechnology, PSG College of Technology, Coimbatore, Tamil Nadu, India

[4]Solar Energy Lab, Department of Chemistry, Thiruvalluvar University, Serkkadu, Vellore-632115, Tamil Nadu, India

[5]Environmental Molecular Microbiology Research (EMMR) Laboratory, Department of Biotechnology, Thiruvalluvar University, Serkadu, Vellore 632115, Tamil Nadu, India.

1. INTRODUCTION

Massive growth of industries and mindless disposal by humans are the prime sources of environmental pollution and its related hazards (Karlapudi et al. 2018). Toxic substances released into the environment are of global concern now. Conventional technologies like incineration, adsorption and excavation are considered less efficient and costly for the removal of pollutants (Bezza and Nkhalambayausi Chirwa 2016). However, the biological based treatment methods have always been more eco-friendly as well as economically feasible alternatives (Selvi et al. 2014, Dadrasnia et al. 2015, Selvi et al. 2015, de la Cueva et al. 2016), since the organic components which account for toxicity are converted to much simpler compounds or H_2O and CO_2 via

*Corresponding author: seljeev@gmail.com Ph: +91-8072642680

biological pathways (Chirwa and Smit 2010). Among various biological approaches, employing biosurfactant producing microbes and/or biosurfactants is considered to be an effective and eco-friendly remediation method in cleaning up the environments contaminated with pollutants. Generally, surfactant-based remedial technologies work on the basis of adsorption behavior, solubilizing capacity, biocompatibility and toxicity of the surfactants. Surfactants are reported to enhance pollutant desorption from soil, thus increasing the bioavailability and promoting bioremediation of organic pollutants. Successful application of cationic, anionic, non-ionic and zwitterionic surfactants have been reported for soil remediation (Mao et al. 2015). Though chemical surfactants are equally advantageous in terms of application prospective, microbial surfactants are highly prefered due to their low toxicity, solubilization capacity, and efficacy in improving the biodegradation process (Usman et al. 2015).

Though there are many studies that have been published in the last 10 years covering different aspects of biosurfactants, this chapter is a compilation of research updates on the role of biosurfactants towards the removal of various emerging contaminants. The latest successful remediation techniques using biosurfactants towards pollutant removal discussed here will surely aid the researchers in understanding about the nature and persistence behaviour of the environmental pollutants and its associated risks due to their existence. Research prospects discussed in this chapter will encourage future researchers to focus on new technological attempts towards the development of sustainable options to treat heavy metals, polycyclic aromatic hydrocarbons (PAHs), pesticides, synthetic dyes etc., contaminated environments.

2. BIOSURFACTANTS

Biosurfactants are structurally diversified surface-active agents, composed of hydrophilic and hydrophobic group that are produced by wide range of microbes (Nurfarahin et al. 2018, Das and Kumar 2018). Chemically, they exist as lipopeptides, lipopolysaccharides, glycolipids, phospholipids, polysaccharide-protein complexes, neutral lipids and fatty acids (Varjani and Upasani 2017). The hydrophilic-lipophilic balance in biosurfactants infers the proportion of the hydrophobic and hydrophilic components in them (Usman et al. 2015). Biosurfactants have been widely exploited commercially for many prospective applications in various fields, such as microbial mediated enhanced oil removal/recovery, bioremediation and biodegradation of pollutant food emulsifiers, moisturizering lotions etc. (Chen et al. 2013, Manickam et al. 2014, Joshi et al. 2016, Varjani and Upasani 2017, Kumari et al. 2018). Due to the amphiphilic nature, the biosurfactants increase the water bioavailability of the hydrophobic water-insoluble materials, thus changing the properties of microbial cell surface. This property makes them function as excellent emulsifiers, dispersing and foaming agents. These are eco-friendly, less toxic, biodegradable, highly selective and active even at extreme pH, temperature, and salinity, along with better foaming characteristics.

The unique features of biosurfactants make them suitable to deploy in bioremediation methods. Depending on their chemical compositions and functional characteristics, they can be classified as high molecular weight bioemulsifiers and low molecular weight biosurfactants. Though both types can form stable emulsions, bioemulsifiers will not reduce interfacial and surface tension, whereas the other one will effectively reduce both the interfacial and surface tension (Uzoigwe et al. 2015).

Compared to chemical based surfactants, bio-based surfactants possess certain distinctive features, like low toxicity, biodegradability, high biocompatibility, efficiency, and on-site production capability (Uzoigwe et al. 2015). They are widely being used towards the remediation of various pollutants in the environment.

3. REMEDIATION OF HEAVY METALS

Heavy metals are noteworthy persistent soil contaminants and regarded as biosphere hazard (Lal et al. 2018). Their entry into the environment is by both natural and anthropogenic means. Natural calamities like volcanic eruptions are known to release heavy metals directly into the environment. However, majorly contributing anthropogenic source processes include alloy production, smelting, mining, welding, and excessive pesticide usage (Segura et al. 2016, Sing and Kumar 2017, Selvi and Aruliah 2018). Remediation of soil contaminated with toxic heavy metals such as chromium, cadmium, lead, zinc etc. is done following the conventional strategies of adsorption, thermal treatment, chlorination, chemical extraction, ion-exchange, membrane separation, etc. (Wuana and Okieimen 2011). But, utilization of microorganisms has been considered preferably for *in situ* treatment in order to reduce the immense cost of the conventional remediation techniques (Pacwa-Płociniczak et al. 2011).

The use of a biosurfactant will stimulate the heavy metals desorption in two ways. The first one is by complexing the free metal ions present in the solution, thus decreasing the interaction of the metal at solution-phase. The second approach is by accumulating the biosurfactants near the solid-solution interfacial region under reduced tension conditions, thus allowing the direct interaction between the the sorbed metal and the biosurfactant (Olaniran et al. 2013). Utilization of biosurfactants has been reported to show promising results in remediation of heavy metal contaminated soil sites. It is well understood that some microorganisms are capable of producing biosurfactants, which are obviously eco-friendly alternatives to chemical based surfactants (Das et al. 2017, Heryani and Putra 2017). The hydrophilic (polar) heads of biosurfactant possess high binding capacity towards heavy metal ions and thus been widely used for treating heavy metal contaminated soil and water (Zhou et al. 2017, Chen et al. 2017). The underlying mechanism by which heavy metal is removed by ionic biosurfactants is depicted in Fig. 1 (Akbari et al. 2018).

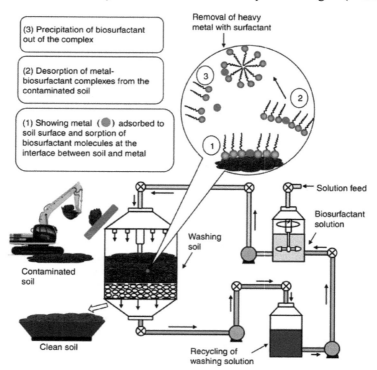

Figure 1   Mechanism of heavy metal removal from contaminated soil using biosurfactants (Source: Akbari et al. 2018).

The following three steps are mainly involved in heavy metal removal from the contaminated soil. Firstly, by washing the soil with biosurfactant solution. Secondly, separation of the electrostatically adsorbed heavy metals from the soil particles/sludge surface, that are confined within the micelle of the biosurfactant. Finally, the recovery of the biosurfactant by membrane separation method.

Figure 2 depicts the use of biosurfactants at different voltage gradients as bio-enhanced electrokinetic remediation of heavy metals and PAHs in dredged marine sediments using a mixture of citric acid (CA), a chelating agent and a surfactant in the processing fluids.

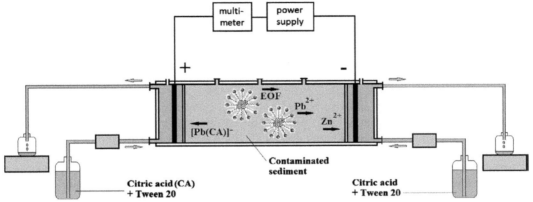

Figure 2 Application of biosurfactants and periodic voltage gradient for enhanced electrokinetic remediation of metals and PAHs in dredged marine sediments (Source: Ammami et al. 2015).

Various operating conditions were tested, resulting in the replacement of Tween 20, a synthetic surfactant by biosurfactants. Increased concentrations of CA seemed to favour both PAH and the metal removal. Likewise, a periodic voltage gradient, a low CA concentration and Tween 20 showed the superior results towards 16 priority PAHs and the heavy metals metals viz. Cd, Pb and Zn removal. The solutions with rhamnolipids (0.028%) and biosurfactant produced by *P. fluorescens* showed promising results. Low electroosmotic flow was noted on using biosurfactants, which reflected in less PAH removal (Ammami et al. 2015).

The applicability of a mixed consortium of acidophilic bacterium and biosurfactant-producing bacterium was reported towards the bioleaching of metals from combustion wastes (Karwowska et al. 2015). They have achieved best bioleaching outcomes with high metal content that were bound to iron oxide, magnesium oxides, carbonates, and also in exchangeable form fractions. The highest metal recovery percentage of 90% was noted for Cu, Ni, and Zn. Ni and Cr, bounded to carbonates were easily removed than Pb and Cd that were less bioleached.

Implementation of bioleaching on heavy metal-contaminated soils with biosurfactant producing bacterial strain, *Burkholderia* sp. Z–90, isolated from vegetable oil source showed the potentiality of various heavy metal removal (Yang et al. 2016). The biosurfactant produced by the bacterial strain was characterized as glycolipid. The removal efficiency for As, Cu, Cd, Mn, Zn, and Pb was noted as 31.6, 24.1, 37.7, 52.2, 44.0 and 32.4%, respectively. On comparison, Cd, Zn and Mn were removed more easily than As, Pb and Cu from the soil. Performance of biosurfactant sourced from *Candida sphaerica* UCP0995 was found to exhibit excellent heavy metal remediation of Pb, Zn and Fe with their observed removal rates of 79%, 90% and 95%, respectively, by Luna et al. (2016). Various combinations of biosurfactant solutions were also tested in their study.

Among various microbial surfactants, rhamnolipid constituting biosurfactants have been explored more in terms of research, owing to their natural degradability nature, low toxicity, high production efficiency, and higher environmental compatibility. They are stable at wide range of pH and temperatures. Rhamnolipid was reported to be more effective, especially on targeted metals and heavy metals that exist in exchangeable, Fe-Mn oxide-bound or carbonate-bound fractions (Chen et al. 2017, Zhou et al. 2017). Abyaneh and Fazaelipoor (2016) evaluated the role of rhamnolipid as a biosurfactant produced by *P. aeruginosa* for chromium(III) removal from aqueous medium by precipitate flotation technique. Ferrous sulphate was used to precipitate Cr(VI) to Cr(III), which was consequently removed by flotation process. The obtained results evidenced the effectiveness of the biosurfactant chromium removal (greater than 95%) with all the factors favouring the flotation process. Rhamnolipid was used as washing agent towards heavy metal removal in river sediments by Chen et al. (2017). Batch studies were done to test the removal efficiency of the biosurfactant. Various parameters, like effects of solution pH, washing time, liquid/solid ratio, rhamnolipid concentration, were studied. The heavy metal speciation was analyzed before and after the washing process in sediment samples. Washing of heavy metals was noted at high pH, high concentration of biosurfactant, and prolonged washing time, thus signifying the potentiality of the rhamnolipid as a washing agent for heavy metal removal.

The combined activity of biosurfactant and complexing agent, tetrasodium of N, N-bis (carboxymethyl) glutamic acid (GLDA) were employed towards enhanced heavy metal removal by electrokinetic (EK) treatment. Rhamnolipid and GLDA were reported to produce high electric output and conductivity during EK treatment. Contrasting trend of lower pH for rhamnolipid at cathode and higher pH for GLDA at anode was noted. However, better heavy metal removal efficiencies were noted with the synergic process of EK combined with chelating agent and biosurfactant (Tang et al. 2017).

Zhou et al. (2017) conducted the column experiments to confirm the capacity of rhamnolipid and saponin to remove the rare earth metals viz. La, Eu, Y, and Ce from the soil. The order of leaching efficiency was noted with 25 g/L of saponin, followed by 10 g/L of saponin and 10 g/L of rhamnolipid, thus proving the great potential of saponin solution towards the leaching of the rare earth metals.

The simultaneous removal of cadmium(II) ($Cd^{+2)}$ from aqueous matrix by micellar-enhanced ultrafiltration (MEUF), along with rhamnolipid, was reported by Verma and Sarker (2017). The process was optimized using the response surface methodology. The effects of experimental conditions such as, feed temperature, stirring speed and transmembrane pressure on permeate flux and Cd^{+2} rejection was studied under the optimal feed conditions. Efficiency of MEUF process was assessed by estimating vesicle binding constants, vesicle loading, and the distribution coefficient for Cd^{+2}. A maximum removal of cadmium (98.8%) was achieved by this process.

The removal of lead from contaminated port sediment using biosurfactant produced from *Bacillus* sp. G1 was reported by Guo et al. (2017). The removal efficiencies of lead were studied in terms of pH, ionic strength and water:solid ratio. Highest lead removal efficiency occurred at pH 2.0. However, the removal of lead was not affected significantly by NaCl. Fourier transform infrared spectra (FTIR) analysis showed the involvement of $-CH_3$ and C=O functional groups. The results of Scanning electron microscopy (SEM) confirmed the stable complex formation between lead ions and the biosurfactant. It was concluded that the biosurfactant effectively removed lead ions from the contaminated port sediment, which could be suggested as a novel option for remediation of lead contaminated sites.

Enhanced heavy metal removal from sludge was demonstrated by pretreating the sludge with rhamnolipid, sophorolipid and saponin using novel enhanced EK technique (Tang et al. 2018). The obtained results showed higher heavy metal removal efficieny in the biosurfactants enhanced EK3, EK4 and EK5 tests than the unenhanced EK1 test and acid-enhanced EK2 test samples. The free cations of heavy metal and metal-complexes (positively charged) electromigrate towards the cathode. Rhamnolipid, sophorolipid and saponin were found to possess binding and chelating characteristics in acid conditions that resulted in enhanced heavy metal removal.

Recently, Shami et al. (2019) reported the efficient removal of cadmium (Cd) from aqueous medium using coal waste activated by rhamnolipid biosurfactant, based on response surface methodology, following CCD experimental design. The highest Cd removal of 99% was observed at pH 9.0 and the absorbent:metal ratio of 125 in <60 min of contact time after the process optimization. Cadmium removal was found to be 48% using non activated coal waste, which proved the potential role of rhamnolipid on coal waste activation. They concluded on considering this efficient, cost effective and easily available adsorbent for treating wastewater contaminated with heavy metals. The proposed mechanisms behind the metal ion adsorption on waste material surface is illustrated in Fig. 3.

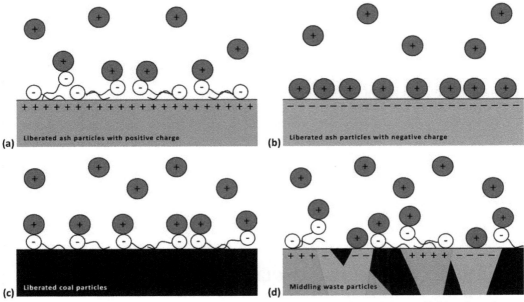

Figure 3   Mechanisms for heavy metal adsorption on the surface of rhamnolipid activated coal waste particles (Source: Shami et al. 2019)

A study on cadmium accumulation in the rice crop during the time of harvest using rhamnolipid, along with Pluronic F127/poly acrylic acid (F127/PAA) hydrogels, was reported by Shen et al. (2019) as shown in Fig. 4.

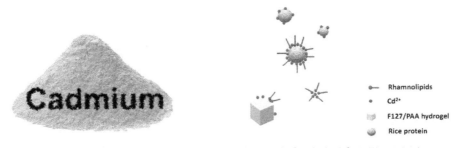

Figure 4   Cadmium removal from rice protein via synergistic treatment of rhamnolipids and F127/PAA hydrogels (Source: Shen et al. 2019)

92% removal of cadmium in rice protein was acheived after two washings with rhamnolipids (0.5 g/L) and F127/PAA (5 g/L) hydrogel. However, the washing process was not found to affect the structure and main constituents of the rice protein. After treatment, a residual cadmium of 0.2 mg/kg was noted, which was well below the accepatable levels. As shown in the figure, the rhamnolipids bound with the Cd from protein and transported it to F127/PAA hydrogel for storage, which was tested for reuse options for upto 4–5 adsorption-desorption cycles. This work reported, for the first time, a green and cost effective approach for cadmium removal from polluted rice grains.

Lal et al. (2018) reported an advanced biological tool for the treatment of heavy metal contaminated soil using rhizobacteria that produces exopolysaccharide and biosurfactant. This was a cost effective, sustainable and promising technique, which could be operative in nature and hold significance over other conventional treatment options.

Though manganese (Mn) is an essential element required for the growth of plants and animals, it can exert toxic effects in humans, plants and animals at high concentrations. Permissible level of Mn for humans is 0.05 mg/L as set by World Health Organization (WHO). Excess Mn concentration was reported to cause anaemia, central nervous system disturbances, Parkinson's and pulmonary disease (Patil et al. 2016, Ramoju et al. 2017, Idress et al. 2018). Ferreira et al. (2019) reported Mn removal from water using emulsion liquid membrane (ELM), made up of chelating agents and the produced biosurfactant *in loco* containing D2EHPA, EDTA, H_2SO_4, and NaCl. The substitution of biosurfactant based membranes to chemical surfactants is due to the need of liquid based membrane to treat wastewater containing toxic metals. Different compositions of ELM of biosurfactant (ELMB) were tested after the optimization by artificial neural network and heuristic optimization using artificial bee colony optimisation approach. The optimization reports were found to show positive results in the simulated Mn removal. The ELMB treatment was found to be economically feasible and fast for the removal of Mn(II) ion from wastewater. The removal of Mn(II) ions was reported to be more than 77% in 2 min and total estimated cost was about 51.62 USD/m^3 of effluent.

4. REMEDIATION OF POLYCYCLIC AROMATIC HYDROCARBONS (PAH)

PAHs are categorised as high priority pollutants because of their mutagenic, carcinogenic, and teratogenic toxic effects. PAH bioremediation is greatly limited due to their high hydrophobicity, low solubility in aqueous matrix, and strong adsorption to soil (Bezza and Chirwa 2017a). Various cleanup techniques have been widely implemented for their removal in the last several decades. Since many PAHs are associated with finely grained and organically rich sediments, they contribute to their contamination, thus making these sediments an ultimate drain-site for these persistant pollutants. Researchers have reported many techniques for estimating the bioavailable fractions of PAHs for remediation of these contaminated sites. However, many recent studies on *in situ* approaches viz. surfactant addition, biostimulation, bioaugmentation, phytoremediation, EK treatment, and using of different sorbents, namely biochar and activated carbon are being used as an effective and long-term solution for remediation of PAH contaminated sediments (Maletić et al. 2019).

As discussed earlier, rhamnolipids, being naturally derived products, can be widely applied over a range of potential commercial applications. Hošková et al. (2015) reported bacterial rhamnolipids produced from *A. calcoaceticus, P. aeruginosa* and *E. asburiae* which were grown as pure culture as well as mixed populations. Homologues of mono and di-rhamnolipid comprising of 1 or 2 saturated/monounsaturated 3–OH fatty acids were found in all the bacterial strains. The rhamnolipids produced by all the bacterial strains were reported to effectively emulsify the crude petroleum compared to synthetic surfactants such as sodium dodecyl sulfate and Tween 80. Rhamnolipids from *E. asburiae* showed good performance towards phenanthrene solubilization. The rhamnolipid mixtures obtained from pure culture and mixed culture strains were proposed to be exploited for remediation of various PAHs.

Blyth et al. (2015) reported on the use of red ash leaves as a biosurfactant source towards the bioremediation of soils contaminated with PAHs. This application can provide an ecofriendly, sustainable, and cost effective substitute to chemical surfactants. Compared to the control, the soil treated with the red ash leaves extracted biosurfactant showed a remarkable decrease in the PAH level in the heavily contaminated soil. On further analysis of the microbial community, the extracted biosurfactant was found to promote the abundancy of the bacterial population, especially the PAH-degrading Gram positive bacteria in the soil, which could serve as a bioagent towards the significant decrease in total PAHs and TPH levels in the soil.

Biosurfactant from *Rhodococcus ruber* IEGM 231 was used for PAH removal in spiked soil containing major petroleum hydrocarbons and heterocycles mixtures that could possibly serve as biopreparation for the replacement of synthetic surfactants (Ivshina et al. 2016). Both the *in situ* produced biosurfactants and the *Rhodococcus* extracted biosurfactants showed almost similar PAH removal. The removal percentage of PAH was found to be varying from 16–69%, with *Rhodococcus* extracted biosurfactant 2.5 times more than the other one. As this biosurfactant could be produced in less time, there is no need of using organic solvents and no requirement of time consuming steps like evaporation and extraction, it was suggested for cost effective industrial applications. A simplified UV-method was used in this study to monitor the residual polyaromatic compounds in the treated soil, that might be used as an effective tool for safe ecological planning to protect and restore PAH contaminated ecosystems.

In general, lipopeptide biosurfactant was reported to exhibit potential application prospects towards the bioremediation of PAH contaminated environment. Bezza and Chirwa (2017b) studied the enhancement of pyrene biodegradation using lipopeptidic biosurfactant from *Paenibacillus dendritiformis* CN5 strain. Batch studies were demonstrated to study the effect of biosurfactant on pyrene biodegradation using a microbial consortium of *P. viridiflava* (49.5%) and *P. nitroreducens* (32.5%) containing lipopeptidic biosurfactant from *P. dendritiformis* CN5 strain in mineral salt medium. The pyrene degradation was enhanced to 83.5% and 67%, respectively, at 600 mg/L and 300 mg/L of lipopeptide at the end of 24 days, than compared to 16% of degradation without lipopeptide. On further increasing the lipopeptide concentration to 900 mg/L, pyrene degradation was found to be reduced to 57%. However, a four-fold increase in the maximum removal rate of pyrene was observed at optimal lipopeptide concentration.

An enhanced biodegradation study of highly contaminated PAHs in aged wood treatment plant soil using lipopeptide biosurfactant extracted from *B. cereus* SPL-4 was reported by Bezza and Chirwa (2017a). Batch experiments showed a significant enhancement in the microcosms augumented with biosurfactant than the surfactant-free controls. Lipopeptide biosurfactant concentration of 0.2% and 0.6% (w/w) showed a maximum PAH removal of 51.2% and 64.1% of 4-ringed structure and 55% and 79% of 5- and 6-ringed structures, respectively. On the contrary, surfactant-free control showed 29% removal of 4-ringed and 25.5% removal of 5- and 6-ringed structured PAHs. The obtained results evidenced that the high biosurfactant dosage had favoured maximum degradation of 5- and 6-ringed PAHs, thus suggesting that the biosurfactant assisted degradation using microbial consortium could be effective while implementing the bioremediation of aged PAH-contaminated soils. Similar studies of lipopeptide biosurfactant (bacterial consortium) assisted remediation of the C_5–C_{11} hydrocarbons from the oily sludge was reported (Chirwa et al. 2017). The molecular structure of the purified biosurfactant was characterised by liquid chromatograph coupled to a tandem mass spectrometry (LC–MS). PAH-degrading potential of phenanthrene enriched isolation of *P. aeruginosa*, W10 from motor oil contaminated soil was reported to utilize vegetable oils and other used oils as their sole carbon and energy source. They reported the production of a biosurfactant BSW10 on olive oil with high stability over salt concentration, temperature, and pH, along with a re-mobilization capability on sorbed oil soil surface. Owing to these charcteristics, they concluded that the potentiality of the strain W10, along with its biosurfactant, can be considered as effective candidates for remediating hydrocarbon polluted sites (Chebbi et al. 2017).

Salamat et al. (2018) reported on the production of biosurfactant from *B. subtilis* isolated from Persian Gulf sediments and employed it towards the reduction of anthracene. A significant reduction of 69.95% was noted after 120 h, which confirmed the anthracene removal ability of the bacterium and its biosurfactant. The remediation potential of lignin and ryegrass, along with the biosurfactant rhamnolipid, was explored by a pot experiment conducted at lab scale using an aged PAH contaminated arable soil (Wu et al. 2018). Treatments involving biostimulation, rhizoremediation, and biosurfactant addition showed promising results towards the detoxification of the aged PAH-contaminated soil. The combination of lignin, ryegrass, and rhamnolipid resulted in the considerable dissipation of PAH in the sampled soil. Lignin and ryegrass was found to alter the composition, microbial abundance, and the network structure which stimulated the rhamnolipid mediated PAH biodegradation in the soil. The findings of this study highlighted the significance of a synchronized microbial activity improvement and the PAH bioavailability.

Recently, on-site remediation of PAHs was demonstrated using the biosurfactant produced from *P. aeruginosa* S5 strain, isolated from coking effluent (Sun et al. 2019). The produced biosurfactant was characterized as glycolipid with a critical micelle concentration (CMC) of 96.5 mg/L, and surface tension was reduced from 72.2 mN/m to 29.6 mN/m. Owing to its CMC value, the biosurfactant was found to possess high surface activity over a wide range of pH from 3.5 to 9.5 and salinity ranging from 0 to 15%. *In situ* treatment of coking wastewater for 15 days with the isolated indigenous *P. aeruginosa* S5 strain showed an enhanced and effective reduction of high molecular weight (HMW) PAHs from 9141.02 µg/L to 5117.16 µg/L, thus considering them as a possible option for in-site remediation technologies.

The role of surfactants viz. rhamnolipid, Tween 80 and dioctyl sodium sulfosuccinate (DOSS) was tested towards the bioremediation of HMW PAHs by solubilization, sorption, followed by benzo (a) pyrene (BaP) biodegradation by *P. aeruginosa* BT-1 strain (Meng et al. 2019). Samples were collected periodically and detected by RP-HPLC attached with fluorescence detector. The highest order of BaP solubilization capacity by the surfactants was achieved using Tween 80, which was followed by Rhamnolipid and DOSS, respectively. Similarly, the highest order of PAH solubility enhancement degree by Tween 80 was seen in BaP, followed by bypyrene and phenanthrene. It was concluded that Tween 80 was more a potent surfactant compared to rhamnolipid from the strain BT-1.

A highly stable bioemulsifier produced by *Rhodococcus erythropolis* OSDS1, along with a microbial consortium of petroleum hydrocarbon biodegraders, was tested for the depletion of crude oil by Xia et al. (2019). It was found that >90% of the initial emulsification capacity was retained even after 168 h. On testing the emulsification activity of the bioemulsifier, the highest emulsification activity was seen with diesel, followed by crude oil/mineral oil, and gasoline. A maximum crude oil depletion of 85.26% at the end of 15 days was noted on treating with mixed bacterial consortium and the biosurfactant, a significantly higher percentage than with the individual strains.

The individual and combined effect of biochar (BC), rhamnolipid (RL) and rhamnolipid modified biochar (RMB) was reported in phytoremediation using a novel plant, *Spartina anglica,* to treat petroleum hydrocarbon-contaminated soil (Zhen et al. 2019). After 60th day of cultivation, the rate of removal of total petroleum hydrocarbons at a concentration of 30 g/kg was noted as 8.6% in unplanted soil, 19.1% in planted soil, 27.7% in planted soil + BC, 32.4% in planted soil + BC + RL and 32.4% in planted soil + RMB. *Spartina anglica* plantation was found to significantly decrease the concentration of tricyclic PAHs and C_8–C_{14} hydrocarbons as compared to unplanted soil. On the other hand, the use of RMB and BC was reported to reduce petroleum hydrocarbon toxicity, thus improving the growth of *Spartina anglica* in terms of height, total chlorophyll content, root vitality and regulated microbial community of rhizosphere. The outcomes of the study clearly established the phyto-mycorrhizal symbiosis of the BC and RMB amended soil.

5.   REMEDIATION OF PESTICIDES

Pesticide remediation has been a matter of concern over the years due to high levels of pollution caused by mindless pesticides usage. Sometimes only 0.1% of pesticide reaches the target species, whereas the remaining reaches the non-target areas. The pesticides application to the crops enhances the yield undoubtedly by minimising the pest attacks, but the pesticide residues cause acute and chronic health impacts in humans and other non-target animals. The pesticide residues can enter the bloodstream directly via inhalation, oral or dermal and indirectly via water and food (EPA 2009). According to WHO (2010) study, out of the total global pesticide use, 45% belongs to organophosphate (OP) compounds, which are reported to create high potential environmental hazards.

Rhamnolipid facilitated enhanced degradation of chlorpyrifos (OP pesticide) using microbial consortium in soil-water system, which was reported earlier by Singh et al. (2016). Five potential bacterial cultures of genus Klebsiella, Pseudomonas, Stenotrophomonas, Bacillus, and Ochrobactrum and their consortium, along with a crude fraction of rhamnolipid biosurfactant system, were found to show an effective degradation of chlorpyrifos and its toxic intermediates. The added biosurfactant was found to improve the pollutant solubility in the aqueous phase by 2–15 folds.

Carbendazim, a widely used fungicide with benzimidazole group, fights against fungal diseases in plants. But, its stability and persistence in soil were reported to cause long-term adverse effects. Acarbendazim degradation study using *Rhodococcus* sp. D-1 bacterium was reported to degrade 98.2% of carbendazim at 200 ppm within 5 days (Bai et al. 2017). Moreover, the effect of biosurfactant rhamnolipid was tested on the extent and degradation rate of carbendazim in batch mode. Rhamnolipid was found to affect the degradation in a concentration-dependent mode. The maximum biodegradation was found to be 97.33% within 2 days at 50 ppm of biosurfactant. Moreover, rhamnolipid also facilitated carbendazim detoxification.

Dichlorodiphenyltrichloroethane (DDT) is a distinct organochlorine pesticide (OCP), commonly used against agricultural insects/pests. Environmental pollution due to DDT has always remained a worldwide issue, owing to its persistence, bioaccumulation, and toxicity to human beings and other species. Wang et al. (2017) reported the use of biosurfactants produced by *Pseudomonas* sp. SB towards DDT-contaminated soil by phytoremediation using two grass species.

Dichlorobenzenes (DCBs) are non-polar and highly volatile organic pollutants, that are used as a disinfectant, pesticide and in the organic chemical synthesis. The United States Environmental Protection Agency (EPA 1988) has listed DCBs as one of the priority pollutants because of their high potential environmental risk and toxicity. Treatment of DCB polluted soils using efficient and eco-friendly remediation technologies like surfactant-enhanced remediation technique has shown promising results. Comparitive studies on demonstration of soil column experiments for the o-DCB and p-DCB degradation using micellar biosurfactant solutions (alkyl polyglycoside, saponin) and Tween 80, a chemical based synthetic surfactant, showed the highest removal efficiency of 80.43% and 76.34% with saponin proceeded by alkyl polyglycoside and Tween 80 (Pei et al. 2017).

The removal of chlorpyrifos, an OP pesticide, was demonstrated using biosurfactants of three isolates belonging to *Pseudomonas* (KX881513) species within 3 days by Shabbier et al. (2018). Instrumental analysis viz. UV-visible, GC–MS and FTIR spectroscopy confirmed the removal of chloropyrifos. The results of the analysis on pesticide uptake showed that chloropyrifos was bound to the cell surface in intracellular components in its native state. Pesticide removal from soil by these bacteria was also noted.

Application of rhamnolipid anionic biosurfactant from bacterium *Lysinibacillus sphaericus* IITR5 1 towards dissolution of hydrophobic chlorinated pesticides was evaluated by Gaur et al. (2019). The bacterium produced 1.6 g/L of biosurfactant with 48% emulsification index and high stability over a wide range of temperature (4°C–100°C), pH (4.0–10.0), and salt concentration (2%–14%). The surface tension was reduced from 72 to 52 N/m. Enhanced dissolution of 7.2, 2.9, and 1.8 folds was noted for γ-hexachlorocyclohexane and α-, β-endosulfan at 90 ppm of rhamnolipid. The bacterium utilized chlorobenzene, benzoic acid, 3- and 4-chlorobenzoic acid as its sole carbon sources. In addition, the biosurfactant also exhibited antimicrobial properties against various bacterial pathogens. The obtained results of the study indicated the importance of rhamnolipid for enhanced dissolution of hydrophobic pesticides.

6. REMEDIATION OF SYNTHETIC DYES

Synthetic dyes are one of the main classes of pollutants found in aqueous system due to indiscriminate release from various industries viz. dyeing, cosmetics, printing, textile, etc. Various conventional methods of dye removal viz. adsorption, coagulation, biological method, chemical oxidation, membrane processes, ozone treatment, photocatalytic degradation processes, etc. have been reported by various researchers (Dasgupta et al. 2015, Brillas and Martínez-Huitle 2015, Holkar et al. 2016). But, these methods are expensive and have some limitations (Yagub et al. 2014), whereas MEUF process has been preferred due to the advantages of high permeate flux, high removal efficiency, and less energy needed than individual membrane process methods (Shah et al. 2016). The solubilization and binding capacity of the surfactant micelles have been used in dye removal from aqueous medium by MEUF process (Samal et al. 2017a).

Saponin is one of the phyto-biosurfactants that has received immense research interest due to its significant functional characteristics, health aids, renewable and eco-friendly nature, biodegradability, and ecologically adaptability (Roy et al. 1997, Schmitt et al. 2014). Solubilization of anionic and cationic dyes viz. eosin yellow and methylene blue in aqueous system was studied using eco-friendly saponin biosurfactant extracted from the pericarp of *Sapindus mukorossi* (fruit) after a detailed characterization study by various instrumentation techniques (Samal et al. 2017a). The solubilization potentiality of dyes in saponin micelle was determined based on solubilization power (SP), partition coefficient (Km) of water and micelle, and Gibbs free energy change of solubilization. The saponin micelle size in aqueous system was noted in the range of 10–11.5 nm, whereas the agglomerated size of saponin was of higher sizes that ranged from 132–235 nm, 390–990 nm, and 5155–8520 nm. The Km of both the dyes was calculated between the aqueous medium and the saponin micelles. Similar study on saponin from the same plant source was employed in MEUF process towards methyl violet removal from wastewater, which was demonstrated in both batch and column mode (Samal et al. 2017b). A maximum methyl violet removal of >99% was achieved in MEUF process using 10 kDa polyethersulfone (PES) membrane with an initial dye concentration of 250 mg/L. In both the studies, the solubilization of the dyes was achieved by the synergised effect of saponin micellar solubilization and dye sorption by saponin agglomerates. The results of these studies explored the potential application of saponin as biosurfactant in many surfactant based processes viz. MEUF, CPE and SER.

Enahnced decolourization of azo dyes was demonstrated with the biosurfactant, BS-L1011, from *P. taiwanensis* L1011 by Liu et al. (2017). The maximum yield of 1.12 g/L of BS-L1011 was noted using 1 g/L yeast extract, 10 g/L D-mannitol as carbon, and 3 g/L urea as nitrogen source. The produced BS-L1011 was found to be highly stable over a wide range of temperatures,

Table 1   Recent updates on application of biosurfactants for the remediation of various pollutants

Application of biosurfactants		*References*
Remediation of Heavy Metals		
Biosurfactant membrane	Removal of Manganese (Mn) from water	Ferreira et al. 2019
Rhamnolipid	Removal of cadmium from contaminated rice protein	Shen et al. 2019
Coal waste activated by rhamnolipid biosurfactant	Removal of cadmium from aqueous solutions	Shami et al. 2019
Rhamnolipid, sophorolipid and saponin	Removal of heavy metal cations	Tang et al. 2018
Micellar-enhanced ultrafiltration (MEUF) with rhamnolipid (RHL)	Removal of cadmium (II) from aqueous solution	Verma and Sarker 2017
Biosurfactant produced from *Bacillus* sp. G1	Removal of lead (Pb) from contaminated port sediment	Guo et al. 2017
Rhamnolipid and Saponin	Removal of the rare earth metals viz. La, Ce, Y and Eu from contaminated soil.	Zhou et al. 2017
Rhamnolipidused as washing agent, effective on targeted metals in exchangeable, carbonate-bound or Fe-Mn oxide-bound fractions	Removal of heavy metals in river sediment, effective on targeted metals in exchangeable, carbonate-bound or Fe-Mn oxide-bound fractions	Chen et al. 2017
Biosurfactant from *Candida sphaerica* UCP0995	Removal of heavy metals viz. Pb, Zn and Fe	Luna et al. 2016
Rhamnolipid as a biosurfactant produced by *Pseudomonas aeruginosa*	Removal of chromium(III) from aqueous solutions	Abyaneh and Fazaelipoor 2016
Glycolipid as biosurfactant produced by bacterial strain *Burkholderia* sp. Z-90	Removal of As, Cu, Cd, Mn, Pb from contaminated soil	Yang et al. 2016
Mixed culture of acidophilic and biosurfactant-producing bacteria	Bioleaching of metals viz. Ni, Cu, Zn, Pb, Cd from combustion wastes	Karwowska et al. 2015
Rhamnolipids biosurfactant produced by *Pseudomonas fluorescens*	Removal of the heavy metals viz. Cd,Pb and Zn.	Ammami et al. 2015
Remediation of Polycyclic Aromatic Hydrocarbons (PAH)		
Biochar (BC) andrhamnolipid (RL)	Phytoremediation of petroleum hydrocarbon-contaminated soil	Zhen et al. 2019
Bioemulsifier produced by *Rhodococcuserythropolis* OSDS1	Improvement of crude oil depletion	Xia et al. 2019
	Biodegradation of benzo (a) pyrene (BaP)	Meng et al. 2019
Biosurfactant-produced by strain *Pseudomonas aeruginosa*S5	*In situ* remediation of polycyclic aromatic hydrocarbons (PAHs)	Sun et al. 2019
Ryegrass and lignin, along with the biosurfactant rhamnolipid	Considerable dissipation of PAH in an aged contaminated arable soil	Wu et al. 2018
Biosurfactant from *Bacillus subtilis*	Removal of anthracene	Salamat et al. 2018
Biosurfactant produced by bacterial consortium	Recovery of the C5-C11 hydrocarbon fraction from oily sludge	Chirwa et al. 2017
Biosurfactant from *Pseudomonas aeruginosa* strain W10	Bioremediation of phenanthrene contaminated sites	Chebbi et al. 2017
Lipopeptide biosurfactant produced by *Bacillus cereus SPL-4*	Degradation of 5 and 6 ring PAHs	Bezza and Chirwa 2017a

Table 4 (Contd...)

Table 1 (Contd...)　Recent updates on application of biosurfactants for the remediation of various pollutants

Application of biosurfactants		References
Remediation of Polycyclic Aromatic Hydrocarbons (PAH)		
Lipopeptide biosurfactant produced by *Paenibacillus dendritiformis* CN5 strain	Enhancement of pyrene biodegradation	Bezza and Chirwa 2017b
Biosurfactant produced by *Rhodococcusruber* IEGM 231	Removal of polycyclic aromatic hydrocarbons in soil	Ivshina et al. 2016
Red ash leaves as a source of biosurfactant	Bioremediation of PAH contaminated soils	Blyth et al. 2015
Rhamnolipids produced by bacteria viz. *Acinetobactercalcoaceticus, Enterobacterasburiae* and *Pseudomonas aeruginosa*	Phenanthrene solubilization and emulsification of crude petroleum	Hošková et al. 2015
Remediation of Pesticides		
Rhamnolipidwas produced from a bacterium *Lysinibacillussphaericus* IITR51	Enhanced dissolution of chlorinated pesticides viz. α-, β-endosulfan, and γ-hexachlorocyclohexane	Gaur et al. 2019
Biosurfactant produced by *Pseudomonas* (KX881513)	Removal of chlorpyrifos, anorgano-phosphorous pesticide	Shabbier et al. 2018
Micellar solutions of biosurfactants viz. saponin, alkyl polyglycoside	Removal efficiencies of o-dichlorobenzene (o-DCB) and p-dichlorobenzene (p-DCB) from contaminated soil	Pei et al. 2017
Biosurfactants produced by *Pseudomonas* sp. SB	Effectively assist the phytoremediation of DDT-contaminated soil by two grass species.	Wang et al. 2017
Biosurfactant produced by *Rhodococcus* sp. D-1	Degradation of carbendazim-	Bai et al. 2017
Rhamnolipid produced by five potential bacteria belonging to genus *Pseudomonas, Klebsiella, Stenotrophomonas, Ochrobactrum* and *Bacillus* and their mixed culture	Degradation of a organophosphate (OP) pesticide (chlorpyrifos)	Singh et al. 2016
Remediation of Synthetic Dyes		
Biosurfactant rhamnolipid produced by *Shewanella putrefaciens* CN32 co-cultured with *Bacillus circulans* BWL1061	Decolourization and detoxification of water-insoluble Sudan dye	Liu et al. 2018
Biosurfactant production from *Pseudomonas taiwanensis* L1011	Decolourization of azo dyes	Liu et al. 2017
Saponin Biosurfactant	Removal of methyl violet from wastewater	Samal et al. 2017b

pH, and salt concentrations and a low CMC value of 10.5 mg/L. Moreover, this study was the first report of the biosurfactant BS-L1011, which accelerated the chemical and biological decolourization of azo dyes in dye bearing wastewater.

Sudan dyes are widely used in various industries, sometimes illegally used as food colouratives to impart orange-red colour. They are categorized as a distinct group of compounds with high structural complexity, recalcitrance, water-insolubility, and toxicity (Xu et al. 2010). Decolourization of these dyes is relatively tough due to its water-insolubility nature. Biological treatment using co-cultured *S. putrefaciens* CN32 and *B. circulans* BW-L1061 and their rhamnolipid biosurfactant was employed towards detoxification and decolourization of the water-insoluble Sudan dye, as reported by Liu et al. (2018). The physicochemical parameters related to dye decolourization were optimized and were found to play an important role in enhanced decolourization of Sudan I. The biosurfactant and the co-cultured bacterial species were found to accelerate the decolourization in a synergistic manner. The highest decolourization percentage of 90.23 was reported within 108 hours. It was suggested that the co-culture technique was found to be effective in treating dye bearing wastewater. Additional microbial toxicity studies showed a significant decrease in Sudan I toxicity towards *E. coli* BL-21 strain and *B. subtilis* 168 strain after decolourization experiments. Table 1 summarizes the recent updates on application of biosurfactants for the remediation of various pollutants.

7. CONCLUSION

Developing of novel remediation strategies that rise upto high standards for cleaning-up the contaminated soil and wastewaters has always seemed to be a complex, challenging and controversial task. Based of the reports from available literature, it is found that technology based on the use of biosurfactant produced from biological sources may be considered as the best suited, successful and cost effective technology for remediation of various pollutants like heavy metals, PAHs, pesticides, synthetic dyes, etc. However, more information is needed towards the exploration of novel biosurfactants from non-microbial biological sources, secretion of biosurfactants, metabolic synthesis route, cell metabolism behind biosurfactant production, and scaling up options. In particular, minimizing the high production cost of biosurfactant is of high interest, in order to apply them in more productive applications. Though high yield can be achieved by fermentation methods using cheap or waste precussors/substrates, improved or enhanced recovery rate and amount of biosurfactants produced needs more research attention. In addition, optimization of fermentation methods with biotechnological and engineering prospects is also much needed.

References

Abyaneh, A.S. and M.H. Fazaelipoor. 2016. Evaluation of rhamnolipid (RL) as a biosurfactant for the removal of chromium from aqueous solutions by precipitate flotation. J. Environ. Manage. 165: 184–187.

Akbari, S., N.H. Abdurahman, R.M. Yunus, F. Fayaz, and O.R. Alara. 2018. Biosurfactants–A new frontier for social and environmental safety: A mini review. Biotech. Res. Innov. 2(1): 81–90.

Ammami, M.T., F. Portet-Koltalo, A. Benamar, C. Duclairoir-Poc, H. Wang and F.L. Derf. 2015. Application of biosurfactants and periodic voltage gradient for enhanced electrokinetic remediation of metals and PAHs in dredged marine sediments. Chemosphere 125: 1–8.

Bai, N.S., S.Wang, R. Abuduaini, M. Zhang, X. Zhu, and Y. Zhao. 2017. Rhamnolipid-aided biodegradation of carbendazim by *Rhodococcus* sp. D-1: Characteristics, products, and phytotoxicity. Sci. Total. Environ. 590-591: 343–351.

Bezza, F.A. and E.M.N. Chirwa. 2016. Biosurfactant-enhanced bioremediation of aged polycyclic aromatic hydrocarbons (PAHs) in creosote contaminated soil. Chemosphere 144: 635–644.

Bezza, F.A and E.M.N. Chirwa. 2017a. The role of lipopeptide biosurfactant on microbial remediation of aged polycyclic aromatic hydrocarbons (PAHs)-contaminated soil. Chem. Eng. J. 309: 563–576.

Bezza, F.A. and E.M.N. Chirwa. 2017b. Pyrene biodegradation enhancement potential of lipopeptide biosurfactant produced by *Paenibacillus dendritiformis* CN5 strain. J. Hazard. Mater. 321: 218–227.

Blyth, W., E. Shahsavari, P.D. Morrison and A.S. Ball. 2015. Biosurfactant from red ash trees enhances the bioremediation of PAH contaminated soil at a former gasworks site. J. Environ. Manage. 162: 30–36.

Brillas, E. and C.A. Martínez-Huitle. 2015. Decontamination of wastewaters containing synthetic organic dyes by electrochemical methods: An updated review. Appl. Catal. B Environ. 166-167: 603–643.

Chebbi, A., D. Hentati, H. Zaghden, N. Baccar, F. Rezgui, M. Chalbi, et al. 2017. Polycyclic aromatic hydrocarbon degradation and biosurfactant production by a newly isolated *Pseudomonas* sp. strain from used motor oil-contaminated soil. Int Biodeter. Biodegr. 122: 128–140.

Chen, Q., M. Bao, X. Fan, S. Liang and P. Sun. 2013. Rhamnolipids enhance marine oil spill bioremediation in laboratory system. Mar. Pollut. Bull. 71(1–2): 269–275.

Chen, W., Y. Qu, Z. Xu, F. He, Z. Chen, S. Huang, et al. 2017. Heavy metal (Cu, Cd, Pb, Cr) washing from river sediment using biosurfactant rhamnolipid. Environ. Sci. Pollut. Res. 24 (19): 16344–16350.

Chirwa, E. and H. Smit. 2010. Simultaneous Cr(VI) reduction and phenol degradation in a trickle bed bioreactor: shock loading response. Chem. Eng. Trans. 20: 55–60.

Chirwa, E.M.N., C.T. Mampholo, O.M. Fayemiwo and F.A. Bezza. 2017. Biosurfactant assisted recovery of the C_5–C_{11} hydrocarbon fraction from oily sludge using biosurfactant producing consortium culture of bacteria. J. Environ. Manage. 196: 261–269.

Dadrasnia, A., N. Shahsavari and I. Salmah. 2015. The top 101 cited articles in environmental clean-up: Oil spill remediation. Global NEST J. 17: 692–700.

Das, A.J., S. Lal, R. Kumar and C. Verma. 2017. Bacterial biosurfactants can be an ecofriendly and advanced technology for remediation of heavy metals and contaminated soil. Int J Environ. Sci. Technol. 14(6): 1343–1354.

Das, A.J. and R. Kumar. 2018. Utilization of agro-industrial waste for biosurfactant production under submerged fermentation and its application in oil recovery from sand matrix. Bioresour. Technol. 260: 233–240.

Dasgupta, J., J. Sikder, S. Chakraborty, S. Curcio and E. Drioli. 2015. Remediation of textile effluents by membrane based treatment techniques: A state of the art review. J. Environ. Manage. 147: 55–72.

De la Cueva, S.C., C.H. Rodríguez, N.O.S. Cruz, J.A.R. Contreras and J.L. Miranda. 2016. Changes in bacterial populations during bioremediation of soil contaminated with petroleum hydrocarbons. Water Air Soil Poll. 227: 1–12.

Environmental Protection Agency (EPA), U.S. 1988. National Pollutant Discharge Elimination System. US Government Printing Office, Washington, DC.

EPA Registering Pesticides. 2009. Available online: http://www.epa.gov/pesticides/regulating/re-gistering/index.htm

Ferreira, L.C., L.C. Ferreira, V.L. Cardoso and U.C. Filho. 2019. Mn(II) removal from water using emulsion liquid membrane composed of chelating agents and biosurfactant produced *in loco*. J. Water Process Eng. 29: 100792.

Gaur, V.K., A. Bajaj, R.K. Regar, M. Kamthana, R.R. Jha, J.K. Srivastava, et al. 2019. Rhamnolipid from a *Lysinibacillus sphaericus* strain IITR51 and its potential application for dissolution of hydrophobic pesticides. Bioresour. Technol. 272: 19–25.

Guo, Y.M., Y.G. Liu, H. Li, A.B. Zheng, X.F. Tan and M.N. Zhang. 2017. Remediation of Pb-contaminated port sediment by biosurfactant from *Bacillus* sp. G1. T. Nonferr. Metal Soc. China 27(6): 1385–1393.

Heryani, H. and M.D. Putra. 2017. Kinetic study and modeling of biosurfactant production using *Bacillus* sp. Electron. J. Biotechn. 27: 49–54.

Hošková, M., R. Ježdík, O. Schreiberová, J. Chudoba, M.M. Sír, A. Cejkova, et al. 2015. Structural and physiochemical characterization of rhamnolipids produced by *Acinetobacter calcoaceticus*, *Enterobacter asburiae* and *Pseudomonas aeruginosa* in single strain and mixed cultures. J. Biotechnol. 193: 45–51.

Holkar, C.R., A.J Jadhav, D.V. Pinjari, N.M Mahamuni and A.B. Pandit. 2016. A critical review on textile wastewater treatments: possible approaches. J. Environ. Manage. 182: 351–366.

Idress, M., S. Batool, H. Ullah, Q. Hussain, M.I. Al-Wabel, M. Ahmad, et al. 2018. Adsorption and thermodynamic mechanisms of manganese removal from aqueous media by biowaste-derived biochars. J. Mol. Liq. 266: 373–380.

Ivshina, I., L. Kostina, A. Krivoruchko, M. Kuyukina, T. Peshkur, P. Anderson, et al. 2016. Removal of polycyclic aromatic hydrocarbons in soil spiked with model mixtures of petroleum hydrocarbons and heterocycles using biosurfactants from *Rhodococcus ruber* IEGM 231. J. Hazard. Mater. 312: 8–17.

Joshi, S.J., Y.M. Al-Wahaibi, S.N. Al-Bahry, A.E. Elshafie, A.S. Al-Bemani, A. Al-Bahri, et al. 2016. Production, characterization, and application of *Bacillus licheniformis* W16 biosurfactant in enhancing oil recovery. Front. Microbiol. 7: 1853.

Karlapudi, A.P., T.C. Venkateswarulu, J. Tammineedi, L. Kanumuri, B.K. Ravuru, V.R. Dirisala, et al. 2018. Role of biosurfactants in bioremediation of oil pollution–A review. Petroleum. 4(3): 241–249.

Karwowska, E., M. Wojtkowska and D. Andrzejewska. 2015. The influence of metal speciation in combustion waste on theefficiency of Cu, Pb, Zn, Cd, Ni and Cr bioleaching in a mixed culture of sulfur-oxidizing and biosurfactant-producing bacteria. J. Hazard. Mater. 299: 35–41.

Kumari, S., R.K. Regar and N. Manickam. 2018. Improved polycyclic aromatic hydrocarbon degradation in a crude oil by individual and a consortium of bacteria. Bioresour. Technol. 254: 174–179.

Lal, S., S. Ratna, O.B. Said and R. Kumar. 2018. Biosurfactant and exopolysaccharide-assisted rhizobacterial technique for the remediation of heavy metal contaminated soil: An advancement in metal phytoremediation technology. Environ. Technol. Innov. 10: 243–263.

Liu, C., Y. You, R. Zhao, D. Sun, P. Zhang, J. Jiang, et al. 2017. Biosurfactant production from *Pseudomonas taiwanensis* L1011 and its application in accelerating the chemical and biological decolorization of azo dyes. Ecotox. Environ. Safe. 145: 8–15.

Liu, W., Y. You, D. Sun, S. Wang, J. Zhu and C. Liu. 2018. Decolorization and detoxification of water-insoluble Sudan dye by *Shewanella putrefaciens* CN32 co-cultured with *Bacillus circulans* BWL1061. Ecotox. Environ. Safe. 166: 11–17.

Luna, J.M., R.D. Rufino and L.A. Sarubbo. 2016. Biosurfactant from *Candida sphaerica* UCP0995 exhibiting heavy metal remediation properties. Process Saf. Environ. 102: 558–566.

Maletić, S.P., J.M. Beljin, S.D. Rončević, M.G. Grgić and B.D. Dalmacija. 2019. State of the art and future challenges for polycyclic aromatic hydrocarbons is sediments: sources, fate, bioavailability and remediation techniques. J. Hazard. Mater. 365: 467–482.

Manickam, N., N.K. Singh, A. Bajaj, R.M. Kumar, G. Kaur, N. Kaur, et al. 2014. *Bacillus mesophilum* sp. nov., strain IITR-54[T], a novel 4-chlorobiphenyl dechlorinating bacterium. Arch. Microbiol. 196(7): 517–523.

Mao, X., R. Jiang, W. Xiao and J. Yu. 2015. Use of surfactants for the remediation of contaminated soils: A review. J. Hazard. Mater. 285: 419–435.

Meng, L., W. Li, M. Bao and P. Sun. 2019. Effect of surfactants on the solubilization, sorption and biodegradation of benzo(a)pyrene by *Pseudomonas aeruginosa* BT-1. J Taiwan Inst. Chem. E 96: 121–130.

Nurfarahin, A.H., M.S. Mohamed and L.Y. Phang. 2018. Culture medium development for microbial-derived surfactants production–An overview. Molecules. 23(5): 1049.

Olaniran, A.O., A. Balgobind and B. Pillay. 2013. Bioavailability of heavy metals in soil: Impact on microbial biodegradation of organic compounds and possible improvement strategies. Int. J. Mol. Sci. 14(5): 10197–10228.

Pacwa-Płociniczak, M., G.A. Płaza, Z. Piotrowska-Seget and S.S. Cameotra. 2011. Environmental applications of biosurfactants: recent advances. Int. J. Mol. Sci. 12: 633–654.

Patil, D.S., M.S. Chavan and J.U.K. Oubagaranadin. 2016. A review of technologies for manganese removal from wastewaters. J. Environ. Chem. Eng. 4(1): 468–487.

Pei, G., Y. Zhu, X. Cai, W. Shi and H. Li. 2017. Surfactant flushing remediation of o-dichlorobenzene and p-dichlorobenzene contaminated soil. Chemosphere 185: 1112–1121.

Ramoju, S.P., D.R. Mattison, B. Milton, D. McGough, N. Shilnilkova, H.J. Clewell, et al. 2017. The application of PBPK models in estimating human brain tissue manganese concentrations. NeuroToxicol. 58: 226–237.

Roy, D., R.R. Kommalapati, S.S. Mandava, K.T. Valsaraj and W.D. Constant. 1997. Soil washing potential of a natural surfactant. Environ. Sci. Technol. 31(3): 670–675.

Salamat, N., R. Lamoochi and F. Shahaliyan. 2018. Metabolism and removal of anthracene and lead by a *B. subtilis*-produced biosurfactant. Toxicol. Reports. 5: 1120–1123.

Samal, K., C. Das and K. Mohanty. 2017a. Eco-friendly biosurfactant saponin for the solubilization of cationic and anionic dyes in aqueous system. Dyes Pigments. 140: 100–108.

Samal, K., C. Das and K. Mohanty. 2017b. Application of saponin biosurfactant and its recovery in the MEUF process for removal of methyl violet from wastewater. J. Environ. Manage. 203 (Part 1): 8–16.

Schmitt, C., B. Grassl, G. Lespes, J. Desbrieres, V. Pellerin, S. Reynaud, et al. 2014. Saponins: a renewable and biodegradable surfactant from its microwaveassisted extraction to the synthesis of monodisperse lattices. Biomacromolecules. 15(3): 856–862.

Segura, F.R., E.A. Nunes, F.P. Paniz, A.C.C. Paulelli, G.B. Rodrigues, G.U.L. Braga, et al. 2016. Potential risks of the residue from Samarco's mine dam burst (Bento Rodrigues, Brazil). Environ. Pollut. 218: 813–825.

Selvi, A., J.A. Salam and N. Das. 2014. Biodegradation of cefdinir by a novel yeast strain, *Ustilago* sp. SMN03 isolated from pharmaceutical wastewater. World J. Microbiol. Biotechnol. 30: 2839–2850.

Selvi, A., D. Das and N. Das. 2015. Potentiality of yeast, *Candida* sp. SMN04 for degradation of cefdinir, a cephalosporin antibiotic: kinetics, enzyme analysis and biodegradation pathway. Environ. Technol. 36: 3112–3124

Selvi, A. and R. Aruliah. 2018. A statistical approach of zinc remediation using acidophilic bacterium via an integrated approach of bioleaching enhanced electrokinetic remediation (BEER) technology. Chemosphere. 207: 753–763.

Shabbier, M., M. Singh, S. Maiti, S. Kumar and S.K. Saha. 2018. Removal enactment of organophosphorous pesticide using bacteria isolated from domestic sewage. Bioresour. Technol. 263: 280–288.

Shah, A., S. Shahzad, A. Munir, M.N. Nadagouda, G.S. Khan, D.F. Shams, et al. 2016. Micelles as soil and water decontamination agents. Chem. Rev. 116: 6042–6074.

Shami, R.B., V. Shojaei and H. Khoshdast. 2019. Efficient cadmium removal from aqueous solutions using a sample coal waste activated by rhamnolipid biosurfactant. J. Environ. Manage. 231: 1182–1192.

Shen, C., S. Tang and Q. Meng. 2019. Cadmium removal from rice protein via synergistic treatment of rhamnolipids and F127/PAA hydrogels. Colloids Surf. B Biointerfaces. 181: 734–739.

Singh, P., H.S. Saini and M. Raj. 2016. Rhamnolipid mediated enhanced degradation of chlorpyrifos by bacterial consortium in soil-water system. Ecotox. Environ. Safe. 134(1): 156–162.

Sun, S., Y. Wang, T. Zang, J. Wei, H. Wu, C. Wei, et al. 2019. A biosurfactant-producing *Pseudomonas aeruginosa* S5 isolated from coking wastewater and its application for bioremediation of polycyclic aromatic hydrocarbons. Bioresour. Technol. 281: 421–428.

Tang, J., J. He, T. Liu, X. Xin and H. Hu. 2017. Removal of heavy metal from sludge by the combined application of a biodegradable biosurfactant and complexing agent in enhanced electrokinetic treatment. Chemosphere 189: 599–608.

Tang, J., J. He, X. Xin, H. Hu and T. Liu. 2018. Biosurfactants enhanced heavy metals removal from sludge in the electrokinetic treatment. Chem. Eng. J. 334: 2579–2592.

Usman, M.M., A. Dadrasnia, K.T. Lim, A.F. Mahmud and S. Ismail. 2015. Application of biosurfactants in environmental biotechnology; remediation of oil and heavy metal. AIMS Bioeng. 3(3): 289–304

Uzoigwe, C., J.G. Burgess, C.J Ennis and P.K.S.M. Rahman. 2015. Bioemulsifiers are not biosurfactants and require different screening approaches. Front. Microbiol. 6: 245.

Varjani, S.J. and V.N. Upasani. 2017. Critical review on biosurfactant analysis, purification and characterization using rhamnolipid as a model biosurfactant. Bioresour. Technol. 232: 389–397.

Verma, S.P. and B. Sarkar. 2017 Rhamnolipid based micellar-enhanced ultrafiltration for simultaneous removal of Cd(II) and phenolic compound from wastewater. Chem. Eng. J. 319: 131–142.

Wang, B., Q. Wang, W. Liua, X. Liu, J. Hou, Y. Teng, et al. 2017. Biosurfactant-producing microorganism *Pseudomonas* sp. SB assists the phytoremediation of DDT-contaminated soil by two grass species. Chemosphere 182: 137–142.

WHO. 2010. International Code of Conduct on the Distribution and Use of Pesticides: Guidelines for the Registration of Pesticides. World Health Organization; Rome, Italy.

Wu, Y., Q. Ding, Q. Zhu, J. Zeng, R. Ji, M.G. Dumont, et al. 2018. Contributions of ryegrass, lignin and rhamnolipid to polycyclic aromatic hydrocarbon dissipation in an arable soil. Soil Biol. Biochem. 118: 27–34.

Wuana, R.A. and F.E. Okieimen. 2011. Heavy metals in contaminated soils: A review of sources, chemistry, risks and best available strategies for remediation. ISRN Ecol. 2011: 1–20.

Xia, M., D. Fu, R. Chakraborty, R.P. Singh and N. Terry. 2019. Enhanced crude oil depletion by constructed bacterial consortium comprising bioemulsifier producer and petroleum hydrocarbon degraders. Bioresour. Technol. 282: 456–463.

Xu, H., H. Thomas, D. Paine, C. Cerniglia and H. Chen. 2010. Sudan azo dyes and Para Red degradation by prevalent bacteria of the human gastrointestinal tract. Anaerobe. 16: 114–119.

Yagub, M.T., T.K. Sen, S. Afroze and H.M. Ang. 2014. Dye and its removal from aqueous solution by adsorption: A review. Adv. Colloid Interfac. Sci. 209: 172–184.

Yang, Z., Z. Zhang, L. Chai, Y. Wang, Y. Liu and R. Xiao. 2016. Bioleaching remediation of heavy metal-contaminated soils using *Burkholderia* sp. Z-90. J. Hazard. Mater. 301: 145–152.

Zhen, M., H. Chen, Q. Liu, B. Song, Y. Wang and J. Tang. 2019. Combination of rhamnolipid and biochar in assisting phytoremediation of petroleum hydrocarbon contaminated soil using *Spartina anglica*. J. Environ. Sci. 85: 107–118.

Zhou, D., Z. Li, X. Luo and J. Su. 2017. Leaching of rare earth elements from contaminated soils using saponin and rhamnolipid bio-surfactant. J. Rare Earths. 35(9): 911–919.

CHAPTER 10

Mycorrhizoremediation:
A Novel Tool for Bioremediation

V. Vijaya Kumar[1*] **and P. Suprasanna**[2]

[1]Natems Sugar Private Limited, III Floor, Plot No. 22, Survey No. 90/1, Trendz Eternity, Green land Colony, Gachibowli Hyderabad - 500032, Telangana, India.
[2]Nuclear Agriculture and Biotechnology Division, Bhabha Atomic Research Centre, Trombay, Mumbai -400085, Maharashtra, India.

1. INTRODUCTION

The world population will rise from the present 6.8 billion to 9.1 billion by the year 2050. To feed this growing population, the overall food production will have to be increased by 70%; for example, the production of cereal crops will have to be increased by additional 1 billion tons and meat production by 200 million tons (FAO 2009). In addition to the many confounding factors that majorly affect food production, land degradation and climate change play a key role in achieving food security. More than 75% of the Earth's land area is degraded and this may increase to 90% by 2050 (Cherlet et al. 2018). Crop yields are reduced globally by 10% due to climate change and land degradation. The most affected are India, China and Sub Saharan Africa (European Commission – Press Release 2018).

Soil is the biologically active matrix on which plants grow and which provides them with physical support, water and nutrients. Apart from acting as a medium for plant growth, various animals and human activities are supported by the soil. It takes 3000 years to replace 1 mm loss of top soil (Laishram et al. 2012, , Baxter and Williamson 1968). The principle components of soil are minerals, organic matter, water and air. The soils vary in their composition and this will impact the water holding capacity of soils, nutrient cycling, chemical reactions and composition of the living organisms in soil. The soil health is also the vital factor in agriculture productivity (Laishram et al. 2012, Nortcliff et al. 2012).

Soil degradation is mainly due to human intervention, use of agrochemicals, chemical waste and industrial activities. There are both short term and long term measures to tackle the problem.

*Corresponding author: vankayalapati99@hotmail.com

For the long term, restoration of the polluted soil, physico-chemical and biological remedial technologies can be used (Jankaite and Vasarevičius 2005). To mitigate the soil pollution and for the better growth of plant species, the ecosystem containing beneficial biota plays a vital, contributory role. The process, which relies upon biological mechanisms to reduce (degrade, detoxify, mineralize or transform) concentration of pollutants to an innocuous state is generally referred to as bioremediation. Through bioremediation, soil organic chemical waste, food processing and chemical plants' effluent, and petroleum refineries' oily sludge contamination can be reduced (Azubuike et al. 2016, Glazer and Nikaido 1995). Bioremediation is generally classified into *in situ* and *ex situ* bioremediation. In case of *in situ* bioremediation, the pollutants are treated at the contaminated site avoiding the excavation of soil. These techniques involve bioventing, biosparging, bioslurping, and phytoremediation (Pilon-Smits 2005). On the contrary, *ex situ* bioremediation involves excavation of pollutants from the polluted sites and transportation to the other sites for treatment. *Ex situ* bioremediation techniques involve biopile, windrows, land farming, and bioreactor (Kumar 2017, Azubuike et al. 2016).

2. PHYTOREMEDIATION

Use of plants for bioremediation is referred to as phytoremediation, and it relies on utilizing plants to remove pollutants from the environment and/or to render them harmless (Liu et al. 2013). Phytoremediation also refers to the use of plants and associated microorganisms to degrade the intensity and lethal effects of pollutants in the nature (Egamberdieva et al. 2016). Phytoremediation is an *in situ* remediation technique (Muthusaravanan et al. 2018) (REF). Phytoremediation is mostly limited to the root zone of the plants. The advantages (Hettiarachchi et al. 2012) and limitations of phytoremediation technology are given below (Table 1). Phytoremediation is further categorized into phytoextraction, phytodegradation, phytostabilization, phytovolatalization, and rhizofiltration (Ozyigit and Dogan 2015).

Table 1 Advantages and limitations of bioremediation

Advantages	*Limitations*
It is an *in situ* remediation technology, where plants are grown at the contaminated site facilitating the removal of contaminants, immobilization or degradation of contaminants.	Phytoremediation will work only when the site contains low levels of pollutants.
It is cost effective compared to conventional technologies.	Depth of root spread may limit the phytoremediation process.
Plants reduce the movement of contaminants in soil (reduced runoff), enhance evapotranspiration rate, and absorption of contaminants by roots.	By consuming the leaves or seeds of plants employed for phytoremediation, by birds, grazers or other animals, the toxic compounds may enter into the food chain.
Phytoremediation with plants enhances the aesthetics of the site, and restores the ecological function to site.	More space and time is required for the successful implementation of phytoremediation.
Phytoremediation uses different mechanisms to treat multiple pollutants simultaneously.	

2.1 Phytoextraction

Phytoextraction, also referred to as phytomining, is a method in which crop plants are grown to accumulate contaminants in their shoots and leaves. These plants after accumulation of contaminants are harvested for removal of contaminants from the site. The harvested biomass volume can be reduced by burning or metals having commercial value can be extracted from the harvested biomass of the plants grown in mining areas (Adams et al. 2000, Robinson et al. 2003, Ghori et al. 2016). The plants that accumulate higher quantities of heavy metals in their aerial parts are called as "hyperaccumulators". Normally, the hyperaccumulators are slow growing and produce less biomass. One of the limitations of phytoextraction is that it is a slow process and it may require 1–20 years, depending on the type and extent of metal contamination for effective removal of metals from the contaminated sites (Thangavel and Subbhuram 2004).

The transition metals which have a mass of above 20 and density above 5 are termed as heavy metals, whose concentration at low levels is toxic to plants and animals. The heavy metals which do not have any physiological function in plants include Arsenic (As), Lead (Pb), Cadmium (Cd), Mercury (Hg), Selenium (Se) and are called non essential elements. The other elements which are essential for the growth and development of plants comprise of Cobalt (Co), Copper (Cu), Iron (Fe), Manganese (Mn), Molybdenum (Mo), Nickel (Ni) and Zinc (Zn) (Rascio and Navari-Izzo 2011). The use of hyperaccumulators was first suggested by Chaney and group (1997) for phytoremediation of sites polluted by metals, while halophytes were also present as useful candidates for phytoremediation (Manousaki and Kalogerakis 2011, Nikalje and Suprasanna 2018). The method of chemically induced phytoextraction uses chelating agents such as EDTA, CDTA, DTPA etc. for chelation with the metals to increase their mobility, facilitating the uptake by the plants (Ghori et al. 2016).

In case of natural phytoextraction process of heavy metals, several examples of hyperaccumulators are available, which include hyperaccumulating plants such as *Helianthus annuus, Pelargonium* sp., *Arabidopsis halleri, Artiplex hamilus*, etc., accumulating more than 400 mg/kg of Cd, *Brassica juncea* accumulating 100 mg/kg of Au, *Betula papyrifera* accumulating above 10,000 mg/kg. Hg., *Aeollanthus biformifolius* accumulating >1000 mg./kg, Co,, *Ipomea alpina* accumulating 12300 mg./kg. Cu, accumulating *Leptospermum scoparium, Pimelea suteri* accumulating upto or above 30,000 mg/kg Cr (Mahmood 2010). In the chemically induced phytoextaction of heavy metals, EDTA is a much widely studied additive. Other amendments such as synthetic aminopolycarboxylic acids (APCAs), diethylene tetraamine pentaacetic acid, nitroloacetates, ammonium isothiocyanate, sodium cyanide, organic acids, chlorides, elemental sulfur, hydrogen peroxide, ammonium fertilizer, etc. play a key role in phytoextraction process (Meers et al. 2008). Sadasivam et al. (2010) reported the phytoextraction of Radiocaesium (^{137}Cs) using ammonium chloride as chemical extractant in a pot culture experiment. Plants such as *Amaranthus viridis* L., *Zea mays* L., *Phaeolus vulgaris* L. and *Helianthus annuus* L. were grown in the presence of ^{137}Cs at 20 Bq/g and the bioaccumulation pattern was studied in the shoots. In the absence of ammonium chloride, the highest bioaccumulation of ^{137}Cs (Bq/g) was recorded in *Amaranthus* (20.46) followed by *Zea mays* (1.63), *Phaseolus* (1.64) and *Helianthus* (1.48), respectively. Similarly, in the presence of ammonium chloride, the highest bioaccumulation was recorded in *Amaranthus* (24.37), followed by *Zea mays* (1.94), whereas *Phaseolus* (1.69) and *Helianthus* (1.69) recorded bioaccumulation of equal quantity of ^{137}Cs. In all the four plants, more bioaccumulation of ^{137}Cs was noted in the presence of ammonium chloride than in the absence of ammonium chloride (Sadasivam et al. 2010).

2.2 Phytodegradation

Phytodegradation, or phytotransformation, comprises of degradation of complex organic pollutants into simple molecules in the soil directly by releasing the enzymes through roots or in the plant tissues by various metabolic activities (Newmann and Reynolds 2004). In this process, the organic contaminants are taken up by the plants through roots and are broken down in the plant tissues into less toxic forms. The microorganisms living in the rhizosphere degrade the organic contaminants in the rhizosphere into less toxic forms. The plant roots secrete various nutrients such as sugars, alcohols and acids required for the growth of soil microorganisms. The bacterial population in the rhizopsphere can be augmented to required levels by controlling the nutrient levels and pH by a process known as biostimulation (Greipssson 2011, Tangahu et al. 2011). The enzymes in the plants such as reductases, dehalogenases and oxygenases break down the contaminants such as trichloroethylene and some herbicides (Galadima et al. 2018). Singh et al. (2004) demonstrated the degradation of cement fixed atrazine and simazine in cement blocks of a long term contaminated soil when mixed with normal soil in 1:1 ratio using the plants rye grass (*Lolium perenne*), tall fescue (*Festuca arundinacae*), Pennisetum (*Pennisetum clandestinum*) and a spring onion (*Allium* sp.). Only *P. clandestinum* survived, while all other plants did not survive after germination or transplantation. Approx. 45% of atrazine and 52% of simizine was degraded by *P. clandestinum* in the contaminated soil, while 22% and 20% of the atrazine and simizine, respectively, degraded in the unplanted soil. Significant increase in dehydrogenase activity and in microbial biomass was observed in soil augmented with *P. clandestinum* compared to the unplanted soil.

2.3 Phytostabilization

Phytostabilization utilizes the plants to reduce the mobility of metals or organic contaminants by immobilizing them by hyperaccumulating plants by root absorption and accumulation or root adsorption and precipitation, which prevents the contaminants from reaching to the foodwebs (Radziemska et al. 2017, Nanthi et al. 2011). Phytostabilization limits the bioavailability of contaminants, reducing their accumulation in the above ground parts. The plant cover in the contaminated sites prevents the soil erosion through its compact root system and prevents the leaching of contaminants reaching the water bodies, underground water and to the other areas through runoff (McIntyre 2003). The plants producing high biomass with extensive root system and low translocation of contaminants from root to shoot are ideal for use in phytostabilization (Alkorta et al. 2010). Phytostabilization is an emerging concept with good potential for field applications with different contamination levels and different soil characteristics such as soil texture, pH, salinity or metal concentrations (Shackira and Puthur 2019).

2.4 Phytovolatalization

Phytovolatalization involves the uptake of organic contaminants such as trichloromethane, tetrachloromethane, tetrachhloroethane etc. and certain metals such as Mercury (Hg), Selinium (Se), Arsenic (As), etc. covert them into volatile forms and releasedinto the atmosphere through transpiration process, either in less toxic forms after metabolic modification or in their native forms (Limmer and Burken 2016). The phytovolatalization process comprises of the interactions between plants and plant-microbe alliance. This alliance converts Hg and Se into less toxic forms such as Hg^0 and di-methyl selinide. The advantage of this method is that toxic contaminants are converted into less toxic forms. The disadvantage of this method is that contaminants like Hg, upon release into the atmosphere, is prone to recycling by precipitation and deposited back into soil, lakes and oceans, where it is converted to methyl mercury by anaerobic bacteria (Etim 2012, Saha et al. 2017, Ozyigit and Dogan 2015).

2.5 Rhizofiltration

Rhizofiltration is the process of elimination of contaminants from the surface water, waste water or from the solution surrounding the roots, by adsorption onto the roots or absorption into the roots (Dushenkov et al. 1995, Adiloğlu 2017). Some of the metals are precipitated by the exudates secreted by the plant roots and become immobile or accumulated on the roots or within the roots. Rhizofiltration works well in the high moisture condition, which is the requirement for the metals to be in solution to be sorbed to the plant roots. Rhizofiltration can be used for removal of contaminants mostly from aquatic environment such as damp soils and ground water / surface waters and is mainly used to treat metals such as Pb, Cd, Cu, Fe, Ni, Mn, Zn, and Cr(VI) (Adams et al. 2000). Both terrestrial and aquatic plants can be used in rhizofiltration process. The plants are grown in water collected from the contaminated site for acclimatization. After acclimatization, the plants are planted back into the contaminated site, where the contaminants are taken up by the plants along with water (Ahalya and Ramachandra 2006).

3. MYCORRHIZOREMEDIATION

The method of using mycorrhizal plants in the phytoremediation of heavy metal contaminated soils is considered as one of the advanced methods of phytoremediation (Meier et al. 2012, Das et al. 2017). This technology employs both plants and mycorrhiza by using different mechanisms to eliminate heavy metals from contaminated soils. Mycorrhiza, which is responsible for the symbiotic association between soil fungus and higher plant roots, include different types: ectomycorrhiza (ECM), endomycorrhiza (AMF-Arbuscular Mycorrhizal Fungus), ectendomycorrhiza, Arbutoid mycorrhiza, Ericoid mycorrhiza, Monotropoid mycorrhiza and Orchidoid mycorrhiza. Plant tolerance is increased by the Ectomycorrhiza and ericoid mycorrhiza. Ericoid mycorrhiza can effectively "detoxify" metal ions and phenolic compounds, which occur in phytotoxic levels in the acidic soils (Leake et al. 1989, Read 1983). The term "Mykorrhizen" was first described by AB Frank to refer to the species in trees of temperate forest and pines in Germany. AM have a very long evolutionary history and one of the oldest associations found on the earth. Their occurrence is described in bryophytes, pteridophytes, many gymnosperms, and maximum angiosperms. (Lakshmanan and Channabasava 2013, Manjhi et al. 2016). Ectomycorrhiza is found in the roots of trees, whereas endomycorrhiza is found in more than 80% of the land plant families. In the roots of gymnosperms and angiosperms, ectomycorrhizal hyphae form Hartig net in between the cells, which increases the surface area for exchange of nutrients between the host and the fungus (Watkinson 2016).

3.1 Endomycorrhiza and Plant Association

The endomycorrhiza enter the host roots through the epidermis, into the cortex region through hypodermis and forms characteristic structures called arbuscules (sites of nutrient exchange) inside the cells for establishing nutrient exchange between host and fungus (Morton et al. 2004, Wipf et al. 2019). Gerdmann and Trappe (1974) observed that in *Glomus radiatus* the vesicles in the roots are like chlamydospores. The authors observed that the "vesicles in senescent roots often become thick walled and convert to Chlamydospores". The AMF increases the uptake of mainly Phosphorus and micronutrients such as calcium, iron, copper, zinc, magnesium, manganese, etc. The presence of AMF increases the absorptive surface of the root by 10 times and the absorption capacity of the immobile elements by 60 times (Srivatsava 1996, Menge 1981). Studies conducted in the past few decades suggest that the mycorrhizal and plant association has been one of the most practiced agronomic methods for augmenting plant health and productivity (Salvioli et al. 2019).

Apart from improving the nutritional status of plans, mycorrhiza (especially AMF) imparts tolerance to abiotic (Aparecida et al. 2019) and biotic stresses. Both ectomycorrhiza and endomycorrhiza (AMF) are used in the remediation process, but AMF is widely used due to its wide occurrence in most of the plant species, while Ectomycorrhiza is present only in woody species (Chibuike 2013, Gomathy et al. 2018). The cell walls of AMF contains cations that can be attached to heavy metals like Cu, Pb, Cd and Cr. AMF secretes glomalin, a glycoprotein that aggregates soil particles, and is known to bind with heavy metals, thereby preventing them from translocating into the aerial parts of the plant. Glomalin is very stable under soil conditions and less degradable. It is tough to dissolve in water and requires highly specific conditions for its extraction such as high temperature (121°C) and citrate buffer at neutral/alkaline pH (Vlček and Pohanka 2020).

Soil is contaminated with heavy metals due to various activities such as industrialization, mining, agriculture, defense activities, etc. At higher concentration, all metals are lethal to plants, humans and other living organisms. The heavy metals interrupt the cellular enzymes, which use various nutritional minerals, thereby disrupting plant growth and development (Aparecida et al. 2019). During the process of phytoremediation, plant–AMF association will help in metabolizing the contaminants and their reduction in the rhizosphere (Hildebrandt et al. 2007, Shi et al. 2019). In this mechanism, the plants secrete phenolics, short chain organic acids, small quantities of proteins and enzymes which stimulate bacterial transformations (enzyme induction), and organic carbon to enhance microbial mineralization rates (substrate enhancement). The plants colonized by AMF show the increased uptake of heavy metals (Miransari 2011, Pathare et al. 2017) and translocate the same from root to shoot (phytoextraction), and in some cases heavy metals are immobilized by AM fungi in soil (phytostabilization). Several transporters have been characterized through transcriptomic analysis in AM-treated roots for their role in metal acquisition (Ferrol et al. 2016). Molecular studies have shed light on the plant-AMF association through activation of genes related to signaling and detoxification pathways (Pathare et al. 2016, Poonam et al. 2017).

3.2 Symbiosis Helps Plants in Field

In a field experiment, Bi et al. (2018) studied the effect of mycorrhiza, at the Daliuta coal mining area located at the transition zone between Loess Plateau in Northern Shaanxi and the Mu Us Desert, on the growth of *Amygdalus pedunculata* inoculated with *Funneliformis mosseae,* and *Rhizophagus intraradices.* The authors reported significant improvement in the shoot and root growth of *Amygdalus pedunculata* Pall. Root colonization was increased considerably after one year of inoculation and the available phosphorus, phosphatase activity and electrical conductivity in the rhizosphere of treated plants was higher than that of controls. AMF inoculation was also found to enhance the bacterial and fungal population in the rhizosphere compared to control.

There have been other studies on the application of AM for phytoremediation of lead (Pb). Yang et al. (2015) studied the effect of AMF species (*Funneliformis mosseae* and *Rhizophagus intraradices*) on the growth, accumulation of Pb, photosynthesis and antioxidant enzyme activities, at various concentrations of lead (Pb) – 0, 500,1000 and 2000 mg/kg soil– in a leguminous tree, *Robinia pseudoacacia* L., under greenhouse conditions. The increase in the biomass and the decrease in the Pb concentration in the leaves was observed in the Mycorrhizal plants than that of non Mycorrhizal plants. Increased Pb concentration was observed in the roots of Mycorrhizal plants at 1000 and 2000 mg/kg soil. Similarly, increase in the shoot height, dry weight of stem, leaf, shoot, and root were observed in Mycorrhizal plants. The mycorrhiza also increased chlorophyll and carotenoid content in *R. pseudoacacia* L. plants. Mycorrhizal plants also showed increase in activities of antioxidant enzymes, and reduced hydrogen peroxide and malondialdehyde (MDA) contents (Yang et al. 2015).

Towards remediation of Lead (Pb) contaminated soils in Saudi Arabia, Abu-Muriefah (2016) studied the effect of *Acacia saligna* plants inoculated with two AM fungi, *Glomus mosse* and *Glomus deserticola,* in a greenhouse experiment by supplementing soil with various concentrations of Pb – 0, 10, 100, 1000 mg/kg soil. After 6 months of inoculation, plants were harvested and analyzed for various parameters such as N, P, K, chlorophyll and uptake of lead in roots and shoots. It was observed that the P, K and chlorophyll contents were higher in AM inoculated treatments compared to control at all the concentrations of lead. N uptake was reduced in *G. mosse* treated plants at 10 mg of Pb (3.02%) treatment than control (3.09%), whereas reduced N uptake was also noticed in *G. deserticola* treated plants at 100 mg/kg Pb (2.15%) compared to control (2.22%). The Pb concent of roots was increased in all treatments except in control in which the accumulation of Pb was reduced by *G. mosse* (130.2 mg/kg) and there was no change in *G. deserticola* treatment (135.5 mg/kg) compared to control (135.5 mg/kg), whereas the shoot accumulation of Pb was reduced in all treatments inoculated with AM fungi.

Channabasava et al. (2015) assessed the effect of AMF in fly ash amended soil in *Paspalum scrobiculatum* L. inoculated with *Rhizophagus fasciculatus* under green house conditions. The soil was amended with fly ash at 0%, 2%, 4% and 6% with or without the addition of *R. fasciculatus*. The plants harvested at 60 days and 90 days showed increase in plant height, root length, shoot dry weight and root dry weight in the treatment of 2% fly ash amended with AMF compared to control, whereas in the other treatments amended with 4% and 6% flys ash, the decrease in the above parameters was observed in both 60 days and 90 days harvested plants. At the same time, an increase in the shoot P, Ca, Mg, Na was recorded in 2% and 4% fly ash amended treatment compared to 0% fly ash in the presence of mycorrhiza at 60 days after sowing (DAS). The K concentration was more in all treatments with AMF inoculation at 60 DAS. At 90 DAS, K content was more in shoots in all treatments with AMF compared to control, whereas more Ca, Mg, Na content was recored in the 2% and 4% fly ash treatment; at 6% fly ash treatment, the concentration of the above elements was redcued in the presence of AMF compared to control. P content was reduced in 4%, 6% fly ash treatment in the presence of AMF. The study indicates that application of AM and fly ash can effectively help in retrieval of low fertile or marginal soils for further cultivation.

Mycoremediation is considered as one of the onsite strategies of remediation of phytotoxic levels of metal contaminants. Abu-Elasoud et al. (2017) studied the impact of Zinc at various concentrations – 50, 100, 200 mg/kg – in wheat *(Triticum aestivum* L. cv. Gemmeza-10) in the presence or absence of mycorrhizal fungus *Funnelliformis geosporum* (Nicol. & Gerd.) Walker & Schüßler. The mycorrhiza improved all the growth parameters studied such as plant height, spike length, grain yield, biological yield, 1000 grain weight, and number of grains per spike at all concentrations of Zinc in Mycorrhiza applied plants compared to controls. The mycorrhiza applied plants matured early compared to non mycorrhizal plants in all the treatments with or without zinc. The accumulation of Zinc was reduced in the presence of mycorrhiza in all the treatments in roots, shoots and grain, suggesting that plant growth can be managed in Zn contaminated sites.

Mycorrhiza can also be effectively used in the reclamation of abandoned fly ash ponds at thermal power stations (Das et al. 2013). Fly ash is produced by burning coal in thermal power stations, which contains ash, dust and soot, which contains metals such as lead, arsenic, cadmium, cobalt, silica, mercury, and other toxic elements. Even though fly ash is being used in cement, bricks, roads, concrete, land reclamation, soil amendment in agriculture, etc., major portion (about 70%) remains unutilized in lagoons and land fills. In India, TERI (The Energy and Resources Institute), New Delhi has developed technology to reclaim fly ash ponds, chloralkali sludge laded sites, and distillery effluent loaded sites by using AMF (Das et al. 2017). Several mycorrhizal strains were isolated from India and abroad, and were tested in green house for adaptability in fly ash. The resistant strains for fly ash were selected and applied to plants in the fly ash dumps, along with organic manure. It was interesting to see that over a period of time, the greyish fly ash dumps were turned into green belt with vegetation (Das et al. 2013).

The chemical waste from alkali and chloride loaded residue create health hazard to human population residing in the locality. The sites are generally characterized with high pH, electrical conductivity and without any plant life. The sludge flies away with high winds and causes skin problems, decolorization of clothes and corrosion of metal structures. The mycorrhiza was applied along with the native microbes isolated and multiplied from the chlor-alkali sludge site gave excellent results in terms of establishment of vegetation and change in the physical chemical properties of soil (Das et al. 2017). Initially, sweet water is used and later the plants sustained with sea water. The change of soil properties encouraged the native grass species established in the reclaimed site.

The effluent generated in distilleries is rich in organics and, when it is stored on ground, it percolates into ground resulting in ground water contamination. The color of ground water gets changed, giving the water foul smell. The organic sludge deposited on the land will have high alkaline pH, with a very high salt content. The distillery effluent was applied in a specially designed landscape with broad ridges and furrows, and mycorrhized trees with high transpiration rates are planted on the furrows, making the site lush green (Das et al. 2017). The mycorrhiza mediated High Rate Transpiration System (HRTS) made the effluent loaded site into a commercially viable area with economic plants. TERI has also reclaimed the hyper saline desert land and phosphogypsum loaded site in an economic way using mycorrhizal fungi (Das et al. 2017).

4. CONCLUSION

Mycorrhiza (AMF) is one of the symbiotic fungi, most often exploited in agriculture for commercial use as biofertilizer as well as bioremediation. Mycorrhizoremediation is an emerging and promising strategy for remediation of heavy metal contaminated sites, effluent sites and other polluted environments. The method is also economically feasible and has potential for applicability at the field level. AMF has been shown to play a role in nutrient exchange, improve soil fertility and help plants to cope with phytotoxic heavy metals and also different environmental stresses in different plants including hyperaccumulator plants. AMF association is also crucial for maintaining beneficial microbial communities around mycorhizosphere. Successful field based exploitation of AMF has been done by some organizations on transforming degraded soils, fly ash ponds and effluent loaded sites into greener patches. These, in the long run, would contribute to sustainable, balanced ecosystem. Further research is needed to understand the mechanism of AM-plant communication through signaling pathways, hormonal modulations and plant micronutrient status in contaminated soils.

Acknowledgments

The author V. Vijaya Kumar is thankful to the Management of Natems Sugar Private Limited, for support and encouragement during the preparation of this chapter.

References

Abu-Elsaoud, A.M., N.A. Nafady and A.M. Abdel-Azeem. 2017. Arbuscular mycorrhizal strategy for zinc mycoremediation and diminished translocation to shoots and grains in wheat. PLoS ONE 12: e0188220. https: //doi.org/10.1371/journal.pone.0188220

Abu-Muriefah, S.S. 2016. The use of *Acacia saligna* inoculated with mycorrhizae in phytoremediation of lead-contaminated soils in the Kingdom of Saudi Arabia. Int. J. Curr. Res. Aca. Rev. 4: 297–309.

Adams, N., D. Carroll, K. Madalinski, S. Rock, T. Wilson and B. Pivetz. 2000. Introduction to phytoremediation. EPA/600/R-99/107. 104 p.

Adiloğlu, S. 2017. Heavy metal removal with phytoremediation, *In*: N. Shiomi (ed.), Advances in Bioremediation and Phytoremediation. IntechOpen, Croatia, pp. 115-126. doi: 10.5772/intechopen.70330.

Ahalya, N. and T.V. Ramachandra. 2006. Phytoremediation: process and mechanisms. J. Ecobiol. 18: 33–38.

Alkorta, I., J.M. Beeerril and C. Garbisu. 2010. Phytostabilization of metal contaminated soils. Rev. Environ. Health. 25: 135–146.

Aparecida, L., F. Vilela and M.V. Barbosa. 2019. Contribution of arbuscular mycorrhizal fungi in promoting cadmium tolerance in plants. *In*: M. Hasanuzzaman M.N.V. Prasad and K. Nahar (eds), Cadmium Tolerance in Plants. Academic Press, UK, pp. 553–586.

Azubuike, C.C., C.B. Chikere and G.C. Okpokwasili. 2016. Bioremediation techniques–Classification based on site of application: Principles, advantages, limitations and prospects. World J. Microbiol. Biotechnol. 32: 180–185.

Baxter, N. and J. Williamson. 1968. Introduction to soils. *In*: Leisa Macartney (ed.), Know Your Soils. Part-1. Mulqeen Printers, Victoria, 28 p.

Bi, Y., Y. Zhang and H. Zou. 2018. Plant growth and their root development after inoculation of arbuscular mycorrhizal fungi in coal mine subsided areas. Int. J. Coal Sci. Technol. 5: 47–53.

Chaney, R.L., M. Malik, Y.M. Li, S.L. Brown, E.P. Brewer, J.S. Angle and A.J.M. Baker. 1997. Phytoremediation of soil metals. Curr. Opinion Biotechnol. 8: 279–284

Channabasava, A., H.C. Lakshman and T. Muthukumar. 2015. Fly ash mycorrhizoremediation through *Paspalum scrobiculatum* L., inoculated with *Rhizophagus fasciculatus*. C. R. Biologies 338: 29–39

Cherlet, M., C. Hutchinson, J. Reynolds, J. Hill, S. Sommer, and G. von Maltitz (eds). 2018. World Atlas of Desertification, Publication Office of the European Union, Luxembourg.

Chibuike, G.U. 2013. Use of mycorrhiza in soil remediation: A review. Sci. Res. Essays 8: 1679–1687.

Das, M., P. Agarwal, R. Singh and A. Adholeya. 2013. A study of abandoned ash ponds reclaimed through green cover development. Int. J. Phytoremediation. 15: 320–329.

Das, M., V.S. Jakkula and A. Adholeya. 2017. Role of mycorrhiza in phytoremediation processes: A review. *In*: A. Varma, R. Prasad and N. Tuteja (eds), Mycorrhiza–Nutrient Uptake, Biocontrol, Ecorestoration. Springer, Cham, Switzerland, pp. 271–286.

Egamberdieva, D., A.A.F. Elsayed and A.J. Teixeira da Silva. 2016. Microbially assisted phytoremediation of heavy metal-contaminated soils. *In*: P. Ahmad (ed.), Plant Metal Interaction. Elsevier, pp. 483–498.

Dushenkov, V., P.B.A. Nanda Kumar, H. Motto and I. Raskin. 1995. Rhizofiltration: The use of plants to remove heavy metals from aqueous streams. Environ. Sci. Technol. 29: 1239–1245.

Etim, E.E. 2012. Phytoremediation and its mechanisms: A review. Int. J. Environ. Bioener. 2: 120–136.

European Commission – Press Release. 2018. New World Atlas of Desertification shows unprecedented pressure on the planet's natural resources. Brussels, 21 June 2018, IP/18/4202.

FAO. 2009. How to Feed the World 2050 -Global Agriculture Towards 2050. High Level Expert Forum, Rome 12–13 October. 4 p.

Ferrol, N., E. Tamayo, and P. Vargas. 2016. The heavy metal paradox in arbuscular mycorrhizas: From mechanisms to biotechnological applications. J. Exp. Bot. 67: 6253–6265.

Galadima, I. Ahmed, S. Mohammed, A. Abubakar and A.A. Deba. 2018. Phytoremediation: A preeminent alternative method for bioremoval of heavy metals from environment. J. Adv. Res. Appl. Sci. Eng. Technol. 10: 59–71.

Gerdemann, J.W. and J.M. Trappe. 1974. The Endogonaceae in the Pacific Northwest. Mycol. Mem. 5: 1–76.

Ghori, Z., H. Iftikhar, M.F. Bhatti, N. Minullah, I. Sharma, A.G. Kazi, et al. 2016. Phytoextraction: The use of plants to remove heavy metals from soil, *In*: P. Ahmad (ed.), Plant Metal Interaction. Elsevier, pp. 385–409.

Glazer, A.N. and H. Nikaido. 1995. Microbial Biotechnology: Fundamentals of Applied Microbiology. New York: Freeman.

Gomathy, M., K.G. Sabarinathan, T.S. Devi and P. Pandiyarajan. 2018. Arbuscular mycorrhizal fungi and glomalin-super glue. Int. J. Curr. Microbiol. App. Sci. 7: 2853–2857.

Greipsson, S. 2011. Phytoremediation. Nat. Edu. Knowl. 3: 7.

Hettiarachchi, G.M., S.C. Agudelo-Arbelaez, N.O. Nelson, Y.A. Mulisa and J.L. Lemunyon. 2012. Phytoremediation protecting the environment with plants. Kansas State University publication. 1–8. www.ksre.ksu.edu

Hildebrandt, U., M. Regvar and H. Bothe. 2007. Arbuscular mycorrhiza and heavy metal tolerance. Phytochem. 68: 139–146.

Jankaite, A. and S. Vasarevičius. 2005. Remediation technologies for soils contaminated with heavy metals, J. Environ. Eng. Landscape Manage. 13(2): 109–113.

Korade, D.L. and M.H. Fulekar. 2009. Development and evaluation of mycorrhiza for rhizosphere bioremediation. J. App.l Biosci. 17: 922–929.

Kumar, V.V. 2017. Mycoremediation: A step toward cleaner environment. *In*: R. Prasad (ed.), Mycoremediation and Environmental Sustainability. Springer International Pubslishing AG, Cham, Switzerland, pp. 117–128.

Laishram, J., K.G. Saxena, R.K. Maikhur and Rao. 2012. Soil quality and soil health: A review. Int. J. Ecol. Environ. Sci. 38(1): 19–37.

Lakshman, C. and A. Channabasava. 2013. Mycorrhizoremediation of mine spoil by using foxtail millet inoculated with *Rhizophagus fasciculatus*: An *ex-situ* solid waste management. Int. J. Curr. Sci. 8: E85–92.

Leake, J.R., G. Shaw and D.J. Read. 1989. The role of ericoid mycorrhizas in the ecology of ericaceous plants. Agric. Ecosystems Environ. 29: 237–250.

Limmer, M. and J. Burken. 2016. Phytovolatilization of Organic Contaminants. Environ Sci Technol. 50(13): 6632–43.

Liu, W.T., J.C. Ni and Q.X. Zhou. 2013. Uptake of heavy metals by trees: Prospects for phytoremediation. Mater. Sci. Forum. 743–744: 768–781.

Mahmood, T. 2010. Phytoextraction of heavy metals– the process and scope for remediation of contaminated soils. Soil Environ. 29: 91–109.

Manjhi, B.K., S. Pal, S.K. Meena, R.S. Yadav, A. Farooqui, H.B. Singh, et al. 2016. Mycorrhizoremediation of nickel and cadmium: A promising technology. Nat. Env. Poll. Tech. 15: 647–652.

Manousaki, E. and N. Kalogerakis. 2011. Halophytes-An emerging trend in phytoremediation. Int. J. Phytorem. 13(10): 959–969.

McIntyre, T. 2003. Phytoremediation of heavy metals from soils. Adv. Biochem. Eng. Biotechnol. 78: 97–123.

Meier, S., F. Borie, N. Bolan and P. Cornejo. 2012. Phytoremediation of metal-polluted soils by arbuscular mycorrhizal fungi. Critical Rev. Environ. Sci. Technol. 42(7): 741–775.

Menge, J.A. 1981. Mycorrhiza agriculture technologies. *In*: C. Elfring (ed.), Innovative Biological Technologies for Lesser Developed Countries-Workshop Proceedings (Washington, DC: US Congress, Office of Technology Assessment, OTA 13P-F-29. July 1985). pp. 383–424. https://princeton.edu./-ota/disk2/1985/8512/851512.PDF.

Meers, E., F.M.G. Tack, S.V. Slycken, A. Ruttens, G.D. Laing, J. Vangronsveld, et al. 2008. Chemically assisted phytoextraction: A review of potential soil amendments for increasing plant uptake of heavy metals. Int. J. Phytoremed. 10: 390–414.

Miransari, M. 2011. Hyperaccumulators, arbuscular mycorrhizal fungi and stress of heavy metals. Biotechnol. Adv. 29(6): 645–53.

Morton, J.B., R.E. Koske, S.L. Stürmer and S.P. Bentivenga. 2004. Mutualistic Arbuscular Endomycorrhizal Fungi, In: G.M. Mueller, G.F. Bills, M.S. Foster (eds), Biodiversity of Fungi: Inventory and Monitoring Methods. Academic Press, NY, pp. 317–336.

Muthusaravanan, S., N. Sivarajasekar, J.S. Vivek, T. Paramasivan, M.U. Naushad, J. Prakashmaran, et al. 2018. Phytoremediation of heavy metals: mechanisms, methods and enhancements. Environ. Chem. Lett. 16: 1339–1343.

Nanthi, S.B., J..H. Park, B. Robinson, R. Naidu and K.Y. Huh. 2011. Phytostabilization: A green approach to contaminant containment. Adv. Agron. 112: 145–204.

Newman, L. and C. Reynolds. 2004. Phytodegradation of organic compounds. Curr. Opin. Biotechnol. 15. 225–230.

Nikalje, G.C. and P. Suprasanna. 2018. Coping with metal toxicity-cues from halophytes. Front. Plant Sci. 9: 777. doi: 10.3389/fpls.2018.00777

Nortcliff, S., H. Hulpke, C.G. Bannick, K. Terytze, G. Knoop, M. Bredemeier, et al. 2012. Soil, 1. Definition, function, and utilization of Soil. ULLMANN'S Encyclopedia of Industrial Chemistry 33: 399–420.

Ozyigit, I.I. and I. Dogan. 2015. Plant-microbe interactions in phytoremediation. *In*: K.R. Hakeem, M.O. Ahmet, R. Mermut and Md. Sabir (eds), Soil Remediation and Plants. Prospects and Challenges. Elsevier, USA, pp. 255–285.

Pathare, V., B.V. Sonawane, S. Srivastava and P. Suprasanna. 2016. Arsenic stress affects the expression profile of genes of 14-3-3 proteins in the shoot of mycorrhiza colonized rice. Physiol. Mol. Biol. Plants. 22: 515–522.

Pathare, V., A. Shukla, S. Srivastava. and P. Suprasanna. 2017. Response of rice-mycorrhizal association to arsenate exposure. Bull. Environ. Sci. Res. 6: 1–6.

Poonam, S. Srivastava, V. Pathare and P. Suprasanna. 2017. Physiological and molecular insights into rice-arbuscular mycorrhizal interactions under arsenic stress. Plant Gene. 11: 232–237.

Pilon-Smits E. 2005. Phytoremediation. Ann. Rev. Plant Biol. 56: 15–39.

Radziemska, M., M.D. Vaverková, and A. Baryła. 2017. Phytostabilization-management strategy for stabilizing trace elements in contaminated soils. Int. J. Environ. Res. Public Health. 14(9): 958. doi:10.3390/ijerph14090958

Rascio, N. and F. Navari-Izzo. 2011. Heavy metal hyperaccumulating plants: How and why do they do it? And what makes them so interesting? Plant Sci. 180: 169–181.

Read, D.J. 1983. The biology of mycorrhiza in the Ericales. Can. J. Bot. 61: 985–1004.

Robinson, B., J.E. Fernández, P. Madejón, T. Marañón, J.M. Murillo, S. Green, et al. 2003. Phytoextraction: an assessment of biogeochemical and economic viability. Plant Soil. 249: 117–125.

Sadhasivam, M., S. Pitchamuthu and V. Ayyavu. 2010. Chemically induced phytoextraction of caesium–137. *In*: 19th World Congress of Soil Science, Soil Solutions for a Changing World 1–6 August 2010, Brisbane, Australia. pp. 39–41.

Saha, J.K., R. Selladurai, M.V. Coumar, M.L. Dotaniya, S. Kundu and A.K. Patra. 2017. Remediation and management of polluted sites. *In*: J.K. Saha, R. Selladurai, M.V. Coumar, M.L. Dotaniya, S. Kundu and A.K. Patra (eds), Soil Pollution—An Emerging Threat to Agriculture. Springer Nature Singapore, Singapore, pp. 317–372.

Salvioli di Fossalunga, A. and M. Novero. 2019. To trade in the field: The molecular determinants of arbuscular mycorrhiza nutrient exchange. Chem. Biol. Technol. Agric. 6, 12 doi:10.1186/s40538-019-0150-7

Science Communication Unit, University of the West of England, Bristol (2013). Science for Environment Policy In-depth Report: Soil Contamination: Impacts on Human Health. Report produced for the European Commission DG Environment. 29 p.

Shackira, A.M. and J.T. Puthur. 2019. Phytostabilization of heavy metals: Understanding of principles and practices. *In*: S. Srivastava, A. Srivastava and P. Suprasanna (eds), Plant-Metal Interactions. Springer, Cham, pp. 263–282.

Shi, W., Y. Zhang, S. Chen, A. Polle, H. Rennenberg and Z.B. Luo. 2019. Physiological and molecular mechanisms of heavy metal accumulation in nonmycorrhizal versus mycorrhizal plants. Plant Cell Environ. 1:1. https://doi.org/10.1111/pce.13471.

Singh, N., M. Megharaj, R.S. Kookana, R. Naidu and N. Sethunathan. 2004. Atrazine and simazine degradation in *Pennisetum* rhizosphere. Chemosphere 56: 257–263.

Srivatsava, D., R. Kapoor, S.K. Srivatsava and K.G. Mukherji. 1996. Vesicular arbuscular mycorrhiza-an overview. *In*: K.G. Mukherji (ed.), Concepts in Mycorrhizal research. Kluwer Academic Publishers, Netherlands, pp. 1–39.

Tangahu, B.V., S.R.S. Abdullah, H. Basri, M. Idris, A. Anuar and M. Mukhlisin. 2011. A review on heavy metals (As, Pb, and Hg) uptake by plants through phytoremediation. Int J. Chem. Eng. Article ID 939161, 31 p. doi:10.1155/2011/939161.

Thangavel, P. and C.V. Subbhuraam. 2004. Phytoextraction: Role of hyperaccumulators in metal contaminated sites. Proc. Indian Natn. Sci. Acad. B70: 109–130.

Vlček, V. and M. Pohanka. 2020. Glomalin – an interesting protein part of the soil organic matter. Soil Water Res. 15: 67–74.

Watkinson, S.C. 2016. Mutualistic symbiosis between fungi and autotrophs. *In*: S.C. Watkinson, L. Boddy, N.P. Money (eds), The Fungi, 3rd Ed. Academic Press, Cambridge, MA, USA, pp. 205–243.

Wipf, D., F. Krajinski, D.V. Tuinen, G. Recorbet and P. Emmanuel. 2019. Trading on the arbuscular mycorrhiza market: From arbuscules to common mycorrhizal networks. New Phytol. 223: 1127–1142.

Yang, Y., X. Han, Y. Liang, A. Ghosh, J. Chen and M. Tang. 2015. The combined effects of Arbuscular Mycorrhizal Fungi (AMF) and Lead (Pb) stress on Pb accumulation, plant growth parameters, photosynthesis, and antioxidant enzymes in *Robinia pseudoacacia* L. PLoS ONE 10: e0145726. doi:10.1371/journal.pone.0145726 1–24.

CHAPTER 11

Agro-ecosystem Bioremediation Mediated by Plant-microbe Associations

**Maryam Bello-Akinosho[1*], Busiswa Ndaba[1],
Ashira Roopnarain[1], Emomotimi Bamuza-Pemu[1],
Rosina Nkuna[2], Haripriya Rama[1], Rasheed Adeleke[3]**

[1]Microbiology and Environmental Biotechnology Research Group,
Agricultural Research Council–Soil, Climate and Water (ARC-SCW),
Arcadia, Pretoria 0001, South Africa.
[2]Institute for the Development of Energy for African Sustainability (IDEAS),
University of South Africa's College of Science,
Engineering and Technology Florida, 1710, South Africa
[3]Unit for Environmental Science and Management, North-West University
(Potchefstroom Campus), Potchefstroom 2520, South Africa.

1. INTRODUCTION

Agro-ecosystems are unwilling sinks for vast varieties of contaminants, which may originate from natural sources but mainly from anthropogenic sources. The anthropogenic sources of agro-ecosystem contaminants could be as a result of on-farm activities or from other off-farm activities such as mining, petroleum exploration and transportation occurring on or near the farm. Contaminants arising from agricultural activities include pesticides, herbicides, chemical fertilizers as well as veterinary medications (Zhang et al. 2018). A list of veterinary pharmaceuticals includes antibiotics, hormonal growth implants, anti-inflammatories and steroids (Sarmah et al. 2006). These agroecosystem contaminants are broadly grouped as organic chemicals and heavy metals (Schaffner et al. 2002, Cherian and Oliveira 2005, Ojuederie and Babalola 2017, Sarwar et al. 2017). Unlike organic contaminants, heavy metals are not

*Corresponding author: BelloakinoshoM@arc.agric.za

metabolised, have tendency to accumulate in living beings, and could be carcinogenic to humans (Ma et al. 2016, Atieh et al. 2017). Although heavy metals are natural constituents of agro-ecosystems because many of them (such as Cobalt, Iron, Manganese, Molybdenum, Zinc, Copper and Nickel) are required in minute quantities by soil organisms and plants for growth and best performance, their indiscriminate production has led to heavy metal pollution in agro-ecosystems. While it is possible to transform heavy metals to less toxic forms through biological processes, they are not completely biodegraded. On the contrary, organic contaminants, although foreign by their nature, could be biodegraded into innocuous forms (Xiong et al. 2010, Megharaj et al. 2011, Meier et al. 2012, Chibuike and Obiora 2014, Dixit et al. 2015).

Contamination of agro-ecosystems has dwindled available arable land and reduced productivity in aquatic agro-ecosystems (Chirwa and Bamuza-Pemu 2010, Shukla et al. 2010, Sarwar et al. 2017). Remediation of contaminated agro-ecosystems is, therefore, imperative to ensure food security for the ever-growing population. Quite a number of strategies and technologies involving the use of chemical, physical and biological treatments to remediate contaminated agro-ecosystems are deployed. The chemical and physical treatments include excavation and confinement of soil in specified landfill or dumpsites, soil washing, vitrification and other thermal treatments (Liu et al. 2018). They, generally, are expensive and could deteriorate the chemical, physical and biological properties of soil (Khan 2005, Gkorezis et al. 2016). Application of biological treatment for remediation, referred to as bioremediation, has received acceptance as an ecosystem-friendly, economical and self-sustaining remediation method. It involves using natural or engineered organisms and/or biological processes to eliminate, attenuate or transform contaminants to less toxic or innocuous substances (Wenzel 2009, Azubuike et al. 2016).

A number of bioremediation approaches are in use and they encompass phytoremediation, land farming, *in situ* microbial remediation, use of bioreactors and composting (Banerjee et al. 1995, Šašek et al. 2003, Robles-González et al. 2008). Biodegradation is a primary mechanism for bioremediation; it is the process of biologically breaking down complex contaminants into simple, less toxic substances (Wenzel 2009, Varjani and Upasani 2016). The utilization of plants in the removal or transformation of environmental contaminants is termed phytoremediation (Salt et al. 1998, Peuke and Rennenberg 2005). More often, many authors define phytoremediation as being assisted by microorganisms, which are innately capable of degrading environmental contaminants (Meier et al. 2012, Gkorezis et al. 2016, Srivastav et al. 2018). These microorganisms proliferate in the presence of plant root exudates and have assured habitat within the rhizosphere facilitated by the presence of plants.

Plants associate with diverse beneficial microorganisms in natural environments as well as in agro-ecosystems (Jacoby et al. 2017) and these associations are useful in many ecosystem functions. The co-operation between plants and microorganisms for biodegradation of contaminants can occur naturally or artificially. Naturally, associations of plants with different phyllospheric and rhizospheric microbes are often encountered and they are quintessential to the growth and optimum performance of the plants as well as the microbes (van der Heijden et al. 2008, Jambon et al. 2018, Adeleke et al. 2019). Furthermore, contaminated environments could trigger mechanisms in plants for the active recruitment of relevant phyllospheric and rhizospheric microbes (Compant et al. 2019).

Artificially, plant-microbe interactions could be engineered with the goal of manipulating the association towards an enhanced beneficial outcome for the plant (Farrar et al. 2014, Quiza et al. 2015). In the case of bioremediation, microorganism that would enhance survival of plants in a contaminated site in addition to being active in the degradation process would be introduced to the plant. Microorganisms often associated with plants for bioremediation purposes include bacteria, fungi, cyanobacteria and algae. The focus of this chapter, therefore, is the associations of plants and rhizospheric microorganisms performing functional roles in bioremediation of contaminated agro-ecosystems.

2. RHIZOSPHERIC PLANT—MICROBE ASSOCIATIONS FOR REMEDIATION

For decades, greater research efforts have been directed to the use of microorganisms in the rhizosphere for bioremediation strategies (Correa-Garcia et al. 2018). Beneficial microorganisms play several roles when in association with plant roots and their association can assist in remediating polluted sites in the agro-ecosystems. In the course of biodegradation of contaminants by plants-microbes interactions, plants provide carbon, a product of photosynthesis, which is transferrable to the roots and soil for subsequent utilization by microbial communities for growth enhancement. Root exudates also provide energy, which is needed by the microbial communities for metabolising contaminants in agro-ecosystems (Mimmo et al. 2014). In addition, microbes assist in vegetation conservation by reducing the detrimental effects of environmental pollutants on plant, thereby improving soil's physical, chemical and biological properties to ensure plant growth and well-being. For instance, in sites contaminated with crude oil, plant growth rate is usually impacted due to the negative effects of inhibition of photosynthetic processes exerted by a complex mixture of hydrocarbons associated with crude oil. The complex mixture comprise of cycloalkanes and other alkanes as well as aromatic compounds, such as polycyclic aromatic hydrocarbons (PAHs) (Frick et al. 1999, Vega-Jarquin et al. 2001, Dominguez-Rosado and Pichtel 2004, Balasubramaniyam 2015). Their inhibitory effects can vary based on plant species as well as physiological behaviour stimulated in response to the contaminants. Bacteria, yeasts and filamentous fungi are able to synthesize an array of surface-active organic compounds such as biosurfactants and bioemulsifiers, which enhance/facilitate bioremediation. There is, therefore, an active involvement of microorganisms in pollutant degradation and improvement of soil's chemical, physical and biological properties, which consequently improve plant growth and enhance remediation of pollutants in contaminated sites by plants. This process is referred to as phytoremediation and is discussed both within the context of plants being directly involved in the remediation as well as plants and microbes associatively carrying out the bioremediation (Kuiper et al. 2004, Gkorezis et al. 2016, Srivastav et al. 2018).

2.1   Phytoremediation

Phytoremediation is a plant-based remediation technology through which environmental contaminants are extracted, sequestered or detoxified. The technology is useful for several pollutant types but more applicable for removing non-biodegradable pollutants such as heavy metals. In agro-ecosystems, phytoremediation involves four major different processes, which are phytoextraction, phytodegradation, phytostabilization and phytovolatilization. Each of these is applicable to specific contaminant and agro-ecosystem type. Phytoextraction relies on the accumulation of heavy metals by harvestable plant parts. After harvesting, plant materials are incinerated and the ashes are treated as hazardous residues. Otherwise, phytomining is employed as a means of recovery of metals from ash residues (Khan et al. 2000, Chen et al. 2018b). The process requires a highly tolerant host plant with large shoots that can tolerate the toxicity of the heavy metals to be accumulated. One such plant is Plantain (*Plantago orbignyana*), which is able to accumulate approximately 1000 mg/kg lead in its shoots (Bech et al. 2012). Phytodegradation is the breakdown or transformation of contaminants by plants and their associated microbes which leads to the contaminant being detoxified. In a study by Park et al. (2011), plants such as field mustard (*Brassica campestris*), tall fescue (*Festuca arundinacea*) and common sunflower (*Helianthus annuus*) were used for successful phytodegradation of total petroleum hydrocarbons (TPH) at degradation efficiencies of 86%, 64%, and 85%, respectively. The enhanced biodegradation performance was attributed to an increase in microbial activities in soils, confirming the role of microbes during degradation. In their study, dehydrogenase enzyme was found to be functional for biological oxidation, which is an indication of microbial-assisted remediation. Phytovolatilization is applicable in the remediation of organically contaminated soil. After uptake by the plant, the organic contaminant is volatilized into the atmosphere (Chatterjee et al. 2013). Phytostabilization is the reduction in contaminant mobility or bioavailability through immobilization or transformation of contaminants to a less toxic substance, thus preventing migration of the contaminant. This is particularly useful to prevent contaminants leaching to groundwater at dumpsites (Bolan et al. 2011).

Phytoremediation without microbial involvement is a relatively slow process (Waigi et al. 2017, Correa-Garcia et al. 2018), which takes several years to reduce metal contents in soil to a safe and acceptable level due to small size and slow growth of several identified metal hyperaccumulator plants. Therefore, to make phytoremediation a viable technology with enhanced efficiency, fast growing and metal tolerant plants with extensive root system, capable of absorbing heavy metals have to be identified. Furthermore, microorganisms with innate ability to metabolise pollutants and capable of associating with the plants should be sought for in order to accelerate the process.

2.2   Microbe-assisted Phytoremediation

It is generally recognized that plant fitness would only be characterized and understood within the plant holobiont concept, where plant plus its intimately associated microbiota is considered a functional whole (Zilber-Rosenberg and Rosenberg 2008, Vandenkoornhuyse et al. 2015). A holistic discussion of phytoremediation would, therefore, be within the context of microbe-assisted phytoremediation, also referred to as rhizoremediation, which is the remediation of contaminants by metabolically active microorganisms through their association with plant roots within the rhizosphere. During rhizoremediation, plant roots exude several organic compounds useable by microbes as nutrient sources of carbon, nitrogen, phosphorous or sulphur in order to multiply in the rhizosphere. These root exudates can also act as inducers of catabolic pathways for biodegradation of different contaminants (Kuiper et al. 2004, El Amrani et al. 2015). The presence of the root exudates is responsible for enhancing the density and activity of rhizospheric microbial communities as well as potentially enhancing the breakdown of hydrocarbons (Correa-Garcia et al. 2018) (Fig. 1). In addition, some root exudates assist in separating organic contaminants from the organic matter in soil, increasing their availability for microbial degradation (Gao et al. 2010). Therefore, biodegradation of contaminants through plant–microbe interactions is more effective in the presence of roots and the effectiveness of rhizoremediation is reduced in compacted soils where roots cannot penetrate contaminated zones (Manschadi et al. 2006).

Rhizoremediation may also incorporate bioaugmentation, where microbes with biodegradation potential are inoculated into contaminated sites for enhanced remediation (Abhilash et al. 2012). Furthermore, using consortia of microorganisms for degradation of contaminants in the soil is considered more effective than individual strains due to the presence of species in the consortium with distinct abilities to use the various intermediates of the biodegradation pathway (Mrozik and Piotrowska-Seget 2010, Shankar et al. 2014, Correa-Garcia et al. 2018, Rahman et al. 2002, Bello-Akinosho 2018). On the contrary, Piakong and Zaida (2018) showed that the use of *Sphingomonas paucimobilis* as a single strain resulted in a more effective degradation of oil sludge in soil by 1.1-fold in comparison to the microbial consortium. During rhizoremediation, either the inoculated microorganism or microbial consortia forms an association with the plant roots because of strong chemotaxis exhibited towards root exudates and/or strong adherence to the root surface (Bashan et al. 2014). A consortium can consist of functional microbes, including plant growth-promoting microorganisms (PGPM) such as nitrogen-fixing rhizobacteria, mycorrhiza-helping bacteria (MHB) and mycorrhizal fungi. All these microbes could benefit the plant, thereby assisting in remediation through plant-microbe interactions for restoring polluted agro-ecosystems (Khan 2005). Plant growth promoting microbes (PGPM) are able to promote plant health by producing phytohormones that are involved in the regulation of plant physiological processes. Plant growth enhancement by PGPM promotes phytoaccumulation or phytodegradation of toxic compounds. Phytohormones produced by PGPM include abscisic acid (ABA), auxins such as indole acetic acid (IAA), gibberellins (GB), cytokinins (CK) and ethylene (Hayat et al. 2010). These phytohormones assist plants in combating abiotic stresses, which include heavy metal contamination, flooding, temperature fluctuations, drought, xenobiotic compounds and salinity in the environment (Fig. 1) (Choudhary et al. 2011, Mishra et al. 2017). The bacterial phytohormone, IAA, is important as a signalling molecule to improve plant immunity and facilitate the biodegradation of contaminants through phytostimulation (Cheynier et al. 2013).

Certain enzymes produced by PGPM also assist in the regulation of phytohormones involved in plant tolerance to abiotic stresses. For example, some PGPM and other soil microorganisms within the rhizosphere produce 1-Aminocyclopropane-1-carboxylase (ACC) deaminase. This enzyme cleaves ACC, which is exuded in high concentrations by plant roots under abiotic stress, into ammonia and α-ketobutyrate, as opposed to being converted into ethylene. High concentrations of ethylene inhibits plant growth; therefore, the cleaving of ACC enables plants to grow under stressed conditions (Santoyo et al. 2016). This plant-microbe interaction is shown to enable plants to grow under stressed conditions as demonstrated by a study performed by Ali et al. (2014), which reported on the ability of the ACC deaminase-containing endophyte, *Pseudomonas migulae* 8R6 to ameliorate tomato plant damage under high salinity levels and promote plant growth. Furthermore, aromatic root exudates with structural similarity to organic contaminants, for instance polychlorinated biphenyls (PCBs), promote the expression of bacterial *bph* genes (Vergani et al. 2017). These genes encode for various enzymes such as dioxygenases and dehydrogenases that are involved in the biodegradation of PCBs to benzoate and pentanoic acid (Brazil et al. 1995). These particular interactions between plants and microbial communities enhance bioremediation of contaminants in the rhizosphere (Fig. 1).

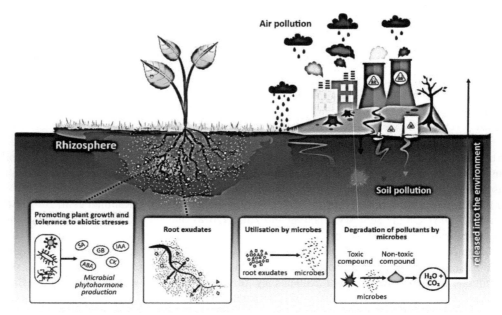

Figure 1   Plant—microbe interactions during rhizoremediation
(Adapted with modifications from Correa-Garcia et al. 2018).

2.2.1　Plant-bacteria Associations for Bioremediation

In harnessing plant-bacteria associations for bioremediation of agro-ecosystems, plant roots are able to alter the physical and chemical conditions (such as improving soil porosity and oxygen bioavailability) in contaminated soil in order to promote biodegradation by microorganisms (Joner et al. 2006). Rhizospheric bacteria act as plant growth promoters, mobilising nutrients for the plant and they are primarily responsible for biodegradation of contaminants in agro-ecosystems (Liu et al. 2007, Lors et al. 2012, Lu et al. 2017). Their abundance, microscopic size and ability to produce extracellular polymeric substances, bioemulsifiers, exoenzymes and biofilms as well as their possession of a variety of metabolic degradation genes make them well suited for biodegradation of organic contaminants (Johnsen and Karlson 2004, Gkorezis et al. 2016). Furthermore, their ability to degrade or transform contaminants facilitate derivation of carbon and/or electrons necessary for their metabolism. Several genera of both Gram-positive and Gram-negative bacteria have been identified and characterized for their ability to utilize different environmental contaminants. For instance, Liu et al. (2007) developed an association of bacterial species *Comamonas* and alfalfa plant (*Medicago sativa* L.) for rhizoremediation of 4-chloronitrobenzene commonly found in wastewater treatment facilities. Furthermore, bispyribac sodium, a herbicide, was degraded by more than 96% in a wheat (*Triticum aestivum*) plantation inoculated with a microbial consortium (Ahmad et al. 2019a,b).

Some metal-tolerant, siderophore-producing, rhizospheric bacteria are also able to facilitate phytoremediation in heavy metal contaminated agro-ecosystems. Bacterial siderophores such as pyoverdine, pyochelin and coelichelin are metal chelators that are able to bind with heavy metal ions to form stable complexes (Rajkumar et al. 2010). The solubility and bioavailability of the metal contaminants increases as a result of complexation. Subsequently, plants are able to take up the metal contaminants more easily within the rhizosphere. Dimkpa et al. (2009) reported

Table 1　Plant-bacteria associations for bioremediation

Plant	*Bacteria*	*Possible mechanism*	*Reference*
Maize (*Zea mays* L.)	*Pseudomonas* sp. UG14Lr	Bacterial metabolism of phenanthrene through enzymatic mineralization in soils was indirectly enhanced by maize	Chouychai et al. (2012)
Canola (*Brassica napus*)	*Pseudomonas asplenii* AC-1	Bacterial production of ACC deaminase and accumulation of copper in plant roots and shoots	Reed and Glick (2005)
Barmultra grass (*Lolium multiflorum*)	*Pseudomonas putida* PCL1444	Enzymatic degradation of naphthalene by bacteria enhanced by Barmultra grass	Kuiper et al. (2001)
Horseradish (*Armoracia rusticana*)	*Bacillus pumilus*	PCB degradation by bacterial metabolism, phytoaccumulation and phytotransformation	Ionescu et al. (2009)
Maize (*Z. mays* L.)	*Pseudomonas aeruginosa* and *Ralstonia metallidurans*	Bacterial production of siderophores as well as enhanced phytoextraction of chromium and lead	Braud et al. (2009)
Winter wheat (*T. aestivum*)	*Sphingobium chlorophenolicum* ATCC 39723	Enzymatic degradation of pentachlorophenol by bacterial mineralization	Dams et al. (2007)
Indian mustard (*Brassica juncea* L.)	*Bacillus subtilis* SJ-101	Bacterial production of IAA and increased phytoextraction of nickel	Zaidi et al. (2006)
Italian ryegrass (*Lolium Multiflorum*)	*Mycobacterium gilvum*	PAH-ring hydroxylation by bacterial dioxygenase was enhanced by ryegrass	Guo et al. (2017)

on the increased plant growth and uptake of solubilized cadmium and iron by sunflower plants (*Helianthus annuus*) resulting from the siderophores produced by *Streptomyces tendae* F4 in the rhizosphere. Other examples of plant-bacteria associations in bioremediation are shown in Table 1.

Endophytic bacteria are also active in bioremediation of agro-ecosystem contaminants. They are afforded nutrients and habitats by their host plants (Arslan et al. 2017). Through cross-talk of signalling molecules, endophytes are able to pass through the plant's innate defence system and colonize the roots (Khare et al. 2018). Reports abound on the beneficial associations between plants and endophytic bacteria for bioremediation (Segura et al. 2009, Afzal et al. 2014). Pollutant-degrading endophytic bacteria are able to protect the plant through mineralization or degradation of toxic pollutants taken up by the plant (Arslan et al. 2017). In addition to protecting the plant from toxic pollutants, endophytic bacteria also enhance phytoremediation capabilities of plants (Table 2) and promote their growth (Ryan et al. 2008, Stępniewska and Kuźniar 2013).

Table 2   Associations of plant and endophytic bacteria for bioremediation

Plant	*Endophytic bacteria*	*Possible mechanism*	*Reference*
Wheat (*T. aestivum*)	*Pseudomonas fluorescens* 2-79 TOM	Trichloroethylene (TCE) biodegradation by plant and bacteria that produce toluene ortho-monooxygenase	Yee et al. (1998)
Yellow lupine (*Lupinus luteus* L.)	*Burkholderia cepacia* G4	Lowered evapotranspiration and degradation of toluene by bacterial toluene ortho-monooxygenase as well as enhanced phytoremediation through decreased phytotoxicity	Barac et al. (2004)
Wheat (*T. aestivum*) and Maize (*Z. mays* L.)	*Enterobacter* sp. 12J1	Pyrene degradation through bacterial production of IAA, which enhanced phytoremediation abilities	Sheng et al. (2008)
White poplar (*Populus alba*) and Weeping willow (*Salix babylonica*)	*Pseudomonas putida* VM1450	2,4-Dichlorophenoxyacetate degradation by bacterial enzymes and enhanced phytoaccumulation of pollutant	Germaine et al. (2006)
Yellow lupin (*Lupinus luteus* L.)	*Burkholderia cepacia* L.S.2.4	Precipitation of nickel by bacteria prevented nickel accumulation in plant roots and shoots	Lodewyckx et al. (2001)
Poplar tree cuttings (*Populus deltoids* x *Populus nigra* hybrid)	*Enterobacter* sp. PDN3	Enhanced phytoremediation of TCE assisted by bacterial metabolism of TCE	Doty et al. (2017)

2.2.2   Plant-fungi Associations for Bioremediation

Filamentous fungi are also potent in biodegradation of different soil contaminants. Soil supports the development of fungal foraging mycelia, which could penetrate soil and rocks. In addition, fungi secrete non-specific exoenzymes (such as peroxidases and laccases) capable of degrading a wide range of organic agro-ecosystem contaminants. Another ability that makes fungi relevant in pollutant degradation is their independence from using pollutants as growth substrates. They, therefore, degrade the pollutant co-metabolically while using other sources of energy. Co-metabolism is a phenomenon in which microbes degrade non-growth substrates. This is often employed in co-metabolic biodegradation by groups of fungi where contaminant degradation is mediated by an enzyme or cofactor produced while metabolising another compound (Wen et al. 2011). Several groups of fungi including the white-rot fungi (WRF) (Pointing 2001, Lee et al. 2014), mycorrhizal fungi (Huang and Wang 2013, Franco et al. 2014) and some species of *Penicillium*, *Aspergillus* and *Trichoderma* (Chávez-Gómez et al. 2003) have been reported with ability to degrade organic contaminants. Perhaps, the most investigated groups of fungi with biodegradation potentials are the WRF and the mycorrhizal fungi (Ellouze and Sayadi 2016, Bosco and Mollea 2019). While other fungi are free-living saprotrophs, mycorrhizal fungi are plants symbionts. Mycorrhizal fungi are a heterogeneous group of fungi that associate with roots of over 90% of plant species, which include forest trees, grasses and food crops (Bonfante and Genre 2010). Both the plants and the fungi benefit from the association. Mycorrhizal fungi are able to mobilise nutrients for plants, while plants provide photosynthesized carbon for fungi in addition to habitat for reproduction. Common classification for mycorrhizae, based on anatomical structures, are the ectomycorrhizae and endomycorrhizae. When the fungus colonizes root intercellular spaces, an ectomycorrhiza is formed. When the fungus develops inside the plant cells, then an endomycorrhiza is formed (Bonfante and Genre 2010).

Arbuscular mycorrhizal fungi (AMF) are a prominent group of endomycorrhizal fungi and they are also known to play a pivotal role in rhizoremediation. They have been reported to improve plant growth in contaminated soils through their capacities to sequester minerals in addition to their defensive role against abiotic stress (Chen et al. 2018a). These capabilities are considered a form of bioremediation (Leyval et al. 2002, Xun et al. 2015). Heavy metal remediation by AMF involves the accumulation and sequestration of such toxic metal ions, which results in the host plant being protected from the pollutant. AMF could also act to convey heavy metals to the host plant in similar manner as mineral nutrients are transferred. This results in heavy metal accumulation in the plant in a process of phytoextraction (Chen et al. 2018a). Furthermore, AMF act to directly uptake and convey water through their fungal hyphae to the host plant in organic contamination sites in order to mitigate the effects of water deficiency and inadequate root gaseous exchange occasioned by the hydrophobic and lipophilic properties of the contaminants (Lenoir et al. 2016). Although some studies have suggested reduction of AMF diversity and colonization in polluted soils, their role after colonization of plant roots in polluted soils is very significant (Debiane et al. 2008, 2009, Driai et al. 2015). Enhanced pollutant degradation has been reported in the rhizosphere of AMF-colonized plants. For instance, Lu and Lu (2015) reported very high PAH dissipation of 93.4% in the combined treatment with tall fescue, AMF and earthworms. Ectomycorrhizal fungi (ECMF) associate with plant root tips and form hyphae, which extensively ramify the roots of plants while growing into the soil in search of nutrients acting as the direct connection between the host plant and microsites of the soil. This feature of ECMF enables them to enhance phytoremediation. For example, Bell et al. (2015) reported that a significant correlation was found between *Sphaerosporella brunnea* (an ECMF that was attracted by planted willow) abundance and total zinc uptaken by willow after four months of planting.

2.2.3 Cyanobacteria and Algae for Bioremediation

Cyanobacteria have been utilised for the bioremediation of an array of environmental pollutants including pesticides, heavy metals, crude oil, phenols and xenobiotics (Singh et al. 2016, Hamouda et al. 2016, Pathak et al. 2018). A factor that contributes to the bioremediation abilities of cyanobacteria is their capability to inhabit a multitude of heavily polluted aquatic environments (El-Bestawy et al. 2007). Their presence in such environments has conferred resistance on them against selected environmental pollutants. Additionally, the decrease in the concentration of the pollutants that cyanobacteria breakdown or accumulate does not result in a reduction in their metabolic activity and viability (El-Bestawy et al. 2007). Cyanobacteria are also promising bioremediating agents due to their ability to fix nitrogen and their photoautotrophic characteristics, which ensure the self-sustainability and adaptability of these microorganisms in low nutrient environments (El-Bestawy et al. 2007, Singh et al. 2016).

Cyanobacteria have exhibited tremendous potential in sustainable agricultural practices due to their application as bioremediating agents of contaminated agro-ecosystems. For example, herbicide-contaminated soil was remediated with the aid of cyanobacterial species such as *Nostoc* sp., *Anaebaena* sp., *Microcystic* sp. and *Spirulina* sp., which have the ability to metabolise glyphosphate herbicides, thereby removing it from contaminated agricultural soils (Forlani et al. 2008). Cyanobacterial mats have also shown potential to remediate soil and water systems that have been contaminated with the chloroacetamide herbicide, acetochlor (El-Nahhal et al. 2013). Cyanobacteria have been utilised for the remediation of insecticide-contaminated environments. For example, organo-phosphorus and organo-chloride insecticides have been successfully removed from polluted aquatic systems by cyanobacterial species such as *Microcystis aeruginosa*, *Synechococcus elongatus* and *Anacystics nidulans* (Vijayakumar 2012). Residues of lindane, another widely used insecticide, have also been removed from agricultural systems by cyanobacterial species such as *Nostoc* sp., *Synechococcus* sp. and *Nodularia* sp. (El-Bestawy et al. 2007). Cyanobacteria such as *Nostoc linckia*, *Spirulina platensis* and *Anaebaena subcylindrica* also play important roles in the bioremediation of sites contaminated with heavy metals (Singh et al. 2016). From all of the above-stated examples, it can be deduced that cyanobacteria may provide an economical and low maintenance method of agro-ecosystem remediation from contaminants such as pesticides, herbicides and metal-contaminated sites. Furthermore, the potential of cyanobacteria to degrade agro-ecosystem contaminants can be improved by genetic engineering approaches (Pathak et al. 2018).

Unlike bacteria, fungi and cyanobacteria, algae are less frequently utilised for agro-ecosystem bioremediation. However, some instances of their bioremediation applications are documented. Algae were used to alleviate heavy metal toxicity in different agro-ecosystems through the introduction of dead or living algal cells to contaminated soil. Dead algal cells remediate heavy metals via biosorption processes, whereas living cells accumulate heavy metals intracellularly (bioaccumulation) (Hassan et al. 2017). The capability of algae to form symbiotic associations with other beneficial organisms also promotes the exploitation of algae-containing consortia in bioremediation of agricultural soils (Hamouda et al. 2016, Pathak et al. 2018). For example, a study conducted by Hamouda et al. (2016) indicated that a consortium of the microalga, *Chlorella kessleri*, and cyanobacterium, *Anabaena oryzae*, enhanced the degradation of crude oil. Mixed consortia convey the advantage of improved functionality due to the combination of functions associated with individual species in the consortia. Moreover, the combination of various species in bioremediation improves the ability of the organisms within the consortium to withstand environmental perturbations and nutrient limitations. The consortium of *Chlorella kessleri* and *Anabaena oryzae* enabled the biodegradation of both aliphatic and aromatic hydrocarbons due to their flexible metabolism. Furthermore, these microorganisms possessed the ability to grow under mixotrophic conditions making them suitable for remediation of polluted agro-ecosystems, which are frequently nutrient deficient and/or low-light environments (Hamouda et al. 2016).

An indirect use of microalgal biomass for bioremediation is its exploitation as a sustainable alternative to chemical pesticides. Excessive use of chemical pesticides contributes to the contamination of agro-ecosystems, hence the gradual shift to biopesticides. The ability of algal species to control agricultural pests is due to the synthesis of secondary metabolites such as antifungals and antibiotics (Costa et al. 2019). The use of microalgae as biopesticides as opposed to excessive use of chemical pesticides may contribute to prevention of agro-ecosystem pollution/ contamination.

3. MECHANISMS INVOLVED IN PLANT—MICROBE REMEDIATION APPROACHES WITHIN THE RHIZOSPHERE

Although bacterial strains can substantially metabolise and hence degrade organic pollutants in bulk soil without plants, the efficiency of this process is enhanced in the rhizosphere, where plant root exudates enhance the abundance of microorganisms. In fact, the rhizosphere houses 10–100 times higher microbial activity than bulk soil (Berendsen et al. 2012, Ho et al. 2017, Hassan et al. 2019). Root exudates enhance mobility of metals and nutrients by methods, which include acidification, intracellular binding, enzymatic electron transfer and indirect stimulation of rhizospheric microbial activities. This, therefore, enhances overall phytoremediation efficiency within the rhizosphere (Ma et al. 2016). Root exudates are also important in attracting the rhizomicrobiome of plants, which ultimately improves adaptation, and survival of plants in contaminated environments.

Plants, by their photoautotrophic nature, do not depend on external organic compounds as carbon and/or energy sources. They are also not evolutionarily under selection pressure to degrade pollutants and do not supposedly have adopted/developed pathways of pollutant degradation but they, nonetheless, detoxify pollutants in certain ways (Gehardt et al. 2009). In fact, a vast diversity of plant enzymes is reported as capable of detoxifying organic pollutants (Schaffner et al. 2002). Detoxification/degradation of organic pollutants could be *in planta* or *ex planta*. *In planta* degradation is intracellular detoxification by the plant itself utilizing enzymes (which include cytochrome P450, glutathione S-tranferase, glycosyltransferases and transporters) to detoxify organic pollutants in stages. Once organic pollutants are taken up from the soil by plant roots mainly via diffusion, they could be further translocated and metabolized or detoxified. The three stages of detoxification are transformation, conjugation and compartmentalization. Transformation involves pollutant hydrolysis, reduction and/or oxidation into less hydrophobic compounds with increased reactivity. Conjugation involves the conjugation of the less hydrophobic compounds to glutathione, amino acids or sugars, while compartmentalization is the stage in which conjugates are sorted into organelles such as the vacuole and/or the cell wall. These stages of *in planta* detoxification are generally termed the 'green-liver' detoxification (Sandermann 1994, Schaffner et al. 2002, Arslan et al. 2017) in semblance to the mammalian liver functions of detoxification of waste materials.

On the other hand, e*x planta* degradation involves pollutant degradation by enzymes secreted by plants or the rhizosphere microbial community that has been greatly influenced by the presence of plants. This is most often referred to as rhizodegradation, in which the microbes or their enzymes play an active role in pollutant degradation (Chen et al. 2013). Many organic contaminants are reported to be completely degraded to innocuous substances by bacteria that metabolise them for their carbon/energy needs. Mechanisms of their degradation involve metabolic and/or enzymatic reactions. Organic contaminants have different structural orientation, which influences mechanism employed by microorganisms for their degradation. Mechanism of degradation for a typical hydrophobic organic contaminant, polycyclic aromatic hydrocarbons (PAHs), will be highlighted here. Polycyclic aromatic hydrocarbons are fused ring aromatic structured, persistent organic pollutants. Aerobic bacterial degradation of PAHs is initiated by the dihydroxylation of the ringed structure, which is catalysed by a ring-hydroxylating dioxygenase (RHD) (Jakoncic et al. 2007). Two oxygen atoms are incorporated into the PAH structure by dioxygenases, targeting two adjacent carbon atoms of the aromatic ring. This leads to opening up of the aromatic ring for biodegradation to proceed and it is a rate-limiting step in the biodegradation of PAHs (Cerniglia and Heitkamp 1989). Dihydroxylation then leads to the formation of transient dioxethanes, which are oxidized to cis-dihydrodiols (Wilson and Jones 1993). A cis-hydrodiol is dehydrated by a dehydrogenase, forming catechol, a dihydroxylated intermediate which is further degraded via meta or ortho cleavage pathways (Aghapour et al. 2013) 2013. Ortho cleavage yields cis-, cis-muconic acid, while meta cleavage yields 2-hydroxymuconic semialdehyde (Aghapour et al. 2013, Gupta et al. 2016)2016. Cis-muconic acid is further degraded to succinate and acetyl-coA, while 2- hydroxymuconic semialdehyde yields acetaldehyde and pyruvate. These are subsequently degraded in the tricarboxylic acid(TCA) cycle to carbon(IV) oxide(CO_2) and water(H_2O) (Fig. 2).

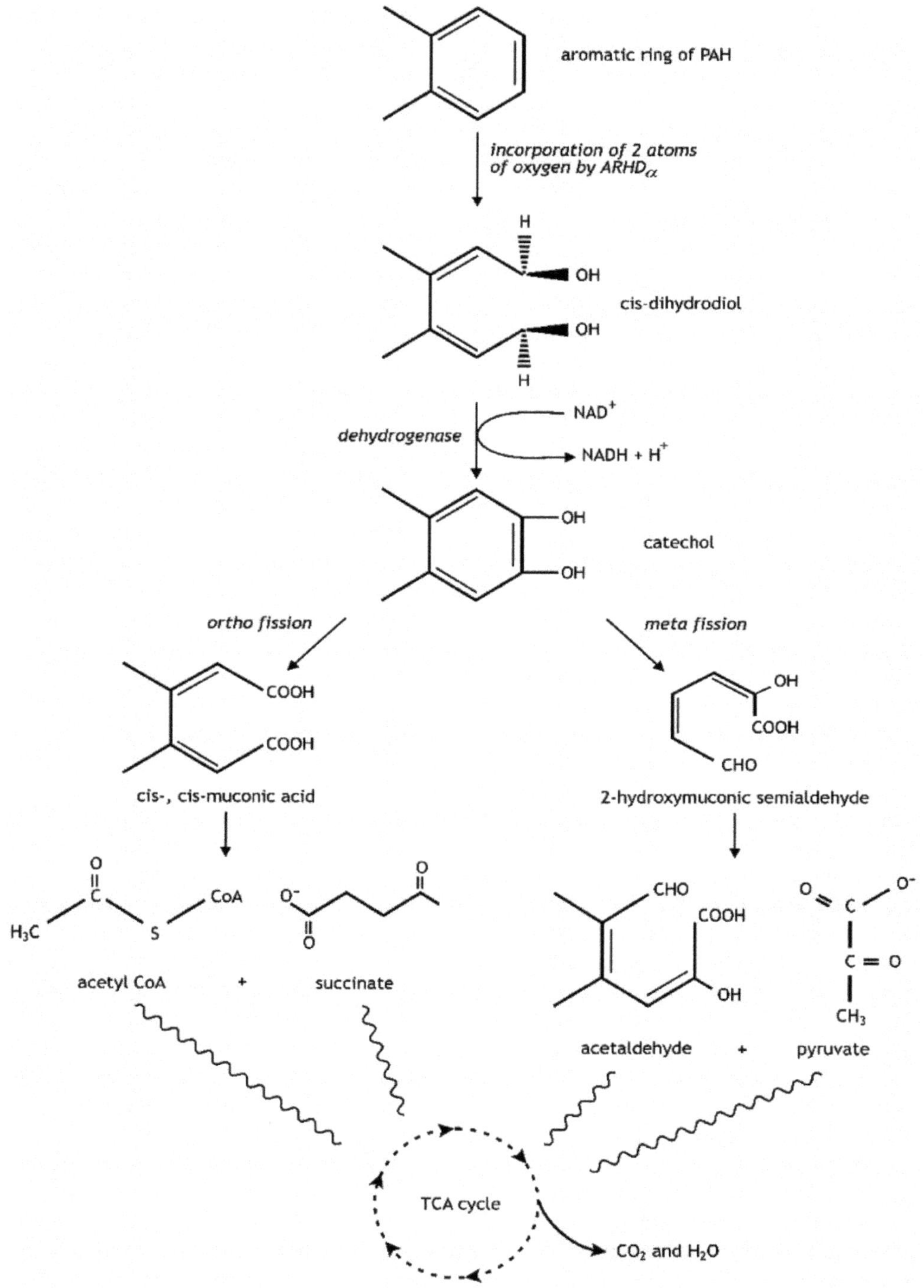

Figure 2   Diagrammatic depiction of degradation products of an aromatic ring of a PAH

While organic contaminants are degraded by microorganisms into metabolic intermediates (useable as primary substrates for their growth) or innocuous compounds such as CO_2 and H_2O, the microbial remediation strategy of heavy metals depends on the metabolizing capabilities of the microorganisms (Dixit et al. 2015). They have developed and adopted different detoxifying mechanisms such as biosorption, bioaccumulation, biotransformation and biomineralization that are exploited for both *in situ* and *ex situ* bioremediation. In a plant-microbe remediation, accumulation ability of the remediating plants becomes relevant. Phytoremediation strategies such as phytoextraction, phytostabilization and phytovolatilization are utilized.

4. CHALLENGES AND PROSPECTS OF PLANT–MICROBE ASSOCIATIONS OF BIOREMEDIATION

Priority is given globally to bioremediation due to its environmental-friendliness, sustainability and cost-effectiveness. Plant–microbe associations in bioremediation of agro-ecosystems is widespread and have proven to be effective and reliable in comparison to remediation strategies with plants or microbes, individually (Chaudhry et al. 2005, Gerhardt et al. 2009, Abhilash et al. 2013, Mahar et al. 2016). Whilst vast research efforts on the use of plant–microbe associations in agro-ecosystem remediation have been carried out in laboratories or greenhouses (Huang et al. 2005, Gerhardt et al. 2009, Hussain et al. 2010, Cook and Hesterberg 2013, Azubuike et al. 2016), plant-microbe associations in remediation are challenged in a couple of ways. Laboratory and greenhouse trials have contributed to a better understanding of the principles and mechanisms involved during bioremediation by plant–microbe interactions. However, replicating the findings of these experiments in the field has become a major challenge (Gerhardt et al. 2009). This challenge is foreseeable because laboratory experiments, unlike field trials, are conducted under controlled conditions such as temperature, humidity and nutrient sources. Difficulty in translating laboratory or greenhouse experiments to successful field trials is also exacerbated by contaminant toxicity and severity which limit plant and microbial growth (Chaudhry et al. 2005). In laboratory experiments, contaminants are introduced in known concentrations followed by thorough mixing to evenly distribute the contaminants, which is not the case in field experiments which is characterized by uneven concentration of the contaminant even if the soil is considerably tilled prior to phytoremediation. Adjudging remediation success, in the light of regulatory standards where success of remediation is frequently measured based on point-by-point assessment across the field, is challenging. Therefore, when a point does not meet the standard, then the whole site is deemed not to have been successfully remediated (Gerhardt et al. 2009).

Another challenge is the non-bioavailability of soil-bound and insoluble contaminants, making them inaccessible to remediating plants and microbes. According to Chaudhry et al. (2005) and Chech (2003), organic contaminants are not readily bioavailable, thereby hindering the bioremediation ability of involved plants and microbes. This is due to sequestration that occurs through sorption and incorporation into soil organic complexes as well as the lack of specific enzymes required for degradation of such compounds (Chech 2003). Thus, bioavailability of the contaminant is important in determining the success or failure of a bioremediation strategy (Azubuike et al. 2016). Furthermore, the correct disposal of plant biomass after phytoremediation of heavy metals is a major concern. Several disposal methods for post phytoremediation plant biomass have been described (Cunningham et al. 1995, Raskin et al. 1997, Mulligan et al. 2001, Sas-Nowosielska et al. 2004). Examples include direct disposal, incineration, ashing as well as liquid extraction. The direct disposal is not a suitable method since it transfers pollutant to another site and requires availability of dumping sites as well as high capital input (Sas-Nowosielska et al. 2004). Incineration has been proposed to be the most feasible and economically acceptable method with environmentally friendly approach (Sas-Nowosielska et al. 2004). In the process of incineration, organic matter is destroyed, whereas metal is released in the form of oxides. These released metals are either entrained in the slag or released to the effluent gases (where metal-containing dust is captured effectively) (Sas-Nowosielska et al. 2004).

To overcome challenges associated with high concentration of contaminants during remediation by plant–microbe-interaction, a multi-process phytoremediation strategy (MPPS) was developed (Huang et al. 2005). The MPPS uses a combination of two different groups of bacteria with contaminant degrading capabilities and plant growth promoting abilities (Huang et al. 2005). The PGPB were responsible for promoting plant growth by providing the plants with nitrogen, phosphorus and other important nutrients (Huang et al. 2005, Rayu et al. 2012). In addition, the PGPB secreted hormones or enzymes responsible for increasing the plant's ability to tolerate the contaminants. These bacteria were introduced to the remediation site along with the selected plants used for contaminant accumulation. The efficiency of plants in accumulating pollutants is limited by toxicity of the pollutant (Powter et al. 2012, Chibuike and Obiora 2014). The plant supported the growth of bacteria by secreting a number of sugars and other important metabolites through the roots, which were utilised by the bacteria. Developing MPPS is a strategy of simulating what happens in natural environment, where the association of plant and rhizospheric microorganisms is functioning such that the limitation of remediation strategies can be overcome. However, this may also be seen as an ecological risk due to introduction of non-native microbes or plant into the environment. One way to avoid this is to use native plants and the associated indigenous microorganisms. This has advantages of sustainability and dual purposefulness of contaminant uptake as well as habitat reconstruction after remediation has been successful (Gerhardt et al. 2009, Bello-Akinosho et al. 2016). Although such multi-process design could not completely overcome challenges of slow rate of contaminant removal and low efficiency in comparison to physical and chemical remediation methods (Zhuang et al. 2007), the environmental benefits offered by plant–microbe remediation make it worthwhile (Juwarkar et al. 2010).

Furthermore, recombinant DNA technology can be used to increase the rate of contaminant removal. For example, the bacteria associated with the plants during microbial associated phytoremediation can be modified genetically to express specific genes that can increase the plant's tolerance to pollutants (Dowling and Dot 2009, Ram 2015). Since plants are autotrophs, their associated bacteria would then express specific genes able to increase the plant's ability to metabolise organic compounds (Dowling and Doty 2009). In essence, genetic modifications are targeted at both in planta and ex planta degradation. The need to optimize rate and reliability of plant-microbe bioremediation is urgent (James and Strand 2009). Therefore, tapping into the potentials of using transgenic plant-microbe associations to remediate contaminants is a way forward. Seemingly good as this is, regulatory restrictions have prevented substantial progress in the field. Furthermore, genetically engineered microbial strains often do not match-up with native microbes in the rhizosphere (Gehardt et al. 2009). The use of metaomics for the prediction of degradation potentials for plant-microbe associations in contaminated environment would also help to identify optimal partnership that would inform bioremediation strategies.

5. CONCLUSION

The polluted agro-ecosystem management by plant-microbe interactions is a promising approach, though there are some challenges which need to be overcome. Consistent attempts are necessitated to comprehend and know how microbes and plants are going to counter the occurrence of unnecessary noxious substances. Lot of research work remains to be carried out for real field investigations and surveys based on lab-scale research experimentations. The field data could help us in commercializing the efficient plant microbe based technology. Further, development of efficient and potential microbes will improve the process of bioremediation. During the process of rhizo-remediation, root exudates from host plant could aid in stimulating the proliferation, survival and microbial action, which consequently result in better and effectual pollutants. The efficient and intricate host root system can help in spreading microbes through soil and aid in penetrating the impervious layers of soil system. Inoculation of contaminant degrading microbes on seeds might be a significant approach to augment the effectiveness of phyto-remediation or biological augmentation process.

References

Abhilash, P.C., J.R. Powell, H.B. Singh and B.K. Singh. 2012. Plant-microbe interactions: novel applications for exploitation in multipurpose remediation technologies. Trends Biotechnol. 30: 416–420.

Abhilash, P.C., R.K. Dubey, V. Tripathi, P. Srivastava, J.P Verma and H.B Singh. 2013. Remediation and management of POPs-contaminated soils in a warming climate: challenges and perspectives. Environ. Sci. Pollut. Res. 20: 5879–5885.

Adeleke, R.A., B. Nunthkumar, A. Roopnarain and L. Obi. 2019. Applications of plant–microbe interactions in agro-ecosystems. *In*: V. Kumar, R. Prasad, M. Kumar and D. Choudhary (eds), Microbiome in Plant Health and Disease. Springer, Singapore, pp. 1–34.

Afzal, M., Q.M. Khan and A. Sessitsch. 2014. Endophytic bacteria: prospects and applications for the phytoremediation of organic pollutants. Chemosphere 117: 232–242.

Aghapour, A.A., G. Moussavi and K. Yaghmaeian. 2013. Biological degradation of catechol in wastewater using the sequencing continuous-inflow reactor (SCR). J. Environ. Health Sci. Eng. 11(1): 3–10.

Ahmad, F., N. Ashraf, Y. Da-Chuan, H. Jabeen, S. Anwar, A.Q. Wahla, et al. 2019a. Application of a novel bacterial consortium BDAM for bioremediation of bispyribac sodium in wheat vegetated soil. J. Hazard. 374: 58–65.

Ahmad, F., N. Ashraf, R.B. Zhou and Y. Da-Chuan. 2019b. Enhanced remediation of bispyribac sodium by wheat (*Triticum aestivum*) and a bispyribac sodium degrading bacterial consortium (BDAM). J. Environ. Manage. 244: 383–390.

Ali, S., T.C. Charles and B.R. Glick. 2014. Amelioration of high salinity stress damage by plant growth-promoting bacterial endophytes that contain ACC deaminase. Plant Physiol. Bioch. 80: 160–167.

Arslan, M., A. Imran, Q.M. Khan and M. Afzal. 2017. Plant–bacteria partnerships for the remediation of persistent organic pollutants. Environ. Sci. Pollut. R. 24: 4322–4336.

Atieh, M.A., Y. Ji and V. Kochkodan. 2017. Metals in the environment: toxic metals removal. Bioinorg. Chem. Appl. Article ID 4309198 https://doi.org/10.1155/2017/4309198

Azubuike, C.C., C.B. Chikere and G.C. Okpokwasili. 2016. Bioremediation techniques–classification based on site of application: Principles, advantages, limitations and prospects. World J. Microbiol. Biotechnol. 32: 1–18.

Balasubramaniyam, A. 2015. The Influence of plants in the remediation of petroleum hydrocarbon-contaminated sites. Pharm. Anal. Chem. Open Access 1: 1–11.

Banerjee, D.K., P.M. Fedorak, A. Hashimoto, J.H. Masliyah, M.A. Pickard and M.R. Gray. 1995. Monitoring the biological treatment of anthracene-contaminated soil in a rotating-drum bioreactor. Appl. Microbiol. Biotechnol. 43: 521–528.

Barac, T., S. Taghavi, B. Borremans, A. Provoost, L. Oeyen, J.V. Colpaert, et al. 2004 Engineered endophytic bacteria improve phyto-remediation of water soluble, volatile, organic pollutants. Nat. Biotechnol. 22: 583–588.

Bashan, Y., L.E. de-Bashan, S.R. Prabhu and J.P. Hernandez. 2014. Advances in plant growth-promoting bacterial inoculant technology: Formulations and practical perspectives (1998–2013). Plant Soil. 378: 1–33.

Bech, J., P. Duran, N. Roca, W. Poma, I. Sánchez, J. Barceló, et al. 2012. Shoot accumulation of several trace elements in native plant species from contaminated soils in the Peruvian Andes. J. Geochem. Explor. 113: 106–111.

Bell, T.H., B. Cloutier-Hurteau, F. Al-Otaibi, M.C. Turmel, E. Yergeau, F. Courchesne, et al. 2015. Early rhizosphere microbiome composition is related to the growth and Zn uptake of willows introduced to a former landfill. Environ. 17: 3025–3038.

Bello-Akinosho, M., R. Makofane, R. Adeleke, M. Thantsha, M. Pillay and G.J. Chirima. 2016. Potential of polycyclic aromatic hydrocarbon-degrading bacterial isolates to contribute to soil fertility. Biomed. Res. Int. 2016: 1–10.

Bello-Akinosho, M. 2018. Biodegradation of polycyclic aromatic hydrocarbons in contaminated soils using a tripartite association of ectomycorrhizal fungi, pine plant and their rhizosphere bacteria. Unpublished doctoral dissertation. University of Pretoria, Pretoria, South Africa.

Berendsen, R.L., C.M. Pieterse and P.A Bakker. 2012. The rhizosphere microbiome and plant health. Trends Plant Sci. 17: 478–486.

Bolan, N.S., J.H. Park, B. Robinson, R. Naidu and K.Y. Huh. 2011. Phytostabilization: A green approach to contaminant containment. *In*: D.L. Sparks (ed.), Advances in Agronomy. Academic Press, pp. 145–204.

Bonfante, P. and A. Genre. 2010. Mechanisms underlying beneficial plant–fungus interactions in mycorrhizal symbiosis. Nat. Commun. 1: 48.

Bosco, F. and C. Mollea. 2019. Mycoremediation in soil. *In*: H. Saldarriaga-Noreña, M.A. Murillo-Tovar, R. Farooq, R.S. Dongre and S. Riaz (eds), Environmental Chemistry and Recent Pollution Control Approaches. IntechOpen, DOI: 10.5772/intechopen.84777. Available from: https://www.intechopen.com/books/environmental-chemistry-and-recent-pollution-control-approaches/mycoremediation-in-soil

Braud, A., K. Jézéquel, S. Bazot and T. Lebeau. 2009. Enhanced phytoextraction of an agricultural Cr-and Pb-contaminated soil by bioaugmentation with siderophore producing bacteria. Chemosphere 74: 280–286.

Brazil, G.M., L. Kenefick, M. Callanan, A. Haro, V. De Lorenzo, D.N. Dowling, et al. 1995. Construction of a rhizosphere pseudomonad with potential to degrade polychlorinated biphenyls and detection of bph gene expression in the rhizosphere. Appl. Environ. Microbiol. 61: 1946–1952.

Cerniglia C.E. and M.A. Heitkamp. 1989. Microbial degradation of polycyclic aromatic hydrocarbons (PAH) in the aquatic environment. *In*: U. Varanasi (ed.), Metabolism of Polycyclic Aromatic Hydrocarbons in the Aquatic Environment. CRC Press, Inc, Boca Raton, Fla, pp. 41–68.

Chatterjee, S., A. Mitra, S. Datta and V. Veer. 2013. Phytoremediation protocols: An overview. *In*: D.K. Gupta (ed.), Plant-Based Remediation Processes. Springer, Berlin, Heidelberg, pp. 1–18.

Chaudhry, Q., M. Blom-Zandstra, S. Gupta and E.J. Joner. 2005. Utilising the synergy between plants and rhizosphere microorganisms to enhance breakdown of organic pollutants in the environment. Environ. Sci. Pollut. R. 12: 34–48.

Chávez-Gómez, B., R. Quintero, F. Esparza-Garcıa, A. Mesta-Howard, F.Z.D. de la Serna, C. Hernández-Rodrıguez, et al. 2003. Removal of phenanthrene from soil by co-cultures of bacteria and fungi pregrown on sugarcane bagasse pith. Bioresour. Technol. 89: 177–83.

Chech, A.M. 2003. Evaluation of the feasibility for in situ bioremediation of mineral oil-contaminated soil. Ph.D. Thesis, University of Arizona, Arizona, USA.

Chen, M., M. Arato, L. Borghi, E. Nouri and D. Reinhardt. 2018a. Beneficial services of arbuscular mycorrhizal fungi–from ecology to application. Front. Plant Sci. 9: 1270–1276.

Chen, Y., Q. Ding, Y. Chao, X. Wei, S. Wang and R. Qiu. 2018b. Structural development and assembly patterns of the root-associated microbiomes during phytoremediation. Sci. Total Environ. 644: 1591–1601.

Chen, J., Q.X. Xu, Y. Su, Z.Q. Shi and F.X. Han. 2013. Phytoremediation of organic polluted soil. J. Bioremed. Biodeg. 4: 132–134.

Cherian, S. and M.M. Oliveira. 2005. Transgenic plants in phytoremediation: Recent advances and new possibilities. Environ. Sci. Technol. 39: 9377–9390.

Cheynier, V., G. Comte, K.M. Davies, V. Lattanzio and S. Martens. 2013. Plant phenolics: recent advances on their biosynthesis, genetics, and ecophysiology. Plant Physiol. Bioch. 72: 1–20.

Chibuike, G.U. and S.C. Obiora. 2014. Heavy metal polluted soils: Effect on plants and bioremediation methods. Appl. Environ. Soil Sci. https://doi.org/10.1155/2014/752708.

Chirwa, E.M.N. and E.E. Bamuza-Pemu. 2010. Investigation of photocatalysis as an alternative to other advanced oxidation processes for the treatment of filter backwash water. Water Research Commission, Gezina, South Africa. Report 1717/1: 10 p.

Choudhary, D.K., K.P. Sharma and R.K. Gaur. 2011. Biotechnological perspectives of microbes in agro-ecosystems. Biotechnol. 33: 1905–1910.

Chouychai, W., A. Thongkukiatkul, S. Upatham, P. Pokethitiyook, M. Kruatrachue and H. Lee. 2012. Effect of corn plant on survival and phenanthrene degradation capacity of *Pseudomonas* sp. UG14Lr in two soils. Int. J. Phytorem. 14: 585–595.

Compant, S., A. Samad, H. Faist and A. Sessitsch. 2019. A review on the plant microbiome: Ecology, functions and emerging trends in microbial application. J. Adv. Res. 19: 29–37.

Cook, R.L. and D. Hesterberg. 2013. Comparison of trees and grasses for rhizoremediation of petroleum hydrocarbons. Int. J. Phytoremediat. 15: 844–860.

Correa-García, S., P. Pande, A. Séguin, M. St-Arnaud and E. Yergeau. 2018. Rhizoremediation of petroleum hydrocarbons: a model system for plant microbiome manipulation. Microb. Biotechnol. 11: 819–832.

Costa, J.A.V., B.C.B. Freitas, C.G. Cruz, J. Silveira and M.G. Morais. 2019. Potential of microalgae as biopesticides to contribute to sustainable agriculture and environmental development. J. Environ. Sci. Health, Part B. 54: 366–375.

Cunningham, S.D., W.R. Berti and J.W. Huang. 1995. Remediation of contaminated soils and sludges by green plants. 3rd International Symposium on In Situ and On-Site Bioreclamation. San Diego, CA, United States, 24–27 April 1995, pp. 183–186.

Dams, R.I., G.I. Paton and K. Killham. 2007. Rhizoremediation of pentachlorophenol by *Sphingobium chlorophenolicum* ATCC 39723. Chemosphere 68: 864–870.

Debiane, D., G. Garçon, A. Verdin, J. Fontaine, R. Durand, A. Grandmougin-Ferjani, et al. 2008. In vitro evaluation of the oxidative stress and genotoxic potentials of anthracene on mycorrhizal chicory roots. Environ. Exp. Bot. 64(2): 120–127.

Debiane, D., G. Garçon, A. Verdin, J. Fontaine, R. Durand, P. Shirali, et al. 2009. Mycorrhization alleviates benzo[a]pyrene-induced oxidative stress in an in vitro chicory root model. Phytochemistry 70: 1421–1427.

Dimkpa, C.O., D. Merten, A. Svatoš, G. Büchel and E. Kothe. 2009. Siderophores mediate reduced and increased uptake of cadmium by *Streptomyces tendae* F4 and sunflower (*Helianthus annuus*), respectively. J. Appl. Microbiol. 107: 1687–1696.

Dixit, R., M. Deepti, P. Kuppusamy, B.S. Udai, S. Asha, S. Renu, et al. 2015. Bioremediation of heavy metals from soil and aquatic environment: An overview of principles and criteria of fundamental processes. Sustainability 7: 2189–2212.

Dominguez-Rosado, E. and J. Pichtel. 2004. Phytoremediation of soil contaminated with used motor oil: II. Greenhouse studies. Environ. Eng. Sci. 21(2): 169–180.

Doty, S.L., J.L. Freeman, C.M. Cohu, J.G. Burken, A. Firrincieli, A. Simon, et al. 2017. Enhanced degradation of TCE on a superfund site using endophyte assisted poplar tree phytoremediation. Environ. Sci. Technol. 51: 10050–10058.

Dowling, D.N. and S.L. Doty. 2009. Improving phytoremediation through biotechnology. Curr. Opin. Biotechnol. 20: 1–3.

Driai, S., A. Verdin, F. Laruelle, A. Beddiar and A.L.H. Sahraoui. 2015. Is the arbuscular mycorrhizal fungus *Rhizophagus irregularis* able to fulfil its life cycle in the presence of diesel pollution? Int. Biodeter. Biodegr. 105: 58–65.

El Amrani, A., A.S. Dumas, L.Y. Wick, E. Yergeau and R. Berthomé. 2015. "Omics" insights into PAH degradation toward improved green remediation biotechnologies. Environ. Sci. Technol. 49: 11281–11291.

El-Bestawy, E.A., A.Z.A. El-Salam and A.E.R.H. Mansy. 2007. Potential use of environmental cyanobacterial species in bioremediation of lindane-contaminated effluents. Int. Biodeter. Biodegr. 59: 180–192.

Ellouze, M and S. Sayadi. 2016. White-rot fungi and their enzymes as a biotechnological tool for xenobiotic bioremediation. *In*: H.E.M. Saleh and R.O.A. Rahman (eds), Management of Hazardous Wastes. InTech, London, pp. 103–120.

El-Nahhal, Y., Y. Awad and J.M. Safi. 2013. Bioremediation of acetochlor in soil and water systems by cyanobacterial mat. Int. J. Geosci. 4: 880–890.

Farrar, K., D. Bryant and N. Cope-Selby. 2014. Understanding and engineering beneficial plant–microbe interactions: Plant growth promotion in energy crops. Plant Biotechnol. J. 12: 1193–1206.

Forlani, G., M. Pavan, M. Gramek, P. Kafarski and J. Lipok. 2008. Biochemical basis for a wide spread tolerance of cyanobacteria to the phosphonate herbicide glyphosate. Plant Cell Physiol. 49: 443–456.

Franco, A.R., A.C. Ferreira and P.M. Castro. 2014. Co-metabolic degradation of mono-fluorophenols by the ectomycorrhizal fungi *Pisolithus tinctorius*. Chemosphere 111: 260–265

Frick, C.M., J.J. Germida and R.E. Farrell. 1999. Assessment of phytoremediation as an in-situ technique for cleaning oil-contaminated sites. Petroleum Technology Alliance Canada, pp. 105a–124a.

Gao, Y., P. Zhou, L. Mao, Y. Zhi, C. Zhang and W. Shi. 2010. Effects of plant species coexistence on soil enzyme activities and soil microbial community structure under Cd and Pb combined pollution. J. Environ. Sci. 22: 1040–1048.

Gerhardt, K.E., X.D. Huang, B.R. Glick and B.M. Greenberg. 2009. Phytoremediation and rhizoremediation of organic soil contaminants: Potential and challenges. Plant Sci. 176: 20–30.

Germaine, K.J., X. Liu, G.G. Cabellos, J.P. Hogan, D. Ryan and D.N. Dowling. 2006. Bacterial endophyte-enhanced phytoremediation of the organochlorine herbicide 2,4-dichlorophenoxyacetic acid. FEMS Microbiol. Ecol. 57: 302–310.

Gkorezis, P., M. Daghio, A. Franzetti, J.D. Van Hamme, W. Sillen, et al. 2016. The interaction between plants and bacteria in the remediation of petroleum hydrocarbons: an environmental perspective. Front. Microbiol. 7: 1836–1841.

Guo, M., Z. Gong, R. Miao, J. Rookes, D. Cahill and J. Zhuang. 2017. Microbial mechanisms controlling the rhizosphere effect of ryegrass on degradation of polycyclic aromatic hydrocarbons in an aged-contaminated agricultural soil. Soil Biol. Biochem. 113: 130–142.

Gupta, G., V. Kumar and A.K. Pal. 2016. Biodegradation of polycyclic aromatic hydrocarbons by microbial consortium: a distinctive approach for decontamination of soil. Soil Sediment Contam. 25(6): 597–623.

Hamouda, R.A.E.F., N.M. Sorour and D.S. Yeheia. 2016. Biodegradation of crude oil by *Anabaena oryzae*, *Chlorella kessleri* and its consortium under mixotrophic conditions. Int. Biodeterior. Biodegrad. 112: 128–134.

Hassan, Z., S. Ali, M. Rizwan, M. Ibrahim, M. Nafees and M. Waseem. 2017. Role of bioremediation agents (bacteria, fungi, and algae) in alleviating heavy metal toxicity. *In*: V. Kumar, M. Kumar, S. Sharma and R. Prasad (eds), Probiotics in Agroecosystem. Springer, Singapore, pp. 517–537.

Hassan, M.K., J.A. McInroy and J.W. Kloepper. 2019. The interactions of rhizodeposits with plant growth-promoting rhizobacteria in the rhizosphere: A review. Agriculture. 9: 142–147.

Hayat, R., S. Ali, U. Amara, R. Khalid and I. Ahmed. 2010. Soil beneficial bacteria and their role in plant growth promotion: A review. Ann. Microbiol. 60: 579–598.

Ho, Y.N., D.C. Mathew and C.C. Huang. 2017. Plant-microbe ecology: interactions of plants and symbiotic microbial communities. *In*: Z. Yousaf (ed.), Plant Ecology–Traditional Approaches to Recent Trends. InTech, Rijeka, Croatia, pp. 517–537.

Huang, X.D., Y. El-Alawi, J. Gurska, B.R. Glick and B.M. Greenberg. 2005. A multi-process phytore-mediation system for decontamination of persistent total petroleum hydrocarbons (TPHs) from soils. Microchem. J. 81: 139–147.

Huang, Y. and J. Wang. 2013. Degradation and mineralization of DDT by the ectomycorrhizal fungi, *Xerocomus chrysenteron*. Chemosphere 92: 760–764.

Hussain, S.T., T. Mahmood and S.A. Malik. 2010. Phytoremediation technologies for Ni++ by water hyacinth. Afr. J. Biotechnol. 9: 8648–8660.

Ionescu, M., K. Beranova, V. Dudkova, L. Kochankova, K. Demnerova, T. Macek, et al. 2009. Isolation and characterization of different plant associated bacteria and their potential to degrade polychlorinated biphenyls. Int. Biodeterior. Biodegrad. 63: 667–672.

Jacoby, R., M. Peukert, A. Succurro, A. Koprivova and S. Kopriva. 2017. The role of soil microorganisms in plant mineral nutrition-current knowledge and future directions. Front. Plant Sci. 8: 1617–1622.

Jakoncic J., Y. Jouanneau, C. Meyer and V. Stojanoff. 2007. The catalytic pocket of the ring-hydroxylating dioxygenase from Sphingomonas CHY-1. Biochem. Biophys. Res. Commun. 352(4): 861–866.

Jambon, I., S. Thijs, N. Weyens and J. Vangronsveld. 2018. Harnessing plant-bacteria-fungi interactions to improve plant growth and degradation of organic pollutants. J. Plant Interact. 13: 119–130.

James, C.A. and S.E. Strand. 2009. Phytoremediation of small organic contaminants using transgenic plants. Curr. Opin. Biotech. 20(2): 237–241.

Johnsen, A. and U. Karlson. 2004. Evaluation of bacterial strategies to promote the bioavailability of polycyclic aromatic hydrocarbons. Appl. Microbiol. Biotechnol. 63: 452–459.

Joner, E.J., C. Leyval and J.V. Colpaert. 2006. Ectomycorrhizas impede phytoremediation of polycyclic aromatic hydrocarbons (PAHs) both within and beyond the rhizosphere. Environ. Pollut. 142: 34–38.

Juwarkar, A.A., S.K. Singh and A. Mudhoo. 2010. A comprehensive overview of elements in bioremediation. Rev. Environ. Sci. Bio. 9: 215–288.

Khan, A.G. 2005. Role of soil microbes in the rhizospheres of plants growing on trace metal contaminated soils in phytoremediation. J. Trace Elem. Med. Biol. 18: 355–364.

Khan, A.G., C. Kuek, T.M. Chaudhry, C.S. Khoo and W.J. Hayes. 2000. Role of plants, mycorrhizae and phytochelators in heavy metal contaminated land remediation. Chemosphere 41: 197–207.

Khare, E., J. Mishra and N.K. Arora. 2018. Multifaceted interactions between endophytes and plant: Developments and prospects. Front. Microbiol. 9: 1–12.

Kuiper, I., E.L. Lagendijk, G.V. Bloemberg and B.J. Lugtenberg. 2004. Rhizoremediation: A beneficial plant-microbe interaction. Mol. Plant-Microbe Interact. 17: 6–15.

Kuiper, I., G.V. Bloemberg and B.J. Lugtenberg. 2001. Selection of a plant-bacterium pair as a novel tool for rhizostimulation of polycyclic aromatic hydrocarbon-degrading bacteria. Mol. Plant-Microbe Interact. 14: 1197–1205.

Lee, H., Y. Jang, Y.S. Choi, M.J. Kim, J. Lee, H. Lee, et al. 2014. Biotechnological procedures to select white rot fungi for the degradation of PAHs. J. Microbiol. Methods 97: 56–62.

Lenoir, I., A. Lounes-Hadj Sahraoui and J. Fontaine. 2016. Arbuscular mycorrhizal fungal assisted phytoremediation of soil contaminated with persistent organic pollutants: A review. Eur. J. Soil Sci. 67(5): 624–640.

Leyval, C., E.J. Joner, C. Del Val and K. Haselwandter. 2002. Potential of arbuscular mycorrhizal fungi for bioremediation. *In:* S. Gianinazzi, H. Schüepp, J.M. Barea and K. Haselwandter (eds), Mycorrhizal Technology in Agriculture: From Genes to Bioproducts. Birkhäuser, Basel, Switzerland, pp. 175–186.

Liu, L., C.Y. Jiang, X.Y. Liu, J.F. Wu, J.G. Han and S.J. Liu. 2007. Plant-microbe association for rhizoremediation of chloronitroaromatic pollutants with *Comamonas* sp. strain CNB-1. Environ. 9: 465–473.

Liu, L., W. Li, W. Song and M. Guo. 2018. Remediation techniques for heavy metal-contaminated soils: principles and applicability. Sci. Total Environ. 633: 206–219.

Lodewyckx, C., S. Taghavi, M. Mergeay, J. Vangronsveld, H. Clijsters and D. van der Lelie. 2001. The effect of recombinant heavy metal resistant endophytic bacteria in heavy metal uptake by their host plant. Int. J. Phytorem. 3: 173–187.

Lors, C., D. Damidot, J.F. Ponge and F. Périé. 2012. Comparison of a bioremediation process of PAHs in a PAH-contaminated soil at field and laboratory scales. Environ. Pollut. 165(Supplement C): 11–17.

Lu, Y.F. and M. Lu. 2015. Remediation of PAH-contaminated soil by the combination of tall fescue, arbuscular mycorrhizal fungus and epigeic earthworms. J. of Hazard. Mater. 285: 535–541.

Lu, H., J. Sun and L. Zhu. 2017. The role of artificial root exudate components in facilitating the degradation of pyrene in soil. Sci. Rep. 7: 7130–7135.

Ma, Y., R.S. Oliveira, H. Freitas, C. Zhang. 2016. Biochemical and molecular mechanisms of plant-microbe-metal interactions: relevance for phytoremediation. Front. Plant Sci. 7: 918–923.

Mahar, A., P. Wang, A. Ali, M.K. Awasthi, A.H. Lahori, Q. Wang, R. Li and Z. Zhang. 2016. Challenges and opportunities in the phytoremediation of heavy metals contaminated soils: A review. Ecotoxicol. Environ. Saf. 126: 111–121.

Manschadi, A.M., J. Christopher, P. deVoil and G.L. Hammer. 2006. The role of root architectural traits in adaptation of wheat to water-limited environments. Funct. Plant Biol. 33: 823–837.

Megharaj, M., B. Ramakrishnan, K. Venkateswarlu, N. Sethunathan and R. Naidu. 2011. Bioremediation approaches for organic pollutants: a critical perspective. Environ. Int. 37(8): 1362–1375.

Meier, S., F. Borie, N. Bolan and P. Cornejo. 2012. Phytoremediation of metal-polluted soils by arbuscular mycorrhizal fungi. Crit. Rev. Env. Sci. Tec. 42(7): 741–775.

Mimmo, T., D. Del Buono, R. Terzano, N. Tomasi, G. Vigani, C. Crecchio, et al. 2014. Rhizospheric organic compounds in the soil-microorganism-plant system: Their role in iron availability. Eur. J. Soil Sci. 65: 629–642.

Mishra, J., R. Singh and N.K. Arora. 2017. Plant growth-promoting microbes: diverse roles in agriculture and environmental sustainability. *In*: V. Kumar, M. Kumar, S. Sharma and R. Prasad (eds), Probiotics and Plant Health. Springer, Singapore, pp. 71–111.

Mrozik, A. and Z. Piotrowska–Seget. 2010. Bioaugmentation as a strategy for cleaning up of soils contaminated with aromatic compounds. Microbiol. 165: 363–375.

Mulligan, C.N., R.N. Yong and B.F. Gibbs. 2001. Remediation technologies for metal-contaminated soils and groundwater: An evaluation. Eng. Geol. 60: 193–207.

Ojuederie, O. and O. Babalola. 2017. Microbial and plant-assisted bioremediation of heavy metal polluted environments: A review. Int. J. Environ Res. Public Health. 14(12): 1504–1511.

Park, S., K.S. Kim, J.T. Kim, D. Kang and K. Sung. 2011. Effects of humic acid on phytodegradation of petroleum hydrocarbons in soil simultaneously contaminated with heavy metals. J. Environ. Sci. 23: 2034–2041.

Pathak, J., P.K. Maurya, S.P. Singh, D.P. Häderand, R.P. Sinha. 2018. Cyanobacterial farming for environment friendly sustainable agriculture practices: innovations and perspectives. Front Env. Sci. Eng. 6: 7–12.

Peuke, A.D. and H. Rennenberg. 2005. Phytoremediation. EMBO Reports 6: 497–501.

Piakong, M.T. and N.Z. Zaida. 2018. Effectiveness of single and microbial consortium of locally isolated beneficial microorganisms (LIBEM) in bioaugmentation of oil sludge contaminated soil at different concentration levels: A laboratory scale. J. Bioremediat. Biodegrad. 9: 2–11.

Pointing, S. 2001. Feasibility of bioremediation by white-rot fungi. Appl. Microbiol. Biotechnol. 57: 20–33.

Powter, C., N. Chymko, G. Dinwoodie, D. Howat, A. Janz, R. Puhlmann, et al. 2012. Regulatory history of Alberta's industrial land conservation and reclamation program. Can. J. Soil Sci. 92: 39–51.

Quiza, L., M. St-Arnaud, M. and E. Yergeau. 2015. Harnessing phytomicrobiome signaling for rhizosphere microbiome engineering. Front. Plant Sci. 6: 507–12.

Rahman, K.S.M., T. Rahman, P. Lakshmanaperumalsamy and I.M. Banat. 2002. Occurrence of crude oil degrading bacteria in gasoline and diesel station soils. J. Basic Microbiol. 42: 284–291.

Rajkumar, M., N. Ae, M.N.V. Prasad and H. Freitas. 2010. Potential of siderophore-producing bacteria for improving heavy metal phytoextraction. Trends Biotechnol. 28: 142–149.

Ram, C. 2015. Advances in Biodegradation and Bioremediation of Industrial Waste. CRC Press.

Raskin, I., R.D. Smith and D.E. Salt. 1997. Phytoremediation of metals: Using plants to remove pollutants from the environment. Curr. Opin. Biotech. 8: 221–226.

Rayu, S., D.G. Karpouzas and B.K. Singh. 2012. Emerging technologies in bioremediation: Constraints and opportunities. Biodegradation. 23: 917–926.

Reed, M.L.E. and B.R. Glick. 2005. Growth of canola (*Brassica napus*) in the presence of plant growth-promoting bacteria and either copper or polycyclic aromatic hydrocarbons. Can. J. Microbiol. 51: 1061–1069.

Robles-González, I.V., F. Fava and H.M. Poggi-Varaldo. 2008. A review on slurry bioreactors for bioremediation of soils and sediments. Microb. Cell Fact. 7: 5–12.

Ryan, R.P., K. Germaine, A. Franks, D.J. Ryan and D.N. Dowling. 2008. Bacterial endophytes: Recent developments and applications. FEMS Microbiol. Lett. 278: 1–9.

Salt, D.E., R.D. Smith and I. Raskin. 1998. Phytoremediation. Annu. Rev. Plant Biol. 49: 643–668.

Sandermann, J.H. 1994. Higher plant metabolism of xenobiotics: The 'green liver' concept. Pharmacogenetics 4(5): 225–241.

Santoyo, G., G. Moreno-Hagelsieb, M. del Carmen Orozco-Mosqueda and B.R. Glick. 2016. Plant growth-promoting bacterial endophytes. Microbiol. 183: 92–99.

Sarmah, A.K., M.T. Meyer and A.B. Boxall. 2006. A global perspective on the use, sales, exposure pathways, occurrence, fate and effects of veterinary antibiotics (VAs) in the environment. Chemosphere 65: 725–759.

Sarwar, N., M. Imran, M.R. Shaheen, W. Ishaque, M.A. Kamran, A. Matloob, et al. 2017. Phytoremediation strategies for soils contaminated with heavy metals: Modifications and future perspectives. Chemosphere 171: 710–721.

Šašek, V., M. Bhatt, T. Cajthaml, K. Malachova and D. Lednicka. 2003. Compost-mediated removal of polycyclic aromatic hydrocarbons from contaminated soil. Arch. Environ. Contam. Toxicol. 44: 0336–0342.

Sas-Nowosielska, A., R. Kucharski, E. Małkowski, M. Pogrzeba, J.M. Kuperberg and K. Kryński. 2004. Phytoextraction crop disposal—An unsolved problem. Environ. Pollut. 128: 373–379.

Schäffner, A., B. Messner, C. Langebartels and H. Sandermann. 2002. Genes and enzymes for in-planta phytoremediation of air, water and soil. Acta Biotechnol. 22: 141–151.

Segura, A., S. Rodríguez-Conde, C. Ramos and J.L. Ramos. 2009. Bacterial responses and interactions with plants during rhizoremediation. Microb. Biotechnol. 2: 452–464.

Shankar, S., C. Kansrajh, M.G. Dinesh, R.S. Satyan, S. Kiruthika and A. Tharanipriya. 2014. Application of indigenous microbial consortia in bioremediation of oil-contaminated soils. Int. J. Environ. Sci. Technol. 11: 367–376.

Sheng, X., X. Chen and L. He. 2008. Characteristics of an endophytic pyrene degrading bacterium of *Enterobacter* sp. 12J1 from Allium macrostemon Bunge. Int. Biodeterior. Biodegrad. 62: 88–95.

Shukla, K.P., N.K. Singh and S. Sharma. 2010. Bioremediation: Developments, current practices and perspectives. Genet. Eng. Biotechnol. J. 3: 1–20.

Singh, J.S., A. Kumar, A.N. Raiand and D.P. Singh. 2016. Cyanobacteria: A precious bio resource in agriculture, ecosystem, and environmental sustainability. Front. Microbiol. 7: 529–534.

Srivastav, A., K.K. Yadav, S. Yadav, N. Gupta, J.K. Singh, R. Katiyar, et al. 2018. Nano-phytoremediation of pollutants from contaminated soil environment: current scenario and future prospects. *In*: A.A. Ansari, S.S. Gill, R. Gill, G.R. Lanza and L. Newman (eds), Phytoremediation. Springer, Cham, Switzerland AG, pp. 383–401.

Stępniewska, Z. and A. Kuźniar. 2013. Endophytic microorganisms-promising applications in bioremediation of greenhouse gases. Appl. Microbiol. Biotechnol. 97: 9589–9596.

van der Heijden, M.G., R.D. Bardgett and N.M. van Straalen. 2008. The unseen majority: Soil microbes as drivers of plant diversity and productivity in terrestrial ecosystems. Ecol. Lett. 11: 296–310.

Vandenkoornhuyse, P., A. Quaiser, M. Duhamel, A. Le Van, A. Dufresne. 2015. The importance of the microbiome of the plant holobiont. New Phytol. 206(4): 1196–1206.

Varjani, S.J. and V.N. Upasani. 2016. Biodegradation of petroleum hydrocarbons by oleophilic strain of *Pseudomonas aeruginosa* NCIM 5514. Bioresour. Technol. 222: 195–201.

Vega-Jarquin, C., L. Dendooven, L. Magana-Plaza, F. Thalasso and A. Ramos-Valdivia. 2001. Biotransformation of hydrocarbon by cells cultures of *Cinchoma robusta* and *Dioscorea* composite. Environ. Toxicol. Chem. 20: 2670–2675.

Vergani, L., F. Mapelli, E. Zanardini, E. Terzaghi, A. Di Guardo, C. Morosini, et al. 2017. Phytorhizoremediation of polychlorinated biphenyl contaminated soils: An outlook on plant-microbe beneficial interactions. Sci. Total. Environ. 575: 1395–1406.

Vijayakumar, S. 2012. Potential applications of cyanobacteria in industrial effluents—A review. J. Bioremed. Biodeg. 3: 1–6.

Waigi, M.G., K. Sun and Y. Gao. 2017. Sphingomonads in microbe-assisted phytoremediation: Tackling soil pollution. Trends Biotechnol. 35: 883–899.

Wen, J., D. Gao, B. Zhang and H. Liang. 2011. Co-metabolic degradation of pyrene by indigenous whiterot fungus *Pseudtrametes gibbosa* from the northeast China. Int. Biodeter. Biodegr. 65: 600–604.

Wenzel, W.W. 2009. Rhizosphere processes and management in plant-assisted bioremediation (phytoremediation) of soils. Plant Soil. 321: 385–408.

Wilson S.C. and K.C. Jones. 1993. Bioremediation of soil contaminated with polynuclear aromatic hydrocarbons (PAHs): A review. Environ. Pollut. 81(3): 229–249.

Xiong, J., L. Wu, S. Tu, J.D. Van Nostrand, Z. He, J. Zhou, et al. 2010. Microbial communities and functional genes associated with soil arsenic contamination and the rhizosphere of the arsenic-hyperaccumulating plant *Pteris vittata* L. Appl. Environ. Microbiol. 76(21): 7277–7284.

Xun, F., B. Xie, S. Liu and C. Guo. 2015. Effect of plant growth-promoting bacteria (PGPR) and arbuscular mycorrhizal fungi (AMF) inoculation on oats in saline-alkali soil contaminated by petroleum to enhance phytoremediation. Environ. Sci. Pollut. R. 22: 598–608.

Yee, D.C., J.A. Maynard and T.K. Wood. 1998. Rhizoremediation of trichloroethylene by a recombinant, root-colonizing *Pseudomonas fluorescens* strain expressing toluene ortho-monooxygenase constitutively. Appl. Environ. Microbiol. 64: 112–118.

Zaidi, S., S. Usmani, B.R. Singh and J. Musarrat. 2006. Significance of *Bacillus subtilis* strain SJ-101 as a bioinoculant for concurrent plant growth promotion and nickel accumulation in *Brassica juncea*. Chemosphere 64: 991–997.

Zhang, L., C. Yan, Q. Guo, J. Zhang and J. Ruiz-Menjivar. 2018. The impact of agricultural chemical inputs on environment: Global evidence from informetrics analysis and visualization. Int. J. Low-Carbon Tech., 13(4): 338–352.

Zhuang, X., J. Chen, H. Shim and Z. Bai. 2007. New advances in plant growth-promoting rhizobacteria for bioremediation. Environ. Int. 33: 406–413.

Zilber-Rosenberg, I. and E. Rosenberg. 2008. Role of microorganisms in the evolution of animals and plants: The hologenome theory of evolution. FEMS Microbiol. Rev., 32(5): 723–735.

CHAPTER 12

Heavy Metal Contamination in Groundwater and Potential Remediation Technologies

Yung Shen Lee[1], Peck Kah Yeow[1],
Tony Hadibarata[1]* and Mohamed Soliman Elshikh[2]

[1]Department of Environmental Engineering, Faculty of Engineering and Science, Curtin University Malaysia, CDT 250, Miri98009, Sarawak, Malaysia.
[2]Department of Botany and Microbiology, College of Science, King Saud University, P.O. Box 2455, Riyadh 11451, Kingdom of Saudi Arabia.

1. INTRODUCTION

Water is the most vital resource on earth, without which the survival of mankind will not be possible. Thus, there will be no living things on earth without the presence of water. Water covers 70% of the world's total surface, but only less than 3% of the water is freshwater that is safe to consume by humans and for other human activities. Water pollution is one of the global environmental challenges, which is a cause of concern for the global society due to the anthropogenic activities and rapid population growth. A study conducted in the United States found that water sample that was extracted from approximately 621 out of 932 public wells tapping contain a high concentration of pollutants that not fulfill the standards of drinking water (Eberts 2014, Megdal 2018).

Other than surface water resources pollution, groundwater pollution is one of the main troubling issues as well. Organic and inorganic pollutants are common contaminants that can be found in groundwater. Groundwater is defined as freshwater underneath the ground surface sourced from rainfall, which seeps into the pores of soil or rocks linking to surface water resources such as rivers, streams and lakes. Groundwater plays an essential role in human daily life and ecosystem as it is clean, fresh and suitable for humans, flora and fauna. In our

*Corresponding author: hadibarata@curtin.edu.my

daily life, humans require water for drinking and household activities. Irrigation, industries, cultivation, restaurants were depended on the clean water resources to function properly. Evidently, approximately 40% of drinking water originates from groundwater and 97% of the rural population in the United States relies on the groundwater for their drinking water needs, where 30 to 40% of the groundwater is used for the cultivation of crops. As for the ecosystem, groundwater ensures the constant level of surface water is maintained by the base flow of lakes and rivers, which protect the habitat of aquatic biodiversity indirectly. Other serious cases such as oil spilling, surface leaching, surface runoff flow to recharge basin and saltwater intrusion are also of major concern (Owa 2013, Mays and Scheibe 2018).

The source of pollution is divided into point source where the pollutants are discharged from one source, and non-point source refers to the contamination caused by various sources such as contamination (chemical substances, excrement of animals and others) runoff from ground surface through the help of rainfall or snow melting to groundwater, lakes, rivers, and others. Point and non-point source pollutants cause groundwater contamination. In general, the contaminants or pollutants refer to the elements or compounds which affect and change the quality of water and bring serious catastrophic effects on the ecosystem. Thus, engineers have to think of ways to treat the critical phenomena of contaminated groundwater around the world by introducing suitable remediation technology to the polluted site. Remediation is a process of removing, treating or preventing pollutants in the media (e.g. atmosphere, soil or water) at the contaminated sites (Kuppusamy et al. 2016, Talabi and Kayode 2019). The remediation technology applied to the contamination site should solve the problem of the remediation site within a range of specified time. Groundwater may be treated by three major technologies such as physical, biological and chemical remediation. These three remediation technologies use more or less the same concept to treat water but application is made through different media. Normally, groundwater remediation technologies are a combination of physical and chemical groundwater remediation or physical and biological groundwater remediation or biological and chemical groundwater remediation. As many countries around the world such as the USA, Africa and China rely on the groundwater to sustain their daily life, groundwater should be treated and purified (Kuppusamy et al. 2016).

Therefore, it is a very common condition that there are many people in the world who do not have proper access to clean and potable water as the demand for clean freshwater is always higher than the supply. Groundwater is one of the most important sources of freshwater that is used in daily human life, agricultural activity and industrial activity throughout the world. However, the rapid development in many countries has caused the quality of groundwater to be unsuitable for human consumption anymore. This is due to the contaminants from the industrial and agricultural sector that have leached or run off into the groundwater causing the groundwater to be polluted (Sawyerr et al. 2017, Liao et al. 2018). In addition, anthropogenic activities such as disposing rubbish into public areas or dumping waste without proper procedure have also caused the groundwater to be polluted. The quality of the groundwater is very important for mankind as the polluted groundwater affects the health conditions of mankind and might cause the outbreak of disease. Long term consumption of contaminated groundwater with high level of heavy metals cause health problems such as difficulties in breathing, heart attack, kidney failure, neurological diseases, hypertension and may even lead to death (Kaur et al. 2017, Nawab et al. 2017).

2.   GROUNDWATER CONTAMINATION

Groundwater is the world's most used water, with an extracted rate currently about 982 km^3/year. It is the main source of water in some developing countries such as China, India, Pakistan, Iran and Bangladesh. In the case study of China, the usage of groundwater has increased to 90% as it provides drinking, industrial, and irrigation water to a huge population. However, it was found that the depletion of groundwater in the arid areas has occurred due to the unsustainable anthropogenic activities in China, which threatens the domestic or industrial water supplies and crop yields. The overexploitation of groundwater also causes several issues, which include ecological damage, degradation of seawater quality, land subsidence and affecting human health. Not only that, it was found that more than 60% of the groundwater in China has the presence of contaminants due to poor groundwater management. A national water quality survey observed that more than one harmful contaminant was found in the laboratory test. Fe, Mn, N and As are the most common contaminants found in the groundwater due to their higher total dissolved solids (TDS) (Hou et al. 2018, Jia et al. 2019).

The anthropogenic activities such as industrial and agricultural activity are the main factors for the groundwater contamination. Groundwater systems are considered as a part of the human nature system. Thus, a comprehensive groundwater system assessment framework has to be established in order to maintain the stability of groundwater and minimize the depletion of groundwater as groundwater depletion can increase the mitigation of contaminants. Moreover, the mining activity also produced the pollutant and become groundwater contaminant; for example, leakage of petroleum hydrocarbon occurs during transportation and the contaminants. The uncontrolled hazardous waste sites can also lead to the contamination of groundwater as the hazardous materials will flow through the soil and into the groundwater. Contaminants were also found in the hydrological cycle, as groundwater is part of the cycle. The groundwater contaminants evaporate into the atmosphere and fall onto the ground surface through the rain. The number of contaminants strongly impacts the distribution of groundwater resources for the long term, wildlife and human health (Aeschbach-Hertig and Gleeson 2012, Jia et al. 2019).

Municipal Solid Waste (MSW) landfills are one of the major contributors and source of the contamination of groundwater due to the wastewater of leachate that consist various toxic organic pollutants (Yang et al. 2006). Previous studies showed that the main pollutants found in the aquatic environment were ammonium, chlorine, total dissolved solids, and chemical oxygen demand (Regadío et al. 2012). Drinking water standards may potentially be exposed to hazardous chemicals posing threats to the overall quality of groundwater at the site. However, even lined landfills pose threats to groundwater. This is because the liners eventually fail and leakage occurs (Sizirici and Tansel 2015). This is due to the mixture of contaminants and leachate with the aquifers. Site contamination in landfills is evident and has been recorded in several developing countries such as Tunis, Malaysia, and India (Marzougui and Ben Mammou 2006). Groundwater contamination resulting from landfills remains a problem and a major concern for developing countries since the 1970s. In China, site contamination in landfills is apparent as the groundwater contamination near those sites has become more and more obvious. The condition of groundwater in six MSW landfills in Beijing was categorized as low quality. Other indications for decreasing quality in groundwater was the high concentrations of some pollutants such as total organic carbon, dissolved methane, ammonia, and other organic pollutants (Li et al. 2008, Eschauzier et al. 2013).

As for the marine environments, the semi-enclosed Gulf remains one of the most physiologically stressful environments as it displays extreme fluctuations in water temperatures as well as elevated salinities. Water temperatures are known to fluctuate at a difference of more than 20°C seasonally. The poor water circulation in the Gulf environment has resulted in a long residence time of water. This is due to anthropogenic activities carried out in sites in the Gulf's vicinity such as oil spill, run-off, and dumping. The water body then received a limited dilution due to the long residence time, ultimately causing the contaminants to dissipate more slowly (Freije and Awadh 2009). Site contamination, especially from the industrial sectors in countries bordering the Gulf, is the main contributor to the major environmental threats faced by the marine environment due to its function as the final repository for both natural and anthropogenic sources. During the last several decades, countries surrounding the Gulf have undergone major growth in population, which led to the development of industrial and economic sectors. This might result in an increase in organic pollutants such as petroleum and heavy metals discharged into the marine environment (Sale et al. 2011).

3. HEAVY METAL POLLUTION

The occurrence of heavy metals in groundwater is frequently associated to anthropogenic activities, where the pollutants may enter into the soil thus contaminating the groundwater. However, automobiles, landfill sites and roadways are now categorized as the biggest causes of inorganic pollutants compared to areas of intensive industry. Groundwater contaminant is one of the major environmental issues that has happened currently and among the different types of contaminants, heavy metals pollution is the most important problem due to their strong toxicity even at low concentration. This contamination is usually caused by heavy metals such as copper, zinc, lead, nickel, mercury, chromium and cadmium that had been leaked from the urban area; therefore, the concentration of heavy metals may be particular in groundwater. Heavy metal is any metalloid element or metal that is poisonous and toxic even at low concentrations and has a relatively high density. It is non-biodegradable and can be found wisely in the Earth's crust. Most of the heavy metals are 4–5 times weightier in terms of specific gravity and makes them a colloidal and dissolved particles in water bodies (Vodela et al. 1997, Chen et al. 2017, Nawab et al. 2017). Table 1 gives toxicity level, anthropogenic sources of heavy metals, as well as their effect on human health.

The anthropogenic origin refers to industrial or domestic effluents, solid waste disposal, and harbor channel dredging, while natural origin refers to volcanism extruded products and eroded minerals in the sediments. Some of the metals such as sodium, potassium, magnesium, and calcium are quite important to sustain life and for normal body functions, while low concentration of molybdenum, iron, cobalt, zinc, manganese, and iron is used as a catalyst during enzyme activities. Although the metal industry is very important to humans, uncontrolled waste disposal of heavy metals can cause large amounts of toxic substance and hazards to humans, animals, plants and ecosystems (Nowierski et al. 2006). Over the last 30 years, heavy metals, is one of the main contributor of pollution to the environment and this is proved by the World Health Organization. Another research showed that serious health problems for humans were detected because of heavy metals (Izah et al. 2016). The toxicity of the heavy metals came from the formation of complexes with protein, where thiol, amine and the carboxylic acid are involved. When these groups bonded with heavy metal, their important enzyme systems are deactivated, thus disturbing the protein development in human body. These changed genetic fragments would not function properly and result in the malfunction or death of the cells. The toxicity of the heavy metals may also form radicals that can oxides the biological molecules and bring a huge impact on the cells. At present, the human food chain is much affected by heavy metal contamination in the environment It is a must to investigate trace metal background levels to understand their effects and distribution on the quality of groundwater and agricultural crop to assess their contamination in groundwater and to evaluate natural concentration variations of heavy metals (Eberts 2014, Izah et al. 2016, Nawab et al. 2017). Case study of high poisonous heavy metal contamination in some countries is shown in Table 2.

Table 1   Source and effect of heavy metals on human health

Toxicity level	Heavy metals	Sources	Harmful effect	References
High poisonous	As	Antipyretics, antiseptics, caustics, antispasmodics, herbicides, fungicides, insecticides, paint, and cosmetics.	Carcinogen, cellular damage and cell death	(Chen and Olsen 2016)
	Cd	Batteries, ceramics, electronic and metal-finishing industries, electroplating industries, pigments, petroleum products, textiles, insecticides, solders, television sets, metallurgical industries	Causing irreversible damage to the renal tubules, which are involved in the mechanisms of nutrient reabsorption	(Pulford and Watson 2003, Rubio et al. 2018) (Clarkson and Magos 2006, Rodrigues et al. 2012)
	Hg	Mining and coal combustion, medical waste	Neurotoxin which impairs brain function, cause damage to kidney and lung	(Zulfiqar et al. 2019)
	Pb	Melting and smelting of ores, exhaust from automobiles, fertilizer, pesticide, additive in pigment and gasoline, urban soil waste	Respiratory diseases and heart, brain and kidney damage	(Etteieb et al. 2020)
	Se	Electronics, paint industry, glass industry, ceramics, metallurgy, chemical industry, and pharmaceutics and in agriculture	Acute toxicity in respiratory, gastrointestinal, cardiovascular difficulties and hypertension, or chronic toxicity.	(Vardhan et al. 2019)
	Zn	Galvanizing iron and steel, construction, siding, apparatus housings, office hardware, heating and ventilation conduits, vehicle and building enterprises for roofing	Diarrhea, liver failure, bloody urine, icterus, kidney failure, stomach cramps, abdominal cramps, epigastric pain	
Moderately poisonous	Co	Metal industry, construction industry, cosmetics, jewelry, and medical exposure	Neurological, cardiovascular and endocrine	(Leyssens et al. 2017)
	Cr	Leather tanning, metallurgy, electroplating and refractory	Ulcerations, dermatitis, and allergic skin reactions	(Almeida et al. 2019)
	Cu	Electronic chips, batteries, cell phones, semiconductors, water pipes, fertilizer industry, pulp, and paper industry, fungicides, insecticides, catalysts, and metal processing products	Epigastric torment, gastrointestinal bleeding, diarrhea, spewing, tachycardia, hematuria, respiratory challenges, hemolytic anemia.	(Vardhan et al. 2019)
	Ni	Mining, smelting, emissions from vehicles, disposal of household, municipal and industrial waste, steel manufacturing and cement industry	Lung irritation, lung inflammation (pneumonia) and emphysema	(Schaumlöffel 2012)
Low poisonous	Ba	Ore mining and refining	Damage to the cardiovascular, renal, respiratory, hematological, nervous, and endocrine systems	(Lu et al. 2019)
	Mn	Mining and metallurgical industries	Manganese neurotoxicity	(Nelson et al. 2018)
	Sr	Thermal power plants, ferrous metallurgy enterprises, non-ferrous metallurgy enterprises, and chemical industry	Destruction of the whole body (general toxic effect)	(Mironyuk et al. 2019)

Table 2 Case study of high poisonous heavy metal contamination in some countries

Heavy metal	Country	Category	Source of contamination	References
As	Italy	Residential	Leaching from water distribution pipes	(Tamasi and Cini 2004)
	Vietnam	Residential	Natural anoxic conditions in the aquifers	(Buschmann et al. 2008)
	Mexico	Industrial, mining, and agricultural	Industrial, mining, and agricultural	(Jane Wyatt et al. 1998)
	India	Industrial, agriculture	Paint, pharmaceutical, fertilizer, and pesticide industries	(Krishna et al. 2009)
Cd	Saudi Arabia	Residential	Corrosion of metal pipes, water tanks, and plumbing systems	(Al-Saleh and Al-Doush 1998)
	Greece	Industrial	Heavy metal activities and industrial effluents	(Simeonov et al. 2003)
	India	Industrial	Industrial wastewater effluents and/or dumping of industrial waste	(Srinivasa Gowd and Govil 2008)
	Egypt	Industrial	Steel, plastic, and battery industries and household plumbing systems	(Mandour and Azab 2011)
Hg	Mexico	Industrial, mining, and agricultural	Industrial, mining, and agricultural	(Jane Wyatt et al. 1998)
	Saudi Arabia	Industry	Desalination water process	(Al-Saleh and Al-Doush 1998)
	Poland	Mining	Industrial effluents, packaging activities of bottled water	(Antoszczyszyn and Michalska 2016)
	China	Mining	Found in demolition bituminous coal landfills Gold mining area	(Li et al. 2009)
Pb	Ghana	Industrial	Battery recycling plant	(Gottesfeld et al. 2018)
	Nigeria	Industrial	Battery recycling plant	(Gottesfeld et al. 2018)
	Australia	Residential	Household drinking water	(Harvey et al. 2016)
	Brazil	Residential	Food	(Vasconcelos Neto et al. 2019)
Zn	Turkey	Industrial	Coal-fired power plant	(Demirak et al. 2006)
	Morocco	Mining	Mining activities	(Bouzekri et al. 2019)
	Zambia	Mining	Mining and smelting	(Kříbek et al. 2019)
	Greece	Mining, Industrial	Heavy metal activities and industrial effluents	(Simeonov et al. 2003)

3.1 Arsenic

Arsenic is considered as an abundant metalloid formed in either soil, water, air and rock. Volcanic explosions, dissolution of the Earth's crust, leaching and runoff are the most common natural sources of arsenic in water. Most of the arsenic pollutants found in groundwater supplies are usually inorganic, while the organic arsenic like arsenobetaine are usually discovered in aquatic organism which as a result cause an increase of arsenic toxification in the human population. Two major factors that currently are the main causes for arsenic contamination in water, air, and soil are combustion of petroleum and the manufacturing of metals as smelting activities are known to be the biggest anthropogenic source of atmospheric contamination (Cumbal et al. 2010, Hamidian et al. 2019). Generally, people tend to get exposed to arsenic through ingestion of food, drinking water and inhalation of air. In South America, it is reported that nearly 15 million civilians regularly drink more than 10 µg/L concentration of water polluted with arsenic (Bundschuh et al. 2012). Arsenic is a very hazardous compound and its exposure may lead to various deadly sicknesses, which include hemolysis, melanosis, gastrointestinal symptoms, bone marrow depression, encephalopathy, polyneuropathy and in severe cases death may occur. Another ecological study based on mortality due to skin cancer has also stated that the concentration of arsenic in drinking water increases the risks of contracting skin cancer (Cumbal et al. 2010, Begum et al. 2015). Not only does arsenic has devastating effects on human health but it also causes damage to the sustainability of the ecological system. As arsenic pollutants are very hazardous substances, when they travel through the ecological system using water as its media, the trees and plants are affected and poisoned by the arsenic through absorption of the water. As a result, when other animals and humans consume the vegetation, they too will be poisoned by the arsenic. A study based on rice paddies has discovered that arsenic pollution has affected the growth of the rice paddies and there are arsenic particles present inside the stems of the rice paddies, which may poison other organisms who consume it. Thus, arsenic pollutants present in groundwater must be treated to ensure safety of human health and sustainability of the ecological system (Bundschuh et al. 2012, Siddiqui et al. 2019).

3.2 Cadmium

Cadmium is a soft silver-white metal, most commonly in the form of complex oxides, carbonates, and sulfides and is usually utilized as a stabilization agent in polyvinyl chloride, various alloys, colorants, and rechargeable nickel-cadmium batteries. Even small amounts of cadmium from smoking are highly toxic to humans, as the lungs absorb cadmium more efficiently than the stomach and other disease such as diarrhea, nausea, and renal dysfunction. A study has shown that long-term exposure to cadmium results in chronic airway inflammation and oxidative stress, which are the symptoms for obstructive pulmonary disease, a state that is distinguished by an ongoing and irremediable airflow limitation. Another research regarding osteoporosis has also discovered that the exposure to cadmium in human body even in low concentration may result in skeletal destruction (Wu and Cao 2010, Idrees et al. 2018, Zhao 2018). Moreover, cadmium groundwater pollution has the ability to destroy the surrounding ecological system, changing and harming local ecosystem by contaminating soil and drinking water. In the aquatic ecosystems, the biochemistry of nutrients and the ecosystem function are greatly affected by the cadmium pollutants. Many marine life is unable to survive as the zooplanktons, which are a food source for some krills and fishes, are unable to live under the cadmium polluted conditions, which cause a massive chain reaction to other marine life. In addition, cadmium is also available inside phosphate fertilizers, which are the main cause of eutrophication (Di Natale et al. 2008, Hayes et al. 2018). Eutrophication is a process whereby there is an excessive amount of nutrients inside a river or large body of water, which causes a growth explosion of algae on top of the body of water. This condition commonly occurs due to excessive fertilizer runoff from agricultural lands and it causes water quality impairment of inland and marine waters, which disrupts the aquatic ecosystems. Many aquatic species are found dead during the eutrophication process as the algae plants consume huge amounts of oxygen in the water during the process, which results in insufficient oxygen for the marine life to survive. Research has pointed out that eutrophication, which was expanding in the Northern Hemisphere lakes, was mainly caused by phosphorus point source pollution coming from fertilizers which contain cadmium pollutants. Hence, cadmium pollutants pose a very big threat to the living organisms, including humans and animals, and the sustainability of the environment (Hayes et al. 2018, Idrees et al. 2018).

Cadmium does not exist in nature as native metal but principally as sulfide ore with the name of greenokite, which is related with zinc sulfide and recovered from refining and smelting of copper, and therefore it is not easy to find cadmium in the natural water. This compound is really useful, which acts as a stabilizer in PVC products, several alloys, color pigment, production of batteries, and an anti-corrosion agent in some metals, but is also present as a pollutant in fertilizers. Soil contaminations can be caused by the excessive application of fertilizer and sewage sludge to farmland, and the emission by industrial activities.. The cadmium in the contaminated area can be uptaken by crop to increase their productivity. (Vodela et al. 1997, Wu and Cao 2010). Men usually have higher daily cadmium intake than women due to higher energy consumption. Moreover, one of the nutritional factors, which is iron status, affects the gastrointestinal absorption of cadmium (Jarup et al. 1998). Cadmium is considered as a toxic chemical compound if its concentration is higher than 0.01 mg/L in irrigation and drinking water. The examples of disease that may be caused by the toxicity of cadmium are destruction of red blood cells, destruction of testicular tissue, kidney damage and high blood pressure. Cadmium is a chemical compound that has carcinogenic properties and long biological half-life, which cause chronic effects as a result of accumulation in the renal cortex and the liver. Therefore, over exposure to high concentration of cadmium will damage the kidney and bring serious damage to human health. The entry of cadmium particles into the human body may cause acute pulmonary edema. (Orisakwe et al. 2006, Wu and Cao 2010). A case study in 1950 showed that cadmium in human body may cause kidney damage. The studies also proved that there will be an additional danger of kidney stones due to the augmented secretion of calcium in urine. Occupationally exposed workers also confirmed this incident in their studies (Friberg 1950, Jarup et al. 1998). Furthermore, a new type of skeletal disease named itai-itai disease had been discovered in Japan in the 1950s and the disease occurred due to long-term high exposure to cadmium. The exposure was actually caused by the groundwater used for irrigation of local rice fields that were polluted by cadmium and this was proved by a few researches in Japan. During recent years, new evidence and data had also shown that skeletal damage is related to the low cadmium exposure, evidenced by low bone mineral density and fractures. Lastly, people that had signs of tubular kidney damage had a higher death rate when exposed to the cadmium environment compared to those people without kidney damage, as shown by the studies conducted in Japan (Nishijo et al. 1995, Jarup et al. 1998).

3.3   Lead

Lead is generally used in food application such as pots which are used for storage and cooking, and the previous generations used lead acetate to sweeten wine. In the last century, lead pollution is the worst due to the widespread use of petrol which has 50% lead. The huge amount of lead emission to air by petrol ignition decreases environmental quality and has adverse impact on living organisms. Occupational exposure to a high concentration of lead occurs in the mining and smelters area. Airborne lead is dumped on ecosystem and can have impact on humans through the food chain. However, in the last few decades, unleaded petrol has been introduced in developed countries and this has helped to reduce the emission of lead to the environment. Long-term exposure to lead in drinking water has become an important concern to the public (Zietz et al. 2007). In previous study, water from 12 boreholes were investigated and showed the maximum concentration of 0.024 mg/L that exceeded the maximum contaminant level. According to the result of the experiment, lead has been known as a cumulative general metabolic poison for human health due to the neurotoxin damage. Therefore, low concentration of lead in the groundwater will also impact human health such as reduced intelligence in children, brain damage and an increase in blood pressure. In some of the cases, lead toxicity may affect the person suffering from consciousness, acute psychosis, prolonged reaction time, reduced ability to understand, memory deterioration, and diminished intellectual capacity in children. Previous research showed that the high level of lead in the blood may impact to human health. The data in this analysis indicated that the increase of 0.48 µmol/l of blood lead level will lead to a decrease of 2 points of IQ (Needleman 1993, Mendelsohn et al. 1998, Zietz et al. 2007).

3.4 Mercury

Mercury is a chemical compound named cinnabar (HgS) and was utilized by pre-historic humans for cave painting. In the current century, mercury is commonly used as a cosmetic supplement to make the skin look brighter and this was known in ancient Greece a long time ago. Liquid mercury is also found in barometers, thermometers, as well as medical instruments for measuring blood pressure and diuretics purpose. According to research, mercury in dental amalgam is a hidden source of global mercury pollution. During the 1970s, mercury concentration in the air in some dental surgeries had reached 20 $\mu g/m^3$ but since then the level has generally dropped to about 1/10 of that concentration. This chemical compound can also act as an electrode in the electrochemical process of manufacturing chlorine in a Chlor-alkali industry. Recently, the exposure of mercury in humans occurred via food. Previous studies showed that mercury is present in the urine, where it is mainly associated to exposure of human to toxic inorganic compounds containing excess mercury (Sallsten et al. 1996, Freije and Awadh 2009, Wu and Cao 2010).

Although mercury is a useful and important compound to human beings, it contains hazardous material that may hugely impact the environment and human health. Paresthesia and numbness in the hands and feet will develop in the beginning stage of mercury infection. The next symptoms will be coordination difficulties, concentric constriction and auditory symptoms. High concentration may lead to death, usually around 1 month after onset of symptoms. The toxicity of the mercury may also result in impairment of speech, movement, vision and hearing and mental disturbance (Weiss et al. 2002). Furthermore, mercury may affect the development of autoimmunity in which a person's immune system attacks its own cells. This can lead to many human diseases such as joint diseases and ailment of the kidneys, circulatory system and neurons. It will cause irreversible brain damage at high concentrations. Besides that, Minamata Bay of Japan is one of the places that had been impacted by the toxicity of mercury. This event occurred in the 1950s, where fish were contaminated by mercury contamination caused by industrial waste. Physically deformed and mentally disturbed babies were born to mothers at this place who were exposed to toxic mercury due to the consumption of polluted fish. The high dietary intake of polluted fish has been hypothesized to increase the risk of coronary heart disease. Another case of mercury poisoning is in the early 1970s where ten thousand Iraq citizens were poisoned after consuming bread made from mercury-polluted grain and a few thousand citizens died due to mercury poisoning. Lastly, a recent case-control study experimented on the relationship between mercury levels in toenails and the risk of coronary heart disease among male health professionals with no previous history of cardiovascular disease. The experimental result showed that the mean mercury level was higher in dentists than in non-dentists, while mercury levels were significantly correlated with fish consumption. (Salonen et al. 1995, Yoshizawa et al. 2002).

4. REMEDIATION TECHNOLOGIES

Recently, a lot of *in situ* and *ex situ* remediation procedures had been established for organic and inorganic contaminant as shown in Figure 1. These methods can be categorized into three treatments: physical, chemical and biological. In general, these techniques employ various mechanisms and demonstrate specific application advantages and limitations. The above statements had shown that the heavy metals present in the groundwater can have a huge impact on humans and the environment through industrial effluents, agricultural runoff, and household uses; therefore, some of the remediation technologies can be implemented to prevent them. There is some equipment that has been developed that can eliminate a massive amount of heavy metals that is frequently found in the groundwater. The newest is nanoparticles as nanosorbents, which is one of the most recommended technology that can be used to remove heavy metals and contaminants from an aqueous solution such as groundwater. The magnetic nanomaterial has some special properties, which are low toxicity, chemical stability, excellent recycling capability and ease of synthesis; therefore, it is very suitable to eliminate heavy metal ions from aquatic environment.

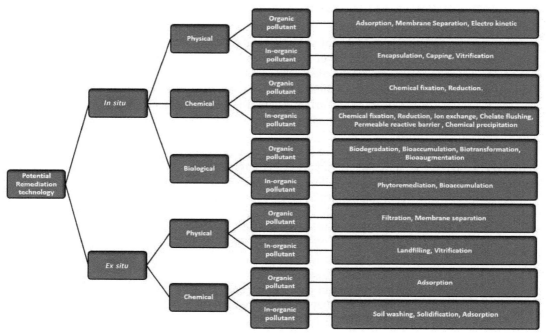

Figure 1   Potential remediation technologies for organic and inorganic contaminants.

Nanoparticles can be used in various fields including bioremediation by researchers from different disciplines such as magnetic resonance imaging, drug delivery, biomedicine, magnetic fluids, catalysis and remediation (Ambashta and Sillanpaa 2010, Jung et al. 2011). Although this technology had been experimental and was developed for a range of different compositions, the success rate of application of this technology is not consistent and particularly depend on the strength of the materials under a range of diverse conditions. After a series of tests and experiments, the best performance of the particles is when the size of the nanoparticles is lowered a critical value, which is around 10–10 nm in the majority of the envisaged applications. However, it also depends on the source material and the surrounding environment of the nanoparticles. This type of technology is generally used in absorbents in many types of scientific communities, which are biological, chemical, environmental science, and even engineering. The magnetic nanoparticle-based absorbents are useful for purification and separation of environmentally and biologically relevant target species that include heavy metals from groundwater with high accuracy and precision (Lu et al. 2007, Girginova et al. 2010).

Furthermore, electrocoagulation is also one of the methods to remove heavy metal from groundwater and this method consists of two electrodes, which are cathode for reduction and anode for oxidation. This method is often used in physiochemical processes such as adsorption, reduction, oxidation, and coagulation. The advantages of electrocoagulation technique are cost-effectiveness and easy applicability for industrial scale (Arroyo et al. 2009). Chromium, a toxic heavy metal that is present in contaminated groundwater, can be eliminated feasibly by this technique. One of the research on removal of chromium from water by Golder and co-workers proved that the adsorption and coagulation are major processes in the removal of chromium during electrocoagulation. In the electrocoagulation system, two electrodes were used to remove chromium compounds from aqueous solution with both monopolar and bipolar configurations and found that the removal rate of chromium is the highest during higher current density (Golder et al. 2007). The previous study showed the iron and aluminum electrodes with monopolar configurations to remove heavy metals from groundwater and that pH value, current density, electrode material and conductivity may affect the efficiency of the process. The result also proved the success of using the iron and aluminum electrode by removing all of the nickel, chromium, and copper in 20 minutes with a 3 pH value and a current density of 10 mA cm^{-2}. This showed that electrocoagulation is a fast and highly efficient method to reduce and remove heavy metals in groundwater compared to other existing techniques; therefore, it is very suitable to be implemented (Akbal and Camcı 2011). The membrane filtration is also an alternative method to remove heavy metals from groundwater. The membranes consist of nano-sized active elements in multifaceted construction that is very useful for this process. Recently, a porous support structure in the homogeneous polymer thin films was used in the reverse osmosis membranes. Partitioning water and dissolved salts between the membrane and the bulk solution, and transport of salts and water across the membrane and the bulk solution depend on the chemical properties of the membrane as well as the physical and chemical structures on the nanoscale to microscale. It also indicated that the distinct modes of material transport across a membrane are related to the dimensions of it. Convective flow through pores will occur when the transport of the water passes through a larger dimension of porous membranes (Nemati et al. 2017, Shariful et al. 2017). Current remediation techniques for heavy metal removal are summarized in Table 3.

Table 3 Current remediation techniques for heavy metal removal

Techniques	Principles	Pollutant	Reference
Air sparging	A technique of injecting air into the target contaminated zone, with the expectation that volatile and semivolatile contaminants will undergo mass transfer (volatilization) from the groundwater to the air bubbles	Fe and Mn	(Pleasant et al. 2014)
Soil flushing	The application of aqueous solutions in the soil to solubilize contaminants	Cd and Pb	(Lee et al. 2011)
Permeable Reactive Barriers (PRBs)	The emplacement of a reactive media perpendicular to the potential trajectory of the contaminated groundwater	Cu	(Komnitsas et al. 2007)
Chemical oxidation	The application of chemical oxidants to the subsurface soil to oxidize and degrade contaminants	Cu, Zn, Pb, Cd, and Cr	(Guo and Zhou 2020)
Carbon absorption	A process by which a solid holds molecule of a gas or liquid or solute as a thin film	Pb, As, and Cd	(Nejadshafiee and Islami 2019)
Bioaugmentation	The use of additional bacterial cultures to speed up the rate of degradation of a contaminant	Pb and Zn	(Sprocati et al. 2012)
Rhizofiltration	The removal of pollutants from environment by adsorption or precipitation on the roots, or for adsorption of contaminants around the root zone	U	(Lee and Yang 2010)
Phytoremediation	A techniques of using living plants to clean up soil, air, and water contaminated with hazardous contaminants	Cd, Ni, Zn, Pb, and Cu	(Doni et al. 2015)

4.1 Physical Remediation

4.1.1 Air Sparging

Due to the limitations of the pump and treat method to remediate sources of groundwater contaminations, *in situ* air sparging (ISAS) has gained popularity due to its easy implementation and economical standings. Air sparging involves the injection of clean pressurized gas into the contaminated zone to treat and mitigate contaminant plumes. Unlike biosparging, its primary function is to increase the amount of oxygen content inside the soils, which in turn, allows aerobic biodegradation. Unfortunately, large areas with contaminations may be deemed ineffective in sparging due to its limited radius of sparging effectiveness. Apart from that, the incorrect air injection pressure could result in the mobilization of free nonaqueous-phase liquid contaminants to other uncontaminated areas (Mohamed et al. 2007, Wang et al. 2007).

4.1.2 Soil Flushing

Applicable to both organic and inorganic pollutants, in-soil flushing involves injecting water or liquid solution into the contaminated area. After the aqueous solution has been circulated through the contaminated zone, the contaminant-solution is then collected and brought for treatment to remove pollutants. This technique is preferable and effective in soil conditions which are permeable and contains little-to-none organic matter and clay content. On the downside, this technology exhibits poor efficiency in soils with low permeability and with pollutants that observe strongly towards the soil surface. In such conditions, it would require many completed flushing cycles to actually significantly reduce levels of pollutant concentrations. Thus, application of this technology would be deemed unsuitable and impractical in terms of the timescales for completed remediation and associated operating costs (Yan et al. 2017, Kwon et al. 2018).

4.2 Chemical Remediation

Chemical remediation technology is a technology applying compound or mixture to damage or change the properties of pollutants chemically, which is different from physical and biological technology. Some of the chemical remediation technology is combined with physical remediation technology as the pollutants have to be pumped from underground to the ground level for treatment or separate certain pollutants from water physically (Liu et al. 2010). The methods stated below are some conventional and popular chemical technologies that are used for groundwater treatment.

4.2.1 Permeable Reactive Barriers (PRBs)

Permeable reactive barriers are on-site technology and placed underground to treat contaminated groundwater. PRB technology utilizing a barrier wall that is made of adsorbents (reactive medium) and reactive media adsorbs contamination in groundwater when passing through it. The criteria of the barrier wall are high permeability, the reactive medium must be more permeable than surrounding media (soil) to induce water flow through the wall naturally and the material of barrier wall depends on the type of contamination. There are two designs for the permeable reactive barriers, which are continuous reactive barrier and funnel and gate permeable reactive barrier. The continuous reactive barrier is in a rectangle shape and high permeability, which is more suitable for uniform flow of contamination. For funnel and gate permeable reactive barrier, the funnel part is low permeability while the gate part is high permeability, which allows the barrier to cover more volume of contaminated groundwater in unsteady flow. Normally, permeable reactive barriers consist of zerovalent iron, which is used to remove pollutant associated with heavy metal (e.g. chromium, uranium or technetium) and acid-mine drainage can be got rid by using organic carbon. The flow rate of contamination water passes through the barrier when adsorption is undergoing. For instance, the thickness of the wall increases as the flow rate of contamination is slow to make sure that contamination in groundwater is removed. The lifetime of permeable reactive barrier depends on the material that has been chosen to treat the groundwater pollution (Obiri-Nyarko et al. 2014, Luo et al. 2016, Beiyuan et al. 2017).

4.2.2 Chemical Oxidation

Remediation through *In situ* Chemical Oxidation (ISCO) involves the injection of oxidants into the subsurface of the soil to eliminate contaminants. Delivery of the oxidants is usually accomplished through direct flushing by vertical groundwater wells, soil mixing, pneumatic mixing and permeation by injection probes. The most commonly implemented oxidants for these applications are potassium permanganate, sodium persulfate, ozone or combination of them. This method is recognized as one of the fastest methods for groundwater remediation (Liu et al. 2014). The advanced oxidation process is also one of the remediation methods to remove or reduce groundwater contamination. Advanced oxidation process is considered as an on-site chemical treatment that uses an oxidizing agent such as hydrogen peroxide H_2O_2 to treat the contaminated region. Other than hydrogen peroxide, other compounds or elements which are able to demonstrate oxidizing agent (hydroxyl radical) characteristics fully can be used as well. Examples for AOP systems that involve either H_2O_2 or ultraviolet light (UV) are H_2O_2/Fe_2^+, $TiO_2/UV/O_2$, $TiO_2/UV/H_2O_2$ and so on. Owing to the advanced technology development, an advanced oxidation process has proved that effectiveness in heavy metal removal increases compared to common treatment methods (e.g. mineralization and degradation). Even though the advanced oxidation process has high efficiency, it is manipulated by some measurable factor. Those measurable factors are surrounding temperature, pH value, the concentration of oxidizing agent and intensity of light (Muruganandham et al. 2014, Guba et al. 2015).

4.2.3 Carbon Absorption

Activated carbon is one of the famous and most frequently applied adsorbent materials that come in different shapes and sizes (pelletized, granular, etc.). Furthermore, it possesses the strongest physical adsorption of any other material known to mankind and is mainly implemented for the removal of unwanted components in chemical, food, wastewater, and groundwater treatment. However, activated carbon is subjected to regeneration or replacement once their adsorption capacity has saturated, whereby deterioration of adsorption performance will occur. However, the adsorptive capacity of the saturated carbon will still decline after regeneration. After a few cycles, the carbon will have to undergo replacement (Yu and Chou 2000, Georgi et al. 2015).

4.3 Biological Remediation

Biological technology that is used to control or remove waste air in the atmosphere, especially volatile organic compounds and toxic air pollutants, is being used widely. As some of the pollutants in groundwater come from the atmosphere through raining, so use of microbes and plants are employed as an important biological remediation tools.

4.3.1 Bioaugmentation

According to Zhang et al. (2016), bioaugmentation is cheaper and eco-friendly in comparison to physico-chemical methods. This approach was developed to improve the process of biodegradation as some pollutants that pose high stability, toxicity or low biodegradability, bioavailability and water solubility tend to be resistant to biodegradation. One of the fundamental benefits of bioaugmentation is that the treatment can be altered to satisfy the needs of eliminating specific pollutants (Zhang et al. 2016). The most commonly appearing organic pollutants are 3-chloroaniline, 4-fluotoaniline, quinoline, pyridine, cyanide and naphthalene. They can be cured by specific bioaugmented bacteria such as *Comamonas testosteroni, Acinetobacter* sp., *Bacillus* sp., *Pseudomonas* sp. and many more together with the bioaugmentation medium. In the United States, the cost of treating groundwater with bioaugmentation method ranges from $30 to $100 per cubic meter. However, the incapability of pumping hydrophobic contaminants that are adsorbed to the aquifers marks up the remediation expenses. Bioaugmentation was modified to include chlorinated ethenes as electron acceptors into the process to undergo dehalorespiration. Another breakthrough was that dehalococcoides microorganisms are observed to be able to dechlorinate and reduce the metabolism of vinyl chloride to ethene (Abeysinghe et al. 2002, Liu et al. 2013). Bioaugmentation is widely applied to degrade chlorinated compounds, such as vinyl chloride and cis-1,2 dichloroethylene, faster and more completely compared to the natural microbial community. Based on research focusing on removal of synthetic dyes from groundwater, vast quantity of azo and anthraquinone components caused by cosmetics and textile industries were detected. In the beginning, the biodegradation method was imposed but anthraquinone-dyes were still detected after the treatment and were found to be toxic and resistant to biodegradation. At the same time, bioaugmentation was implemented with *Sphingomonas xenophaga,* where it has successfully removed anthraquinone-dyes from the wastewater. In general, bioaugmentation can undertake reductive dichlorination of chlorinated ethenes to the safe by-product ethene with the aid of dihydrocodeine. On the financial aspect, expenses on clean-up can be reduced as bioaugmentation is a sustainable technology that is capable of utilizing renewable materials such as soy oil, molasses, and lactate with the least energy expenditure. Besides, bioaugmentation can be extensively practiced in different aquifers as well as serious chlorine contaminated groundwater (Weyens et al. 2009, Fang et al. 2013, Payne et al. 2013).

4.3.2 Rhizofiltration

Rhizofiltration (commonly known as phytofiltraton) is the disposal of plant roots of contaminants in wastewater, extracted groundwater or any other contaminated aqueous solution. The root environment may explicitly produce biogeochemical conditions that might leads to the precipitations of contaminants onto the roots and into the water body itself. Phytoextraction and rhizofiltration are quite similar due to their accumulation of contaminants in or on the plant itself. However, rhizofiltration differs from phytoextraction due to the different accumulation points of the plant as accumulation only occurs on the root or portion of the plant for rhizofiltration. Apart from that, rhizofiltration is also distinct from phytoextraction in that the contaminants are located in the water rather than embedded in the soil itself. Furthermore, the contaminant removal from the rhizofiltration itself is done through the harvesting of the vegetation roots and is disposed of later on (Galal et al. 2018).

4.3.3 Phytoremediation

In the current era, various methods have been introduced to mitigate and eliminate the effects of contaminants in the environment. Phytoremediation is an effective and cheap technology option that is able to remove hazardous metal pollutants from soil and water. Unlike other conventional technologies, it aims to use selected plants to clean contaminated environments from hazardous pollutants to provide a greener environment. Vegetation have grown very specific and efficient mechanisms that are able to extract main element. Phytoremediation can be subcategorized into few different forms such as phytoextraction, phytostabilization, rhizofiltration, etc. Phytoremediation is a potential technology to be implemented for the remediation of a hazardous waste zone. The success of this remediation technique greatly depends on the environmental conditions (humidity, wind direction, air temperature, etc.). Apart from that, vegetation growth should be monitored for optimum pH value, nutrient levels and appropriate water content in the soil (Pulford and Watson 2003).

4.3.4 Phytoextraction

Phytoextraction mainly refers to the uptake of a contaminant in the aboveground portion of a plant. Once the contaminant has subsequently accumulated enough, the plant is then harvested and disposed of. Furthermore, organic and nutrient contaminants are generally not considered to be taken up by the plant as the following contaminants can be metabolized by the plant itself. The effectiveness of phytoextraction is limited to the low solubility of the metals and the sorption capacity of metals to soil particles. Soluble hazardous metals in extracted groundwater can also be cleaned through this method and if necessary, rhizofiltration as well (Ma et al. 2013).

4.3.5 Phytostabilization

Phytostabilization is the use of vegetation control to contain soil contaminants, through chemical, physical and biological modification of the contaminated soil. Contaminant migration is hindered by precipitation, complexation, metal valence reduction, adsorption, and accumulation by the roots of the plant itself. Up to date, phytostabilization has only been focused on hazardous metals with mercury and lead being potential candidates for phytostabilization. The advantages of this method are soil removal, disposal of hazardous contaminants is not required, associated cost to site activities is considered less than other remediation technologies and the ecosystem restoration is further enhanced as well. The disadvantages of phytostabilization are the long period of maintenance before the vegetation can be self-sustaining in the long run. It is to ensure re-release of the contaminants and the leaching possibilities are minimized (Lorestani et al. 2013).

5. CRITERIA TO SELECT SUITABLE TECHNOLOGY

Nowadays, pollutants are being spread widely in our environment. To sustain our life, different types of remediation technologies were introduced a century ago. Before implementation of technology, the professionals involved have a lot of workload. The expenses (e.g. capital cost, analysis cost, operational cost and maintenance cost), effect of secondary pollution, effectiveness of remediation technology, life span of technology, impact on social and economic development and policy from the government are criteria required to be considered before installation (An et al. 2017). Hence, to install a remediation technology is a difficult task. First, researchers the geographical location, and the type of pollutants which need to be remediated and classify them. Then, engineers design or select a suitable technology no matter the physical, biological, chemical or combination method for sustainability assessment. Next, the determination of geographical location and weather of contamination site, type of soil and depth of groundwater are essential criteria that need to be considered before selecting the proper technology (Ye et al. 2019).

Groundwater remediation technology from a physical aspect is used to extract, inject, transport or deliver contaminants as well as separate contamination from groundwater for further treatment. For instance, vapor air stripping, pump, and treat, dual-phase vacuum extraction and air stripping come under physical remediation technology. Apart from that, the greatest feature of physical remediation technology is that it is suitable to install *in situ* or *ex situ* or combine with chemical and biological remediation to have higher efficiency in groundwater treatment. This might due to the geographical location of contamination site, type of soil and depth of contamination. The combination of biological and chemical remediation technology is difficult due to the reaction between microorganisms and chemical compounds. Air stripping is an effective method to reduce volatile organic compounds from wastewater or groundwater. However, the application of the air stripping method is limited to volatile organic compounds (Ghoreyshi et al. 2014). Soil such as clays and silts are low permeable, which is not suitable for pump and treat installation. Next, remediation technology from chemical aspect are permeable reactive barriers and advanced oxidation process. The technologies are high in efficiency but they are with some limitations. The contamination must flow through the permeable barrier wall to have a reaction with it in order to purify the polluted groundwater. The water which does not pass through is considered as leakage and the purified water gets mixed with polluted groundwater again. Thus, the treatment is considered failure. All the criteria must be included in the calculation to prevent failure. Other than that, some chemical compounds will have an effect on secondary pollution to the environment. The main objective of remediation technology is to remove or degrade the contamination but not contaminate the environment again (An et al. 2017).

In addition, biological treatment technology is not a comprehensive technology, as some of them are not fully developed; even so, some countries are still using it. This is because the benefit of biological remediation technology is very attractive. As biological treatment does not require extra material such as catalyst or chemical compound, only microorganisms or bacteria are used to enhance the process of remediation. The technology require low operating and capital costs compared to physical and chemical remediation technology. This is because the technology utilize the media from the environment such as microorganisms and bacteria. However, there is some limitation for biological techniques such as bioremediation, which is suitable to remove biodegradable pollutants only. Although there are a lot of advantages, the pollutants that can be remediated are very limited, unlike chemical and physical remediation technology. Apart from that, physical or chemical treatments are rapid and cost efficient if compared to the biological treatment. On-site technology is cheaper than off-site technology. On-site technology is suitable for a small region of remediation and shallow aquifer, whereas offsite technology is more suitable for large remediation areas and deep aquifer. The expenses of off-site technology are much higher than on-site technology as it require transport cost and excavation fee (Guba et al. 2015, Kuppusamy et al. 2016, Ye et al. 2019).

6.   CONCLUSION

Groundwater is an important natural resource that is commonly used for commercial, domestic, agricultural and industrial purposes all over the world, but it is not clean and pure in recent years as it acquires contaminants from its surroundings and those arising from animals, humans and other biological activities. Human activities such as heavy use of fertilizer and pesticide, unplanned industrialization, excessive use of groundwater and uncontrolled industrial waste disposal are also responsible for the groundwater contamination. The presence of heavy metals in groundwater is usually related to human industrial activities, where vertical displacement and leaching of heavy metals may happen in the soil profile, thus contaminating the groundwater. The toxicity of heavy metals that are present in groundwater pose a major threat to the environment and human beings, thus posing a serious challenge to environmental engineers due to the treatment of wastewater effluents prior to discharge into nearby water sources such as groundwater. To prevent pollution or contamination from deteriorating, major technologies such as physical, chemical or biological have to keep improving. Most of the conventional methods are unable to remove the heavy metals in groundwater, so advanced technology such as air stripping, permeable reactive barriers and so on are introduced to treat water more efficiently. Engineers and researchers should identify and analyze the pollutant in underground water before appropriate technology is applied for water treatment. Physical groundwater remediation is more recommended as it suitable for *in situ* and *ex situ* and can also combine with biological or chemical remediation technology, although the expenses will be higher. In conclusion, humans should be aware about the importance of the environment to us and to biodiversity and take action on it before the pollution is irreparable. This is because water is one of the basic needs of humans to survive and the huge population relies on groundwater to have clean drinking water as the water above ground is polluted because of human activities. Many ecosystems and species depend on groundwater for some or all of their water supply.

References

Abeysinghe, D.H., D.G. De Silva, D.A. Stahl and B.E. Rittmann. 2002. The effectiveness of bioaugmentation in nitrifying systems stressed by a washout condition and cold temperature. Water Environ. Res. 74(2): 187–199.

Aeschbach-Hertig, W. and T. Gleeson. 2012. Regional strategies for the accelerating global problem of groundwater depletion. Nat. Geosci. 5(12): 853–861.

Akbal, F. and F. Camcı. 2011. Copper, chromium and nickel removal from metal plating wastewater by electrocoagulation. Desalination 269(1): 214–222.

Al-Saleh, I. and I. Al-Doush I. 1998. Survey of trace elements in household and bottled drinking water samples collected in Riyadh, Saudi Arabia. Sci. Total Environ. 216(3): 181–192.

Almeida, J.C., C.E.D. Cardoso, D.S. Tavares, R. Freitas, T. Trindade, C. Vale and E. Pereira. 2019. Chromium removal from contaminated waters using nanomaterials – A review. TrAC, Trends Anal. Chem. 118: 277–291.

Ambashta, R.D. and M. Sillanpaa. 2010. Water purification using magnetic assistance: A review. J. Hazard. Mat. 180(1–3): 38–49.

An, D., B. Xi, J. Ren, Y. Wang, X. Jia, C. He and Z. Li. 2017. Sustainability assessment of groundwater remediation technologies based on multi-criteria decision making method. Resour. Conser. Recycling. 119: 36–46.

Antoszczyszyn, T. and A. Michalska. 2016. The potential risk of environmental contamination by mercury contained in Polish coal mining waste. J. Sustain. Mining. 15(4): 191–196.

Arroyo, M.G., V. Perez-Herranz, M.T. Montanes, J. Garcia-Anton and J.L. Guinon. 2009. Effect of pH and chloride concentration on the removal of hexavalent chromium in a batch electrocoagulation reactor. J. Hazard. Mater. 169(1-3): 1127–1133.

Begum, M., J. Horowitz and M.I. Hossain. 2015. Low-dose risk assessment for arsenic: A meta-analysis approach. Asia Pac. J. Public Health. 27(2): Np 20–35

Beiyuan, J., D.C. Tsang, A.C. Yip, W. Zhang, Y.S. Ok and X.D. Li. 2017. Risk mitigation by waste-based permeable reactive barriers for groundwater pollution control at e-waste recycling sites. Environ. Geochem. Health. 39(1): 75–88

Bouzekri, S., M.L. El Hachimi, N. Touach, H. El Fadili, M. El Mah and E.M. Lotfi. 2019. The study of metal (As, Cd, Pb, Zn and Cu) contamination in superficial stream sediments around of Zaida mine (High Moulouya-Morocco). J. African Earth Sci. 154: 49–58.

Bundschuh, J., M.I. Litter, F. Parvez, G. Román-Ross, H.B. Nicolli, J.S. Jean, et al. 2012. One century of arsenic exposure in Latin America: A review of history and occurrence from 14 countries. Sci. Total Environ. 429: 2–35.

Buschmann, J., M. Berg, C. Stengel, L. Winkel, M.L. Sampson, P.T.K Trang and P.H. Viet. 2008. Contamination of drinking water resources in the Mekong delta floodplains: Arsenic and other trace metals pose serious health risks to population. Environ. Int. 34(6): 756–764.

Chen, A.Y.Y. and T. Olsen. 2016. Chromated copper arsenate-treated wood: a potential source of arsenic exposure and toxicity in dermatology. Int. J. Women's Dermatol. 2(1): 28–30.

Chen, P., Y. Ruan, S. Wang, X. Liu and B. Lian. 2017. Effects of organic mineral fertiliser on heavy metal migration and potential carbon sink in soils in a karst region. Acta Geochimica. 36(3): 539–543.

Clarkson, T.W. and L. Magos. 2006. The toxicology of mercury and its chemical compounds. Crit. Rev. Toxicol. 36(8): 609–662.

Cumbal, L., P. Vallejo, B. Rodriguez and D. Lopez. 2010. Arsenic in geothermal sources at the north-central Andean region of Ecuador: concentrations and mechanisms of mobility. Environ. Earth Sci. 61(2): 299–310.

Demirak, A., F. Yilmaz, A. Levent and N. Ozdemir. 2006. Heavy metals in water, sediment and tissues of Leuciscus cephalus from a stream in southwestern Turkey. Chemosphere 63(9): 1451–1458.

Di Natale, F., M. Di Natale, R. Greco, A. Lancia, C. Laudante and D. Musmarra. 2008. Groundwater protection from cadmium contamination by permeable reactive barriers. J. Hazard. Mat. 160(2–3): 428–434.

Doni, S., C. Macci, E. Peruzzi, R. Iannelli and G. Masciandaro. 2015. Heavy metal distribution in a sediment phytoremediation system at pilot scale. Ecol. Eng. 81: 146–157.

Eberts, S.M. 2014. If groundwater is contaminated, will water from the well be contaminated? Ground Water. 52(Suppl 1): 3–7.

Eschauzier, C., K.J. Raat, P.J. Stuyfzand and P. De Voogt. 2013. Perfluorinated alkylated acids in groundwater and drinking water: Identification, origin and mobility. Sci. Total Environ. 458–460: 477–485.

Etteieb, S., S. Magdouli, M. Zolfaghari and S. Brar. 2020. Monitoring and analysis of selenium as an emerging contaminant in mining industry: A critical review. Sci. Total Environ. 698: 134–139.

Fang, F., H. Han, Q. Zhao, C. Xu and L. Zhang. 2013. Bioaugmentation of biological contact oxidation reactor (BCOR) with phenol-degrading bacteria for coal gasification wastewater (CGW) treatment. Biores. Technol. 150: 314–320.

Freije, A. and M. Awadh. 2009. Total and methylmercury intake associated with fish consumption in Bahrain. Water Environ. J. 23(2): 155–164.

Friberg, L. 1950. Health hazards in the manufacture of alkaline accumulators with special reference to chronic cadmium poisoning: A clinical and experimental study. Acta Med. Scand. Suppl. 240: 1–12.

Galal, T.M., E.M. Eid, M.A. Dakhil and L.M. Hassan. 2018. Bioaccumulation and rhizofiltration potential of *Pistia stratiotes* L. for mitigating water pollution in the Egyptian wetlands. Int. J. Phytoremed. 20(5): 440–447.

Georgi, A., A. Schierz, K. Mackenzie and F.D. Kopinke. 2015. Colloidal activated carbon for in-situ groundwater remediation – Transport characteristics and adsorption of organic compounds in water-saturated sediment columns. J. Contam. Hydrol. 179: 76–88.

Ghoreyshi, A.A., H. Sadeghifar and F. Entezarion. 2014. Efficiency assessment of air stripping packed towers for removal of VOCs (volatile organic compounds) from industrial and drinking waters. Energy. 73: 838–843.

Girginova, P.I., A.L. Daniel-Da-Silva, C.B. Lopes, P. Figueira, M. Otero, V.S. Amaral, et al. 2010. Silica coated magnetite particles for magnetic removal of Hg^{2+} from water. J. Colloid Intrf. Sci. 345(2): 234–240.

Golder, A.K., A.N. Samanta and S. Ray. 2007. Removal of Cr^{3+} by electrocoagulation with multiple electrodes: bipolar and monopolar configurations. J. Hazard. Mat. 141(3): 653–661.

Gottesfeld, P., F.H. Were, L. Adogame, S. Gharbi, D. San, M.M. Nota, et al. 2018. Soil contamination from lead battery manufacturing and recycling in seven African countries. Environ. Res. 161: 609–614.

Guba, S., V. Somogyi and E. Szabóné Bárdos. 2015. Groundwater remediation using biological and photocatalytic methods. Hung. J. Ind. Chem. 43(1): 30–43.

Guo, J. and Y. Zhou. 2020. Transformation of heavy metals and dewaterability of waste activated sludge during the conditioning by Fe^{2+}-activated peroxymonosulfate oxidation combined with rice straw biochar as skeleton builder. Chemosphere 238: 124–128.

Hamidian, A.H., N. Razeghi, Y. Zhang and M. Yang. 2019. Spatial distribution of arsenic in groundwater of Iran, a review. J. Geochem. Explor. 201: 88–98.

Harvey, P.J., H.K. Handley and M.P. Taylor. 2016. Widespread copper and lead contamination of household drinking water, New South Wales, Australia. Environ. Res. 151: 275–285.

Hayes, F., D.J. Spurgeon, S. Lofts and L. Jones. 2018. Evidence-based logic chains demonstrate multiple impacts of trace metals on ecosystem services. J. Environ. Manage. 223: 150–164.

Hou, D., G. Li and P. Nathanail. 2018. An emerging market for groundwater remediation in China: Policies, statistics, and future outlook. Front. Environ. Sci Eng. 12(1): 16–21.

Idrees, N., B. Tabassum, E.F. Abd Allah, A. Hashem, R. Sarah and M. Hashim. 2018. Groundwater contamination with cadmium concentrations in some West U.P. Regions, India. Saudi J. Biol. Sci. 25(7): 1365–1368.

Izah, S.C., I.R. Inyang, T.C.N. Angaye and I.P. Okowa. 2016. A Review of heavy metal concentration and potential health implications of beverages consumed in Nigeria. Toxics. 5(1): 11–16.

Jane Wyatt, C., C. Fimbres, L. Romo, R.O. Méndez and M. Grijalva. 1998. Incidence of heavy metal contamination in water supplies in Northern Mexico. Environ. Res. 76(2): 114–119.

Jarup, L., M. Berglund, C.G. Elinder, G. Nordberg and M. Vahter. 1998. Health effects of cadmium exposure—A review of the literature and a risk estimate. Scand. J. Work Environ. Health. 24(Suppl 1): 1–51.

Jia, X., D. O'connor, D. Hou, Y. Jin, G. Li, C. Zheng, et al. 2019. Groundwater depletion and contamination: Spatial distribution of groundwater resources sustainability in China. Sci. Total Environ. 672: 551–562.

Jung, J.H., J.H. Lee and S. Shinkai. 2011. Functionalized magnetic nanoparticles as chemosensors and adsorbents for toxic metal ions in environmental and biological fields. Chem. Soc. Rev. 40(9): 4464–4474.

Kaur, T., R. Bhardwaj and S. Arora. 2017. Assessment of groundwater quality for drinking and irrigation purposes using hydrochemical studies in Malwa region, southwestern part of Punjab, India. App. Water Sci. 7(6): 3301–3316.

Komnitsas, K., G. Bartzas, K. Fytas and I. Paspaliaris. 2007. Long-term efficiency and kinetic evaluation of ZVI barriers during clean-up of copper containing solutions. Miner. Eng. 20(13): 1200–1209.

Kříbek, B., I. Nyambe, V. Majer, I. Knésl, M. Mihaljevič, V. Ettler, et al. 2019. Soil contamination near the Kabwe Pb–Zn smelter in Zambia: Environmental impacts and remediation measures proposal. J. Geochem. Explor. 197: 159–173.

Krishna, A.K., M. Satyanarayanan and P.K. Govil. 2009. Assessment of heavy metal pollution in water using multivariate statistical techniques in an industrial area: A case study from Patancheru, Medak District, Andhra Pradesh, India. J. Hazard. Mater. 167(1): 366–373.

Kuppusamy, S., T. Palanisami, M. Megharaj, K. Venkateswarlu and R. Naidu. 2016. In-Situ remediation approaches for the management of contaminated sites: a comprehensive overview. Rev. Environ. Contam. Toxicol. 236: 11–15.

Kwon, M.J., E.J. O'loughlin, B. Ham, Y. Hwang, M. Shim and S. Lee. 2018. Application of an in-situ soil sampler for assessing subsurface biogeochemical dynamics in a diesel-contaminated coastal site during soil flushing operations. J. Environ. Manage. 206: 938–948.

Lee, M. and M. Yang. 2010. Rhizofiltration using sunflower (*Helianthus annuus* L.) and bean (*Phaseolus vulgaris* L. var. vulgaris) to remediate uranium contaminated groundwater. J. Hazard. Mater. 173(1): 589–596.

Lee, Y.C., E.J. Kim, D.A. Ko and J.W. Yang. 2011. Water-soluble organo-building blocks of aminoclay as a soil-flushing agent for heavy metal contaminated soil. J. Hazard. Mater. 196: 101–108.

Leyssens, L., B. Vinck, C. Van Der Straeten, F. Wuyts and L. Maes. 2017. Cobalt toxicity in humans— A review of the potential sources and systemic health effects. Toxicol. 387: 43–56.

Li, C.P., G.X. Li, Y.M. Luo and Y.F. Li. 2008. Fuzzy mathematics-based groundwater quality evaluation of six MSW landfills in Beijing. Huan Jing Ke Xue. 29(10): 2729–2735.

Li, P., X.B. Feng, G.L. Qiu, L.H. Shang and Z.G. 2009. Mercury pollution in Asia: A review of the contaminated sites. J. Hazard. Mater. 168(2): 591–601.

Liao, F., G. Wang, Z. Shi, X. Huang, F. Xu, Q. Xu, et al. 2018. Distributions, sources, and species of heavy metals/trace elements in shallow groundwater around the Poyang Lake, East China. Exposure Health 10(4): 211–227.

Liu, L., S.P. Hu, Y.X. Chen and H. Li 2010. Feasibility of washing as a remediation technology for the heavy metals-polluted soils left by chemical plant. Ying Yong Sheng Tai Xue Bao, 21(6): 1537–1541.

Liu, C.S., M.C. Kuo, C.Y. Su, Y.C. Chen, W.C. Cheng, C.Y. Chou, et al. 2013. A bacteria injection scheme for in situ bioaugmentation. J. Environ. Sci. Health, Part A 48(9): 1079–1085.

Liu, H., T.A. Bruton, F.M. Doyle and D.L. Sedlak. 2014. In situ chemical oxidation of contaminated groundwater by persulfate: Decomposition by Fe(III)- and Mn(IV)-containing oxides and aquifer materials. Environ. Sci. Technol. 48(17): 10330–10336.

Lorestani, B., N. Yousefi, M. Cheraghi and A. Farmany. 2013. Phytoextraction and phytostabilization potential of plants grown in the vicinity of heavy metal-contaminated soils: A case study at an industrial town site. Environ. Monit. Assess. 185(12): 10217–10223.

Lu, A.H., E.L. Salabas and F. Schuth. 2007. Magnetic nanoparticles: Synthesis, protection, functionalization, and application. Angew. Chem. Int. Ed. Engl., 46(8): 1222–1244.

Lu, Q., X. Xu, L. Liang, Z. Xu, L. Shang, J. Guo, et al. 2019. Barium concentration, phytoavailability, and risk assessment in soil-rice systems from an active barium mining region. Appl. Geochem., 106: 142–148.

Luo, X., H. Liu, G. Huang, Y. Li, Y. Zhao and X. Li. 2016. Remediation of arsenic-contaminated groundwater using media-injected permeable reactive barriers with a modified montmorillonite: Sand tank studies. Environ. Sci. Poll. Res. Int. 23(1): 870–877.

Ma, Y., M. Rajkumar, Y. Luo and H. Freitas. 2013. Phytoextraction of heavy metal polluted soils using *Sedum plumbizincicola* inoculated with metal mobilizing *Phyllobacterium myrsinacearum* RC6b. Chemosphere 93(7): 1386–1392.

Mandour, R.A. and Y.A. Azab. 2011. The prospective toxic effects of some heavy metals overload in surface drinking water of Dakahlia Governorate, Egypt. The Int. J. Occup. Env. Med. 2(4): 245–253.

Marzougui, A. and A. Ben Mammou. 2006. Impacts of the dumping site on the environment: Case of the Henchir El Yahoudia Site, Tunis, Tunisia. C. R. Geosci. 338(16): 1176–1183.

Mays, D.C. and T.D. Scheibe. 2018. Groundwater contamination, subsurface processes, and remediation methods: Overview of the special issue of water on groundwater contamination and remediation. Water. 10(12): 170817 https://doi.org/10.3390/w10121708

Megdal, S.B. 2018. Invisible water: The importance of good groundwater governance and management. npj Clean Water. 1(1): 15

Mendelsohn, A.L., B.P. Dreyer, A.H. Fierman, C.M. Rosen, L.A. Legano, H.A. Kruger, et al. 1998. Low-level lead exposure and behavior in early childhood. Pediatrics, 101(3): E10 DOI: 10.1542/peds.101.3.e10

Mironyuk, I., T. Tatarchuk, H. Vasylyeva, M. Naushad and I. Mykytyn. 2019. Adsorption of Sr(II) cations onto phosphated mesoporous titanium dioxide: Mechanism, isotherm and kinetics studies. J. Environ. Chem. Eng. 7(6): 103430–103435.

Mohamed, A.M., N. El-Menshawy and A.M. Saif. 2007. Remediation of saturated soil contaminated with petroleum products using air sparging with thermal enhancement. J. Environ. Manage. 83(3): 339–350.

Muruganandham, M., R.P.S. Suri, S. Jafari, M. Sillanpaa, G.J. Lee, J.J. Wu, et al. 2014. Recent developments in homogeneous advanced oxidation processes for water and wastewater treatment. Int. J. Photoenergy, 2014: 21–25.

Nawab, J., S. Khan, M.A. Khan, H. Sher, U.U. Rehamn, S. Ali, et al. 2017. Potentially toxic metals and biological contamination in drinking water sources in chromite mining-impacted areas of Pakistan: A comparative study. Expos. Health 9(4): 275–287.

Needleman, H.L. 1993. The current status of childhood low-level lead toxicity. Neurotoxicology 14(2-3): 161–166.

Nejadshafiee, V and M.R. Islami. 2019. Adsorption capacity of heavy metal ions using sultone-modified magnetic activated carbon as a bio-adsorbent. Mat. Sci. Eng. 101: 42–52.

Nelson, M., T. Adams, C. Ojo, M.A. Carroll and E.J. Catapane. 2018. Manganese toxicity is targeting an early step in the dopamine signal transduction pathway that controls lateral cilia activity in the bivalve mollusc *Crassostrea virginica*. Comp. Biochem. Physiol. C Toxicol. Pharmacol. 213: 1–6.

Nemati, M., S.M. Hosseini and M. Shabanian. 2017. Novel electrodialysis cation exchange membrane prepared by 2-acrylamido-2-methylpropane sulfonic acid: Heavy metal ions removal. J. Hazardous Mat. 337: 90–104.

Nishijo, M., H. Nakagawa, Y. Morikawa, M. Tabata, M. Senma, K. Miura, et al. 1995. Mortality of inhabitants in an area polluted by cadmium: 15 year follow up. Occup. Environ. Med. 52(3): 181–184.

Nowierski, M., D.G. Dixon and U. Borgmann. 2006. Lac Dufault sediment core trace metal distribution, bioavailability and toxicity to *Hyalella azteca*. Environ. Pollu. 139(3): 532–540.

Obiri-Nyarko, F., S.J. Grajales-Mesa and G. Malina. 2014. An overview of permeable reactive barriers for in situ sustainable groundwater remediation. Chemosphere 111: 243–259.

Orisakwe, O.E., I.O. Igwilo, O.J. Afonne, J.M. Maduabuchi, E. Obi and J.C. Nduka. 2006. Heavy metal hazards of sachet water in Nigeria. Arch. Environ. Occup. Health. 61(5): 209–213.

Owa, F.D. 2013. Water pollution: Sources, effects, control and management. Mediterr. J. Soc. Sci. 4(8): 65–68.

Payne, R.B., S.K. Fagervold, H.D. May and K.R. Sowers. 2013. Remediation of polychlorinated biphenyl impacted sediment by concurrent bioaugmentation with anaerobic halorespiring and aerobic degrading bacteria. Environ. Sci. Technol. 47(8): 3807–3815.

Pleasant, S., A. O'donnell, J. Powell, P. Jain and T. Townsend. 2014. Evaluation of air sparging and vadose zone aeration for remediation of iron and manganese-impacted groundwater at a closed municipal landfill. Sci. Total Environ. 485-486: 31–40.

Pulford, I.D. and C. Watson. 2003. Phytoremediation of heavy metal contaminated land by trees— A review. Environ. Int. 29(4): 529–540.

Regadío, M., A.I. Ruiz, I.S. De Soto, M. Rodriguez Rastrero, N. Sánchez, M.J. Gismera, et al. 2012. Pollution profiles and physicochemical parameters in old uncontrolled landfills. Waste Manage. 32(3): 482–497.

Rodrigues, S.M., B. Henriques, A.T. Reis, A.C. Duarte and P.F.M. Pereira. 2012. Hg transfer from contaminated soils to plants and animals. Environ. Chem. Lett. 10(1): 61–67.

Rubio, C., Paz, S., E. Tius, A. Hardisson, A.J. Gutierrez, D. Gonzalez-Weller, et al. 2018. Metal contents in the most widely consumed commercial preparations of four different medicinal plants (Aloe, Senna, Ginseng, and Ginkgo) from Europe. Biol. Trace Elem. Res. 186(2): 562–567.

Sale, P.F., D.A. Feary, J.A. Burt, A.G. Bauman, G.H. Cavalcante, K.G. Drouillard, et al. 2011. The growing need for sustainable ecological management of marine communities of the Persian Gulf. Ambio 40(1): 4–17.

Sallsten, G., J. Thoren, L. Barregard, A. Schutz and G. Skarping. 1996. Long-term use of nicotine chewing gum and mercury exposure from dental amalgam fillings. J. Dent. Res. 75(1): 594–598.

Salonen, J.T., K. Seppanen, K. Nyyssonen, H. Korpela, J. Kauhanen, M. Kantola, et al. 1995. Intake of mercury from fish, lipid peroxidation, and the risk of myocardial infarction and coronary, cardiovascular, and any death in eastern Finnish men. Circulation. 91(3): 645–655.

Sawyerr, H.O., A.T. Adeolu, A.S. Afolabi, O.O. Salami and B.K. Badmos. 2017. Impact of dumpsites on the quality of soil and groundwater in satellite towns of the federal capital territory, Abuja, Nigeria. J Health Pollut. 7(14): 15–22.

Schaumlöffel, D. 2012. Nickel species: Analysis and toxic effects. J. Trace Elem. Med Biol. 26(1): 1–6.

Shariful, M.I., S.B. Sharif, J.J.L. Lee, U. Habiba, B.C. Ang and M.A. Amalina. 2017. Adsorption of divalent heavy metal ion by mesoporous-high surface area chitosan/poly (ethylene oxide) nanofibrous membrane. Carbohydr. Polym. 157: 57–64.

Siddiqui, S.I., M. Naushad and S.A. Chaudhry. 2019. Promising prospects of nanomaterials for arsenic water remediation: A comprehensive review. Process Saf. Environ. Prot. 126: 60–97.

Simeonov V., J.A. Stratis, C. Samara, G. Zachariadis, D. Voutsa, A. Anthemidis, et al. 2003. Assessment of the surface water quality in Northern Greece. Water Res. 37(17): 4119–4124.

Sizirici, B. and B. Tansel. 2015. Parametric fate and transport profiling for selective groundwater monitoring at closed landfills: A case study. Waste Manage. 38: 263–270.

Sprocati, A.R., C. Alisi, F. Tasso, P. Marconi, A. Sciullo, V. Pinto, et al. 2012. Effectiveness of a microbial formula, as a bioaugmentation agent, tailored for bioremediation of diesel oil and heavy metal co-contaminated soil. Process Biochem. 47(11): 1649–1655.

Srinivasa Gowd, S. and P.K. Govil. 2008. Distribution of heavy metals in surface water of Ranipet industrial area in Tamil Nadu, India. Environ. Monit. Assess. 136(1): 197–207.

Talabi, A.O. and T.J. Kayode. 2019. Groundwater pollution and remediation. J. Water Resource Prot. 11(1): 19–23.

Tamasi, G. and R. Cini. 2004. Heavy metals in drinking waters from Mount Amiata (Tuscany, Italy). Possible risks from arsenic for public health in the Province of Siena. Sci. Total Environ. 327(1): 41–51.

Vardhan, K.H., P.S. Kumar and R.C. Panda. 2019. A review on heavy metal pollution, toxicity and remedial measures: Current trends and future perspectives. J. Mol. Liq. 290: 111197.

Vasconcelos Neto M.C.D., T.B.C. Silva, V.E.D. Araújo and S.V.C.D. Souza. 2019. Lead contamination in food consumed and produced in Brazil: Systematic review and meta-analysis. Food Res. Int. 126: 108671.

Vodela, J.K., S.D. Lenz, J.A. Renden, W.H. Mcelhenney, B.W. Kemppainen. 1997. Drinking water contaminants (arsenic, cadmium, lead, benzene, and trichloroethylene). 2. Effects on reproductive performance, egg quality, and embryo toxicity in broiler breeders. Poult Sci. 76(11): 1493–1500.

Wang, Z.Q., Q. Wu, Z.G. Zou, H. Chen, X.C. Yang and J.C. Zhao. 2007. Study on the groundwater petroleum contaminant remediation by air sparging. Huan Jing Ke Xue, 28(4): 754–760.

Weiss, B., T.W. Clarkson and W. Simon. 2002. Silent latency periods in methylmercury poisoning and in neurodegenerative disease. Environ. Health Perspect. 110(Suppl 5): 851–854.

Weyens, N., D. Van Der Lelie, K. Artois, K. Smeets, S. Taghavi, L. Newman, et al. 2009. Bioaugmentation with engineered endophytic bacteria improves contaminant fate in phytoremediation. Environ. Sci. Technol. 43(24): 9413–9418.

Wu, G.H. and S.S. Cao. 2010. Mercury and cadmium contamination of irrigation water, sediment, soil and shallow groundwater in a wastewater-irrigated field in Tianjin, China. Bull. Environ. Contam. Toxicol. 84(3): 336–341.

Yan, X., Q. Liu, J. Wang and X. Liao. 2017. A combined process coupling phytoremediation and in situ flushing for removal of arsenic in contaminated soil. J. Environ. Sci. (China) 57: 104–109.

Yang, Y., M. Watanabe, X. Zhang, J. Zhang, Q. Wang and S. Hayashi. 2006. Optimizing irrigation management for wheat to reduce groundwater depletion in the piedmont region of the Taihang Mountains in the North China Plain. Agric. Water Manage. 82(1): 25–44.

Ye, J., X. Chen, C. Chen and B. Bate. 2019. Emerging sustainable technologies for remediation of soils and groundwater in a municipal solid waste landfill site—A review. Chemosphere 227: 681–702.

Yoshizawa, K., E.B. Rimm, J.S. Morris, V.L. Spate, C.C. Hsieh, D. Spiegelman, et al. 2002. Mercury and the risk of coronary heart disease in men. N. Engl. J. Med. 347(22): 1755–1760.

Yu, J.J. and S.Y. Chou. 2000. Contaminated site remedial investigation and feasibility removal of chlorinated volatile organic compounds from groundwater by activated carbon fiber adsorption. Chemosphere 41(3): 371–378.

Zhang, S., Z. Hou, X.M. Du, D.M. Li and X.X. Lu. 2016. Assessment of biostimulation and bioaugmentation for removing chlorinated volatile organic compounds from groundwater at a former manufacture plant. Biodegradation 27(4-6): 223–236.

Zhao, F.J. 2018. Soil and human health. Eur. J. Soil Sci. 69(1): 158–158.

Zietz, B.P., J. Lass and R. Suchenwirth. 2007. Assessment and management of tap water lead contamination in Lower Saxony, Germany. Int. J. Environ. Health Res. 17(6): 407–418.

Zulfiqar, U., M. Farooq, S. Hussain, M. Maqsood, M. Hussain, M. Ishfaq, et al. 2019. Lead toxicity in plants: Impacts and remediation. J. Environ. Manage. 250: 109557. doi: 10.1016/j.jenvman.2019.109557

Application of Plant-microbe Interactions in Contaminated Agro-ecosystem Management

Leila El-Bassi

[1]Laboratory of Wasstewater and Environment,
Center of Water Research and Technologies, Borj Cedria Ecopark-Tunisia.

1. INTRODUCTION

Increasing industrialization, mechanized agriculture practices, and the needs of population have released huge amounts of chemical contaminants in the environment (Arslan et al. 2015). The FAO (2009) reported that by 2050, the agricultural production will have to increase by 70% to respond the world's agriculture need (Altieri 2004). The agricultural practices, relying on the extensive use of pestcides and agrochemical fertilizers, also lead to environmental disturbances (Paul and Lade 2014). Previous research underlined the consequences of organic and metallic contaminants on soils (Pinedo et al. 2013, Teng et al. 2014), plants (Copaciu et al. 2016), and humans (Cachada et al. 2012, Malchi et al. 2014).

Under starving or limiting conditions, plants have developed an array of biological and structural mechanisms to combat stresses. Recently, plant-microbe association has gained greater attention because of the beneficial effects for the crop growth and fertility of the agro-ecosystems. Plant-microbe associations include a wide range of interdependent mechanisms, from obligately symbiotic to purely saprotrophic microorganisms (Fester et al. 2014).

In this chapter, a summary of the significance of plant-microbe interactions in the agro-ecosystem will be discussed. Therefore, their eventual use as biofertilizer, biocontrol agent, phytoremedior and biopesticide as well as the risk and the limitation of their applications, is reported.

---

*Corresponding author: l.elbassi@gmail.com

2. MICRO-ORGANISMS IN DEPOLLUTION OF CONTAMINATED ENVIRONMENTS

Depollution of contaminated environments using microbes is widely considered as an eco-freindly method for biotransformation of contaminants and intensive research is being conducted to improve its performance (Abhilash et al. 2012, Van Aken et al. 2010). The environmental clean-up using microorganisms involves the bioconversion of organic xenobiotics into carbon dioxide, water and a byproduct resulting from the specific selected microorganisms' activities (Hlihor et al. 2017a,b, Sarkar 2018). The microorganisms' action implies principally the trans-formation of xenobiotics as part of their metabolic processes.

Bioremediation may be functional under aerobic or anerobic conditions and it was involved in the stabilization and the mineralization of organic pollutants. Anaerobic microbes have shown their efficiency to degrade or transform recalcitrant polluants like polychlorinated biphenyls, polyaromatic hydrocarbons, dechlorinate trichloroethylene and chloroform (Passatore et al. 2014, Hlihor et al. 2017c). However, under aerobic conditions, environmental bacteria such as *Rhodococcus, Pseudomonas, Alcaligenes, Sphingomonas*, and *Mycobacterium* were able to degrade pesticides and hydrocarbons (Vergani et al. 2017, El-Bassi et al. 2010, 2009).

2.1 Bioremediation with Engineered Microbes

In the 1990, the genetic engineering was applied to enhance the bioremediation yield and promote the microbial efficiency when applied to clean-up the contaminated environments. Previous studies have shown that inoculating a genetically modified *Pseudomonas* sp. B13 strain significantly enhances the degradation of chlorotoluene and its substitutes (Brinkmann and Reineke 1992, Abril et al. 1989). *Deinococcus radiodurans*, first characterized as radiation-tolerant microbe, was genetically modified and oriented for toluene degradation (Lang and Wullbrandt 1999, Ezezika and Singer 2010).

However, many attempts to apply and commercialize this technique have failed due to several factors such as public acceptability, formal approval before being released in natural environment (soil, water) and associated risk evaluation.

2.2 Bioremediation Application Using Recombinant Technology

The metabolism of microbial communities used for bioremediation can be quite slow despite its efficiency in eliminating hazardous pollutants (Liu et al. 2017). The advances in biotechnological techniques such as recombinant DNA methods or natural gene transfer, and bacterial engineering could be performed with the aim to promote the secretion of specific enzymes involved in the biotransformation of toxic organic substances (Pandotra et al. 2018).

Germaine et al. (2009) reported that the naphthalene degradation rate increased to 40% after inoculation of VM1441 (pNAH7) on pea plants (Germaine et al. 2009). A natural endophyte *Burkholderia cepacia* G4 recombined with a plasmid (pTOM) encoding enzymes promote the biotransformation of toluene in the yellow lupine plant (Barac et al. 2004).

Four *Pseudomonas* strains, showing oil biodegradation performances, have genes carried on bacterial extra-chromosomal plasmids, which codes for degradation enzymes synthesis were detected. Previous studies (Chebbi et al. 2017, Gao et al. 2017), showed that isolation of involved plasmids and their introduction in *Pseudomonas* spp., promote the oil degradation efficiencies by 10–100 times compared to wild strains.

PCBs, forming a class of organic aromatic compound, are used worldwide for a variety of applications. PCBs accumulating in the soil can be harmful for human and environmental health. It was reported that the biphenyl dioxygenase enzyme, produced by *Burkholderia cepacia* strain LB400, *C. testosterone* B-365, and *Rhodococcus globerulus* P6 (Barriault et al. 2002) is able to degrade PCBs in presence of oxygen (Ang et al. 2005). Genetically modified strains of *Pseudomonas fluorescens* F113rifbph and *Pseudomonas fluorescens* F113:1180 strains positively impacted the degradation capacity of PCBs compared with wildtype rhizospheric bacteria (Brazil et al. 1995, Villacieros et al. 2005). Furthermore, *in situ* applications using the genetically modified strain *Pseudomonas fluorescens* F113:1180 were successful (Villacieros et al. 2005).

The introduction of the plasmid PC3 coding for hydrolytic para-dechlorination gene or the plasmid E43 coding for oxygenolytic ortho-dechlorination gene into *Comamonas testosterone* strain VP44 was able to cometabolite 95% of ortho and para-chlorobiphenyls (Hrywna et al. 1999).

3. PHYTOREMEDIATION OF ORGANIC CONTAMINANTS

Phytoremediation is a set of applications using plants for environmental detoxification to remove organic compound and metallic contaminants from soils and water (Schnoor et al. 1995, Raskin et al. 1997, Yan et al. 2020). It has been considered as an essential tool to decontaminate soil, water and air by extracting, sequestering contaminants, detoxifying and/or hyperaccumulating pollutants (Heinekamp and Willey 2007). Plants metabolize the xenobiotic compounds through three steps pathway including activation, conjugation and compartmentalization, which involves mostly enzymes such as cytochrome P450, N-glucosyltransferase and glutathione S-transferase (Vergani et al. 2017).

Phytoremediation can have different forms: phytoextraction, phytostabilization, rhizofiltration, phytodegradation, phytorestauration and phytovolatilization (Ozyigit and Dogan 2015).

Phytoextraction acts by absorption of heavy metals by roots and their transportation to upper parts of plants, decreasing soil metal concentrations (Zhang et al. 2010, Ali et al. 2013). In phytoextraction processes, several plants were considered; for instance, alfalfa (*Medicago sativa*) and maize (*Zea mays*) for Hg, Ni and Pb decontamination (EPA 2000), sunflower (*Helianthus annuus*) for Cs and Sr (Adler 1996), and Indian mustard (Brassica juncea) for As, B, Cd, Cr(VI), Cu, Ni, Pb, Se, Sr and Zn (Salido et al. 2003).

Phytostabilization process uses the ability of the plant to immobilize the metal contaminant and reduces its hazard (Arthur et al. 2005). The phytostabilization acts by sorption, precipitation and complexation. The most commonly known plants used during this process are *Eucalyptus urophylla* and *Eucalyptus saligna* used for Zn stabilization (Magalhães et al. 2011); *Vigna unguiculata* used for Pb and Zn complexation (Kshirsagar and Aery 2007); *Sorghum* sp. used for Cd, Cu Ni, Pb and Zn phytostabilization (Jadia and Fulekar 2008).

Under phytovolatilization, the pollutants are uptaken and released by transpiration under their initial form or after transformation (Susarla et al. 2002, Ali et al. 2013). For instance, Pilon-Smits et al. (1999) and Arthur et al. (2005) reported the Iris-leaved rush (*Juncus xiphioides*), cattail (*Typha latifolia*) and club-rush (*Scirpus robustus*) to be Selinium (Se) phytoremediator, which perform by phytovolatilization (Pilon-Smits et al. 1999, Arthur et al. 2005).

4. POTENTIAL OF PLANT ASSOCIATED MICROORGANISMS IN BIOREMEDIATION

Soil indigenous microorganisms can assimilate organic xenobiotics as carbon source, sulfur source and other energy sources, with the presence of corresponding substrates. Naturally, plants get along with indigenous microbes in the same ecosystem and their association significantly impact the sustainable agriculture (Table 1) (Newton et al. 2010). To make an effective communication and mutualistic symbiosis between microbes and plants, various signals were secreted (Badri et al. 2009).

Table 1 Plant-endophytic microbe associations for organic pollutant bioremediation in soils

Case study	Bacteria	Associated plant	Pollutant
Siciliano and Germida 1999	*Pseudomonas aeruginosa* (R75)	*Elymus dauricus*	Chlorobenzoic acids
Germaine et al. 2009	*Pseudomonas putida*	*Pisum sativum*	Naphtalene
Weyens et al. 2010a	*Pseudomonas putida* (W619-TCE)	*populus trichocarpa*	Trichloroethylene
Afzal et al. 2012	*Pseudomonas* sp. ITRI53	*Lolium mutiforum*	TPH
Germaine et al. 2006	*Pseudomonas putida* (VM1450)	Pisum sativum	Pseticide-2,4-D
Becerra-Castro et al. 2013	*Sphingomonas* sp. (D4)	Cytisusstriatus	Hexachlorocyclohexane
Eevers et al. 2016	*Sphingomonas* sp. *Stenotrophomonas* sp.	*Cucurbita pepo*	DDE and DDT
Barac et al.2004	*Burkholderia* (G4)	*Lupinus luteus*	VOCs and toluene
Wang et al. 2010	*Burkholderia cepacia* (FX2)	*Zea mays*	Toluene
Weyens et al. 2013	*Bulkholderia* sp. (HU001)	*salix* spp.	Toluene
Kang et al. 2012	*Enterobacter* sp. (PDN3)	*Populus trichocarpa*	BTEX
Sheng et al. 2008	*Enterobacter* sp. (12J1)	*Triticum aestivum* *Zea Mays*	Pyrene
Yousaf et al. 2011	*Enterobacter ludwigii* strain	*Medicago sativa* *Lotus carniculatus* *Lolium multiforum*	TPH
Mannisto et al. 2001	*Herbaspirillum* sp. (K1)	*Triticum aestivum*	PCBs, TCP
Jabeen et al. 2016	*Mesorhizobium* sp. (HN3yfp)	*Lolium multiforum*	Chlorpyrifos
Weyens et al. 2010b	*Bacillus cepacia* (VM1468)	*Populus trichocarpa*	TCE
Yousaf et al. 2010	*Pantoea* sp. Strain (ITS10)	*Lolium multiforum*	TPH
Ho et al. 2009	*Achromobacter xylosoxidans*	*Phragmites australis,* *Ipomoea aquatica* *Vetiveria zizanioides*	Phenol and Catechol

4.1 Plant-microbe Interaction as Biofertilizers

Plant-microbe association plays an important role in maintaining a sustainable agriculture because of their use as biofertilizers in the agro-ecosystems and their effects on promoting plant growth. The biofertilizer effect of the plant-microbe occurs via different mechanisms, mainly, improving nitrogen uptake process (Khan 2005, Sahin et al. 2004), promoting the bioavailability of organic and inorganic phosphate (Bashan and de-Bashan 2002), and inducting the production of growth regulators such as indole acetic acid (IAA), cytokinins, and gibberellins (Glick 1995, Marques et al. 2010). It was prevously reported that the genera belonging to azorhizobium, bradyrhizobium, mesorhizobium, rhizobium and sinorhizobium have shown an efficient biofertilization effect (Singh et al. 2019).

Nitrogen fixation by plant-microbe symbiosis: Nitrogen is the key nutrient required to promote plant growth. The N is actively involved in mostly all biochemical and physiological mechanisms in the plant (Krapp 2015). N_2 is considered as the key element directly influencing the crop productivity (Sainju 2013). However, it is worth mentioning that an excessive use of nitrogenous chemical fertilizers leads to huge environmental impacts, from ecosystem sustainability to terrestrial and aquatic eutrophication and acidification (Galloway et al. 2008, Yang et al. 2014).

Biological Nitrogen Fixation (BNF) represents the main way for N_2 fixation and contributes globally 180×10^6 tonnes of fixed N provided every year. About 80% of BNF need for plant growth is the result of interactive association of plant-microbe and the rest is provided by free-living systems (Graham 1988). Many reports have been interested in the plant-microbe association for N_2 fixation; *Azospirillum* sp. establishes symbiotic associations with Zea mays (De Salamone et al. 1996); *Bacillus polymyxa* with *Triticum aestivum* (Omar et al.1996); *Herbaspirillum* sp. with *Oryza sativa* (James et al. 2002) and *Azotobacter aceae* with *Fagopyrum esculentum* (Bhattacharyya and Jha 2012).

Phosphate solubilization: Phosphorus is one of the key elements in the main metabolic processes of the plant like photosynthesis, respiration, energy transfer, signal transduction, and enzymes biosynthesis (Anand et al. 2016). 95% of the phosphate present in the soils is insoluble and immobilized and not so accessible to plants. Phosphate solubilizing bacteria operates by enhancing the bioavailability of phosphorus and its transformation as monobasic ($H_2PO_4^-$) and dibasic ($HPO4_2^-$).

Phosphate solubilizing bacteria working in association witn plants belong to genera Bacillus, Burkholderia, *Enterobacter, Arthrobacter, Pseudomonas, Beijerinckia, Microbacterium, Erwinia, Rhizobium, Mesorhizobium, Serratia, Rhodococcus,* and *Flavobacterium* (Oteino et al. 2015). Successful associations between microbes and plants promoting the P solubilization have been previously reported like *Azotobacter chroococcum* and *Triticuma estivum* (Bhattacharyya and Jha 2012) and between Rhizobium Leguminosarum and Phaseolus vulgaris (Ahemad and Kibret 2014).

Siderophore production: Under iron-starving conditions, siderophores produced by microorganisms are able to improve the iron uptake capacity of plants (Li et al. 2016b). Generally, iron exists under the ferric ions form (Fe^{3+}), hardly assimilated by plants (Ammari and Mengel 2006). The association of microbe-plants is used to overcome this iron starvation by siderophores production. Bacterial strains like *Pseudomonas putida* have the ability to use siderophores produced by different microbes to assimilate iron in limiting conditions and to enhance the iron level in the soil (Rathore 2014).

Up to date, more than 500 siderophores were determined and 270 among them have a known chemical structure (Hider and Kong 2010). Beneduzzi et al. (2012) demonstrated that the ferric-siderophore complex enhances the iron uptake potential in the presence of metal like nickel and cadmium (Beneduzzi et al. 2012). However, considerable research should be done to determine the bacterial ability to produce siderophores.

4.2   Plant-microbe Interaction in Phytoremediation

The phytoremediation can be defined as a process by which the use of plants, including grasses and trees, to sequester, remove, transform or destroy hazardous contaminants from soil, water and air, is performed (Prasad and Freitas 2003). Phytoremediation can operate under different forms and specific mechanisms could be used on each of phytoremediation process (Ozyigit and Dogan 2015).

However, a practical and cost effective technology using the beneficial interaction between plants and microbial flora for environmental remediation has gained importance (Table 2) (Salt et al. 1998).

Table 2   Plant-microbe associations in phytoremediation

Plant	Microbe	Heavy metal	Associated effect	Reference
Brassica Juncea, Brassica napus, Solanum lycopersicum	*Kluyvera ascorbata*	Zn, Ni, Pb	Promote ACC deaminase and IAA	Burd et al. (2000)
Sorghum vulgare	*Glomus mossea, Glomus claroideum, Glomus caledonium*	Cu	Promoted uptake rate of Cu	Gonzalez-Chavez et al. (2002)
Zea mays	*Scutellospora gilmorei, Gigaspora decipiens, Glomus caledonium, Gigaspora margarita*	Cu, Zn, Pb, Cd	Promoted the extraction efficiencies of heavy metals; increase of P uptake	Wang et al. (2007)
Thlaspi praecox	*Rhodotorula aurantiaca Phialophora verrucosa, Rhizoctonia* sp., *Penicillium* sp.	Zn, Cd, Pb	Incresead metalic adsorption by plants	Pongrac et al. (2009)
Chrysopogon zizanioides	*Glomus mosseae*	Pb	Enhanced metalic adsorption by plant	Punaminiya et al. (2010)
Oenothera picensis	*Glomus claroideum*	Cu	Increased heavy metal uptake by plant	Meier et al. (2011)
Cloezia Artensis, Alphitonia neocaledonica	*Glomus etunicatum*	Ni	Increased P uptake and heavy metal uptake	Amir et al. (2013)
Polygonum pubescens	*Rahnella* sp. JN6	Cd, Pb, Cu, and Zn	ACC deaminase	He et al. (2013)
Solanum nigrum	*Pseudomonas* sp. Lk9	Cd, Zn, Cu, and Cr	Enhanced organic acid production	Chen et al. (2014)
Alyssum pintodasilvae	*Arthrobacter nicotinovorans* SA40	Ni	Increased P uptake; enhanced IAA and siderophore production	Cabello-Conejo et al. (2014)
Cymbopogon citratus	*Rhizophagus clarus*	Pb	Improved the production of essential oils	Lermen et al. (2015)
Pinus densiora, Quercus variabilis	*Laccaria laccata, Cenococcum geophilum, Pisolithus* sp.	Cu	Decreased heavy metal accumulation in shoots	Zong et al. (2015)
Vigna radiata	*Exiguobacterium* sp.	As	Increased P uptake; enhanced indole-3-acetic acid and exopolysaccharide (EPS) production	Pandey and Bhatt (2016)
Sorghum sudanense	*Enterobacter* sp. K3-2	Cu	Improved IAA, siderophores, ACC deaminase production	Li et al. (2016a)
Raphanus sativus	*Stenotrophomonas* sp.CIK-517Y and *Bacillus* sp. CIK-516	Ni	Enhanced IAA and 1-aminocyclopropane-1-carboxylate deaminase potentials	Akhtar et al. (2018)

4.3 Plant-microbe Interaction: Symbiosis versus Pathogenesis

Pathogenic microorganisms represent a chronic and major threat to sustainable agriculture (Prasad et al. 2019). Regular use of pesticides has caused pathogen resistance. The plant-microbe association can be regarded as an alternative to replace the excessive use of pesticides.

Numerous mechanisms are involved in biocontrol action as an important tool used in sustainable agriculture (Vimal et al. 2012). The biocontrol action operates by direct action via the production of antibiotics, enzymes (chitinases, lipases, proteases, etc.), and siderophores, or indirect action by competing with the pathogen for habitation (Lugtenberg and Kamilova 2009). (Table 3).

Pseudomonas and *Bacillus* are known to act through their antibiosis mechanisms (Jayaprakashvel and Mathivanan 2011).

Table 3 Plant-microbe associations in pathogens control

Plant	*Microbe*	*Pathogens*	*Resulting effect*	*References*
Hordeum vulgare	*Glomus mosseae*	*Gaeumannomyces graminis*	Pathogens inhibition caused by mycorrhize colonization	Khaosaad et al. (2007)
Lycopersicon esculentum	*Glomus mosseae, Glomus intraradices*	*Phytophthora nicotianae*	Resistance to pathogens	Lioussanne et al. (2009)
Phaseolus vulgaris	*Glomus mosseae, Glomus intraradices, Glomus clarum*	*Fusarium solani*	Defensive enzyme activities	Al-Askar and Rashad (2010)
Lycopersicon esculentum	*Glomus mosseae*	*Meloidogyne incognita, Pratylenchus penetrans*	Reducing disease development	Vos et al. (2012)
Helianthus tuberosus	*Trichoderma harzianum, Glomus clarum*	*Sclerotium rolfsii*	Limitations in disease incidence	Sennoi et al. (2013)
Inula conyza, Conyza canadensis, Solidago virgaurea, Solidago gigantean Senecio vernalis, Senecio inaequidens	*Glomus etunicaTum, Glomus claroideum, Glomus intraradices, Glomus mosseae, Glomus geosporum*	*Pythium ultimum*	Delay disease appearance	Fabbro and Prati (2014)
Nicotiana tabacum	*Trichoderma harzianum, Glomus mosseae*	*Ralstonia solanacearum*	Improve plant resistance and plant growth	Yuan et al. (2016)

4.4 Plant-microbe Interaction: Biocontrol Agent

The application of the plant-microbe interaction as control agent aims to prevent the plant disease in order to limit the high doses of chemical pesticides and fungicides. As biocontrol agent, a multitude of bacterial and final genera were used such us: *Pseudomonas, Bacillus, Streptomyces, Fusarium* and *Gliocladium* (Kloepper et al. 2004, Bakker et al. 2014, Kumar et al. 2017).

The biocontrol agents generally act by limiting the growth and the toxic potential of diseases or pathogenic organisms. The direct effect of biocontrol agent occurs via the nutrients' competition, degradation of enzymes or toxic secondary metabolites, or inducing the resistance systems in the host plant (Xu et al. 2014, Hao et al. 2014, Singh and Siddiqui 2015).

Table 4   Plant-microbe interaction as biocontrol agents against plant disease

Biocontrol agent	Plant	Disease	Reference
Bacillus amyloliquefaciens strain IN937a, *Bacillus pumilus* strain SE34, *Bacillus subtilis* strain IN937b	Cucumber	Cucumber mosaic virus	Zehnder et al. (2000)
Bacillus subtilis 937b, *Bacillus pumilus* SE34, *Bacillus amyloliquefaciens* 937a	Tomato	Tomato mottle virus	Murphy et al. (2000)
B. pumilus strain INR7	Cucumber	Bacterial wilt	Zehnder et al. (2001)
Pseudomonas sp.	White clover Medicago	Acyrthosiphon Kondoi	Kempster et al. (2002)
Streptomyces marcescens strain	Tobacco	Blue mold	Zhang et al. (2002)
Pseudomonas syringae	Tomato	Bacterial speck of tomato	Bashan and e-Bashan (2002)
Bacillus cereus strains	Tomato	Foliar diseases	Silva et al. (2004)
Ralstonia solanacearum	Tomato	Bacterial wilt of tomato	Guo et al. (2004)
Alternaria triticina	Wheat	Leaf blight of wheat	Siddiqui and Singh (2005)
Bacillus spp. strains	Bell pepper	Blight of bell pepper	Jiang et al. (2006)
Bacillus subtilis strain	Cucumber, Pepper	Soilborne pathogens	Chung et al. (2008)
Burkholderia strains	Maize	Maize rot	Hernandez-Rodriguez et al. (2008)
Azospirillum strains	Rice	Rice blast	Naureen et al. (2009)
Pseudomonas fluorescens strain	Banana	Banana bunchy	Kavino et al. (2010)
Bacillus amyloliquefaciens and *Bacillus subtilis*	Crops	Infection and fungal diseases	Siahmoshteh et al. (2017)

5.   APPLICATION OF PLANT-MICROBE INTERACTIONS

Currently, it has been well demonstrated that the association of plant-microbe promotes the crop productivity via various modes of actions. However, this association is highly dependant on environmental factors, which affect the micro-organisms' growth and consequently their impact on the plants. The environmental factors could range from the indigenous microbial diversity to soil specificities and climate conditions (Gupta et al. 2015).

On sustainable agriculture approach, the applications of plant-microbe associations are mainly concerned with the treatment of the young plants by potential bacterial or fungal strains as biocontrol agents, as biofertilization methods, are also carried out to clean-up organic polluted environment. The recent advances in biological tools tend to use molecular and sub-molecular units of micro-organisms (Bourras et al. 2015). Furthermore, the beneficial plant-microbe interactions leading to the overexpression of secondary metabolites such us antibiotics and volatile oils could be used for medical and economical benefits (Maag et al. 2015).

Plant-microbe interaction studies should emphasize on selecting potential microbes and use multidisciplinary research advances in agro-biotechnology, molecular biology, and nanobiotechnology to provide optimal formulations and opportunities.

5.1 Risks and limitations of Plant-microbe Associations

Although microbial use in bioremediation to remove environmental pollutants is known to be an eco-freindly approach and cost-effective technique, many microbial organisms have indesirable impacts.

For instance, the strain *Burkholderia cepacia,* which has considerable bioremediation effect on toxic nitro-compounds' biotransformation, presents multiple antibiotic resistances. Furthermore, the *Pseudomonas* strain, involved in toluene biodegradation, pump out various antibiotics and biocides (Davison 2005). Thus, such bacteria may increase or produce more virulent progeny.

Furthermore, the use of genetic engeneering and microbial recombinant in bioremediation might have an effect on ecosystem disruption after the degradation of the pollutant (Khan et al. 2016, Prakash et al. 2011).

To control these risks, several possibilities were considered like transposition vectors without antibiotic-resistant genes (Chamekh, 2015).

6. CONCLUSION

Bioremediation and urgent environmental clean-up is highly recommended to maintain a sustainable agro-ecosystem. Taking into consideration the beneficial aspects of plant-microbe interactions, it is evident that it could be an attractive approach for sustainable agriculture. Apart its cost effictiveness and eco-freindly aspect, this association has proved its vital role in soil fertility and crop productivity as biofertilizer, biocontrol and bioremedation agent. However, this symbiosis is highly dependent on various factors like climate change, soil specificities and indigenous bacterial communities that potentially impact the efficiency of the corresponding plant-microbe interaction. In this approach, supplementary research have to be performed to better understand the mechanisms involved with the aim to find the best plant-microbe interaction and to overcome its limits.

References

Abhilash, P.C., J.R. Powell, H.B Singh and B.K Singh. 2012. Plant–microbe interactions: Novel applications for exploitation in multipurpose remediation technologies. Trends Biotechnol. 30: 416–420.

Abril, M.A., C. Michan, K.N. Timmis and J.L. Ramos. 1989. Regulator and enzyme specificities of the TOL plasmid-encoded upper pathway for degradation of aromatic hydrocarbons and expansion of the substrate range of the pathway. J. Bacteriol. 171: 6782–6790.

Adler, T. 1996. Botanical cleanup crews. Sci. News. 150: 42–43.

Afzal, M., S. Yousaf, T.G. Reichenauer and A. Sessitsch. 2012. The inoculation method affects colonization and performance of bacterial inoculant strains in the phytoremediation of soil contaminated with diesel oil. Int. J. Phytoremed. 14: 35–47.

Ahemad, M. and M. Kibret. 2014. Mechanisms and applications of plant growth promoting rhizobacteria: current perspective. J. King Saud Univ. Sci. 26: 120.

Akhtar, M.J., S. Ullah, I. Ahmad, A. Rauf, S.M. Nadeem, M.Y. Khan, et al. 2018. Nickel phytoextraction through bacterial inoculation in *Raphanus sativus*. Chemosphere 190: 234–242.

Al-Askar, A.A. and Y.M. Rashad. 2010. Arbuscular mycorrhizal fungi: A biocontrol agent against common bean Fusarium root rot diseases. Plant Pathol. 9: 31–38.

Ali, H., E. Khan and M.A. Sajad. 2013. Phytoremediation of heavy metals-concepts and applications. Chemosphere 91: 869–881.

Altieri, M.A. 2004. Linking ecologists and traditional farmers in the search for sustainable agriculture. Front. Ecol. Environ. 2(1): 35–42.

Amir, H., A. Lagrange, N. Hassaine and Y. Cavaloc. 2013. Arbuscular mycorrhizal fungi from New Caledonian ultramafic soils improve tolerance to nickel of endemic plant species. Mycorrhiza 23: 585–595.

Ammari, T. and K. Mengel. 2006. Total soluble Fe in soil solutions of chemically different soils. Geoderma 136(3): 876–885.

Anand, K., B. Kumari and M.A. Mallick. 2016. Phosphate solubilizing microbes: An effective and alternative approach as biofertilizers. Int. J. Pharma. Pharmaceut. Sci. 8(2): 37–40.

Ang, E.L., H. Zhao and J.P. Obbard. 2005. Recent advances in the bioremediation of persistent organic pollutants via biomolecular engineering. Enzym. Microb. Technol. 37: 487–496.

Arslan, M., A. Imran, Q.M. Khan and M. Afzal. 2015. Plant-bacteria partnerships for the remediation of persistent organic pollutants. Environ. Sci. Pollut. Res. 24(5): 4322–4336.

Arthur, E.L., Pamela J. Rice, Patricia J. Rice, T.A. Anderson, S.M. Baladi, K.L.D. Henderson, et al. 2005. Phytoremediation—An overview. Crit. Rev. Plant Sci. 24: 109–122.

Badri, D.V., T.L. Weir, D. van der Lelie and J.M. Vivanco. 2009. Rhizosphere chemical dialogues: Plant-microbe interactions. Curr. Opin. Biotechnol. 20(6): 642–650.

Bakker, P.A., L. Ran and J. Mercado-Blanco. 2014. Rhizobacterial salicylate production provokes headaches. Plant Soil. 382(1-2): 1–16.

Barac, T., S. Taghavi, B. Borremans, A. Provoost, L. Oeyen, J.V. Colpaert, et al. 2004. Engineered endophytic bacteria improve phytoremediation of water-soluble, volatile, organic pollutants. Nat. Biotechnol. 22: 583–588.

Barriault, D., M.M. Plante and M. Sylvestre. 2002. Family shuffling of a targeted *bphA* region to engineer biphenyl dioxygenase. J. Bacteriol. 184: 3794–3800.

Bashan, Y. and L.E. de-Bashan. 2002. Protection of tomato seedlings against infection by *Pseudomonas syringae* pv. tomato by using the plant growth-promoting bacterium *Azospirillum brasilense*. Appl. Environ. Microbiol. 68(6): 2637–2643.

Becerra-Castro, C., P.S. Kidd, B. Rodríguez-Garrido, C. Monterroso, P. Santos-Ucha and A. Prieto-Fernández. 2013. Phytoremediation of hexachlorocyclohexane (HCH)-contaminated soils using *Cytisus striatus* and bacterial inoculants in soils with distinct organic matter content. Environ. Pollut. 178 : 202–210.

Beneduzi, A., A. Ambrosini and L.M. Passaglia. 2012. Plant growth-promoting rhizobacteria (PGPR): Their potential as antagonists and biocontrol agents. Genet. Mol. Biol. 35(4): 1044–1051.

Bhattacharyya, P.N. and D.K. Jha. 2012. Plant growth-promoting rhizobacteria (PGPR): Emergence in agriculture. World J. Microbial. Biotechnol. 28: 1327–1350.

Bourras, S., T. Rouxel and M. Meyer. 2015. *Agrobacterium tumefaciens* gene transfer: How a plant pathogen hacks the nuclei of plant and nonplant organisms. Phytopathol. 105 (10) : 1288–1301.

Brazil, G.M., L. Kenefick, M. Callanan, A. Haro, V. De Lorenzo, D.N. Dowling, et al. 1995. Construction of a rhizosphere pseudomonad with potential to degrade polychlorinated biphenyls and detection of bph gene expression in the rhizosphere. Appl. Environ. Microbiol. 61: 1946–1952.

Brinkmann, U. and W. Reineke. 1992. Degradation of chlorotoluenes by in vivo constructed hybrid strains: problems of enzyme specificity, induction and prevention of metapathway. FEMS Microbiol. Lett. 96: 81–87.

Burd, G.I., D.G. Dixon and B.R. Glick. 2000. Plant growth-promoting bacteria that decrease heavy metal toxicity in plants. Can. J. Microbiol. 46(3): 237–245.

Cabello-Conejo, M.I., C. Becerra-Castro, A. Prieto-Fernández, C. Monterroso, A. Saavedra-Ferro, M. Mench, et al. 2014. Rhizobacterial inoculants can improve nickel phytoextraction by the hyperaccumulator *Alyssum pintodasilvae*. Plant Soil. 379(1-2): 35–50.

Cachada, A., P. Pato, T. Rocha-Santos, E.F. da Silva and A.C. Duarte. 2012. Levels, sources and potential human health risks of organic pollutants in urban soils. Sci. Total Environ. 430: 184–192.

Chamekh, M. 2015. Genetically engineered bacteria in gene therap hopes and challenges. *In*: D. Hashad (ed.), Gene Therapy-Principles and Challenges. InTech, Croatia, pp. 145–162.

Chebbi, A., D. Hentati, H. Zaghden, N. Baccar, F. Rezgui, M. Chalbi, et al. 2017. Polycyclic aromatic hydrocarbon degradation and biosurfactant production by a newly isolated *Pseudomonas* sp. strain from used motor oil-contaminated soil. Int. Biodeterior. Biodegrad. 122: 128–140.

Chen, L., S. Luo, X. Li, Y. Wan, J. Chen and C. Liu. 2014. Interaction of Cd hyperaccumulator *Solanum nigrum* L. and functional endophyte *Pseudomonas* sp. Lk9 on soil heavy metals uptake. Soil Biol. Biochem. 68: 300–308.

Chung, S., H. Kong, J.S. Buyer, D.K. Lakshman, J. Lydon, S.D. Kim, et al. 2008. Isolation and partial characterization of Bacillus subtilis ME488 for suppression of soilborne pathogens of cucumber and pepper. Appl. Microbiol. Biotechnol. 80(1): 115–123.

Copaciu, F., O. Opriş, Ü. Niinemets and L. Copolovici. 2016. Toxic influence of key organic soil pollutants on the total flavonoid content in wheat leaves. Water Air Soil Pollut. 227(6): 196. doi:10.1007/s11270-016-2888-x.

Davison, J. 2005. Risk mitigation of genetically modified bacteria and plants designed for bioremediation. J. Ind. Microbiol. Biotechnol. 32(11-12): 639–650.

De Salamone G, I.E., J. Dobereiner, S. Urquiaga and R.M. Boddey. 1996. Biological nitrogen fixation in *Azospirillum* strain-maize genotype associations as evaluated by the ^{15}N isotope dilution technique. Biol:Fertil:Soils 23: 249–256.

Eevers, N., J.R. Hawthorne, J.C. White, J. Vangronsveld and N. Weyens. 2016. Exposure of *Cucurbita pepo* to DDE-contamination alters the endophytic community: A cultivation dependent vs a cultivation independent approach. Environ. Pollut. 209: 147–154.

El-Bassi, L., N. Shinzato, T. Namihira, H. Oku and T. Matsui. 2009. Biodegradation of thiodiglycol, a hydrolyzate of the chemical weapon yperite, by benzothiophene-desulfurizing bacteria. J. Hazard. Mat. 167: 124–127.

El-Bassi, L., H. Iwasaki, H. Oku, N. Shinzato and T. Matsui. 2010. Biotransformation of benzothiazole derivatives by the *Pseudomonas putida* strain HKT554. Chemosphere 81: 109–113.

EPA. 2000. Introduction to phytoremediation. Environmental Protection Agency (EPA) Report EPA/600/R-99/107. US Environmental Protection Agency, Cincinnati.

Ezezika, O.C. and P.A. Singer. 2010. Genetically engineered oil-eatingmicrobes for bioremediation: Prospects and regulatory challenges. Technol. Soc. 32: 331–335.

Fabbro, C.D. and D. Prati. 2014. Early responses of wild plant seedlings to arbuscular mycorrhizal fungi and pathogens. Basic Appl Ecol. 15: 534–542.

Fester, T., J. Gielber, L.Y. Wick, D. Scholsser and M. Kastner. 2014. Plant–microbe interactions as drivers of ecosystem functions relevant for the biodegradation of organic contami nants. Curr. Opin. Biotechnol. 27: 168–175.

Galloway, J.N., A.R. Townsend, J.W. Erisman, M. Bekunda, Z. Cai, J.R. Freney, et al. 2008. Transformation of the nitrogen cycle: Recent trends, questions, and potential solutions. Science 320: 889–892.

Gao, G., A.S. Clare, C. Rose and G.S. Caldwell. 2017. Ulva rigida in the future ocean: Potential for carbon capture, bioremediation, and biomethane production. GCB Bioenergy 10, 39–51, doi:10.1111/gcbb.12465

Germaine, K.J., E. Keogh, D. Ryan and D.N. Dowling. 2009. Bacterial endophyte-mediated naphthalene phytoprotection and phytoremediation. FEMS Microbiol. Lett. 296: 226–234.

Germaine, K.J., X. Liu, G.C. Cabellos, J.P. Hogan, D. Ryan and D.N. Dowling. 2006. Bacterial endophyte-enhanced phytoremediation of the organochlorine herbicide 2,4-dichlorophenoxyacetic acid: Bacterial endophyte-enhanced phytoremediation. FEMS Microbiol. Ecol. 57: 302–310.

Glick, B.R. 1995. The enhancement of plant growth by freeliving bacteria. Can. J. Microbiol. 4: 1109–1114.

Gonzalez-Chavez, C., J. D'Haen, J. Vangronsveld and J.C. Dodd. 2002. Copper sorption and accumulation by the extraradical mycelium of different *Glomus* spp. (arbuscular mycorrhizal fungi) isolated from the same polluted soil. Plant Soil 240: 287–297.

Guo, J.H., H.Y. Qi, Y.H. Guo, H.L. Ge, L.Y. Gong, L.X. Zhang, et al. 2004. Biocontrol of tomato wilt by plant growth-promoting rhizobacteria. Biol. Control. 29(1): 66–72.

Gupta, G., S.S. Parihar, N.K. Ahirwar, S.K. Snehi and V. Singh. 2015. Plant growth promoting Rhizobacteria (PGPR): Current and future prospects for development of sustainable agriculture. J. Microbiol. Biochem. Tech. 7: 96–102.

Graham, P.H. 1988. Principles and Application of Soil Microbiology, Prentice Hall, Upper Sadele River, pp. 322–345.

Hao, K., J.Y. Liu, F. Ling, X.L. Liu, L. Lu, L. Xia, et al. 2014. Effects of dietary administration of *Shewanella haliotis* D4, *Bacillus cereus* D7 and *Aeromonas bivalvium* D15, single or combined, on the growth, innate immunity and disease resistance of shrimp, *Litopenaeus vannamei*. Aquaculture 428: 141–149.

He, H., Z. Ye, D. Yang, J. Yan, L. Xiao, T. Zhong, et al. 2013. Characterization of endophytic *Rahnella* sp. JN6 from Polygonum pubescens and its potential in promoting growth and Cd, Pb, Zn uptake by *Brassica napus*. Chemosphere 90(6): 1960–1965.

Heinekamp, Y. and N. Willey. 2007. Using real-time polymerase chain reaction to quantify gene expression in plants exposed to radioactivity. *In*: N. Willey (ed.), Phytoremediation: Methods and Reviews. Humana Press, Totowa, pp. 59–70.

Hernandez-Rodriguez, A., M. Heydrich-Pérez, Y. Acebo-Guerrero, M.G. Velazquez-Del Valle and A.N. Hernandez-Lauzardo. 2008. Antagonistic activity of Cuban native rhizobacteria against *Fusarium verticillioides* (Sacc.) Nirenb. in maize (*Zea mays* L.). Appl. Soil Ecol. 39(2): 180–186.

Hider, R.C. and X. Kong. 2010. Chemistry and biology of siderophores. Nat. Prod. Rep. 27(5): 637–657.

Hlihor, R.M., L.C. Apostol and M. Gavrilescu. 2017a. Environmental bioremediation by biosorption and bioaccumulation: principles and applications. *In*: N.A. Anjum, S.S. Gill and N. Tuteja (eds), Enhancing Cleanup of Environmental Pollutants. Springer International Publishing, pp. 289–315.

Hlihor, R.M., M. Gavrilescu, T. Tavares, L. Favier and G. Olivieri. 2017b. Bioremediation: An overviewon current practices, advances, and new perspectives in environmental pollution treatment. Biomed. Res. Int. 2017.

Hlihor, R.M., H. Figueiredo, T. Tavares and M. Gavrilescu. 2017c. Biosorption potential of dead and living Arthrobacter viscosus biomass in the removal of Cr (VI): Batch and column studies. Process. Saf. Environ. Prot. 108: 44–56.

Ho, Y.N., C.H. Shih, S.C. Hsiao and C.C. Huang. 2009. A novel endophytic bacterium, *Achromobacter xylosoxidans*, helps plants against pollutant stress and improves phytoremediation. J. Biosci. Bioeng. 108: 9–17.

Hrywna, Y., T.V. Tsoi, O.V. Maltseva, J.F. Quensen and J.M. Tiedje. 1999. Construction and characterization of two recombinant bacteria that grow on ortho-and parasubstituted chlorobiphenyls. Appl. Environ. Microbiol. 65: 2163–2169.

Jabeen, H., S. Iqbal, F. Ahmad, M. Afzal and S. Firdous. 2016. Enhanced remediation of chlorpyrifos by ryegrass (*Lolium multiflorum*) and a chlorpyrifos degrading bacterial endophyte *Mezorhizobium* sp. HN3. Int. J. Phytoremediat. 18: 126–133.

Jadia, C.D. and M.H. Fulekar. 2008. Phytoremediation: The application of vermicompost to remove zinc, cadmium, copper, nickel and lead by sunflower plant. Environ. Eng. Manag. J. 7: 547–558.

James, E.K., P. Gyaneshwar, N. Mathan, W.L. Barraquio, P.M. Reddy, P.P.M. Iannetta, et al. 2002. Infection and colonization of rice seedlings by the plant growth promoting bacterium *Herbaspirillum seropedicae* Z67. Mol. Plant Microbe. Interact. 15: 894–906.

Jayaprakashvel, M. and N. Mathivanan. 2011. Management of plant diseases by microbial metabolites. *In*: D.K. Maheshwari (ed.), Bacteria in Agrobiology: Plant Nutrient Management. Springer, Berlin, Heidelberg, pp. 237–265.

Jiang, Z.Q., Y.H. Guo, S.M. Li, H.Y. Qi and J.H. Guo. 2006. Evaluation of biocontrol efficiency of different *Bacillus* preparations and field application methods against Phytophthora blight of bell pepper. Biol. Control 36(2): 216–223.

Kang, J.W., Z. Khan and S.L. Doty. 2012. Biodegradation of trichloroethylene by an endophyte of hybrid poplar. Appl. Environ. Microbiol. 78: 3504–3507.

Kavino, M., S. Harish, N. Kumar, D. Saravanakumar and R. Samiyappan. 2010. Effect of chitinolytic PGPR on growth, yield and physiological attributes of banana (*Musa* spp.) under field conditions. Appl. Soil Ecol. 45(2): 71–77.

Kempster, V.N., E.S. Scott and K.A. Davies. 2002. Evidence for systemic, cross-resistance in white clover (*Trifolium repens*) and annual medic (*Medicago truncatula* var *truncatula*) induced by biological and chemical agents. Biocon. Sci. Technol. 12(5): 615–623.

Khan, A.G. 2005. Role of soil microbes in the rhizospheres of plants growing on trace metal contaminated soils in phytoremediation. J. Trace Elem. Med. Biol. 18: 355–364.

Khan, S., M.W. Ullah, R. Siddique, G. Nabi, S. Manan, M. Yousaf, et al. 2016. Role of recombinant DNA technology to improve life. Int. J. Genom. Article ID 2405954 https://doi.org/10.1155/2016/2405954

Khaosaad, T., J.M. Garcia-Garrid, S. Steinkellner and H. Vierheilig. 2007. Take-all disease is systemically reduced in roots of mycorrhizal barley plants. Soil Biol. Biochem. 39: 727–734.

Kloepper, J.W., C.M. Ryu and S. Zhang. 2004. Induced systemic resistance and promotion of plant growth by *Bacillus* spp. Phytopathol. 94(11): 1259–1266.

Krapp, A. 2015. Plant nitrogen assimilation and its regulation: A complex puzzle with missing pieces. Curr. Opin. Plant. Biol. 25: 115–122.

Kshirsagar, S. and N.C. Aery. 2007. Phytostabilisation of mine waste: Growth and physiological responses of *Vigna unguiculata* (L.) Walp. J. Environ. Biol. 2: 651–654.

Kumar, A., H. Verma, V.K. Singh, P.P. Singh, S.K. Singh, W.A. Ansari, et al. 2017. Role of *Pseudomonas* sp. in sustainable agriculture and disease management. *In*: V.S. Meena, P.K. Mishra, J.K. Bisht, A. Pattanayak (eds), Agriculturally Important Microbes for Sustainable Agriculture. Springer, Singapore, pp. 195–215.

Lang, S. and D. Wullbrandt. 1999. Rhamnose lipids–biosynthesis, microbial production and application potential. Appl. Microbiol. Biotechnol. 51(1): 22–32.

Lermen, C., F. Morelli, Z.C. Gazim, P.A.D. Silva, J.E. Goncalves, D.C. Dragunski, et al. 2015. Essential oil content and chemical composition of *Cymbopogon citratus* inoculated with arbuscular mycorrhizal fungi under different levels of lead. Ind. Crop. Prod. 76: 734–738.

Li, Y., Q. Wang, L. Wang, L.Y. He and X.F. Sheng. 2016a. Increased growth and root Cu accumulation of *Sorghum sudanense* by endophytic *Enterobacter* sp. K3-2: implications for *Sorghum sudanense* biomass production and phytostabilization. Ecotoxicol. Environ. Saf. 124: 163–168.

Li, M., G.J. Ahammed, C. Li, X. Bao, J. Yu, C. Huang, et al. 2016b. Brassinosteroid ameliorates zinc oxide nanoparticles-induced oxidative stress by improving antioxidant potential and redox homeostasis in tomato seedling. Front. Plant Sci. 7: 615 doi: 10.3389/fpls.2016.00615.

Liu, S., N. Qureshi and S.R. Hughes. 2017. Progress and perspectives on improving butanol tolerance. World J. Microbiol. Biotechnol. 33(3): 51–57.

Lioussanne, L., M. Jolicoeur and M. St-Arnaud. 2009. Role of the modification in root exudation induced by arbuscular mycorrhizal colonization on the intraradical growth of *Phytophthora nicotianae* in tomato. Mycorrhiza 19: 443–448.

Lugtenberg, B. and F. Kamilova. 2009. Plant-growth-promoting rhizobacteria. Annu. Rev. Microbiol. 63: 541–556.

Malchi, T., Y. Maor, G. Tadmor, M. Shenker and B. Chefetz. 2014. Irrigation of root vegetables with treated wastewater: evaluating uptake of pharmaceuticals and the associated human health risks. Environ. Sci. Technol. 48: 9325–9333.

Männistö, M.K., M.A. Tiirola and J.A. Puhakka. 2001. Degradation of 2, 3, 4, 6-tetrachlorophenol at low temperature and low dioxygen concentrations by phylogenetically different groundwater and bioreactor bacteria. Biodegradation 12: 291–301.

Maag, D., M. Erb, T.G. Köllner and J. Gershenzon. 2015. Defensive weapons and defense signals in plants: Some metabolites serve both roles. Bioessays 37(2): 167–174.

Magalhães, M.O.L., N. Sobrinho, F.S. dos Santos and N. Mazur. 2011. Potential of two species of eucalyptus in the phytostabilisation of a soil contaminated with zinc. Rev. Ciênc. Agron. 42: 805–812.

Marques, A.P.G.C., C. Pires, H. Moreira, A.O.S.S. Rangel and P.M.L. Castro. 2010. Assessment of the plant growth promotion abilities of six bacterial isolates using *Zea mays* as indicator plant. Soil Biol. Biochem. 42: 1229–1235.

Meier, S. 2011. Contribution of metallophytes/arbuscular mycorrhizal symbiosis to the phytoremediation processes of copper contaminated soils. Ph.D. Thesis, Universidad de la Frontera.

Murphy, J.F., G.W. Zehnder, D.J. Schuster, E.J. Sikora, J.E. Polston and J.W. Kloepper, 2000. Plant growth-promoting rhizobacterial mediated protection in tomato against Tomato mottle virus. Plant Dis. 84(7): 779–784.

Naureen, Z., A.H. Price, F.Y. Hafeez and M.R. Roberts. 2009. Identification of rice blast disease-suppressing bacterial strains from the rhizosphere of rice grown in Pakistan. Crop Prot. 28(12): 1052–1060.

Newton, A.C., B.D. Fitt, S.D. Atkins, D.R. Walters and T.J. Daniell. 2010. Pathogenesis, parasitism and mutualism in the trophic space of microbe-plant interactions. Trends Microbiol. 18(8): 365–373.

Omar, M.N.A., N.M. Mahrous and A.M. Hamouda. 1996. Evaluating the efficiency of inoculating some diazatrophs on yield and protein content of 3 wheat cultivars under graded levels of nitrogen fertilization. Ann. Agric. Sci. 41: 579–590.

Oteino, N., R.D. Lally, S. Kiwanuka, A. Lloyd, D. Ryan, K.J. Germaine, et al. 2015. Plant growth promotion induced by phosphate solubilizing endophytic *Pseudomonas* isolates. Front. Microbiol. 6: 745–752.

Ozyigit, I.I. and I. Dogan. 2015. Plant-microbe interactions in pytoremediations. *In*: K. Rehman, H. Muhammad, S. Münir, Ö. Ahmet, R. Mermut (eds), Soil Remediation and Plants. Academic Press, pp. 255–285.

Pandey, N. and R. Bhatt. 2016. Role of soil associated Exiguobacterium in reducing arsenic toxicity and promoting plant growth in *Vigna radiata*. Eur. J. Soil. Biol. 75: 142–150.

Pandotra, P., M. Raina, R.K. Salgotra, S. Ali, Z.A. Mir, J.A. Bhat, et al. 2018. Plant-bacterial partnership: A major pollutants remediation approach. *In*: M. Oves, M.Z. Khan, I.M.I. Ismail (eds), Modern Age Environmental Problems and their Remediation. Springer, Cham, pp. 169–200.

Passatore, L., S. Rossetti, A.A. Juwarkar and A. Massacci. 2014. Phytoremediation and bioremediation of polychlorinated biphenyls (PCBs): State of knowledge and research perspectives. J. Hazard. Mater. 278: 189–202.

Paul, D. and H. Lade. 2014. Plant-growth-promoting rhizobacteria to improve crop growth in saline soils: A review. Agron. Sustain. Dev. 34: 737–752.

Pilon-Smits, E.A.H., M.P. de Souza, G. Hong, A. Amini, R.C. Bravo, S.T. Payabyab, et al. 1999. Selenium volatilisation and accumulation by twenty aquatic plant species. J. Qual. 28: 1011–1018.

Pinedo, J., R. Ibáñez, J.P.A. Lijzen and A. Irabien. 2013. Assessment of soil pollution based on tota petroleum hydrocarbons and individual oil substances. J. Environ. Manag. 130: 72–79.

Pongrac, P., S. Sonjak, S.K. Vogel-Miku, P. Kump, M. Nečemer and M. Regvar, 2009. Roots of metal hyperaccumulating population of *Thlaspi praecox* (Brassicaceae) harbour arbuscular mycorrhizal and other fungi under experimental conditions. Int. J. Phytoremediat. 11: 347–359.

Prakash, D., S. Verma, R. Bhatia and B.N. Tiwary. 2011. Risks and precautions of genetically modified organisms. ISRN Ecol. 2011: 1–13.

Prasad, M., R. Srinivasan, C. Manoj, M. Choudhary and L.K. Jat. 2019. Plant growth promoting rhizobacteria (PGPR) for sustainable agriculture: Perspectives and challenges. *In*: A.K. Singh, A. Kumar, P.K. Singh (eds), PGPR Amelioration in Sustainable Agriculture. Elsevier Inc., pp. 129–157.

Prasad, M.N.V. and H.M.D. Freitas. 2003. Metal hyperaccumulation in plants-biodiversity prospecting for phytoremediation technology. Electron. J. Biotechn. 6: 285–321.

Punaminiya, P., R. Datta, D. Sarkar, S. Barber, M. Patel and P. Das. 2010. Symbiotic role of Glomus mosseae in phytoextraction of lead in vetiver grass (*Chrysopogon zizanioides* L.). J. Hazard Mater. 177: 465–474.

Raskin, I., R.D. Smith and D.E. Salt. 1997. Phytoremediation of metals: Using plants to remove pollutants from the environment. Curr. Opin. Biotechnol. 8: 221–226.

Rathore, P. 2014. A review on approaches to develop plant growth promoting rhizobacteria. International J. Recent Sci. Res. 5: 403–407.

Sahin, F., R. Cakmakci and F. Kanta. 2004. Sugar beet and barley yields in relation to inoculation with N2-fixing and phosphate solubilizing bacteria. Plant Soil 265: 123–129.

Sainju, U.M. 2013. Tillage, cropping sequence, and nitrogen fertilization influence dryland soil nitrogen. Agron. J. 105: 1253–1263.

Salt, D.E., R.D. Smith and I. Raskin. 1998. Phytoremediation. Annu. Rev. Plant Physiol. 49: 643–668.

Salido, A.L., K.L. Hasty, J.M. Lim and D.J. Butcher. 2003. Phytoremediation of arsenic and lead in contaminated soil using Chinese Brake Ferns (*Pteris vittata*) and Indian mustard (*Brassica juncea*). Int. J. Phytoremediat. 5: 89–103.

Sarkar, S.K. 2018. Phytoremediation of Trace Metals by Mangrove Plants of Sundarban Wetland. *In*: S.K. Sarkar (ed.), Trace Metals in a Tropical Mangrove Wetland. Springer, Singapore, pp. 209–247.

Schnoor, J.L., L.A. Light, S.C. McCutcheon, N.L. Wolfe and L.H. Carreia. 1995. Phytoremediation of organic and nutrient contaminants. Environ. Sci. Technol. 29(7): 318A–323A.

Sennoi, R., N. Singkham, S. Jogloy, S. Boonlue, W. Saksirirat, T. Kesmala, et al. 2013. Biological control of southern stem rot caused by *Sclerotium rolfsii* using *Trichoderma harzianum* and arbuscular mycorrhizal fungi on Jerusalem artichoke (*Helianthus tuberosus* L.). Crop Prot. 54: 148–153.

Sheng, X., X. Chen and L. He. 2008. Characteristics of an endophytic pyrene-degrading bacterium of *Enterobacter* sp. 12J1 from *Allium macrostemon* Bunge. Int. Biodeterior. Biodegrad. 62: 88–95.

Siahmoshteh, F., Z. Hamidi-Esfahani, D. Spadaro, M. Shams-Ghahfarokhi and M. Razzaghi-Abyaneh. 2017. Unraveling the mode of antifungal action of *Bacillus subtilis* and *Bacillus amyloliquefaciens* as potential biocontrol agents against aflatoxigenic *Aspergillus parasiticus*. Food Control 89: 300–307.

Siciliano, S.D. and J.J. Germida. 1999. Enhanced phytoremediation of chlorobenzoales in rhizosphere soil. Soil. Biol. Biochem. 31: 299–305.

Siddiqui, Z.A. and L.P. Singh. 2005. Effects of fly ash and soil micro-organisms on plant growth, photosynthetic pigments and leaf blight of wheat. J. Plant Dis. Protect. 112(2): 146–155.

Singh, N. and Z.A. Siddiqui. 2015. Effects of *Bacillus subtilis*, *Pseudomonas fluorescens* and *Aspergillus awamori* on the wilt-leaf spot disease complex of tomato. Phytoparasitica 43(1): 61–75.

Singh, P.P., A. Kujur, A. Yadav, A. Kumar, S.K. Singh and B. Prakash. 2019. Mechanisms of plant-microbe interactions and its significance for sustainable agriculture. *In*: A.K. Singh, A. Kumar, P.K. Singh (eds), PGPR Amelioration in Sustainable Agriculture. Woodhead Publishing, Sawston, pp. 17–39.

Silva, H.S.A., R. da Silva Romeiro, D. Macagnan, B. de Almeida Halfeld-Vieira, M.C.B Pereira and A. Mounteer. 2004. Rhizobacterial induction of systemic resistance in tomato plants: non-specific protection and increase in enzyme activities. Biol. Control. 29(2) : 288–295.

Susarla, S., V.F. Medina and S.C. McCutcheon. 2002. Phytoremediation: An ecological solution to organic chemical contamination. Ecol. Eng. 18 : 647–658.

Teng, Y., J. Wu, S. Lu, Y. Wang, X. Jiao and L. Song. 2014. Soil and soil environmental quality monitoring in China: a review. Environ. Int. 69: 177–199.

Van Aken, B., P.A. Correa and J.L. Schnoor. 2010. Phytoremediation of polychlorinated biphenyls: new trends and promises. Environ. Sci. Technol. 44: 2767–2776.

Vergani, L., F. Mapelli, E. Zanardini, E. Terzaghi, A. Di Guardo, C. Morosini, et al. 2017. Phyto-rhizoremediation of polychlorinated biphenyl contaminated soils: An outlook on plant-microbe beneficial interactions. Sci. Total Environ. 575: 1395–1406.

Villacieros, M., C. Whelan, M. Mackova, J. Molgaard, M. Sanchez-Contreras, J. Lloret, et al. 2005. Polychlorinated biphenyl rhizoremediation by *Pseudomonas fluorescens* F113 derivatives, using a *Sinorhizobium meliloti nod* system to drive *bph* gene expression. Appl. Environ. Microbiol. 71: 2687–2694.

Vimal, S.R., J.S Singh, N.K Arora and S. Singh. 2017. Soil-Plant-Microbe Interactions in Stressed Agriculture Management: A Review. Pedosphere 27(2): 177–192.

Vos, C.M., A.F. Tesfahun, B. Panis, D.D. Waele and A. Elsen. 2012. Arbuscular mycorrhizal fungi induce systemic resistance in tomato against the sedentary nematode *Meloidogyne incognita* and the migratory nematode *Pratylenchus penetrans*. Appl. Soil Ecol. 61: 1–6.

Wang, Y., H. Li, W. Zhao, X. He, J. Chen, X. Geng, et al. 2010. Induction of toluene degradation and growth promotion in corn and wheat by horizontal gene transfer within endophytic bacteria. Soil Biol. Biochem. 42: 1051–1057.

Wang, F.Y., X.G. Lin and R. Yin. 2007. Effect of arbuscular mycorrhizal fungal inoculation on heavy metal accumulation of maize grown in a naturally contaminated soil. Int. J. Phytore-mediat. 9: 345–353.

Weyens, N., S. Croes, J. Dupae, L. Newman, D. van der Lelie, R. Carleer, et al. 2010a. Endophytic bacteria improve phytoremediation of Ni and TCE co-contamination. Environ. Pollut. 158: 2422–2427.

Weyens, N., S. Truyens, J. Dupae, L. Newman, S. Taghavi, D. van der Lelie, et al. 2010b. Potential of the TCE-degrading endophyte *Pseudomonas putida* W619-TCE to improve plant growth and reduce TCE phytotoxicity and evapotranspiration in poplar cuttings. Environ. Pollut. 158: 2915–2919.

Weyens, N., K. Schellingen, B. Beckers, J. Janssen, R. Ceulemans, D. van der Lelie, et al. 2013. Potential of willow and its genetically engineered associated bacteria to remediate mixed Cd and toluene contamination. J. Soils Sediments 13: 176–188.

Xu, Z., R. Zhang, D. Wang, M. Qiu, H. Feng, N. Zhang, et al. 2014. Enhanced control of cucumber wilt disease by *Bacillus amyloliquefaciens* SQR9 by altering the regulation of its DegU phosphorylation. Appl. Environ. Microbiol. 80(9): 2941–2950.

Yan, A., Y. Wang, S.N. Tan, M.L. Mohd Yusof, S. Ghosh and Z. Chen. 2020. Phytoremediation: A promising approach for revegetation of heavy metal-polluted land. Front. Plant Sci. 11:359. doi: 10.3389/fpls.2020.00359

Yang, B., H.Y. Ma, X.M. Wang, Y. Jia, J. Hu, X. Li, et al. 2014. Improvement of nitrogen accumulation and metabolism in rice (*Oryza sativa* L.) by the endophyte *Phomopsis liquidambari*. Plant Physiol. Bioch. 82: 172–182.

Yousaf, S., K. Ripka, T.G. Reichenauer, V. Andria, M. Afzal and A. Sessitsch. 2010. Hydrocarbon degradation and plant colonization by selected bacterial strains isolated from Italian ryegrass and birdsfoot trefoil: Bacterial hydrocarbon degradation and plant colonization. J. Appl. Microbiol. 109: 1389–1401.

Yousaf, S., M. Afzal, T.G. Reichenauer, C.L. Brady and A. Sessitsch. 2011. Hydrocarbon degradation, plant colonization and gene expression of alkane degradation genes by endophytic *Enterobacter ludwigii* strains. Environ. Pollut. 159: 2675–2683.

Yuan, S.F., M.Y. Li, Z.Y. Fang, Y. Liu, W. Shi, B. Pan, et al. 2016. Biological control of tobacco bacterial wilt using *Trichoderma harzianum* amended bio-organic fertilizer and the arbuscular mycorrhizal fungi *Glomus mosseae*. Biol Control. 92: 164–171.

Zehnder, G.W., C. Yao, J.F. Murphy, E.R Sikora and J.W. Kloepper. 2000. Induction of resistance in tomato against cucumber mosaic cucumovirus by plant growth promoting rhizobacteria. Biocont. 45(1): 127–137.

Zehnder, G.W., J.F. Murphy, E.J. Sikora and J.W. Kloepper. 2001. Application of rhizobacteria for induced resistance. Eur. J. Plant Pathol. 107(1): 39–50.

Zhang, S., A.L. Moyne, M.S. Reddy and J.W. Kloepper. 2002. The role of salicylic acid in induced systemic resistance elicited by plant growth-promoting rhizobacteria against blue mold of tobacco. Biol. Control. 25(3): 288–296.

Zhang, B.Y., J.S. Zheng and R.G. Sharp. 2010. Phytoremediation in engineered wetlands: Mechanisms and applications. *In*: Z. Yang, and B. Chen (eds), International Conference on Ecological Informatics and Ecosystem. Procedia Environ. Sci. pp. 1315–1325.

Zong, K., J. Huang, K. Nara, Y.H. Chen, Z.G. Shen and C.L. Lian. 2015. Inoculation of ectomycorrhizal fungi contributes to the survival of tree seedlings in a copper mine tailing. J. For. Res. 20: 493–500.

CHAPTER 14

Heavy Metals, Hydrocarbons, Radioactive Materials, Xenobiotics, Pesticides, Hazardous Chemicals, Explosives, Pharmaceutical Waste and Dyes Bioremediation

Elżbieta Wołejko*, Agata Jabłońska-Trypuć,
Andrzej Butarewicz and Urszula Wydro

[1]Białystok University of Technology, Department of Chemistry, Biology and Biotechnology,
Wiejska Str 45E, Białystok 15-351, Poland.

1. INTRODUCTION

Soil and water are the main elements of the natural environment in which important biochemical processes occur. They are necessary for the proper development of microorganisms and plant growth (Oleszczuk 2007). Unfortunately, soil and water are constantly being degraded due to the technological progress of the recent decades, which has contributed to the increase of this phenomenon. Because of the scale of this problem in the industrialized society, immediate action is required in order to reduce pollutants occurring in the environment (Zhang et al. 2014). The terrestrial and aquatic ecosystems may contain many harmful compounds in their matrix, for example: heavy metals (Mishra et al. 2017), polychlorinated biphenyls (PCBs), polycyclic aromatic hydrocarbons (PAHs), halogenated organic compounds (AOX), polychlorinated dibenzodioxins (PCDD) and polychlorinated dibenzenefurans (PCDFs), commonly known as

*Corresponding author: e.wolejko@pb.edu.pl

furans and dioxins (Oleszczuk 2007), explosives as well as pharmaceutical and dye wastes. These substances can cause contamination of many hectares of land and kilometres of rivers or drinking water collection points. The effects of contamination may appear after a few dozens of years, as these compounds are slowly deposited in organisms in low doses (Thapa et al. 2012). It should be remembered that humans can be poisoned by dangerous substances indirectly, i.e. by eating products, be they plant or animal, in which they are found and can result in an increase in occurrences of cancer or genetic diseases.

In order to remove xenobiotics or hazardous chemicals from the natural environment, it is possible to subject them to various remediation processes using physical, chemical and biological techniques. Chemical and physical methods can irreversibly affect the properties of soils, destroying biodiversity and influencing the proper growth and development of plants (Jha et al. 2015). Moreover, these methods require large financial expenditure. Therefore, bioremediation is considered to be effective, eco-friendly, non-invasive, financially beneficial and socially acceptable as an alternative to the ecosystem, which can be used instead of physical or chemical methods interfering with it (Fig. 1). According to Kaplan and Kitts (2004), three types of soil bioremediation can be distinguished: natural bioremediation, biostimulation and bioaugmentation. Natural bioremediation involves the use of natural biodegradation processes conducted by various groups of microorganisms (Vidali 2001). During the biostimulation process, the growth and activity of soil microorganisms should be stimulated through their proper nutrition and oxygen supply. In turn, bioaugmentation will consist in supplying different groups of microorganisms to soil, where the number of native microorganisms is too low to conduct proper reclamation. Such bacteria can be isolated from contaminated soil and multiplied in bioreactors. It is also possible to use specially selected "mix of microorganisms" originating from various areas where they are well distributed by xenobiotics. Bouchez et al. (2000) claim that biostimulation and bioaugmentation may be utilized simultaneously to increase microbiological and physicochemical environment. Like any method, bioremediation has advantages (Fig. 1) as well as limitations. The limitations of its use may be related to the type of contamination, adaptation of microorganisms, seasonal adjustment activity of microorganisms under the influence of changing environmental conditions, depth of root penetration, solubility and availability of impurities (Thapa et al. 2012).

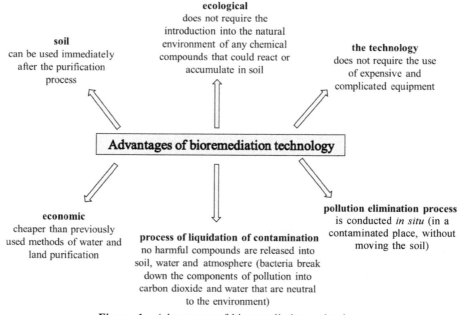

Figure 1 Advantages of bioremediation technology.

To effectively remove all harmful compounds from soil and water, using one method may not be sufficient. Therefore, scholars need to have an in-depth knowledge of various techniques for removing pollution from the environment. However, to use these bioremediation methods skilfully without adversely affecting the environment, one should become well-acquainted with the sources of pollution and their degradation products and it is necessary to determine how they enter the environment and biochemical cycles and to learn about the pathways of pollutants' degradation by microorganisms. Moreover, one ought to analyse how the introduction of microorganisms influences flora, fauna and humans living there.

2. BIOREMEDIATION OF XENOBIOTICS

The term "xenobiotics" is not precise because it refers to any chemical foreign to living organisms which occur in the natural environment as a result of human activity. In the environment, in natural conditions, such substances either do not exist or their content is very low. According to Gribble (2004), some insects, for example *Amblyomma americanium* and grasshopper, produce sex pheromone, 2,6-dichlorophenol and 2,5-dichlorophenol, respectively. Xenobiotics include plant protection products applied on a large scale in agriculture, gardening and vegetable growing, synthetic detergents in cleaning and washing agents, components of tobacco smoke, PAHs, PCBs, PCDDs, etc. (Sinha et al. 2009).

If present in the environment in low doses, xenobiotics favour the development of microorganisms. Moreover, the substances tend to remain in soil at higher doses for months or years (Jha et al. 2015). More toxic xenobiotics, after being introduced into the environment, are often located there because microorganisms at the beginning cannot handle them. Yet, it does not mean that their loss is never observed because not all compounds of anthropogenic origin are resistant to microbial degradation. When it comes to decomposition of such compounds, microorganisms first use simple substances readily transported by cellular shields and directly incorporated into the main metabolic pathways (Sinha et al. 2009, Jha et al. 2015). However, as Gribble (2004) reports, processes of xenobiotic transformation do not usually bring any measurable benefits to microorganisms, except for the removal of potentially toxic substances from the environment or at least a reduction of their toxicity as a result of changing the structure of the molecule. Table 1 presents various microorganisms which have the ability to remove xenobiotic compounds from environment.

Biotransformation of xenobiotics involving microorganisms can occur both in the absence or presence of oxygen (Cao et al. 2009). In most cases, oxygen is involved in the first reactions of xenobiotic transformations, regardless of whether their chemical structure is aromatic or aliphatic (Sinha et al. 2009). However, as suggested by Cao et al. (2009), hydroxylation reactions of these compounds seem to be of key importance in their transformation processes, and sometimes can be the limiting stage for xenobiotics metabolism by microorganisms. The major hydroxylation reactions involve oxygenase enzymes, mainly dioxygenase or monooxygenase (Ullrich and Hofrichter 2007). Moreover, through oxidation of aliphatic xenobiotics, carboxylic acids are formed, which may be the central indirect metabolite participating in the fatty acid transformation cycle in the microbial cell. In turn, the distribution of xenobiotics with an aromatic structure involves the transformation of the xenobiotic to one of the key indirect metabolites, such as procatechuic acid, catechol, hydroquinone or gentisic acid (Cao et al. 2009). Their common feature is the presence of two hydroxyl groups located either in the *para* or *ortho* position. If the hydroxyl group is in the structure of the compound which undergoes microbiological biodegradation, monooxygenase involved in the transformation of this compound introduces one of the oxygen atoms into the aromatic ring, which reduces other to water (Ullrich and Hofrichter 2007). Vaillancourt et al. (2006) report that the situation is different when the structure of the aromatic compound has

Table 1 Degradation of xenobiotic compounds by microorganisms

Rhizosphere microorganisms degrading xenobiotic	Xenobiotic compounds	References
Bacteria		
Achromobacter sp.	Glyphosate (H), carbazole	Sviridov et al. 2015 Salam et al. 2014
Acinetobacter sp.	Atrazine (H), naphthalene, Acenaphthylene; Acenaphthene; Fluorene; Phenanthrene; Fluoranthene;Benzo(a)anthracene; Chrysene; Benzo(b)fluoranthene; Benzo(k) fluoranthene; Indeno(1,2,3cd)pyrene; Dibenzo(a,h)anthracene; Benzo(g,h,i)perylene	Popov et al. 2005 Haritash and Kaushik 2009
Alcaligenes faecalis	Atrazine (H), Arylacetonitrils	Siripattanakul et al. 2009
Arthrobacter sp.	Glyphosate (H), Carbofuran, Parathion, isoproturon	Sviridov et al. 2015
Aspergillus Niger	cypermethrin	Qin et al. 2010
Bacillus cereus ZH-3	Fenpropathrin (I)	Liu et al. 2015
Bacillus licheniformis JTC-3	Carbendazim (F)	Panda et al. 2018
Bacillus megaterium,	Atrazine (H)	Siripattanakul et al. 2009
Bacillus polymyxa	Metribuzine,	Jha et al. 2015
Bacillus sp. ISTDS2	Cypermethrin (I)	Chen et al. 2010
Bacillus sp. SG2	Cypermethrin (I)	Chen et al. 2010
Bacillus subtilis	Chloropyrifos, Metribuzine, Lindane, Imidacloprid	Jha et al. 2015
Brevibacterium aureum DG-12	Cyfluthrin (I)	Chen et al. 2010
Dehalospirilum multivorans	dichlorodiphenyl tetrachloroethane (DDT)	Chaudhry and Chapalamadugu 1991
Escherichia coli	Diethylphosphate, chlorferon	Ha et al. 2009
Flavobacterium sp.	Parathion, Pentachlorophenol,	Zhang et al. 2014 Rabus et al. 2016
Janibacter sp.	carbazole, fluorene, diphenyl ether, dibenzo-p-dioxin	Yamazoe et al. 2004
Kelbsiella plenticola	Imidacloprid	Jha et al. 2015
Klebsiella ornithinolytica	Atrazine (H)	Siripattanakul et al. 2009
Lysinibacillus sphaericus FLQ-11-1	Cyfluthrin (I), benzothiophene, ethanethiol and dichloromethane	Hu et al. 2014 Wan et al. 2010
Mycobacterium sp.	Pyrene, Fluorene; Phenanthrene; endosulfan, Benzo(a)anthracene; Benzo(a)piren,	Sarma et al. 2011 Haritash and Kaushik 2009
Mycobacterium sp. LB501T	Anthracene	van Herwijnen et al. 2003
Mycobacterium sp. SD-4	Carbendazim (F)	Zhang et al. 2017
Mycobacterium vanbaalenii PYR-1	Pyrene, biphenyl, naphthalene, anthracene, fluoranthene, 1-nitropyrene, phenanthrene, benzo-[a]pyrene, benz[a]anthracene, and 7,12-dimethylbenz[a]anthracene	Kim et al. 2007
Ochrobactrum tritici pyd-1	Fenpropathrin (I)	Wang et al. 2011
Penibacillus	PAHs	Su et al. 2016
Proteus vulgaris	Chloropyrifos, Metribuzine, Imidacloprid	Jha et al. 2015
Pseudomonas aeruginosa	Atrazine (H), DDT, Endosulfan	Popov et al. 2005 Barragán et al. 2007

Table 1 Contd...

Table 1 Contd... Degradation of xenobiotic compounds by microorganisms

Rhizosphere microorganisms degrading xenobiotic	*Xenobiotic compounds*	*References*
Pseudomonas putida	benzene, toluene, p-xylene, phenols, Dinitrotoluene, Trifluralin	Lee et al. 1995 El-Naas et al. 2010
Pseudomonas putida VM1450	2,4-D (H)	Kumar et al. 2014 Germaine et al. 2006
Pseudomonas putida	beta-cypermetryny, deltametryny i cyhalotryny	Bhatt et al. 2019
Pseudomonas sp.	Indeno(1,2,3cd)pyrene; Dibenzo(a,h) anthracene; Benzo(g,h,i)perylene, Acenaphthylene; Fluorene; Phenanthrene; Fluoranthene; Benzo(a)anthracene; Chrysene; Benzo(b)fluoranthene; Phenol Benzo(k) fluoranthene; 2,4,5-T Diazinon	Su et al. 2016
Pseudomonas sp. P3	Benzo(a)pyrene	Zhang et al. 2014
Pseudomonas striata	Chloropyrifos, Metribuzine, Imidacloprid	Jha et al. 2015
Rhodococcus sp.	Fluorene; Phenanthrene; Fluoranthene; Benzo(a)anthracene; Chrysene; Benzo(b) fluoranthene; Benzo(k)fluoranthene; Atrazine (H), Pentachlorophenol (PCP), naphthalene, Indeno(1,2,3cd)pyrene; Dibenzo(a,h) anthracene; Benzo(g,h,i)perylene, triazines	Popov et al. 2005 Haritash and Kaushik 2009
Sphingomonas paucimobilis	Pyrene	Habe and Omori 2003
Sphingomonas paucimobilis B90A	gamma- and alpha- HCH	Pal et al. 2005
Staphylococcus sp. strain PN/Y	phenanthrene	Mallick et al. 2007
Stenotrophomonas sp. P1	Benzo(a)pyrene	Zhang et al. 2014
Streptomyces sp.	Glyphosate (H), Atrazine (H)	Sviridov et al. 2015 Popov et al. 2005
Fungi		
Aspergillus flavus	DDT, carbazole	Lobastova et al. 2004
Aspergillus paraceticus	DDT	Mitra et al. 2001
Botrytis cinerea	Linuron (H), metroburon (H)	Ortiz-Hernandez et al. 2013
Fusarium oxysporum	DDT	Mitra et al. 2001
Candida tropicalis	Phenol	Adav et al. 2007
Penicillium chrysogenum CLONA2	2,4-D	Bhosle and Thore 2016
Phlebia acanthocystis	Aldrin, Dieldrin	Xiao et al. 2011
Phlebia brevispora	Aldrin, Dieldrin	Xiao et al. 2011
Phlebia aurea	Aldrin, Dieldrin	Xiao et al. 2011
Trichoderma viride	β-cyfluthrin, DDT, endosulfan	Saikia and Gopal 2004 Senthilkumar et al. 2011

no hydroxyl substituents. Then, it is required to introduce two hydroxyl groups into the ring and this transition is catalysed by hydroxylating dioxygenase.

During the decomposition of xenobiotics, when oxygen is limited or it lacks, there may be an accumulation of intermediate products of decomposition of organic substances, often adversely affecting organisms located in such an environment and indirectly harmful to humans

(Sinha et al. 2009). For example, benzo(a)pyrene is biotransformed by microorganisms and, as a result, derivative products containing epoxide moieties are formed. They easily come into association with nucleic acid nitrogen bases. According to Kebamo et al. (2015), biotransformation comprises the following categories: reduction, oxidation, hydrolysis, condensation, isomerization, introduction of functional groups, and formation of new carbon bonds.

The most difficult to degrade are xenobiotics, especially polycyclic compounds and substances containing halogen, nitro, cyano and sulfonic substituents as well as heavy metals (Cao et al. 2009). There are several reasons for this state of affairs. First of all, some compounds have been introduced into the environment relatively recently and the majority of microorganisms is not able to transform them. Secondly, these substances are often highly toxic disrupting the basic physiological processes of microorganisms. Thirdly, many of them are poorly water-soluble, which makes it difficult to transport them inside the cell at various depths in soil. Then, they are gradually released as a result of natural biological and physicochemical processes, poisoning the environment locally. Therefore, it is important to monitor the population of bacteria, which effectively contribute to the biotransformation processes of xenobiotics in a polluted environment. Such monitoring can be conducted using traditional methods or molecular techniques (Gribble 2004, Sinha et al. 2009). However, it should be remembered that both approaches have drawbacks: traditional methods are very labour-intensive and it takes long to obtain the results, and molecular techniques are very expensive. Moreover, using mutagenesis also offers many opportunities to increase microorganisms' degradation properties, which is important in the case of impurities not susceptible to microbial degradation (Thapa et al. 2012). The knowledge of degradation mechanisms facilitates the selection of the most effective strains and allows for increasing the efficiency of the process of contamination degradation (Kumar and Gopal 2015).

3. BIOREMEDIATION OF PESTICIDES

Among all types of xenobiotics, one of the main threats to the environment and human health are pesticides. In the recent years, pesticide use has increased due to the rapid population growth and a greater demand for high-quality food. As indicated by Javorekova et al. (2010), efforts are still being taken to successfully detoxify pesticides in soil by using already existing microflora or supplementing it with a specific microbiological culture. Laffin et al. (2010) show that there are still some lacunae in the knowledge on how microorganisms biodegrade many active substances of pesticides in soil. As reported by Yu et al. (2016), the retention time of such substances is often expressed by the half-life of biologically active ingredients and fillers present in the preparations used. According to Oleszczuk (2007), the scope and pace of all pesticide transformations depend on their chemical structure and concentration, type and number of microorganisms capable of their degradation or transformation, as well as the physicochemical properties of the environment in which the pesticides appear or accumulate. Such investigations are still desirable because pesticides can release more toxic metabolites than primary compounds into the environment. Most research focuses only on active substances and does not take into account the stabilizers, emulsifiers and auxiliaries present in pesticides, which may also adversely affect environmental biocoenosis. There are multidirectional transformations of such compounds in the environment, and identifying all possible intermediates is difficult or actually impossible (Sassolas et al. 2012).

The most frequently isolated microorganisms found in pesticides-contaminated environments include bacteria of the following strains: *Acetobacter, Arthrobacter, Pseudomonas, Bacillus, Brevibacterium, Flavobacterium, Methylococcus, Klebsiella*, etc. (Table 1) (Badawi et al. 2009, Hussain et al. 2011). Several fungal species have some potential of pesticides' residues degradation, like *Penicillium, Candida, Pleurotus, Trichoderma, Fusarium, Rhodotorula, Phaenerochaete* and *Aspergillus* (Saikia and Gopal 2004, Senthilkumar et al. 2011).

2,4-Dichlorophenoxyacetic acid (2,4-D) is one of the most widespread herbicides (McGuinness and Dowling 2009). 2,4-D salts are easily absorbed by the roots of plants and translocated into meristematic tissues of roots and shoots, where the compound acts as a plant hormone, causing their uncontrolled growth. Mobility of 2,4-D in soil often leads to pollution of surface and groundwater. Although the herbicide is biodegradable, it may persist in soil and water for a longer period of time (Germaine et al. 2006). The ability of endophytic bacteria to degrade 2,4-D has been demonstrated in an experiment conducted by Germaine et al. (2006). In these studies, the endophytic *Pseudomonas putida* VM1450 strain, derived from internal poplar tissues (*Populus deltoids*), was introduced into peas (*Pisum sativum*). The inoculated plants were treated with 2,4-D and it was found that strain VM1450 actively colonized internal tissues of the plants. The inoculated plants were characterized by a greater ability to remove 2,4-D from the soil and they did not accumulate this herbicide in their tissues (Germaine et al. 2006).

Badawi et al. (2009) claim that the same bacteria and fungi can break down different groups of pesticides. For example, *Arthrobacter globiformis* D47, *Sphingobium* sp. YBL1 and *Arthrobacter* sp. N$_2$ degrade phenylurea herbicides such as diuron, isoproturon, chlorotoluron and fluometuron (Sun et al. 2011). In turn, among the fungi decomposing isoproturon, chlorotoluron and diuron, there are *Rhizoctonia solani, Bjerkandera adusta, Oxysporus* sp. and *Mortierella* sp. Gr4 (Badawi et al. 2009). However, as observed by El-Sebai et al. (2007) and Hussain et al. (2011), *Methylopila* sp. TES and *Sphingomonas* sp. SH can mineralize isoproturon and its metabolites, but they are not able to decompose other pesticides related to phenylurea herbicides.

Among technologies of bioremediation *in situ*, the most used by researchers include biostimulation, bioaugmentation, bioreactors, bioventing, composting and land farming which are implemented to reduce, degrade, eliminate, and/or transform the pesticides in soil (Singh 2008).

4. BIOREMEDIATION OF HYDROCARBONS

Bioremediation of petroleum pollutants in the environment depends on both abiotic factors (concentration and type of pollution, physicochemical properties of contaminated soil, content of organic and biogenic compounds, e.g. nitrogen and phosphorus, temperature, oxygen content, humidity, reaction), and biotic ones such as the quantitative and qualitative composition of microorganisms in soil. Aromatic hydrocarbons (PAHs) are ubiquitous, and toxic for the environment. Among PAHs, the components of petroleum products are the most burdensome and at the same time dangerous for the environment and living organisms (Chambers et al. 2018). It is related to their resistance to degradation and to their accumulation in terrestrial and aquatic ecosystems (Maliszewska-Kordybach et al. 2008). As noted in the studies by Saleem (2016), such petroleum compounds can be removed using physical, chemical and/or biological methods. Each of these techniques can be employed *in situ*, i.e. in the place of contamination, or *ex situ*, in which the soil needs to be transported to the point of its treatment (Thapa et al. 2012). Unfortunately, most of the proposed solutions are still at an experimental stage because their use is limited in practice and the knowledge of PAH-degrading bacteria populations *in situ* is not sufficient.

In soil, petroleum products can come in various forms: as hydrocarbons dissolved in water, those floating on the surface of the soil solution or as pollutants adsorbed on soil particles. Using chemical bonds, hydrocarbons are combined with the organic components of humus and, therefore, PAHs are mainly instanced in the upper levels of soils rich in humus substances (Hajabbasi 2016). Soil resistance to degradation increases with the content of colloids and sorption capacity. The fastest-purifiable soils in biological processes are mainly those made of mineral materials with a small amount of humus. Moreover, Chen et al. (2010) point out that the soils with a high content of humus or clay minerals have a large soil sorption capacity because oil-derivative compounds are absorbed in them and can be more easily decomposed, processed and collected by various microorganisms. To increase the rate of degradation of pollutants, soil tends to be additionally enriched with nutrients necessary for microbial growth.

As reported by Mrozik and Piotrowska-Seget (2010), several approaches have been proposed for PAHs bioremediation in soil, which consist in increasing augmenting microorganisms adapted to heightened contaminant concentrations and/or producing surfactants to enhance bioavailability of PAHs. The most frequently determined microorganisms found in oil-contaminated environments include bacteria of the following genera: *Acetobacter, Pseudomonas, Cornynebacterium,* and fungi: *Aspergillus, Candida, Trichoderma* and *Mortierella* (Liste and Felgentreu 2006).

Noteworthy are *Bacillus* bacteria because these strains have the ability to produce biosurfactants and degrade oil in soil (Souza et al. 2015). For example, the introduction of *B. subtilis* A1 into contaminated soil decreased the concentration of hydrocarbons in soil by 87% after 7 days (Parthipan et al. 2017). In addition, many *Pseudomonas* bacteria have enzymes involved in the degradation of aliphatic and aromatic hydrocarbons (Weyens et al. 2009, Pawlik et al. 2017). For this reason, these strains are used as biovaccines that, in assisted phytoremediation, increase the efficiency of the transformation of oil-derived impurities (Sessitsch et al. 2013, Sun et al. 2014).

Biodegradation of petroleum derivatives is a multi-stage process, taking place under both aerobic and anaerobic conditions, involving many different groups of microorganisms, often with a synergistic effect. One of the methods of controlling *in situ* biodegradation of areas contaminated with petroleum derivatives is to monitor the concentration of so-called key metabolites (Pawlik et al. 2017). Benzoyl-CoA, which does not belong to synthetic compounds, may be a specific marker for the catabolism of toluene, xylene and ethylbenzene, coupled with the sulphate respiration of microorganisms (Gabrielson et al. 2003). In anaerobic conditions, microorganisms can decompose hydrocarbons like benzene, toluene, ethylbenzene, monoaromatic hydrocarbons (BTEX), hexadecane and naphthalene, but their distribution is 4 times slower than in the aerobic environment (Sun et al. 2011).

Researchers pay special attention to endophytic bacteria which can promote the growth and development of plants, consequently increasing their biomass in soil contaminated with petroleum substances. They can also affect the bioavailability of organic pollutants in soil (Weyens et al. 2009). Many endophytic bacteria are able to grow in the presence of PAHs, and some can degrade these compounds, using them as a source of carbon and energy. Examples comprise endophytic *Pseudomonas putida* PD1 and *Pseudomonas* sp. Ph6 degrading phenanthrene (Hajabbasi 2016) or strains *Stenotrophomonas* sp. P1 and *Pseudomonas* sp. P3 isolated from Canadian canon (*Conyza canadensis*) and cloverleaf (*Trifolium pratense* L.) capable of degrading naphthalene, phenanthrene, fluorene, pyrene and benzo(a)pyrene. Zhang et al. (2014) show that the inoculation of plants and soil with a bacterial consortium containing *Bacillus subtilis* strain J4 AJ (capable of degradation of diesel fuel) and strain *Pseudomonas* sp. U-3 (producing a bio-surfactant effectively reducing surface tension) promote the removal of diesel oil from the environment. In their studies, scholars observed that biosurfactants secreted by bacteria may also be carriers of microbial material, thanks to which the bioremediation process also takes place inside the sorbent. The use of metabolic cycles of microorganisms can therefore lead to final transformation products, i.e. CO_2 and H_2O (Aidberger et al. 2005, Zhang et al. 2014). Moreover, degradation of phenanthrene by *Acinetobacter* sp., *Flavobacterium* sp., *Pseudomonas* sp., *Rhodococcus* sp. and *Stretomyces* flavovirens produces 2,4-hydrosyphenantrene, 1,2-dihydroxynaphthalene and phthalates, while 3,4-dihydroxyfluoren, 3,4 -dihydroxycoumarin, 1-indanone and salicylate were identified during biotransformation of fluorene (Zhang et al. 2014).

The literature provides a number of studies on the biodegradation processes of BTEX and PAHs in soil using fungi. Sašek et al. (2003) conducted research on the use of the fungus

Cladophialophoria sp. for the mineralization of BTEX, which showed that such degradation proceeds with higher dynamics for toluene, ethylbenzene and m-xylene than for benzene. Only the introduction of bacteria of the genus *Rhodococcus* sp. significantly accelerated the process of benzene mineralization. Metabolic profiles and the inhibitory nature of substrate interaction indicate that ethylbenzene, toluene, and m-xylene are degraded in the side chain by the same monooxygenase enzyme (Wiesche et al. 2003).

Wiesche et al. (2003) and Sašek et al. (2003) conducted PAH biodegradation processes with biopreparation based on indigenous microorganisms enriched with fungi of the genus *Pleurotus ostreatus* and *Dichomitus squalens*. Research with the above fungi showed that they have the ability to produce ligninoline enzymes responsible for biodegradation of PAHs with a higher number of rings (5–6) in soil (Ahn et al. 2010). Enrichment of microorganisms with fungi of the genus *Pleurotus ostreatus* increases the biodegradation efficiency of pentacyclic PAHs (benzo(a)pyrene (BaP), benzo(a)anthracene (BaA) and dibenzo(a, h)anthracene (DaA)) as compared to the action of a biopreparation only based on the bacteria themselves (Wiesche et al. 2003). In addition, by using bioaugmentation with a biopreparation containing bacterial cultures *Pseudomonas putida* enriched with fungi of the genus *Pleurotus ostreatus* and *Irpex lacteus*, the content of individual PAHs was reduced by 66% over a period of 10 weeks (Li et al. 2007).

The interaction between fungi and bacteria is beneficial for the process of mineralization of petroleum hydrocarbons. The above mentioned mushrooms have a metabolic capacity of biodegradation of BTEX and PAH, similar in many respects to that of bacteria. Therefore, fungi should not be ignored in the development of effective bioremediation strategies (Wiesche et al. 2003).

Moreover, research is being conducted using biofilters technology where both bacteria and fungi strains are applied. However, some strains of fungi are able to stay longer on the biofilter as compared to bacteria (Husaini et al. 2008). For example, *Exophiala oligospermum* placed into the perlite deposit biodegraded toluene in a stable way for several months (Esteves et al. 2004), while spores of *Scedosporium apiospermum* introduced into the vermiculite deposit increased greatly the speed of toluene purification at high loads for 2 months of continuous operation of the deposit (Kennes and Veiga 2004). Therefore, one should further explore the possibility of inoculation of biofilters with strains of fungi in bioremediation technology. The focus should not be placed only on bacteria, as usually is the case, but rather on filamentous mycelium fungi that offer greater prospects than yeast fungi. Therefore, fungi should not be ignored in the development of effective bioremediation strategies (Wiesche et al. 2003).

5. BIOREMEDIATION OF EXPLOSIVES

According to the literature data, explosives occurring in environment as well as their transformation products can affect microorganisms, plants and higher organisms in a toxic or mutagenic way (Lewis et al. 2004, Thijs et al. 2014). The production and processing of ammunition led to serious environmental pollution with compounds such as 2,4,6-trinitrotoluene(TNT), hexahydro-1,3,5-trinitro-1,3,5-triazine (RDX), pentaerythritol tetranitrate (PETN), triamino-trinitrobenzene (TATB), 2,4,6,8,10,12-heksanitro-2,4,6,8,10,12-heksaazaizowurcytanu (HNIW) or octahydro-1,3, 5,7-tetranitro-1, 3,5,7-tetrazocine (HMX) (Ndibe et al. 2018). These toxic and mutagenic compounds are characterized by high durability and resistance to chemical agents, biological oxidation and hydrolysis.

Studies conducted by many researchers suggest that there are microorganisms and fungi that are able to break down explosives (Boopathy 2000, Thijs et al. 2014, Anasonye et al. 2015). As noted in the studies by Van Aken et al. (2004), *Methylobacterium* sp. BJ001 strain showed the ability to totally degrade TNT, RDX and HMX after 55 days. Thijs et al. (2014) used a consortium of bacteria isolated from the rhizosphere of sycamore maple (*Acer pseudoplatanus*) growing in soil contaminated with TNT. It was found that the consortium is able to efficiently transform TNT into hydroxylamine and amino-dinitrotoluene and support the development of plants growing there (Ziganshin et al. 2010). Table 2 presents various microorganisms and fungi which can degrade explosive compounds.

Table 2 Degradation of explosive compounds by various microorganisms

Rhizosphere microorganisms degrading explosives	Explosives compounds	References
Bacteria		
Stenotrophomonas maltophilia OK-5, *Pseudomonas fluorescens* I-C, *Bacillus* sp. SF, *Morganella morganii* B2, *Myriophyllum brasiliense*, *Pseudomonas aeruginosa*, *Pseudomonas putida* KP-T201, *Pseudomonas pseudoalcaligenes* JS52, *Yarrowia lipolytica* AN-L15	TNT	Anasonye et al. 2015 Ndibe et al. 2018 Park et al. 2003 Ziganshin et al. 2010
Methylobacterium sp. BJ001, *Rhodococcus, Microbacterium* sp. and *Gordonia* sp. KTR9, *Clostridium bifermentans, Corynebacterium, Rhodococcus* DN22, *Stenotrophomonas maltophilia* PB1, *Morganella morganii* strain B2, *Rhodococcus rhodochrous* 11Y	RDX	Van Aken et al. 2004 Singh et al. 2012
Methylobacterium sp. BJ001, *Clostridium bifermentans* HAW-1, *Caldicellulosiruptor owensensis*	HMX	Van Aken et al. 2004 Nejidat et al. 2008
Enterobacter cloacae PB2, *Agrobacterium radiobacter*	PETN	Singh et al. 2012
Pseudomonas sp. FA1, *Rhodococcus, Ochrobactrum, Mycobacterium, Rulstonia*	HNIW	Singh et al. 2012
Fungi		
Gymnopilus luteofolius, Agaricus aestivalis, Agrocybe praecox, Clitocybeodora, Kuehneromyces mutabilis, Mycena galericulata, Phanerochaete velutina, Physisporinus rivolosus, Stropharia rugosoannulata, Phanerochaete chrysosporium, Phlebia radiata, Bjerkandera adusta DSM 3375	TNT	Anasonye et al. 2015
Aspergillus niger, Rhodotorula, Bullera, Acremonium, Penicillium, Phanerochaete chrysosporium	RDX	Singh et al. 2012 Bhatt et al. 2006
Phanerochaete chrysosporium	HMX	Nejidat et al. 2008
Phanerochaete chrysosporium, Irpex lacteus, Clostridium sp. EDB2, *Phanerochaete chrysosporium,*	HNIW	Singh et al. 2012 Bhushan et al. 2005 Singh et al. 2012

Moreover, literature data indicate that bioremediation technologies can be used to effectively remove explosives from soil and water by using microorganisms, fungi and plants which give promising results in their detoxification (Anasonye et al. 2015, Ndibe et al. 2018). Bioremediation methods of removing explosives, such as using bio-reactors inoculated with sludge, land farming and composting, have been developed. The research conducted by Boopathy (2000) shows that for soil slurry, the reactor used molasses as a co-substrate for bacterial growth and one observed effective biodegradation of TNT and its metabolites within a period of 3 months. For *in situ* cleaning of soil contaminated with TNT, one can also use perforated organic glass tubes that are filled with inoculum in the form of wood chips inhabited by fungi like *Gymnopilus luteofolius, Kuehneromyces mutabilis, Mycena galericulata, Phanerochaete velutina, Physisporinus rivolosus* and *Stropharia rugosoannulata*. Such studies were performed in laboratory and field conditions by Anasonye et al. (2015): inoculum tubes were placed vertically in the ground to the depth of contamination, the growing mycelium filled the spaces between the tubes and the secreted enzymes degraded TNT at the level of 70 ÷ 80%. As the authors suggest, this model can also be used for other types of pollution in the environment, such as metals, pesticides, PAHs, etc.

6. OCCURRENCE AND TOXICITY OF HEAVY METALS AND METALLOIDS

Natural sources of heavy metals occurring in the environment are geochemical processes such as weathering of rocks, and volcanic eruption (Oves et al. 2016). Anthropogenic sources of heavy metals include metallurgy, in particular iron metallurgy, mining, glass and cement industry as well as combustion of fossil fuels, heat and power plants and local boiler houses. Heavy metals also enter the environment due to the use of artificial fertilizers, pesticides and sewage sludge (Yan-de et al. 2007, Wei and Zhu 2008, Memon and Schröder 2009, Oves et al. 2016). Furthermore, a wide range of heavy metals originate from transportation, in particular from wear on brakes, tyres and other vehicle parts and fuel combustion (Padmavathiamma and Li 2012).

According to Oves et al. (2016) and Etesami (2018), heavy metals, depending on their concentration and occurence, may stimulate living organisms or be toxic for them. Metals necessary for proper growth and various metabolic processes in animal, human and plant organisms exhibit stimulating effects. Metals such as chromium (Cr), copper (Cu), iron (Fe), manganese (Mn), magnesium (Mg), molybdenum (Mo) and zinc (Zn) in specific concentrations play an important role in the development and functioning of organisms; however, in excessive concentrations (higher than physiological concentration), they can be harmful. Toxic effects are also shown by heavy metals in a low concentrations whose presence is not necessary for the body, like nickel (Ni), lead (Pb), cadmium (Cd), mercury (Hg) and arsenic (As) (Etesami 2018).

Heavy metal toxicity in humans is reflected by changes in the enzymatic activity, protein synthesis and disturbances of adenosine triphosphate (ATP) production due to damage of cell membranes and membranes of cellular organelles, such as lysosomes, mitochondria and nuclei. Toxicity of heavy metals can also cause disturbance of basic physiological processes, including mineral management. Acute and chronic poisoning with heavy metals in humans and animals may cause damage to the digestive, respiratory, nervous, circulatory, haematopoietic and renal systems. Some heavy metals have teratogenic and embryotoxic effects (Padmavathiamma and Li 2007, Oves et al. 2016). Janssen et al. (2003) reported that heavy metals penetration through biological membranes can cause their bioaccumulation and reduction of excretion from the organism, which depends on the chemical form, the stability of the compound and its solubility in water and fats, pH and the presence of other substances in the environment.

As reported by Foti et al. (2017), high concentrations of Cu, Ni and Zn are the most toxic to plants. Toxicity of heavy metals in plants causes disturbances in the functioning of cell membranes, photosynthetic and mitochondrial electron transport as well as inactivation of enzymes of basic cell metabolism. It ultimately leads to a decrease in the cell energy balance, mineral management disorders and growth reduction.

7. BASIC MECHANISMS, STRATEGIES AND IMPORTANCE OF RHIZOMICROBIOME IN HEAVY METALS AND METALLOIDS BIOREMEDIATION

Rhizosphere is the soil zone surrounding the roots of plants and characterized by a high microbiological activity. It may contain about 10^{11} microbial cells per 1 g of roots and it is called rhizomicrobiome (Berendsen et al. 2012). Rhizobia are also important plant growth promoting (PGP) microbes which care for plant health (Mendes 2013, Nazir et al. 2016).

Rhizospheric microbes also play a crucial role in heavy metals remediation and the process involving them is referred to as rhizoremediation (Mishra et al. 2017). Pires et al. (2017) reported that rhizobacteria used in remediation of soil contaminated with heavy metals results from the ability of microbes to detoxify some heavy metals, their capacity to promote plant growth and to accumulate metals in the cell. According to Ma et al. (2016), plant-growth promoting microorganisms (PGPMs) belong to the microorganisms, which contribute to the remediation in the most effective way and includes plant growth-promoting bacteria (PGPB) and arbuscular mycorrhizal fungi (AMF). The most frequently detected rhizosphere microorganisms that are capable of bioremediation of heavy metals comprise *Firmicutes, Proteobacteria* and *Actinobacteria* (Pires et al. 2017). As reported by Narendrula-Kotha and Nkongolo (2017), among the fungal population, *Ascomycota* and *Basidiomycota* are the most abundant in heavy metals-contaminated soil.

Mishra et al. (2017) have mentioned that detoxification and remediation of heavy metals by rhizospheric microorganisms depend on their availability. On the other hand, rhizobia may immobilize metal in the environment due to complexation, precipitation etc. In addition, they can also mobilize heavy metals by altering properties of soil such as pH and redox reaction, while supporting uptake of metals by plants in the phytoremediation process. Additionally, metal solubilization mechanism is used in bioleaching (Hietala and Roane 2009).

It should be highlighted that the transformation of metals by microbes tends to be associated with metal resistance mechanisms. In practice, the removal of heavy metals from heavily contaminated areas using bioremediation methods relies primarily on the immobilization of heavy metals by microorganisms. It mainly occurs under the influence of biosorption and intracellular accumulation, precipitation and binding by various polymers of microbial origin, biotransformation (methylation, demethylation, volatilization, oxidation or reduction, physical sequestration) and alcalization (Hietala and Roane 2009, Etesami 2018). Examples of the application rhizospheric microorganisms in bioremediation of heavy metals are presented in Table 3.

Biosorption is a physico-chemical process involving the binding of metal ions through cellular covers or more precisely, through their negative charge such as –SH, –OH and –OOH functional groups. This process may include one process or a combination of, for instance, adsorption, chelation, ion exchange reactions, microprecypitation and complexation (Pokethitiyook and Poolpak 2016).

Complexation of metals consists in binding metal cations with two polymers hydroxyl groups of cell covers (exopolysaccharides, lypopolysaccharides, proteins, carbohydrates with various functional groups) (Hietala and Roane 2009). Biosorption is a passive absorption in which heavy metals are absorbed by inactive or dead microbe cells or by organic material from biological sources (organic wastes, sewage sludge) wherein this process usually occurs in the wall of cell. Intracellular accumulation relies on intracellular transport of metals/metalloids, which are accumulated through the cell membrane and the cell metabolic cycle (Etesami et al. 2018). Metals/metalloids may occur in the cell as complexes with phytochelatins or metallothioneins. Furthermore, they may be associated with organic polyphosphates. Gutierrez-Corona et al. (2016) reported that intracellular bioaccumulation is an "active uptake" which takes place only in living cells, depends on metabolism and needs energy for transport of metals. The accumulation of metals by bacteria may take the intracellular sequestration form in which metals bind with intracellular compartments such as vacuole or special proteins – bacterial metallothioneins (MTs) and metallo-chaperones (Etesami 2018). Some plant growth-promoting rhizobacteria (PGPR) may produce MTs which can bind metals like cytoplasmic proteins (Blindauer et al. 2002).

The study conducted by Ma et al. (2015) show that *Bacillus* sp. SC2b isolated from the rhizosphere of *Sedum plumbizincicola* has potential ability to conduct biosorption and biomobilization of Cd, Pb and Zn. Besides, it might serve as a future biofertilizer, which has a capability of solubilizing phosphate as well as producing indole-3-acetic acid (IAA) and siderophores. According to Zaidi et al. (2006), *Bacillus subtilis* SJ-101 strain shows high biosorption capacity for Ni and ability to protect *Brassica juncea* against toxic concentration of nickel in soil. Similar effects were observed for Ni resistant *Bacillus megaterium* SR28C in the studies conducted by Rajkumar et al. (2013). The tested strain bound considerable concentration of nickel in the resting cells and, in addition, exhibited the plant growth promoting features.

Table 3 Rhizospheric microorganisms involved in heavy metal bioremediation

Metals	Example of rhizo-microorganism	Mechanisms	References
Metalloids			
As	*Bacillus aryabhattai* AS6	Bioaccumulation	Ghosh et al. 2018
	Bacillus sp, Geobacillus sp.	Oxidasing	Majumder et al. 2013
	Pseudomonas stutzeri KC	Precipitation	Zawadzka et al. 2006
	Gram-positive strain (ASI-1)	Volatilization	Meyer et al. 2007
Se	*Stenotrophomonas maltophilia*	Reduction	Di Gregorio et al. 2005
	Pseudomonas stutzeri KC	Precipitation	Zawadzka et al. 2006
	Gram-positive strain (ASI-1)	Volatilization	Meyer et al. 2007
Te	*Pseudomonas stutzeri* KC	Precipitation	Zawadzka et al. 2006
Heavy metals			
Cd	*Pseudomonas fluorescens* *Microbacterium oxydans*		Piccirillo et al. 2013
	Bacillus sp. SC2b s	Biosorption	Ma et al. 2015
	Klebsiella planticola	Precipitation	Sharma et al. 2000
Cr	*Pseudomonas stutzeri* KC	Precipitation	Zawadzka et al. 2006
	Pseudomonas aeruginosa OSG41	Reduction	Oves et al. 2013
	Cellulosimicrobium cellulans	Reduction	Chatterjee et al. 2009
Hg	Gram-positive strain (ASI-1)	Volatilization	Meyer et al. 2007
Ni	*Bacillus subtilis* SJ-101	Biosorption	Zaidi et al. 2006
	Bacillus megaterium SR28C	Biosorption	Rajkumar et al. 2013
Pb	*Bacillus* sp. SC2b s	Biosorption	Ma et al. 2015
	Providencia vernicola SJ2A	Sequestration	Sharma et al. 2017
	Gram-positive strain (ASI-1)	Volatilization	Park et al. 2011
Zn	*Pseudomonas fluorescens* *Microbacterium oxydans*		Piccirillo et al. 2013
	Bacillus sp. SC2b s	Biosorption	Ma et al. 2015
	Anabaena PCC 7120, *Pseudomonas aeruginosa*, *Pseudomonas putida*	Intracellular sequestration	Blindauer et al. 2002
Radionuclides			
U	*Geobacter* sp.	Reduction	Newsome et al. 2014
	Serratiaa sp.	Precipitation	
	Pseudomonas sp.	Bioaccumulation	
I	*Mycobacterium chubuense*	Volatilization (methylation)	Amachi et al. 2003
Sr	*Bacillus pasteurii*	Precipitation	Cuthbert et al. 2012

Biotransformation of metals/metalloids by rhizomicrobes using redox reactions leads to a reduction of their toxicity or obtaining non-toxic forms. Oves et al. (2013) demonstrated usefulness of inoculation with chromium tolerant *Pseudomonas aeruginosa* OSG41. The strain had capability of reducing toxic hexavalent chromium to non-toxic and immobile Cr^{3+}. A similar effect was described for *Cellulosimicrobium cellulans* in a study conducted by Chatterjee et al. (2009). In the same vein, Di Gregorio et al. (2005) discussed the ability to reduce Se(IV) to Se(0) shown by *Stenotrophomonas* sp. strain isolated from the root system of *Astragalus bisulcatus*.

The reduction effect of rhizospheric microorganisms against radionuclides elements (uranium) has also been observed (Finneran et al. 2002). In turn, Majumder et al. (2013) presented oxidative effect in which *Bacillus* sp. and *Geobacillus* sp. participate and contribute to transforming toxic As^{3+} to less toxic As^{5+}.

Biotransformation of metal into a gaseous state is called volatilization. This phenomenon is observed in the case of metalloids such as As, Se and Te as well as Hg and Pb. It has been reported that Gram-positive strain (ASI-1) is able to transform metal(loid)s' ions into a gaseous state due to the addition of methyl groups (methylation). Amachi et al. (2003) have observed that some soil bacteria has the capability of reducing radioactive iodine by means of methylation. On one hand, volatilization is a microorganism strategy of detoxification of the microenvironment from metals; on the other hand, some volatile products have higher mobility and toxicity than the input substances (Meyer et al. 2007).

Some rhizosphere microbes can accumulate metals in the precipitation process (Mishra et al. 2017, Etesami 2018). This can be favoured by the production of acid phosphates by some bacteria, as well as by strong alkalisation of the environment as a result of denitrification processes. As reported by Zawadzka et al. (2007), many heavy metals and metalloids (Pb, Hg, Cd and As) can create insoluble precipitates with pyridine-2,6-bis(thiocarboxylic acid) (PDTC) produced by *Pseudomonas stutzeri* KC. Selenium and tellurium may also be reduced by the mentioned strain and can be bound with siderophores by precipitation (Zawadzka et al. 2006).

Chelating metal ions by microbes living in rhizosphere is possible due to the release of natural organic agents such as organic acid anions, biosurfactants and siderophores. These substances can reduce the metal bioavailability and toxic effects. The release of these substances by rhizomicrobes involved in environmental detoxification from heavy metals support this process and promote plant growth. The participation of biosuractants, siderophores and indole acetic acids (IAA) in bioremediation of heavy metals is described by Zawadzka et al. (2007), Ma et al. (2015), and Adhikary et al. (2019). Furtheremore, Zawadzka et al. (2006) drew attention to the fact that siderophores such as PDTC can support metal bioremediation by participating in metal reduction, complexation, selective precipitation and solubilization processes. The surface-active biomolecules (biosurfactants) can also contribute to decreasing metals mobility. For instance, *Pseudomonas aeruginosa* produces rhamnolipids, whereas *Bacillus subtilis* has a capability of releasing surfactin, which causes an increase in the metals' resistance (Mulligan et al. 2001). It should be noted that biosurfactant complexation can increase metal mobility and this technique is often applied in soil-washing technologies (Hietala and Roane 2009).

Generally, some of the described microbiological processes may either cause an increase in mobility (acidification, chelatation) or immobility (precipitation, complexation) of inorganic elements in the environment (Ma et al. 2016). Both mechanisms are being used to enhance phytoremediation – phytoextraction or phytostabilization. Additionally, rhisospheric micro-organisms (bacteria, fungi) introduced to metal-polluted soil promote plant growth and act as biofertilisers which enhance phytoremediation efficiency.

8. BIOREMEDIATION OF RADIOACTIVE MATERIALS

The development of civilization and growing dependence on nuclear power results in the increase in the production of radioactive waste. Therefore, the use of microorganisms for biodegradation of radioactive materials seems to be justified. As suggested by Reena et al. (2012), microorganisms can influence radionuclides' mobility, solubility and bioavailability occurring in the terrestrial and aquatic ecosystems. The biological effect and sensitivity of microorganisms to radiation depend on, among others, the type of radiation, its intensity, exposure time, species or strains of the microorganism, their growth phase, metabolic state and also on the culture medium and environmental conditions (Lloyd and Renshaw 2005). Kumar et al. (2007) indicated that the half-lives of radionuclides such as Uranium (U), Neptunium (Np), Plutonium (Pu), Americium (Am), Technetium (Tc), Cesium (Cs) and Strontium (Sr) depending on their isotopes may vary from 4.47×10^9 to 29 years.

According to the literature data, exposure of microorganisms to radiation can induce radiation tolerance in select strains – often producing dyes with antioxidant properties, e.g. melanin. Such a situation took place, for example, after the accident in 1986 in Ukraine in the region of the damaged Chernobyl nuclear power plant (Lloyd and Renshaw 2005). By means of laboratory methods, *E. coli* strains with extremely high resistance to radiation were isolated and selected by repeated exposures to increasing doses of ionizing radiation (Confalonieri and Sommer 2005).

Radiation–resistant microorganisms can potentially be used for bioremediation of radioactive waste, including by bioadsorption, bioaccumulation of radioisotopes by cells or bioprecipitation, e.g. of uranium, using genetically modified *Deinococcus radiodurans* strains. However, resistance to radiation does not mean complete insensitivity to radioisotopes (Kulkarni et al. 2013). The isotopes present in the environment tend to be associated with various heavy metals, which can also influence *Deinococcus radiodurans*' activity – a sensitive bacterium to these bonds (Bauermeister et al. 2011).

Deinococcus radiodurans can be used for, for instance, the bioremediation of radioactive waste not only from soils but also from waters. Genetically recombined strains of the bacterium in question equipped with alkaline phosphatase or non-specific acid phosphatase precipitate with uranium phosphate ions (also adsorbed on the surface of cells) (Kulkarni et al. 2013).

Radiation resistance of *Deinococcus radiodurans* does not mean full tolerance to radioisotopes (Egas et al. 2014). The main reason for *Deinococcus radiodurans* high resistance to radiation seems to be not so much the prevention of DNA degradation, but rather the rapid and effective ability to remove serious damage in the genome. Bacteria that also show high resistance to radiation include *Methylobacterium radiotolerans*, *Rubrobacter radiotolerans* and cyanobacteria of the genus *Chroococcidiopsis* (Billi et al. 2000, Egas et al. 2014).

In addition, various species of fungi can be successfully used for bioremediation of areas contaminated with radioactive waste (Kulkarni et al. 2013). As suggested by Dighton et al. (2008), some species of fungi can also be used as biomarkers (bioindicators) for the degree of radioactive contamination of soil. Examples of such organisms include *Chaetomium aureum* and *Purpureocillium liliacinus* occurring at a high level of pollution (3.7×10^6–3.7×10^8 Bq/kg), *Acremonium strictum* and *Arthrinium phaeospermum* – at a medium level (3.7×10^3–3.7×10^5 Bq/kg), *Myrothecium roridum* and *Metarhizium anisioplia* – at a low level ($<3.7 \times 10^2$ Bq/kg) (Dighton et al. 2008).

9.   BIOREMEDIATION OF PHARMACEUTICALS

Pharmaceuticals are widely used in many areas, including medicine, veterinary medicine, industry, agriculture and everyday life. Due to the very intensive and widespread use, as well as the weakness of the traditional wastewater treatment process, they are becoming increasingly common in the environment. The division into classes by their functions and main properties is presented in Table 4 (Barra Caracciolo et al. 2015, Wang and Wang 2016).

A huge number of different active compounds are currently used in various areas and by many industries, especially for the prevention and treatment of numerous diseases. Widespread detection methods allow for a better estimation of their presence in the environment, which subsequently has engendered a significant scientific and regulatory issue (Zuccato et al. 2010, Barra Caracciolo et al. 2015). A significantly large proportion of pharmaceuticals in

Table 4 Classification, function and main properties of pharmaceutical contaminants most frequently detected in the environment

Group of compounds	Main functions	Representative compound name	CAS	Molecular weight	Structure	Analytical methods of detection
Antibiotics	Kill bacteria	Tylosin	1401-69-0	916.1		HPLC, LC-MS, LC-MS/MS
Nonsteroidal anti-inflammatory drugs	Reduce pain and inflammation	Ibuprofen	15687-27-1	206.3		HPLC, LC-MS, GC-MS, LC-MS/MS
Antidepressants	Cure physical dysfunctions	Fluxetine	59333-67-4	309.3		LC-MS/MS
Anticonvulsants	Improve mood disorders	Carbamazepine	298-46-4	236.3		HPLC,GC-MS, LC-MS/MS
Antineoplastics	Kill or control cancer cells	Epirubicin	56420-45-2	543.5		LC-MS/MS

Hormones	Regulation of metabolism, homeostasis maintaining	Estriol	50-27-1	288.4	HPLC
Lipid regulators	Regulation of cholesterol and triglycerides level in blood	Gemfibrozil	25812-30-0	250.3	HPLC, LC-MS/MS
Beta-blockers	Inhibit adrenalin and noradrenalin	Propranolol	525-66-6	259.3	HPLC, GC-MS, LC-MS/MS

unchanged form or as chemically active metabolites flow into the wastewater treatment plants. If pharmaceuticals are not effectively removed in the wastewater treatment process, they are an important source of exposure to the above-mentioned compounds. Factors causing the pharmaceuticals to enter the agricultural soils are primarily the use of sewage sludge as manure and crop fertilizers. From soils, they can then penetrate into groundwater and be detected even in coastal waters (Boxall 2004, Dolliver and Gupta 2008, Bottoni et al. 2010, Lapworth et al. 2012, McEneff et al. 2014). Increasingly, there are concerns related to the potential risk to human health of eating food and water contaminated with pharmaceutical residues present in agricultural soils and groundwater. The results of ecotoxicological studies indicate that human pharmaceuticals are not usually characterized by acute toxicity to aquatic organisms because in the environment they usually occur in low concentrations. However, long-term exposure to even low doses of pharmaceuticals causes adverse effects on living organisms. This is due to the fact that pharmaceuticals and their residues constitute a very specific type of environmental pollution because they are synthesized in such a way that they are bioactive in very low concentrations. Their targets are usually macromolecules that occur in various forms of life. Therefore, even extremely low, environmental concentrations of compounds from this group of pollutants can be potentially dangerous causing sub-lethal effects on non-target organisms (Santos et al. 2009, Franzellitti et al. 2013). At present, a significant lack of literature data regarding toxicological effects of above mentioned compounds on soil and water organisms is being observed. However, some research data indicate that pharmaceutical residues may influence endocrine system and cause pathogen resistance, e.g. in case of antibiotics. It was also shown that they have deleterious effect on not only specific microorganisms, but also on microbial communities' functionality (Marti et al. 2013).

The most important factors taken into account when determining the concentrations of environmental exposure are the amount of consumption, the degree of release into the environment and persistence in the environment. Stability in the environment is associated with various factors, such as physicochemical properties, and environmental factors (e.g. light, temperature, pH), but most importantly the presence and activity of microorganisms. Biodegradation of pharmaceuticals, but also of other chemicals on metabolic pathways, involves the use of a chemical as a source of energy, carbon, nitrogen or other nutrients (Barra Caracciolo et al. 2015). Microbiological biodegradation is currently one of the most important ways of eliminating organic pollutants in the environment. The methods that are used in the area of microbial biodegradation have many undoubted advantages, including low costs. Microorganisms can remove environmental contaminants by consuming them for their metabolic functions and in some cases different microorganisms can cooperate to remove specific contaminants (Wang and Wang 2016).

9.1 Biodegradation of Antibiotics on Natural Microbial Communities

Antimicrobial pharmaceuticals are of concern as chemical substances which inhibit growth or eliminate viruses, bacteria and fungi. The term antibiotics mainly refers to active compounds that act against bacteria. This group includes synthetic, semi-synthetic compounds and substances of natural origin derived from fungi or bacteria. Currently, there are about 250 registered antibiotics used in both medicine and veterinary medicine (Kümmerer 2003). In recent years, there has been a widespread concern related to the fact that the release of antibiotics into the environment through urban wastewater and agricultural wastewater may promote the development of resistance in bacteria to most commonly used antibiotics. In most countries around the world, governments now recognise that increasingly common antibiotic resistance is a priority in public health. Guidelines developed in the European Union (EU) suggest a very prudent and reasonable use of antimicrobials in both medicine and veterinary medicine only in a state of higher necessity (European Union 2011). It should be mentioned that the key point in pharmaceuticals biodegradation is the production of a specific enzyme by the microorganisms, which subsequently is able to decompose chemical compound. Antibiotics belong to the group of compounds which cannot stimulate microorganisms to enzyme production, leading to their poor biodegradability. Therefore, pure bacterial cultures are not efficient in antibiotics biodegradation. However, literature data indicate that the Gram-positive strain *Microbacterium* sp. C448 grows using sulfamethasine as the only carbon source, mineralizing the benzyl portion of the molecule and secreting the pyrimidine portion as the final product (Topp et al. 2013).

9.2   Biodegradation of Nonsteroidal Anti-inflammatory Drugs (NSAID)

Drugs from the group of NSAID are the most frequently detected pain killers in surface water in an unchanged or slightly modified form. NSAIDs are detected in different areas of the environment in concentrations in the range of nanograms to micrograms per litre (Wu et al. 2012). Their high chemical stability causes problems with their elimination in sewage treatment plants, which may explain their presence in tap water with which they are unintentionally consumed by consumers. Bioremediation processes for the aforementioned group of drugs are a promising alternative to chemical methods of their removal, which in turn are often associated with the generation of large amounts of free radicals in the course of complex chemical reactions and generate high costs (Zhang et al. 2013). Among the microorganisms capable of decomposing and transforming NSAID, currently several species of fungi are mentioned, the most often being *Penicillium* sp., *Trametes versicolor*, *Cunninghamella elegans*, *C. echinulata*, *C. blakesleeana*, *Beauveria bassiana*, *Phanerochaete chrysosporium*, *Ph. sordida*, *Actinoplanes* sp., *Bjerkandera* sp. R1, *Bj. adusta*, *Irpex lacteus*, and *Ganoderma lucidum* (Wojcieszyńska et al. 2014, Rodarte-Morales et al. 2012). For instance naproxen, which is commonly found both in surface and groundwater and whose photolytic degradation generates photoproducts more toxic than a parent molecule, can be degraded by the fungus *Trametes versicolor* in a few hours process (Marco-Urrea et al. 2010). According to the literature data, other fungi species like *Bjerkandera* sp. R1 and *Bj. adusta*, require longer time, e.g. seven days, for complete naproxen removal (Rodarte st al. 2011). On the other hand, there are a few literature data regarding NSAID removal by different bacterial strains. Among them, *Pseudomonas*, *Nocardia*, *Rhodococcus*, *Sphingomonas*, *Patulibacter* and *Stenotrophomonas* were mentioned (Wu et al. 2012). For instance, paracetamol can be degraded by *Delftia tsuruhatensis*, *Pseudomonas aeruginosa* and *Stenotrophomonas*, which use drug as a sole carbon and energy source. Above mentioned strains use different mechanisms and a variety of enzymes in biodegradation process.

9.3   Biodegradation of Antidepressants

Antidepressants, as other pharmaceuticals, are chemically stable and biologically active compounds, which can pollute the aquatic environment by inappropriate elimination in wastewater treatment plants, irresponsible disposal of expired drugs by patients and metabolic excretion (Minagh et al. 2009). There is significant lack of data regarding their biodegradation owing to natural microbial communities. One of the antidepressants is fluoxetine, which is characterized by very high toxicity against aquatic organisms. In natural surface waters where processes of natural microbial degradation occur, the half-life of this drug is estimated at 6–10 days (Nödler et al. 2014).

9.4   Biodegradation of Anticonvulsants

Selected drugs from the group of anticonvulsants, such as carbamazepine, are characterized by very stable structure, which results in their poor biodegradability. Except that they can be detected in surface and ground water, they were also found to accumulate in soil being highly refractory to microbial degradation (Maeng et al. 2011). However, they can be degraded in the presence of an extra substrate, e.g. glucose. According to Santosa et al. (2012), two unidentified basidiomycete can degrade carbamazepine in line with glucose (Santosa et al. 2012). Elsewhere, Gauthier et al. (2010) claimed that *Rhodococcus rhodochrous* and the fungus *Aspergillus niger* have the ability to degrade carbamazepine in liquid media also containing glucose.

9.5 Biodegradation of Blood Lipid Regulators

In the group of drugs that lower blood lipids, two subgroups can be distinguished with different mechanisms of action: statins and fibrates. Statins are the most commonly consumed drugs, but they rarely occur as environmental contaminants because they are mainly excreted as metabolites. Fibrates and their derivatives are drugs that inhibit the production of very low density lipoprotein (VLDL) and reduce plasma triglyceride levels and are also among the most commonly detected drugs in European waters (Gracia-Lor et al. 2012). Literature data from a few years ago reported that gemfibrozil is not biodegradable (Stumpf et al. 1999) However, it turned out that the fungus *Cunninghamella elegans* ATCC 9245 has the ability to break down the above-mentioned drug as a result of the hydroxylation process (Kang et al. 2009).

10. BIOREMEDIATION OF DYES

In recent years, the threat of dye contamination has reached the highest level, which is due to the fact that they are one of the most commonly used chemicals in many industries, especially in the textile industry. Dyes have many different application also in area of drugs, detergents and cosmetics. Group of dyes can be divided into many different sub-groups according to their chemical structures (acridine, anthraquinone, chromophoric, azin, and nitroso dyes), origin (natural and synthetic) and applications (vat dyes, dispersive dyes, and azoic colours). Non-ionic (disperse dyes), anionic (direct, acid and reactive dyes), and cationic forms (basic dyes) can be also distinguished (Vikrant et al. 2018). Textile dyes can be removed from the environment using selected microorganisms and biocatalysts; however, it should be taken into account that each of the methods has its own advantages and disadvantages associated with deinking efficiency, usefulness and working ability (depending on factors such as pH, temperature and dye concentration). Selected microorganisms are capable of synthesizing and releasing enzymes that can remove or neutralize harmful pollutants from the environment.

Literature data indicate that the colour of dyes reflects the presence of conjugated systems in chromophores. In addition to chromophores, certain dyes contain auxochromes (such as hydroxyl groups, amino groups and carboxylic acids), which are associated with colour change and with the solubility of compounds in various solvents (Xiang et al. 2016). The decomposition and thus discolouration of textile dyes occurs as a result of chemical reactions that break electronic bonds in chromophores. In the processes of effective dye degradation, both the pure cultures of bacteria, mixed cultures of bacteria and the combination of bacteria with fungi and algae are being used. Azo dyes belong to the group of the oldest industrially synthesized organic compounds and are probably the most dominant dyes present in industrial wastewater. The breakdown of the above dyes by bacteria involves the cleavage of azo bonds by various enzymes, mainly azoreductase and laccase, which are produced during the early stationary stage of microbial growth (Brüschweiler and Merlot 2017). For the decomposition of azo dyes *Pseudomonas entomophila* BS1, *Pseudomonas rettgeri* strain HSL1 and *Pseudomonas specie* SUK1 are being used (Lade et al. 2015). Literature data also indicate that among the different strains of *Penicillium oxalicum* fungi, the SAR-3 strain is capable of biodegradating azo dyes. *Candida, Magnusiomyces* and *Aspergillus niger* turned out to be also effective in azo dyes biodegradation (Mahmoud et al. 2017). Among other microorganisms, algae and cyanobacteria should be also mentioned as microorganisms able to the elimination of textile dyes. The following are mentioned in the scientific literature: *Anabaena flos-aquae* UTCC64, *Cosmarium, Chlorella vulgaris, Chlorella pyrenoidosa Lyngbya lagerlerimi, Nostoc linckia, Oscillatoria rubescens, Elkatothrix viridis* and *Volvox aureus Phormidium autumnale* UTEX1580, and *Synechococcus* sp. PCC7942 (El-Sheekh et al. 2009, Dellamatrice et al. 2017).

11. CONCLUSION

Using rhizospheric microorganisms in bioremediation of contaminated soil seems to be very useful and promising technique; however, it requires further research and improvement. In contrast to physicochemical methods, the application of biological methods for removing pollutants from the environment is economical and eco-friendly (Kumar and Gopal 2015). Furthermore, the capacity of rhizomicrobiota to detoxification and removal of toxic contaminants from the environment is used in the bioaugmentation technique. Contaminat-resistant microorganisms may enter into different environmental elements in order to improve the remediation efficiency. The colonization of bacteria introduced into the new environment is no longer a serious problem, but it is still observed that few introduced microorganisms survive. Therefore, the main future perspective is searching for strains that can remove and/or detoxify many contaminants at the same time (simultaneous removal of contaminants by microorganisms) (Zhang et al. 2019). Moreover, for a more efficient bioremediation process, it is advisable to combine different groups of microorganisms (e.g. bacteria, fungi and earthworms) to form a complex which support each other at various stages of pollution detoxification and removal (Mahohi and Raiesi 2019).

References

Adav, S.S., M.Y. Chen, D.J. Lee and N.Q. Ren. 2007. Degradation of phenol by aerobic granules and isolated yeast Candida tropicalis. Biotechnol. Bioeng. 96(5): 844–852.

Adhikary, A., R. Kumar, R. Pandir, P. Bhardwaj, R. Wusirika and S. Kumar. 2019. *Pseudomonas citronellolis*; A multi-metal resistant and potential plant growth promoter against arsenic(V) stress in chickpea. Plant Physiol. Biochem. 142: 179–192.

Ahn, D.W., S.S. Kim, S.J. Han and B.I. Kim. 2010. Characteristics of electrokinetic remediation of unsaturated soil contaminated by heavy metals-I: Experimental study. Int. J. Offshore Polar Eng. 20(2): 140–146.

Aidberger, H., M. Hasinger, R. Braun and A.P. Loibner. 2005. Potential of preliminary test methods to predict biodegradation performance of petroleum hydrocarbons in soil. Biodegradation 16: 115–125,

Amachi, S., M. Kasahara, S. Handa, Y. Kamagata and H. Shinoyama. 2003. Microbial participation in iodine volatilization from soils. Environ. Sci. Technol. 37: 3885–3890.

Anasonye, F., E. Winquist, M. Rasanen, J. Kontro, K. Bjorklof, G. Vasilyeva, et al. 2015. Bioremediation of TNT contaminated soil with fungi under laboratory and pilot scale conditions. Int. Biodeterior. Biodegradation 105: 7–12.

Badawi, N., S. Ronhede, S. Olsson, B.B. Kragelund, A.H. Johnsen, O.S. Jacobsen, et al. 2009. Metabolites of the phenylurea herbicides chlorotoluron, diuron, isoproturon and linuron produced by the soil fungus *Mortierella* sp. Environ. Pollut. 157(10): 2806–2812.

Barra Caracciolo, A., E. Topp and P. Grenni. 2015. Pharmaceuticals in the environment: Biodegradation and effects on natural microbial communities. A review. J. Pharm. Biomed. Anal. 106: 25–36.

Barragán, H.B., C. Costa, J. Peralta, J. Barrera, F. Esparza and R. Rodríguez. 2007. Biodegradation of organochlorine pesticides by bacteria grown in microniches of the porous structure of green vean coffee. Int. Biodeterior. Biodegr. 59: 239–244.

Bauermeister, A., R. Moeller, G. Reitz, S. Sommer and P. Rettberg. 2011. Effect of relative humidity on *Deinococcus radiodurans*' resistance to prolonged desiccation, heat, ionizing, germicidal, and environmentally relevant UV radiation. Microb. Ecol. 61: 715–22.

Berendsen, R.L., C.M. Pieterse and P.A. Bakker. 2012. The rhizosphere microbiome and plant health. Trends Plant Sci. 17(8): 478–486.

Bhatt, M., J.S. Zhao, A. Halasz and J. Hawari. 2006. Biodegradation of hexahydro-1, 3, 5-trinitro-1,3,5-triazine by novel fungi isolated from unexploded ordnance contaminated marine sediment. J. Ind. Microbiol. Biotechnol. 33: 850–858.

Bhatt, P., Y. Huang, H. Zhan and S. Chen. 2019. Insight into microbial applications for the biodegradation of pyrethroid insecticides. Front Microbiol. 10: 1778.

Bhosle, N.P. and A.S. Thore. 2016. Biodegradation of the herbicide 2,4-D by some fungi. Am. Eurasian J. Agric. Environ. Sci. 16(10): 1666–1671.

Bhushan, B., A. Halasz and J. Hawari. 2005. Biotransformation of CL-20 by a dehydrogenase enzyme from *Clostridium* sp. EDB2. Appl. Microbiol. Biotechnol. 69: 448–455.

Billi, D., E.I. Friedmann, K.G. Hoffer, M.G. Caiola and R. Ocampo-Friedman. 2000. Ionizing-radiation resistance in dessication-tole-rant cyanobacterium *Chroococcidiopsis*. Appl. Environ. Microbiol. 66: 1489–1492.

Blindauer, C.A., M.D. Harrison, A.K. Robinson, J.A. Parkinson, P.W. Bowness, P.J. Sadler, et al. 2002. Multiple bacteria encode metallothioneins and SmtA-like zinc fingers. Mol. Microbiol. 45(5): 1421–1432.

Boopathy, R. 2000. Bioremediation of explosives contaminated soil. Int. Biodeterior. Biodegr. 46(1): 29–36.

Bottoni P., S. Caroli and A. Barra Caracciolo. 2010. Pharmaceuticals as priority water contaminants. Toxicol. Environ. Chem. 92: 549–565.

Bouchez, T., D. Patureau, P. Dabert, S. Juretschko, J. Doré, J., P. Delgenès, et al. 2000. Ecological study of a bioaugmentation failure. Environ. Microbiol. 2: 179–190.

Boxall, A.B.A. 2004. The environmental side effects of medication-how are human and veterinary medicines in soils and water bodies affecting human and environmental health? EMBO Rep. 5: 1110–1116.

Brüschweiler, B.J. and C. Merlot. 2017. Azo dyes in clothing textiles can be cleaved into a series of mutagenic aromatic amines which are not regulated yet. Regul. Toxicol. Pharmacol. 88(Suppl C): 214–226.

Cao, B., K. Nagarajan and K.C. Loh. 2009. Biodegradation of aromatic compounds: Current status and opportunities for biomolecular approaches. Appl. Microbiol. Biotechnol. 85(2): 207–28.

Chambers, D.M., C.M. Reese, L.G. Thornburg, E. Sanchez, J.P. Rafson, B.C. Blount, et al. 2018. Distinguishing petroleum (crude oil and fuel) from smoke exposure within populations based on the relative blood levels of benzene, toluene, ethylbenzene, and xylenes (BTEX), styrene and 2,5-dimethylfuran by pattern recognition using artificial neural networks. Environ. Sci. Technol. 52: 308–316.

Chatterjee, S., G.B. Sau and S.K. Mukherjee. 2009. Plant growth promotion by a hexavalent chromium reducing bacterial strain, *Cellulosimicrobium cellulans* KUCr3. World J. Microbiol. Biotechnol. 25: 1829–1836.

Chaudhry, G.R. and S. Chapalamadugu. 1991. Biodegradation of halogenated organic compounds. Microbiol. Rev. 55: 59–79.

Chen, K.F., Ch.M. Kao, Ch. Chen, R.Y. Surampalli and M.S. Lee. 2010. Control of petroleum hydrocarbon contaminated ground water by intrinsic and enhanced bioremediation. J. Environ. Sci. 22: 846–871.

Confalonieri, F. and S. Sommer. 2005. Bacterial and archaeal resistance to ionizing radiation. J. Phys. Conf. Ser. 261: 1–15.

Cuthbert, M.O., M.S. Riley, S. Handley-Sidhu, J.C. Renshaw, D.J. Tobler, V.R. Phoenix, et al. 2012. Controls on the rate of ureolysis and the morphology of carbonate precipitated by *S. pasteurii* biofilms and limits due to bacterial encapsulation. Ecol. Eng. 41: 32–40.

Dellamatrice, P.M., M.E. Silva-Stenico, L.A.B.D. Moraes, M.F. Fiore and R.T.R. Monteiro. 2017. Degradation of textile dyes by cyanobacteria Braz. J. Microbiol. 48(1): 25–31.

Di Gregorio, S., S. Lampis and G. Vallini. 2005. Selenite precipitation by a rhizospheric strain of *Stenotrophomonas* sp. isolated from the root system of *Astragalus bisulcatus*: A biotechnological perspective. Environ. Int. 31(2): 233–241.

Dighton, J., T. Tugay and N. Zhdanova. 2008. Fungi and ionizing radiation from radionuclides. FEMS Microbiol. Lett. 281: 109–120.

Dolliver, H. and S. Gupta. 2008. Antibiotic losses in leaching and surface runoff from manure-amended agricultural land. J. Environ. Qual., 37: 1245–1253.

Egas, C., C. Barroso, H.J.C. Froufe, J. Pacheco, L. Albuqerque and M.S. Costa. 2014. Complete genome sequence of the radiation-resi-stant bacterium *Rubrobacter radiotolerans* RSPS-4. Stand. Geno-mics Sci. 9: 1062–1075.

El-Naas, M.H., S. Al-Zuhair and S. Makhlouf. 2010. Continuous biodegradation of phenol in a spouted bed bioreactor (SBBR). Chem. Eng. J. 160: 565–570.

El-Sebai, T., B. Lagacherie, G. Soulas and F. Martin-Laurent. 2007. Spatial variability of isoproturon mineralizing activity within an agricultural field: Geo-statistical analysis of simple physicochemical and microbiological soil parameters. Environ. Pollut. 145(3): 680–690.

El-Sheekh, M.M., M.M. Gharieb and G.W. Abou-El-Souod. 2009. Biodegradation of dyes by some green algae and cyanobacteria Int. Biodeterior. Biodegrad. 63(6): 699–704.

Esteves, E., M.C. Veiga and C. Kennes. 2004. Fungal biodegradation of toluene in gas-phase biofilters. *In*: W. Verstraete (ed.), European Symposium on Environmental Biotechnology, Taylor & Francis Group, London, pp. 337–340.

Etesami, H. 2018. Bacterial mediated alleviation of heavy metal stress and decreased accumulation of metals in plant tissues: Mechanisms and future prospects. Ecotox. Environ. Safe. 147: 175–191.

European Union 2011. Communication from the Commission to the European Parliament and the Council, Action Plan Against the Rising Threats from Antimicrobial Resistance. COM (accessed May 2019) http://ec.europa.eu/health/antimicrobial_resistance/key_documents/ index_en.htm

Finneran, K.T., M.E. Housewright and D.R. Lovley. 2002. Multiple influences of nitrate on uranium solubility during bioremediation of uranium-contaminated subsurface sediments. Environ. Microbiol. 4: 510–516.

Foti, L., F. Dubs, J. Gignoux, J.C. Lata, T.Z. Lerch, J. Mathieu, et al. 2017. Trace element concentrations along a gradient of urban pressure in forest and lawn soils of the Paris region (France). Sci. Total Environ. 598: 938–948.

Franzellitti, S., S. Buratti, P. Valbonesi and E. Fabbri. 2013. The mode of action (MOA) approach reveals interactive effects of environmental pharmaceuticals on *Mytilus galloprovincialis*. Aquat. Toxicol. 140–141: 249–256.

Gabrielson, J., I. Kühna, P. Colque-Navarro, M. Hart, A. Iversen, D. Mc Kenzie, et al. 2003. Microplate-based microbial assay for risk assessment and (eco)toxic fingerprinting of chemicals. Analytica Chimica Acta 485: 121–130.

Gauthier H., V. Yargeau and D.G. Cooper. 2010. Biodegradation of pharmaceuticals by *Rhodococcus rhodochrous* and *Aspergillus niger* by co-metabolism. Sci. Total Environ. 408: 1701–1706.

Germaine, K.J., X. Liu, G.G. Cabellos, J.P. Hogan, D. Ryan and D.N. Dowling. 2006. Bacterial endophyte-enhanced phytoremediation of the organochlorine herbicide 2,4-dichlorophenoxyacetic acid. FEMS Microbiol. Ecol. 57: 302–310.

Ghosh, P.K., T.K. Maiti, K. Pramanik, S.K. Ghosh, S. Mitra and T.K. De. 2018. The role of arsenic resistant *Bacillus aryabhattai* MCC3374 in promotion of rice seedlings growth and alleviation of arsenic phytotoxicity. Chemosphere. 213: 611–612.

Gracia-Lor E., J.V. Sancho, R. Serrano and F. Hernández. 2012. Occurrence and removal of pharmaceuticals in wastewater treatment plants at the Spanish Mediterranean area of Valencia. Chemosphere 87: 453–462.

Gribble, G.W. 2004. Amazing organohalogens. Am. Sci. 92: 342.

Gutierrez-Corona, J.F., P. Romo-Rodriguez, F. Santos-Escobar, A.E. Espino-Saldana and H. Hernandez-Escoto. 2016. Microbial interactions with chromium: Basic biological processes and applications in environmental biotechnology. World J. Microbiol. Biotechnol. 32: 191.

Ha, J., C.R. Engler and J. Wild. 2009. Biodegradation of coumaphos, chlorferon and diethylthiophosphate using bacteria immobilized in Ca-alginate gel beads. Bioresour. Technol. 100: 1138–1142.

Habe, H. and T. Omori. 2003. Genetics of polycyclic aromatic hydrocarbon metabolism in diverse aerobic bacteria. Biosci. Biotechnol. Biochem. 67(2): 225–43.

Hajabbasi, A.M. 2016. Importance of soil physical characteristics for petroleum hydrocarbons phytoremediation: A review. Afr. J. Environ. Sci. Technol. 10(11): 394–405.

Haritash, A.K. and C.P. Kaushik. 2009. Biodegradation aspects of polycyclic aromatic hydrocarbons (PAHs): A review. J. Hazard. Mater. 169: 1–15.

Hietala, K.A. and T.M. Roane. 2009. Microbial remediation of metals in soils. *In.* A. Singh, R.C. Kuhad, O.P. Ward (eds), Advances in Applied Bioremediation, Soil Biology. Springer, Berlin, Heidelberg, pp. 201–220.

Hu, G.P., Y. Zhao, F.Q. Song, B. Liu, L. Vasseur, C. Douglas, et al. 2014. Isolation, identification and cyfluthrin-degrading potential of a novel *Lysinibacillus sphaericus* strain, FLQ-11-1. Res. Microbiol. 165(2): 110–118.

Husaini, A., H.A. Roslan, K.S.Y. Hii and C.H. Ang. 2008. Biodegradation of aliphatic hydrocarbon by indigenous fungi isolated from used motor oil contaminated sites. World J. Microbiol. Biotechnol. 24(12): 2789–2797.

Hussain, S., M. Devers-Lamrani, N. El Azhari and F. Martin-Laurent. 2011. Isolation and characterization of an isoproturon mineralizing *Sphingomonas* sp. strain SH from a French agricultural soil. Biodegradation 22: 1637–1650.

Javorekova, S.N., I. Svrcbreve and J. Makova. 2010. Influence of benomyl and prometryn on the soil microbial activities and community structures in pasture grasslands of Slovakia. J. Environ. Sci. Health B 45: 702–709.

Jensen, J.K., P.E. Holm, J. Nejrup, M.B. Larsen and O.K. Borggaard. 2009. The potential of willow for remediation of heavy metal polluted calcareous urban soils. Environ. Pollut. 157: 931–937.

Jha, S.K., J. Paras and H.P. Sharma. 2015. Xenobiotic degradation by bacterial enzymes. Int. J. Curr. Microbiol. App. Sci. 4(6): 48–62.

Kang, S.I., S.Y. Kang, R.A. Kanaly, E. Lee, Y. Lim and H.G. Hur. 2009. Rapid oxidation of ring methyl groups is the primary mechanism of biotransformation of gemfibrozil by the fungus *Cunninghamella elegans*. Arch. Microbiol. 191: 509–517.

Kaplan, C. and C. Kitts. 2004. Bacterial succession in a petroleum land treatment unit. Appl. Environ. Microbiol. 70(3): 1777–1786.

Kebamo, S., S. Tesema and B. Geleta. 2015. The role of biotransformation in drug discovery and development. J. Drug Metab. Toxicol. 6: 196–202.

Kennes, C. and M.C. Veiga. 2004. Fungal biocatalysts in the biofiltration of VOC-polluted air. J. Biotechnol. 113: 305–319.

Kim, S.J., O. Kweon, R.C. Jones, J.P. Freeman, R.D. Edmondson and C.E. Cerniglia. 2007. Complete and integrated pyrene degradation pathway in *Mycobacterium vanbaalenii* PYR-1 based on systems biology. J. Bacteriol. 189: 464–472.

Kulkarni, S., A. Ballal and S.K. Apte. 2013. Bioprecipitation of uranium from alkaline waste solutions using recombinant *Deinococcus radiodurans*. J. Hazard. Mater. 262: 853–861.

Kumar, R.A.J., S. Singh and O.V. Singh. 2007. Bioremediation of radionuclides: emerging technologies. OMICS: Int. J. Integr. Biol. 11(3): 295.

Kumar, B.L. and D.V.R.S. Gopal. 2015. Effective role of indigenous microorganisms for sustainable environment. 3 Biotech. 5: 867–876.

Kümmerer, K. 2003. Significance of antibiotics in the environment. J. Antimicrob. Chemother. 52: 5–7.

Lade, H., A. Kadam, D. Paul, S. Govindwar. 2015. Biodegradation and detoxification of textile azo dyes by bacterial consortium under sequential microaerophilic/aerobic processes. Excli J. Exp. Clin. Sci. 14: 158–174.

Laffin, B., M. Chavez and M. Pine. 2010. The pyrethroid metabolites 3-phenoxybenzoic acid and 3-phenoxybenzyl alcohol do not exhibit estrogenic activity in the MCF-7 human breast carcinoma cell line or sprague-dawley rats. Toxicol. 267: 39–44.

Lapworth, D.J., N. Baran, M.E. Stuart and R.S. Ward. 2012. Emerging organic contaminants in groundwater: A review of sources, fate and occurrence Environ. Poll. 163: 287–303.

Lee, J., K. Jung, S. Choi and H. Kim. 1995. Combination of the tod and the tol pathways in redesigning a metabolic route of *Pseudomonas putida* for the mineralization of a benzene, toluene, and p-xylene mixture. Appl. Environ. Microbiol. 61: 2211–2217.

Lewis, T.A., D.A. Newcombe and R.L. Crawford. 2004. Bioremediation of soils contaminated with explosives. J. Environ. Manage. 70: 291–307.

Li, X., P. Li, X. Lin, C. Zhang, Q. Li and Z. Gong. 2007. Biodegradation of aged polycyclic aromatic hydrocarbons (PAHs) by microbial consortia in soil and slurry phases. J. Hazard. Mater. 150(1): 21–26.

Liste, H. and D. Felgentreu. 2006. Crop growth, culturable bacteria, and degradation of petrol hydrocarbons in a long term contaminated field solid. Appl. Soil Ecol. 31: 43–52.

Liu, J., W. Huang, H. Han, C. She and G. Zhong. 2015. Characterization of cell-free extracts from fenpropathrin-degrading strain *Bacillus cereus* ZH-3 and its potential for bioremediation of pyrethroid-contaminated soils. Sci. Total Environ. 523: 50–58.

Lloyd, J.R. and J.C. Renshaw. 2005. Bioremediation of radioactive waste: Radionuclide-microbe in interactions in laboratory and field-scale studies. Curr. Opin. Biotechnol. 16: 254–260.

Lobastova, T.G., G.V. Sukhodolskaya, V.M. Nikolayeva, B.P. Baskunov, K.F. Turchin and M.V. Donova. 2004. Hydroxylation of carbazoles by *Aspergillus flavus* VKM F-1024. FEMS Microbiol. Lett. 235(1): 51–56.

Ma, Y., R.S. Oliveira, L. Wu, Y. Luo, M. Rajkumar, I. Rocha and H. Freitas. 2015. Inoculation with metal-mobilizing plant-growth-promoting rhizobacterium *Bacillus* sp. SC2b and its role in rhizoremediation. J. Toxicol. Env. Heal. Part A. 78(13-14): 931–944.

Ma, Y., R.S. Oliveira, H. Freitas and C. Zhang. 2016. Biochemical and molecular mechanisms of plant-microbe-metal interactions: Relevance for phytoremediation. Front. Plant Sci. 7.

Maeng, S.K., S.K. Sharma, C.D.T. Abel, A. Magic-Knezev and G.L. Amy. 2011. Role of biodegradation in the removal of pharmaceutically active compounds with different bulk organic matter characteristics through managed aquifer recharge: Batch and column studies. Water Res. 45L: 4722–4736.

Mahmoud, M.S., M.K. Mostafa, S.A. Mohamed, N.A. Sobhy and M. Nasr. 2017. Bioremediation of red azo dye from aqueous solutions by *Aspergillus niger* strain isolated from textile wastewater. J. Environ. Chem. Eng. 5(1): 547–554.

Mahohi, A. and F. Raiesi. 2019. Functionally dissimilar soil organisms improve growth and Pb/Zn uptake by *Stachys inflata* grown in a calcareous soil highly polluted with mining activities. J. Environ. Manage. 247: 780–789.

Majumder, A., K. Bhattacharyya, S. Bhattacharyya and S.C. Kole. 2013. Arsenic-tolerant, arsenite-oxidising bacterial strains in the contaminated soils of West Bengal. India. Sci. Total Environ. 463: 1006–1014.

Maliszewska-Kordybach, B., B. Smreczak, A. Klimkowicz-Pawlas and H. Terelak. 2008. Monitoring of the total content of polycyclic aromatic hydrocarbons (PAHs) in arable soils in Poland. Chemosphere 73: 1284–1291.

Mallick, S., S. Chatterjee and T.K. Dutta. 2007. A novel degradation pathway in the assimilation of phenanthrene by Staphylococcus sp. strain PN/Y via meta-cleavage of 2-hydroxy-1-naphthoic acid: Formation of trans-2,3-dioxo-5-(2'-hydroxyphenyl)-pent-4-enoic acid. Microbiol. 153: 2104–2115.

Marco-Urrea, E., M. Perez-Trujillo, P. Blanquez, T. Vicent and G. Caminal. 2010. Biodegradation of the analgesic naproxen by *Trametes versicolor* and identification of intermediates using HPLC-DAD-MS and NMR. Bioresour. Technol. 101: 2159–2166.

Marti R., A. Scott, Y.C. Tien, R. Murray, L. Sabourin, Y. Zhang, et al. 2013. Impact of manure fertilization on the abundance of antibiotic-resistant bacteria and frequency of detection of antibiotic resistance genes in soil and on vegetables at harvest Appl. Environ. Microbiol. 79: 5701–5709.

Mc Eneff, G., L. Barron, B. Kelleher, B. Paull and B. Quinn. 2014. A year-long study of the spatial occurrence and relative distribution of pharmaceutical residues in sewage effluent, receiving marine waters and marine bivalves. Sci. Total Environ. 476-477: 317–326.

McGuinness, M. and D. Dowling. 2009. Plant-associated bacterial degradation of toxic organic compounds in soil. Int. J. Environ. Res. Public Health. 6: 2226–2247.

Memon, A.R. and P. Schröder. 2009. Implications of metal accumulation mechanisms to phytoremediation. Environ. Sci. Pollut. Res. 16: 162–175.

Mendes, R. 2013. The rhizosphere microbiome: Significance of plant beneficial, plant pathogenic, and human pathogenic microorganisms. FEMS Microbiol. Rev. 37: 634–663.

Meyer, J., A. Schmidt, K. Michalke and R. Hensel. 2007. Volatilisation of metals and metalloids by the microbial population of an alluvial soil. Syst. Appl. Microbiol. 30: 229–238.

Minagh, E., R. Hernan, K. O'Rourke, F.M. Lyng and M. Davoren. 2009. Aquatic ecotoxicity of the selective serotonin reuptake inhibitor sertraline hydrochloride in a battery of freshwater test species. Ecotoxicol. Environ. Saf. 72: 434–440.

Mishra, J., R. Singh and N.K. Arora. 2017. Alleviation of heavy metal stress in plants and remediation of soil by rhizosphere microorganisms. Front. Microbiol. 8: 1706.

Mitra, J.1., P.K. Mukherjee, S.P. Kale and N.B. Murthy. 2001. Bioremediation of DDT in soil by genetically improved strains of soil fungus *Fusarium solani*. Biodegrad. 12(4): 235–245.

Mrozik, A. and Z. Piotrowska-Seget. 2010. Bioaugmentation as a strategy for cleaning up of soils contaminated with aromatic compounds. Microbiol. Res. 165: 363–375.

Mulligan, C.N., R.N. Yong and B.F. Gibbs. 2001. Heavy metal removal from sediments by biosurfactants. J. Hazard. Mater. 85: 111–125.

Narendrula-Kotha, R. and K.K. Nkongolo. 2017. Microbial response to soil liming of damaged ecosystems revealed by pyrosequencing and phospholipid fatty acid analyses. PLoS ONE. 12(1): e0168497.

Nazir, N., N.K. Azra, M.Y. Zargar, I. Khan, D. Shah, J.A. Parray, et al. 2016. Effect of root exudates on rhizosphere soil microbial communities. J. Res. Dev. 16: 88–95.

Ndibe, T., B. Benjamin, W. Eugene and J. Usman. 2018. A review on biodegradation and biotransformation of explosive chemicals. European J. Eng. Res. Sci. 3(11): 58–65.

Nejidat, A., L. Kafka, Y. Tekoah and Z. Ronen. 2008. Effect of organic and inorganic nitrogenous compounds on RDX degradation and cytochrome P-450 expression in *Rhodococcus* strain YH1. Biodegrad. 19: 313–320.

Newsome, L, K. Morris and J.R. Lloyd. 2014. The biogeochemistry and bioremediation of uranium and other priority radionuclides. Chem. Geol. 363: 164–184.

Nödler, K., D. Voutsa and T. Licha. 2014. Polar organic micropollutants in the coastal environment of different marine systems Mar. Pollut. Bull. 85: 50–59.

Oleszczuk, P. 2007. Organic pollutants in sewage sludge-amended soil. Part II fate of contaminants in soils. Ecol. Chem. Eng. 14(2): 185.

Ortiz-Hernández, M.L., E. Sánchez-Salinas, E. Dantán-González and M.L. Castrejón-Godínez. 2013. Pesticide biodegradation: mechanisms, genetics and strategies to enhance the process. *In:* R. Chamy and F. Rosenkranz (eds), Biodegradation— Life of Science. InTech, Croatia, pp. 251–287.

Oves, M., M.S. Khan and A. Zaidi, 2013. Chromium reducing and plant growth promoting novel strain *Pseudomonas aeruginosa* OSG41 enhance chickpea growth in chromium amended soils. Eur. J. Soil Biol. 56: 72–83.

Oves, M., M. Saghir Khan, A. Huda Qari, M. Nadeen Felemban and T. Almeelbi. 2016. Heavy metals: Biological importance and detoxification strategies. J. Bioremediat. Biodegrad. 7: 334.

Padmavathiamma, P.K. and L.Y. Li. 2007. Phytoremediation technology: Hyper-accumulation metals in plants. Water Air Soil Pollut. 184: 105–126.

Padmavathiamma, P.K. and L.Y. Li. 2012. Rhizosphere influence and seasonal impact on phytostabilisation of metals—A field study. Water Air Soil Pollut. 223: 107–124.

Pal, R., S. Bala, M. Dadhwal, M. Kumar, G. Dhingra, O. Prakash, et al. 2005. Hexachlorocyclohexane-degrading bacterial strains *Sphingomonas paucimobilis* B90A, UT26 and Sp$^+$, having similar lin genes, represent three distinct species, *Sphingobium indicum* sp. nov., *Sphingobium japonicum* sp. nov. and *Sphingobium francense* sp. nov., and reclassification of [*Sphingomonas*] chungbukensis as *Sphingobium chungbukense* comb. nov. Int. J. Syst. Evol. Microbiol. 55: 1965–1972.

Panda, J., T. Kanjilal and S. Das. 2018. Optimized biodegradation of carcinogenic fungicide Carbendazim by *Bacillus licheniformis* JTC-3 from agro-effluent. Biotechnol. Res. Innov. 2: 45–57.

Park, C., T.H. Kim, S. Kim, S.W. Kim, J. Lee and S.H. Kim. 2003. Optimization for biodegradation of 2,4,6-trinitrotoluene (TNT) by *Pseudomonas putida*. J. Biosci. Bioeng. 95: 567–571.

Park, J.H., N. Bolan, M. Megharaj and R. Naidu. 2011. Isolation of phosphate solubilizing bacteria and their potential for lead immobilization in soil. J. Hazard. Mater. 185: 829–836.

Parthipan, P., E. Preetham., L.L. Machuca, P.K.S.M. Rahman, K. Murugan and A. Rajasekar. 2017. Biosurfactant and degradative enzymes mediated crude oil degradation by bacterium *Bacillus subtilis* A1. Front. Microbiol. 8, doi 10.3389/fmicb.2017.00193.

Pawlik, M., B. Cania, S. Thijs and J. Vangronsveld. 2017. Hydrocarbon degradation potential and plant growth-promoting activity of culturable endophytic bacteria of *Lotus corniculatus* and Oenothera biennis from a long-term polluted site. Environ. Sci. Pollut. Res. 24: 19640–19652.

Piccirillo, C., S.I.A. Pereira, A.P.G.C. Marques, R.C. Pullar, D.M. Tobaldi, M.E. Pintado and P.M.L. Castro. 2013. Bacteria immobilisation on hydroxyapatite surface for heavy metals removal. J. Environ. Manage. 121: 87–95.

Pires, C., A.R. Franco, S.I.A. Pereira, I. Henriques, A. Correia, N. Magan, et al. 2017. Metal(loid)-contaminated soils as a source of culturable heterotrophic aerobic bacteria for remediation applications. Geomicrobiol. J. 34(9): 760–768.

Pokethitiyook, P. and T. Poolpak. 2016. Biosorption of heavy metal from aqueous solutions. *In*: A.A. Ansari, S.S. Gill, R. Gill, G.R. Lanza and L. Newman (eds), Phytoremediation: Management of Environmental Contaminants. Springer, Switzerland, pp. 113–114.

Popov, V.H., P.S. Cornish, K. Sultana and E.C. Morris. 2005. Atrazine degradation in soils: The role of microbial communities, atrazine application history, and soil carbon. Aust. J. Soil Res. 43: 861–871.

Qin, K., L.S. Zhu and J.H. Wang. 2010. Screening and degradation characteristics of cypermethrin degrading fungi. Chin J. Environ. Eng. 4: 950–954.

Rabus, R., M. Boll, J. Heider, R.U. Meckenstock, W. Buckel, O. Einsle, et al. 2016. Anaerobic microbial degradation of hydrocarbons: from enzymatic reactions to the environment. J. Mol. Microbiol. Biotechnol. 26: 5–28.

Rajkumar, M., Y. Ma and H. Freitas. 2013. Improvement of Ni phytostabilization by inoculation of Ni resistant *Bacillus megaterium* SR28C. J. Environ. Manage. 128: 973–980.

Reena, R., M.C. Majhi, A. Kumar Arya, R. Kumar and A. Kumar. 2012. BioRadBase: A database for bioremediation of radioactive waste. Afr. J. Biotechnol. 11(35): 8718–8721.

Rodarte-Morales, A.I., G. Feijoo, M.T. Moreira and J.M. Lema. 2012. Biotransformation of three pharmaceutical active compounds by the fungus *Phanerochaete chrysosporium* in a fed batch stirred reactor under air and oxygen supply. Biodegrad. 23: 145–156.

Saikia, N. and M. Gopal. 2004. Biodegradation of b-cyfluthrin by fungi. J. Agric. Food Chem. 52: 1220–1223.

Salam, L.B., M.O. Ilori, O.O. Amund, M. Numata, T. Horisaki and H. Nojiri. 2014. Carbazole angular dioxygenation and mineralization by bacteria isolated from hydrocarbon-contaminated tropical African soil. Environ. Sci Pollut. Res. Int. 21(15): 9311–24.

Saleem, H. 2016. Plant-bacteria partnership: Phytoremediation of hydrocarbons contaminated soil and expression of catabolic genes. Bull. Environ. Stud. 1: 18–28.

Santos, J., I. Aparicio, M. Callejón and E. Alonso. 2009. Occurrence of pharmaceutically active compounds during 1-year period in wastewaters from four wastewater treatment plants in Seville (Spain). J. Hazard. Mater. 164: 1509–1516.

Santosa, I.J., M.J. Grossmana, A. Sartorattob, A.N. Ponezib and L.R. Durranta. 2012. Degradation of the recalcitrant pharmaceuticals carbamazepine and 17α-ethinylestradiol by ligninolytic fungi Chem. Eng. 27: 169–174.

Sarma, S.J., K. Pakshirajan and B. Mahanty. 2011. Chitosan-coated alginate-polyvinyl alcohol beads for encapsulation of silicone oil containing pyrene: A novel method for biodegradation of polycyclic aromatic hydrocarbons. J. Chem. Technol. Biotechnol. 86: 266–272.

Sašek, W., T. Cajthaml and M. Bhatt. 2003. Use of fungal technology in soil remediation: A case study. Water Air Soil Pollut. 3: 5–14.

Sassolas, A., B. Prieto-Simón and J.L. Marty. 2012. Biosensors for pesticide detection: New trends. Am. J. Anal. Chem. 3: 210–232.

Senthilkumar, S., A. Anthonisamy, S. Arunkumar and V. Sivakumari. 2011. Biodegradation of methyl parathion and endosulfan using *Pseudomonas aeruginosa* and *Trichoderma viridae*. J. Environ. Sci. Eng. 53(1): 115–122.

Sessitsch A., M. Kuffner, P. Kidd, J. Vangronsveld, W. Wenzel, K. Fallmann, et al. 2013. The role of plant-associated bacteria in the mobilization and phytoextraction of trace elements in contaminated soils. Soil Biol. Biochem. 60: 182–194.

Sharma, P.K., D.L. Balkwill, A. Frenkel and M.A. Vairavamurthy. 2000. A new *Klebsiella planticola* strain (Cd-1) grows anaerobically at high cadmium concentrations and precipitates cadmium sulfide. Appl. Environ. Microbiol. 66: 3083–3087.

Sharma, J., K. Shamim, S.K. Dubey and R.M. Meena. 2017. Metallothionein assisted periplasmic lead sequestration as lead sulfite by *Providencia vermicola* strain SJ2A. Sci. Total Environ. 579: 359–365.

Singh, D.K. 2008. Biodegradation and bioremediation of pesticide in soil: Concept, method and recent developments. Indian J. Microbiol. 48: 35–40.

Singh, B., J. Kaur and K. Singh 2012. Microbial remediation of explosive waste. Critical Nat. Rev. Microbiol. 38(2): 152–167.

Sinha, S., P. Chattopadhyay, I. Pan, S. Chatterjee, P. Chanda, D. Bandyopadhyay, et al. 2009. Microbial transformation of xenobiotics for environmental bioremediation. Afr. J. Biotechnol. 8(22): 6016–6027.

Siripattanakul, S., W. Wirojanagud, J.M. McEvoy, F.X. Casey and E. Khan. 2009. Atrazine removal in agricultural infiltrate by bioaugmented polyvinyl alcohol immobilized and free *Agrobacterium radiobacter* J14a: A sand column study. Chemosphere 74: 308–313.

Souza R., A. Ambrosini, L.M.P. Passaglia. 2015. Plant growth-promoting bacteria as inoculants in agricultural soils. Genet. Mol. Biol. 38: 401–419.

Stumpf, M., T.A. Ternes, R.D. Wilken, S.V. Rodrigues and W. Baumann. 1999. Polar drug residues in sewage and natural waters in the state of Rio de Janeiro. Brazil Sci. Total Environ. 225: 135–141.

Su, J., W. Ouyang, Y. Hong, D. Liao, S. Khan and H. Li. 2016. Responses of endophytic and rhizospheric bacterial communities of salt marsh plant (*Spartina alterniflora*) to polycyclic aromatic hydrocarbons contamination. J. Soil. Sediment. 16: 707–715.

Sun, M., D. Fu, Y. Teng, Y. Shen, Y. Luo, Z. Li and P. Christie. 2011. In situ phytoremediation of PAH-contaminated soil by intercropping alfalfa (*Medicago sativa* L.) with tall fescue (*Festuca arundinacea* Schreb.) and associated soil microbial activity. J. Soils Sediments. 11: 980–989.

Sun, K., J. Liu, L. Jin and Y. Gao. 2014. Utilizing pyrene-degrading endophytic bacteria to reduce the risk of plant pyrene contamination. Plant Soil. 374: 251–262.

Sviridov, A.V., T.V. Shushkova, I.T. Ermakova, E.V. Ivanova, D.O. Epiktetov and A.A. Leontievsky. 2015. Microbial degradation of glyphosate herbicides (Review). Appl. Biochem. Microbiol. 51: 188–195.

Thapa, B., A.K.C. Kumar and A. Ghimire. 2012. A review on bioremediation of petroleum hydrocarbon contaminants in soil. J. Sci. Eng. Technol. 8: 164–170.

Thijs, S., P. Van Dillewijn, W. Sillen, S. Truyens, M. Holtappels, J. D'Haen, et al. 2014. Exploring the rhizospheric and endophytic bacterial communities of *Acer pseudoplatanus* growing on a TNT-contaminated soil: Towards the development of a rhizocompetent TNT-detoxifying plant growth promoting consortium. Plant Soil. 385: 15–36.

Topp, E., R. Chapman, M. Devers-Lamrani, A. Hartmann, R. Marti, F. Martin-Laurent, et al. 2013. Accelerated biodegradation of veterinary antibiotics in agricultural soil following long-term exposure, and isolation of a sulfamethazine-degrading *Microbacterium* sp. J. Environ. Qual. 42: 173–178.

Ullrich, R. and M. Hofrichter. 2007. Enzymatic hydroxylation of aromatic compounds. Cell. Mol. Life Sci. 64: 271–93.

Vaillancourt, F.H., J.T. Bolin and L.D. Eltis. 2006. The ins and outs of ring-cleaving dioxygenases. Crit. Rev. Biochem. Mol. Biol. 41: 241–267.

Van Aken, B., J.M. Yoon and J.L. Schnoor. 2004. Biodegradation of nitrosubstituted explosives 2,4,6-trinitrotoluene, hexahydro-1,3,5-trinitro-1,3,5-triazine, and octahydro-1,3,5,7-tetranitro-1,3,5-tetrazocine by a photosymbiotic *Methylobacterium* sp. associated with poplar tissues (*Populus deltoids × nigra* DN34). Appl. Environ. Microbiol. 70: 508–517.

van Herwijnen, R., D. Springael, P. Slot, H.A.J. Govers and J.R. Parsons. 2003. Degradation of Anthracene by *Mycobacterium* sp. strain LB501T proceeds via a novel pathway, through o-phthalic acid. Appl. Environ. Microbiol. 69: 186–190.

Vidali, M. 2001. Bioremediation: An overview. Pure Appl. Chem. 73: 1163–1172.

Vikrant, K., B.S. Giri, N. Raza, K. Roy, K.H. Kim, B.N. Rai, et al. 2018. Recent advancements in bioremediation of dye: Current status and challenges. Bioresour. Technol. 253: 355–367.

Wan S.G., G.Y. Li, T.C. An, B. Guo, L. Sun, L. Zu, et al. 2010. Biodegradation of ethanethiol in aqueous medium by a new *Lysinibacillus sphaericus* strain RG-1 isolated from activated sludge. Biodegrad. 21: 1057–66.

Wang, B., Y. Ma, W. Zhou, J. Zheng, J. Zhu, J. He, et al. 2011. Biodegradation of synthetic pyrethroids by *Ochrobactrum tritici* strain pyd-1. World J. Microbiol. Biotechnol. 27: 2315–2324.

Wang, J. and S. Wang. 2016. Removal of pharmaceuticals and personal care products (PPCPs) from wastewater: A review. J. Environ. Manage. 182: 620–640.

Wei, S. and Q. Zhu. 2008. Trace elements in agro-ecosystems. *In*: M.N.V. Prasad (ed.), Trace elements as Contaminants and Nutrients-consequences in Ecosystems and Human Health. Wiley, New Jersey, USA, pp. 55–80.

Weyens N., D. van der Lelie, T. Artois, K. Smeets, S. Taghavi, L. Newman, et al. 2009. Bioaugmentation with engineered endophytic bacteria improves phytoremediation. Environ. Sci. Technol. 43: 9413–9418.

Wiesche, C., R. Martens and F. Zadrazil. 2003. The effect of interaction between white-root fungi and indigenous microorganisms on degradation of polycyclic aromatic hydrocarbons in soil. Water Air Soil Pollut. 3: 73–79.

Wojcieszyńska, D., D. Domaradzka, K. Hupert-Kocurek and U. Guzik. 2014. Bacterial degradation of naproxen-undisclosed pollutant in the environment. J. Environ. Manage. 1: 157–161.

Wu, S., L. Zhang and J. Chen. 2012. Paracetamol in the environment and its degradation by microorganism Appl. Microbiol. Biotechnol. 96: 875–884.

Xiang, X., X. Chen, R. Dai, Y. Luo, P. Ma, S. Ni, et al. 2016. Anaerobic digestion of recalcitrant textile dyeing sludge with alternative pre-treatment strategies. Bioresour. Technol. 222(Suppl C): 252–260.

Xiao, P., T. Mori, I. Kamei, H. Kiyota, K. Takagi and R. Kondo. 2011. Novel metabolic pathway of organochlorine pesticides dieldrin and aldrin by the white rot fungi of the genus *Phlebia*. Chemosphere 85: 218–224.

Yamazoe, A., O. Yagi and H. Oyaizu. 2004. Biotransformation of fluorene, diphenyl ether, dibenzo-p-dioxin and carbazole by *Janibacter* sp. Biotechnol. Lett. 26: 479.

Yan-de, J., H. Zhen-li, Y. Xiao-e. 2007. Role of soil rhizobacteria in phytoremediation of heavy metal contaminated soils. J. Zhejiang Univ. Sci. B. 8(3): 192–207.

Yu, R., Q. Liu, J. Liu, Q. Wang and Y. Wang. 2016. Concentrations of organophosphorus pesticides in fresh vegetables and related human health risk assessment in Changchun, Northeast China. Food Control. 60: 353–360.

Zaidi, S., S. Usmani, B.R. Singh and J. Musarrat. 2006. Significance of *Bacillus subtilis* strain SJ 101 as a bioinoculant for concurrent plant growth promotion and nickel accumulation in *Brassica juncea*. Chemosphere 64: 991–997.

Zawadzka, A.M., R.L. Crawford and A.J. Paszczynski. 2006. Pyridine-2,6-bis(thiocarboxylic acid) produced by *Pseudomonas stutzeri* KC reduces and precipitates selenium and tellurium oxyanions. Appl. Environ. Microbiol. 72: 3119–3129.

Zawadzka, A.M., R.L. Crawford and A.J. Paszczynski. 2007. Pyridine-2,6-bis(thiocarboxylic acid) produced by *Pseudomonas stutzeri* KC reduces chromium(VI) and precipitates mercury, cadmium, lead and arsenic. BioMetals. 20: 145–158.

Zhang, L., J. Hu, R. Zhu, Q. Zhou and J. Chen. 2013. Degradation of paracetamol by pure bacterial cultures and their microbial consortium. Appl. Microbiol. Biotechnol. 97: 3687–3698.

Zhang, X., X. Liu, Q. Wang, X. Chen, H. Li, J. Wei, et al. 2014. Diesel degradation potential of endophytic bacteria isolated from *Scirpus triqueter*. Int. Biodeterior. Biodegrad. 87: 99–105.

Zhang, Y., H. Wang, X. Wang, B. Hu, C. Zhang, W. Jin, et al. 2017. Identification of the key amino acid sites of the carbendazim hydrolase (MheI) from a novel carbendazim-degrading strain *Mycobacterium* sp. SD-4. J. Hazard. Mater. 331: 55–62.

Zhang, C., Y. Tao, S. Li, J. Tian, T. Ke, S. Wei, et al. 2019. Simultaneous degradation of trichlorfon and removal of Cd(II) by *Aspergillus sydowii* strain PA F-2. Environ. Sci. Pollut. R. 26(26): 26844–26854.

Ziganshin, A.M., R.P. Naumova, A.J. Pannier and R. Gerlach. 2010. Influence of pH on 2,4,6-trinitrotoluene degradation by *Yarrowia lipolytica*. Chemosphere 79: 426–433.

Zuccato, E., S. Castiglioni, R. Bagnati, M. Melis and R. Fanelli. 2010. Source, occurrence and fate of antibiotics in the Italian aquatic environment. J. Hazard. Mater. 179(1–3): 1042–1048.

CHAPTER 15

Bioremediation of Hydrocarbons and Classic Explosives:
An Environmental Technology Removing Hazardous Wastes

Ahmed Ibrahim Jessim

Center for Research and Evaluation, Directorate of Treatment and Disposal of Chemical, Biological and Military Hazardous Wastes, Ministry of Science and Technology, Iraq.

1. INTRODUCTION

Bioremediation is a term refers that to a process used for treatment of contaminated soils, waters, clays and any subsurface material via modifications of environmental conditions in order to stimulate the growth of microorganisms to degrade the target pollutants (Azubuike et al. 2016). Before 1919, when bioremediation was identified, the term biotechnology has been coined by Karl Ereky when referring to the use of living organisms or their products to benefit the human health and environment (Judge 2004). Simply, bioremediation is using microorganisms in order to degrade the pollutants (Abatenh et al. 2017). In many cases of bioremediation, studies have shown that they have less cost, efficiency benefits and are more sustainable than other conventional remediation techniques (Paniagua-Michel and Rosales 2015, Guo et al. 2019). Bioremediation can also be applied in those areas where we cannot reach without drilling, for example, hydrocarbons' leaks or spills in deep. Bioremediation costs are lower than other conventional methods because there is no use of costly chemicals and physical equipments, and microorganisms adapt themselves in the surrounding environmental conditions and use hydrocarbon as a source of carbon and energy (Azubuike et al. 2016). Bioremediation is also used to treat different types of wastes, including wastewater (Fei et al. 2018), solid wastes (Leslie Grady et al. 1999), and industrial wastes (Kowalczyk et al. 2015). Since the last two decades, bioremediation has been adopted globally and is used as a reliable approach (Semple et al. 1999). Various microorganisms having bioremediation potential such as bacteria, fungi and some species of blue-green algae are being used to degrade hydrocarbons and other organic compounds (Sorkhoh et al. 1995).

In bioremediation process, the microbes convert the toxic or organic compounds into water, CO_2 and other non-toxic or less toxic compounds. Single microbe can alone metabolize a narrow range of hydrocarbons by means of cellular metabolism, but the consortium of these have an extended enzymatic range for better biodegradation. The potential biodegrader microbial community are generally isolated from hydrocarbon polluted soils (Olajire and Essien 2014, Ambrazaitienė et al. 2013). All oil spills adversely affect the particular ecosystem, resulting in accumulation of toxic compounds, and the soil micro as well macro flora and fauna being negatively affected (Gumuscu and Tekinay 2013). The toxic wastes of explosives are also being remediated by different strains of microbes, chiefly by bacteria (Cooper and Kurowski 1997). There are about 60 highly explosive compounds, which were synthesized and developed to use for military purposes (Lin et al. 2013, Morley et al. 2006). Due to routine military activities, the natural soil suffers from accumulation of explosives and it goes on accumulating. In due course of time, the explosive residues can find their ways to the groundwater and may spread with wind to different areas (Phelan et al. 2002, Hamoudi-Belarbi et al. 2018). From these toxic explosive materials, Hexahydro-1,3,5-trinitro-1,3,5, triazine (RDX), 2,4,6-Trinitrotoluene, globally considered as the most known explosive material (TNT), and 1,3,5,7-tetranitro-1,3,5,7-tetrazine (HMX) are the main culprits for soil toxicity. These toxic compounds are heat stable, have high density and high persistence (Cooper and Kurowski 1997, Lin et al. 2013). The environmental pollution with hydrocarbons and explosive materials has become a global phenomenon and all efforts are needed to remediate to that extent which is less toxic (Hamoudi-Belarbi et al. 2018, Trigo et al. 2009). To ensure the success of bioremediation, we have to develop those microbes which can proliferate and survive the toxicity of pollutants and harsh environment conditions (Adams et al. 2015). The organic materials can be remediated biologically, but the volatile organic compounds (VOCs) are subjected to remediation under *in situ*. It has also been shown that the duration of some bioremediation operations is relatively longer, depending on the quality and concentration of pollutants. Moreover, the efficiency of microbes plays a significant role in bioremediation, since environmental factors do play their own role (Mishra et al. 2001).

2. BIOREMEDIATION OF HYDROCARBON CONTAMINATED SOILS

Soils all over the world play an important role in ecosystems. Microbes in soil carry out several functions- these minute creatures can adjust the flow of energy, and the recycling of elements; in addition to that, they play a pivotal role in the growth and development of plants (Thomas et al. 1992). Drilling operations and exploration lead to polluted soils with oily compounds; these organic compounds must be treated *in situ* so that further contamination of ecosystem can be regulated (Azubuike et al. 2016). The organic polluted sites can be reverted to natural conditions by using potential bacteria, fungi and micro algal species. These organisms have shown to convert toxic organic compounds into nontoxic substances. During this process of remediation, water and CO_2 are also produced (Maruthi et al. 2013). Several modern studies have established new trends of biodegradation by isolating microbes from contaminated sites. These microbes were tested for degrading several different contaminants and then releasing such microbes *in situ* for efficient bioremediation (Peng et al. 2015, Shahab et al. 2017). Furthermore, some species, for example, the strain *Enterobacter* sp. CMG 457, were identified as having resistance to organic hydrocarbons like aromatic paraffin and pesticides, and also exhibited tolerance against 3 mM concentration of heavy metals like $CuSO_4$, $CdCl_2$ and $CrCl_2$ (Agnes 2017). Therefore, such bacterial strains can be employed in bioremediation process of hydrocarbons as well as heavy metals; further, these microbes can be improved for bioremediation using genetic engineering techniques (Olajire and Essien 2014).

According to many scientific reports and projects, there are a lot of species and strains of bacteria in many places in the world; for example, in China from oil field isolated *Acidobacteria*, *Actinobacteria*, *Bacteriocytes*, *Chlorogenic*, *Planctomycetes* and *Proteobacteria* (Cooper and Kurowski 1997). Other strains of bacteria were also isolated from oily soils and they were identified as *Aeromicrobiu*, *Brevibacterium*, *Burkholderia*, *Gordonia* and *Mycobacterium*. These bacterial strains were tested for bioremediation process and showed good ability to remediate hydrocarbons. In another study, strains of bacteria *Bacillus subtilis* SA7, *Pseudomonas aeruginosa* SA3 sp. and *Citrobacter* SB9, two species of microalgae *Aphanocaps* and *Chlorella minutissimma*, and the species of fungus *Amorphoteca*, *Candida*, *Graphium*, *Neosartorya*, *Pichia*, *Talaromyces* and *Yarrowia* have shown the ability to remediate petrol (Omojevwe and Ezekiel 2016). The positive and synergistic approach of all different groups of microorganisms such as bacteria, fungi and algae in hydrocarbon polluted areas of soils and aquatic environments was due to their enzymatic ability to remediate hydrocarbons (Jahangeer and Kumar 2013). Under field conditions of hydrocarbons' bioremediation, an addition of nutrients as a catalyst results in gradual increase in growth. A ratio of Carbon: (C): Nitrogen (N): Phosphorus (P) at 120:10:1 is necessary to ensure the growth of bacteria for the purpose of raising the efficiency of biological treatment (Peng 2015).

3. MECHANISMS OF BIOREMEDIATION OF HYDROCARBONS

3.1 Aerobic Bioremediation of Hydrocarbons

Many species of microorganisms are distributed widely in the environment and owing to their versatile metabolic ability, they can proliferate easily in wide range of different environmental conditions. This nutritional and environmental diversity of microorganisms can be exploited for biodegradation of pollutants, and is therefore useful in bioremediation. (Tang et al. 2007). The biological degradation of hyrocarbons in the ecosystem is a multifaceted course, whose qualitative and quantitative aspects rely on the type and quantity of hydrocarbons present. Local clime, seasonal variation conditions, and the types of native microbial population decides the bioremediation process (Strong and Burgess 2008).

The quick and complete biodegradation of the common organic contaminants is carried out under aerobic conditions. Figure 1 depicts the chief process of hydrocarbons' aerobic degradation (Foght 2008). Early intra cellular attack on organic pollutants is a process of oxidation and the incorporation as well as activation of oxygen molecule is an enzymatic chief reaction carried out by peroxidases and oxygenases. Degradation pathways on periphery convert organic contaminants one by one into intermediates of pivotal intermediary metabolism, for example, the TCA cycle. Cell biomass biological synthesis takes place from the pivotal precursor metabolites, e.g. acetyl-CoA, succinate, pyruvate. Sugars needed for numerous biological syntheses and growth are made by the process of gluconeogenesis.

3.2 Anaerobic Bioremediation of Hydrocarbons

Anaerobic bioremediation is a slow process. There are three supposed processes of anaerobic bioremediation of hydrocarbons, as shown in Fig. 2. The square brackets denote a postulated intermediate product, broken arrows signify a multiple of an enzymatic step, and an open arrows imply further substrate metabolism.

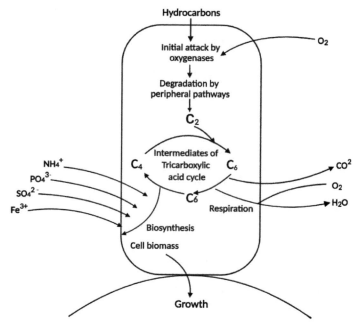

Figure 1 The chief principle of aerobic bioremediation of hydrocarbons using microbes. Adapted from (Foght 2008).

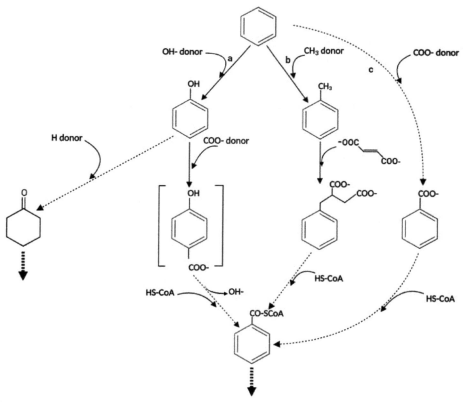

Figure 2 Three supposed anaerobic bioremediation of hydrocarbons, which they proposed for benzene. Adapted from (Foght 2008).

(a) Hydroxylation, to form phenol cyclohexanone or *p*-hydroxybenzen and benzoyl-CoA. The hydroxyl donor under methanogenic conditions is postulated to be H_2O, or under nitrate-reducing conditions with *Dechloromonas aromatica*

(b) Alkylation, to form toluene, which is followed by fumarate addition to the form of benzyl succinate and benzoyl-CoA. The methyl donor may be methyl tetrahydrofolate, S-adenosyl methionine or cobalamin protein.

(c) Carboxylation, to form benzoate (possibly through more than one enzymatic step) and benzoyl-CoA. The carboxyl donor isn't likely bicarbonate but may be derived from benzene, as shown in Fig. 2 (Foght 2008).

4. TYPES OF BIOREMEDIATION

There are several methods which can be used in the field of bioremediation, the most commonly used being bioventing (Lee et al. 2001), landfarming (Kuo et al. 2012), bioreactors (Quijano et al. 2010), composting (Ouyang et al. 2005), phytoremediation (Ziarati et al. 2019) and biostimulation (Silva-Castro et al. 2012). It is possible that bioremediation can occur on itself, but it can also be supported and stimulated by adding microbes called bio-stimulation. Addition of microbes in a site for remediation of heavy metals does not result in metal bioremediation, since microbes do not bioremedite metal, but they just adsorb or absorb it; this is true in case of metals like Pb, Cd and Hg. Therefore, phytoremediation is the best approach to remediate the metal contaminated soils (Stals et al. 2010). Plants have proven to be useful in phytoremediation at such conditions because these have the ability to absorb, transport and accumulate heavy metals within their parts (Key et al. 2008). Later, plants are harvested, incinerated and metals or their salts are recovered from plant ashes.

4.1 Bioreactors

This method can be carried out *ex situ* and *in situ*, and this is considered as one of the best bioremediation method for contaminated soils with intractable pollutants (difficult to degrade) under controlled environmental conditions. This is an ecofriendly approach and its outputs are within environmental parameters, which can be discharged into aquatic streams or open sites (Kulshreshtha et al. 2014, de L. Rizzo et al. 2010). In case of water contaminated with heavy metals, for efficient remediation, this method can be used (Kuyukina et al. 2003). Generally, there are two types of bioreactors which are used for bioremediation purposes 1) Column reactor 2) Slurry reactor (Kuyukina et al. 2003, Prakash et al. 2015). The reactors can be modified, depending on quality and quantity of polluted soil or water, but the principle remains the same in all shapes and sizes.

4.2 Biostimulation

Biostimulation implies the amendment of the particular natural conditions to stimulate existing microbes which are able to carry out bioremediation. This is generally done by addition of various rate limiting nutrients (Adams et al. 2015, Azubuike et al. 2016). The rate limiting nutrients and electron acceptors generally used are phosphorous (P), nitrogen (N), and oxygen (O_2) or carbon (Cunningham and Philp 2000, Kanissery and Sims 2011).

4.3 Biopilling

Biopilling is an ecofriendly method of bioremediating the polluted soils with intractable pollutants under controlled environmental conditions, where final output can be recycled and thrown, and they are within environmental parameters (Germaine et al. 2015). Biopilling is one of the preferred bioremediation method widely used in most oily countries such as Canada (Snelgrove 2010). This method is a combination of two techniques, landfarming and composting. In this method, piles can be prepared as geometric shapes in a way that can compress the air into them. Biopilling methods are used to treat surface soils polluted with petroleum hydrocarbons. They can also control pollutants via leaching and volatilization (Mary Kensa 2011). Biopiles can be used to remediate large amounts of polluted soils and is carried out along the pile from the bottom to the top. This method is up to 98% efficient, depending on the type of pollutants and the microorganisms involved (Cristorean et al. 2016).

4.4 Bioventing

There are some systems which are not able to remediate pollutants due to several reasons. The problem was studied from all directions to find a viable solution to enhance the efficiency of remediations. To solve this problem, bioventing is one of the methods that can be used *in situ* (Lee and Swindoll 1993). Bioventing uses microorganisms for the bioremediation of organic pollutant in soils and groundwater (Garima and Singh 2014). Bioventing process enhances the activity of indigenous microorganisms and stimulates their natural activity for biodegradation process. In bioventing, air is provided or oxygen is added to the existing soil microflora (Gracía Frutos et al. 2010). This process uses low flow rates to supply only enough oxygen to maintain microbial activity in vadose or unsaturated zone. Bioventing primarily helps in decomposing the residual hydrocarbons or contaminats that have been absorbed, but also contributes to the decomposition of volatile organic compounds (VOCs) as vapours move slowly through bioactive soil (Azubuike et al. 2016). This method is characterized by its superiority over other methods, and it has the ability to remediate many types of organic pollutants in relatively short periods. Taking into consideration the processing time, continuous monitoring and daily sampling analysis should be carried out in order to complete the target in given time (Yu et al. 2013).

4.5 Composting

This method is used in *ex situ*, where microbes (bacteria, fungi and actinomycetes) are activated to consume different types of solid pollutants. Hydrocarbon pollutants are decomposed using soil microorganisms, under aerobic and anaerobic conditions (Prakash et al. 2015). This method is also used *in situ* for treating various contaminants having chemical, biological, and military waste origins (Pavel and Gavrilescu 2008). This method is also used to remediate polluted soils with oily hydrocarbons by mixing polluted soil with clean soil in the ratio 20:80 and also with enough quantity of cow or sheep dung. Mixture of above is prepared in the longitudinal piles of 2 m length, 1 m breadth and 30 cm height on a flat hard ground. The mixture is covered with dark plastic sheets to ensure the activity of aerobic (on surface) and anaerobic (deep inside) microbes. Once a week sprayed with water to maintain wetness and stirred once a month for a period ranging from 3–6 months, results of the analysis using Gas Chromatography (GC) showed an efficiency of this method, with remediation of 98% of oily hydrocarbons (Pavel and Gavrilescu 2008).

4.6 Landfarming

It is an *ex situ* technique of waste management process, which is carried out in biotreatment cells or in upper zone of soils. Polluted sediments, soils and sludges are transported to the sites of landfarming. Here the transported material is incorporated in soil surface and turned or mixed periodoically to nicely aerate the mixture (Lukić et al. 2017). To increase the efficiency of bioremediation, some nutrients and minerals are also added to increase the bio-efficiency (Ghazali et al. 2004). For optimum activity of microbes, water is regularly added to maintain a proper level of moisture (Thomas et al. 1992). Sometimes, if required, additives may be added such as organic fertilizers to improve the bio-content specification (Rubinos et al. 2007). It should be also noted that oxygen movement in soil promotes aerobic bacterial activity, which enhances decomposition of organic compounds (Neria et al. 2015).

4.7 Phytoremediation

Phytoremediation is known as a biological technology which uses the plants for remediation of hydrocarbons or other contaminants to enhance the process of degradation and removing of pollutants in polluted soil or groundwater (Padmavathiamma and Li 2007). Phytoremediation is not a costly technique for large sites with shallow levels of pollutants' residue with organic nutrients or metal pollutants, (Schnoor et al. 1995, Vidali 2001). The advantages of phytoremediation is that it is a cost effective approach, can be implemented and maintained easily, is environmental friendly, aesthetically pleasing, and reduces land filled wastes and harvestable plant material, from which metals or contaminants can be recovered. The disadvantages are that it is climate-dependent, has long remediation times, its effects to food web might be unknown, pollutants ultimate fate may be mysterious and results are variable (Longly 2007). Use of phytoremediation as a polishing or secondary approach for *in situ* remediation mitigates land disturbances, gets rid of the transportation problem and other accountability costs linked with off site remediation and disposing (Schnoor et al. 1995)

5. BIOREMEDIATION OF CONTAMINATED SOILS WITH CLASSIC EXPLOSIVES

In the early 20th century, modern activities witnessed new types of organic material; about 60 or more highly explosive compounds were synthesized and developed for military use. The most widely used of these explosives probably are Hexahydro-1,3,5-trinitro-1,3,5triazine (RDX), 2,4,6-Trinitrotoluene, the most known explosive material globally (TNT), and 1,3,5,7-tetranitro-1,3,5,7-tetrazosine (HMX). These chemical compounds are heat stable, can be kept for long times and have high density (Das and Chandran 2011, Lin et al. 2013). In this chapter, I would like to review biodegradation of (TNT) in detailed manner, whose pollution represents a global phenomenon. The traces of TNT are generally found in soil and groundwater due to its extensive production and use in many military activities (Shemer et al. 2015, Rosenbaltt et al. 1991). Pollution of water and soils with explosives depends on their production, manufacture, loading of ammunition items, and their waste disposal. The highest extent of explosives' contamination is generally related to improper waste water management part, particularly waste water release methods, which are generally employed by explosive mechanized plants and their units where waste water is released to the unlined ponds or lagoons or natural water stream, thus resulting in indirect pollution of groundwater and soil (Pennington and Brannon 2002). In fact, it is natural that soils can be polluted with explosive materials in military training areas due to the accumulation of residues, which is accumulating day by day. The groundwater pollution with explosive residues depends upon the spreading of the surface layer of soil, which comes in contact with natural precipitation, solubility of the inorganic and organic compounds and on dissolution kinetics (Morley et al. 2006, Phelan et al. 2002). A perfect example of explosive material which has been used extensively is 2,4,6-Trinitrotoluene (TNT). TNT is solid, yellow coloured, odourless compound (CRREL 2006). This is manufactured in laboratories and factories using nitric acid (HNO_3) and sulphuric acid (H_2SO_4) with toluene ($C_6H_5-CH_3$) (ATSDR 2007).

During early 1916, TNT was manufactured to be used in First and Second World War, which was manufactured in government factories and commercially employed in military ammunitions (Steen 2006). Since TNT has been used extensively during war, therefore, its chemicals have been added to the soil and water in ecosystem. The polluted sites with (TNT) may contain up to 10 gm kg^{-1} of TNT in sediments of soil and reach up to 100 mg L^{-1} in aquatic system. At the same time, TNT and its evolved products are very noxious and mutagenic agents for prokaryotic and eukaryotic organisms. Therefore, such polluted sites need urgent attention for remediation of hazardous chemical explosives for safety and quality of ecosystem. It has been estimated that nearly 3320 locations in Germany alone need restitution of a green and clean environment. Numerous physical and chemical techniques have been developed and tried to deal with TNT polluted soils, but all techniques are not cost effective. To remediate TNT, activated charcoal has often been tried in contaminated surface and groundwater but left over carbon substances accumulated as another type of waste, which becomes another problem to resolve (Clause 2014). TNT soluble phase has some sludge part which contains a higher residue amount of TNT (67.8 mg L^{-1}), which results in environment toxicity. This sludge toxicity adversely affects root germination and development. TNT is converted into 4-amino-2,6-DNT and 2-amino-4,6-DNA with decreasing amount of nitrogenous aromatic compounds after 21 days. Bioremediation experiments could be the assessment criterion for suitable treatment methodologies for these explosive substances. Inoculation of suitable bacterial consortium from contaminated soils may enhance the biodegradation process, since the microbes are harmonious with substance to be treated. TNT solid phase prevents the microbial action; therefore, its conversion into the aquatic phase increases its biodegradation process. Enzymes of the microbes are mainly responsible for TNT biodegradation (Nehrenheim et al. 2013). TNT derivatives' partial transitions in the ecosystem are highly hazardous to organisms such as fishes and other aquatic animals; moreover, their presence negatively influences the fungus *Phanerochaete chrysosporium* at 20 ppm or higher concentrations (Hawari et al. 1999). Recently, numerous research works have suggested that this chemical may result in adverse health effects on some environmental organisms such as *Salmonella typhimurium*, algal and fungal cells, plants, invertebrates, some vertebrates and also human beings. Other studies conducted on animals such as *Rattus, Mus musculus, Canis lupus* and *Rana temporaria*, with respect to TNT and its derivatives, resulted in mutilated embryos, cell toxicity and mutation (Maeda et al. 2006, Kalderis et al. 2011, Kuperman et al. 2015, Habineza et al. 2016). Wide ranges of bacteria, fungi and plants have been studied to determine the effectual *in situ* TNT biodegradation technology. Fungi can tolerate higher aromatic nitrogenous explosives' levels, but they cannot tolerate harsh environmental conditions such as higher temperatures, high alkalinity and acidity (Gumuscu and Tekinay 2013). Aerobic bacterial species possess potential to transform TNT into non-toxic substances (Smets et al. 2007). *Bacillus* sp. YRE1 has been used on charcoal and polystyrene surface for direct biodegradation of TNT; it was observed that it degraded TNT up to 94% under acidic conditions at pH (5–7) (Ullah et al. 2010). It was also observed that isolated bacterial species from the TNT contaminated soils can give astonishing results in terms of bioremediation by analyzing the TNT residues in the media (Rahal and Moussa 2011). There are several bacterial strains having potential of treating this noxious substance in environment. Strain of *Achromobacter spanius* STE 11 exhibited TNT bioremediation. Absolute removal of 100 mg L^{-1} TNT was observed under aerobic conditions within 20 h by STE11 isolate. In this biological conversion course, TNT was converted to 4-aminodinitrotoluene (49 mg L^{-1}), 2-aminodinitrotoluene (16 mg L^{-1}), 2,4-dinitrotoluene (7 mg L^{-1}) and 2,6-dinitrotoluene (3 mg L^{-1}) as main metabolites (Gumuscu and Tekinay 2013). Analysis indicated that in the cell biomass, 24.77 mg L^{-1} nitrogen accumulated from TNT, which indicates that this bacterial strain can employ TNT as its solitary source of N. Biodegradation of TNT was observed between pH 4.0–8.0 and temperature range of 4–43°C, but the highest degradation was observed at pH 6.0–7.0 and temperature of 30°C.

6. TNT BIODEGRADATION

There are diverse type of microbes which possess the capability to degrade TNT, like bacteria such as *Bacillus cereus* (Gumuscu and Tekinay 2013, Gh and Moussa 2011), *Clostridium* sp., *Desulfovibrio* sp. and *Methanococcus* sp. (Rosenblatt et al. 1991), and also *Pseudomonas* (Habineza et al. 2016), and *Achromobacter spanius* STE 11 (Gumuscu and Tekinay 2013, Gh and Moussa 2011). Interestingly, *Corynebacterium* and *Sphingomonas sanguinis* have also been reported to degrade TNT (Gh and Moussa 2011, Litake et al. 2005). Further, *Salmonella typhimurium* has been reported to biodegrade TNT (Gumuscu and Tekinay 2013, Van Aken and Agathos 2001). Partial degradation of TNT has also been reported by some fungal species such as *Phanerochaete chrysosporium*. On the other hand, research works have shown that around 91 fungal strains, which belong to 32 species isolated from decomposing garbage wood, belong to class Basidiomycetes, and *Saprophytic micromycetes* especially have good capability to transform TNT to another form, Monoaminodinitrotoluenes (ADNT). These fungi can produce higher amount of ADNT from TNT. Two fungal strains, *Clitocybula dusenii* TMb12 and *Stropharia rugosa-annulata* DSM11372 (Gumuscu and Tekinay 2013, Funk et al. 1995), and also *Phanerochaete chrysosporium,* have been classified as white rot species (Gh and Moussa 2011). Moreover, fungi belonging to basidiomycetes such as *Agaricus aestivalis, Agrocybe praecox, Clitocybe odora, Stropharia, Shewanella putrefaciens, Aspergillus, Alternaria, Penicillium, Trichoderma,* and the acid tolerant yeasts *Yarrowia lipolytica* AN-L15 have been reported to biodegrade TNT (Litake et al. 2005).

6.1 Bioremediation of TNT under Anaerobic Conditions

Bioremediation of TNT under anaerobic conditions can degrade nitroaromatic compounds in soil completely; this is done *ex situ* to degrade the TNT in polluted soil. In this case, the reactor of 23 m^3 is incubated with polluted soil mixed with TNT (50:50), along with some mud and water. Some amount of (1–2%) carbon starch, buffered with phosphate, was also added to maintain the pH near neutral. Microbes from native soil can also be used under controlled conditions as anaerobic for 1–2 days. It was observed that microbes degraded 3000 mg of TNT per kg^{-1} of soil in 11 weeks, though the process was slow as it was more of safety. The reason for slow processing could be the low ambient temperature, and higher amount of clay in soils polluted with higher amount of other contaminants in soil with higher TNT concentrations (Park et al. 2003). Generally, two types of bioreactors are used: first is soil column reactor, and second is slurry bioreactor. Both require living microbes but second one requires more water during the period of remediation (Gh and Moussa 2011).

6.2 Biodegradation of TNT under Aerobic Condition

In the majority of studies described till date, aerobic bacteria likely convert TNT molecule by reducing one or two nitro groups to hydroxyl amino or amino groups and produce various isomers of amino nitroaromatic compounds. These compounds later on usually build up in the culture medium with no further metabolism. There are reports that partly reduced TNT forms react among themselves in the presence of oxygen to form recalcitrant azoxy tetra nitrotoluenes (Haïdour and Ramos 1996), which leads to higher rate of mutations, even more than TNT (George et al. 2001). These conversion reactions eliminate TNT but yield in extremely recalcitrant products that are not utilized by most of the microbes that produce them (Abraham et al. 2001, Lee et al. 2002).

6.3   Biological Composting

Biocomposting is believed to be a very beneficial approach to biodegrade contaminated soils with non-metabolized TNT. Several projects have been carried out in this field, where they have removed TNT from soils. In biological composting, contaminated soil is mixed with same amount of additional material; these materials are generally agricultural by-products (straw, hay, wood chips, clover, corn stalks, plant leaves). Aerobic-anaerobic fertilization and pneumatic air-fertilization system are considered to be highly proficient in treating TNT contaminated soils. These conditions convert the substance into amine and diammonium nitrate toluene during aerobic phase, and airy process removes the transformed products, probably by the equivalent correlation with the soil (Esteve-Núñez et al. 2001). This technique is used *ex situ*. Here, live microbes are activated to biodegrade liquid and solid contaminated materials, and it is very effective for removing TNT, RDX and PAH (Semple et al. 2001).

6.4   Biopilling

Biological pilling or biopilling technique is considered as more attractive for biological degradation of dangerous materials like TNT, and each contaminant requires little modification on the fundamental technique. Biopilling approach is economically efficient owing to the less cost for biopilling combination parts. Moreover, less operation cost of the technique employed in process of biodegradation makes it more fascinating and adoptable. In case of TNT, this is mainly biodegraded on site, using bioreactors, which are efficient and eliminate the transportation cost of contaminated soils (Van Deuren et al. 2002, Aktaş 2013).

6.5   Phytoremediation of TNT

Phytoremediation is a technique used *in situ*, which uses efficient plants to uptake pollutants from soils. This approach is considered the most appropriate for field sites, where other techniques of biodegradation options are not available; this technique is also cost effective. Deep rooted trees, grasses, legumes and aquatic plants, all are used for bioremediation purpose. This technique has been employed to biodegrade wide range of inorganic and organic pollutants and also explosive materials like TNT. Here, plants efficiently remove contaminants from ground water and reservoirs, and microbes associated with roots also improve the plant efficiency (Litake et al. 2005).Though these approaches are somehow slow, but the presence of efficient microbes helps in consuming organic compounds such as TNT explosives, since rhizospheric microbe and plant roots work synergistically (Schulz-Berendt 2000).

7.   MECHANISMS OF TNT DEGRADATION

The foremost obstacle in microbial degradation of TNT is the balanced organization of three nitro groups. These groups makes a great electron shortfall state at the aromatic ring. This also mostly causes the withdrawal and release of electrons along the whole structure. The usage of TNT as a source of energy and carbon source is very difficult for microorganisms. The oxidative biodegradation leads to destruction or transformation of TNT to different forms, but does not lead to absolute mineralization. (Glick 2003, Serrano-González et al. 2018). As compared to nitrogen atom, oxygen is more electronegative; therefore, there is polarization of N–O bond. Partially positive charge on the N atom, in association with its higher electronegativity, makes the nitro group simply reducible. Among living organisms, nitro group's reduction on aromatic rings is widely distributed (Schnoor 2000).

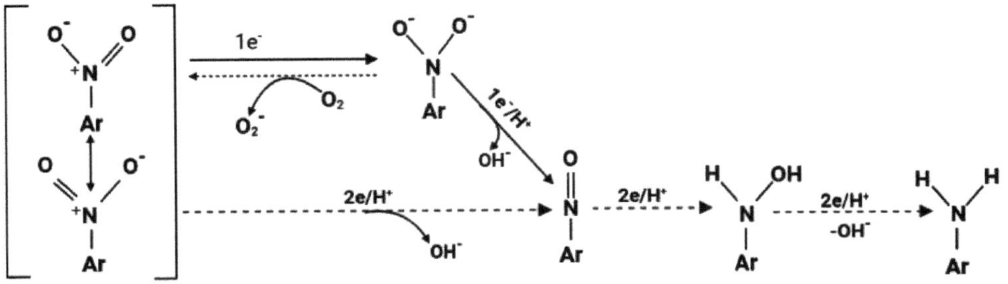

Figure 3 Reduction mechanisms of nitro groups in nitro aromatic compounds. First stage in reduction of nitro group can be attained through one-electron transfer (shown in solid lines) or two electron transfer (shown in dashed line). First mechanism manufactures a nitro anion radical that can react with O_2 to produce superoxide radicals and the original nitro aromatic compound through a futile cycle (shown in dotted line). The electron density at the nitro groups makes them easily reducible, which was noted during TNT decay by one electron or two electron transfers, produces an aromatic amine and hydroxylamine (Roldán et al. 2008).

This reduction is referred to as type first, this reaction type is independent of O_2 molecule, and no radicals are formed. The nitrite reductase enzymes are able to carry out this electronic duplex donation by reduced pyridine nucleotides. This also includes other reduction reactions carried out by enzymes that reduce nitro aromatic compounds. The main enzymes include dihydrolipic amide dehydrogenase, aldehyde oxidase, xanthine oxidase, diaphorases, CO dehydrogenase and hydrogenases. The nitro group could also be reduced *in vitro* via single-electron transfer, which forms a nitro anion radical. This radical is an alleged intermediate in reducing a nitro group to a nitroso group. But it can react with O_2 molecule to generate a superoxide anion and modify the original nitro aromatic compound. The enzymes which carry out this reaction are known as O_2-sensitive (type II) nitro reductases and are observed in bacteria such as *Escherichia coli* and *Clostridium* spp. (Roldán et al. 2008). Abiotic reduction could also happen for nitro groups to the corresponding amines in aquifers, soils and sediments. There are numerous probable electron donors in natural system, such as sulphur and iron reducing microbial species, and also in natural organic materials, which may be reduced from nitro aromatic compounds through biological degradation. TNT reacts with siloxane surface of clays to create covalent groups. Alive organisms play a significant role in these biological processes but they are yet to be explored to understand properly (Gh and Moussa 2011, Spain 1995).

8. CONCLUSION

Whatever have been discussed above, I can say that bioremediation plays an important role in reducing or mitigating organic pollutants. But this reduction or removal relies on the type of organic or inorganic substances. Also, the type of groups, their concentration and location in ecosystem, type of microbes involved and environmental conditions decide their fate. Nitro aromatic explosives, such as 2,4,6-trinitrotoluene (TNT), also pose some problems, as mentioned in earlier text, but efficient microbes do transform it into less hazardous form. In general, microbes can ease decontamination of environment by many means, such as volatilization, immobilization binding, reduction, oxidation, and finally biodegrading pollutants. The process of bioremediation is attractive because of several reasons, such as ecofriendliness, cost effectiveness and in most cases no accumulation of secondary compounds, which pose another environmental issue. Numerous molecular, genetic, and metagenomic engineering techniques have enhanced the progress of bioremediation. Here, purposely made or created microbes are employed for various decontamination processes, which in most of the cases perform as per expectation.

9. FUTURE ASPECTS

An enhancement in our knowledge about enzymes, which are accountable for biological degradation of structure, pathways, and other functional processes, will help us in biodegrading the xenobiotics as well as other recalcitrant compounds. An understanding of the physiology and genetics of potential microorganisms is going to be very effective to assess and augment the process of biodegradation. For this reason, having an up to date database about the environmental molecular evaluation results of polluted and bioremediated sites should help us further to solve the problem of decontamination. In this chapter, focus on bioremediation principles, various process, microbes involved in various applications, and with special reference to TNT were discussed. It is very important that we must have a basic knowledge and good understanding of an organism's biodegradative potential under varied environmental conditions. Further, an understanding of metabolomics of biodegradation process will certainly pave way for a 'clean' and 'green' environment, which is the need of the hour.

References

Abatenh, E., B. Gizaw, Z. Tsegaye and M. Wassie, 2017. Application of microorganisms in bioremediation-Review. J. Environ. Microbiol. 1(1): 02–09.

Abraham, E.N., A.A. Caballero and J.L. Ramos. 2001. Biological degradation of 2,4,6-trinitrotoluene. Microbiol Mol. Biol. Rev. 65(3): 335–352. doi:10.1128/MMBR.65.3.335-352.2001

Adams, G.O., P.T. Fufeyin, S.E. Okoro and I. Ehinomen. 2015. Bioremediation, biostimulation and bioaugmentation: A Review. Int. J. Environ. Bioremed. Biodegrad. 3(1): 28–39.

Agnes, M.N. 2017. Screening, Isolation and Characterization of Hydrocarbonoclastic Bacteria from Oil Contaminated Soils. MSc. Thesis University of Nairobi, p. 65.

Aktaş, F. 2013. Bioremediation techniques and strategies on removal of polluted environment. J. Eng. Res. Appl. Sci. 2(1): 107–115.

Ambrazaitienė, A., V. Jakubauskaitė, M. Zubrickaitė and D. Karčauskienė. 2013. Biodegradation activity in the soil contaminated with oil products. Zemdirbyste-Agric. 100(3): 235–242.

ATSDR. 2007. Agency for Toxic Substances and Disease Registry. U.S. Department of Health and Human Services. https://www.atsdr.cdc.gov/toxprofiles/tp124.pdf

Azubuike, C.C., C.B. Chikere and G.C. Okpokwasili. 2016. Bioremediation techniques-classification based on site of application: Principles, advantages, limitations and prospects. Review. World J. Microbiol. Biotechnol. 32(180): 1–18.

Clause, H. 2014. Microbial degradation of 2,4,6-trinitrotoluene in vitro and in natural environments. *In*: S.N. Singh (ed.), Biological Remediation of Explosive Residues, Environmental Science and Engineering. Springer International Publishing, pp. 15–38.

Cooper, P.W. and S.R. Kurowski. 1997. Chemistry of explosives. *In*: Introduction to the Technology of Explosives. Wiley-VCH Inc, New York, pp. 1–38.

Cristorean, C., V. Micle and I.M. Sur. 2016. A critical analysis of ex-situ bioremediation technologies of hydrocarbon polluted soils. ECOTERRA – J. Environ. Res. Protec. 13(1): 17–29.

CRREL. 2006. Cold Regions Research and Engineering Laboratory. Conceptual Model for the Transport of Energetic Residues from Surface Soil to Groundwater by Range Activities. J.L. Clausen, N. Korte, M. Dodson, J. Robb, S. Rieven, ERDC/CRREL TR-06-18

Cunningham C.J. and J.C. Philp. 2000. Comparison of bioaugmentation and biostimulation in ex-situ treatment of diesel contaminated soil. Land Contam. Reclam. 8(4): 261–269.

Das, N. and P. Chandran. 2011. Microbial degradation of petroleum hydrocarbon contaminants: An overview. Biotechnol. Res. Int. 2011: 13, Article ID 941810. https://doi.org/10.4061/2011/941810

de L. Rizzo, C.A., R. daM. dos Santos, R.L.C. dos Santos, A.U. Soriano, C.D. da Cunha, A.S. Rosado, et al. 2010. Petroleum-contaminated soil remediation in a new solid phase bioreactor. J. Chem. Technol. Biotechnol. 85: 1260–1267.

Esteve-Núñez, A., A. Caballero and J.L. Ramos. 2001. Biological degradation of 2,4,6-trinitrotoluene. Microbiol. Mol. Biol. Rev. 65(3): 335–352.

Fei, F., Z. Wen, S. Huang and D. De Clercq. 2018. Mechanical biological treatment of municipal solid waste: Energy efficiency, environmental impact and economic feasibility analysis. J. Cleaner Prod. 178: 731–739.

Foght, J. 2008. Anaerobic biodegradation of aromatic hydrocarbons: pathways and prospects. J. Mol. Microbiol. Biotechnol. 15: 93–120.

Funk, S.B., D.L. Crawford, R.L. Crawford, G. Mead and W. Davis-Hoover. 1995. Full-scale anaerobic bioremediation of trinitrotoluene (TNT) contaminated soil. Appl. Biochem. Biotechnol. 51(1): 625–633.

García Frutos, F.J., O. Escolano, S. García, M. Babín and M.D. Fernández. 2010. Bioventing remediation and ecotoxicity evaluation of phenanthrene-contaminated soil. J. Hazard Mater. 15: 183(1–3): 806–813.

Garima, T. and S.P. Singh. 2014. Application of bioremediation on solid waste management: A review. J. Bioremed. Biodeg. 5(6): 1–8.

George S.E., G. Huggins-Clark and L.R. Brooks. 2001. Use of a *Salmonella* microsuspension bioassay to detect the mutagenicity of munitions compounds at low concentrations. Mutat. Res. 490: 45–56.

Germaine, K.J., J. Byrne, X. Liu, J. Keohane, J. Culhane, R.D. Lally, et al. 2015. Ecopiling: A combined phytoremediation and passive Biopilling system for remediating hydrocarbon impacted soils at field scale. Front. Plant Sci. 5(756): 1–6. doi: 10.3389/fpls.2014.00756

Gh, R.A. and L.A. Moussa. 2011. Degradation of 2,4,6-trinitrotoluene (TNT) by soil bacteria isolated from tnt contaminated soil. Austral J. Basic. Appl. Sci. 5(2): 8–17.

Ghazali, F.M., R.N. Zaliha, R.N.Z. Abdul Rahman, A.B. Salleh and M. Basri. 2004. Biodegradation of hydrocarbons in soil by microbial consortium. Int. Biodeter. Biodegrad. 54: 61–67.

Glick, B.R. 2003. Phytoremediation: synergistic use of plants and bacteria to clean up the environment. Biotechnol. Adv. 21: 383–393.

Gumuscu, B. and T. Tekinay. 2013. Effective biodegradation of 2,4,6-trinitrotoluene using a novel bacterial strain isolated from TNT-contaminated soil. Int. Biodeterior. Biodegrad. 85: 35–41.

Guo, Z., Y. Sun, S.Y. Pan and P.C. Chiang. 2019. Integration of green energy and advanced energy-efficient technologies for municipal wastewater treatment plants. Int. J. Environ. Res. Public Health 16(1282): 1–29.

Habineza, A., J. Zhai, T. Mai, D. Mmereki and T. Ntakirutimana. 2016. Biodegradation of 2,4,6-trinitrotoluene (TNT) in contaminated soil and microbial remediation options for treatment. Periodica. Polytechnica. Chem. Eng. 61(3): 171–187.

Haïdour. A. and J.L. Ramos. 1996. Identification of products resulting from the biological reduction of 2,4,6-trinitrotoluene, 2,4-dinitrotoluene and 2,6-dinitrotoluene by *Pseudomonas* sp. Environ. Sci. Technol. 30: 2365–2370.

Hamoudi-Belarbi, L., S. Hamoudi, K. Belkacemi, L. Bendifallah and M. Khodja. 2018. Bioremediation of polluted soil sites with crude oil hydrocarbons using carrot peel wastes. Environs. 5(124): 1–12.

Hawari, J., A. Halasz, S. Beaudet, L. Paquet, G. Ampleman and S. Thiboutot. 1999. Biotransformation of 2,4,6-trinitrotoluene with phanerochaete chrysosporium in agitated cultures at pH 4.5. Appl. Environ. Microbiol. 65(7): 2977–2986.

Jahangeer, and V. Kumar. 2013. An Overview on microbial degradation of petroleum hydrocarbon contaminants. Int. J. Eng. Tech Res. 1(8): 34–37.

Judge, L.R. 2004. Biotechnology: Highlights of the science and law shaping the industry. Santa Clara HTLJ 20(1): 79–93.

Kalderis, D., A.L. Juhasz, R. Boopathy and S. Comfort. 2011. Soils contaminated with explosives: Environmental fate and evaluation of state-of-theart remediation processes (IUPAC Technical Report). Pure Appl. Chem. 83(7): 1407–1484.

Kanissery, R.G. and G.K. Sims. 2011. Biostimulation for the enhanced degradation of herbicides in soil: Review article. Appl. Environ. Soil Sci. 2011: 10. Article ID 843450.

Key, S., K.-C.M. Julian and P.M.W. Drake. 2008. Genetically modified plants and human health. J. R. Soc. Med. 101: 290–298.

Kowalczyk, A., T.J. Martin, O.R. Price, J.R. Snape, R.A. van Egmond, C.J. Finnegan, et al. 2015. Refinement of biodegradation tests methodologies and the proposed utility of new microbial ecology techniques. Ecotoxicol. Environ. Safety. 111: 9–22.

Kulshreshtha, A., R. Agrawal, M. Barar and S. Saxena. 2014. A review on bioremediation of heavy metals in contaminated water. IOSR J. Environ. Sci. Toxicol. Food Technol. 8(7): 44–50.

Kuo, Y.C., S.Y. Wang, C.M. Kao, C.W. Chen and W.P. Sung. 2012. Using enhanced landfarming system to remediate diesel oil-contaminated soils. Appl. Mech. Mater. 121: 554–558.

Kuperman R.G., R.T. Checkai, M. Simini and C.T. Phillips. 2015. Toxicity determinations for five energetic materials, weathered and aged in soil, to the Collembolan Folsomia candida. U.S Army Research, Development and Engineering Command ECBC-TR-1273-A Report.

Kuppusamy, S., T. Palanisami, M. Megharaj, K. Venkateswarlu and R. Naidu. 2016. Ex-Situ remediation technologies for environmental pollutants: a critical perspective. Rev. Environ. Contam. Toxicol. 236: 117–192.

Kuyukina, M.S., I.B. Ivshina, M.I. Ritchkova, J.C. Philp, C.J. Cunningham and N. Christofi. 2003. Bioremediation of crude oil contaminated soil using slurry-phase biological treatment and land farming techniques. Soil Sed. Contam. 12(1): 85–99.

Lee, M.D and C.M. Swindoll. 1993. Bioventing for *in situ* remediation. Hydrological Sci. 38(4): 273–282.

Lee, J.Y., C.H. Lee, K. Lee and S.I. Choi. 2001. Evaluation of soil vapour extraction and bioventing for a petroleum contaminated shallow aquifer in Korea. Soil Sed. Contam. 10: 439–458.

Lee, M.S., H.W. Chang, H.Y. Kahng, J.S. So and K.H. Oh. 2002. Biological removal of explosive 2,4,6-trinitrotoluene by *Stenotrophomonas* sp. OK-5 in bench-scale bioreactors. Biotechnol. Bioprocess Eng. 7 (2): 105–111.

Leslie Grady Jr., C.P., G.T. Daigger and H.C. Lim. 1999. Biological wastewater treatment. Handbook, 2nd Ed. Marcel Dekker, New York, pp. 8–10.

Lin, K.S., K. Dehvari, M.J. Hsien, P.J. Hsu and H. Kuo. 2013. Degradation of TNT, RDX, and HMX explosive wastewaters using zero-valent iron nanoparticles. Propellants Explos. Pyrotech. 38: 78–790.

Litake, G.M., S.G. Joshi and V.S. Ghole. 2005. TNT biotransformation potential of the clinical isolate of Salmonella typhimurium-potential ecological implications. Indian J. Occup. Environ. Med. 9(1): 29–34.

Longly, K. 2007. The feasibility of poplars for phytoremediation of TCE contaminated groundwater: A cost-effective and natural alternative means of groundwater treatment. Master Thesis of Environmental Studies The Evergreen State College. pp. 18–74.

Lukić, B., A. Panico, D. Huguenot, M. Fabbricino, E.D. van Hullebusch and G.A. Esposito. 2017. A review on the efficiency of land farming integrated with composting as a soil remediation treatment. Environ. Technol. Rev. 6(1): 94–116.

Maeda, T., K. Kadokami and H.I. Ogawa. 2006. Characterization of 2,4,6-trinitrotoluene (TNT)-metabolizing bacteria isolated from TNT polluted soils in the yamada green zone, Kitakyushu, Japan. J. Environ. Biotechnol. 6(1): 33–39.

Maruthi, Y.A., K. Hossain and S. Thakre. (2013). *Aspergillus flavus*: A potential bioremediator for oil contaminated soils. European J. Sustain. Develop. 2(1): 57–66.

Mary Kensa, V. (2011). Bioremediation—An overview. J. Ind. Poll. Control. 27(2): 161–168.

Mishra, S., J. Jyot, R.C. Kuhad and B. Lal. 2001. *In situ* bioremediation potential of an oily sludge-degrading bacterial consortium. Curr. Microbiol. 43: 328–335.

Morley, M.C., H. Yamamoto, G.E. Speitl Jr and J. Clausen. 2006. Dissolution kinetics of high explosives particles in a saturated sandy soil. J. Contam. Hydrol. 85: 141–158.

Nehrenheim, E., O. Muter, M. Odlare, A. Rodriguez, G. Cepurnieks and V. Bartkevics. 2013. Toxicity assessment and biodegradation potential of water-soluble sludge containing 2,4,6-trinitrotoluene. Water Sci. Technol. 68(8): 1707–1714.

Neira, J., M. Ortiz, L. Morales and E. Acevedo. 2015. Oxygen diffusion in soils: Understanding the factors and processes needed for modelling. Chilean J. Agric. Res. 75(1): 35–45.

Olajire, A.A. and J.P. Essien. 2014. Aerobic degradation of petroleum components by microbial consortia. J. Petrol Environ. Biotechnol. 5(5): 1–22.

Omojevwe, E.G. and F.O. Ezekiel. 2016. Microalgal-bacterial consortium in polyaromatic hydrocarbon degradation of petroleum-based effluent. J. Bioremed. Biodegrad. 7: 359. doi:10.4172/2155-6199.1000359

Ouyang, W., H. Liu, V. Murygina, Y. Yu, Z. Xiu and S. Kalyuzhnyi. 2005. Comparison of bioaugmentation and composting for remediation of oily sludge: A field-scale study in China. Process Biochem. 40: 3763–3768.

Padmavathiamma, P.K. and L.Y. Li. 2007. Phytoremediation of technology: Hyper-accumulation metals in plants. Water Air Soil Pollut. 184: 105–126.

Paniagua-Michel, J. and A. Rosales. 2015. Marine bioremediation-a sustainable biotechnology of petroleum hydrocarbons biodegradation in coastal and marine environments. J. Bioremed. Biodegrad. 6(273): 1–6.

Park, C., T.H. Kim, A. Kim, J. Lee and S.W. Kim. 2003. Bioremediation of 2,4,6-trinitrotoluene contaminated soil in slurry and column reactors. J. Biosci. Bioeng. 96(5): 429–433.

Pavel, L.V. and M. Gavrilescu. (2008). Overview of ex-situ decontamination techniques for soil cleanup. Environ. Eng. Manage. J. 7(6): 815–834.

Peng, S. 2015. The nutrients, total petroleum hydrocarbons and heavy metal contents in the sea water of Bohai bay China: Temporal – spatial variations, sources, pollution statuses and ecological risks. Marine Poll. Bull. 95(1): 445–451.

Peng, M., X. Zi, and Q. Wang. 2015. Bacterial community diversity of oil-contaminated soils assessed by high throughput sequencing of 16S rRNA Genes. Int. J. Environ. Res. Public Health 12: 12002–12015.

Pennington, J.C. and J.M. Brannon. 2002. Environmental fate of explosives. Thermochimica Acta. 384(1–2): 163–172.

Phelan, J.M., J.V. Romero, J.L. Barnett and D.R. Parker. 2002. Solubility and dissolution kinetics of composition B explosive in water. Sandia National Laboratories, P.O. Box 5800, Albuquerque, NM 87185–0719.

Prakash, V., S. Saxena, A. Sharma, S. Singh and S.K. Singh, 2015. Treatment of oil sludge contamination by composting. J. Bioremed. Biodeg. 6(3): 1–6.

Quijano, G., J. Rocha-Rios, M. Hernandez, S. Villaverde, S. Revah, R. Munoz, et al. 2010. Determining the effect of solid and liquid vectors on the gaseous interfacial area and oxygen transfer rates in two-phase partitioning bioreactors. J. Hazard. Mater. 175: 1085–1089.

Roldán, M.D., E. Pérez-Reinado, F. Castillo and C. Moreno-Vivián. 2008. Reduction of polynitroaromatic compounds: The bacterial nitroreductases. FEMS Microbiol. Rev. 32: 474–500.

Rosenblatt, D.H., E.P. Burrows, W.R. Mitchell and D.L. Parmer. 1991. Organic explosives and related compounds. *In*: O. Huntzinger (ed.), The Handbook of Environmental Chemistry-Antropogenic Compounds, Vol 3. Part G. Springer-Verlag, Berlin, pp. 133–145.

Rubinos, D., R. Villasuso, S. Muniategui, M. Barral and F. Díaz-Fierros. (2007). Using the land farming technique to remediate soils contaminated with hexachlorocyclohexane isomers. Water Air Soil Pollu. 181: 385–399.

Schnoor, J.L., L.A. Licht, S.C. McCutcheon, N.L. Wolfe and L.H. Carriera. 1995. Phytoremediation of organic and nutrient contaminants. Environ. Sci. Technol. 29(7): 318A–23A.

Schnoor, J.L. 2000. Degradation by plants-phytoremediation. *In*: H.J. Rehm and G. Reed (eds), Biotechnology. WILEY-VCH Verlag GmbH, USA, Canada, pp. 372–384.

Schulz-Berendt, V. 2000. Bioremediation with heap technique. *In*: H.J. Rehm and G. Reed (eds), Biotechnology. WILEY-VCH Verlag GmbH, USA, Canada, pp. 320–328.

Semple, K.T., B.J. Reid and T.R. Fermor. 2001. Review Impact of composting strategies on the treatment of soils contaminated with organic pollutants. Environ. Poll. 112: 269–283.

Semple, K.T., R.B. Cain and S. Schmidt. 1999. Biodegradation of aromatic compounds by microalgae. Mini-review. FEMS Microbiol. Lett. 170: 291–300.

Serrano-González, M.Y., R. Chandra, C. Castillo-Zacarias, F. Robledo-Padilla, M. de J. Rostro-Alanis and R. Parra-Saldiva. 2018. Biotransformation and degradation of 2,4,6-trinitrotoluene by microbial metabolism and their interaction. Defence Technol. 14(2): 151–164.

Shahab, S., I. Shafi and N. Ahmed. 2017. Indigenous oil degrading bacteria: Isolation, screening and characterization. Nat. J. Health Sci. 2(3): 100–105.

Shemer, B., N. Palevsky, S. Yagur-kroll and S. Belkin. 2015. Genetically engineered microorganisms for the detection of explosives' residues. Review. Front. Microbiol. 6: 1–7.

Silva-Castro, G., L. Santa Cruz-Calvo, I. Uad, C. Perucha, J. Laguna, J. Gonzalez-Lopez and C. Calvo. 2012. Treatment of diesel-polluted clay soil employing combined biostimulation in microcosms. Int. J. Environ. Sci. Technol. 9: 535–542.

Smets, B.F., H. Yin and A. Esteve-Nunez. 2007. TNT biotransformation: when chemistry confronts mineralization. Appl. Microbiol. Biotechnol. 76: 267–277.

Snelgrove, J. 2010. Biopile: bioremediation of petroleum hydrocarbon contaminated soils from a sub-arctic site. Thesis of Master of Engineering. Department of Civil Engineering and Applied Mechanics. McGill University, Montreal, pp. 1–6.

Sorkhoh, N.A., R.H. Al-Hasan, M. Khanafer and S.S. Radwan. 1995. Establishment of oil-degrading bacteria associated with cyanobacteria in oil-polluted soil. J. Appl. Bacteriol. 78: 194–199.

Spain, J.C. 1995. Biodegradation of nitroaromatic compounds. Ann. Rev. Microbiol. 49: 523–555.

Stals, M., R. Carleer, G. Reggers, S. Schreurs and J. Yperman. 2010. Flash pyrolysis of heavy metal contaminated hardwoods from phytoremediation: characterization of biomass, pyrolysis oil and char/ ash fraction. J. Anal. Appl. Pyrol. 89: 22–29.

Steen, K. 2006. Technical Expertise and U.S. Mobilization, 1917–18: High explosives and war gases. *In*: R. Macleaod and J.A. Johnson (eds), Frontline and Factory: Comparative Perspectives on the Chemical Industry at War, 1914–1924. Springer Nature, Switzerland, pp. 103–122.

Strong, P.J. and J.E. Burgess. 2008. Treatment methods for wine-related and distillery wastewaters: A review. Biorem. J. 12(2): 70–87.

Tang, C.Y., Q.S. Criddle, C.S. Fu and J.O. Leckie. 2007. Effect of flux (transmembrane pressure) and membranes properties on fouling and rejection of reverse osmosis and nanofiltration membranes treating perfluorooctane sulfonate containing wastewater. J. Environ. Sci. Tech. 41(6): 2008–2014.

Thomas, J.M., R.L. Raymond, J.T. Wilson, R.C. Loehr and C.H. Ward. 1992. Bioremediation. *In*: J. Lederberg (ed.), Encyclopedia of Microbiology. Academic Press, Inc. San Diego, CA, pp. 369–385.

Trigo, A., A. Valencia and I. Cases. 2009. Systemic approaches to biodegradation. FEMS Microbiol. Rev. 33: 98–108.

Ullah, H., A.A. Shah, F. Hasan and Abdull Hameed. 2010. Biodegradation of trinitrotoluene by immobilized *Bacillus* sp. YRE1. Pak. J. Bot. 42(5): 3357–3367.

Van Aken, B. and S.N. Agathos. 2001. Biodegradation of nitro-substituted explosives by white rot fungi. Adv. Appl. Microbiol. 48: 1–77.

Van Deuren, J., T. Lloyd, S. Chhetry, L. Raycharn and J. Peck. 2002. Remediation Technologies Screening Matrix and Reference Guide, Federal Remediation Technologies Roundtable, 4.

Vidali, M. 2001. Bioremediation. An overview. Pure Appl. Chem. 73(7): 1163–1172.

Yu, L., M. Han and F. He. 2013. A review of treating oily wastewater. Arabian J. Chem. 10: 1913–1922.

Ziarati, P., M. El-Esawi, B. Sawicka, K. Umachandran, A. Mahmoud, B. Hochwimmer, et al. 2019. Investigation of prospects for phytoremediation treatment of soils contaminated with heavy metals. Rev. J. Med. Discov. 4(32): 1–16.

CHAPTER 16

Soil Bioremediation and Sustainability

Nour Sh. El-Gendy[1,2,3*] and Hussein N. Nassar[1,2,3,4]

[1]Department of Processes Design and Development, Egyptian Petroleum Research Institute (EPRI), Nasr City, Cairo, Egypt, PO 11727.
[2]Center of Excellence, October University for Modern Sciences and Arts (MSA), 26 July Mehwar Rd. Intersection with Wahat Rd., 6th of October City, Egypt, PO 12566.
[3]Nanobiotechnology Program, Faculty of Nanotechnology for Postgraduate Studies, Cairo University, Sheikh Zayed Branch Campus, Sheikh Zayed City, Giza, Egypt, PO 12588.
[4]Department of Microbiology, Faculty of Pharmacy, October University for Modern Sciences and Arts (MSA), 26 July Mehwar Rd. Intersection with Wahat Rd., 6th of October City, Egypt, PO 12566.

1. INTRODUCTION

Biotreatment of oil polluted soils is known to be cost effective and ecofriendly, in contrast to the worldwide applied physical and chemical techniques. However, it exemplifies a challenging domain. Soil oil pollution affects the carbon-nitrogen percentage at the contaminated location, as petroleum is mainly composed of carbon and hydrogen. Oil contamination decreases soil porosity and permeability (Tara et al. 2014). Moreover, it negatively impacts the soil flora and fauna (Gan et al. 2018) and drastically affects the crops' germination and yields. The suffocation of plants also occurs by prohibition of air by oil and enervation of oxygen by enhanced microbial action, intervention of soil-water-plant associations and contagion from sulfides and additional manganese produced during the degradation of polluting petro-hydrocarbons. However, oil pollution usually affects the abundance and diversity of soil microbial population (i.e. bacteria, fungi and actinomycetes), which will consequently affect the rate of contaminant biodegradation and biotreatment of contaminated soils (Gan et al. 2018). This can be attributed to the chemical constituents of petroleum itself. This is considered as incomplete nourishment for microbial

*Corresponding author: nourepri@yahoo.com

communities, as it has hydrocarbons, the main source of energy and carbon, but insufficient concentrations of nitrogen, phosphorus and other nutrients required for microbial growth. Thus, the outdrawing of large amounts of organic carbon inclines to a hasty reduction of accessible inorganic nutrients, and thus lowers the biotransformation process, since extremely high C/P or C/N ratio, or both, have negative impact on microbial diversity and activity (Liu et al. 2018). Vidali (2001) reported the nutritional requirement of C/P and C/N to be approximately 10:1 and 30:1, respectively. Most frequently employed soil clean up standard for total petroleum hydrocarbons (TPH) is 100 mg kg^{-1}, though the guidelines and standard range from background concentrations is up to 10,000 mg kg^{-1} TPH in soil, depending on the countries and region (Sari et al. 2019).

2. BIOSTIMULATION

Biostimulation is the stimulation of the indigenous microbes to consume the pollutant hydrocarbons as carbon source. This occurs by altering certain factors such as moisture content, temperature, pH and/or addition of nutrients and surfactants, whereas bioaugmentation involves the addition of microorganisms, nutrients and sometimes surfactants into the contaminated area in order to attain precise and foreseeable biodegradation. There are mainly two strategies: *in situ* and *ex situ* bioremediation. The former is much cheaper, as it omits excavation of polluted soil, its transportation for remediation and engineering for applied equipment, but it takes a longer time than *ex situ* strategy. Schirmer et al. (2000) noted that the half-lives of compounds in the field tend to be 4–10 times longer than in laboratory as *ex situ* provides more consistency in homogenizing, screening, monitoring and treatment of waste.

Mostly, biostimulation is the worldwide recommended and endorsed approach (Sabate et al. 2004, Mrayyan and Battikhi 2005), specifically when applying inexpensive and/or costless, widely existing and sustainable nutrients sources, such as molasses, corn steep liquor, etc. (El-Gendy and Farah 20011, Soliman et al. 2014, Godleads et al. 2015). Biostimulant enhances the rate of bioremediation via two routes: (1) as nutrients source, especially nitrogen and phosphorous, which enrich the hydrocarbon degrading microorganisms (Shahi et al. 2016), (2) as biosurfactant source, increasing the bioavailability of the poorly soluble recalcitrant hydrocarbons (Yi and Crowley 2007). Nevertheless, an adapting period of about ten or less days up to few weeks is required by endogenous microbial degraders, in the course of non-remarkable occurrence of pollutants' degradation (Sharma et al. 2014). Moreover, severe toxicity from biostimulant would happen upon immediate solubilization of concentrated nutrients, causing severe damage to the endogenous microbial community. Accordingly, the fine-tuning of C/N/P percentage is so imperative to accomplish a fruitfully effective biostimulation treatment. The carbon-nitrogen-phosphorus (C:N:P) of 100:10:1 is the commonly recommended reported value for applicable biotreatment of hydrocarbon polluted soil via biostimulation. Yet, this value can be varied according to the degree and age of pollution and soil type. In biostimulation process, polluted soil composting, addition of organic matter such as manures, farm yard waste, and wastes from food processing are frequently added to supplement the organic material in soil (Ferrari et al. 2019). However, most of the research concerning the application of composting of polluted soils is *ex situ* biotreatment (Antizar-Ladislao et al. 2004, Joo et al. 2008). But, in recent times, some *in situ* applications have been published. Some case studies recommended 2:1 (polluted soil to biostimulant, e.g. compost and/or sewage sludge) for enhancing the biodegradation rate (Admon et al. 2001, Hwang et al. 2001, Namkoong et al. 2002). Nevertheless, applying fresh compost usually attains improved recuperation outcomes than when applying mature compost. This is attributed to the abundance of biological available organic carbon in the fresh biostimulating compost, which promotes the sorption of hydrophobic pollutants and acts as a carbon source for endogenous microbial population. Moreover, fresh compost increases the metabolic activity of microorganisms, consequently increasing the secretion of extracellular enzymes and biosurfactants, and accelerating the biotreatment process (Tsui and Roy 2007). Aged refuse from landfill is known to be rich in organic matter and essential nutrients are required for microbial growth. It is characterized by large specific surface area and high porosity and moisture withholding. It is also self pH buffering and can be considered as a bulking agent and source of microbial biomass, which is well adapted for being habituating severe and tough conditions of landfills. Consequently, such microbes are characterized by high biodegrading capabilities for labile and recalcitrant xenobiotics and different organic pollutants. Thus, aged refuse can be applied as a cost effective and sustainable amendment for bioremediation of oil polluted soil (Abbasian et al. 2016, Liu et al. 2018, Chen et al. 2019). Sewage sludge alone has been reported as a biostimulant for bioremediation of lubricant oil-polluted soil (Agamuthu et al. 2013).

Petroleum is composed of hydrocarbons (alkanes, alkenes, and aromatics) and non-hydrocarbon fractions (resins and asphaltenes) (Jian et al. 2011). Usually, *n*-alkanes of odd carbon numbers C15–C19 and C25–C35 indicate marine phytoplankton and higher land plant leaf waxes, respectively. However, shorter chain *n*-alkanes (odd and even) indicate petroleum contamination. Thus, carbon preference index (CPI), which is the ratio of odd to even carbon numbers with CPI < 1 indicates petroleum contamination, while CPI > 1 indicates highly weathered contaminated area with biogenic addition (Scholz-Böttcher et al. 2009). Petroleum components differ in their vulnerability to biodegradation and have been commonly classified from the most degradable to the most recalcitrant ones as follows: *n*-alkanes > branched alkanes > low-molecular weight aromatics > cyclic alkanes > high molecular weight- polynuclear aromatic hydrocarbons > polar compounds > petroleum biomarkers compounds (El-Gendy et al. 2014, Macaulay and Rees 2014). Moreover, the larger the size of compound and the greater the alkylation and substitution, the firmer will be the steric hindrance, and the lesser the water solubility, thus lower bioavailability and interaction between the enzymes' active sites and compound occurs, with a concomitant decrease in bioremediation rate (Bressler and Gray 2003, Pasumarthi et al. 2013). Occasionally, biodegradation of higher molecular weight hydrocarbons into low molecular weight and short chain hydrocarbons occurs (El-Gendy and Farah 2011). Not only this, but the biodegradation of lower fractions enhanced the biodegradation of higher ones. Leahy and Colwell (1990) described the asphaltene biodegradation as a consequence of a co-oxidation process, since asphaltenes contain the necessary elements for microbial enrichments, i.e. the carbon, hydrogen, sulfur, nitrogen, and oxygen. Consequently, when microorganisms reach their full metabolic prospective and activities, they can use asphaltene as a source of energy and carbon, and/or degrade them via co-metabolism and/or co-oxidation (Pineda-Flores and MestaHoward 2001). Soliman et al. (2014) observed 4.3 and 31% asphaltenes biodegradation percentage with a concomitant 7.3 and 56% maltenes biodegradation percentage in natural attenuation and CSL-biostimulation treatments of oily sludge polluted soil on microcosm level, respectively.

Due to the hydrophobic nature of polyaromatic hydrocarbons (PAHs), they are characterized by slow biotransformation, as they are highly sorbed on soil particles. However, low-molecular weight PAHs with two and/or three aromatic rings are comparatively more soluble, volatile and biodegradable than the higher molecular weight PAHs. But the aqueous solubility plays an important factor. For example, chrysene and pyrene both are 4-membered ring PAHs, but pyrene is more biodegradable than chrysene. Chrysene is insoluble, but pyrene water solubility reaches 0.135 mg/L. Addition of co-substrate that supports microbial growth enables the degradation of these recalcitrant compounds by co-metabolism. Moreover, addition of such co-substrates enhances the bioremediation under high salinity conditions. In a CSL biostimulation on oily sludge contaminated soil, PAHs were reduced by 47%, whereas approximately 72% and 43% removal occurred for anthracene and pyrene, respectively, with a significant observed biodegradation for 5-membered aromatic rings compound benzo(k)flouranthene, recording 67%. However, slightly lower biodegradation percentage was recorded for benzo(b)flouranthene

(35%), benzo[a]pyrene (23%) and the six-membered ring compounds benzo(ghi)perylene and indeno(1,2,3-cd)pyrene, recording 21% and 18%, respectively (Soliman et al. 2014).

Moreover, PAHs, as example of lipophilic compounds, have a tranquilizer mode of toxicity and interrelate with lipophilic components of the bacterial cytoplamitic membranes, negatively impacting their permeability and structure. Consequently, heavy metals' contaminants can easily penetrate into the microbial cells and intensely damage their functions (Shen et al. 2006). Thus, one of the important factors upon applying biostimulation process is to also overcome the toxicity of heavy metals.

The stability of the co-substrate is very important for the success of the biostimulation process. As stable as it would be, much desorption of pollutant would occur and consequently would be more available for microbial attack. That was very obvious in a study performed by Sayara et al. (2011), where the biostimulation of soil contaminated with PAHs mixture – fluorene, phenanthrene, anthracene, fluoranthene, pyrene, benzo(a)anthracene, and chrysene – as performed using stable compost of municipal solid wastes and non-stable rabbit food pellets, in the ratio of 1:0.25 (soil:biostimulant, w/w). A bulking agent consisting of wood chips was also introduced in a ratio of 1:1 (v/v) to ensure aerobic conditions and tap water was provided to retain the water holding capacity (50–60%), whereas 80% and 71% removal of such pollutants occurred within 30 days, respectively.

Reaching sustainable and cost effective biotreatment processes represents a worldwide major goal. This chapter summarizes examples of valorization of abundant, readily available and inexpensive different organic wastes into stable, self-emulsifier and buffering biostimulant and a source for required nutrients and microbial biodegraders to be applied for enhancement of sustainable, ecofriendly and cost effective bioremediation processes of petroleum hydrocarbons' polluted soil.

3. ANIMAL AND FOOD WASTES AS BIOSTIMULANT

To make the bioremediation process more cost effective and increase its sustainability, food waste compost has been applied as biostimulant for bioremediation of oil polluted soil. In a study reported by Hara et al. (2013), the TPH decreased from 8300 to 2300 ppm, with 25–30%, 30–35% and 37–42% reduction of alkanes, aromatics and resins, within 74 days of incubation, respectively, applying food waste compost. Residual mushroom compost of *Agaricus bisporus* as a source of coarse laccase enzyme for oxidation of phenolic compounds has been reported (Trejo-Hernandez et al. 2001). In one more study by Lau et al. (2003), spent mushroom compost has been successfully applied for bioremediation of PAH-polluted soil and expressed 82% PAH-degradation efficiency. As such, mushroom derived composts mitigated the toxic compounds, and reduced the required amounts of enzymes, microbes, and nutrients that stimulate the bioremediation process. Abioye et al. (2009) reported the biostimulation of crude oil polluted soil with mellon shells, which showed 30% higher biodegradation efficiency than the unstimulated soil within 28 days, while banana peels showed approximately 39% higher biodegradation efficiency than control soil within 56 days of incubation (Romanus et al. 2015). A comparative study by Abioye et al. (2012) for biostimulation of 5% and 15% (w/w) lubricant oil polluted soil with 10% (w/w) banana skin (BS), brewery spent grain (BSG), or spent mushroom compost (SMC) on bench scale, over a period of 84 days has been reported. Generally, the low polluted soil showed rapid and better bioremediation than high polluted soil, expressing 79% and 92% biological degradation of used lubricating oil, whereas the BSG expressed the best biostimulant activity at low and high pollutant concentrations, recording 92% and 55% biodegradation efficiency, respectively, as it had the highest N and P content that consequently led to the presence of highest population of hydrocarbon degrading microorganisms. Moreover, it expressed the highest germination index of lettuce seed. Lin et al. (2012) proved the efficiency of indigenous thermophiles in the bioremediation of 26,315 mg/kg diesel oil polluted soil by composting with a mixture of food wastes from fruit and vegetable markets, residences and schools, mature compost and saw dust with a C/N ratio of 32 and water holding capacity of 50–60%. The temperature raised gradually to 73°C and then decreased to reach 30°C at maturation stage, with concomitant increase of pH reaching 8.5 then decreased again to 7.1, indicating compost maturity and 90% TPH removal within 30 days of incubation. The gas chromatography with flam ionization detector (GC/FID) proved that most of the degradation occurred within the thermophilic period, the first 10 days, with the predominance of the thermophilic isolate *Pseudoxanthomonas* sp. In a study by Rhbal et al. (2014), a mixture of food wastes (equal amounts of carrots, cucumber, lettuce, onions, potatoes and tomatoes) from a restaurant were used for preparing a compost to biostimulate a crude oil polluted soil. The compost composed of (w/w) food waste (3%), sawdust (38%), leaves (17%), grass (27%) and wheat straw (14%) with total C:N (40–50). Three applications have been studied: polluted soil/compost ratios (60/40, 70/30, and 80/20), the water holding capacity maintained at 60–70%, the aeration and oxygen level were maintained by periodic mixing and the samples were weekly taken within a period of 12 weeks for analysis. The increase in pH values by the end of applied biotreatment indicated the maturity of the compost and better TPH removal, recording 8.7, 7.9 and 7.6, respectively. Not only this, but monitoring the change of temperature during the biotreatment processes showed an increase, reaching the maximum 54°C, 49°C, and 41°C, respectively, within 40 days of incubation and then gradually decreased to ambient temperature, indicating the stabilization of the compost. Such increase in temperature was beneficial as it lowers the viscosity of petroleum pollutants and consequently increases their motilities and interactions with microbial populations, thus enhancing the rate of biodegradation. Even the increase of the total kjeldahl nitrogen (TKN) reaching 6.2, 5.1 and 4.2 mg/kg within 5 weeks, 6 weeks and 4 weeks, respectively, it then decreased reaching nearly the same value 3.2 mg/kg at the end of incubation period. That proved the efficient TPH removal in the first treatment as the nitrogen is very important for the microbial activity, whereas the C/N ratio decreased from 30.7, 35.4 and 32.9 to 16, 16.4 and 18, respectively. This coincided with the results of TPH measured by gas chromatography-mass spectrometer (GC/MS), which recorded 96%, 78% and 64%, respectively. It is worth mentioning that the microbial population in such soil/compost treatment was of the most reported popular strains for petroleum hydrocarbons' degradation: *Pseudomonas*, *Bacillus*, *Kelbsiella*, *Serratia*, *Klyvera* and *Escherichia coli*. Water extract of carrot peels and carob kibbles were used as biostimulants of crude oil polluted soil (Hamoudi–Belarbi et al. 2018). The pH and water content were within the recommendable range for maximum biodegradation performance, pH 8.5–7.6 and 51–54%, respectively. The former expressed better efficiency and that was attributed to its higher nitrogen and phosphorous content, which concomitantly increased the endogenous microbial population and the ability of carrot to release linoleic acid, which increases the solubility of hydrocarbon pollutants and its bioavailability.

Biochar is another source for soil amendment, which is produced from bone meal and animal carcasses under oxygen limited conditions, and is characterized by large amounts of calcium phosphate, higher ash, N, P, K, S and micronutrients contents, as compared to the biochars produced from wood products which have relatively lower carbon content. However, 1% (w/w) biochar obtained from slow pyrolysis of birch wastes at 450°C has been applied as biostimulant for bioremediation of oil polluted soil, proving its action as a source of organic matter and inorganic nutrients for microbes and as adsorbent for the toxic pollutants decreasing its soil toxicity (Galitskaya et al. 2016). The main advantages of biochar is that they act as bulking agent, providing soil with some nutrients, and due to its high porosity and large specific surface area, it can act as an agent for retaining nutrients and their transformation when supplied with fertilizer or biostimulant. It improves the cation exchange capacity, water holding capacity, aeration and prevention of soil compaction, thus increasing the nutrients' supply rate and consequently the biodegradation of oil pollutants, even under low temperatures (Karppinen et al. 2017). In another study, meat and bone meal (MBM), which is a byproduct of the rendering industry has been successfully used as a biostimulant for bioremediation of soil polluted by diesel oil, keeping the pH of the soil under treatment within the recommended pH for remediation, i.e. pH 7 to 7.5 (Liu et al. 2019). MBM is a rich source of nutrients: it contains approximately 30% C, 10% Ca, 8% N, 5% P and also K, Mg and O. Its C:N and N:P ratios range between 3 to 4 and 0.5 to 2, respectively (Jeng et al. 2004).

Animal wastes can be also used as amendments for enriching soil hydrocarbon degrading microbial indigenous population. Wellman et al. (2001) and Lee et al. (2008) reported the usage of manure for enriching soil bioremediation. A comparative study for applying different animal manure for bioremediation of soil polluted with 10% (w/w) mixture of gasoline, kerosene and diesel oil has been performed by Agarry et al. (2010). The specific degradation rate constants calculated from the first order kinetic model equation showed that the efficiency of the biostimulant ranked in the following decreasing order: poultry manure > piggery manure > goat manure > NPK fertilizer with TPH % removal of 73%, 63%, 50%, and 39% after 4 weeks of incubation, respectively. The enrichment efficiency of the indigenous microbial population followed the same trend. Pig manure was applied for bioremediation of oily sludge polluted soil (Liu et al. 2010). The water holding capacity, TPH and PAHs degraders increased at the end of treatment and the concentrations of TPH, alkanes, aromatics, resin and asphaltenes decreased from 240.5 ± 8.9, 115.0 ± 4.8, 88.0 ± 2.5, 35.2 ± 7.3 and 10.4 ± 0.5 g/kg to 100.6 ± 12.6, 30.3 ± 7.3, 36.5 ± 5.8, 29.1 ± 4.4 and 7.4 ± 0.9, respectively, after 360 days. After the biotreatment process, the biotoxicity test was performed based on the decrease in light emission by *Photobacterium phosphoreum* T3, and a sharp increase in EC50 occurred, proving the effectiveness of biotreatment to reduce soil pollution.

Agamuthu et al. (2013) reported the application of cow dung, which showed 94% bioremediation of 10% (w/w) lubricant oil polluted soil, with rapid degradation rate of 0.2086 days^{-1} and 3.32 days half-life. The first-order kinetic model was applied in a comparative study at the microcosm level for bioremediation of Bonny Light crude oil polluted soil (10% w/w) using melon shell, ground nut shell, bean shell, cassava peels, pig manure and cattle dung, alone and/or in mixtures, as biostimulants, over a period of 42 days, relative to NPK fertilizer and non-amended microcosm, i.e. the negative control (Agarry et al. 2013). The biodegradation rate constants (k) and half-life times ($t_{1/2}$) revealed, in general, the superiority of the microcosms amended with food and animal wastes alone and/or in mixture, compared to the negative control and NPK amended microcosms. The best one was pig dung amended one, followed by the microcosm amended with pig dung and cassava peels, recording TPH removal 96.62% and 94.86%, respectively. Umanu and Babaden (2013) proved the good efficiency of non-sterile poultry droppings as a biostimulant to enhance the bioremediation of kerosene polluted soil as it acted not only as a source of nutrients but source of hydrocarbon degraders— *Bacillus* sp., *Alcaligenes faecalis*, *Serratia* sp., *Pseudomonas aeruginosa*, *Aspergillus niger*, *Penicillium chrysogenum*, and *Candida* sp. In another study, the first order kinetic model was applied also to evaluate the biostimulation treatment of crude oil polluted soil by inorganic fertilizer (NPK), cow dung (CD) and palm kernel husk ash (PKHA) individually and in a mixture in the ratio 1:1 (NPK:CD and CD:PKHA) over a period of 40 days (Ofoegbu et al. 2015). The best biodegradation efficiency (84.62%), with the highest rate 0.042 days^{-1} and lowest half-life of 16.5 days, was recorded in biostimulation by NPK:CD, while the lowest biodegradation efficiency (59.45%), with the lowest rate 0.021 days^{-1} and lowest half-life of 33 days, was recorded in biostimulation by PKHA. In another comparative study performed by Obiakalaije et al. (2015) for biostimulation of soil polluted by 53,966.60 mg/kg crude oil using goat manure, poultry droppings and cow dung, the main hydrocarbon degrading bacteria were found to be *Bacillus* sp, *Micrococcus* sp., *Escherichia coli*, *Arthrobacter* sp., *Citrobacter* sp.,

Pseudomonas sp., *Alicagenes* sp., *Flavobacterium* sp., *Corynebacterium* sp., and *Aeromonas* sp., while the main fungal isolates were *Fusarium* sp., *Candida* sp., *Aspergillus* sp., *Rhodotorula* sp., *Mucor* sp., *Penicillum* sp., and *Rhizopus* sp. Moreover, there were significant differences at the $p < 0.05$ in the decrease of total petroleum hydrocarbons' content for all the amended microcosms relative to the unamended one, recording 70.7%, 78.6% and 87.1% for soil biostimulated by cow dung, poultry droppings and goat manure, respectively, while only 32.1% in the control, with a significant increase in total heterotrophic microbial population and insignificant changes in pH relative to the unamended soil, within 28 days of incubation. This was attributed to the presence of considerable amounts of nitrogen, phosphorous and indigenous microbial populations in amended animal wastes. Cow dung and goat manure were also used as biostimulants in bioremediation of diesel oil contaminated soil (Williams and Amaechi 2017). Ogbeh et al. (2019) applied response surface methodology based on Box-Behnken-Design to optimize and study the interactive effects of poultry manure, cow manure and inorganic nitrogen-phosphorus-potassium (NPK) fertilizer on the bioremediation of spent-engine oil polluted sandy loam soil over a period of six weeks. Two second-order quadratic regression models have been predicted for bioremoval of total petroleum hydrocarbons (TPH) and improvement of the total soil porosity (TSP). Based on those models, 66.92% and 52.65% were achieved using 125.0 g/kg cow manure, 100.0 g kg^{-1} poultry manure and 10.5 g kg^{-1} NPK fertilizer for degradation of TPH and enhancement of TSP, respectively. Sari et al. (2019) reported the application of a mixture of yard waste and rumen residue (3:1 w/w) as nutrient sources in composting process to remediate polluted silty-clay loam soil in Indonesia with TPH content of 2,153.33 mg kg^{-1}, that reached 6,974.58 mg kg^{-1} within 150 days of incubation, with the predominance of *Bacillus* sp.

4. AGRO-INDUSTRIAL WASTES AS BIOSTIMULANT

Molasses is a readily available and cheap by-product of sugar industries from sugarcane or sugar beet. Molasses is a rich source of different microbial essential nourishment elements; its C, N, P, Na and K contents are 64, 6, 0.3, 0.33, and 5.5 (wt.%), respectively. It is characterized by a high carbohydrates content of approximately 50% and, to a lesser extent, non-nitrogenous compounds, (e.g. citric acid, oxalic acid) the recorded ratio of 2–8% (wt.%). Molasses has no furfural, which is toxic to most of the fermenting microorganisms. The ash content reaches approximately 11% wt.% and acts as a source of mineral elements. Molasses is also rich in valuable compounds such as vitamins, approximately 0.7% calcium and have substantial amounts of trace minerals like copper (2.2 ppm), zinc (3.91 ppm), manganese (4.74 ppm), iron (78.37 ppm), and magnesium (1370 ppm) (El-Gendy et al. 2013). Moreover, molasses is also characterized by high biochemical oxygen demand (BOD) in the range (40000–60000 mg/L) and chemical oxygen demand (COD) concentrations in the range (80000–120000 mg/L). Corn steep liquor (CSL) is also a readily available and cheap byproduct of starch industry. It is a rich source of protein, lipids, carbohydrates, vitamins, and amino acids and acts as an emulsifier and buffering agent. Its carbon : nitrogen : phosphorus content can reach up to 6.8:1.06:2.14 (% w/w) (El-Gendy and Farah 2011, Henkel et al. 2012, El Mahdi et al. 2016). The efficient nourishment composition of the aforementioned sustainable resources, together with being readily available and inexpensive, mark them as recommendable candidates for being used in microbial culture medium and/or microbial nutrient supplements in different bioprocesses.

Adebusoye et al. (2010) reported the application of cassava steep liquor which is composed of 20.3 mg/L protein, 33.4 g/L soluble starch and 0.765 mg/L phosphate, for biological remediation of diesel oil polluted soil that removed 98% of 1,102.3 mg diesel oil/kg soil, with complete degradation of the pristane and phytane, within 35 days of incubation. The predominant microbial population was found to belong to the bacterial strains that can utilize, survive in and tolerate toxic pollutants, for example; *Lactobacillus* sp., *Clostridium* sp., *Streptomyces* sp., *Pseudomonas* sp., *Bacillus* sp., *Flavobacterium* sp., *Corynebacterium* sp., *Nocardia* sp., *Candida* sp., *Saccharomyces* sp., *Rhodotorula* sp., *Aspergillus* sp., *Penicilium* sp. and *Geotrichum* sp. The treated soil was then planted with maize and the germination and growth profiles of maize seed plants were good, proving the retrieval of the oil-polluted soil. El-Gendy and Farah (2011) used corn steep liquor (CSL) as a biostimulant for being widely abundant and cost effective source of nutrients, co-substrate and emulsifier, in biotreatment of petroleum contaminated soil at microcosm level. The CSL contains the required carbon, nitrogen and phosphorus nutrients for microbial enrichment, 6.8:1.06:2.14 (% w/w), respectively, in the form of carbohydrates, protein and lipids. The degradation of the total petroleum hydrocarbons followed hockey stick model, where it expressed two distinct stages: an initial rapid biodegradation stage (for 21 days), followed by a slow biodegradation stage. They ascribed the fast degradation phase to the well adaptive native indigenous microbial soil community for the pollutants as well as hydrocarbon loss via volatilization. El-Gendy and Farah (2011) proved that CSL acted as an emulsifier, increasing the bioavailability of the total petroleum hydrocarbons. Furthermore, El-Gendy and Farah (2011) related the degradation rate to the petroleum sequestration in soil particles, which affects the availability of the hydrocarbons to the microbial cells. Within the slow biodegradation stage, the desorption rate of petroleum from soil particles was the limiting factor rather than the bacterial activity. It was noticed in that study that the microbial progression reached its stationary stage within 21 days, through the addition of CSL as biostimulant twice during the biotreatment process, at zero-time and after 14 days. That was associated with the decline in biodegradation rate and the reduction in the ratio of the total resolvable peaks to the recalcitrant unresolved peaks (TRP/UCM), as detected by gas-chromatography analysis. That was explained by the enrichment and enhancement of the enzymatic activities of the endogenous microbial population via the biodegradation of the easily amendable and labile linear, open-chain hydrocarbons and low-molecular weight PAHs within the initial rapid 21 days biodegradation stage. However, upon the removal of such labile types of pollutants, the enriched microbial community would have been enforced with the metabolism of the other persistent, obstinate petroleum pollutants, for example, the high molecular weight PAHs and more refractory long-chain hydrocarbons and petroleum biomarkers, with a consequent production of some noxious byproducts, which might have obstructed the microbial capabilities and activities. Kinetic modeling proved that the first order model described the biodegradation rate of most of the total petroleum hydrocarbons (TPH) and most of the petroleum components, which were the total normal- and iso- alkanes, the total naphthenes and cyclo-alkanes, the pristane (Pr 2, 6, 10, 14-tetramethylpentadecane) and phytane (Ph 2, 6, 10, 14-tetramethylhexadecane) as examples of conformist petroleum biomarkers, the short chain hydrocarbons ($nC10$–$nC14$ and $nC15$–$nC19$), and the moderate chain hydrocarbons ($nC25$–$nC29$ and $nC30$–$nC34$). Nevertheless, the second order model was the best in describing the biodegradation of some of the n-alkanes, which were $nC18$, $nC20$–$nC24$ and $nC35$–$nC42$. The half-life time ($t_{1/2}$) calculated from the applied kinetic models has been used to estimate the susceptibility of the petroleum pollutants for biodegradation, which was found to be: short-chain n-alkanes > long-chain n-alkanes > iso-alkanes > naphthenes and cyclo-alkanes. El-Gendy and Farah (2011) observed that the chain length governs the biodegradation rate. The short-chain n-alkanes degraded faster than the long chain ones and the half-life time of nC_{17} was lower than that of nC_{18}. Moreover, the least recorded $t_{1/2}$ was of nC_{10}–nC_{14} due to the abiotic loss via volatilization of such short chain compounds, in addition to biotic metabolism of such labile hydrocarbons by endogenous microbial community. However, pristane expressed the longest half-life time and it was even higher than that of phytane, proving the recalcitrance nature of pristane. Nevertheless, the half-life times of the rest of the studied petroleum pollutants did not express a definite inclination. Seklemova et al. (2001) stated that at elevated concentrations of pollutants, the patterns of microbial metabolism of such pollutants are not definite. Recently, Obayori et al. (2015) proved that CSL acted as an excellent biostimulant in a microcosm for biological remediation of petro-hydrocarbons polluted soils and the predominant aboriginal microbial population was found to be from the hydrocarbon degraders *Rhodococcus* sp., *Pseudomonas* sp., *Bacillus* sp., and *Corynebacterium* sp. 57.1% of TPH was removed at the rate of 1.33 mg/kg/day within 21 days and 78% of TPH at the rate of 3.52 mg/kg/day, with a decrease of nC_{17}/Pr and nC_{18}/Ph ratios from 1.298 and 1.153 to 0.182 and 0.182 by the end of incubation after 42 days, respectively.

Biostimulation of petroleum contaminated soil sample by peanut hull powder removed 38% of TPH contamination (Xu and Liu 2010). The first order kinetic model equation has been also applied to estimate the biotreatment competences of dissimilar biological stimulation routes for used lubricant oil contaminated soil (5 and 15% w/w) by means of adding variable biowastes: spent mushroom compost, brewery spent grain and banana skin as biostimulants instead of inorganic fertilizers (Abioye et al. 2012). Regardless of the applied biostimulant, the toxicity of high pollutants' concentrations towards the indigenous microbial habitats was proved and the bioremediation efficiency of low-polluted soil was higher than that of high-polluted one. The maximum biodegradtion efficacy of 92% was noted for 5% polluted soil biostimulated by 10% brewery spent grain, which was related to the constituents of the brewery spent grain itself compared with the other applied biostimulants, as brewery spent grain had the highest nitrogen and phosphorous contents, the most important limiting nutrients for successful bioremediation. It had also the highest moisture content, which empowering it to natively have some microbial communities that synergistically impacted the pollutant biodegradation. The recorded biodegradation rates were 0.4361, 0.410 and 0.3100 days^{-1}, for 5% polluted soil, biostimulated by brewery spent grain, banana skin and spent mushroom compost, respectively.

Soliman et al. (2014) accomplished biostimulation study on an Egyptian persistent and extremely oily sludge contaminated soil (53,100 mg/kg) with corn steep liquor (CSL) on microcosm level for 180 days; there was an incessant rise in microbial growth up to 130 days. That was attributed to the well adaptation of endogenous microbial habitats to the physicochemical criteria of the contaminated location and type of pollutants. Yet, a diminution in microbial population occurred thereafter, and was accredited to the decrease of the easily degradable hydrocarbons' content with time. The observed low microbial biodiversity, Gram positive *Micrococcus lutes* RM1, *Brevibacterium* sp. RM4 and *Curtobacterium* sp. RM5, with the numerousness of *M. lutes* RM1, was related to the elevated contamination and the low nutrients-nourishment sandy soil nature. A highly statistical significant difference was observed for total petroleum hydrocarbons' bioremoval rates in biostimulation microcosms, relative to the natural attenuation one ($p < 0:001$ at $\alpha = 0.5$, 95% confidence level), with a recorded total petroleum hydrocarbon biodegradation percentage of 44, and 6%, respectively. The biostimulation process was best fitted to the first-order kinetic model equation. However, the natural attenuation was better described by the second-order kinetic model equation. The biodegradation rates of oily-sludge components in biostimulation microcosms were ranked as: saturates > aromatics > resins > asphaltenes. That recorded a total removal of 78%, 47%, 42%, and 31%, after 180 days, respectively. The recorded biodegradation rate constants were 0.0004 kg/mg/day and 0.0032 days^{-1}. The initial total petroleum hydrocarbons' biodegradation rates were 21.24 and 169.92 mg/kg/days. The half-life time were 72.2 and 9 days, in natural attenuation and biostimulation microcosms, respectively. Consequently, it was predicted that to achieve total petroleum hydrocarbons content of about 500 mg/kg, it needs approximately 5,932 and 717 days biotreatment times, respectively.

Dadrasnia and Agamuthu (2013a, b) observed that the agro-industrial wastes – potato skin, soy cake and tea leaf – have prospective biostimulating behavior in the biotreatment of diesel oil polluted soil. Hamzah et al. (2014) proved the biostimulation activities of oil palm empty fruit bunch and sugarcane bagasse in a successful biotreatment of oil polluted soil. The advance of slow relief biostimulant ball using natural supplies has been also reported as a promising enhancer for the endogenous anaerobic microbial habitats in coastal sediments (Subha et al. 2015). The benefits of applying slow release biostimulant ball are longer acting times, reduced toxicity, and amplified cost effectiveness. Moreover, the direct bioavailability of such biostimulant to the indigenous microbial community would lead to lower energy consumption in the absorption step. Thus, an enhancement in microbial growth and enzymatic activities occurs with a concomitant increase in contaminant bioremoval rate (Subha et al. 2015).

5. LIGNOCELLULOSIC WASTES AS BIOSTIMULANT

Different plant residues and lignocellulosic wastes have been reported for biostimulation of hydrocarbon polluted soils. However, it is important to be sure that they stimulate and enrich the hydrocarbons' degrading microorganisms and not the microbial community that are not involved in the degradation of the target pollutants, as has been reported by application of wheat straw in creosote contaminated soil (Hultgren et al. 2009). Moreover, it is also important to be used the lignocellulosic wastes as a co-substrate that enhances the degradation process and not as a preferred source of carbon and energy sources for the degrading organisms than the targeted pollutants. Similarly, Adetutu et al. (2012) documented the application of sawdust and pea straw to biostimulate the microbes in soil contaminated by ^{14}C-hexadecane.

However, many studies report the positive impact of such plant residues and lignocellulosic wastes on the biostimulation process. It can act as absorber for organic pollutants via the lignin biopolymer, which also prevents the movement of these pollutants to groundwater. Moreover, the cellulose and hemicellulose polysaccharides act as co-substrates, enriching the soil microflora and enhancing the biodegradation process. These natural wastes themselves can act as a source of microorganisms which would contribute in the biodegradation process. Not only these, but they would improve the physical properties of the polluted soil (i.e. aeration, moisture, nutrition and structural properties), thus enhancing the pollutants' biodegradation (Fig. 1). Lignocellulosic wastes and crop residues such as wheat straw, sawdust, oil palm empty fruit bunch, rice husk and sago waste can act also as bulking agents, which reduce soil bulk density, act as co-substrate, and increase soil porosity and oxygen diffusion that stimulate the microbial activity (Rhykerd et al. 1999, Molina-Barahona et al. 2004, Omosiowho 2014, Kota et al. 2014, Alotaibi et al. 2018).

Barathi and Vasudevan (2003) reported the degradation of 94% of contaminating aliphatic hydrocarbons in a biostimulation of crude oil polluted soil using 20% wheat bran, which supported nutrients in the soil and increased aeration. In another study by Shahsavari et al. (2013), pea straw led to 83% total petroleum hydrocarbons (TPH) removal from 10,000 mg/kg polluted soil, with an obvious increase in petrogenic hydrocarbon-utilizing microorganisms, within 90 days incubation period at room temperature (27–30°C) and water holding capacity of 60–70%. Water hyacinth (*Eichhornia crassipes*) has been reported as an efficient biostimulant in bioremediation of heavy crude oil polluted soil (Udeh et al. 2013). Guinea corn shaft has been used as biostimulant for crude oil polluted soil, expressing approximately 90% removal of 5% (w/w) crude oil polluted soil within 56 days of incubation period (Romanus et al. 2015). Yelebe et al. (2015) proved that a mixture of palm bunch ash and wood ash acts as a good biostimulant for bioremediation of petroleum hydrocarbons polluted soil.

Moringa Oleifera seed cake has been successfully used for biostimulation treatment of five oil sludge polluted soil samples with different total petroleum hydrocarbons (TPH) and polyaromatic hydrocarbons' (PAHs) contents ranged between (48063.61–293846 mg/kg) and 189.93–4255.87 mg/kg), respectively, which degraded to 652.58–61.29 and 5.66–1.66 mg/kg, respectively, within 90 days of incubation (Uwem et al. 2015). Adams et al. (2017) proved that rice husk is better for bioremediation of oil polluted soil than chicken manure and a mixture of them. However, in another comparative study by Adams et al. (2018), poultry manure and sorghum husk in a ratio of 1:1 expressed higher biostimulation activity and TPH removal in soil polluted with petroleum hydrocarbons and heavy metals than those treated with individual biostimulant.

But, soil treated with sorghum husk alone expressed the highest removal of different heavy metals, 96.1%, 97.5%, 100% and 99.3% reduction in nickel, zinc, lead and copper, respectively, at the end of incubation period of 30 days. The soil pH in all the applied treatments ranged between 7.13 and 7.92, which is suitable for the microbial growth and the soil electrical conductivity was also within the acceptable range 130–2320 μS/cm.

Lignocellulosic wastes have been also used for bioremediation of heavy metals' polluted soil (Hidayah and Mangkoedihardjo 2010).

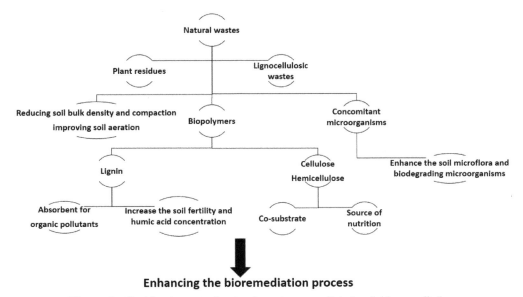

Figure 1 Positive impact of natural wastes on polluted soil bioremediation.

6. BIOSURFACTANTS FROM AGRO-INDUSTRIAL WASTES

Bioaccessibility can be enhanced by numerous chemical- or bio- solubilizers and/or surfactants, which form micelles that improve the release and microbial availability of pollutants in soil and solution (Dave et al. 2014). Biosurfactants are more effective than chemical surfactants and they can solely decrease heavy metal toxicity in contaminated areas and promote biodegradation of crude oil without the need of adding fertilizers, reducing the expenses of the applied biotreatment route and minimizing the dilution or wash-away obstacles that would occur upon the application of conventional water-soluble fertilizers (Thavasi et al. 2011). In recent studies, biosurfactants have been proved to have antimicrobial and anti-adhesive activities against pathogenic microorganisms (Janek et al. 2013). Moreover, unlike chemical surfactants, biosurfactants are less toxic and partially biodegradable, characterized by low critical micelles concentration (CMC) and low interfacial tension in aqueous solutions, tolerable to changes in temperature and pH, high emulsification and solubilization indexes (Santos et al. 2016). Yet, in-spite of the aforementioned benefits of biosurfactants, they are still not competitive in the market with their synthetic equivalents due to practical causes and expensive manufacturing costs, especially regarding substrates, that account for about 10 to 30% of the total production expenses and the complexity of the purification step (Rocha e Silva et al. 2014, Santos et al. 2019). Thus, its price ranges between 2 to 3 US$/kg and it is estimated to be 20–30% higher than their chemically synthesized equivalents (Sarubbo et al. 2015). According to Henkel et al. (2012) and Andrade Silva et al. (2014), the culture media cost up to 50% of the total manufacturing expenses of biosurfactants. Thus, applying readily available and costless agro-industrial wastes, e.g. ground-nut oil refinery residue, sugarcane vinasse, corn steep liquor, sugarcane molasses, sugar-beet molasses, waste frying oil, cheese whey waste, cassava wastewater, glycerol as by-product of biodiesel production, etc., for production of biosurfactants would overcome that obstacle (Guerra-Santos et al. 1984, Person and Molin 1987, Banat et al. 2010, Rocha e Silva et al. 2014, Hashemi et al. 2016, Colin et al. 2017, Santos et al. 2018, 2019).

Gudiña et al. (2015a) reported the production of 3.2 g/L biosurfactant by *Pseudomonas aeruginosa* strain using 10% (v/v) CSL and 10% (v/v) molasses. Upon supplementing the cultivation medium with 10% (v/v) CSL, 2.0 mM $FeSO_4$, 0.2 mM $MnSO_4$, and 0.8 mM $MgSO_4$, 4.8 g/L biosurfactant was produced using *Bacillus subtilis* #573, which was 3.6 times higher than the yield in the medium without heavy metals' supplements (Gudiña et al. 2015b). In an older study by Patel and Desai (1997), *P. aeruginosa* strain produced 240 mg/L biosurfactant using 7% (v/v) molasses and 0.5% (v/v). Different agro-industrial wastes have been applied in the production processes of rhamnolipids, including orange peel (George and Jayachandran 2009), wastes from the sunflower-oil production process (Benincasa and Accorsini 2008), waste frying oil (Haba et al. 2003), and molasses (Aparna et al. 2012). Moreover, solid state fermentation has been applied for production of biosurfactants using solid agro-industrial wastes (Amani et al. 2010, Geshev et al. 2010).

The response surface methodology (RSM) based on three 2^2 full factorial design followed by a 2^2 central composite design was applied to optimize and study the effect of two inexpensive agro-industrial wastes – soybean oil from refinery, and corn steep liquor (CSL) from corn industry – by *Candida sphaerica* UCP 0995 (Luna et al. 2011). Thus, a second order polynomial equation was predicted describing the dependency of surface tension on the studied wastes and the lowest surface tension 25.25±0.21 mN/m was achieved at optimum 8.63% (v/v) and 8.80% (v/v), respectively. That recovered 95% of motor oil polluting sandy soil sample, recommending the usage of such surfactant in enhanced oil recovery and/or soil bioremediation. However, in another study, the RSM based on central composite rotatable design (CCRD) of experiments was applied to optimize the production of glycolipid biosurfactant by *Pantoea* sp. in submerse fermentation process using corn steep liquor, vegetable fat and juice from the pineapple peel, which revealed optimum values of 5% (v/v), 2% (v/v) and 25% (v/v), respectively, with maximum reduction of surface tension from 69.67 mN/m to 30.00 mN/m (Gomes de Almeida et al. 2015). The biosurfactant produced by *Candida sphaerica* in distilled water, supplemented with 9% ground-nut oil refinery residue and 9% corn steep liquor, was reported to stimulate the biotreatment of motor oil contaminated sandy soil enriched with sugar cane molasses to more than 95% removal, applying twice its CMC value (Luna et al. 2017). In another study, Amaro da Silva et al. (2018) produced biosurfactant by *Candida sphaerica* and *Bacillus cereus* in distilled water, supplemented with 5% oil soy residue and 2.5% corn steep liquor, and in mineral medium, supplemented with 2% residual soybean oil, which significantly enhanced the bioremediation of motor oil polluted sandy soil enriched with cane molasses, removing more than 90% of the pollutant within 15 days and 30 days, respectively. The enhancement of bioremediation rate applying biosurfactants was attributed to the intensification of the surface area of hydrophobic water-insoluble contaminants and upsurge of the bioavailability of hydrophobic substrates (Jadhav et al. 2013). In another study, *Candida sphaerica* produced 4.5 g/L biosurfactant in a mineral medium framed with a ground-nut oil refinery residue CSL (Sobrinho et al. 2008), while *Pseudomonas aeruginosa* was reported to produce 8 g/L biosurfactant in mineral medium amended with glycerol and CSL (Khopade et al. 2012).

The full 2^2 factorial design was applied to optimize the biosurfactant production using the fungal isolate *Cunninghamella echinulate* (Andrade Silva et al. 2014). 4 g/L biosurfactant was produced with CMC of 20 g/L at optimum concentrations of 4% (w/v) corn steep liquor (CSL), 3% (w/v) soybean oil waste (SOW) and 7% (w/v) NaCl, with a decrease in surface tension from

72 to 36 mN/m and good emulsification index (%E_{24}) of 80% and 85% for engine and burnt engine oils, while 65% and 70% for soybean and canola oils, respectively. These properties were considerably stable over a wide range of temperatures (0–100°C), pH (2–12) and salinities (2–12% NaCl). The 2^2 full factorial design (FFD) was also applied to study the interactive effect of cassava wastewater (CW) obtained from a food industry and corn waste oil (CWO) from a restaurant on the produced biosurfactant by *Serratia marcescens* UCP/WFCC 1549 (Montero-Rodríguez et al. 2015). The lowest surface tension of 27.8 mN/m was obtained using 6% CW and 7.5% CWO, with oil dispersion rate of 78%, and %E_{24} of 60% for diesel and engine oil and 72.7% for burned engine oil, which were stable over a wide range of temperatures 0–120°C, pH 2–12 and salinity 2–12% NaCl. Moreover, that biosurfactant removed 88.27% and 73.70% of burned engine oil from beach sand and mangrove sediment, respectively, recommending its application in bioremediation processes. In another study, the 2^2 factorial design was used to optimize the concentration of CSL from corn wet-processing industry and crude glycerol (CG) from biodiesel production from cotton oil and study their interactive effects on producing biosurfactant by using the fungus *Rhizopus arrhizus* UCP 1607 (Pele et al. 2019). The optimum concentrations of 3% CG and 5% CSL produced 1.74 g/L glycoprotein biosurfactant, that reduced the surface tension of water from 72 to 28.8 mN/m and was characterized by a low CMC of 1.7% and high emulsification index %E_{24}, that ranged between 50% and 79.4% for different petroleum products (hexadecane, gasoline, diesel oil, kerosene, motor oil and burnt motor oil). The produced surfactant was also characterized by high stability and maintained its surface tension reduction value over a wide range of temperature (0–100°C), pH (2–12) and salinity 2–10%). Moreover, it removed 79.4% of the diesel steeped in a marine soil sample. A yield of 10 g/L ecofriendly lipopeptide based biosurfactant was produced by *Bacillus methylotrophicus* UCP1616 using CSL and sugarcane molasses (3% each), within 144 h (Chaprão et al. 2018), which removed 70%, 63% and 25% of motor oil adsorbed onto marine rocks, *natura* sand and sandy soil, respectively. It also expressed non-inhibitory potentials on seed germination or elongation of the roots of cabbage plant. Moreover, it promoted the microbial growth of autochthonous microorganisms throughout the applied motor oil biodegradation process. Thus, it was recommended for bioremediation processes for both petroleum hydrocarbons marine and terrestrial polluted areas. Moreover, the surfactant expressed good reduction in surface tension, reaching 29 mN/m, with high stability over a wide range of temperatures (0–120°C), pH (2–12) and salinity (0–12% NaCl). However, the %E_{24} decreased with the increase of temperature, pH and salinity from 100% at pH2, 0°C and 0% NaCl to 55.5%, 35% and 10% at 120°C, 12% NaCl and pH12, respectively. That proved the findings of Campos et al. (2013), which is that a good surfactant does not guarantee its performance as a good emulsifier. That was explained by different factors. (i) the variation in pH, which would cause some changes in the biosurfactant, and consequently empowering more or less interaction with the motor oil as a function of the changes in its composition and/or structure. (ii) The intensification of NaCl would lead to a weaker development of the oil-water-biosurfactant emulsion complex due to the affinity of NaCl towards the water molecules, reduces the efficiency of the biosurfactant and the inequity of that complex. (iii) While the elevation of temperature would lower the viscosity of the motor oil, and consequently, weakening the interaction between the biosurfactant and oil. In another study by Santos et al. (2019), a full 2^4-factorial design was applied to optimize the factors affecting the yield of biosurfactant produced by *Streptomyces* sp. DPUA1566 using soybean waste frying oil and corn steep liquor as agro-industrial wastes. 1.9 g/L biosurfactant was produced in a medium containing 10 g/L soybean waste frying oil and 20 g/L corn steep liquor, as C and N sources, respectively, at optimum physicochemical factors of pH 8.5, 150 rpm, 28°C and air saturation of 80%. The produced lipoprotein biosurfactant was efficient over a wide ranges of temperature, pH and salt concentrations and decreased the water surface tension from 72 to 28 mN/m, with a critical micelle concentration of 0.08%. The produced biosurfactant expressed

considerably high emulsification indices (E_{24}) with different hydrocarbons and oils: soybean oil (60%), Cariocar brasiliense oil (65%), motor oil and waste motor oil (100%), canola oil (35%) and *n*-hexadecane (30%). The main component of the biosurfactant was bioelan. The toxicity test was also performed for the produced biosurfactant to ensure its safe environmental application and it was proved to have no toxic effects against vegetable seeds or brine shrimp. Thus, it can be applied in bioremediation field or also in food, pharmaceutical and cosmetic industries.

7. CONCLUSIONS AND FUTURE ASPECTS

The global market for bioremediation technologies is steadily growing. This chapter provides comprehension about the different biostimulants obtained from different types of bioorganic wastes - lignocellulosic wastes, animal wastes, food wastes and agro-industrial wastes - and their applications into biostimulation treatment of petroleum hydrocarbon polluted soil as they are rich sources of microbial hydrocarbon degraders, readily biodegradable organic matter and nutrients and water. It also covers the approaches for enhancing the biotreatment efficiency via the composting by biowastes to stimulate the biodegradation of organic pollutants in soil. Moreover, this chapter covers also the production of biosurfactant from such wastes and their applications in bioremediation of petroleum hydrocarbons polluted soil to improve the bioavailability of hydrophobic pollutants to the endogenous microbial populations. The chapter emphasized on the importance of effect of the biochemical composition of such bioorganic wastes on the proficiency of the biotreatment. Since the organic content of those bioorganic wastes, their biochemical origin and macronutrient contents such as total nitrogen and phosphorus affect the soil properties, causing an increase in soil aggregate stability, moisture content, water infiltration and hydraulic conductivity, decreasing the need for adding expensive bulking agents and also adjust and/or maintain the soil pH within the favorable range for microbial activity (pH 6–8). Thus, such bioorganic wastes also have a vital role in enhancing the endogenous microbial progression and its catabolic capabilities and enzymatic activities, which are vital for partial and/or complete pollutants' mineralization.

However, more research should be done to shorten the process time as much as possible to make it more feasible in real applied field. The effect of biostimulation applying such bioorganic wastes for bioremediation of persistent petroleum hydrocarbons biomarkers, high molecular weight polyromantic hydrocarbons, polyromantic heterocyclic compounds, asphaltenes and heavy metals still need to be investigated, especially for *in situ* treatment. Nevertheless, some studies have proved that the addition of bioorganic wastes has negative impact on the bioremediation process. Thus, the nourishing balance of C:N:P proportion in the soil as well in the applied organic biostimulant and/or biowaste is very important. The synergetic effect between the soil endogenous microbial population and those in the amended biostimulant is also important to be investigated and studied.

Furthermore, research is still required on the production of biosurfactants, which can act as an efficient surfactant as well as an efficient emulsifier using sustainable resources of abundant, readily available and inexpensive biowastes. Research is also needed on the stability of such sustainable produced biosurfactants as well as bioemulsifiers over a wide range of pH, temperatures and salinities to widen their application on different polluted sites having different physicochemical characteristics.

Moreover, the periodic monitoring and/or evaluation of soil toxicity during and after the application of such biotreatment is very important to assure the success of the applied bioremediation process and its eco-safety, as it is recommended to reuse the biotreated and/or bioreclaimed soil for cultivation to overcome the worldwide shortage of arable land and the desertification problem.

References

Abbasian, F., R. Lockington, M. Megharaj and R. Naidu. 2016. The biodiversity changes in the microbial population of soils contaminated with crude oil. Curr. Microbiol. 72(6): 663–670.

Abioye, O.P., A. Abdul Aziz and P. Agamuthu. 2009. Stimulated biodegradation of used lubricating oil in soil using organic wastes. Malaysian J. Sci. 28(2): 127–133.

Abioye, O.P., P. Agamuthu and A.R. Abdul Aziz. 2012. Biodegradation of used motor oil in soil using organic waste amendments. Biotechnol. Res. Int. 2012: Article ID 587041. doi: 10.1155/2012/587041.

Adams, F.V., A. Niyomugabo and O.P. Sylvester. 2017. Bioremediation of crude oil contaminated soil using agricultural wastes. Procedia. Manuf. 7: 459–464.

Adams, F.V., M.F. Awode and B.O. Agboola. 2018. Effectiveness of sorghum husk and chicken manure in bioremediation of crude oil contaminated soil. *In*: N. Shiomi (ed.), Advances in Bioremediation and Phytoremediation. InTech Open Publisher, pp. 99–113. http://dx.doi.org/10.5772/intechopen.71832.

Adebusoye, S.A., M.O. Ilori, O.S. Obayori, G.O. Oyetibo, K.A. Akindele and O.O. Amund. 2010. Efficiency of cassava steep liquor for bioremediation of diesel oil-contaminated tropical agricultural Soil. Environmentalist. 30: 24–34.

Adetutu, E.M., A.S. Ball, J. Weber, S. Aleer, C.E. Dandie and A.L. Juhasz. 2012. Impact of bacterial and fungal processes on ^{14}C-hexadecane mineralization in weathered hydrocarbon contaminated soil. Sci. Total. Environ. 414: 585–591.

Admon, S., M. Green and Y. Avinimelech. 2001. Bioremediation kinetics of hydrocarbons in soil during land treatment of oily sludge. Bioremediation. 5: 193–209.

Agamuthu, P., Y.S. Tan and S.H. Fauziah. 2013. Bioremediation of hydrocarbon contaminated soil using selected organic wastes. Procedia. Environ. Sci. 694–702.

Agarry, S.E., C.N. Owabor and R.O. Yusuf. 2010. Bioremediation of soil artificially contaminated with petroleum hydrocarbon oil mixtures: Evaluation of the use of animal manure and chemical fertilizer. Bioremdiat. J. 14(4): 189–195.

Agarry, S.E., M.O. Aremu and O.A. Aworanti. 2013. Kinetic modelling and half-life study on enhanced soil bioremediation of Bonny light crude oil amended with crop and animal-derived organic wastes. J. Petrol. Environ. Biotechnol. 4: 137. doi:10.4172/2157-7463.1000137.

Alotaibi, H.S., A.R. Usman, A.S. Abduljabbar, Y.S. Ok, A.I. Al-Faraj, A.S. Sallam, et al. 2018. Carbon mineralization and biochemical effects of short-term wheat straw in crude oil contaminated sandy soil. Appl. Geochem. 88: 276–287.

Amani, H., M.H. Sarrafzadeh, M. Haghighi and M.R. Mehrnia. 2010. Comparative study of biosurfactant producing bacteria in MEOR applications. J. Pet. Sci. Eng. 75: 209–14.

Amaro da Silva, I., A.H.M. Resende, N.M.P. Rocha e Silva, P.P.F. Brasileiro, J.D.P. de Amorim, J. Moura de Luna, et al. 2018. Application of biosurfactants produced by *Bacillus cereus* and *Candida sphaerica* in the bioremediation of petroleum derivative in soil and water. Chem. Eng. Trans. 64: 553–558.

Andrade Silva, N.R., M.A.C. Luna, A.L.C.M.A. Santiago, L.O. Franco, G.K.B. Silva, P.M. de Souza, et al. 2014. Biosurfactant-and-bioemulsifier produced by a promising *Cunninghamella echinulata* isolated from caatinga soil in the northeast of Brazil. Int. J. Mol. Sci. 15: 15377–15395.

Antizar-Ladislao, B., J.M. Lopez-Real and A.J. Beck. 2004. Bioremediation of polycyclic aromatic hydrocarbon (PAH) contaminated waste using composting approaches. Crit. Rev. Environ. Sci. Technol. 34: 249–289.

Aparna, A., G. Srinikethan and H. Smitha. 2012. Production and characterization of biosurfactant produced by a novel *Pseudomonas* sp. 2B. Colloids Surf. B 95: 23–29.

Banat, I.M., A. Franzetti, I. Gandolfi, G. Bestetti, M.G. Martinotti, L. Fracchia, et al. 2010. Microbial biosurfactants production, applications and future potential. Appl. Microbiol. Biotechnol. 87: 427–444.

Barathi S. and N. Vasudevan. 2003. Bioremediation of crude oil contaminated soil by bioaugmentation of *Pseudomonas fluorescens* NS1. J. Environ. Sci. Health A. 38: 1857–1866.

Benincasa, M. and F.R. Accorsini. 2008. *Pseudomonas aeruginosa* LBI production as an integrated process using the wastes from sunflower-oil refining as a substrate. Bioresour. Technol. 99: 3843–3849.

Bressler, D.C. and M.R. Gray. 2003. Transport and reaction processes in bioremediation of organic contaminants. 1. Review of bacterial degradation and transport. Int. Chem. React. Eng. 1(1). https://doi.org/10.2202/1542-6580.1027.

Campos, J.M., T.L.M. Stamford, L.A. Sarubbo, J.M. Luna, R.D. Rufino and I.M. Banat. 2013. Microbial biosurfactants as additives for food industries. Biotechnol. Prog. 29: 1097–1108.

Chaprão, M.J., R.C.F. Soares da Silva, R.D. Rufino, J.M. Luna, V.A. Santos and L.A. Sarubbo. 2018. Production of a biosurfactant from *Bacillus methylotrophicus* UCP1616 for use in the bioremediation of oil-contaminated environments. Ecotoxicol. 27: 1310–1322.

Chen, F., X. Li, Q. Zhu, J. Ma, H. Hou and S. Zhang. 2019. Bioremediation of petroleum-contaminated soil enhanced by aged refuse. Chemosphere. 222: 98–105.

Colin, V.L., N. Bourguignon, J.S. GoÂmez, K.G. de Carvalho, M.A. Ferrero and M.J. Amoroso. 2017. Production of surface active compounds by a hydrocarbon-degrading *Actinobacterium*: presumptive relationship with lipase activity. Water Air Soil Pollut. 228: 454–559.

Dadrasnia, A. and P. Agamuthu. 2013a. Dynamics of diesel fuel degradation in contaminated soil using organic wastes. Int. J. Environ. Sci. Technol. 10: 769–778.

Dadrasnia, A. and P. Agamuthu. 2013b. Potential biowastes to remediate diesel contaminated soils. Global NEST J. 15(4): 474–484.

Dave, B.P., C.M. Ghevariya, J.K. Bhatt, D.R. Dudhagara and R.K. Rajpara. 2014. Enhanced biodegradation of total polycyclic aromatic hydrocarbons (TPAHs) by marine halotolerant *Achromobacter xylosoxidans* using Triton X-100 and β-cyclodextrin—A microcosm approach. Mar. Pollut. Bull. 79: 123–129.

El-Mahdi, A.M., H. Abdul Aziz, S.S. Abu Amr, N.Sh. El-Gendy and H.N. Nassar. 2016. Isolation and characterization of *Pseudomonas* sp. NAF1 and its application in biodegradation of crude oil. Environ. Earth Sci. 75: 380–391.

El-Gendy, N.Sh. and J.Y. Farah. 2011. Kinetic modeling and error analysis for decontamination of different petroleum hydrocarbon components in biostimulation of oily soil microcosm. Soil Sediment Contam. 20(4): 432–446.

El-Gendy, N.Sh., H.R. Ali, M.M. El-Nady, S.F. Deriase, Y.M. Moustafa and M.I. Roushdy. 2014. Effect of different bioremediation techniques on petroleum biomarkers and asphaltene fraction in oil polluted sea water. Desalin Water Treat. 52(40–42): 7484–7494.

El-Gendy, N.Sh., H.R. Madian and S.S. Abu Amr. 2013. Design and optimization of a process for sugarcane molasses fermentation by *Saccharomyces cerevisiae* using response surface methodology. Int. J. Microbiol. 2013: 9. Article ID 815631. http://dx.doi.org/10.1155/2013/815631.

Ferrari, D.G., J. Pratscher and T.J. Aspray. 2019. Assessment of the use of compost stability as an indicator of alkane and aromatic hydrocarbon degrader abundance in green waste composting materials and finished composts for soil bioremediation application. Waste Manag. 95: 365–369.

Galitskaya, P., L. Akhmetzyanova and S. Selivanovskaya. 2016. Biochar-carrying hydrocarbon decomposers promote degradation during the early stage of bioremediation. Biogeosciences. 13: 5739–5752.

Gan, L., J.P. Wang and Q.S. Wu. 2018. Bacterial diversity change in oil-contaminated soils in Jianghan oilfield via a high-throughput sequencing technique. Biotechnol. 17: 128–134.

George, S. and K. Jayachandran. 2009. Analysis of rhamnolipid biosurfactants produced through submerged fermentation using orange fruit peelings as sole carbon source. Appl. Biochem. Biotechnol. 158: 694–705.

Gesheva, V., E. Stackebrandt and V. Tonkova. 2010. Biosurfactant production by halotolerant *Rhodococcus fascians* from Casey station, Wilkes Land, Antarctica. Curr. Microbiol. 61: 112–17.

Godleads, O.A., T.F. Prekeyi, E.O. Samson and E. Igelenyah. 2015. Bioremediation, biostimulation and bioaugmentation: A review. Int. J. Environ. Bioremediat. Biodegr. 3(1): 28–39.

Gomes de Almeida, F.C., T. Alves de Lima e Silva, I. Garrard, L.A. Sarubbo, G. Maria de Campos-Takaki and E.B. Tambourgi. 2015. Optimization and evaluation of biosurfactant produced by *Pantoea* sp. using pineapple peel residue, vegetable fat and corn steep liquor. J. Chem. Chem. Eng. 9: 269–279.

Gudiña, E.J., A.I. Rodrigues, E. Alves, M.R. Domingues, J.A. Teixeira and L.R. Rodrigues. 2015a. Bioconversion of agro-industrial by-products in rhamnolipids toward applications in enhanced oil recovery and bioremediation. Bioresour. Technol. 177: 87–93.

Gudiña, E.J., E.C. Fernandes, A.I. Rodrigues, J.A. Teixeira and L.R. Rodrigues. 2015b. Biosurfactant production by *Bacillus subtilis* using corn steep liquor as culture medium. Front Microbiol. 6. Article 59. doi: 10.3389/fmicb.2015.00059.

Guerra-Santos, L., O. Käppeli and A. Fiechter. 1984. *Pseudomonas aeruginosa* biosurfactant production in continuous culture with glucose as carbon source. Appl. Environmen. Microbiol. 48(2): 301–305.

Haba, E., A. Pinazo, O. Jauregui, M.J. Espuny, M.R. Infante and A. Manresa. 2003. Physicochemical characterization and antimicrobial properties of rhamnolipids produced by *Pseudomonas aeruginosa* 47T2 NCBIM 40044. Biotechnol. Bioeng. 81: 316–322.

Hamoudi-Belarbi, L.S. Hamoudi, K. Belkacemi, L. Nouri, L. Bendifallah and M. Khodja. 2018. Bioremediation of polluted soil sites with crude oil hydrocarbons using carrot peel waste. Environments. 5: 124. doi:10.3390/environments5110124.

Hamzah, A., C.-W. Phan, P.-H. Yong and N.H. Mohd Ridzuan. 2014. Oil palm empty fruit bunch and sugarcane bagasse enhance the bioremediation of soil artificially polluted by crude oil. Soil Sediment Contam. 23: 751–62.

Hara, E.M., M. Kurihara, N. Nomura, T. Nakajima and H. Uchiyama. 2013. Bioremediation field trial of oil-contaminated soil with food waste compost. J. JSCE. 1: 125–132.

Hashemi, S.Z., J. Fooladi, G. Ebrahimipour and S. Khodayari. 2016. Isolation and identification of crude oil degrading and biosurfactant producing bacteria from the oil contaminated soils of Gachsar. Appl. Food Biotechnol. 3(2): 83–89.

Henkel, M., M.M. Müller, J.H. Kügler, R.B. Lovaglio, J. Contiero, C. Syldatk, et al. 2012. Rhamnolipids as biosurfactants from renewable resources: Concepts for next-generation rhamnolipid production. Process Biochem. 47: 1207–1219.

Hidayah, W.R. and S. Mangkoedihardjo. 2010 Rice husk for bioremediation of chromium(VI) polluted soil. Int. J. Acad. Res. 2: 35–38.

Hultgren, J., L. Pizzul, Md. P. Castillo and U. Granhall. 2009. Degradation of PAH in a creosote contaminated soil: A comparison between the effects of willows (*Salix viminalis*), wheat straw and a nonionic surfactant. Int. J. Phytoremed. 12: 54–66.

Hwang, E.Y., W. Namkoong and J.S. Park. 2001. Recycling of remediated soil for effective composting of diesel-contaminated soil. Comput. Sci. Util. 9(2): 143–148.

Jadhav V.V., A. Yadav, Y.S. Shouche., S. Aphale, A. Moghe and S. Pillai. 2013, Studies on biosurfactant from *Oceanobacillus* sp. BRI 10 isolated from Antarctic seawater. Desalination. 318: 64–71.

Jain, P.K., V.K. Gupta, R.K. Gaur, M. Lowry, D.P. Jaroli and U.K. Chauhan 2011. Bioremediation of petroleum oil contaminated soil and water. Res. J. Environ. Tox. 5(1): 1–26.

Janek, T., M. Lukaszewicza and A. Krasowska. 2013. Identification and characterization of biosurfactants produced by the Arctic bacterium *Pseudomonas putida* BD2. Colloids Surf. B. 110: 379–386.

Jeng, A.S., T.K. Haraldsen, N. Vagstad and A. Grønlund. 2004. Meat and bone meal as nitrogen fertilizer to cereals in Norway. Agr. Food Sci. 13(3): 268–275.

Joo, H.S., P.M. Ndegwa, M. Shoda and C.G. Phae. 2008. Bioremediation of oil-contaminated soil using *Candida catenulate* and food waste. Environ. Pollut. 156: 891–896.

Karppinen, E.M., K.J. Stewart, R.E. Farrell and S.D. Siciliano. 2017. Petroleum hydrocarbon remediation in frozen soil using a meat and bonemeal biochar plus fertilizer. Chemosphere. 173: 330–339.

Khopade, A., B. Ren, X.Y. Liu, K. Mahadik, L., Zhang and C. Kokare. 2012. Production and characterization of biosurfactant from marine *Streptomyces* species B3. Desalination. 285: 198–204.

Kota, M.F., A.A.S.A. Hussaini, A. Zulkharnain and H.A. Roslan. 2014. Bioremediation of crude oil by different fungal genera. Asian J. Plant Biol. 12(1): 11–18.

Lau, K.L., Y.Y. Tsang and S.W. Chiu. 2003. Use of spent mushroom compost to bioremediate PAH-contaminated samples. Chemosphere. 52(9): 1539–1546.

Leahy, J.G. and R.R. Colwell. 1990. Microbial degradation of hydrocarbons in the environment. Microbiol. Rev. 54(3): 305–315.

Lee, S.H., B.I. Oh and J.G. Kim. 2008. Effect of various amendments on heavy mineral oil bioremediation and soil microbial activity. Bioresour. Technol. 99: 2578–2587.

Lin, C., D.S. Sheu, T.C. Lin, C.M. Kao and D. Grasso. 2012. Thermophilic biodegradation of diesel oil in food waste composting processes without bioaugmentation. Environ. Eng. Sci. 29(2): 117–123.

Liu, Q., Q. Li, N. Wang, D. Liu, L. Zan, L. Chang, et al. 2018. Bioremediation of petroleum-contaminated soil using aged refuse from landfills. Waste Manag. 77: 576–585.

Liu, W., Y. Luo, Y. Teng, Z. Li and L.Q. Ma. 2010. Bioremediation of oily sludge-contaminated soil by stimulating indigenous microbes. Environ. Geochem. Health. 32: 23–29.

Liu, X., V. Selonen, K. Steffen, M. Surakka, A.L. Rantalainen, M. Romantschuk, et al. 2019. Meat and bone meal as a novel biostimulation agent in hydrocarbon contaminated soils. Chemosphere. 225: 574–578.

Luna, J.M., B.G.A. Lima, M.I.S. Pinto, P.P.F. Brasileiro, R.D. Rufino and L.A. Sarubbo. 2017. Application of *Candida sphaerica* biosurfactant for enhanced removal of motor oil from contaminated sand and seawater. Chem. Eng. Trans. 57: 565–570.

Luna, J.M., R.D. Rufino, C.D.C. Albuquerque, L.A. Sarubbo and G.M. Campos-Takaki. 2011. Economic optimized medium for tensio-active agent production by *Candida sphaerica* UCP0995 and application in the removal of hydrophobic contaminant from sand. Int. J. Mol. Sci. 12: 2463–2476.

Macaulay, B.M. and D. Rees. 2014. Bioremediation of oil spills: A review of challenges for research advancement. Ann. Environ. Sci. 8: 9–37.

Molina-Barahona, L., R. Rodriguez-Vázquez, M. Hernández-Velasco, C. Vega-Jarquín, O. Zapata-Pérez, A. Mendoza-Cantú, et al. 2004. Diesel removal from contaminated soils by biostimulation and supplementation with crop residues. Appl. Soil Ecol. 27: 165–175.

Montero-Rodríguez, D., R.F.S. Andrade, D.L.R. Ribeiro, D. Rubio-Ribeaux, R.A. Lima, H.W.C. Araújo, et al. 2015. Bioremediation of petroleum derivative using biosurfactant produced by *Serratia marcescens* UCP/WFCC 1549 in low-cost medium. Int. J. Curr. Microbiol. App. Sci. 4(7): 550–562.

Mrayyan, B. and M.N. Battikhi. 2005. Biodegradation of total organic carbons (TOC) in Jordanian petroleum sludge. J. Hazard. Mater. 120: 127–134.

Namkoong, W., E. Hwang, J. Park and J. Choi. 2002. Bioremediation of diesel-contaminated soil with composting. Environ. Pollut. 119: 23–31.

Obayori, O.S., L.B. Salam, W.T. Anifowoshe, Z.M. Odunewu, O.E. Amosu and B.E. Ofulue. 2015. Enhanced degradation of petroleum hydrocarbons in corn-steep-liquor-treated soil microcosm. Soil and Sed. Contam. 24(7): 731–743.

Obiakalaije, U.M., O.A. Makinde and E.R. Amakoromo. 2015. Bioremediation of crude oil polluted soil using animal waste. Int. J. Environ. Bioremediat. Biodegr. 3(3): 79–85.

Ofoegbu, R.U., Y.O.L. Momoh and I.L. Nwaogazie. 2015. Bioremediation of crude oil contaminated soil using organic and inorganic fertilizers. Pet. Environ. Biotechnol. 6(1): 1–6. doi:10.4172/2157-7463.1000198.

Ogbeh, G.O., T.O. Tsokar and E. Salifu. 2019. Optimization of nutrients requirements for bioremediation of spent-engine oil contaminated soils. Environ. Eng. Res. 24(3): 484–494.

Omosiowho, U.E. 2014. Comparative analysis of composting and landfarming as bioremediation techniques in hydrocarbon degradation. Intl. J. Sci. Env. Tech. 3(6): 1977–1995.

Pasumarthi, R., S. Chandrasekaran and S. Mutnuri 2013. Biodegradation of crude oil by *Pseudomonas aeruginosa* and *Escherichia fergusonii* isolated from the Goan coast. Mar. Pollut. Bull. 76: 276–282.

Patel, R.M. and A.J. Desai, A.J. 1997. Biosurfactant production *by Pseudomonas aeruginosa* GS3 from molasses. Lett. Appl. Microbiol. 25: 91–94.

Pele, M.A., D.R. Ribeaux, E.R. Vieira, A.F. Souza, M.A.C. Luna, D.M. Rodríguez, et al. 2019. Conversion of renewable substrates for biosurfactant production by *Rhizopus arrhizus* UCP 1607 and enhancing the removal of diesel oil from marine soil. Electron. J. Biotechnol. 38: 40–48.

Person, A. and G. Molin. 1987. Capacity for biosurfactant production of environmental *Pseudomonas* and *Vibrionaceae* growing on carbohydrates. Appl. Microbiol. Biotechnol. 26: 439–442.

Pineda-Flores, G. and A.M. Mesta-Howard. 2001. Petroleum asphaltenes: Generated problematic and possible biodegradation mechanisms. Rev. Latinoam. Microbiol. 43: 143–150.

Rhbal, H., S.M. Safi, M. Terta, M. Arad, A. Anouzla and M. Hafid. 2014. Hydrocarbons and green waste elimination through composting process. Int. J. Environ. Monit. Anal. 2(3): 13–22.

Rhykerd, R.L., B. Crews, K.J. McInnes and R.W. Weaver. 1999. Impact of bulking agents, forced aeration, and tillage on remediation of oil contaminated oil. Bioresour. Technol. 67(3): 279–285.

Rocha e Silva, N.M.P., R.D. Rufino, J.M. Luna, V.A. Santos and L.A. Sarubbo. 2014. Screening of *Pseudomonas* species for biosurfactant production using low-cost substrates. Biocat. Agric. Biotechnol. 3: 132–139.

Romanus, A.A., E.F. Ikechukwu, A.S. Patrick, U. Goddey and O. Helen. 2015. Efficiency of plantain peels and guinea corn shaft for bioremediation of crude oil polluted soil. J. Microbiol. Res. 5(1): 31–40.

Sabate, J., M. Vinas and A.M. Solanas 2004. Laboratory-scale bioremediation experiments on hydrocarbon-contaminates soils. Int. Biodeter. Biodegr. 52: 19–25.

Santos, A.P.P., M.D.S. Silva, E.V.L. Costa, R.D. Rufino, V.A. Santos, C.S. Ramos, et al. 2018. Production and characterization of a biosurfactant produced by *Streptomyces* sp. DPUA1559 isolated from lichens of the amazon region. Brazilian. J. Med. Biol. Res. 51. e6657. https://doi.org/10.1590/1414-431X20176657.

Santos, D.K.F., R.D. Rufino, J.M. Luna, V.A. Santos and L.A. Sarubbo. 2016. Biosurfactants: Multifunctional biomolecules of the 21st century. Int. J. Mol. Sci, 17: 401. doi:10.3390/ijms17030401.

Santosa, E.F., M.F.S. Teixeirab, A. Convertic, A.L.F. Portoa and L.A. Sarubbo. 2019. Production of a new lipoprotein biosurfactant by *Streptomyces* sp. DPUA1566 isolated from lichens collected in the Brazilian Amazon using agroindustry wastes. Biocatal. Agric. Biotechnol. 17: 142–150.

Sari, G.L., Y. Trihadiningrum and Ni'matuzahroh. 2019. Bioremediation of petroleum hydrocarbons in crude oil contaminated soil from wonocolo public oilfields using aerobic composting with yard waste and rumen residue amendments. J. Sustain. Dev. Energy Water Environ. Syst. 7(3): 482–492.

Sarubbo L.A., J.M. Luna and R.D. Rufino. 2015. Application of a biosurfactant produced in low-cost substrates in the removal of hydrophobic contaminants. Chem. Eng. Trans. 43: 295–300.

Sayara, T., E. Borràs, G. Caminal, M. Sarrà and A. Sánche. 2011. Bioremediation of PAHs-contaminated soil through compo sting: Influence of bioaugmentation and biostimulation on contaminant biodegradation. Int. Biodeter. Biodegr. 65: 859–865.

Schirmer, M., J. Molson, E. Frind and J. Barker. 2000. Biodegradation modeling of a dissolved gasoline plume applying independent laboratory and field parameters. J. Conatm. Hydrol. 46: 339–374.

Scholz-Böttcher, B.M., S. Ahlf, F. Vázquez-Gutiérrez and J. Rullkötter. 2009. Natural *vs.* anthropogenic sources of hydrocarbons as revealed through biomarker analysis: A case study in the southern Gulf of Mexico. B. Soc. Geol. Mex. 61(1): 47–56.

Seklemova, E., A. Pavlova and K. Kovacheva 2001. Biostimulation based biodegradation of diesel fuel: Field demonstration. Biodegradation. 12: 311–316.

Shahi, A., S. Aydin, B. Ince and O. Ince. 2016. Reconstruction of bacterial community structure and variation for enhanced petroleum hydrocarbons degradation through biostimulation of oil contaminated soil. Chem. Eng. J. 306: 60–66.

Shahsavari, E., E.M. Adetutu, P.A. Anderson and A.S. Ball. 2013. Plant residues-A low cost, effective bioremediation treatment for petrogenic hydrocarbon-contaminated soil. Sci. Total. Environ. 443: 766–774.

Sharma, P., J. Singh, S. Dwivedi and M. Kumar. 2014. Bioremediation of oil spill. J. Biosci. Technol. 5(6): 571–581.

Shen, G., Y. Lu and J. Hong. 2006. Combined effect of heavy metals and polycyclic aromatic hydrocarbons on urease activity in soil. Ecotox. Environ. Safe. 63: 474–480.

Sobrinho, H.B.S., R.D. Rufino, J.M. Luna, A.A. Salgueiro, G.M. Campos-Takaki, L.F.C. Leite, et al. 2008. Utilization of two agroindustrial by-products for the production of a surfactant by *Candida sphaerica* UCP0995. Process Biochem. 43: 912–917.

Soliman, R.M., N.Sh. El-Gendy, S.F. Deriase, L.A. Farahat and A.S. Mohamed. 2014. The evaluation of different bioremediation processes for Egyptian oily sludge polluted soil on a microcosm level. Energ. Source. Part A. 36(3): 231–24.

Subha, B., Y.C. Song and J.H. Woo. 2015. Optimization of biostimulant for bioremediation of contaminated coastal sediment by response surface methodology (RSM) and evaluation of microbial diversity by pyrosequencing. Mar. Pollut. Bull. 98: 235–246.

Tara, N., M. Afzal, T.M. Ansari, R. Tahseen, S. Iqbal and Q.M. Khan. 2014. Combined use of alkane-degrading and plant growth-promoting bacteria enhanced phytoremediation of diesel contaminated soil. Int. J. Phytorem. 16(12): 1268–1277.

Thavasi, R., S. Jayalakshmi and I.M. Bana. 2011. Effect of biosurfactant and fertilizer on biodegradation of crude oil by marine isolates of *Bacillus megaterium*, *Corynebacterium kutscheri* and *Pseudomonas aeruginosa*. Bioresour. Technol. 102: 772–778.

Trejo-Hernandez, M.R., A.R. Lopez-Munguia and Q. Ramirez. 2001. Residual compost of *Agaricus bisporus* as a source of crude laccase for enzymatic oxidation of phenolic compounds. Process Biochem. 36: 635–639.

Tsui, L. and W.R. Roy. 2007. Effect of compost age and composition on the atrazine removal from solution. J. Hazard. Mater. B 139(1): 79–85.

Udeh N.U., I.L. Nwaogazie and Y. Momo. 2013. Bio-remediation of a crude oil contaminated soil using water hyacinth (*Eichhornia crassipes*). Adv. Appl. Sci. Res. 4(2): 362–369.

Umanu, G. and M.F. Babade. 2013. Biological degradation of kerosene in soil amended with poultry droppings. Int. J. Adv. Biol. Res. 3(2): 254–259.

Uwem, U.M., D.M. Sunday, A.F. Wosilat and U.U. Jonah. 2019. Bioremediation of petroleum sludge impacted soils using agro-waste from Moringa seed. Sci. J. Anal. Chem. 7(1): 1–12.

Vidali, M. 2001. Bioremediation: An overview. Pure Appl. Chem. 73(7): 1163–1172.

Wellman, D.E., A.L. Ulery, M.P. Barcellona and S. Puerr-Auster. 2001. Aimal waste-enhanced degradation of hydrocarbon-contaminated soil. Soil Sed. Contam. 10: 511–523.

Williams, J.O. and V.C. Amaechi, 2017. Bioremediation of hydrocarbon contaminated soil using organic wastes as amendment. Curr. Stud. Compar. Edu. Sci. Technol. 4(2): 89–99.

Xu, Y. and M. Liu. 2010. Bioremediation of crude oil-contaminated soil: comparison of different biostimulation and bioaugmentation treatments. J. Hazard. Mater. 183(1–3): 395–401.

Yelebe, Z.R., R.J. Samuel and B.Z. Yelebe. 2015. Kinetic model development for bioremediation of petroleum contaminated soil using palm bunch and wood ash. Int. J. Eng. Sci. Invention. 4(5): 40–47.

Yi, H. and D.E. Crowley. 2007. Biostimulation of PAH degradation with plants containing high concentrations of linoleic acid. Environ. Sci. Technol. 41: 4382–4388.

CHAPTER 17

Nitrogen Cycle Bacteria in Agricultural Soils:
Effects of Nitrogen Fertilizers, Heavy Metals, Pesticides and Bioremediation Approaches

**Guillermo Bravo, Paulina Vega-Celedón,
Constanza Macaya, Ingrid-Nicole Vasconez and Michael Seeger***

Laboratorio de Microbiología Molecular y Biotecnología Ambiental,
Departamento de Química & Centro de Biotecnología Daniel Alkalay Lowitt,
Universidad Técnica Federico Santa María, Avenida España 1680, Valparaíso, Chile.

1. INTRODUCTION

Soil plays an essential role in life. It provides the habitat for diverse animals, plants, protists and microorganisms. Soil is the physical place where diverse biogeochemical cycles occur. Phenomena such as soil erosion, nutrient depletion, low organic matter content, changes in pH, pollution and high variability in rainfall distribution generate nutrient deficiency, affecting soil fertility and disrupting biogeochemical cycles (Hai et al. 2009, Hernández et al. 2011, Altimira et al. 2012, Rütting et al. 2018).

Nitrogen is a key element in the productivity of soils and its availability is critical for agriculture. Nitrogen fertilization has allowed to increase crop yield and productivity. However, the massive application of nitrogen compounds showed negative effects on the environment: water pollution resulting from nitrate leaching, ammonia emission, generation of greenhouse gases (N_2O and NO) and ecosystems' eutrophication near agricultural sites (Küstermann et al. 2010, Rütting et al. 2018). In addition, heavy metals pollution and the high use of pesticides, such as herbicides, fungicides and insecticides, negatively affect the nitrogen microbial communities in agricultural soils (Hernández et al. 2011, Altimira et al. 2012, Walia et al. 2014, Thiour–Mauprivez et al. 2019). Therefore, the global trend is to reduce the applications of nitrogen compounds and

*Corresponding author: michael.seeger@usm.cl, michael.seeger@gmail.com

pesticides, and to optimize the natural nitrogen cycle (Hai et al. 2009). The aims of this report are the analyses of the effects of fertilization, heavy metals and pesticides on the microbial communities associated with the nitrogen cycle in agricultural soils, and the bioremediation strategies of agricultural soils polluted with heavy metals and pesticides.

2. NITROGEN CYCLE IN AGRICULTURAL SOILS

The nitrogen cycle in soil includes nitrogen fixation, nitrification, denitrification, and ammonification processes (Fig. 1). However, from an agricultural perspective, nitrogen balance in soils is influenced by biological nitrogen fixation, nitrification and denitrification (Mosier et al. 2004).

Atmosphere

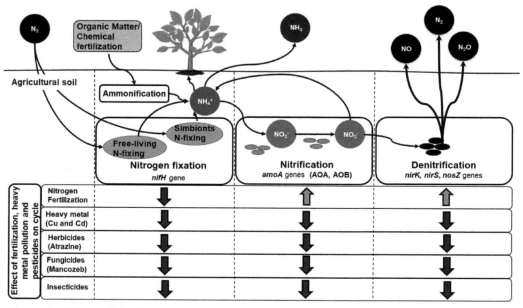

Figure 1   Effects of fertilization, heavy metals and pesticides on nitrogen fixation, nitrification and denitrification in agricultural soils.
(The effects on the key genes of nitrogen fixation (nitrogenase encoding *nifH* gene), nitrification (ammonia monooxygenase encoding *amoA* gene) and denitrification (nitrite reductase encoding *nirK* and *nirS* genes; nitrous oxide reductase encoding *nosZ* gene) are illustrated.)

Nitrogen fixation   The main source of nitrogen is N_2 gas in the atmosphere. N_2 is highly stable and can only be used by a limited group of archaea and bacteria, incorporating it into the cell in a process known as nitrogen fixation. Nitrogen fixation is carried out by microorganisms, which may be archaea, free-living bacteria or bacteria in symbiosis with plant roots (Franche et al. 2008). Nitrogen fixation is carried out in the absence of oxygen by aerobic, facultative anaerobic and strict anaerobic organisms. N_2 gas is reduced to ammonia by a multienzyme complex called nitrogenase (Bravo 2017, Rütting et al. 2018).

In agricultural soils, the presence of nitrogen fixing free-living microorganisms or symbionts has been used as an indicator of soil quality (Filip 2002). Nitrogen fixing microorganisms contribute to the productivity of plants; therefore, their number are relevant in agricultural systems (Hai et al. 2009). Free-living microorganisms play an important role in the nitrogen fixation in various ecosystems, since on average they are responsible for fixing 2–3 kg N ha^{-1} year^{-1} (Mao et al. 2011).

Nitrification Nitrification is an essential process in the nitrogen cycle in soils, which involves the biological oxidation of ammonia via nitrite to nitrate in the presence of oxygen by bacteria and archaea (Hernández et al. 2011). Several enzymes participate in the oxidation of reduced nitrogen compounds. The transmembrane enzyme ammonia monooxygenase oxidizes ammonia to hydroxylamine. NH_2OH is subsequently oxidized by hydroxylamine oxidoreductase to nitrite (Hernández et al. 2011). Due to the high solubility of nitrate in agricultural systems, nitrification may cause negative effects, generating losses in crop production, and causing water eutrophication. It has been estimated that nitrification produces worldwide losses of 37 Tg of N year^{-1} in soil (Mosier et al. 2004).

Denitrification Denitrification is a process in which nitrate is reduced successively to NO, N_2O and N_2 gases. Nitrate reduction by anaerobic microorganisms uses nitrate as electron acceptor. Nitrate reduction starts with the action of the transmembrane nitrate reductase, whose product is nitrite. Nitrite is then reduced to nitric oxide by nitrite reductase. Nitric oxide is successively reduced to gaseous nitrogen (Rütting et al. 2018). Denitrification can be harmful to soils. It has been shown that 17 Tg of N are lost worldwide every year from soils and 15 Tg at the agroecological level due to the microbial denitrifying activity (Mosier et al. 2004, Mao et al. 2011).

Ammonification Ammonification is the process whereby organic N is mineralized to ammonium (NH_4^+) by soil microorganisms. The oxidation of organic nitrogen (e.g., urea, amino acid, nucleic acid, uric acid) releases ammonium and produces energy. Ammonium is assimilated by plants and soil microorganisms (Mosier et al. 2004).

Multiple microorganisms are involved in the nitrogen cycle. In recent decades, tools based on metagenomic approach and q-PCR of genes associated with the nitrogen cycle have allowed the *in situ* study of the nitrogen cycle in agricultural soils. Marker genes frequently used are nitrogenase encoding *nifH* gene that is associated to nitrogen fixation, ammonia monooxygenase encoding *amoA* gene that is involved in nitrification (*amoA* AOA for archaea and *amoA* AOB for bacteria), nitrite reductase encoding *nirK* and *nirS* genes and nitrous oxide reductase encoding *nosZ* gene involved in denitrification (Fig. 1) (Hai et al. 2009, Jung et al. 2011, Ouyang et al. 2018, Pereg et al. 2018).

3. FERTILIZERS AND THEIR EFFECTS ON THE NITROGEN CYCLE

Nitrogen fertilization is a standard practice for improving crop growth and productivity. The nitrogen availability in crops depends on microbial activity. Nitrogen fertilization affects the structure of nitrogen fixing, nitrification and denitrification microorganisms (Fig. 1) (Hai et al. 2009, Morgante et al. 2010, Hernández et al. 2011). Metagenomic studies based on the analysis of the *nifH* gene show discrepancies in the effects of nitrogen fertilization on nitrogen fixation. Jung et al. (2011) described that nitrogen fixation processes are inhibited with increasing nitrogen availability in soils, affecting the abundance of nitrogen fixing microorganisms. The effects of nitrogen fertilization on nitrogen fixing bacteria is dependent on crop type, pH, temperature, type of fertilization (inorganic or organic) and organic matter (Fig. 1) (Ouyang et al. 2018, Pereg et al. 2018). Hai et al. (2009) reported a decrease in the number of nitrogen-fixing microbes during the application of urea and compost in agricultural soils. Ouyang et al. (2018) indicated that inorganic and organic nitrogen fertilization had no effects on the abundance of the *nifH* gene using metagenomic analysis on crop. Pereg et al. (2018) showed that organic fertilization strongly increases the *nifH* gene, whereas fertilizers rich in inorganic ammonia and nitrate showed lower increase of this gene.

Nitrogen fertilization (organic or inorganic) significantly increases the number of microorganisms associated to nitrification (AOA *amoA* and AOB *amoA* genes) and denitrification (*nirK*, *nirS* and *nosZ* genes), which is independent of the crop type (Hai et al. 2009, Jung et al. 2011, Ouyang et al. 2018, Pereg et al. 2018). The increase in both nitrification and denitrification gene abundances under nitrogen fertilization suggests a high rate of nitrogen loss and an inefficient use of nitrogen fertilizer by the crops (Fig. 1) (Mosier et al. 2004, Ouyang et al. 2018).

4. HEAVY METALS AND THEIR EFFECTS ON THE NITROGEN CYCLE

Anthropogenic activities and the wide use of heavy metals as biocides cause soil pollution (Seshadri et al. 2015). The main heavy metals that pollute agricultural soils are copper (Cu), cadmium (Cd), mercury (Hg), zinc (Zn) and lead (Pb) (Altimira et al. 2012, Bravo 2017, Li et al. 2019, Bravo et al. 2020). Heavy metals cause negative effects on living organisms through mechanisms of oxidative stress, disruption of the integrity of cell membranes, enzymatic inhibition by disruption of the active sites, interference with the energy transport system, and replacement of metallic centre of metalloporphyrins, among other effects (Rojas et al. 2011, Ahmad et al. 2012, Gómez-Sagasti and Marino 2015, Ojuederie and Babalola 2017). Heavy metals cause dramatic changes in soil microbial communities and their activities (Ahmad et al. 2012, Altimira et al. 2012). The effects of three main heavy metals such as copper, cadmium and mercury on the nitrogen cycle will be analyzed.

4.1 Copper

Copper is an essential micronutrient for all living organisms. However, in agriculture Cu at higher concentration acts as a broad-spectrum biocide to control plant pathogenic bacteria, fungi, oomycetes, invertebrates and algae (Lamichhane et al. 2018). Therefore, copper-based products lead to an accumulation of Cu in soils, which is associated to a reduction of the microbial community in soils (Seshadri et al. 2015, Ojuederie and Babalola 2017). Copper produces changes in microbial abundance and community structure (Altimira et al. 2012).

Agricultural soil nitrogen cycle microbial communities are affected by copper pollution. Copper has strong inhibitory effects on nitrogen-fixing microorganisms. Copper pollution decreases growth and enzymatic activities of *Rhizobium* and free-living nitrogen fixing bacteria (Ahmad et al. 2012). This heavy metal reduces the abundance of the *amoA* (AOA, AOB) genes, inhibiting nitrification processes in agricultural soils (Figure 1) (Liu et al. 2014, Simonin et al. 2018). However, different effects of copper on nitrifying microorganisms have been reported. Simonin et al. (2018) showed a decrease in the abundance of the *nirS* gene in agricultural soil treated with copper biocide. In contrast, Liu et al. (2014) reported that the abundance of the *nirK* gene in rice paddies was not affected by copper pollution.

4.2 Cadmium

Cadmium (Cd) is a heavy metal that is not involved in any biological processes. Cadmium is part of the top 10 chemicals of major public health concern according to the World Health Organization (Alviz-Gazitua et al. 2019). Cd interacts with photosynthetic, respiratory and nitrogen metabolism in plants, causing wilting and chlorosis, which may cause poor growth, low biomass and death (Zhang et al. 2014, Gómez-Sagasti and Marino 2015). Phosphate fertilizers and farmyard manures are the main cadmium sources in agricultural soils (Afzal et al. 2019).

Due to its high toxicity, Cd causes serious effects on the nitrogen cycle in agricultural soils. Cadmium pollution inhibits symbiosis establishment, nodule formation and symbiotic nitrogen fixation in legume (Gómez-Sagasti and Marino 2015). Cd strongly decreases the abundance of the nitrification *amoA* genes (AOA and AOB) and bacterial denitrification *nirS, nirK* and *nosZ* genes in agricultural soils (Fig. 1) (Afzal et al. 2019).

4.3   Mercury

Mercury is a highly toxic heavy metal for macroorganisms and microorganisms (Rojas et al. 2011). Hg is distributed at low concentration in air, water and soils due to natural cycles and human activities (Branco et al. 2012, Tagliaferro et al. 2015). Mercury(II) is a highly toxic ion with high affinity to sulfhydryl, thioester and imidazole groups of proteins, inactivating proteins (Tagliaferro et al. 2015). Mercury pollution in agricultural soils cause negative effects on plants. Hg replaces the central magnesium atom of chlorophyll, interrupting photosynthesis (Patra and Sharma 2000). In microorganisms, mercury causes inhibition of the microbial activity and reduces the diversity of microbial communities (Ranjard et al. 2006).

In agricultural soils, the effects of mercury pollution on the nitrogen cycle microorganisms have been reported. Hg reduces the abundance of nitrogen-fixing bacteria (*nifH*) and bacteria involved in nitrification (*amoA*) (Bravo 2017). Liu et al. (2010) reported no significant differences in the number of *amoA* gene copies in long-term polluted soils exposed to various mercury concentrations. Specific microorganisms have adapted to mercury pollution by acquiring genetic determinants and physiological changes such as saturation-rigidification of cell membrane, production of stress proteins, and elimination of toxic compounds using efflux pump (Rojas et al. 2011, Altimira et al. 2012, Madrova et al. 2018). The effects of mercury on soil denitrifying bacteria have not been reported. However, a strong reduction of the *nirK* gene abundance caused by mercury pollution has been observed in aquatic systems (Mosier and Francis 2010).

4.4   Bioremediation of Heavy Metals

Traditional heavy metal agricultural soil treatment strategies are based on metal immobilization by an increase of the organic matter content and application of lime for altering soil pH (Lamichhane et al. 2018). However, these methods showed low selectivity, are applicable only on a range of contaminant concentrations and may generate toxic sludge (Fuentes et al. 2014). Bioremediation is a sustainable technology based on the capability of bacteria, fungi, archaea and plants to remove, degrade or transform compounds or elements into less toxic forms (Orellana et al. 2018).

Phytoremediation is based on the capability of plants to accumulate heavy metals, allowing the extraction of heavy metals for the clean-up of soils (Lamichhane et al. 2018). Hyperaccumulators have been widely used for the remediation of heavy metals-polluted soils (Muthusaravanan et al. 2018). Plant species such as *Jatropha curcas, Zea mays, Thlaspi caerulescens, Helianthus annuus, Brassica juncea,* and *Nicotiana tabacum* have been applied for copper and cadmium phytoremediation.

Bacterial heavy metals' bioremediation is based on aerobic reduction, biotransformation, biosorption, bioleaching, bioaccumulation and biomineralization (Wagner-Döbler 2003, Bravo 2017, Bravo et al. 2020). Copper and cadmium have been removed from soils by biosorption using the bacterium *Cupriavidus metallidurans* CH34 (Diels et al. 1999, Alviz-Gazitua et al. 2019). *Desulfovibrio desulfuricans* is capable to transform heavy metals such as Cd into insoluble forms, reducing its toxicity (Ojuederie and Babalola 2017). Mercury bioremediation is based on the reduction of Hg^{2+} to gaseous Hg^0 by mercury-resistant bacteria (Rojas et al. 2011). Bravo (2017) established a bioremediation process using *C. metallidurans* MSR33 of a mercury-polluted agricultural soil in a rotary drum reactor, reaching up to 70% of mercury removal in soil.

The effects of heavy metal bioremediation on nitrogen cycle microorganisms in soil have been scarcely reported. Soil bioremediation processes influence the dynamics of the microbial communities (Morgante et al. 2010, Fuentes et al. 2014, 2016). Mercury bioremediation of an agricultural soil by *C. metallidurans* strain MSR33 increased the communities of nitrogen fixing bacteria and nitrifying bacteria (Bravo 2017).

Sustainable agriculture requires bioremediation of heavy metal from soils and balanced biogeochemical cycles (Hernández et al. 2011, Fuentes et al. 2014, Orellana et al. 2018).

5. PESTICIDES AND THEIR EFFECTS ON THE NITROGEN CYCLE

Pesticides including herbicides, fungicides, and insecticides are widely applied in the management and control of undesirable weeds, microorganisms and insects in agriculture. Although the pesticide industry delivers immediate benefits, protecting the agricultural production, the negative effects on the ecosystems and wild crop biodiversity are of increasing concern. The persistence of pesticides and their metabolites negatively impacts macroorganisms, as well as the abundance and activity of non-target microorganisms involved in the nitrogen cycle (Hernández et al. 2011, Singh and Jauhari 2017).

5.1 Herbicides

The application of herbicides is widely used in the management practices of crop production in modern agriculture (Singh and Jauhari 2017). Globally, more than eight thousand species of weeds injure crops, causing losses of 13% (Zhang et al. 2011, 2018). There are 9 main herbicide categories: phenoxies, triazines, amides, carbamates, dinitroanilines, urea derivates, sulfonyl ureas, bipiridils and uracils (Zhang 2018). Diverse mechanisms of action have been reported: acetyl-CoA carboxylase inhibitors, acetolactate synthase or acetohydroxy acid synthase (AHAS) inhibitors, photosystem II inhibitors, photosystem I inhibitors, protoporphyrinogen oxidase (PPG oxidase or protox) inhibitors, carotenoid biosynthesis inhibitors, enolpyruvylshikimate-3-phosphate (EPSP) synthase inhibitors, glutamine synthetase inhibitors, dihydropteroate synthetase inhibitors, mitosis inhibitors, cellulose inhibitors, oxidative phosphorylation uncouplers, fatty acid and lipid biosynthesis inhibitors, synthetic auxins and auxin transport inhibitors (Forouzesh et al. 2015). The modes of action of many herbicides are well defined and the targeted enzyme in weeds is known. Interestingly, some of these targets are also present in a range of non-target organisms, including soil microorganisms (Thiour-Mauprivez et al. 2019). Glyphosate and atrazine herbicides affect also non-target organisms. Bacteria play a crucial role in nitrogen cycling in the soil ecosystem, including nitrification and denitrification processes. Herbicides' application affect nitrification and denitrification bacteria (Hernández et al. 2011, Mertens et al. 2018, Tyler and Locke 2018). The countries with the highest herbicide consumption are USA, Brazil and Argentina, exceeding 203, 153 and 116 thousand tonnes on average from 1990 to 2017, respectively (http://www.fao.org/faostat/en/#data). Three of the most commonly used agricultural herbicides by USA are glyphosate, atrazine and 2,4-dichlorophenoxyacetic acid (2,4-D) (Meyer and Scribner 2009, Fernández-Cornejo et al. 2014), which belong to a not classified, triazine and phenoxy categories of herbicides (Sherwani et al. 2015, Zhang 2018). These types of herbicides will be further discussed.

Glyphosate   Glyphosate is the most prevalent herbicide used in the world due to its broad-spectrum targets (Sherwani et al. 2015). Glyphosate or N-(phosphonomethyl) glycine (Fig. 2) is the aminophosphonic acid derivate of the natural amino acid glycine that has an amphoteric and zwitterion structure containing three functional groups: phosphonate, amino and carboxylic acid (Jayasumana et al. 2014). Glyphosate affects the EPSP synthase, a key enzyme in the

Type of Pesticide	General Classification	Mechanism of action	Active compound
Herbicide	-	Enolpyruvylshikimate-3-phosphate synthase inhibitor	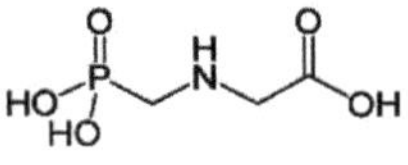 Glyphosate
Herbicide	Triazine	Photosystem II inhibitor	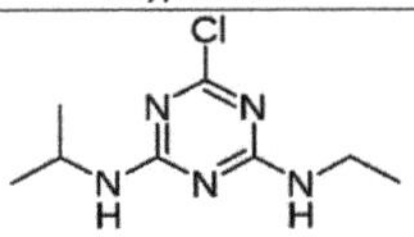 Atrazine
Herbicide	Phenoxy	Synthetic auxin	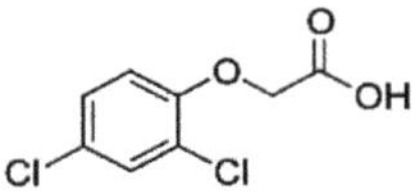 2,4-dichlorophenoxyacetic acid (2,4-D)
Fungicide	Dithiocarbamate	Multi-site activity	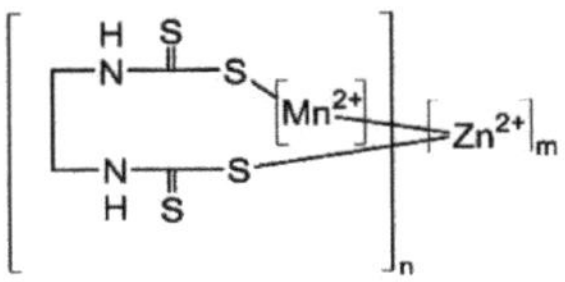 Mancozeb
Fungicide	Benzimidazole	Microtubules formation inhibitor	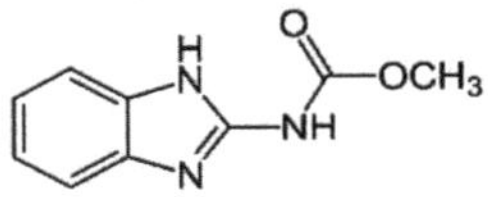 Carbendazim
Fungicide	Triazole	Ergosterol biosynthesis inhibitor	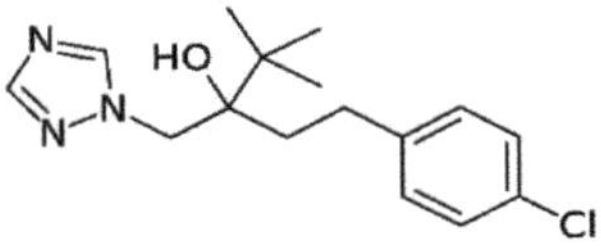 Tebuconazole
Insecticide	Neonicotinoid	Acetylcholinesterase inhibitor	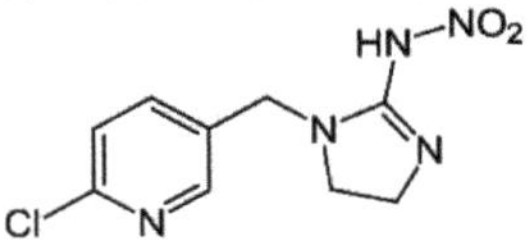 Imidacloprid
Insecticide	Neonicotinoid	Acetylcholinesterase inhibitor	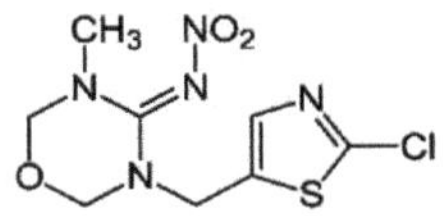 Thiamethoxam
Insecticide	Pyrethrin derivatives	Closure of voltage-gated sodium channel inhibitor	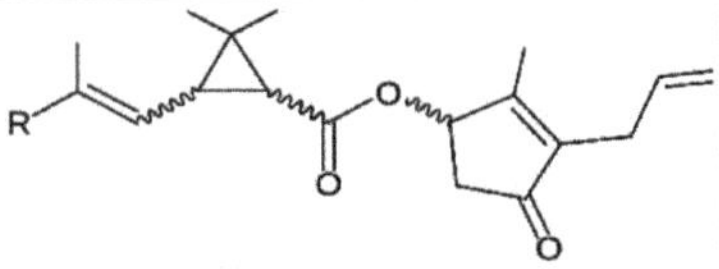 Pyrethroids

Figure 2. Classification, mechanism of action and active compound structure of some of the most commonly used agricultural herbicides, fungicides and insecticides.

shikimate pathway in plants, fungi and bacteria (Thiour-Mauprivez et al. 2019), affecting the biosynthesis of the aromatic amino acids tryptophan, tyrosine, and phenylalanine (Sherwani et al. 2015). Glyphosate belongs to the groups G and 9 in HRAC and WSSA classifications systems, respectively (Forouzesh et al. 2015).

Triazines *s*-Triazine herbicides (e.g. atrazine, simazine and terbutylazine) are applied worldwide for the control of weeds in agricultural and forest soils (Flores et al. 2009, Seeger et al. 2010, Hernández et al. 2011). The chemical structures of triazine herbicides are permutations of alkyl substituted 2,4-diamines of chlorotriazine. Atrazine (6-chloro-N^2-ethyl-N^4-(propan-2-yl)-1,3,5-triazine-2,4-diamine) (Fig. 2) is the most studied triazine herbicide (Barr and Buckley 2011). Atrazine inhibits the photosystem II, which is located in the thylakoid membrane of plants, algae and cyanobacteria (Thiour-Mauprivez et al. 2019). Atrazine belongs to the groups C_1 and 5 in HRAC and WSSA classifications systems, respectively (Forouzesh et al. 2015).

Phenoxies Phenoxy herbicides are synthetic analogues of auxin plant growth hormones (Vega-Celedón et al. 2016), which include 2,4-dichlorophenoxyacetic acid (2,4-D), 2,4,5-trichlorophenoxyacetic acid (2,4,5-T), 2-methyl-4-chlorophenoxyacetic acid (MCPA) and methylchlorophenoxypropionic acid (MCPP). These compounds have a similar molecular structure comprising an aromatic ring with a carboxylic side chain (Jayakody et al. 2015). 2,4-D is the most studied phenoxy herbicide; however, its effect on microorganisms has not been reported (Thiour-Mauprivez et al. 2019) (Fig. 2). 2,4-D belongs to the groups O and 4 in HRAC and WSSA classifications systems, respectively (Forouzesh et al. 2015).

5.2 Herbicides and their Effects on the Nitrogen Cycle

Glyphosate influences the soil microflora and their interactions with plants (Mertens et al. 2018). Several studies in genetic resistant (GR) soybean systems contaminated with glyphosate reported a reduction of the nitrogen fixation, nitrogenase enzyme activity and inhibition of the growth of rhizobia isolated from root nodules (Zablotowicz and Reddy 2004, Means et al. 2007, Zobiole et al. 2012, Fan et al. 2017, Pires Bomfim et al. 2017). Metagenomic studies on rhizosphere soil after glyphosate treatment of a GR soybean reveal a decrease of nitrogen-fixing bacteria *Bradyrhizobium* and *Rhizobium*, and an increase of *Burkholderia*. Nitrogenase *nifH*, *nifD* and *nifK* genes significantly decreased, whereas nodulation related *nodB*, *nodC* and nirK genes increased (Lu et al. 2018). An increase in the number of nodules has been reported in green gram inoculated with *Bradyrhizobium* with low concentrations of glyphosate, but a reduction of the nodulation was observed at high glyphosate concentration (Zaidi et al. 2005). Glyphosate and atrazine decreased nitrification in sugarcane, inhibiting both AOA and AOB *amoA* genes and decreasing denitrifying *narG*, *nirK*, *nirS* and *nosZ* genes, which reduces N_2O emissions (Zhang et al. 2018). Studies on corn polluted with atrazine and simazine demonstrated an inhibition of nitrification (Somda et al. 1990). Simazine strongly inhibited the nitrification processes in fertilized agricultural soil, affecting the AOB community (Hernández et al. 2011). Simazine inhibited *Nitrosobacteria* and specifically *Nitrosospira* species (Hernández et al. 2011). Reduced nodulation was observed in green gram inoculated with *Bradyrhizobium* in presence of atrazine (Khan et al. 2006). Long-term 2,4-D application reduced fungal and bacterial populations (Rai et al. 1992). Auxin and 2,4-D induce nodular outgrowths (Bhat et al. 2015). Treatments in wheat with 2,4-D and *Azospirillum* sp. resulted in an increased colonization, nodulation and nitrogenase activity post treatment (Zeman et al. 1992, Katupitiya et al. 1995, Elanchezhian and Panwar 1997). Treatments on maize with *Azorhizobium* and 2,4-D caused nodulation and the induction of nitrogenase activity (Saikia et al. 2006). Low concentrations of 2,4-D promote a high nitrogenase activity in *Stenotrophomonas maltophilia* Sb16, while high concentration inhibit nitrogenase activity (Nahi et al. 2016).

5.3 Bioremediation of Herbicides

The use of herbicides not only affects the crop growth and development, but also the quality of crops. The herbicides may pollute the food chain, threatening human health. Hence, it is crucial to develop strategies to minimize the pesticide residues in soil, water contamination and the toxicity of herbicides (Singh and Jauhari 2017).

Two major degradation pathways in glyphosate-degrading microorganisms involves the conversions of glyphosate into amidomethylphosphonic acid (AMPA) and sarcosine. Specific *Achromobacter, Agrobacterium, Arthrobacter, Bacillus, Flavobacterium, Geobacillus, Providencia, Ochrobactrum* and *Pseudomonas* strains have the AMPA pathway (Zhan et al. 2018, Xu et al. 2019). *Achromobacter, Alcaligenes, Arthobacter, Bacillus, Enterobacter, Lysinibacillus, Ochrobactrum, Pseudomonas, Rhizobium* and *Streptomyces* strains have the sarcosine pathway (Zhan et al. 2018, Pérez Rodríguez et al. 2019). Glyphosate and its breakdown product AMPA are classified as probably human carcinogens (Van Bruggen et al. 2018). Therefore, the bioremediation of glyphosate via sarcosine is the most promising catabolic pathway. Microorganisms with the sarcosine oxidase gene, such as the plant growth-promoting bacterium (PGPB) *Lysinibacillus sphaericus,* are capable of degrading glyphosate through the Carbon-Phosphorus (C–P) pathway without leading to AMPA production. Bioremediation with *L. sphaericus* in a potato crop soil sprayed with glyphosate showed a 79% degradation of glyphosate in soil with minimal AMPA production and increased ammonium (Pérez Rodríguez et al. 2019).

s-Triazine-degrading bacterial strains belonging to *Pseudomonas, Arthrobacter, Chelatobacter, Agrobacterium, Alcaligenes, Clavibacter, Ralstonia, Rhizobium, Rhodococcus, Stenotrophomonas, Pseudaminobacter* and *Nocardiodes* genera have been characterized (Hernández et al. 2008a, b, Govantes et al. 2009, Seeger et al. 2010). *Pseudomonas* sp. ADP is a model strain for *s*-triazine biodegradation studies. The enzymes of the upper pathway are encoded by the *atzA, atzB* and *atzC* genes and the enzymes of the lower pathway are encoded by the *atzDEF* operon (Martínez et al. 2001). Combined bioaugmentation and biostimulation strategies with *Pseudomonas* sp. ADP and citrate or succinate in atrazine-contaminated soil showed 80% atrazine removal after 13 days (Silva et al. 2004). A consortium composed by *Arthrobacter* sp. AK_YN10, *Pseudomonas* sp. AK_AAN5 and *Pseudomonas* sp. AK_CAN1 was capable of degrading 80% atrazine after 14 days in soil mesocosms (Sagarkar et al. 2014). A mixed bacterial consortium significantly enhances the atrazine degradation rates in contaminated soils (61%) compared to control (12%) after 30 days (Dehghani et al. 2013). *Pseudomonas* sp. MHP41 was successfully applied for bioremediation in simazine-contaminated agricultural soil, with 89% removal after 27 days (Morgante et al. 2010, 2012).

2,4-D catabolic pathways have been elucidated in diverse microorganisms including *Cupriavidus necator* JMP134 and *Pseudomonas* strains. The *tfdABCDEF* genes encode the enzymes of this pathway that funnel into the tricarboxylic acid cycle (Kumar et al. 2016). Bioaugmentation with *Cupriavidus* sp. CY-1 showed an almost complete removal of 2,4-D from common soil and forest soil after 7 days, without disturbing the indigenous microorganisms (Chang et al. 2015). Soil bioaugmentation with *Novosphingobium* sp. DY4 degraded 95% of the herbicide 2,4-D after 5–7 days. The herbicide application significantly altered soil bacterial community structure; however, bioaugmentation with strain DY4 produced only minor changes in the microbial structure (Dai et al. 2015).

5.4 Fungicides

The major soil microbial communities are represented by fungi and bacteria, while in diverse soils archaea are also important (Prosser et al. 2007). Environmental conditions such as soil properties, regional climate, and anthropogenic activity can affect the structure and function of soil microbial communities (Richter et al. 2018). In order to control plant pathogens and soil-borne diseases, the use of fungicides has become common in agriculture (Ullah and Dijkstra 2019). The application of these pesticides controls the abundance and activity of microbial communities, affecting both targeted and non-targeted microorganisms (Milenkovski et al. 2010) and, therefore, the nitrogen-cycle.

In order to assure crop production and avoid pest injuries, agricultural use of fungicides is critically high; by Italy, France and Brazil are the highest consumers of these compounds (http://www.fao.org/faostat/en/#data). Among the main organic fungicides, dithiocarbamates, benzimidazoles, triazoles and diazoles, diazines, morpholines, seed treatment fungicides, and disinfectants are the most used globally (Zhang 2018). Eleven mechanisms of action for fungicides have been described and listed by the Fungicide Resistance Action Committee (FRAC). These mechanisms of action may act on the nucleic acids metabolism, cytoskeleton and motor proteins, respiration, amino acid and protein synthesis, signal transduction, membrane integrity and function, melanin synthesis in cell wall, sterol biosynthesis in membranes, cell wall biosynthesis, host plant defense induction, and in some cases, a multi-site activity (https://www.frac.info/docs). Three of the most important categories of organic fungicides in agriculture will be analyzed.

Dithiocarbamates Dithiocarbamates (DTCs) are organosulfur ligands used as fungicides for the first time during 1940s. The dithiocarbamate ($N-CS_2$) is the functional group of these compounds (Odularu and Ajibade 2019). DTCs can be classified into three main groups: dimethyl dithiocarbamates (e.g. ziram, thiram), propylene-bis-dithiocarbamates (e.g. propineb), and ethylene-bis-dithiocarbamates (e.g. mancozeb, maneb) (Kakitani et al. 2017). Mancozeb, thiram and maneb are the most used dithiocarbamate-based products in agriculture.

Mancozeb (MCZ) is a manganese and zinc ethylene-bis-dithiocarbamate fungicide and bactericide (Fig. 2), classified in the multi-site action group (group M) by FRAC (https://www.frac.info/docs, Kakitani et al. 2017). The fungicidal activity of MCZ derives from the bisisothiocyanate sulfide and ethylene bisisothiocyanate compounds generated during MCZ exposure to water and UV light. These toxic compounds interfere with sulphydryl groups of fungal enzymes, which affect cytoplasmic and mitochondrial biochemical processes (Gullino et al. 2010, Yang et al. 2019).

Benzimidazoles Benzimidazoles (BZDs) are heterocyclic aromatic compounds introduced as broad-spectrum systemic fungicides in the 1960s. BZDs have a benzene ring and an imidazole group (Heneberg et al. 2018). Carbendazim, benomyl, fuberidazole and thiabendazole are BDZs derivatives used as fungicides (https://www.frac.info/docs). Carbendazim is the most used fungicide derivative (Dong et al. 2017, Zhou et al. 2016).

Carbendazim (CBZ) is a methyl 2-benzimidazole carbamate (Fig. 2) with a broad-spectrum activity widely used in agriculture to prevent fungal diseases in cereals, vegetables, fruits and field crops (Zikos et al. 2015). It has been proposed that the systemic fungicide CBZ binds to fungi β-tubulin genes, delimiting microtubules formation and stopping hyphal growth (Zhou et al. 2016, Heneberg et al. 2018).

Triazoles Azoles are nitrogen containing heterocycles with several biological properties including anti-microbial activities (Zhang et al. 2017). Triazole compounds are widely used in agriculture. Tebuconazole, epoxiconazole, propiconazole, bromuconazole, and difenoconazole are the most studied fungicides due to their toxicity, bioaccumulation and persistence (Crini et al. 2017).

Tebuconazole (TBZ) is a widely used systemic fungicide (Youness et al. 2018). Its structure is (RS)-1-*p*-chlorophenyl-4,4-dimethyl-3-(1H-1,2,4-triazol-1-ylmethyl)-pentan-3-ol (Fig. 2). The fungicidal activity of TBZ is based on its capability to target specifically the 14α-demethylase (CYP51), which is a regulatory enzyme of the ergosterol biosynthetic pathway. Ergosterol plays an essential role in fungal membrane formation, altering its structure and functions such as permeability and fluidity (Desmyttere et al. 2019).

5.5 Fungicides and their Effects on the Nitrogen Cycle

The accumulation of fungicides in soils can be caused by a thin plant canopy, over-application, or application followed by rainfall or irrigation (Man and Zucong 2009). Fungicide accumulation may produce a high environmental risk in soil ecosystems due to changes in the soil microbiota composition and reduction of its function (Dos Santos et al. 2017). The impact of the pesticides MCZ, CBZ and TBZ on nitrogen cycle has been reported. MCZ ($\leq$2000 ppm) inhibited nitrification and reduced nitrifying bacteria and ammonification rates (Walia et al. 2014). In grassland soil microcosms (34% moisture), a single application of MCZ inhibited the production of N_2O and NO by nitrification and denitrification after 48 hours in 55% and 96%, respectively (Kinney et al. 2005). Sandy loam soils' microcosms exposed to MCZ transiently reduced the archaeal and bacterial *amoA* gene expression, recovering basal expression after 20 days (Feld et al. 2015). CBZ and MCZ in combination with chlorpyrifos and monocrotophos insecticides enhanced nitrification in red and black soil and increased ammonification in groundnut soils (Srinivasulu et al. 2012). CBZ combined with urea decreased *Nitrosomonas* populations and also other autotrophic and heterotrophic nitrifying bacteria in soils (Ding et al. 2019). In agricultural silt loam soil, a CBZ treatment caused no significant shifts in urease activity after 7 days (Shao and Zhang 2017). Lower nitrate reductase activity was found in a peanut crop with soil treated with TBZ, although ammonification and nitrification increased with TBZ treatment (Saha et al. 2016). Inhibition of nitrification was observed in a short-term mesocosm of soil treated with TBZ. However, recovery of nitrifying microorganisms was observed with the decrease of TBZ (Muñoz-Leoz et al. 2011).

5.6 Bioremediation of Fungicides

Fungicides are conformed by chemical compounds of diverse structures with different degradation rates (Ahlawat et al. 2010). Microbial metabolism is probably the most important natural process in pesticide degradation (Salunkhe et al. 2014). Microbial bioremediation of MCZ, CBZ and TBZ pesticides has been reported. *Trichodema* sp. degraded 18% MCZ after 6 days. A consortia of *Trichoderma* sp., *Aspergillus* sp. and the bacterial isolate B–I degraded MCZ in soil conditions (100–50 μg g^{-1}) after 15 days (Ahlawat et al. 2010). The three *Trichoderma* species *T. atroviride*, *T. viride* and *T. harzianum* degraded after 5 days 21%, 47% and 85% of CBZ (200 ppm), respectively (Sharma et al. 2016). *Aspergillus niger* and *A. flavus*, which were isolated from pesticide-polluted water, degraded after 5 days 23% and >50% MCZ, respectively (Aimeur et al. 2016). *Pseudomonas* sp. and *Arthrobacter* sp., which were isolated from wheat rhizosphere, have shown to be promising candidates for bioremediation of MCZ and CBZ (100 ppm) (Waghmare et al. 2018). *Stenotrophomonas* sp. that was isolated from *Coriandrum sativum* L. rhizosphere degraded 68% CBZ (250 μg mL^{-1}) after 21 days (Dos Santos et al. 2017). A consortium of *Streptomyces albogriseolus* and *Brevibacillus borstelensis* strains achieved 97% CBZ degradation after 12 h (Ridhima and Anil 2016).

Bioremediation assays on contaminated soil with CBZ and cadmium were carried out under greenhouse conditions for 180 days using a hyperaccumulator plant (*Sedum alfredii*) combined with carbendazim-degrading bacterial strains (*Bacillus subtilis, Paracoccus* sp., *Flavobacterium* sp. and *Pseudomonas* sp.). The plant-bacteria interaction reached a higher degradation of CBZ (83%) than treatments using *S. alfredii* (65%) and the carbendazim-degrading bacterial strains individually (68%) (Xiao et al. 2013). *Bacillus subtilis* strains increased 20-fold the degradation of CBZ (1 g L^{-1}) sprayed in grapes berries (Salunkhe et al. 2014).

Serratia marcescens strain B1 degraded 95% of TBZ (200 mg L^{-1}) in 8 days in mineral salt medium. In sterilized soil inoculated with B1 strain, 97% of TBZ was degraded in 30 days, while the control soil achieved a 70% degradation (Wang et al. 2018). *Bacillus* sp. strains 3B6 and 29B3 inoculated in liquid minimum medium with TBZ as sole carbon source were capable of degrading 50–60% TBZ (100 µM). The degradation of TBZ by *Bacillus* sp. 3B6 includes the hydroxylation of the *t*-butyl group or the alpha carbon of the aromatic ring (Youness et al. 2018).

5.7 Insecticides

Insecticides are agents used to control and kill insect populations, which are widely used in agriculture, horticulture and forestry (Gupta et al. 2019). In addition, insecticides are used to control vectors associated with human diseases such as malaria, dengue fever, yellow fever and Lyme disease (Hinckley et al. 2016, Ranson 2017). This type of pesticide comprises an extensive number of compounds that can be classified according to the type of contact, origin and chemical nature. Receptors, ion channels and enzymes are the most common target molecules (Casida and Durkin 2013). There are diverse insecticide categories: chlorinated hydrocarbons, organophosphates, carbamates, pyrethroids, botanical products, seed treatment insecticides, and other insecticides (Zhang 2018). According to their activity, these compounds can be classified in contact (i) and systemic (ii) insecticides. Contact insecticides act in direct contact with insects, and usually, are inorganic compounds like sulphur, arsenate, copper and fluorine (Villaverde et al. 2017). In the case of organic chemical compounds, they may be synthetically (Jeanmart et al. 2016) or naturally produced (e.g. pyrethrum, neem oil) (Mahmood et al. 2016). Systemic insecticides are absorbed by the plant and distributed throughout its tissues (e.g. plant stem, leaves, roots), making them toxic for insects that feed on these tissues. They are commonly applied to roots, foliage or injected into the tree trunks. According to their chemical classes, three major groups are organophosphates and carbamates, phenyl-pyrazoles and pyrethrins, and neonicotinoids (Simon-Delso et al. 2015).

Organophosphates and Carbamates Organophosphates (OPs) belong to a large class of phosphorus-based pesticides, usually ester, amide or thiol with derivatives of phosphoric acid, with alkyl or aryl groups as R moieties (Fig. 2). This type of insecticide, that is mainly applied as a food crop insecticide, possesses a lethal mechanism of action, inhibiting the acetylcholinesterase (AChE) enzyme by its phosphorylation, resulting in an accumulation of acetylcholine (ACh) that binds to muscarinic and nicotinic receptors (Vale and Lotti 2015). OPs persist in the environment only for relatively short periods of time under alkaline conditions (Sawyer et al. 2003). Carbamates (CMs) are N-substituted esters of carbamic acid, with R_1-NR_2-(C=O)O–R_3 as general formula (Fig. 2), where R_1, R_2, and R_3 are a methyl, a benzimidazole, and an aromatic moiety, respectively (Ufarté et al. 2017). CMs are used in agriculture for more than 120 different crops and ornamental plants. CMs are effective against 160 harmful insects. Carbamates are inhibitors of acetylcholinesterase enzyme, causing an overstimulation of the nervous system (Tiwari et al. 2019).

Pyrethrins Pyrethrins are natural insecticides. Pyrethrin I is the most important compound due to its efficacy and abundance (Ujihara 2019). Pyrethrin I possesses an ester structure, composed by a substituted cyclopentenolone and cyclopropanecarboxylic acid. In contrast to OPs and CMs, the mechanism of action of pyrethrins consists in modifying the kinetics of the voltage-sensitive sodium channel (Soderlund et al. 2002). Actually, pyrethrin I is formulated from several unstable partial structures (e.g. trialkyl-substituted double bond, cyclopentenolone). Therefore, the use of these pesticides has been limited.

Neonicotinoids Neonicotinoids are the most widely used insecticides despite their toxicity (Skandrani et al. 2006). In 2010, these systemic insecticides' production was estimated to be 20,000 tonnes, with presence in more than 120 countries (CCM International 2011, Simon-Delso et al. 2015). These compounds disrupt the neuronal transmission in the central nervous system of invertebrates by mimicking the action of neurotransmitters, binding with a high specific affinity to the acetylcholine receptor (nAChR) of insects (Stygar et al. 2013). Important crop pests, such as sap-feeding insects *Aphidae* (aphids), and *Chrysomelidae* (among others, western corn rootworm) can be controlled by neonicotinoids (Tomizawa and Casida 2005, Elbert et al. 2008). Regarding the chemical structure, the main commercial neonicotinoids contained three structural components: (i) a N-heterocyclyl-methyl moiety, (ii) a heterocyclic or acyclic spacer, and (iii) a N-nitroimine, nitromethylene, or N-cyanoimine pharmacophore (Casida 2011). These compounds are translocated easily into the xylem of plants and exhibit high translaminar movement (Canadian Council of Ministers of the Environment 2007). The first-generation neonicotinoids, imidacloprid (IMI) and thiacloprid (THI), that belong to the chloronicotinyl subclass, and thiamethoxam (TMX), a second-generation neonicotinoid that belongs to the thianicotinyl subclass, are the most used insecticides in the world (Jeschke et al. 2011, Tomizawa and Casida 2010, Zhou et al. 2013, Ge et al. 2014).

5.8 Insecticides and their Effects on the Nitrogen Cycle

Insecticides may directly or indirectly contact non-target soil microorganisms, affecting their activity and abundance, which is crucial to maintain soil fertility (Singh and Singh 2006). Some insecticides (e.g. organophosphates, pyrethrins) suppress nitrogen-fixing bacteria, producing lower crop yields and delayed growth (Potera 2007). High application of organochlorine pesticides to alfalfa (*S. meliloti*) crops generate the disruption of chemical signaling between the host plant and N-fixing rhizobia bacteria, inhibiting the *nod* gene expression, which is required for the initiation of nodulation (Fox et al. 2007). The widespread use of imidacloprid in agricultural soil represents a high risk to honeybees, causing colony collapse disorder (Whitehorn et al. 2012). Imidacloprid is a persistent pesticide in soil, exhibiting a half-time of up to 229 days (Scorza et al. 2004). Organic matter content, pH, temperature and insecticide concentration affect the imidacloprid persistence in soils (Florese-Céspedes et al. 2002, Cycoń and Piotrowska-Seget 2015). A significant decrease in nitrifying and N_2-fixing bacteria accompanied by a reduction in the nitrate soil concentration was observed in pristine sandy loam soil loaded with two IMI dosages (1 and 10 mg kg soil^{-1}). In contrast, denitrifying bacteria exhibit more tolerance to imidacloprid (Tenuta and Beauchamp 1996, Cycoń et al. 2013, Cycoń and Piotrowska-Seget 2015). In thiamethoxam treated soils, a significant decrease of aerobic nitrifying (e.g, *Azotobacter vinellandi, A. chrococcum*), denitrifying and ammonifying bacteria were reported (Filimon et al. 2015).

5.9 Bioremediation of Insecticides

The use of insecticides is not restricted to agricultural fields. Several toxic metabolites of insecticides may persist in the environment as pollutants. Bioremediation of insecticides in soil has been scarcely studied; however, microorganisms capable of degrading these compounds have been reported.

The persistence of imidacloprid in the field may be up to 180 days in non-agricultural soil (Oi 1999). Plants may increase the dissipation rate of imidacloprid, with a half-live range of 42 to 129 days (Liu et al. 2006). Sorption of this insecticide and its metabolites in soil organic carbon reduces its degradation (Beck et al. 1995). IMI removal could be carried out by hydrolysis, photodegradation and biodegradation processes (Zaror et al. 2009, Tang et al. 2012). Enrichment culture assays on imidacloprid-polluted soils showed a decrease of 43% IMI in three weeks. The imidacloprid-degrading bacterial strain PC-21 was capable of degrading 37 to 58% imidacloprid in TSB medium (Anhal et al. 2007).

For thiamethoxam (TMX), three degradation pathways have been described: (i) demethylation to desmethyl-TMX, (ii) nitro reduction to nitrosoimino and (iii) oxadiazine ring cleavage to generate another commercial insecticide, clothianidin (Karmakar et al. 2009, Casida 2011, Zhou et al. 2013). *Bacillus amyloliquefaciens* IN937a, *B. pumilus* SE34 and *B. subtilis* FZB24 degraded

22% TMX after 3 days in TSB medium (Zhou et al. 2013). Additionally, *Pseudomonas* sp. 1G is capable of degrading 70% TMX after 14 days in mineral salt medium (Pandey 2009). At field scale, TMX degradation releases nitrogen into soil, influencing soil microbial community structure, especially bacteria involved in the nitrogen cycle. The nitrogen-fixing bacterium *Ensifer adhaerens* TMX-23 is capable of degrading TMX (Zhou et al. 2013). This strain possesses three TMX catabolic pathways (Karmakar et al. 2006, Casida 2011). The strain TMX-23 is capable of transforming the N-nitroimino group ($=N-NO_2$) to N-nitrosoimino ($=N-NO$) and urea metabolites, degrading 81% TMX by the nitro reduction pathway (Zhou et al. 2014). Interestingly, strain TMX-23 is able to grow on TMX and to produce salicylic acid, which is a phytohormone involved in plant defense (Ford et al. 2010).

Stenotrophomonas maltophilia CGMCC1.1788 is capable of degrading 70% THI by thiazolidine ring hydroxylation (Zhao et al. 2009). The N_2-fixing bacterium *Ensifer meliloti* CGMCC 7333 is capable of transforming 91% of thiacloprid (Ge et al. 2014). The degradation involves the hydrolysis of the cyanoimine group, generating amide metabolites. Thus, *E. meliloti* CGMCC 7333 is an interesting candidate for THI bioremediation processes.

6. CONCLUSION

Agricultural activities influence biogeochemical cycles. Nitrogen cycle in agricultural soil is altered by fertilization, heavy metal pollution and the application of pesticides. The microbial communities associated with nitrogen fixation and nitrification are highly sensitive to heavy metals and pesticides. There is scarce information on the effects of heavy metals and pesticides on denitrification, which should be the focus of future studies.

Further studies are required on the effects of bioremediation of heavy metals and pesticides on the nitrogen cycle. However, bioremediation of heavy metals and pesticides in soil allows to mitigate the negative effects caused by these contaminants on the nitrogen cycle. Therefore, bioremediation is an attractive technology for the restoration of the nitrogen cycle in agricultural soil and for sustainable agriculture.

References

Afzal, M., M. Yu, C. Tang, L. Zhang, N. Muhammad, H. Zhao, et al. 2019. The negative impact of cadmium on nitrogen transformation processes in a paddy soil is greater under non-flooding than flooding conditions. Environ. Int. 129: 451–460.

Ahlawat, O.P., P. Gupta, S. Kumar, D.K. Sharma and K. Ahlawat. 2010. Bioremediation of fungicides by spent mushroom substrate and its associated microflora. Indian J. Microbiol. 50: 390–395.

Ahmad, E., A. Zaidi, M. Saghir Khan and M. Oves. 2012. Heavy metal toxicity to symbiotic nitrogen-fixing microorganism and host legumes. *In*: A. Zaidi, P. Wani and M. Khan (eds), Toxicity of Heavy Metals to Legumes and Bioremediation. Springer, Vienna, Austria, pp. 29–44.

Aimeur, N., W. Tahar, M. Meraghni, N. Meksem and O. Bordjiba. 2016. Bioremediation of pesticide (mancozeb) by two *Aspergillus* species isolated from surface water contaminated by pesticides. J. Chem. Pharm. 9: 2668–2670.

Altimira, F., C. Yáñez, G. Bravo, M. González, L. Rojas and M. Seeger. 2012. Characterization of copper-resistant bacteria and bacterial communities from copper polluted agricultural soils of central Chile. BMC Microbiol. 12: 193.

Alviz-Gazitua, P., S. Fuentes-Alburquenque, L.A. Rojas, R. Turner, N. Guiliani and M. Seeger. 2019. The response of *Cupriavidus metallidurans* CH34 to cadmium involves inhibition of the initiation of biofilm formation, decrease in intracellular c-di-GMP levels, and a novel metal regulated phosphodiesterase. Front. Microbiol. 10: 1499.

Anhalt, J.C., T.B. Moorman and W.C. Koskinen. 2007. Biodegradation of imidacloprid by an isolated soil microorganism. J. Environ. Sci. Health. B. 42: 509–514.

Beck, A., S.C. Wilson, R.E. Alcock and K.C. Jones. 1995. Kinetic constraints on the loss of organic chemicals from contaminated soils: Implications for soil-quality limits. Crit. Rev. Sci. Tec. 25: 1–43.

Bhat, S.V., S.C. Booth, S.G.K. McGrath and T.E.S. Dahms. 2015. *Rhizobium leguminosarum* bv. *viciae* 3841 adapts to 2,4-dichlorophenoxyacetic acid with "auxin-like" morphological changes, cell envelope remodeling and upregulation of central metabolic pathways. PLoS ONE 10: e0123813.

Branco, V., P. Ramos, J. Canario, J. Lu, A. Holmgren and C. Carvalho. 2012. Biomarkers of adverse response to mercury: histopathology versus thioredoxin reductase activity. J. Biomed. Biotechnol. 2012: 1–9.

Bravo, G. 2017. Bioremediation of mercury polluted agricultural soils using *Cupriavidus metallidurans* MSR33 in a rotary drum reactor and its effects on microorganisms involved in the nitrogen cycle. Ph.D. thesis, Universidad Técnica Federico Santa María, Pontificia Universidad Católica de Valparaíso, Valparaíso, Chile.

Bravo, G., Vega-Celedón, P., Gentina, J.C. and Seeger, M. 2020. Effects of mercury II on *Cupriavidus metallidurans* strain MSR33 during mercury bioremediation under aerobic and anaerobic conditions. Processes 8(8): 893.

Barr, D.B. and B. Buckley. 2011. In vivo biomarkers and biomonitoring reproductive and developmental toxicity. *In*: R.C. Gupta (ed.), Reproductive and Developmental Toxicity. Academic Press/Elsevier, Amsterdam, pp. 253–265.

Canadian Council of Ministers of the Environment. 2007. Canadian water quality for the protection of aquatic life: Imidacloprid. *In*: Canadian Council of Ministers of the Environment (eds), Canadian environmental quality guideline, Winnipeg, Canada, pp. 1–8.

Casida, F. 2011. Neonicotinoid metabolism: Compounds, substituents, pathways, enzymes, organisms, and relevance. J. Agr. Food Chem. 59: 2923–2931.

Casida, J.E. and K.A. Durkin. 2013. Neuroactive insecticides: Targets, selectivity, resistance, and secondary effects. Annu. Rev. Entomol. 58: 99–117.

Chang, Y.C., M.V. Reddy, H. Umemoto, Y. Sato, M. Kang, Y. Yajima, et al. 2015. Bio-augmentation of *Cupriavidus* sp. CY-1 into 2,4-D contaminated soil: microbial community analysis by culture dependent and independent techniques. PLoS ONE 10: e0145057.

Crini, G., A. Exposito Saintemarie, S. Rocchi, M. Fourmentin, A. Jeanvoine, L. Millon, et al. 2017. Simultaneous removal of five triazole fungicides from synthetic solutions on activated carbons and cyclodextrin-based adsorbents. Heliyon 3: e00380.

Cycoń, M. and Z. Piotrowska-Seget. 2015. Biochemical and microbial soil functioning after application of the insecticide imidacloprid. J. Environ. Sci. 27: 147–158.

Cycoń, M., A. Zmijowska, M. Wójcik and Z. Piotrowska-Seget. 2013. Biodegradation and bioremediation potential of diazinon-degrading *Serratia marcescens* to remove other organophosphorus pesticides from soils. J. Environ. Manage. 117: 7–16.

Dai, Y., N. Li, Q. Zhao and S. Xie. 2015. Bioremediation using *Novosphingobium* strain DY4 for 2,4-dichlorophenoxyacetic acid-contaminated soil and impact on microbial community structure. Biodegrad. 26: 161–170.

Dehghani, M., S. Nasseri and H. Hashemi. 2013. Study of the bioremediation of atrazine under variable carbon and nitrogen sources by mixed bacterial consortium isolated from corn field soil in Fars province of Iran. J. Environ. Public Health 2013: 973165.

Desmyttere, H., C. Deweer, J. Muchembled, K. Sahmer, J. Jacquin, F. Coutte, et al. 2019. Antifungal activities of *Bacillus subtilis* lipopeptides to two *Venturia inaequalis* strains possessing different tebuconazole sensitivity. Front. Microbiol. 10: 2327.

Ding, H., X. Zheng, J. Zhang, Y. Zhang, J. Yu and D. Chen. 2019. Influence of chlorothalonil and carbendazim fungicides on the transformation processes of urea nitrogen and related microbial populations in soil. Environ. Sci. Pollut. R. 26: 31133–31141.

Dong, Y., L. Yang and L. Zhang. 2017. Simultaneous electrochemical detection of benzimidazole fungicides carbendazim and thiabendazole using a novel nanohybrid material-modified electrode. J. Agr. Food Chem. 65: 727–736.

Diels, L., M. De Smet, L. Hooyberghs and P. Corbisier. 1999. Heavy metals bioremediation of soil. Mol. Biotechnol. 12: 149–158.

Dos Santos, J., I. Batista, H. Santos, J. Silva, P. Costa, A. Ghelfi, et al. 2017. Biodegradation of the fungicide carbendazim by bacteria from *Coriandrum sativum* L. rhizosphere. Acta Sci. Biol. 39: 71–77.

Elanchezhian, R. and J.D.S. Panwar. 1997. Effects of 2,4-D and *Azospirillum brasilense* on nitrogen fixation, photosynthesis and grain yield in wheat. J. Agron. Crop Sci. 178: 129–133.

Elbert, A., M. Haas, B. Springer, W. Thielert and R. Nauen. 2008. Applied aspects of neonicotinoid uses in crop protection. Pest Manag. Sci. 64: 1099–1105.

Fan, L., Y. Feng, D.B. Weaver, D.P. Delaneya, G.R. Wehtje and G. Wang. 2017. Glyphosate effects on symbiotic nitrogen fixation in glyphosate-resistant soybean. Appl. Soil Ecol. 121: 11–19.

Feld, L., M.H. Hjelmsø, M.S. Nielsen, A.D. Jacobsen, R. Rønn, F. Ekelund, et al. 2015. Pesticide side effects in an agricultural soil ecosystem as measured by *amoA* expression quantification and bacterial diversity changes. PLoS ONE. 10: e0126080.

Fernández-Cornejo J., R. Nehring, C. Osteen, S. Wechsler, A. Martin and A. Vialou. 2014. Pesticide use in U.S. Agriculture: 21 selected crops, 1960–2008. Economic Information Bulletin Number 124 Washington (DC): USDA Economic Research Service.

Filimon, M.N., S.O. Voia, R. Popescu, G. Dumitrescu, L. Ciochina, M. Mituletu, et al. 2015. The effect of some insecticides on soil microorganisms based on enzymatic and bacteriological analyses. Rom. Biotechnol. Lett. 20: 10439–10447.

Filip, Z. 2002. International approach to assessing soil quality by ecologically-related biological parameters. Agr. Ecosyst. Environ. 88: 169–174.

Flores, C., V., Morgante., M., González, R., Navia and M. Seeger. 2009. Adsorption studies of the herbicide simazine in agricultural soils of the Aconcagua valley, central Chile. Chemosphere 74(11): 1544–1549.

Flores-Céspedes, F., E. González-Pradas, M. Fernández-Pérez, M. Villafranca-Sánchez, M. Socias-Viciana and M.D. Urena-Amate. 2002. Effects of dissolved organic carbon on sorption and mobility of imidacloprid in soil. J. Environ. Qual. 31: 880–888.

Franche, C., K. Lindström and C. Elmerich. 2008. Nitrogen-fixing bacteria associated with leguminous and non-leguminous plants. Plant Soil 321: 35–59.

Ford, K.A., J.E. Casida, D. Chandran, A.G. Gulevich, R.A. Okrent, K.A. Durkin, et al. 2010. Neonicotinoid insecticides induce salicylate-associated plant defense responses. P. Natl. A. Sci. 107: 17527–17532.

Forouzesh, A., E. Zand, S. Soufizadeh and S. Samadi Foroushani. 2015. Classification of herbicides according to chemical family for weed resistance management strategies—an update. Weed Res. 55: 334–358.

Fox, J., J. Gulledge, E. Engelhaupt, M. Burow and J. McLachlan. 2007. Pesticides reduce symbiotic Efficiency of nitrogen-fixing rhizobia and host plants. P. Natl. A. Sci. USA 104: 10282–10287.

Fuentes, S., V. Méndez, P. Aguila and M. Seeger. 2014. Bioremediation of petroleum hydrocarbons: Catabolic genes, microbial communities, and applications. Appl. Microbiol. Biotechnol. 98: 4781–4789.

Fuentes, S., B. Barra, G. Caporaso and M. Seeger. 2016. From rare to dominant: A fine tuned soil bacterial bloom during petroleum hydrocarbon bioremediation. Appl. Microbiol. Biotechnol. 82: 888–896.

Ge, F., LY. Zhou, Y. Wang, Y. Ma, S. Zhai, Z.H. Liu, et al. 2014. Hydrolysis of the neonicotinoid insecticide thiacloprid by the N_2-fixing bacterium *Ensifer meliloti* CGMCC 7333. Int. Biodeter. Biodegr. 93: 10–17.

Govantes, F., O. Porrúa, V. García-González and E. Santero. 2009. Atrazine biodegradation in the lab and in the field: Enzymatic activities and gene regulation. Microb. Biotechnol. 2: 178–185.

Gómez-Sagasti, M. and D. Marino. 2015. PGPRs and nitrogen-fixing legumes: A perfect team for efficient Cd phytoremediation? Front. Plant Sci. 6: 81. doi:10.3389/fpls.2015.00081

Gullino, M.L., F. Tinivella, A. Garibaldi, G.M. Kemmitt, L. Bacci and B. Sheppard. 2010. Mancozeb: Past, present, and future. Plant Dis. 94: 1076–1087.

Gupta, R., I. Miller, J. Malik, R. Doss, W.D. Dettbarn and D. Milanovic. 2019. Insecticides. *In*: C. Gupta (ed.), Biomarkers in Toxicology. Academic Press, Kentucky, United States, pp. 389–407.

Hai, B., N.H. Diallo, S. Sall, F. Haesler, K. Schauss, M. Bonzi, et al. 2009. Quantification of key genes steering the microbial nitrogen cycle in the rhizosphere of sorghum cultivars in tropical agroecosystems. Appl. Environ. Microbiol. 75: 4993–5000.

Heneberg, P., J. Svoboda and P. Pech. 2018. Benzimidazole fungicides are detrimental to common farmland ants. Biol. Conserv. 221: 114–117.

Hernández, M., V., Morgante, M., Ávila, P., Villalobos, P., Miralles., M., González, et al. 2008a. Novel *s*-triazine-degrading bacteria isolated from agricultural soils of central Chile for herbicide bioremediation. Electron. J. Biotechn. 11(5): 1–7.

Hernández, M., P. Villalobos, V. Morgante, M. González, C. Reiff, E. Moore, et al. 2008b. Isolation and characterization of a novel simazine-degrading bacterium from agricultural soil of central Chile, *Pseudomonas* sp. MHP41. FEMS Microbiol. Lett. 206: 184–190.

Hernández, M., Z. Jia, R. Conrad and M. Seeger. 2011. Simazine application inhibits nitrification and changes the ammonia-oxidizing bacterial communities in a fertilized agricultural soil. FEMS Microbiol. Ecol. 78: 511–519.

Hinckley, A., J. Meek, J. Ray, S. Niesobecki, N. Connally, K. Feldman, et al. 2016. Effectiveness of residential acaricides to prevent Lyme and other tick-borne diseases in humans. J. Infect. 214: 182–188.

Jayakody, N., E.C. Harris and D. Coggon. 2015. Phenoxy herbicides, soft-tissue sarcoma and non-Hodgkin lymphoma: a systematic review of evidence from cohort and case control studies. Brit. Med. Bull. 114: 75–94.

Jayasumana, C., S. Gunatilake and P. Senanayake. 2014. Glyphosate, hard water and nephrotoxic metals: are they the culprits behind the epidemic of chronic kidney disease of unknown etiology in Sri Lanka? Int. J. Environ. Res. Public Health 11: 2125–2147.

Jeanmart, S., A. Edmunds, C. Lamberth and M. Pouliot. 2016. Systemic approaches to the 2010–2014 new agrochemicals. Bioorg. Med. Chem. 24: 317–341.

Jeschke, P., R. Nauen, M. Schindler and A. Elbert. 2011. Overview of the status and global strategy for neonicotinoids. J. Agric. Food Chem. 59: 2897–2908.

Jung, J., J. Yeom, J. Kim, J. Han, H.L. Soo, H. Park, et al. 2011. Change in gene abundance in the nitrogen biogeochemical cycle with temperature and nitrogen addition in Antarctic soils. Res. Microbiol. 162: 1018–1026.

Kakitani, A., T. Yoshioka, Y. Nagatomi and K. Harayama. 2017. A rapid and sensitive analysis of dithiocarbamate fungicides using modified QuEChERS method and liquid chromatography–tandem mass spectrometry. J. Pestic. Sci. 42: 145–150.

Karmakar, R., S.B. Singh and G. Kulshrestha. 2006. Persistence and transformation of thiamethoxam, a neonicotinoid insecticide, in soil of different agroclimatic zones of India. Bull. Environ. Contam. Toxicol. 76: 400–406.

Karmakar, R., R. Bhattacharya and G. Kulshrestha. 2009. Comparative metabolite profiling of the insecticide thiamethoxam in plant and cell suspension culture of tomato. J. Agr. Food. Chem. 57: 6369–6374.

Katupitiya, S., P.B. New, C. Elmerich and I.R. Kennedy. 1995. Improved N_2 fixation in 2,4-D treated wheat roots associated with *Azospirillum lipoferum*: Studies of colonization using reporter genes. Soil Biol. Biochem. 27: 447–452.

Khan, M.S, P. Chaudhry, P.A. Wani and A. Zaidi. 2006. Biotoxic effects of the herbicides on growth, seed yield, and grain protein of greengram. J. Appl. Sci. Environ. Mgt. 10: 141–146.

Kinney, C.A., K.W. Mandernack and A.R. Mosier. 2005. Laboratory investigations into the effects of the pesticides mancozeb, chlorothalonil, and prosulfuron on nitrous oxide and nitric oxide production in fertilized soil. Soil Biol. Biochem. 37: 837–850.

Kumar, A., N. Trefault and A.O. Olaniran. 2016. Microbial degradation of 2,4-dichlorophenoxyacetic acid: Insight into the enzymes and catabolic genes involved, their regulation and biotechnological implications. Crit. Rev. Microbiol. 42: 194–208.

Küstermann, B., O. Christen and K.L. Hülsbergen. 2010. Modelling nitrogen cycles of farming systems as basis of site- and farm-specific nitrogen management. Agr. Ecosyst. Environ. 135: 70–80.

Lamichhane, J., E. Osdaghi, F. Behlau, J. Köhl, J.B. Jones and J. Aubertot. 2018. Thirteen decades of antimicrobial copper compounds applied in agriculture: A review. Agron. Sustain. Dev. 38: 28.

Li, M., L. Ren, J. Zhang, L. Luo, P. Qin, Y. Zhou, et al. 2019. Population characteristics and influential factors of nitrogen cycling functional genes in heavy metal contaminated soil remediated by biochar and compost. Sci. Total Environ. 651: 2166–2174.

Liu, Y., Y. Liu, Y. Ding, J. Zheng, T. Zhou, G. Pan, et al. 2014. Abundance, composition and activity of ammonia oxidizer and denitrifier communities in metal polluted rice paddies from south China. PLoS ONE. 9: e102000.

Liu, W., W. Zheng, Y. Ma and K.K. Liu. 2006. Sorption and degradation of imidacloprid in soil and water. J. Environ. Sci. Heal. B. 41: 623–534.

Liu, Y., Y. Zheng, J. Shen, L. Zhang and J. He. 2010. Effects of mercury on the activity and community composition of soil ammonia oxidizers. Environ. Sci. Pollut. R. 17: 1237–1244.

Lu, G.H., X.M. Hua, J. Cheng, Y.L. Zhu, G.H. Wang, Y.J. Pang, et al. 2018. Impact of glyphosate on the rhizosphere microbial communities of an EPSPS-transgenic soybean line ZUTS31 by metagenome sequencing. Curr. Genomics. 19: 36–49.

Madrova P., T. Vetrovsky, M. Omelka, M. Grunt, Y. Smutna, D. Rapoport, et al. 2018. A short-term response of soil microbial communities to cadmium and organic substrate amendment in long-term contaminated soil by toxic elements. Front. Microbiol. 9: 2807.

Mahmood, I.S. Imadi, K. Shazadi, A. Gul and K. Hakeem. 2016. Effects of pesticides on enviroment. *In*: K. Hakeem, M. Akhar and S. Abdullah (eds), Plant, Soil and Microbes. Springer, Cham, Germany, pp. 253–269.

Man, L. and C. Zucong. 2009. Effects of chlorothalonil and carbendazim on nitrification and denitrification in soils. J. Environ. Sci. 21: 458–467.

Mao, Y., A.C. Yannarell and R.I. Mackie. 2011. Changes in N-transforming archaea and bacteria in soil during the establishment of bioenergy crops. PLoS ONE. 6: e24750.

Martínez, B., J. Tomkins, L.P. Wackett, R. Wing and M.J. Sadowsky. 2001. Complete nucleotide sequence and organization of the atrazine catabolic plasmid pADP-1 from *Pseudomonas* sp. strain ADP. J. Bacteriol. 183: 5684–5697.

Means, N.E., R.J. Kremer and C. Ramsier. 2007. Effects of glyphosate and foliar amendments on activity of microorganisms in the soybean rhizosphere. J. Environ. Sci. Heal. B. 42: 125–132.

Mertens, M., S. Höss, G. Neumann, J. Afzal and W. Reichenbecher. 2018. Glyphosate, a chelating agent-relevant for ecological risk assessment? Environ. Sci. Pollut. Res. Int. 25: 5298–5317.

Meyer, M.T. and E.A. Scribner. 2009. The evolution of analytical technology and its impact on water-quality studies for selected herbicides and their degradation products in water. *In*: S. Ahuja (ed.), Handbook of Water Purity and Quality. Academic Press, New York, USA, pp. 289–313.

Milenkovski, S., E. Bååth, P.E. Lindgren and O. Berglund. 2010. Toxicity of fungicides to natural bacterial communities in wetland water and sediment measured using leucine incorporation and potential denitrification. Ecotoxicology 19: 285–294.

Morgante V., C. Flores, X. Fadic, M. Gonzalez, M. Hernández, F. Cereceda-Balic, et al. 2012. Influence of microorganisms and leaching on simazine attenuation in an agricultural soil. J. Environ. Manage. 95: s300–s305.

Morgante, V., A. López-López, C. Flores, M. González, B. González, V. Vásquez, et al. 2010. Bioaugmentation with *Pseudomonas* sp. strain MHP41 promotes simazine attenuation and bacterial community changes in agricultural soils. FEMS Microbiol. Ecol. 71: 114–126.

Mosier, A., J.K. Syers and F. Jean. 2004. Nitrogen fertilizer: An essential component of increased food, feed, and fiber production. *In*: A. Mosier, J.K. Syers and F. Jean (eds), Agriculture and the Nitrogen Cycle: Assessing the Impacts of Fertilizer use on Food Production and the Environment. Island Press. Washington D.C, USA, pp. 3–15.

Mosier, A. and C. Francis. 2010. Denitrifier abundance and activity across the San Francisco bay estuary. Env. Microbiol. Rep. 2: 667–676.

Muñoz-Leoz, B., E. Ruiz-Romera, I. Antigüedad and C. Garbisu. 2011. Tebuconazole application decreases soil microbial biomass and activity. Soil Biol. Biochem. 43: 2176–2183.

Muthusaravanan, S., N., Sivarajasekar, J.S. Vivek, T. Paramasivan, M. Naushad, J. Prakashmaran, et al. 2018. Phytoremediation of heavy metals: Mechanisms, methods and enhancements. Environ. Chen. Lett. 16: 1339–1359.

Nahi, A., R. Othman, D. Omar and M. Ebrahimi. 2016. Effects of selected herbicides on growth and nitrogen fixing activity of *Stenotrophomonas maltophilia* (Sb16). Pol. J. Microbiol. 65: 377–382.

Odularu, A.T. and P.A. Ajibade. 2019. Dithiocarbamates: Challenges, control, and approaches to excellent yield, characterization, and their biological applications. Bioinorg. Chem. Appl. 2019: 1–15.

Oi, M. 1999. Time dependent sorption of imidacloprid in two different soils. J. Agric. Food Chem. 47: 327–332.

Ojuederie, O. and O. Babalola. 2017. Microbial and plant-assisted bioremediation of heavy metal polluted environments: A review. Int. J. Environ. Res. Public Health. 14: 1504.

Orellana, R., C. Macaya, G. Bravo, F., Dorochesi, A. Cumsille, R. Valencia, et al. 2018. Living at the frontiers of life: Extremophiles in Chile and their potential for bioremediation. Front. Microbiol. 9: 2309.

Ouyang, Y., S. Evans, M. Friesen and L. Tiemann. 2018. Effect of nitrogen fertilization on the abundance of nitrogen cycling genes in agricultural soils: A meta-analysis of field studies. Soil Biol. Biochem. 127: 71–78.

Pandey, G., S. Dorrian, R. Russell and J. Oakeshott. 2009. Biotransformation of the neonicotinoid insecticides imidacloprid and thiamethoxam by *Pseudomonas* sp. 1G. Biochem. Biophys. Res. 380: 710–714.

Patra, M. and A. Sharma. 2000. Mercury toxicity in plants. Bot. Rev. 66: 379–422.

Pereg, L., A. Morugán-Coronado, M. McMillana and F. García-Orenes. 2018. Restoration of nitrogen cycling community in grapevine soil by a decade of organic fertilization. Soil Till. Res. 179: 11–19.

Pérez Rodríguez, M., C. Melo, E. Jiménez and J. Dussán. 2019. Glyphosate bioremediation through the sarcosine oxidase pathway mediated by *Lysinibacillus sphaericus* in soils cultivated with potatoes. Agriculture 9: 217–221.

Pires Bomfim, N.C., B.C. Pereira Costa, L. Anjos Souza, G. Costa Justino, L. Ferreira Aguiar and L. Santos Camargos. 2017. Glyphosate effect on nitrogen fixation and metabolization in RR soybean. J. Agric. Sci. 9: 114–121.

Potera, C. 2007. Agriculture: pesticides disrupt nitrogen fixation. Environ. Health Perspect. 115: A579.

Prosser, J.I., B.J.M. Bohannan, T.P. Curtis, R.J. Ellis, M.K. Firestone, R.P. Freckleton, et al. 2007. The role of ecological theory in microbial ecology. Nat. Rev. Microbiol. 5: 384–392.

Rai, J.P.N. 1992. Effects of long-term 2,4-D application on microbial populations and biochemical processes in cultivated soil. Biol. Fertil. Soils 13: 187–191.

Ranjard, L., L. Lignier and R. Chaussod. 2006. Cumulative effects of short-term polymetal contamination on soil bacterial community structure. Appl. Environ. Microbiol. 72: 1684–1687.

Ranson, H. 2017. Current and future prospects for preventing malaria transmission via the use of insecticides. Cold Spring Harb. Perspec. Med. 7: a0226823.

Richter, A., I. Schöning, T. Kahl, J. Bauhus and L. Ruess. 2018. Regional environmental conditions shape microbial community structure stronger than local forest management intensity. Forest Ecol. Manag. 409: 250–259.

Ridhima, A. and K.S. Anil. 2016. Bioremediation of carbendazim, a benzimidazole fungicide using *Brevibacillus borstelensis* and *Streptomyces albogriseolus* together. Curr. Pharm. Biotechno. 17: 185–189.

Rojas, L.A., C. Yáñez, M. González, S. Lobos, K. Smalla and M. Seeger. 2011. Characterization of the metabolically modified heavy metal-resistant *Cupriavidus metallidurans* strain MSR33 generated for mercury bioremediation. PLoS ONE. 6: e17555.

Rütting, T., H. Aronsson and S. Delin. 2018. Efficient use of nitrogen in agriculture. Nutr. Cycl. Agroecosyst. 110: 1–5.

Saha, A., A. Pipariya and D. Bhaduri. 2016. Enzymatic activities and microbial biomass in peanut field soil as affected by the foliar application of tebuconazole. Environ. Earth Sci. 75: 558.

Saikia, S.P., V. Jain and G.C. Srivastava. 2006. Effect of 2,4-D and inoculation with *Azorhizobium caulinodans* on maize. Acta Agron. Hung. 54: 121–125.

Sagarkar, S., A. Nousiainen, S. Shaligram, K. Björklöf, K. Lindström, K.S. Jørgensen, et al. 2014. Soil mesocosm studies on atrazine bioremediation. J. Environ. Manage. 139: 208–216.

Salunkhe, V.P., I.S. Sawant, K. Banerjee, P.N. Wadkar, S.D. Sawant and S.A. Hingmire. 2014. Kinetics of degradation of carbendazim by *B. subtilis* strains: Possibility of *in situ* detoxification. Environ. Monit. Assess. 186: 8599–8610.

Sawyer, C., P.L. McCarty and G.F. Parkin. 2003. Chemistry for environmental engineering and science. Pesticides 279–288.

Scorza, R.P., J.H. Smelt, J.J. Boesten, R.F. Hendriks and S.E. Van der Zee. 2004. Vadose zone processes and chemical transport: preferential flow of bromide, bentazon, and imidacloprid in a Dutch clay soil. J. Environ. Qual. 33: 1473–1486.

Seeger, M., M. Hernández, V. Méndez, B. Ponce, M. Córdova and M. González. 2010. Bacterial degradation and bioremediation of chlorinated herbicides and biphenyls. J. Soil Sci. Plant Nutr. 10: 320–332.

Seshadri B., N. Bolan and R. Naidu. 2015. Rhizosphere-induced heavy metal(loid) transformation in relation to bioavailability and remediation. J. Soil Sci. Plant Nut. 15: 524–548.

Shao, H. and Y. Zhang. 2017. Non-target effects on soil microbial parameters of the synthetic pesticide carbendazim with the biopesticides cantharidin and norcantharidin. Sci. Rep. 7: 5521.

Sharma, P., M. Sharma, M. Raja, D. Singh and M. Srivastava. 2016. Use of *Trichoderma* spp. in biodegradation of carbendazim. Indian J. Agr. Sci. 86: 59–62.

Sherwani, S.I., I.A. Arif and H.A. Khan. 2015. Modes of action of different classes of herbicides. *In*: A. Price, J. Kelton and L. Sarunaite (eds), Herbicides: Physiology of Action and Safety. InTech, Rijeka, Croatia, pp. 165–186.

Silva, E., A.M. Fialho, I. Sá-Correia, R.G. Burns and L.J. Shaw. 2004. Combined bioaugmentation and biostimulation to cleanup soil contaminated with high concentrations of atrazine. Environ. Sci. Technol. 38: 632–637.

Simon-Delso, N., V. Amaral-Rogers, L. Belzunces, J. Chagnon, C. Downs, L. Furlan, et al. 2015. Systemic insecticides (neocotinoids and fipronil): Trends, uses, mode of action and metabolites. Environ. Sci. Pollut. Res. 22: 5–34.

Simonin, M., A. Cantarel, A. Crouzet, J. Gervaix, J. Martins and A. Richaume. 2018. Negative effects of copper oxide nanoparticles on carbon and nitrogen cycle microbial activities in contrasting agricultural soils and in presence of plants. Front. Microbiol. 9: 3102.

Singh, J. and D. Singh. 2006. Ammonium, nitrate and nitrite nitrogen and nitrate reductase enzyme activity in groundnut (*Arachis hypogaea* L.) fields after diazinon, imidaclopid and lindane treatments. J. Environ. Sci. Health B. 41: 1305–1318.

Singh, S.N. and N. Jauhari. 2017. Degradation of atrazine by plants and microbes. *In*: S.N. Singh (ed.), Microbe-induced degradation of pesticides. Springer International Publishing, Gewerbestrasse, Switzerland, pp. 213–225.

Skandrani, D., Y. Gaubin, B. Beau, J.C. Murat, C. Vincent and F. Croute. 2006. Effect of selected insecticides on growth rate and stress protein expression in cultured human A549 and SH-SY5Y cells. Toxicol. In Vitro 20: 1378–1386.

Soderlund, D., J. Clark, L. Sheets, L. Mullin, V. Piccirillo, D. Sargent, et al. 2002. Mechanisms of pyrethroid neurotoxicity: Implications for cumulative risk assessment. Toxicology 171: 3–59.

Somda, Z.C., H.A. Mills and S.C. Phatak. 1990. Nitrapyrin, terrazole, atrazine, and simazine influence on nitrification and corn growth. J. Plant Nutr. 13: 1179–1193.

Srinivasulu, M., G.J. Mohiddin, K. Subramanyam and V. Rangaswamy. 2012. Effect of insecticides alone and in combination with fungicides on nitrification and phosphatase activity in two groundnut (*Arachis hypogeae* L.) soils. Environ. Geochem. Health 34: 365–374.

Stygar, D., K. Michalczyk, B. Dolezych, M. Makonieczny, P. Migula, M. Zaak, et al. 2013. Digestive enzymes activity in subsequent generations of *Cameraria ohridella* larvae harvested from horse chestnut trees after treatment with imidacloprid. Pestic. Biochem. Phys. 105: 5–12.

Tagliaferro, L., A. Officioso, S. Sorbo, A. Basile and C. Manna. 2015. The protective role of olive oil hydroxytyrosol against oxidative alterations induced by mercury in human erythrocytes. Food Chem. Toxicol. 82: 59–63.

Tang, J., X. Huang and X. Huang. 2012. Photocatalytic degradation of imidacloprid in aqueous suspension of TiO$_2$ supported on H-ZSM-5. Environ. Earth Sci. 66: 441–445.

Tenuta, M. and E. Beauchamp. 1996. Denitrification following herbicide application to a grass sward. Can. J. Soil Sci. 76: 15–22.

Thiour-Mauprivez, C., F. Martin-Laurent, C. Calvayrac and L. Barthelmebs. 2019. Effects of herbicide on non-target microorganisms: Towards a new class of biomarkers?. Sci. Total Environ. 684: 314–325.

Tiwari, B., S. Kharwar and D.N. Tiwari. 2019. Pesticides and rice agriculture. *In*: A.K. Mishra, D.N. Tiwari and A.N. Rai (eds), Cyanobacteria. Academic Press, India, pp. 303–325.

Tomizawa, M. and J. Casida. 2005. Neonicotinoid insecticide toxicology: Mechanisms of selective action. Annu. Rev. Pharmacol. Toxicol. 45: 247–268.

Tomizawa, M. and J. Casida. 2010. Neonicotinoid insecticides: Highlights of a symposium on strategic molecular designs. J. Agric. Food Chem. 59: 2883–2886.

Tyler, H.L. and M.A. Locke. 2018. Effects of weed management on soil ecosystems. *In*: N.E. Korres, N.R. Burgos and S.O. Duke (eds), Weed Control: Sustainability, Hazards, and Risks in Cropping Systems Worldwide. CRC Press, New York, USA, pp. 32–62.

Ufarté, L., E. Laville, S. Duquesne, D. Morgavi, P. Robe, C. Kloop, et al. 2017. Discovery of carbamate degrading enzymes by functional metagenomics. PloS ONE 12: e0189201.

Ujihara, K. 2019. The history of extensive structural modifications of pyrethroids. J. Pestic. Sci. 44: 215–224.

Ullah, M. and F. Dijkstra. 2019. Fungicide and bactericide effects on carbon and nitrogen cycling in soils: A meta-analysis. Soil Systems. 3: 23.

Vale, A. and M. Lotti. 2015. Organophosphorus and carbamate insecticide poisoning. *In*: M. Aminoff, F. Boller and D. Swaab (eds), Handbook of Clinical Neurology. Elsevier, Amsterdam, Netherlands, pp. 149–168.

Van Bruggen, A.H.C., M.M. He, K. Shin, V. Mai, K.C. Jeong, M.R. Finckh, et al. 2018. Environmental and health effects of the herbicide glyphosate. Sci. Total Environ. 616–617: 255–268.

Vega-Celedón, P., H. Canchignia, M., González, M. Seeger. 2016. Biosynthesis of indole-3-acetic acid and plant growth promoting by bacteria. Cultivos Tropicales 37: 31–37.

Villaverde, J., B. Sevilla-Morán, C. López-Goti, P. Sadín-España and J. Alonso-Prados. 2017. An overview of nanopesticides in the framework of European legislation. *In*: A. Mihai (ed.), New Pesticides and Soil Sensors. Academic Press, Bucharest, Romania.

Waghmare, S., S. Khandare and M. Ingale. 2018. Isolation and screening of pesticide resistant rhizobacteria from wheat (*Triticum aestivum*) rhizosphere soil. Int. Arch. App. Sci. Technol. 9: 62–69.

Walia, A., P. Mehta, S. Guleria, A. Chauhan and C.K. Shirkot. 2014. Impact of fungicide mancozeb at different application rates on soil microbial populations, soil biological processes, and enzyme activities in soil. Sci. World J. Article ID 702909 https://doi.org/10.1155/2014/702909

Wagner-Döbler, I. 2003. Pilot plant for bioremediation of mercury-containing industrial wastewater. Appl. Microbiol. Biotechnol. 62: 124–133.

Wang, X., X. Hou, S. Liang, Z. Lu, Z. Hou, X. Zhaong, et al. 2018. Biodegradation of fungicide tebuconazole by *Serratia marcescens* strain B1 and its application in bioremediation of contaminated soil. Int. Biodeter. Biodegr. 127: 185–191.

Whitehorn, P., S. O'Connor, F. Wackers and D. Goulson. 2012. Neonicotinoid pesticide reduces bumble bee colony growth and queen production. Science 336: 351–352.

Xiao, W., H. Wang, T. Li, Z. Zhu, J. Zhang, Z. He, et al. 2013. Bioremediation of Cd and carbendazim co-contaminated soil by Cd-hyperaccumulator *Sedum alfredii* associated with carbendazim-degrading bacterial strains. Environ. Sci. Pollut. R. 20: 380–389.

Xu, B., Q.J. Sun, J.C. Lan, W.M. Chen, C.C. Hsueh and B.Y. Chen. 2019. Exploring the glyphosate-degrading characteristics of a newly isolated, highly adapted indigenous bacterial strain *Providencia rettgeri* GDB 1. J. Biosci. Bioeng. 128: 80–87.

Yang, L.N., M.H. He, H.B. Ouyang, W. Zhu, Z.C. Pan, Q.J. Sui, et al. 2019. Cross-resistance of the pathogenic fungus *Alternaria alternata* to fungicides with different modes of action. BMC Microbiol. 19: 205.

Youness, M., M. Sancelme, B. Combourieu and P. Besse-Hoggan. 2018. Identification of new metabolic pathways in the enantioselective fungicide tebuconazole biodegradation by *Bacillus* sp. 3B6. J. Hazard. Mater. 351: 160–168.

Zablotowicz, R.M. and K.N. Reddy. 2004. Impact of glyphosate on the *Bradyrhizobium japonicum* symbiosis with glyphosate-resistant transgenic soybean. J. Environ. Qual. 33: 825–831.

Zaidi, A., M.S. Khan and P.Q. Rizvi. 2005. Effect of herbicides on growth, nodulation and nitrogen content of greengram. Agron. Sustain. Dev. 25: 497–504.

Zaror, C.A., C. Segura, H. Mansilla, M.A. Mondaca and P. González. 2009. Detoxification of waste water contaminated with imidacloprid using homogeneous and heterogeneous photo-Fenton processes. Water Pract. Technol. 4(1): wpt2009010

Zeman, A.M.M., T. Tchan, C. Elmerich and I.R. Kennedy. 1992. Nitrogenase activity in wheat seedlings bearing para-nodules induced by 2,4-dichlorophenoxyacetic acid (2,4-D) and inoculated with *Azospirillum*. Res. Microbiol. 143: 847–855.

Zhan, H., Y. Feng, X. Fan and S. Chen. 2018. Recent advances in glyphosate biodegradation. Appl. Microbiol. Biotechnol. 102: 5033–5043.

Zhang, F., X. Wan and Y. Zhong. 2014. Nitrogen as an important detoxification factor to cadmium stress in poplar plants. J. Plant Interact. 9: 249–258.

Zhang, M., W. Wang, L. Tang, M. Heenan and Z. Xu. 2018. Effects of nitrification inhibitor and herbicides on nitrification, nitrite, and nitrate consumptions and nitrous oxide emission in an Australian sugarcane soil. Biol. Fertil. Soils 54: 697–706.

Zhang, S., Z. Xu, C. Gao, Q.C. Ren, L. Chang, Z.S. Lv, et al. 2017. Triazole derivatives and their anti-tubercular activity. Eur. J. Med. Chem. 138: 501–513.

Zhang, W.J., F.B. Jiang and J.F. Ou. 2011. Global pesticide consumption and pollution: with China as a focus. Proc. Int. Acad. Ecol. Environ. Sci. 1: 125–144.

Zhang, W. 2018. Global pesticide use: Profile, trend, cost/benefit and more. Proc. Int. Acad. Ecol. Environ. Sci. 8: 1–27.

Zhao, Y.J., Y.J. Dai, C.G. Yu, J. Luo, W.P. Xu, J.P. Ni, et al. 2009. Hydroxylation of thiacloprid by bacterium *Stenotrophomonas maltophilia* CGMCC1.1788. Biodegradation 20: 761–768.

Zhou, G., Y. Wang, S. Zhai, F. Ge, Z.H. Liu, Y.J. Dai, et al. 2013. Biodegradation of the neocotinoid insecticide thiamethoxam by the nitrogen fixing and plant-growth-promoting rhizobacterium *Ensifer adhaerens* strain TMX-23. Appl. Microbiol. Biotechnol. 97: 4065–4074.

Zhou, G., Y. Wang, Y. Ma, S. Zhai, L. Zhou, Y. Dai, et al. 2014. The metabolism of neonicotinoid insecticide thiamethoxam by soil enrichment cultures, and the bacterial diversity and plant growth-promoting properties of the cultured isolates. J. Environ. Sci. Heal. B. 49: 381–390.

Zhou, Y., J. Xu, Y. Zhu, Y. Duan and M. Zhou. 2016. Mechanism of action of the benzimidazole fungicide on *Fusarium graminearum*: Interfering with polymerization of monomeric tubulin but not polymerized microtubule. Phytopathology 106: 807–813.

Zikos, C., A. Evangelou, C.E. Karachaliou, G. Gourma, P. Blouchos, G. Moschopoulou, et al. 2015. Commercially available chemicals as immunizing haptens for the development of a polyclonal antibody recognizing carbendazim and other benzimidazole-type fungicides. Chemosphere 119: S16–S20.

Zobiole, L.H.S., R.J. Kremer, R.S. Oliveira and J. Constantin. 2012. Glyphosate effects on photosynthesis, nutrient accumulation, and nodulation in glyphosate-resistant soybean. J. Plant Nutr. Soil Sci. 175: 319–330.

CHAPTER 18

Bioremediation of Toxic Metals from Wastewater for Water Security

Praveen Solanki[1*], Neha Khanna[2], ML Dotaniya[3], Maitreyie Narayan[4], Shiv Singh Meena[5], RK Srivastava[4] and S Udayakumar[6]

[1]Krishi Vigyan Kendra, Bankhedi, Hoshangabad-461 990, Madhya Pradesh, India
[2]Department of Agricultural Chemistry & Soil Science,
Dr. B. R. Ambedkar University, Agra- 282 004, India
[3]ICAR- Directorate of Rapeseed- Mustard Research, Sewar, Bharatpur- 321 303, India
[4]Department of Environmental Science, GBPUA&T, Pantnagar-263 145, India
[5]Department of Soil Science, GBPUA&T, Pantnagar-263 145, India
[6]Department of Agronomy, Horticulture & Plant Science, South Dakota State University,
Brooking, South Dakota, S.D. 57007, USA.

1. INTRODUCTION

Expansion and increase in population, industrialization and urbanization are the major culprits of the environmental pollution (Kiziloglu et al. 2008, Klay et al. 2010, Kothari et al. 2012, Lei et al. 2013, Solanki et al. 2018a, Dotaniya et al. 2020, Meena et al. 2020). Among pollution of various environmental segments, water pollution is crucial one which threatens the whole living biosphere including soil pollution or soil contamination (Solanki and Debnath 2014, Kaboosi 2016, Khaliq et al. 2017). Rapid extraction of natural resources for different developmental activities, and further soil deposition and/or land filling of remaining material after processing and purification are the greatest culprits of various soil contaminations (Dotaniya et al. 2014d, Kalavrouziotis et al. 2008, Solanki et al. 2017a). In the past two decades, a large deposition of toxic materials were done into various aquatic bodies, which is responsible for the reduction and degradation/pollution of the water across the globe (Khanna and Solanki 2014, Narayan et al. 2018a, Solanki et al. 2018b). Mainly, these contaminants are of two types, organic and inorganic.

---

*Corresponding author: praveen.solanki746@gmail.com

The major scientific challenge is how to degrade or detoxify both organic and inorganic material simultaneously in a cost effective and eco-friendly manner to protect the functionary system of the ecosystem, which supports the entire life system on the earth (Carrier et al. 2001, CPHEEO 2012, de Lima et al. 2013). Much of the work has been completed in many developed countries viz. USA, Japan, UK, Canada, Australia and others; however, in India, the condition is at risk now and, hence, there is urgent need to develop a bioremediation tool for decontamination/ detoxification of wastewater for its further secondary applications and water security as well (CPCB 2015, Solanki et al. 2017b).

Microorganisms' based bioremediation of wastewater as well as contaminated soil is an emerging green technology for wastewater remediation to reduce its contamination level within permissible limits for various applications and to discharge on the land/soil (Agarwal and Joshi 2010, Khatoon et al. 2017, Solanki et al. 2017c, d). Bioremediation is an intervention based on the available biological diversity for the mitigation or complete elimination of toxic and/or noxious effects of various environmental pollutants. It works on the principle of biochemical cycling (Hanumantha et al. 2011, Abhilash et al. 2012, Abdelaziz et al. 2014). The bioremediation at the source of contamination is called *in situ* bioremediation, whereas deliberate relocation of the wastewater and/or contaminated soil to enhance the biocatalysis is called *ex situ* bioremediation (Ahmed 2007, ABRI 2009, Garcia-Hernandez et al. 2017). Most of the microorganisms are capable of detoxifying a particular type of contaminant, whereas some of the efficient microorganisms are able to detoxify more than one toxic compound (Hernandez et al. 2009, Jais et al. 2017). Some of the microorganisms are fortified with bio agents to enhance their efficiency for bioremediation of wastewater and contaminated soil (Agarry and Ogunleye 2012, Clemente et al. 2012, Meena et al. 2017a, b, Solanki et al. 2017c). Some variety of natural, modified transgenic, and associated microorganisms with mychoriza or rhizosphere are extraordinarily active in the bioremediation of contaminated toxic wastewater and for the cleaning up of various toxic pollutants by immobilising or removing them (Ahmed 2007, Anand et al. 2013, Akash et al. 2016, Solanki et al. 2018c). Diverse microbes such as fungi and its oxidative enzymes play a crucial role in decontamination and detoxification of xenobiotic compounds from the wastewater (Madera-Parra et al. 2013, Dotaniya et al. 2018a). Thus, bioremediation in combination with phytoremediation, rhizodegradation and rhizofiltration significantly contribute to reduce the fate of toxic and hazardous contaminants from the various environmental segments (Schroeder and Schwitzguebel 2004, Dowling and Doty 2009, Weyens et al. 2009, Ma et al. 2011, Narayan et al. 2018b).

2. CHARACTERIZATION OF VARIOUS WASTEWATER

Characteristics of wastewater is one of the most important parameter since it gives information about the various beneficial as well as harmful contents of the wastewater, and therefore, based on the contents, various types of reuse as well as different kinds of remediation techniques can be applied (Dotaniya et al. 2014e, Solanki et al. 2018b, Dotaniya et al. 2019a). It is also observed that the characteristics of wastewater are that they highly fluctuate and vary from time to time and also depend on the type of product to be manufactured (Solanki et al. 2018a). Even the quantity of wastewater also fluctuates and varies from hours to hours, day to day and season to season, which creates hurdle to prepare a systemic blueprint for proper management of the wastewater (Solanki et al. 2017a). Shankhwar et al. (2016) have reported the characteristics of domestic wastewater collected from the local residential area of GB Pant University of Agriculture and Technology, Pantnagar, India and reported the annual average characteristics of wastewater as pH of 7.68, chemical oxygen demand of 277.5 (mg/L), bio-chemical oxygen demand of 58.5 (mg/L), total solids of 713 (mg/L), total dissolved solids of 439.89 (mg/L), total suspended solids of 273.11 (mg/L), total carbon of 208.4 (mg/L), total potassium of 3.6 (mg/L), total nitrogen of 38.26 (mg/L) and total phosphorus was 8.31 (mg/L). The characteristics of wastewater at Pantnagar University were also analysed by Solanki et al. (2018c) and reported the pH of 8.86, electrical conductivity of 552 (μS cm^{-1}), chemical oxygen demand of 380 (mg/L), biochemical oxygen demand of 50.4 (mg/L), total solids of 425 (mg/L), total dissolved solids of 289 (mg/L), total suspended solids of 145 (mg/L), total nitrogen of 36.5 (mg/L), total phosphorus of 7.42 (mg/L), total potassium of 2.30 (mg/L) and sodium of 8.84 (mg/L).

3. ENVIRONMENTAL RISK DUE TO WASTEWATER

In the recent decade, various toxic effects of different heavy metals on the living as well as non-living environment are a cause of serious concern to the many scientific bodies and researchers across the globe (Dotaniya et al. 2018b, c). Since wastewater and/or industrial effluents consist various toxic pollutants as well as disease causing microorganisms (Pipalde and Dotaniya 2018), it could lead to many environmental risks and contamination of different environmental segments such as soil, water, air and living biosphere (Arnot 2009, Solanki 2014, Manu and Thalla 2017). The contamination of water segment of environment is very crucial, since remaining segments totally depend on it (Dotaniya et al. 2014a, Bala et al. 2016, Narayan et al. 2018a). Long-term application of effluents, building up of various heavy metals and different toxic contaminants into the water based food chain and food web have been the major concerns in recent years (Cifuentes et al. 2000, Beltrao et al. 2014, Dotaniya et al. 2016b, 2018d, e). Various heavy/toxic metals are introduced into the environment by different process viz. steel and iron processing, tannery, electroplating, isotopes, colouring agents and many other anthropogenic activities (Espinoza-Quinones et al. 2008, Bhattacharya et al. 2013, Elbana et al. 2014). Heavy metal concentration limits the germination and growth of the crop plants (Dotaniya et al. 2014a, b, Meena et al. 2015). Various physico-chemical processes such as ion exchange, chemical precipitation, adsorption on activated carbon, reverse osmosis and electrical deposition are used for removing different kinds of heavy metals from the aqueous wastewater stream (Kim 1981, Basta 1983, Zhou et al. 1999, Acar and Malkoc 2004).

4. HUMAN HEALTH AND WASTEWATER

Contamination of water with various pollutants is an emerging issue in the developing countries like India. Furthermore, this contaminated water is a key factor for the degradation of the entire aquatic environmental segment (Dotaniya et al. 2016a, b). Bioaccumulation of toxic metals into plant biomass and their harvestable yield leads to many serious health diseases in the human beings (Fantroussi et al. 2006, Zhuang et al. 2009, Massaquoi et al. 2015, Dotaniya et al. 2017c, 2019d, e). Nowadays, water contamination/pollution is one of the serious threats being faced by human beings, since most of the vegetarian diets are produced from the soil segments of the environments which is being irrigated with polluted water. Contaminated water is the major culprit for many health diseases in human beings as well as animals due to contamination of food chains and food webs with different kinds of toxic substances (Dotaniya et al. 2014f, 2017a, Migahed et al. 2017). However, the type of disease completely depends upon the particular toxic metal (Nabulo et al. 2010, Shakir et al. 2017). Irrigation of the field or edible crops with wastewater as compared to floricultural crops is more vulnerable to the human health, since most of the flowers are used for worship only (Mojida et al. 2010, Hanjra et al. 2012, Hu et al. 2013, Blasi et al. 2016, Dotaniya et al. 2016b). Large scale deposition of contaminated material on the soil leads to not only polluted soil but also rising health risk of both human and environment

(Brejova et al. 2004, Bunluesin et al. 2004, Dotaniya et al. 2017b, Dotaniya and Pipalde 2018). There is strong evidence that the cocktail of various environmental contaminants is responsible for global epidemic of cancer as well as many degenerative diseases in human beings (Muchuweti et al. 2006, Mulbry et al. 2008, Shammi et al. 2016). Presence of heavy metals' concentration in crop field affects the soil microbial dynamics, plant nutrient supplying dynamics and plant showed nutrient deficiency (Dotaniya and Meena 2013, 2017, Singh et al. 2017, Dotaniya et al. 2019c). In such types of situations, farmers are forced to apply more amount of plant nutrient and further pollute the soil (Meena et al. 2019, Solanki et al. 2019).

5. POTENTIAL AREAS FOR WASTEWATER APPLICATION

The quantitative generation of wastewater is proportional to the increasing trend of population and industrialization (Meers et al. 2010, Judit et al. 2011). The developmental activities such as urbanization, housing, lacking of sewage treatment plant, etc. are triggering the quantitative generation of wastewater (Pettersson and Lavieille 2007, Mekki et al. 2013). Furthermore, the generation of wastewater is positively related with development; therefore, developed countries are producing more quantity of wastewater, while developing countries are producing less quantity of wastewater (Rahman et al. 2012, Rai et al. 2015).

Recycling of wastewater on the agricultural land could be one of best option to transform a problematic land to a healthy land (Shuval and Fattal 2003, Singh et al. 2012). Soil with an acidic pH could be restored by wastewater irrigation since it contains more organic materials, which leads to balancing the pH of soil and also developing a biofilm surrounding the root hair zone, which provides nourishment to the root for sustaining the plant life (Tove et al. 2009, Vasudevan et al. 2011a, b, Dotaniya et al. 2019b). Apart from this, wastewater also contains macro and micro nutrients such as nitrogen, phosphorus, potassium, calcium, magnesium, zinc, iron, copper, sulphur, nickel, etc., which supports the plant life and further sustain animal life and human beings as well, since these trace elements are acquired directly or indirectly by all living things throughout their lives (Zaidi 2007, Velho et al. 2012). Addition of wastewater consisting organic matter, and various macro- and micro-nutrients leads to physical as well as textural transformation of soil by improving physical structure of soil particles, water holding capacity (WHC), cohesion and adhesion property of soil, etc., all of which promote biological growth of planted vegetation (Zema et al. 2012). Addition of organic carbon through application of crop residue enhances the concentration of plant nutrients in soil solution (Dotaniya and Datta 2014, Dotaniya et al. 2014b, c, 2015) as well as reduces the heavy metal concentration by forming different humus metal complexes in soil (Dotaniya et al. 2016a, 2018a).

Additional benefit of wastewater application is the slow releasing nature of nutrients to the planted vegetation, which makes it more economically viable and efficient as compared to the mineral and chemical fertilizers (Adrover et al. 2012, Biswas et al. 2015). Its ready availability in huge quantity across the country again makes it a good and prominent substitute for mineral fertilizers as an emerging soil conditioner (Becerra-Castro et al. 2015).

6. BIOREMEDIATION OF WASTEWATER

The bioremediation of contaminated/polluted water (with various kinds of heavy metals and toxic pollutants) means the remediation of wastewater using different potential micro-organisms and plants who are capable enough to degrade, decompose, detoxify and bio-accumulate various environmental pollutants into their harvestable biomass (Scott et al. 2004, Zhang et al. 2014, Ravanipour et al. 2015). Most of the microphytes as well as some macrophytes are capable of accumulating different kinds of toxic pollutants into their harvestable body parts by the process called phytoremediation (Kumar et al. 2013). Phytoremediation is a cost effective and innovative technology that remediates contamination from the environmental segments in a sustainable manner. The phytoremedial potential of the plant can be assessed using its bio-concentration factor (BCF, the ratio of concentration of a particular heavy metal into root and wastewater) and translocation factor (TF, the ratio of concentration of a particular heavy metal into shoot and root treated with wastewater) (Song et al. 2006, Soda et al. 2012). Less translocation factor (< 1, concentration of a particular heavy metal is more in root as compared to shoot) of the phytoremediation plant is considered a good feature for the safe disposal of the plant since the root (high concentration) proportion of the plant is very less as compared to the shoot (less concentration). However, the bio-concentration factor value more than 1000 is considered as significant phytoremediation potential for a particular plant in removing a particular toxic metal (Zayed et al. 1998, Teodorescu and Gaidau 2008).

7. FUTURE PROSPECTS

The quantitative generation of wastewater will increase by many folds in the upcoming future, since the generating quantity of wastewater is positively correlated with the population exploitation, industrialization and urbanization. In this challenging era of development, the generation of wastewater is a common phenomenon, which is more hectic to control for the management authority as well; therefore, this could be the potential key area for the future research and development study.

8. CONCLUSION

Recycling of quantitative generated wastewater is one of the greatest challenges for the scientific community across the globe to mitigate the crop irrigation water requirement in the areas where freshwater is very crucial and scared. Furthermore, generation of wastewater is increasing due to developmental activities, along with urbanization and industrialization. The manurial value of huge generated wastewater could be one of the potential options to cope up with freshwater scarcity and cost-effective management of wastewater, particularly in peri-urban areas of many developing countries. In many developing countries, the application of wastewater for irrigation of the agricultural and horticultural crops is very common in practice to harvest the wastewater as well as nutrients present in it. However, long-term application of wastewater to the agricultural land for various crop productions could lead to accumulation of significant quantity of different kinds of toxic substances including heavy metals as well. Furthermore, contamination of food chain and food web occurs with the heavy metal and toxic substances which cause human health diseases. Therefore, it is a prime need to remediate wastewater contamination as well as wastewater prior to its application for sustainable crop production without compromising environmental and human health. Microorganisms based bioremediation of highly contaminated wastewater and soils could play a crucial role in effective management of wastewater for reducing the metal toxicity and further human health risk in a cost effective and eco-friendly manner.

Acknowledgment

Authors thank G.B. Pant University of Agriculture and Technology (GBPUA&T), Pantnagar and also University Grant Commission (UGC) for providing financial support and an academic platform to do this work.

References

Abdelaziz, A.E., G.B. Leite, M.A. Belhaj and P.C. Hallenbeck. 2014. Screening microalgae native to Quebec for wastewater treatment and biodiesel production. Bioresour Technol. 157: 140–148.

Abhilash, P.C., H.B. Singh, J.R. Powell and B.K. Singh. 2012. Plant-microbe interactions: Novel applications for exploitation in multi-purpose remediation technologies. Trend Biotechnol. 30: 416–420.

ABRI, 2009. Advancing the Blue Revolution Initiative, pilot project reuse of treated wastewater in agriculture in Meknes, Morocco. Available at http://pdf.usaid.gov/pdf_docs/Pnadp168.pdf. Accessed on 02 Oct. 2018.

Acar, F.N. and E. Malkoc. 2004. The removal of chromium (VI) from aqueous solutions by *Fagus orientalis* L. Bioresour Technol. 94: 13–15.

Adrover, M., E. Farrus, G. Moya and J. Vadell. 2012. Chemical properties and biological activity in soils of Mallorca following twenty years of treated wastewater irrigation. J. Environ. Manag. 95: S188–S192.

Agarry, S.E. and O.O. Ogunleye. 2012. Box-behnken design application to study enhanced bioremediation of soil artificially contaminated with spent engine oil using biostimulation strategy. Int. J. Energy Environ. Eng. 3: 31–38.

Agarwal, A. and H. Joshi. 2010. Application of nanotechnology in the remediation of contaminated groundwater: A short review. Recent Res. Sci. Technol. 2: 51–57.

Ahmed, M.T. 2007. Life cycle assessment a decision making tool in wastewater treatment facilities. *In*: M.K. Zaidi (ed.), Wastewater Reuse–Risk Assessment Decision Making and Environmental Security. Springer, Dordrecht, pp. 305–314.

Akash, M.S., D.S. Samuel and U.H. Rolf. 2016. Mass balance assessment for six neonicotinoid insecticides during conventional wastewater and wetland treatment: Nationwide reconnaissance in United States wastewater. Environ. Sci. Technol. 50(12): 6199–6206.

Anand, K.G., K.V. Sanjeet, K. Khushboo and K.V. Rajesh. 2013. Phytoremediation using aromatic plants: A sustainable approach for remediation of heavy metals polluted sites. Environ. Sci. Technol. 47(18): 10115–10116.

Arnot, J.A. 2009. Mass balance model for chemical fate bioaccumulation exposure and risk assessment. *In*: L.I. Simeonov and M.A. Hassanien (eds), Exposure and Risk Assessment of Chemical Pollution-Contemporary Methodology NATO Science for Peace and Security Series C: Environmental Security. Springer, Dordrecht. pp. 69–91.

Bala, J.D., J. Lalung, A.A. Al-Gheethi and I. Norli. 2016. A review on biofuel and bioresources for environmental applications. *In*: M. Ahmad, M. Ismail and S. Riffat (eds), Renewable Energy and Sustainable Technologies for Building and Environmental Applications. Springer, Switzerland, pp. 205–225.

Basta, N. 1983. Getting the metal out of spent plating baths. Chem. Eng. 90: 22–23.

Becerra-Castro, C., A.R. Lopes, I. Vaz-Moreira, E.F. Silva, C.M. Manaia and O.C. Nunes. 2015. Wastewater reuse in irrigation: A microbiological perspective on implications in soil fertility and human and environmental health. Environ Int. 75: 117–135.

Beltrao, J., P.J. Correia, M. Costa, P. Gamito, R. Santos and J. Seita. 2014. The influence of nutrients turfgrass responds to treated wastewater application under several saline condition and irrigation regimes. Environ. Progress. 1: 105–111.

Bhattacharya, P., S. Ghosh and A. Mukhopadhyay. 2013. Combination technology of ceramic microfiltration and biosorbent for treatment and reuse of tannery effluent from different streams: Response of defence system in *Euphorbia* sp. Int. J. Recycl. Org. Waste Agricult. 2: 19 http://www.ijrowa.com/content/2/1/19.

Biswas, G.C., A. Sarkar, H. Rashid, M.H. Shohan, M. Islam and Q. Wang. 2015. Assessment of the irrigation feasibility of low-cost filtered municipal wastewater for red amaranth (*Amaranthus tricolor* L. cv. Surma). Int. Soil Water Conserv. Res. 3(3): 239–252.

Blasi, B., C. Poyntner, T. Rudavsky, F.X. Prenafeta-Boldu, S. de-Hoog, H. Tafer, et al. 2016. Pathogenic yet environmentally friendly? Black fungal candidates for bioremediation of pollutants. Geomicrobiol. J. 33(3–4): 308–317.

Brejova, B., D.G. Brown and T. Vinar. 2004. The most probable labeling problem in HMMs and its application to bioinformatics. *In*: I. Jonassen and J. Kim (eds), Algorithms in Bioinformatics WABI 2004 Lecture Notes in Computer Science, vol 3240. Springer, Berlin Heidelberg, pp. 426–437.

Bunluesin, S., M. Kruatrachue, P. Pokethitiyook, G.R. Lanza, E.S. Upatham and V. Soonthornsarathool. 2004. Plant screening and comparison of *Ceratophyllum demersum* and *Hydrilla verticillata* for cadmium accumulation. Bull. Environ. Contam. Toxicol. 73: 591–598.

Carrier, G., M. Bouchard, R.C. Brunet and M. Caza. 2001. A toxicokinetic model for predicting the tissue distribution and elimination of organic and inorganic mercury following exposure to methyl mercury in animals and humans II Application and validation of the model in humans. Toxicol. Appl. Pharmacol. 171: 50–60.

Cifuentes, E., M. Gomez, U. Blumenthal, M.M. Tellez-Rojo, I. Romieu, G. Ruiz-Palacios, et al. 2000. Risk factors for giardia intestinalis infection in agricultural villages practicing wastewater irrigation in Mexico. Am. J. Trop. Med. Hyg. 62(3): 388–392.

Clemente, R., D.J. Walker, T. Pardo, D. Martinez-Fernandez and M.P. Bernal. 2012. The use of a halophytic plant species and organic amendments for the remediation of a trace elements contaminated soil under semi-arid conditions. J. Hazard. Mater. 223–224.

CPCB, 2015. Central pollution control board directions under section 18(1)(b) of the Water (Prevention and Control of Pollution) Act 1974 regarding treatment and utilization of sewage File No A-19014/43/06-MON http://cpcbnicin/Mghlya_swg_18(1)(b)_2015pdf.

CPHEEO, 2012 Manual on sewerage and sewage treatment part A: Engineering final draft central public health and environmental engineering organisation Ministry of Urban Development New Delhi http://wwwcmampcom/CP/FDocument/ManualonSewerageTreatmentpdf.

de Lima, M.A.B., L.O. de Franco, P.M. de Souza, A.E. do Nascimento, C.A.A. da Silva, R.C.C. de Maia, et al. 2013. Cadmium tolerance and removal from *Cunninghamella elegans* related to the polyphosphate metabolism. Int. J. Mol. Sci. 14: 7180–7192.

Dotaniya, M.L. and V.D. Meena. 2013. Rhizosphere effect on nutrient availability in soil and its uptake by plants—A review. Proc. Natl. Acad. Sci. India Sec. B: Biol. Sci. 85(1): 1-12.

Dotaniya, M.L. and S.C. Datta. 2014. Impact of bagasse and press mud on availability and fixation capacity of phosphorus in an Inceptisol of north India. Sugar. Tech. 16(1): 109–112.

Dotaniya, M.L., H. Das and V.D. Meena. 2014a. Assessment of chromium efficacy on germination, root elongation, and coleoptile growth of wheat (*Triticum aestivum* L.) at different growth periods. Environ. Monit. Assess. 186: 2957–2963.

Dotaniya, M.L., S.C. Datta, D.R. Biswas and K. Kumar. 2014b. Effect of organic sources on phosphorus fractions and available phosphorus in Typic Haplustept. J. Ind. Soc. Soil Sci. 62(1): 80–83.

Dotaniya, M.L., S.C. Datta, D.R. Biswas, H.M. Meena and K. Kumar. 2014c. Production of oxalic acid as influenced by the application of organic residue and its effect on phosphorus uptake by wheat (*Triticum aestivum* L.) in an Inceptisol of north India. Natl. Acad. Sci. Lett. 37(5): 401–405.

Dotaniya, M.L., V.D. Meena and H. Das. 2014d. Chromium toxicity on seed germination, root elongation and coleoptile growth of pigeon pea (*Cajanus cajan*). Legume Res. 37(2): 225–227.

Dotaniya, M.L., J.K. Saha, V.D. Meena, S. Rajendiran, M.V. Coumar, S. Kundu, et al. 2014e. Impact of tannery effluent irrigation on heavy metal build up in soil and ground water in Kanpur. Agrotechnol. 2(4): 77.

Dotaniya, M.L., J.K. Meena, V.D. Jajoria and D.K. Rathor. 2014f. Chromium pollution: A threat to environment. Agric. Rev. 35(2): 153–157.

Dotaniya, M.L., S.C. Datta, D.R. Biswas, H.M. Meena, S. Rajendiran and A.L. Meena. 2015. Phosphorus dynamics mediated by bagasse, press mud and rice straw in inceptisol of north India. Agrochimica. 59(4): 358–369.

Dotaniya, M.L., S.C. Datta, D.R. Biswas, C.K. Dotaniya, B.L. Meena, S. Rajendiran, et al. 2016a. Use of sugarcane industrial byproducts for improving sugarcane productivity and soil health—A review. Intl. J. Recyc. Org. Waste. Agric. 5(3): 185–194.

Dotaniya, M.L., V.D. Meena, K. Kumar, B.P. Meena, S.L. Jat, M. Lata, et al. 2016b. Impact of biosolids on agriculture and biodiversity. Today and Tomorrow's Printer and Publisher, New Delhi, pp. 11–20.

Dotaniya, M.L. and B.P. Meena. 2017. Rhizodeposition by plants: A boon to soil health. *In*: R. Elanchezhian, A.K. Biswas, K. Ramesh and A.K. Patra (eds), Advances in Nutrient Dynamics in Soil Plant System for Improving Nutrient use Efficiency. New India Publishing Agency, New Delhi, India, pp. 207–224.

Dotaniya, M.L., V.D. Meena, S. Rajendiran, M.V. Coumar, J.K. Saha, S. Kundu, et al. 2017a. Geo-accumulation indices of heavy metals in soil and groundwater of Kanpur, India under long term irrigation of tannery effluent. Bull. Environ. Conta. Toxic. 98(5): 706–711.

Dotaniya, M.L., S. Rajendiran, V.D. Meena, J.K. Saha, M.V. Coumar, S. Kundu, et al. 2017b. Influence of chromium contamination on carbon mineralization and enzymatic activities in Vertisol. Agric. Res. 6(1): 91–96.

Dotaniya, M.L., S. Rajendiran, M.V. Coumar, V.D. Meena, J.K. Saha, S. Kundu, et al. 2017c. Interactive effect of cadmium and zinc on chromium uptake in spinach grown on Vertisol of Central India. Intl. J. Environ. Sci. Tech. 15(2): 441–448.

Dotaniya, M.L. and J.S. Pipalde. 2018. Soil enzymatic activities as influenced by lead and nickel concentrations in a Vertisol of Central India. Bull. Environ. Conta. Toxic. 101(3): 380–385.

Dotaniya, M.L., S. Rajendiran, C.K. Dotaniya, P. Solanki, V.D. Meena, J.K. Saha, et al. 2018a. Microbial assisted phytoremediation for heavy metal contaminated soils. *In*: V. Kumar, M. Kumar and R. Prasad (eds), Phytobiont and Ecosystem Restitution. Springer Singapore, Switzerland AG, pp. 295–317.

Dotaniya, M.L., S. Rajendiran, V.D. Meena, M.V. Coumar, J.K. Saha, S. Kundu, et al. 2018b. Impact of long-term application of sewage on soil and crop quality in Vertisols of central India. Bull. Environ. Conta. Toxic. 101: 779–786.

Dotaniya, M.L., V.D. Meena, S. Rajendiran, M.V. Coumar, A. Sahu, J.K. Saha, et al. 2018c. Impact of long-term application of Patranala sewage on carbon sequestration and heavy metal accumulation in soils. J. Indian Soc. Soil Sci. 66(3): 310–317.

Dotaniya, M.L., V.D. Meena, J.K. Saha, S. Rajendiran, A.K. Patra, C.K. Dotaniya, et al. 2018d. Environmental impact measurements: Tool and techniques. *In*: L.O. Martinez, L.O. Kharissova and B. Kharisov (eds), Handbook of Ecomaterials. Springer, Cham, pp. 183–201.

Dotaniya, M.L., N.R. Panwar, V.D. Meena, C.K. Dotaniya, K.L. Regar, M. Lata, et al. 2018e. Bioremediation of metal contaminated soils for sustainable crop production. *In*: V.S. Meena (ed.), Role of Rhizospheric Microbes in Soil. Springer, India. pp. 143–173.

Dotaniya, M.L., V.D. Meena and B.L. Meena. 2019a. Use of wastewater for sustainable agriculture. *In*: R.S. Meena (ed.), Sustainable Agriculture, Scientific Publishers, India, pp. 71–87.

Dotaniya, M.L., J.K. Saha, S. Rajendiran, M.V. Coumar, V.D. Meena, S. Kundu, et al. 2019b. Chromium toxicity mediated by application of chloride and sulphate ions in vertisol of Central India. Environ. Monit. Assess. 191: 429.

Dotaniya, M.L., K. Aparna, C.K. Dotaniya, M. Singh and K.L. Regar. 2019c. Role of soil enzymes in sustainable crop production. *In*: M. Khudus (ed.), Enzymes in Food Biotechnology. Springer International, London, pp. 569–589.

Dotaniya, M.L., J.S. Pipalde, R.C. Jain, S. Rajendiran, M.V. Coumar, J.K. Saha, et al. 2019d. Can lead and nickel interaction affect plant nutrient uptake pattern in spinach (*Spinacia oleracea*)? Agric. Res. doi.org/10.1007/s40003-019-00428-4.

Dotaniya, M.L., J.K. Saha, S. Rajendiran, M.V. Coumar, V.D. Meena, H. Das, et al. 2019e. Reducing chromium uptake through application of calcium and sodium in spinach. Environ. Monitor. Assess. 191: 754–762.

Dotaniya, M.L., C.K. Dotaniya, P. Solanki, V.D. Meena and R.K. Doutaniya. 2020. Lead contamination and its dynamics in soil-plant system. *In*: D. Gupta, S. Chatterjee and C. Walther (eds), Lead in Plants and the Environment. Radionuclides and Heavy Metals in the Environment. Springer, Cham, pp. 83–88.

Dowling, D.N. and S.L. Doty. 2009. Improving phytoremediation through biotechnology. Curr. Opin. Biotechnol. 20: 204–206.

Elbana, T., N. Bakr, F. Karajeh, Q. El and E.D. Dia. 2014. Treated wastewater utilization for agricultural irrigation in Egypt. *In*: Proceedings of the National Conference on Water Quality: Challenges and Solutions. National Research Centre-Cairo, Egypt, pp. 35–46. April 29th, 2014. http://bio-nrc.jimdo.com/conferences/.

Espinoza-Quinones, F.R., S. Antonio-da, A. Marcia, M. Soraya, N. Aparecido, N.M. Nayara, et al. 2008. Chromium ions phytoaccumulation by three floating aquatic macrophytes from a nutrient medium. World J. Microbiol. Biotechnol. 24: 3063–3070.

Fantroussi, S., S.N. Agathos, D.H. Pieper, R. Witzig, B. Cámara, L. Gabriel-Jürgens, et al. 2006. Biological assessment and remediation of contaminated sediments. *In*: D. Reible and T. Lanczos (eds) Assessment and Remediation of Contaminated Sediments Nato Science Series: IV Earth and Environmental Science, vol 73. Springer, Dordrecht, pp. 179–238.

Garcia-Hernandez, M.A., J.F. Villarreal-Chiu and M.T. Garza-Gonzalez. 2017. Metallophilic fungi research: an alternative for its use in the bioremediation of hexavalent chromium. Int. J. Environ. Sci. Technol. 14: 2023–2038. https://doi.org/10.1007/s13762-017-1348-5.

Hanjra, M.A., J. Blackwella, G. Carrc, F. Zhangd and T.M. Jacksona. 2012. Wastewater irrigation and environmental health: Implications for water governance and public policy. Int. J. Hygiene Environ. Health. 215: 255–269.

Hanumantha, R.P., K.R. Ranjith, B. Raghavan, V. Subramanian and V. Sivasubramanian. 2011. Application of phycoremediation technology in the treatment of wastewater from a leather processing chemical manufacturing facility. Water Sa. 37(1): 7–14.

Hernandez, J.P., L.E. de-Bashan, D.J. Rodriguez, Y. Rodriguez and Y. Bashan. 2009. Growth promotion of the freshwater microalga *Chlorella vulgaris* by the nitrogen-fixing plant growth-promoting bacterium *Bacillus pumilus* from arid zone soils. Eur. J. Soil Biol. 45: 88–93.

Hu, J., F. Wu, S. Wu, Z. Cao, X. Lin and M.H. Wong. 2013. Bioaccessibility, dietary exposure and human risk assessment of heavy metals from market vegetables in Hong Kong revealed with an in vitro gastrointestinal model. Chemosphere. 91(4): 455–461.

Jais, N.M., R.M.S.R. Mohamed, A.A. Al-Gheethi and H. Amir. 2017. The dual roles of phycoremediation of wet market wastewater for nutrients and heavy metals removal and microalgae biomass production. Clean Techn. Environ. Policy. 19: 37–52.

Judit, L., K. Mirjam, K. Jonas, S.M. Christa and S. Nele. 2011. Multiple criteria decision analysis reveals high stakeholder preference to remove pharmaceuticals from hospital wastewater. Environ. Sci. Technol. 45(9): 3848–3857.

Kaboosi, K. 2016. The assessment of treated wastewater quality and the effects of mid-term irrigation on soil physical and chemical properties (case study: Bandargaz-treated wastewater). Appl. Water Sci. 7: 2385–2396.

Kalavrouziotis, I.K., P. Robolas, P.H. Koukoulakis and A.H. Papadopoulos. 2008. Effects of municipal reclaimed wastewater on the macro and micro elements status of soil and of *Brassica oleracea* var. Italica, and *Brassica oleracea* var. Gemmifera. Agric. Water Manage. 95: 419–26.

Khaliq, S.Z.A., A. Al-Busaidi, M. Ahmed, M. Al-Wardy, H. Agrama and B.S. Choudri. 2017. The effect of municipal sewage sludge on the quality of soil and crops. Int. J. Recycl. Org. Waste Agricult. 6: 289–299.

Khanna, N. and P. Solanki. 2014. Role of Agriculture in the Global Economy. Agrotechnol. 2(4): 221–227.

Khatoon, H., P. Solanki, M. Narayan, L. Tewari and J.P.N. Rai. 2017. Role of microbes in organic carbon decomposition and maintenance of soil ecosystem. Int. J. Chem. Studies. 5(6): 1648–1656.

Kim, B.M. 1981. Treatment of metal containing wastewater with calcium sulfide. AIChE Symp. Ser. 77: 39–48.

Kiziloglu, F.M., M. Turan, U. Sahin, Y. Kuslu and A. Dursun. 2008. Effects of untreated and treated wastewater irrigation on some chemical properties of cauliflower (*Brassica olerecea* L. var. botrytis) and red cabbage (*Brassica olerecea* L. var. rubra) grown on calcareous soil in Turkey. Agric. Water Manage. 95: 716–724.

Klay, S., A. Charef, L. Ayed, B. Houman and F. Rezgui. 2010. Effect of irrigation with treated wastewater on geochemical properties (saltiness, C, N and heavy metals) of isohumic soils (Zaouit Sousse perimeter, Oriental Tunisia). Desalination 253(1–3): 180–187.

Kothari, R., V.V. Pathak, V. Kumar and D.P. Singh. 2012. Experimental study for growth potential of unicellular alga *Chlorella phrenoidosa* on dairy wastewater: An integrated approach for treatment and biofuel production. Biores. Technol. 116: 466–470.

Kumar, N., K. Bauddh, S. Kumar, N. Dwivedi, D.P. Singh and S.C. Barman. 2013. Accumulation of metals in weed species grown on the soil contaminated with industrial waste and their phytoremediation potential. Ecolog. Engg. 61: 491–495.

Lei, W., Z. Zhen-Yong, Z. Ke and T. Chang-Yan. 2013. Reclamation and utilization of saline soils in arid north western china: a promising halophyte drip-irrigation system. Environ. Sci. Technol. 47(11): 5518–5519.

Ma, Y., M.N.V. Prasad, M. Rajkumar and H. Freitas. 2011. Plant growth promoting rhizobacteria and endophytes accelerate phytoremediation of metalliferous soils. Biotechnol. Adv. 29: 248–258.

Madera-Parra, C.A., E.J. Pena-Salamanca, M.R. Pena, D.P.L. Rousseau and P.N.L. Lens. 2013. Phytoremediation of landfill leachate with *Colocasia esculenta*, *Gynerum sagittatum* and *Heliconia psittacorum* in constructed wetlands. Int. J. Phytorem. 17(1–6): 16–24.

Manu, D.S. and A.K. Thalla. 2017. Artificial intelligence models for predicting the performance of biological wastewater treatment plant in the removal of Kjeldahl Nitrogen from wastewater. Appl. Water Sci. 7: 3783–3791.

Massaquoi, L.D., H. Ma, X.H. Liu, P.Y. Han, S. Zuo, Z. Hua, et al. 2015. Heavy metal accumulation in soils, plants, and hair samples: An assessment of heavy metal exposure risks from the consumption of vegetables grown on soils previously irrigated with wastewater. Environ. Sci. Pollut. Res. 22: 18456–18468.

Meena, V.D., M.L. Dotaniya, J.K. Saha and A.K. Patra. 2015. Antibiotics and antibiotic resistant bacteria in wastewater: Impact on environment, soil microbial activity and human health. African J. Microbiol. Res. 9(14): 965–978.

Meena, S.S., D.C. Kala, P. Solanki and V. Sarode. 2017a. Effect of rice husk biochar, carpet waste, FYM and PGPR on chemical properties of soil. Int. J. Curr. Microbiol. Appl. Sci. 6(5): 2287–2292.

Meena, S.S., J. Yadav, D.K. Singhal, P. Solanki, D.C. Kala and V. Kumar. 2017b. The assessment of rice husk biochar, carpet waste, FYM and PGPR on nutrient uptake of mungbaean (*Vigna radiata* L.). Ecol. Environ. Conserv. 23(2): 859–865.

Meena, V.D., M.L. Dotaniya, J.K. Saha, B.P. Meena, H. Das and A.K. Patra. 2019. Sustainable C and N management under metal-contaminated soils. *In*: R. Datta, R. Meena, S. Pathan and M. Ceccherini (eds), Carbon and Nitrogen Cycling in Soil. Springer, Singapore, pp. 293–336.

Meena, V.D., M.L. Dotaniya, J.K. Saha, H. Das and A.K. Patra. 2020. Impact of lead contamination on agroecosystem and human health. *In*: D. Gupta, S. Chatterjee and C. Walther (eds), Lead in Plants and the Environment. Radionuclides and Heavy Metals in the Environment. Springer, Cham, pp. 78–86. doi.org/10.1007/978-3-030-21638-2_4.

Meers, E., S. van Slycken, K. Adriaensen, A. Ruttens, J. Vangronsveld, G. Du Laing, et al. 2010. The use of bio-energy crops (*Zea mays*) for "phytoattenuation" of heavy metals on moderately contaminated soils: A field experiment. Chemosphere 78: 3541–3547.

Mekki, A., A. Dhouib and S. Sayadi. 2013. Review: Effects of olive mill wastewater application on soil properties and plants growth. Int. J. Recycl. Org. Waste Agricult. 2: 15–19.

Migahed, F., A. Abdelrazak and G. Fawzy. 2017. Batch and continuous removal of heavy metals from industrial effluents using microbial consortia. Int. J. Environ. Sci. Technol. 14: 1169–1180.

Mojida, M.A., G.C.L. Wyseureb, S.K. Biswasa and A.B.M.Z. Hossaina. 2010. Farmers' perceptions and knowledge in using wastewater for irrigation at twelve peri-urban areas and two sugar mill areas in Bangladesh. Agric. Water Manage. 98: 79–86.

Muchuweti, M., J.W. Birkett, E. Chinyanga, R. Zvauya, M.D. Scrimshaw and J.N. Lister. 2006. Heavy metal content of vegetables irrigated with mixtures of wastewater and sewage sludge in Zimbabwe: Implication for human health. Agric. Ecosyts. Environ. 112: 41–48.

Mulbry, W., S. Kondrad, C. Pizarro and E. Kebede-Westhead. 2008. Treatment of dairy manure effluent using freshwater algae: Algal productivity and recovery of manure nutrients using pilot-scale algal turf scrubbers. Bioresour. Technol. 99: 8137–8142.

Nabulo, G., S.D. Young and C.R. Black. 2010. Assessing risk to human health from tropical leafy vegetables grown on contaminated urban soils. Sci. Total Environ. 408(22): 5338–5351.

Narayan, M., P. Solanki and R.K. Srivastava. 2018a. Treatment of sewage (domestic wastewater or municipal wastewater) and electricity production by integrating constructed wetland with microbial fuel cell. *In*: I. Zhu (ed.), Sewage, IntechOpen, DOI: 10.5772/intechopen.75658.

Narayan, M., P. Solanki, A.K. Rabha and R.K. Srivastava. 2018b. Treatment of pulp and paper industry effluent and electricity generation by constructed wetland coupled with microbial fuel cell (CW-MFC). J. of Pharmaco. and Phytoche. 7(6): 493–498. http://www.phytojournal.com/archives/2018/vol7issue6/PartI/7-5-534-855.pdf.

Pettersson, T. and D. Lavieille. 2007. Evolution on pollutant removal efficiency in storm water ponds due to changes in pond morphology. *In*: G.M. Morrison and S. Rauch (eds), Highway and Urban Environment Alliance for Global Sustainability. Bookseries, vol 12. Springer, Dordrecht, pp. 429–439.

Pipalde, J.S. and M.L. Dotaniya. 2018. Interactive effects of lead and nickel contamination on nickel mobility dynamics in spinach. Intl. J. Environ. Res. 12(5): 553–560. doi.org/10.1007/s41742-018-0107-x.

Rahman, A., J.T. Ellis and C.D. Miller. 2012. Bioremediation of domestic wastewater and production of bioproducts from microalgae using waste stabilization ponds. J. Bioremed. Biodegrad. 3(6): 6199.

Rai, U.N., A.K. Upadhyay, N.K. Singh, S. Dwivedi and R.D. Tripathi. 2015. Seasonal applicability of horizontal sub-surface flow constructed wetland for trace elements and nutrient removal from urban wastes to conserve Ganga River water quality at Haridwar, India. Ecol. Eng. 81: 115–122.

Ravanipour, M., R.R. Kalantary, A. Mohseni-Bandpi, A. Esrafili, M. Farzadkia and S. Hashemi-Najafabadi. 2015. Experimental design approach to the optimization of PAHs bioremediation from artificially contaminated soil: application of variables screening development. J. Environ. Health Sci. Eng. 13: 22.

Schroeder, P. and J.P. Schwitzguebel. 2004. New cost action launched: Phytotechnologies to promote sustainable land use and improve food safety. J. Soils Sed. 4(3): 205.

Scott, C.A., N.I. Faruqui and L. Raschid-Sally. 2004. Wastewater use in irrigated agriculture: management challenges in developing countries. *In*: C.A. Scott, N.I. Faruqui and L. Raschid-Sally (eds), Wastewater Use in Irrigated Agriculture: Confronting the Livelihood and Environmental Realities. IWMI/IDRC-CRDI/CABI, Wallingford, UK, pp. 1–10.

Shakir, E., Z. Zahraw, A. Hameed and M.J. Al-Obaidy. 2017. Environmental and health risks associated with reuse of wastewater for irrigation. Egyptian J. Petrol. 26: 95–102.

Shammi, M., M.A. Kashem, M.M. Rahman, M.D. Hossain, R. Rahman and M.K. Uddin. 2016. Health risk assessment of textile effluent reuses as irrigation water in leafy vegetable *Basella alba*. Int. J. Recycl. Org. Waste Agricult. 5: 113–123.

Shankhwar, A.K., S. Ramola, T. Mishra and R.K. Srivastava. 2016. Grey water pollutant loads in residential colony and its economic management. Renewables 2: 5. https://doi.org/10.1186/s40807-014-0005-6.

Shuval, H. and B. Fattal. 2003. Control of pathogenic microorganisms in wastewater recycling and reuse in agriculture. *In*: D. Mara and N. Horan (eds), The Handbook of Water and Wastewater Microbiology. Elsevier, pp. 241–262.

Singh, P.K., P.B. Deshbhratar and D.S. Ramteke. 2012. Effects of sewage wastewater irrigation on soil properties, crop yield and environment. Agric. Water Manage. 103: 100–104.

Singh, V.S., S.K. Meena, J.P. Verma, A. Kumrar, A. Aeron, P.K. Mishra, et al. 2017. Plant beneficial rhizospheric microorganism (PBRM) strategies to improve nutrients use efficiency: A review. Ecol. Eng. 107: 8–32.

Soda, S., T. Hamad, Y. Yamaoka, M. Ike, H. Nakazato, Y. Saeki, et al. 2012. Constructed wetlands for advanced treatment of wastewater with a complex matrix from a metal-processing plant: Bioconcentration and translocation factors of various metals in *Acorus gramineus* and *Cyperus alternifolius*. Ecol. Eng. 39: 63–70.

Solanki, P. 2014. Effect of Sewage Sludge on Marigold and Golden Rod. Thesis Acharya N.G. Ranga Agricultural University, Rajendranagar, Hyderabad, pp. 172. http://krishikosh.egranth.ac.in/handle/1/75355.

Solanki, P. and P. Debnath. 2014. Role of biosolids in sustainable development. Agrotechnol. 2(4): 220. http://dx.doi.org/10.4172/2168-9881.S1.008.

Solanki, P., M. Narayan and R.K. Srivastava. 2017a. Effectiveness of domestic wastewater treatment using floating rafts a promising phyto-remedial approach: A review. J. Appl. Nat. Sci. 9(4): 1931–1942.

Solanki, P., M. Narayan, S.S. Meena and R.K. Srivastava. 2017b. Floating raft wastewater treatment system: A review. Inter. J. of Pure Appl. Microbiol. 11: 1113–1116.

Solanki, P., S.S. Meena, M. Narayan, H. Khatoon and L. Tewari. 2017c. Denitrification process as an indicator of soil health. Int. J. Curr. Microbiol. Appl. Sci. 6: 2645–2657.

Solanki, P., S.H.K. Sharma, B. Akula and J. Reddy. 2017d. Sewage sludge and its impact on soil property. Environ. Ecol. 35(4C): 3186–3195.

Solanki, P., M. Narayan, S.S. Meena, R.K. Srivastava, M.L. Dotaniya and C.K. Dotaniya. 2018a. Phytobionts of wastewater and restitution. *In*: V. Kumar, M. Kumar and R. Prasad (eds), Phytobiont and ecosystem restitution. Springer, Singapore, pp. 231–243.

Solanki, P., M. Narayan, A.K. Rabha and R.K. Srivastava. 2018b. Assessment of cadmium scavenging potential of *Canna indica* L. Bull. Environ. Contam. Toxicol. 101(4): 446–450.

Solanki, P., A.K. Rabha, M. Narayan and R.K. Srivastava. 2018c. Relative comparison for phytoremediation potential of Canna and Pistia for wastewater recycling. Environ. Ecol. 36(1A): 316–320.

Solanki, P., M.L. Dotaniya, N. Khanna, S. Udayakumar, C.K. Dotaniya, S.S. Meena, et al. 2019. Phycoremediation of industrial effluents contaminated soils. *In*: J.S. Singh (ed.), New and Future Developments in Microbial Biotechnology and Bioengineering. Elsevier, Netherlands, pp. 245–258.

Song, Y.F., B.M. Wilke, X.Y. Song, P. Gong, Q.X. Zhou and G.F. Yang. 2006. Polycyclic aromatic hydrocarbons (PAHs), polychlorinated biphenyls (PCBs) and heavy metals (HMs) as well as their genotoxicity in soil after long-term wastewater irrigation. Chemosphere 65: 1859–1868.

Teodorescu, M. and C. Gaidau. 2008. Facts contradictions and possible improvement actions for hazardous wastewater management a case study. *In*: P. Hlavinek, O. Bonacci, J. Marsalek and I. Mahrikova (eds), Dangerous Pollutants (Xenobiotics) in Urban Water Cycle. Springer, Dordrecht, pp. 267–278.

Tove, A.L., C.A. Alfredo, I.L. Rik, M.M. Eggen and L. Judit. 2009. Source separation: Will we see a paradigm shift in wastewater handling? Environ. Sci. Technol. 43(16): 6121–6125.

Vasudevan, P., P. Griffin, A. Warren, A. Thapliyal, R.K. Srivastava and M. Tandon. 2011a. Localized domestic wastewater treatment: part II- irrigation potential in Indian scenario. J. Scien. Ind. Res. 70: 595–600.

Vasudevan, P., P. Griffin, A. Warren, A. Thapliyal and M. Tandon. 2011b. Localized domestic wastewater treatment: Part I—constructed wetlands (an overview). J. Scien. Ind. Res. 70: 583-594.

Velho, V.F., R.A. Mohedano, P.B. Filho and R.H.R. Costa. 2012. The viability of treated piggery wastewater for reuse in agricultural irrigation. Int. J. Recycl. Org. Waste Agricult. 1: 10–15.

Weyens, N., D. Van, D. Lelie, S. Taghavi and J. Vangronsveld. 2009. Phytoremediation: Plant-endophyte partnerships take the challenge. Curr. Opin. Biotechnol. 20: 248–254.

Zaidi, M.K. 2007. Environmental aspects of wastewater reuse. *In*: Zaidi, M.K. (ed), Wastewater Reuse-Risk Assessment Decision Making and Environmental Security NATO Science for Peace and Security Series. Springer, Dordrecht, pp. 357–366. doi: 101007/9781402060274_36.

Zayed, A., S. Gowthaman and N. Terry. 1998. Phytoremediation of trace elements by wetland plants: 1 Duckweed. J. Environ. Qual. 27: 715–721.

Zema, D.A., B. Bombino, S. Andiloro and S.M. Zimbone. 2012. Irrigation of energy crops with urban wastewater: Effects on biomass yields, soils and heating values. Agric. Water Manage. 115: 55–65.

Zhang, Z., C. Wang, J. Li, B. Wang, J. Wu, Y. Jiang, et al. 2014. Enhanced bioremediation of soil from Tianjin, China, contaminated with polybrominated diethyl ethers. Environ. Sci. Pollut. Res. 21: 14037–14046.

Zhou, P., J.C. Huang, A.W.F. Li and S. Wei. 1999. Heavy metal removal from wastewater in fluidized bed reactor. Water Res. 33: 1918–1924.

Zhuang, P., M.B. McBride, H. Xia, N. Li and Z. Li. 2009. Health risk from heavy metals via consumption of food crops in the vicinity of Dabaoshan mine, South China. Sci. Total Environ. 407(5): 1551–1561.

CHAPTER 19

Plant-microbe Interaction in Attenuation of the Toxic Waste in the Ecosystem

Mary Isabella Sonali and Veena Gayathri Krishnaswamy*

Department of Biotechnology, Stella Maris College, Chennai, Tamil Nadu, India.

1. INTRODUCTION

The microbiome present in soil interacts with the plant and forms an organization for the reduction of toxic waste present in the soil. The process aided by microorganisms is a slow and sometimes an incomplete process, which reduces the amount toxic substances; this process is called bioremediation. Endophytes are a group of plant-associated bacteria that have the potential to degrade the toxic substances/metabolite present in the soil which could lead to phytoremediation.

Bioaccumulation has led to a serious threat to the food chain. The contamination increases to several folds as it goes up the food chain. Another contamination that occurs in terrestrial environments is pesticide contamination. Polycyclic Aromatic Hydrocarbons (PAH), petroleum hydrocarbon, chemical fertilizers, and heavy metals are some of the common anthropogenic compounds present in the ecosystem, which contribute to environmental burden (Bhatia and Kumar 2011, Rhind 2009). The anthropogenic compounds usually present in the soil are mostly present in combined form, for example, PAH contamination along with heavy metals. Due to this, the nature of the soil varies in physico-chemical characters, mineralogy and as a result, they have different grades of fertility (Zhou and Song 2004, Ebrahimi et al. 2018).

During pedogenesis, the weathered soil consists of kaolinite (potassium), oxides and other additional nutrients that are suitable for agricultural use. The fertility of the soil is affected by several factors such as chemical and physical properties, climate and rainfall (Karmakar et al. 2016). A plant needs 18 essential nutrients along with water, air, light and optimum temperature

*Corresponding author: veenagayathri2018@gmail.com

for its growth in varying amounts. These essential nutrients are categorized into two different groups such as macronutrients (primary and secondary) and micronutrients, (Brady and Weil 1999). Tables 1 and 2 show the different roles of micronutrients and macronutrients pertaining to the plant group.

Table 1 The role of macronutrients in the plant growth

Macronutrient	Role
Nitrogen	Chlorophyll, Synthesis of protein
Phosphorus	Photosynthesis
Potassium	Enzyme activity, Synthesis of starch and sugar
Calcium	Cell growth and components of cell wall
Magnesium	Enzyme activation
Sulphur	Synthesis of amino acids and protein

Table 2 The role of micronutrients in the plant growth

Micronutrients	Role
Boron	Reproduction
Chlorine	Root growth
Copper	Enzyme activation
Iron	Photosynthesis
Manganese	Enzyme activation
Sodium	Water movement
Zinc	Components of auxin and enzymes
Molybdenum and Cobalt	Nitrogen fixation
Nickel	Nitrogen liberation
Silicon	Rigidity of cell wall

In recent years, various techniques have been used to clean the environment in order to maintain the fertility of soil and crop productivity. One of the common biological methods used is bioremediation, which involves microorganisms to degrade the toxic waste. The bioremediation technique gives a permanent solution and is cost-effective. The general component and characteristics of bioremediation have three aspects:

• Microbial systems
• Type of contaminant
• Geological and chemical condition at the contaminated site.

Bioremediation using microbes is effective and more satisfactory in today's world as it relies on the enzymes produced by the microorganism. Microorganisms such as bacteria, fungi, and algae react with the anthropogenic pollutants using their metabolites and break them into less toxic form (Dangi 2018). Another technique is plant-based remediation, which is termed as phytoremediation in which plants help in the removal of toxic compounds present in the soil contaminants. Phytoremediation involves techniques such as phytoextraction, phytostabilization, phytodegradation, phytovolatilization and rhizoremediation (Praveen 2019). Table 3 shows different types of phytoremediation processes that exist.

Microbial based phytoremediation is of two types: bacteria based and mycorrhizal fungi. Bacteria secrete some growth regulators and chelating agents that enhance the plant adaptability. AM enhances the tolerance to the host plant and heavy metal and affects absorption, transportation, and accumulation of heavy metals. The efficiency of the phytoremediation process can be increased by the addition of chelating, acidifying agents, bypassing electric current in the soil. The efficacy can also be increased for the GMOs by the application of organic chemicals and fertilizers and GMOs can exhibit high biodegradation capacity (Rostami and Abooalfazl 2019). Recombinant DNA technology is one of the evolving technologies in the recent years, which can be utilized for endophytic and rhizospheric bacteria, which are useful in the degradation of anthropogenic components (Bhatia and Kumar 2011).

Table 3 Different types of phytoremediation processes (Praveen 2019)

Types	Meaning	Contaminants	Reference
Phytoextraction	In this type of phytoremediation, the translocation of pollutants takes place initially in roots from soil or water to the aerial parts of the plant. This involves phytoextraction including: (1) chelate aided phytoextraction to mobilize and enhance the pollutants' uptake. (2) Continuous phytoextraction is based on the inherent potential of the plant	Heavy metals	Sekara et al. (2005), Salt et al. (1995)
Phytostabilisation (phytoimmobilization)	In this, plant stabilizes or immobilizes the contaminants in soil itself. Contaminants which get immobilized get adsorbed, precipitated with changes in the contaminants in the rhizosphere.	Heavy metals	Singh (2012), Yoon et al. (2006)
Phytodegradation	In phytodegradation, the anthropogenic compounds are degraded in the rhizospheric region through the microbial activities. It mainly constitutes plant enzymes which have the capacity to degrade or the organic compounds.	Fuel and solvents	Alkorta and Garbisu (2001)
Phytovolatilization	In phytovolatilization, the organic compounds and the heavy metals get absorbed in the rhizospheric region, which gets converted into volatile form and is released through the stomatal opening. It is a temporary solution as the volatile form of the substance can get back into the environment through precipitation.	Heavy metals like mercury and selenium	Ghosh and Singh (2005)
Rhizoremediation	In rhizoremediation, the breakdown of contaminants takes place in the rhizospheric region with the involvement of the microbes associated with the plants. The rhizospheric region is enriched with substances which have the capability to attract or stimulate the microbiome availability, which aids in the degradation of the pollutants.	Various types of contaminants (heavy metals, pesticides, petroleum products, fly ash, herbicides, etc.)	Kuiper et al. (2004)

Biodegradation is an attenuation of chemical substances by the biological process of the microorganism. It plays a key role in degrading the toxic compounds present in the soil and it happens at a high rate; hence, the toxicity can be reduced rapidly, while the anthropogenic compounds which are persistent will exhibit the toxic effect for a very long time. Biodegradation of soil and water depends upon the microbes present in the soil and many other environmental factors such as aerobic and anaerobic conditions. Bioremediation is divided into two categories: *in situ* and *ex situ* bioremediation. In *in situ* bioremediation, the process of degradation takes place at the contaminated site whereas in *ex situ* bioremediation the treatment process is carried out at a different place. The contaminants are excavated from the contaminated site and treated at another site. The *in situ* technique involves physical, chemical and biological methods and the large volume of contaminants can be treated. It is less tedious, less expensive and releases the pollutants in less amount compared to *ex situ* bioremediation, but it leaves behind a high concentration of residual at the same site. *Ex situ* techniques are easy to control, and the degradation process is very quick and various types of contaminants can be treated (Speight 2017a).

2. EXISTENCE OF ANTHROPOGENIC COMPOUNDS IN THE ECOSYSTEM

Pollution in the ecosystem increases day by day due to the constant increase of anthropogenic compounds. The bacterial metabolic pathways are usually interconnected with each other, and disturbance happens in the metabolic pathways due to globalization. The most commonly occurring anthropogenic compounds are PAH, PCB, BTEX, TCE, heavy metals, chemical fertilizers, and insecticides; along with these, there is an addition of carbon nanotubes, graphene and oxide nanoparticles, which creates negative impact on the environment causing risk to human health and disturbance in the biodiversity (Dangi 2018).

The chemical contaminants that are degraded can also be independent, which can make the environment suitable or adaptive to the chemical. The life cycle of the anthropogenic compound must not create any health hazard issue or environmental problems, disturbance to human beings it can impact the well-being of the human at the least adverse effect (Giebler et al. 2013). The existence of these anthropogenic compounds in the environment can cause toxicity and imbalance in the food chain and the web. The main focus of this heading is the effect of various anthropogenic compounds to the soil health and its impact on human health.

2.1 Synthetic Organic Compounds

The organic compounds and metals are hydrophobic and lipophilic and neither of the two can be degraded readily; they get accumulated in the environment and animal tissues as toxic compounds. This does not create a toxic effect at the concentration below which they are toxic. However, these anthropogenic compounds can create exerted development in animals at their early stages. It causes disturbance in the estrogen cycle in female animals, which disrupts the reproductive system which has a high possibility to create unbalance between the predator and prey (Rhind 2009). Few synthetic organic compounds are listed below:

2.1.1 Dioxins

Dioxins are found everywhere in the ecosystem. It is related to polychlorinated dibenzo-p-dioxins (PCDDs) Lorenzo et al. 2017. Dioxins are never intentionally produced, even though it causes a huge impact on human health causing diabetes, infertility, carcinoma and hormonal imbalance, and high levels of dioxins lead to chloracne (acne-like lesions in the upper parts of the body). The most commonly found dioxins are 2, 3, 7, 8-tetrachlorodibenzo-p-dioxin (TCDD). The origin of dioxins is from the chlorine which is released during the bleaching of pulp, combustion process and other synthesis related to chlorine. Dioxin is a chemical that consists of two carbon rings, which are chlorinated and held together by a double bond of oxygen atoms, and like most other synthetic organochlorines, it is extremely persistent in the environment. It is believed that inhabitants of the United States are carriers of dioxins linked cancer (Houlihan 2005).

2.1.2 Pesticides

Organochlorine pesticides (OCPs) have been categorized into persistent organic pollutants (POPs), bioaccumulative, and toxic. Pesticides comprise of insecticides, herbicides, weedicides, molluscicides, and fungicides. Higher levels of pesticides can cause respiratory tract and reproductive system related diseases, which are more prominent in humans (Seung-Kyu 2019).

2.1.3 Polycyclic Aromatic Hydrocarbons (PAHs)

PAH are the compounds which contain aromatic rings above two. They only contain compounds of carbon and hydrogen and are made up of multiple organic rings with certain modifications in the electron delocalization. The simplest form of polyaromatic hydrocarbons is the naphthalene (two aromatic rings), followed by phenanthrene (three aromatic rings). Apart from these, other compounds linked with PAH are phenanthrene, fluorine, 4-chlorophenol, naphthalene nonylphenol, phenol and 4-tert-octylphenol (Fig. 1). PAHs are mostly described as carcinogens and mutagens. In general, PAH is stable and recalcitrant in the soil and less easy to degrade than many other organic compounds (Seo et al. 2009).

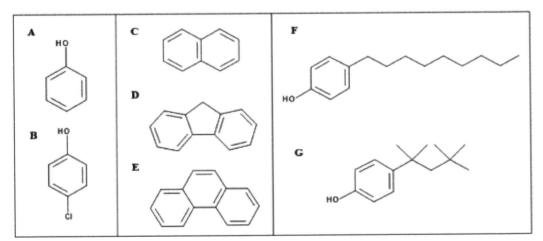

Figure 1   Structure of A) Phenol B) 4-chloro phenol C) Naphthalene D) Fluorene E) Phenanthrene F) Nonylphenol G) 4-tert-octylphenol (Chang 2019).

2.1.4 Phthalates

Phthalates are extensively used as plasticizers, which can increase their flexibility, longevity, durability and reduce their opacity. It is usually esters of phthalic acid. They are mainly blended with polyvinyl chloride (PVC) to make its nature soft. The phthalates give the fragrance of new car interiors. Apart from plasticizers, phthalates are also used in the food industry about their high content of fat such as cheese, milk, and margarine. It is reported that the phthalates have high potential to induce disorders in the reproductive system (Kay et al. 2013).

2.1.5 Heavy Metals

Metals with a density of more than 5 g/cm^3 are generally categorized as heavy metals that have high density and atomic weight as well as atomic numbers. Copper, chromium, mercury, lead, nickel, and zinc are often considered to be heavy metals and cause heavy metal contamination.

The heavy metals to an extent can cause deformity in the components of living cells such as nuclei, lysosome, endoplasmic reticulum and few enzymes which are involved in major processes like metabolism, repairing system and detoxification and can also interfere with the DNA synthesis and nuclear protein, which can cause deformity in the DNA strands and causes carcinogenesis leading to apoptosis (Beyersmann and Hartwig 2008).

2.2 Inorganic Compounds

These compounds are more frequently found to affect the sites which are affected by the anthropogenic compounds and are made up of possibly toxic metals and metalloids, e.g. barium, nickel, vanadium, cadmium, and arsenic (Vargas-García et al. 2012, Li et al. 2017, Ye et al. 2017). Any form of energy emissions also comes under inorganic pollution; it has a higher tendency to get into the environment by its dispersion and recycling capability. Table 4 shows the sources of the inorganic contaminants.

Table 4   Various sources of inorganic contaminants

Inorganic contaminants	*Major sources*
Carbon dioxide and carbon monooxide	Combustion emissions
Hydroxyl and hydrogen peroxide	Photochemistry
Mineral and soil dust	Dust emission
Nitric acid	Gas phase chemistry
Radionuclides	Nuclear accidents
Volcanic ash and sulphur dioxide	Volcanic emission
Mercury and fly ash	Combustion emissions
Ammonium salts, nitrate salts and nitrite salts	Farms and factories

Inorganic pollution usually occurs in the place which is incapable of degrading the anthropogenic pollutants and personnel working with inorganic compounds have a lack of knowledge of how it should be treated and disposed of. There is an inorganic chemical pollutant that is persistently found in the environment, and can enter the ecosystem via gas, liquid or solid phase. It becomes remotely available and gets lodged and redeposited at different regions or proximities (Speight 2017b). Combustion of fossil fuels can also emit inorganic pollutants such as crude oil, natural gases and different types of fuel and coal (Speight 2014, 2017b)

2.2.1   Arsenic

Arsenic and arsenic-related compounds (trivalent arsenite and pentavalent arsenate) are found at low concentrations everywhere in the environmental matrices. The organic form of the arsenic is often the methylated metabolites (MMA, DMA and trimethylarsine oxide). Arsenic enters the environment from natural phenomena such as soil erosion and volcanic eruptions and majorly due to anthropogenic activities. It is formulated along with agricultural compounds, wood preservatives, and dye industry. It is also used in animal husbandry to eradicate tapeworms in sheep and cattle. It causes apoptosis and leukemia and inhibits DNA repair causing chromosomal aberrations in human cells as well as rodents' cells (Yedjou and Tchounwou 2007, Patlolla and Tchounwou 2005).

2.2.2   Chromium

Chromium(Cr) naturally occurs in two forms: chromium(II) and chromium(VI). Cr is released mainly from refractory, metal and chemical industries. Chromium is mostly released in the hexavalent form, which is a man-made form usually found in the environment due to extensive anthropogenic activity and is classified as a carcinogen. The toxicity of chromium is classified from low toxicity to high toxicity depending upon its valence state (Velma et al. 2009). Inhaling high levels of hexavalent chromium can irritate the nose lining and cause nose ulcers in animals. It can cause diseases related to the gastrointestinal tract, respiratory tract, kidney inflammation and damage in the male reproductive system (Hegazy et al. 2016).

2.2.3   Lead

Lead occurs in the natural environment due to anthropogenic activities such as combustion of fossils and ore mining, which contribute to a high level of lead in the ecological system (Gabby 2003). Lead gains entry into the human system as dust particles or aerosols via inhalation and ingestion of food contaminated with lead can cause bioaccumulation. The physiological status of humans can influence the absorption capacity of lead; the highest bioaccumulation takes place in the kidney, liver and soft tissues like brain and heart (Flora 2008). Accumulation of lead in the central nervous system (CNS) can lead to symptoms like headache, irritability, memory loss, dullness and poor attention span (Flora et al. 2012).

3.   FACTORS AFFECTING SOIL FERTILITY AND PLANT GROWTH

Among all the biotic and abiotic factors, the growth and variety of plant species is mainly dependent on the microbial communities of the soil ecosystem (Fierer 2017). The bioavailability of carbon, nitrogen and phosphorus can alter the traits of plant and is also interdependent on the microbiome of the soil (De Deyn et al. 2008). The soil which is contaminated with the anthropogenic activity such as heavy metals and ore mining land constitute an altered plant species diversity (Borymski et al. 2018). In mine tailing, the plant traits (life span, photosynthesis and morphological structures) are associated with the microbiome mediated process (Grigulis et al. 2013, Legay et al. 2017, Navarro et al. 2018). The plant microbe interaction is the driving component/system for the communities of plant; however, it is still not identified whether a plant-soil feedback mechanism is present. It is understood that the feedback mechanism can communicate the biotic and abiotic components in the ecosystem, availability of nutrients in the soil and the different plant species. Accessibility of nutrients for the plant depends on their physical structure and their chemical properties and the nutrient acquisition is believed to be an important driver of PSF (Guochen et al. 2018).

3.1   Soil Temperature

The temperature of the soil is the key determinant of physical, chemical and physiological reactions and, most importantly, for the activity of soil microbes present in the soil. The soil temperature directly affects the physiology of the microbes and indirectly exerts factors such as nutrient and substrate diffusion and water activity. The next factor is soil moisture about 80% of the net radiation is used to evaporate water, 5% for photosynthesis and 15% tends to warm the soil. The quality and the quantity of the rhizosphere also depends upon the temperature. The microbes present in the soil are classified according to the temperature in Table 5.

Table 5   Classification of bacteria based on temperature

Environmental class	Temperature range (°)	Optimum growth
Psychrophile	−5–20	15
Mesophile	15–45	37
Thermophile	40–70	60
Hyperthermophile	65–95	85

3.2 Climate

Climate comprises two key features, temperature and rainfall, which indirectly affect the soil fertility and its productivity. The climate varies from region to region and thus the soil fertility, as the amount of moisture content in the soil also varies. Temperature also affects plant diversity as it also influences the microbiome of the soil and the organic matter content. The arid region usually lacks the topsoil, which provides mechanical support to the plant; hence, the arid region has sparse plant diversity (Sherchan et al. 2005). Apart from temperature and rainfall, greenhouse gases also alter the vegetation of the plant. Greenhouse gases are usually emitted during the combustion of fuel, industrialization, urbanization, and deforestation. These greenhouse gases get into the atmosphere, cause a drastic decrease in plant diversity and affect the temperature of the environment (Pathak et al. 2003).

3.3 Soil Texture

Soil texture refers to the various particles or granules present in the soil. It consists of various proportions of granule particles with varying size present in the soil. Based on the size, the granules are classified into different types (Rowell 1994). The different sizes of granules are listed in Table 6.

Table 6 Different particles with regard to their size

Particle	Size
Clay	<0,002mm
Slit	0–0.5mm
Sand	0.05–2mm

Clay, which is of diameter less than 0.002 mm, has the highest water holding and nutrient acquisition capacity. Soil with 30% of clay is a favourable condition for the plant growth, when other factors are also favourable.

3.4 Water Retention Capacity

The capacity of the soil to retain the water content and the capability of the plant for the uptake of water from the soil is called water retention capacity. The water accessibility for the plant is dependent on the depth of the root and the total water content between field capacity and wilting percentage of each layer invented by the root (Brady and Weil 1999). The water holding capacity of the soil depends on the organic matter, soil texture and structure and bulk density. These characters can affect the root depth.

3.5 Soil pH

Soil pH is an important and key property for the presence of microbial communities (Bohn 2001). Soil pH varies from acidity to alkalinity on a pH scale of 0 to 14, whereas pH 7 is neutral. The values above the neutral areas are alkaline and the value below the neutral pH is acidic pH. However, it has been experimented and proved that the plants can tolerate pH ranging from 5.5–6.5, which is the middle pH range (Blake and Goulding 2002). The availability of nutrients like calcium, magnesium, potassium, nitrogen, and sulphur, which are macronutrients, is generally found in strong acidic pH. On the other hand, ferrous, manganese, zinc, copper and cobalt are found in the soil which has low pH but higher concentration can be toxic to the plants (Bohn 2001).

4. ROLE OF MICROBES IN THE DEGRADATION OF TOXIC WASTE

An increase in anthropogenic activities leading to environmental contamination has gained awareness worldwide. Bioremediation is one of the safer methods which could create a cleaner environment. It is mainly chosen because it is cost-effective for the removal of contaminants from the contaminated sites which contain an extensive range of pollutants. Nature's innate recycling mechanism involves certain types of microorganisms, which have the potential to bio transform high toxic metals into a less toxic form, which is utilized as their energy source. Removing the pollutants from the environment in a workable way is the main idea for choosing bioremediation; the role of microorganism in biodegradation of contaminants has intensified in recent years. Reduction of the pollutant toxicity present in the aquatic and land environment is carried out using biological organisms such as fungi and bacteria, which also play a key role in promoting the plant growth in the contaminated sites. Certain plant growth-promoting rhizomicrobes (PGPR) including *bacillus, pseudomonas, mycobacterium*, etc. help in augmented nutrient uptake, along with higher phosphate and nitrogen content of the plants. These rhizobacteria may also lead to metal mobilization and increase metal uptake by some plant species leading to microbe assisted phytoremediation of an environmentally polluted site. Microorganisms act as fairy-tale in the bioremediation of contaminated sites and for degradation purposes (Banerjee et al. 2018).

During pedogenesis, the weathered materials contain heavy metals which are considered as natural pollutants, thus causing risks to human health, plants, animals and the ecosystem. Heavy metal industries landfill and lead-based paints are the main source for heavy metal contamination. On the other hand, application of fertilizers, animal-based manures, biosolids (sewage sludge), different compost measures, pesticides, coal combustion residues in the form of fly ash, effluent discharge from petrochemical industries, etc. can lead to increase in the pollutants in the soil (Roychowdhury et al. 2019). The biological method is extensively used as it is cost-effective and environmentally friendly and it brings a sustainable way for the eradication of the contaminants (Banerjee et al. 2018).

4.1 Microorganism and Pollutants

Rhizopheric and endophytic microorganisms are found in rhizospheres and the stem, which effectively degrade anthropogenic compounds present in the contaminated soil. The most commonly and efficiently used methods are composting, bioremediation, biodegradation, and bio-transformation. Microorganisms used in the degradation are *Chorella vulgaris, Corynebacterium* sps., *Scenedesmus platydiscus, Streptoccoccocus., Bacillus* sps., *Staphylococcus* sps., etc. Biodegradation of hydrocarbons takes place in the presence of oxygen (aerobic) as well as in the absence of oxygen (anaerobic) but effective degradation takes place in the absence of oxygen (anaerobic) (Mondal and Palit 2019). Plant growth-promoting rhizobacteria are found in the rhizospheres and they help in plant growth stimulation (Saharan and Nehra 2011). They also help in phytoremediation of heavy metals (Glick 2010). Rhizobacteria can uptake the contamination from the soil, especially heavy metals involving organic acid and biosurfactant production, which help to minimize the toxicity in the root region (Wu et al. 2006). Table 7 shows different bacterial strains which could degrade different pollutants.

Fungi also have the ability to degrade like bacteria and the organic matter (OM). Fungi can thrive at low pH and low moisture, which are suitable conditions for degradation of organic matter. Fungi are considered to degrade natural polymeric compounds, most efficiently with the production of multienzymes complexes by the extracellular membranes. Table 8 shows the different fungal strains which could degrade the different pollutants.

Table 7   Bacteria and the pollutants' degradation

Microorganism	Pollutant	Reference
P. alcaligenes P. mendocina and *P. putida*, *P. veronii*, *Achromobacter*, *Flavobacterium*, *Acinetobacter*, *Species of Staphylococcus*, *Shigella*, *Escherichia*, *Klebsiella* and *Enterobacter*.	Hydrocarbon	Safiyanu et al. (2015)
Species of Bacillus, *Stenotrophomonas* and *Staphylococcus*	Dichlorodiphenyltrichloroethane (DDT)	
Pseudomonas putida	Benzene and xylene	Safiyanu et al. (2015)
Bacillus coagulans, *Pseudomonas cepacia*, *Serratia ficaria* and *Citrobacter koseri*.	Crude oil	Kehinde and Isaac (2016)
Listeria denitrificans and *Nocardia atlantica*	Dyes from textile industries	Hassan et al. (2013)
Species of Rhodococcus	Polychlorinated biphenyl	Rajan (2005)
Pseudomonas sps., *Micrococcus luteus* and *Proteus vulgaris*	Plastics	Priyanka and Archana (2011)

Table 8   Fungi and the pollutants' degradation

Fungi	Pollutants	Reference
Trichosporon cutaneum	PAH	Mörtberg and Neujahr (1985)
Candida tropicalis, *Candida lipolytica* and *Candida ernobii*	Diesel	De Cássia et al. (2007)
Candida methanosorbosa	Aniline	Mucha et al. (2010)
Microfungi	PAH, biphenyl and pesticides	Fritsche and Hofrichter (2008)

Table 9 shows the different algal strains which could degrade the different pollutants.

Table 9   Algae and the Pollutant

Algae	Pollutants	Reference
Prototheca zopfi	Aromatic compounds	Walker et al. (1975)
Scenedesmus platydiscus, *S. quadricauda* and *Chlorella vulgaris*	PAH	Wang and Chen (2006)

Certain plant growth promoting rhizobacteria PGPR) including Bacillus, Pseudomonas, Mycobacterium, etc. help in increased nutrient uptake, along with higher N and P content of the plants. These rhizomicrobes may also lead to mobilization of the metal and increase metal uptake by some plant species leading to microbe assisted phytoremediation of an environmentally polluted site (Roychowdhury et al. 2019).

5.   TYPES OF BIODEGRADATION

5.1   Biodegradation of Xenobiotic Compounds

Anthropogenic inorganic and organic pollutants are discrete throughout the atmosphere and in different spheres of the atmosphere, which tend to transform into another compound that may be toxic, less toxic and not toxic to flora and fauna. Xenobiotics are man-made or artificial compounds present in the ecosystem. Xenobiotics cannot be recognized by the naturally occurring microbes and therefore do not enter the common metabolic pathway. The main degrading microbes are bacteria and fungi. The selection of xenobiotics degrading bacteria and their adaptation to the xenobiotic contaminated environment is the key factor for detoxification of the environment. Xenobiotics gain entry into the ecosystem by i) pharmaceutical industry

ii) chemical industry iii) bleaching and paper industry, which expels chlorinated organic compounds into the ecosystem iv) mining industry, and v) intensive agriculture (synthetic fertilizers, herbicides, and pesticides). Mostly bacillus is involved in the degradation processes, namely Actinomycetes, *Pseudomonas, Nocardia, Klebsiella, Azotobacter* and *Flavobacterium*, and fungi species such as *Candida, Thielavia, Penicillium, Thermomyces* and *Ganoderma*. Microbes possess the ability to degrade polycyclic aromatic hydrocarbons (PAHs), which are a persistent constituent of petrochemical wastes, organonitrogen compounds such as nitrotoluenes and chlorinated compounds like pentachlorophenol, polychlorinated biphenyl, and chlorinated dioxin-like compounds. A single bacterium can have the ability to degrade one or more compounds; it may also possess specialized enzymes and metabolic pathways such as chlorobiphenyls and chlorobenzenes (Singh 2017). The most common pathway involved in the degradation of polyaromatic compounds is the β-ketoadipate pathway, which provides utilization of primary substrates which is present in both bacteria and fungi. Primary substrates are converted into protocatechuic acid catechols; finally, the end products obtained by this pathway are two aliphatic products such as succinate and acetyl-Co A (Ghosal et al. 2016).

5.2 Biodegradation of Plastics

Plastic is one of the recalcitrant compounds present in the ecosystem as it comprises of 80% of plastics found in agricultural land, landfills and water bodies (Rummel et al. 2017). Microorganisms can degrade plastics through the production of enzymes by degrading the long polymers, further these degraded polymers act as carbon and energy sources of microbes Enzymes produced by the microorganisms act on polyethylene terephthalate (PET) and polyurethane (PUR). Enzyme PETase produced by the microorganism's hydrolyze the plastic into monomers (Danso et al. 2019). There is a wide range of plastics to be degraded; fungi, which could degrade PHB polyesters, are *Fusarium, Penicillium, Cryptococcus, Aspergillus,* and *Rhizopus*. Polyethylene adipate is biodegraded by *Penicillum* and *Aspergillus*. PLA (polylactic acid) is degraded by *Tritirachium album* and *Penicillum roqueforti* (Ghosh et al. 2013). *Aspergillus niger* has the ability to degrade plastics made up of polyvinyl alcohol (Mogil'nitskii et al. 1987). Styrene degradation is observed in *Pseudomonas, Xanthobacter, Rhodococcus* and *Corynebacterium* (Danso et al. 2019).

5.3 Aerobic Degradation

In aerobic degradation, oxygen is supplemented to the soil to increase the vitality of the indigenous bacterial strains as it is considered to be the growth limiting factor for the bacterial strains, which could degrade hydrocarbon. By adding dissolved oxygen, biodegradation is accelerated up to 10–100 times. Dissolved oxygen, which is obtained from the natural sources, gets exhausted quickly in the presence of petroleum hydrocarbons; thus, it is untreated and oxygen depleted aquifers are slow. Low to moderate levels of contaminants can be treated. The most commonly treated compounds are BTEX, PAHs, TPH, MTBE and TBA (McGregor and Vakili 2019). In cellular respiration, oxygen gets oxidized to sugars and fats to obtain energy. Prior to this process, the glucose molecules get broken into small molecules, which enter mitochondria to take part in aerobic respiration. With the help of oxygen, the small molecules are broken down into water, CO_2 and energy (polimernet.com). Bacteria that are involved in degradation are various species of *Pesudomonas, Rhodococcus, Mycobacterium, Burkhelderia* and *Alcaligens eutrophus* (Srivastava and Kumar 2019).

5.4 Anaerobic Degradation

Anaerobic degradation takes place in the absence of oxygen and, more predominantly, when anaerobes are dominant over aerobes. Landfill biodegradation takes in the anaerobic condition through the digestion process. This is majorly used in the wastewater treatment of sludge as it reduces the large volume of input material. It reduces the risk of landfill gas getting emitted into the environment. It produces fertilizers and renewable energy such as methane and CO_2, which can be utilized for the production of biogas. Bacteria hydrolyze to produce carbohydrates to utilize them as a food source. Acetogenic bacteria are a group of bacteria that can convert proteins (amino acids) into hydrogen, ammonia, carbon dioxide, and organic acid; further, it is converted into acetic acid. Methanogens are a group of bacteria that can produce methane and CO_2 utilizing the products obtained from acetogenins. *E. coli* can take part in aerobic, anaerobic and fermentative respiration using fumarate and nitrites as electron acceptors (Sims and Kanissery 2019).

6. PHYTOREMEDIATION: A GREEN TECHNOLOGY —AN OVERVIEW

Soil pollution is a threat around the globe, which leads to human illness, soil infertility, crop productivity, biodiversity misshape and loss of natural resources. Plants are involved in cleaning up pollutants from the environment by an association with rhizospheric microbes, which is termed as phytoremediation. It is an environmentally friendly technique that provides a sustainable outcome. It can be applied to small to large contaminated sites for phytoremediation of the soil (Pandey et al. 2019). The high concentration of heavy metals was first found to be accumulated in the leaves of *Viola calaminaria* and *Thlaspi caerulescens* (Baumann 1885). Fragrant flowers that are inedible and which can tolerate stress conditions are used in phytoremediation as they do not possess a vegetative body that could affect the food chain, which belongs to the families of Lamiaceae, Geraniaceae, Poaceae, and Asteraceae. It is mostly used in industries that manufacture perfumes, cosmetics, toiletries and insect repellents (Pandey and Singh 2015). *Calamagrostis epigejos,* which is an aromatic wild grass of Serbia, is used in *in situ* phytoremediation of fly ash (Mitrovic et al. 2008); in India, *Cynodon dactylon* and *Vetriveria zizanioides* are used to phytoremediate the heavy metals (Pandey et al. 2015, Das et al. 2013). In mining, contaminated soil *Cymbopogon flexuosus* and *Vetriveria zizanioides* are used in the phytoremediation technique (Srivastava et al. 2014). *V. zianioides* is used in the phytoremediation of asbestos mining waste dumps (Kumar and Maiti 2015). *Lavandula vera* is used for phytoremediation of heavy metals like lead, cadmium and zinc (Angelova et al. 2015).

6.1 Mechanism of Phytoremediation Process

The mechanism of phytoremediation was reported in PAH, pesticides, toluene, PCB, and benzene in various ways such as the interaction of microbes in association with the rhizosphere, phytodegradation, and phytoremediation of organic pollutants in the root region (Stephenson and Black 2014). In phytoremediation, the pollutants are reduced, catabolized and converted into less toxic form. It is well illustrated by "green liver model", which explains how the process of phytoremediation takes place with the help of enzymes at molecular level, but still it is a research subject which is used to understand the xenobiotics transformation and manipulation (Sandermann 1994). Since the last decade, phytoremediation is a hot topic and its molecular and biological mechanisms and the various strategies used in improving phytoremediation with relevance to engineering techniques are still under research. Phytoremediation is not a new technique, it is an upcoming technique.

6.2 Types of Phytoremediation

In general, phytoremediation includes six methods: phytoextraction, phytostabilization, phytodegradation, phytovolatilizationan, and rhizoremediation. It is a new technique which is very cost effective in remediating the areas which are contaminated into less toxic form. Figure 2 shows different pytoremediation that takes place in the plant.

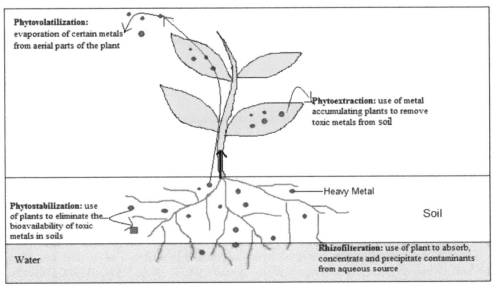

Figure 2   Types of phytoremediation (Adapted from Kushwaha et al. 2015).

6.3 Phytosequestration

Phytosequestration is also termed as phytostabilization. In this process, the remediation takes place in the root surface or plant exudates are formed, which get released into the root which is present closer and the contaminants get sequestered, immobilized or precipitated. Plant varieties that are tolerant of metal can be used to decrease the contaminations. This technique is suitable for the remediation of cadmium, zinc, chromium, copper, and arsenic. In extremely acidic metal contaminated soil, it has been investigated that using perennial ryegrass (*Lolium perenne*), soil heavy metals can be sequestered. The plant which is chosen for phytosequestration must possess tolerance against contaminants, high ability of root biomass and high retention root capacity. The most common species used are *Juncus* sps., Lavandula luisierra and Rumex induratus (Anawar et al. 2011).

6.4 Phytoextraction

Phytoextraction or immobilization is employed to remove heavy metal contamination from the soil. These biological techniques have fewer side effects than physical and chemical techniques. It is the main and most promising technique as both *in situ* or *ex situ* treatment for the removal of contaminated soils, sediments, and water. It is best suitable for the removal of hydrocarbons, heavy metals, and radionuclides. The characteristics of a plant that is chosen for phytoextraction includes the ability of high accumulation of contaminants, high biomass, and high growth rate. Also, hyperaccumulators like *Thlaspi caerulescens* (Brassicaceae family), *Pteris vittata*, *Noccaea caerulescens* and *Arabidopsis halleri* accumulate heavy metals effectively (Panesar et al. 2019). Phytoextraction is predominantly used to remove the heavy metal contaminations.

It is reported that *Sedum alfredii* and *Alyssum bertolonii* can accumulate high levels of cadimum and nickel (Deng et al. 2007, Kramer 2010).

6.5 Phytovolatilization

Plants that are involved in phytovolatilization can absorb the volatile compounds via their roots and transpire it on the metabolites produced via their leaves. Poplar trees have been shown to volatilize 90% of TCE from the roots. Heavy metals are reported to be volatilized into gas form by *Arabidopsis thaliana* and *Brassica juncea* (Ghosh and Singh 2005). The most common plants used in phytovolatilization are *Salix* and *Populus*.

6.6 Phytodegradation

Phytodegradation is used for the degradation of contaminants in the soil and underground water. This process doesn't need microorganisms and it is the most advantageous aspect of this method. Plant tissue takes up the contaminants present in the soil. It gets metabolized (using enzymes) or biotransformed. It can occur in any part of the plant such as stem, root, and leaves. It is used to degrade insecticides, PCBs and herbicides. Cannas can detoxify xenobiotics (Solanki et al. 2018).

6.7 Rhizodegradation

In rhizodegradation, the hydrocarbon are broken down by the microorganisms present in the root. Secondary metabolites are produced which aid in the breakdown of contaminants present in the soil. Rhizodegradation is employed to degrade TCE, TCA, PAH, BTEX, insecticides, PCB, toulene and PCP (Aybar et al. 2015). The common plants used in rhizodegradation are *Typha latifolia, Medicago sativa* and *Morus rubra* (Panesar et al. 2019).

6.8 Rhizofiltration

Contaminants hold on tightly to the roots of the plants or get absorbed in the root in the rhizofiltration method. This method is effectively used to remove radionuclides present in the groundwater and wastewater contaminations. It can also be used to remove heavy metals like copper, zinc, lead, and chromium from the soil. Sunflower is reported to rhizofiltrate radioactive compounds (Panesar et al. 2019). Apart from plants, bacteria can also be used in the removal of selenium such as *Stenotrophomonas maltophilia* and *Bacillus mycoides* (Vallini et al. 2005).

7. EFFECTIVE PLANT GROWTH REGULATORS IN IMPROVING PHYTOREMEDIATION

Plant growth regulators are efficient in the removal of auxins, cytokinins, gibberellins, and salicylic acid. It is influenced by environmental stress and the physiological state of the plant. Each plant growth regulator has a significant role on the plant – auxins help in cellular growth (Ali et al. 2013, Israr and Sahi 2008), and cytokinins increases the adsorption process of the heavy metals (Cassina et al. 2011). Cytokinins reduce the heavy metals' side effect in sunflowers and improve the plant to uptake the water and contaminants from the soil (Tassi et al. 2008). Gibberellin also acts effectively in water stress plants (Tuna et al. 2008). A combination of Ca and gibberellin increases the enzyme in a plant which aids in higher chlorophyll content, which can reduce the negative effects of Ni toxicity (Siddiqui et al. 2011). Salicylic acid brings down the copper toxicity of the plant, which leads to Cu phytoremediation (Afrousheh et al. 2015).

In *Helianthus annus*, cytokinins help in the accumulation of heavy metals (lead and zinc) in the stem and leaves, which can also contribute to increase in transpiration (Rostami and Azhdarpoor 2019, Tassi et al. 2008). Jasmonates acts as a messenger molecule to activate pathways involved in heavy metal stress condition in plants (Poschenrieder et al. 2008).

8. TECHNIQUES TO BOOST PHYTOREMEDIATION EFFICACY

Techniques are needed to improve phytoremediation efficiency to remediate high levels of accumulation. Ethylene diamine tetra acetic Acid (EDTA) is a chelating agent which increases the solubility of heavy metals which could increase the phytoremediation capacity of the plant (Bareen 2012, Nowack et al. 2006). The chelating agent is used in tough condition and when the accessibility of metal content is low in the soil, it helps in the absorption of metals. EDTA acts in the liquid form forming a soil-metal complex that transmits heavy metals to the aerial parts of the plant (Lasat 2002, Farid et al. 2013). The application of an electric field in the soil increases the heavy metal absorption capacity of the root (Kim et al. 2005). But it causes negative effects on the plant such as a change in the genetic material and physical characters. Apart from an external application, pH also contributes to the metal solubility, and low pH increases the solubility of metal in the soil. GMOs are also used in the phytoremediation of heavy metals, which could increase the efficacy of absorption. Bacteria can also degrade the contaminants present in the soil with the production of enzymes, and heavy metals are degraded by the production of chelating and acidifying agents by the bacteria which could possess high surface volume, enabling them to be used as a microbial chelating agent. Bacteria can also influence the pH of the soil, making it favourable for the degradation of the contaminants (Rostami and Azhdarpoor 2019). Plant growth-promoting hormones such as IAA, jasmonates, ethylene, gibberellins, and cytokinins can also increase the efficiency of phytoremediation with minimal side effects of the pollutants present in the plant.

9. ADVANCES IN PHYTOREMEDIATION TECHNOLOGY

Phytoremediation is a thriving field of lively research with its novelty and cost-effectiveness; being eco-friendly, aesthetic, and efficient, phytoremediation can be applied onsite, with remediation by solar-driven technology (Ali et al. 2013). Many researchers have used plant-assisted biosolids or microbes to enhance the efficacy of the phytoremediation process in removing heavy metals from contaminated sites (Kim et al. 2010, Rajkumar et al. 2012). Natural plants and genetically engineered plants with positive outcomes improve the remediation of heavy metals from contaminated soil (Chanu and Gupta 2016, Gomes et al. 2016). The analysis of the plants' ability to tolerate and to accumulate metals and polyaromatic hydrocarbons (PAH), including their bioconcentration factors and correlations between pollutant types, suggests that *Pteris vittata* and *Pteris cretica* are suitable for the joint remediation of arsenic and PAHs; *Boehmeria nivea* for the simultaneous phytoremediation of lead, arsenic, and PAHs; and *Miscanthus floridulus* for the phytoremediation of copper and PAHs pollution. *Ricinus communis* plant is used for phytoremediation of metal-polluted sites as well as in phytoremediation of fly ash (Olivares et al. 2013, Pandey 2013). *Eucalyptus globulus* was found to phytoremediate cadmium polluted soil, which was subjected to nearly 30 years of e-waste disposal (Luo et al. 2015). The potential application of plant enzymes to enhance phytoremediation is greatly recognized by current world biotechnologists. Some of the challenges of phytoremediation, however, could be overcome through the development of the genetically engineered plants as well as endophytic microbes equipped with the property of overexpressing the enzymes competent of contaminant degradation (Tripathi et al. 2020).

10. REMEDIATION METHODS

Environment pollution has become a big threat, and bioremediation can play a vital role in cleaning up the contaminated site. Bioremediation is a process of remediating the environment with the help of microbes. Remediation strategies, such as chemical and physical methods, are not enough to diminish pollution problems because of the continuous production of novel recalcitrant pollutants due to anthropogenic activities. As physical and chemical approaches are tedious, bioremediation using microbes is acceptable and it is an eco-friendly and socially acceptable alternative to conventional remediation approaches. Many microorganisms with a bioremediation potential have been isolated and characterized but, in many cases, cannot completely degrade the targeted pollutant or are ineffective in situations with mixed contaminations (Dangi et al. 2018). Plants and microbes including bacteria, fungi, and algae have a high neutralizing ability in remediating the soil. Bioremediation using microbes is more acceptable, which mainly relies on the enzymes produced by them and takes part in the metabolic pathways. These microorganisms attack the contaminant and degrade them completely or convert them into less harmful products (Sivaperumal et al. 2017). Bioremediation using microorganisms has resulted in successes as well as failures. In the cases of failure of this method, the causes are generally time-consuming, low adaptability, lack of competitiveness of the microbes, and low bioavailability to the target pollutants (Singh et al. 2017). Autochthonous (indigenous) microorganisms present in the contaminated environments play a vital role in solving most of the challenges associated with biodegradation and bioremediation of contaminated sites (Verma and Jaiswal 2016). Bioremediation techniques can be categorized as *ex situ* techniques and *in situ* techniques. Pollutant nature, depth and degree of pollution, type of environment, location, cost, and environmental policies are some of the selection criteria that are considered while choosing any bioremediation technique (Frutos et al. 2012, Smith et al. 2015).

10.1 *Ex situ* Bioremediation Technique

Ex situ bioremediation is a process where contaminated soil or water is excavated from the environment and subsequently transporting them to another site for treatment. *Ex situ* bioremediation can use bioreactors and nutrients can be added to speed up the breakdown of anthropogenic pollutants (Christopher 2016).

10.1.1 Biopile

Biopile-mediated bioremediation involves the piling of excavated polluted soil, followed by nutrient amendment, and sometimes aeration is provided to enhance the bioremediation process by increasing microbial activities. The basic pile system consists of a treatment bed, an aeration unit, an irrigation/nutrient unit, and a leachate collection unit (Fig. 3). The advantage of this particular *ex situ* technique is increasingly being well-thought-out due to its constructive features including cost-effectiveness, which enables effective biodegradation on the condition that nutrient, temperature, and aeration are adequately provided and controlled (Whelan et al. 2015).

   The suppleness of biopile allows remediation time to be shortened as the heating system can be integrated into biopile design to increase microbial activities and contaminant availability thus increasing the rate of biodegradation (Aislabie et al. 2006). Furthermore, heated air can be injected into biopile design to deliver air and heat in cycles to facilitate enhanced bioremediation. It is also studied that humidified biopile had a very low final total petroleum hydrocarbon (TPH) concentration compared to heated and passive biopiles as a result of optimal moisture content, reduced leaching, and negligible volatilization of less degradable contaminants (Whelan et al. 2015). Soil biopiles can be up to 20 feet high. Biopile can be covered with plastic to control leachate runoff, evaporation leachate and volatilization of compounds as well as to enhance solar heating. Treatment time of biopile is typically 3 to 6 months, after which the excavated soil material is either returned to its original location or disposed of. Biopile treatment has been applied in the treatment of non-chlorinated VOCs, fuel-contaminated soil, chlorinated VOCs, semi-volatile organic compounds (SVOCs), and pesticides can also be treated, but effectiveness against each varies.

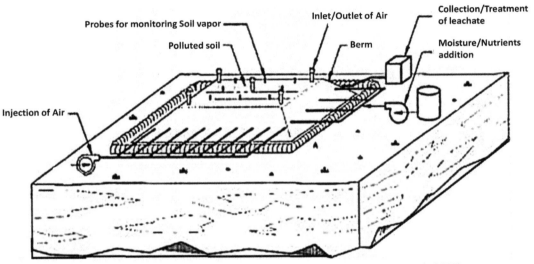

Figure 3 Construction of biopile (adopted from Menendez-Vega et al. 2007).

10.1.2 Windrows

Windrows is one of the *ex situ* technique which relies on the periodic turning of piled contaminated soil to enhance bioremediation by increasing degradation activities of indigenous and/or transient hydrocarbonoclastic bacteria present in polluted soil. The intermittent turning of polluted soil, together with the addition of water, boosts up aeration, and uniformly distributes contaminants, nutrients, and microbial degradative activities, thus enhancing the rate of bioremediation, which can be accomplished through assimilation, biotransformation and mineralization of pollutants. When compared to biopile treatment, windrows showed a higher rate of hydrocarbon removal; however, the higher efficiency of the windrow towards hydrocarbon removal depends on the soil type, which was reported to be more friable (Coulon et al. 2010). The employment of windrow treatment has been implicated in CH4 (greenhouse gas) emission due to intensification of the anaerobic zone within piled polluted soil, which usually occurs following reduced aeration (Hobson et al. 2005). Phytoremediation is an upcoming field in the research as it is cost-effective and eco-friendly (Ali et al. 2013). Plant-assisted biosolids are used or microbes are employed to increase the efficacy of phytoremediation (Kim et al. 2010, Rajkumar et al. 2012). GMOs and wild variety plants have reported being positive in remediating metal contaminated soil (Chanu and Gupta 2016, Gomes et al. 2016). The ability of *Pteris vittata* and *Pteris cretica* to tolerate and remediate hydrocarbons and heavy metals with their bioconcentration proved to be the best remediation. On the other hand, *Boehmeria nivea* remediates Pb, Ar and PAHs, and *Miscanthus floridulus* remediates Cu and PAHs. *Ricinus communis* is used for phytoremediation of metal-polluted sites as well as in phytoremediation of fly ash disposal site (Olivares et al. 2013, Pandey 2013). *Eucalyptus globulus* can remediate e-waste disposal contaminated with Cd which is 30 years old (Luo et al. 2015). Apart from these enzymes secreted by the plants, microbes also play a vital role in phytoremediation of the contaminated soil and it is greatly recognized in the field of biotechnology (Tripathi et al. 2020).

10.1.3 Bioreactor

The bioreactor is an *ex situ* technique which contains a closed vessel which contains the raw materials and is converted into the desired products using biological reactions. There are different types of bioreactors used for different operations. It can support the growth of cell mimicking the natural environment. There are different parameters used in the bioreactor for the effective reduction of bioremediation time, important being the restricted bioaugmentation, the addition of nutrients, pollutant bioavailability in large amount and limited mass transfer for effective bioremediation. Benzene, VOCs and BTEX can be remediated in the bioreactor. GMOs can be used for bioremediation processes and can be destroyed after the process is over; however, biosurfactant cannot be treated in the bioreactor (Mustafa et al. 2015). The bioreactor cannot treat a very large amount of contaminants; therefore, the volume is restricted, and bioreactor-based bioremediation technique involves more capital and manpower (Philp and Atlas 2005). Another factor is that it requires monitored conditions; if it is not maintained properly, then the microbial activities decrease causing a decrease in the bioremediation process. But the main problem with the bioreactor-based bioremediation is the cost involved in pilot-scale bioremediation.

10.1.4 Farming

Land farming is one of the simplest *ex situ* bioremediation technique, which is cost-effective and requires fewer operations. It is widely used in the disposal of oil sludge, drill cuttings, and petroleum wastes. It uses clay at the bottom to prevent the pollutants from contaminating the water table (Kumar and Yadav 2018). Contaminants are tiled periodically for aeration; when it is tilled on the site, then it is called *in situ* treatment, otherwise it is an *ex situ* technique. The bioremediation process is carried out by the indigenous microbes present in the soil. While tilling the land takes place, nutrients can be added into the soil, which can enhance the microbial activities to enhance the bioremediation process. In the absence of nutrients, it is found that heterotrophic bacteria take over remediating the diesel contaminants (Silva-Castro et al. 2015). Contaminants are degraded, transformed, and immobilized by biological processes by the action of microorganisms with pH, aeration and moisture content being monitored. There are a few disadvantages such as large working space and high capital. The major advantage of this method is that it does not need any preliminary testing of the land, is less laborious and has a faster remediation process (Christopher 2016).

11. *IN SITU* BIOREMEDIATION

In *In situ* bioremediation method, the contaminants are treated at the contaminated site and not excavated from the site, thus the soil structure is not disturbed. It requires less capital and the cost required for the excavation process is saved in this method. There are a few enhanced bioremediation methods which require nutrients' supply and aeration such as biosparging, bioventing, and phytoremediation, while others are carried out without any enhancement. It is ideal for the removal of hydrocarbons, heavy metals, PCBs and dyes. Some of the *in situ* bioremediation methods are listed below.

11.1 Bioventing

Bioventing is an *in situ* bioremediation technique which takes place in aerobic condition. In this technique, the indigenous microbes are supplemented with oxygen (vadose) zone and nutrients to enhance bioremediation process. Low air is injected to sustain the microbial activity using wells. It is best suitable for the soil contaminated by petroleum hydrocarbons, non-chlorinated solvents, pesticides, wood preservatives, and other organic chemicals (Christopher 2016, Surajit and Hirak 2014).

11.2   Biosparging

Biosparging technique is similar to bioventing technique; in this, aeration is injected into the saturated zone at the subsurface level to stimulate the growth of indigenous microorganisms, which enhances the degradation of the pollutants present in the soil. The efficacy of remediation in biosparging depends on the permeability of the soil to the pollutants and its degradation capacity. Biosparging system is easy to install with flexible construction and the cost for the installation of the unit is less. It is ideally used for the remediation of petroleum pollutants from the water (Surajit and Hirak 2014).

11.3   Bioaugmentation

Bioaugmentation is a process which involves the addition of microorganisms or GMOs into the subsurface of the soil to speed up the degradation process and it reduces the time taken for the entire process. It is ideally used for the bioremediation of wastewater to restart the bioreactors of the activated sludges. It is also used for the complete degradation of PCBs and TCEs into chlorine and ethylene compounds (Surajit and Hirak 2014).

12.   ENHANCEMENT IN THE REDUCTION OF ENVIRONMENTAL TOXIC WASTES

When agents or active controls are introduced into the soil to reduce toxicity more effectively, this is called enhanced natural attenuation (ENA). Enhanced natural attenuation refers to when nutrient packages, catalysts and other microorganisms are added to the soil to enhance and speed up the bioremediation process. The use of enhanced natural attenuation is to mainly control the leachate which is produced during the remediation process (USEPA-SAB 2001, NRC 2000). The attenuation process requires two partners' contaminants such as soil and water system. The successful interaction between soil and water promotes attenuation. Giving more concentration only to one partner alone (contaminants) can deny the opportunity of the other partner or exploit it completely. The microorganism can resist the metal toxicity and convert heavy metals of toxic form into the less toxic form which is in negligible amount (Foulkes et al. 2016, Battista 1997, White et al. 1999). Apart from bacteria, algae also can degrade heavy metals and organic compounds by establishing a symbiotic relationship with the aerobic bacteria in the soil or through biosorption mechanism (Ungureanu et al. 2015, Majumder et al. 2015). Peat moss and fungus encapsulated withinpolysulfone have been reported to remediate heavy metals from the soil (Ahmad et al. 2019). The external appliance of electricity creates electro-migration and electro-osmosis, which contribute in the remediation of non-ionic pollutants from the soil. Chemical, physical, and biological methods are the commonly used methods; the combination of biological methods with electro-migration is a successful method at small scale or laboratory level experiments, but it is still under the developing stage.

13.   CONCLUSION

This chapter mainly focused on the interaction between the plant and the soil microbe in the attenuation of the toxic waste present in the soil, to maintain a sustainable environment. Soil remediation method is one of the upcoming methods which needs a lot of attention from the researchers. Rhizobacteria interaction increases the bioremediation processes and creates a less toxic environment. This chapter had a special emphasis on bioremediation technology and the phytoremediation process and it also talked about the anthropogenic compounds present in the soil.

References

Afrousheh, M., M. Shoor Mahmud, A. Tehranifar and V.R. Safari. 2015. Phytoremediation potential of copper contaminated soils in *Calendula officinalis* and effect of salicylic acid on the growth and copper toxicity. Int. Lett. Chem. Phys. Astron. 50: 159–163.

Ahmad, S., A. Pandey, V.V. Pathak, V.V. Tyagi and R. Kothari. 2019. Phycoremediation: Algae as eco-friendly tools for the removal of heavy metals from wastewaters *In*: R.N. Bharagava and G. Saxena (eds), Bioremediation of Industrial Waste for Environmental Safety. Springer, India, pp. 53–76.

Aislabie, J., J.S. David and M.F. Julia. 2006. Bioremediation of hydrocarbon-contaminated polar soils. Extremophiles 10(3): 171–179.

Ali, T., M. Sajid, K.M. Yahya, A. Ana, H.M. Baqir, A.H. Naeem, et al. 2013. Phytoremediation of cadmium contaminated soil by auxin assisted bacterial inoculation. Asian. J. Agric. Biol 1: 79e84.

Alkorta, I. and C. Garbisu. 2001. Phytoremediation of organic contaminants in soils. Bioresour. Technol. 79: 273–276.

Anawar, H.M., M.C. Freitas, N. Canha and I.S. Regina. 2011. Arsenic, antimony, and other trace element contamination in a mine tailings affected area and uptake by tolerant plant species. Environ. Geochem. Health. 33: 353–362.

Angelova, V.A., D.F. Grekov, V.K. Kisyov and K.I. Ivanov. 2015. Potential of lavender (*Lavandula vera* L.) for phytoremediation of soils contaminated with heavy metals. Int. J. Biol. Biomol. Agric. Food Biotechnol. Eng. 9:(5) 522–529.

Aybar, M., A. Bilgin and B. Sağlam. 2015. Removing heavy metals from the soil with phytoremediation. Artvin Çoruh Univ. J. Nat. Hazards Environ. 1: 59–65.

Banerjee, A., M.K. Jhariya, D.K. Yadav and A. Raj. 2018. Micro-remediation of metals: A new frontier in bioremediation. *In*: C. Hussain (ed.), Handbook of Environmental Materials Management. Springer, Cham, pp. 1–36. doi:10.1007/ 978-3-319-58538-3_10-1

Bareen, F.E. 2012. Chelate assisted phytoextraction using oilseed *Brassicas*. *In*: N. Anjum, I. Ahmad, M. Pereira, A. Duarte, S. Umar and N. Khan (eds), The Plant Family Brassicaceae. Environmental Pollution, vol 21. Springer, Dordrecht, pp. 289–311.

Battista, J.R. 1997. Against all odds: The survival strategies of *Deinococcus radiodurans*. Ann. Rev. Microbiol. 51: 203–224.

Baumann, A., 1885. Das Verhalten von Zinksatzengegen Pflanzen und im Boden. Landwirtsch. Vers.-Statn. 31: 1–53.

Beyersmann, D. and A. Hartwig. 2008. Carcinogenic metal compounds: recent insight into molecular and cellular mechanisms. Arch. Toxicol. 82(8): 493–512.

Bhatia, D. and M.D. Kumar. 2011. Plant-microbe interaction with enhanced bioremediation. Res. J. Biotech. 6(4): 72–76.

Blake, L. and K.W.T Goulding. 2002. Effects of atmospheric deposition, soil pH and acidification on heavy metal contents in soils and vegetation of semi-natural ecosystems at rothamsted experimental station, UK. Plant Soil. 240(2): 235–251.

Bohn, H.L., B.L. McNeal and G.A. O'Connor. 2001. Soil Chemistry, 3rd Ed. p. 307.

Borymski, S., M. Cycoń, M. Beckmann, L.A.J. Mur and Z. Piotrowska-Seget. 2018. Plant species and heavy metals affect biodiversity of microbial communities associated with metal-tolerant plants in metalliferous soils. Front. Microbiol. 9: Article 1425. https://doi.org/ 10.3389/fmicb.2018.01425.

Brady, N.C. and R.R. Weil. 1999. The Nature and Properties of Soils, 12th Ed. Prentice Hall Publishers, London.

Cassina, L., E. Tassi, E. Morelli, L. Giorgetti, R. Damiano, R.L. Chaney, et al. 2011. Exogenous cytokinin treatments of a Ni hyperaccumulator, *Alyssum murale*, grown in a serpentine soil: implications for phytoextraction. Int. J. Phytorem. 13: 90e101.

Chang, L.W., L. Magos and T.F.L. Boca Raton. 1996. Toxicology of Metals. CRC Press, USA.

Chang, Y.C. 2019. Microbial Biodegradation of Xenobiotic Compounds. CRC Press, Taylor & Francis.

Chanu, L.B. and A. Gupta. 2016. Phytoremediation of lead using *Ipomoea aquatica* Forsk. in hydroponic solution. Chemosphere. 156: 407–411.

Christopher, C., A.C.B. Chikere and G.C. Okpokwasili. 2016. Bioremediation techniques–classification based on site of application: Principles, advantages, limitations and prospects. World J. Microbiol. Biotechnol. 32(11): 180–185.

Coulon, F., M. Al Awadi, W. Cowie, D. Mardlin, S. Pollard, C. Cunningham, et al. 2010. When is a soil remediated? Comparison of biopiled and windrowed soils contaminated with bunker-fuel in a full-scale trial. Environ. Pollut. 158: 3032–3040.

Dangi, A.K., B. Sharma, R.T. Hill and P. Shukla 2018. Bioremediation through microbes: Systems biology and metabolic engineering approach. Critical Rev. Biotechnol. 39(1): 79–98.

Danso, D., J. Chow and W.R. Streit. 2019. Plastics: Microbial degradation, environmental and biotechnological perspectives. Appl. Environ. Microbiol. 85(19): e01095-19

Das, M., P. Agarwal, R. Singh and A. Adholeya. 2013. A study of abandoned ash ponds reclaimed through green cover development. Int. J. Phytorem. 15(4): 320–329.

De Cássia, M.R., C.S. De Souza, G.E. De Barros, B.R. Lovaglio, E.C. Lopes, F. Vieira, et al. 2007. Biodegradation of diesel oil by yeasts isolated from the vicinity of Suape Port in the state of Pernambuco–Brazil. Brazilian Arch. Biol. Technol. 50: 147–152.

De Deyn, G.B., J.H.C. Cornelissen and R.D Bardgett. 2008. Plant functional traits and soil carbon sequestration in contrasting biomes. Ecol. Lett. 11: 516–531.

Deng, D.M., W.S. Shu, J. Zhang, H.L. Zou, Z. Lin, Z.H. Ye, et al. 2007. Zinc and cadmium accumulation and tolerance in populations of *Sedum alfredii*. Environ. Pol. 147(2): 381–386. doi:10.1016/j.envpol. 2006.05.024

Ebrahimi T., M. Yousefi, N. Shariatifar, A. Mortazavian, A.M. Mohammadi, A. Khrshidian, et al. 2018. *In vitro* removal of polycyclic aromatic hydrocarbons by lactic acid bacteria. J App. Microbiol. 126(3): 954–964.

Farid, M.A., S. Shakoor, M. Bilal, A. Bharwana, S. Rizvi, H. Ehsan, et al. 2013. EDTA assisted phytoremediation of cadmium, lead and zinc. Int. J. Agron. Plant Prod. 4(11): 2833–2846.

Fierer, N. 2017. Embracing the unknown: Disentangling the complexities of the soil microbiome. Nat. Rev. Microbiol. 15: 579–590.

Flora, G., D. Gupta and A. Tiwari. 2012. Toxicity of lead: A review with recent updates. Interdiscip. Toxicol. 5(2): 47–58.

Flora, S.J.S. M. Mittal and A. Mehta. 2008. Heavy metal induced oxidative stress & its possible reversal by chelation therapy. Indian J. Med. Res. 128: 501–523

Foulkes, J.M., K. Deplanche, F. Sargent, L. Macaskie and J.R. Lloyd. 2016. A novel aerobic mechanism for reductive palladium biomineralization and recovery by *Escherichia coli*. Geomicrobiol. J. 33(3–4): 230–236.

Fritsche, W. and M. Hofrichter. 2008. Aerobic degradation by microorganisms. *In*: H.-J. Rehm and G. Reed (eds), Biotechnology Set, 2nd Ed. Wiley-VCH Verlag GmbH, Weinheim, pp. 144–167. doi: 10.1002/ 9783527620999.ch6m

Frutos, F.J.G., R. Pérez, O. Escolano, A. Rubio, A. Gimeno, M.D. Fernandez, et al. 2012. Remediation trials for hydrocarbon-contaminated sludge from a soil washing process: Evaluation of bioremediation technologies. J. Hazard. Mater. 199: 262–271.

Gabby, P.N. 2003. Lead, environmental defense, alternatives to lead-acid starter batteries, pollution prevention fact sheet. Battry Alts. Available at http:www.cleacarcam paign.org/factsheet.

Ghosal, D., S. Ghosh, T.K. Dutta and Y. Ahn. 2016. Current state of knowledge in microbial degradation of polycyclic aromatic hydrocarbons (PAHs): A review. Front. Microbiol. 7: 1369. doi: 10.3389/fmicb. 2016.01369.

Ghosh, M. and S.P. Singh. 2005. A review on phytoremediation of heavy metals and utilization of its byproducts. Appl. Ecol. Environ. Res. 3: 1–18.

Ghosh, S.K., S. Pal and S. Ray. 2013. Study of microbes having potentiality for biodegradation of plastics. Environ. Sci. Pollut. Res. Int. 20(7): 4339–4355.

Giebler, J., Y.W. Lukas, A. Chatzinotas and H. Harms. 2013. Alkane degrading bacteria at the soil litter interface: comparing isolates with T-RFLP-based community profiles. FEMS Microbiol. Ecol. 86(1): 45–58.

Glick, B.R. 2010. Using soil bacteria to facilitate phytoremediation. Biotechnol. Adv. 28: 367–374.

Gomes, M.A., R.A. Hauser-Davis, A.N. de Souza and A.P. Vitória. 2016. Metal phytoremediation: General strategies, genetically modified plants and applications in metal nanoparticle contamination. Ecotoxicol. Environ. Safety. 134: 133–147.

Grigulis, K., S. Lavorel, U. Krainer, N. Legay, C. Baxendale, M. Dumont, et al. 2013. Relative contributions of plant traits and soil microbial properties to mountain grassland ecosystem services. J. Ecol. 101: 47–57.

Guochen, K.P., H. Lambers, P. Kardol, B.L. Turner, D.A. Wardle and E. Laliberté. 2018 Biotic and abiotic plant-soil feedback depends on nitrogen acquisition strategy and shifts during long-term ecosystem development. J. Ecol. 107: 142–153.

Hassan, M.M., M.Z. Alam and M.N. Anwar. 2013. Biodegradation of textile azo dyes by bacteria isolated from dyeing industry effluent. Int. Res. J. Biological. Sci. 2(8): 27–31.

Hegazy, R., A. Salama, D. Mansour and A. Hassan. 2016. Renoprotective effect of lactoferrin against chromium-induced acute kidney injury in rats: Involvement of IL-18 and IGF-1 inhibition. PLoS One. 11(3): e0151486. doi: 10.1371/journal.pone.0151486

Hobson, A.M., J. Frederickson and N.B. Dise. 2005. CH_4 and N_2O from mechanically turned windrow and vermicomposting systems following in-vessel pre-treatment. Waste. Manag. 25: 345–352.

Houlihan, J. 2005. Pollution Gets Personal in Human Body Burden of Synthetic Toxic Chemicals conference. 3rd National Report on Human Exposure to Environmental Chemicals. Pub. No: 05-0570. Seattle.

Israr, M. and Sahi S.V. 2008. Promising role of plant hormones in translocation of lead in *Sesbania drummondii* shoots. Environ. Pollut. 153: 29e36.

Karmakar, R., I. Das, D. Dutta and A. Rakshit. 2016. Potential effects of climate change on soil properties: A review. Sci. Int. 4(2): 51–73.

Kay, V.R., C. Chambers and W.G. Foster. 2013. Reproductive and developmental effects of phthalate diesters in females. Crit. Rev. Toxicol. 43(3): 200–219.

Kehinde, F.O. and S.A. Isaac. 2016. Effectiveness of augmented consortia of *Bacillus coagulans*, *Citrobacter koseri* and *Serratia ficaria* in the degradation of diesel polluted soil supplemented with pig dung. African J. Microbiol. Res. 10(39): 1637–1644.

Kim, K.R. and G. Owens. 2010. Potential for enhanced phytoremediation of landfills using biosolids: A review. J. Environ. Manag. 91: 791–797.

Kim, S.S., K.J. Hwan and H.S. Jae. 2005. Application of the electrokinetic-Fenton process for the remediation of kaolinite contaminated with phenanthrene. J. Hazard Mater. 118: 121e131.

Krämer, U. 2010. Metal hyperaccumulation in plants. Ann. Rev. Plant Biol. 61(1), 517–534.

Kuiper, I., E.L. Lagendijk, G.V. Bloemberg and B.J. Lugtenberg. 2004. Rhizoremediation: A beneficial plant-microbe interaction. Mol. Plant Microbe Interact. 17:(1): 6–15.

Kumar, A. and S.K. Maiti. 2015. Effect of organic manures on the growth of *Cymbopogon citratus* and *Chrysopogon zizanioides* for the phytoremediation of chromite-asbestos mine waste: A Pot Scale Experiment. Int. J. Phytoremed. 17: 437–447.

Kumar, R. and P. Yadav. 2018. Novel and cost-effective technologies for hydrocarbon bioremediation. *In*: V. Kumar, M. Kumar and R. Prasad (eds), Microbial Action on Hydrocarbons. Springer-Singapore, pp. 543–565.

Kushwaha, A., R. Rani, S. Kumar and A. Gautam. 2015. Heavy metal detoxification and tolerance mechanisms in plants: Implications for phytoremediation. Environ. Rev. 24(1): 39–51.

Lasat, M.M. 2002. Phytoextraction of toxic metals. J. Environ. Qual. 31: 109e120.

Legay, N., C. Baxendale, K. Grigulis, U. Krainer, E. Kastl, M. Schloter, et al. 2017. Bioremediation mechanisms of combined pollution of PAHs and heavy metals by bacteria and fungi: A mini review. Bioresour. Technol. 224: 25–33.

Li, M., H. Liu, G. Geng, C. Hong, F. Liu, Y. Song, et al. (2017). Anthropogenic emission inventories in China: A review. Natl. Sci. Rev. 4(6): 834–866.

Lorenzo, V., F. Mapelli, E. Zanardini, E. Terzaghi, D. Guardo, A. Morosini, et al. 2017. Phyto-rhizoremediation of polychlorinated biphenyl contaminated soils: An outlook on plant-microbe beneficial interactions. Sci. Total Environ. 575: 1395–1406.

Luo, Y., Y. Ma, R.S. Oliviera, F. Nai, M. Rajkumar, I. Rocha, et al. 2015. The hyperaccumulator *Sedum plumbizincicola* harbors metal-resistant endophytic bacteria that improve its phytoextraction capacity in 1206 multi-metal contaminated soil. J. Environ. Manag. 156: 62–69.

Majumder, S., S. Gupta and S. Raghuvanshi. 2015. Removal of dissolved metals by bioremediation. *In*: S.K. Sharma (ed.), Heavy Metals in Water: Presence, Removal and Safety. RSC, pp. 44–56.

McGregor, R. and F. Vakili. 2019. The in situ treatment of BTEX, MTBE, and TBA in saline groundwater. Remediation. 29(4): 107–116.

Menendez-Vega, D., J.L.R. Gallego, A.I. Pelaez, G.F. de Cordoba, J. Moreno, D. Munoz and J. Sanchez. 2007. Engineered *in situ* bioremediation of soil and groundwater polluted with weathered hydrocarbons. Eur. J. Soil Biol. 43(5–6): 310–321.

Mitrovic, M., P. Pavlovic, D. Lakusic, L. Djurdjevic, B. Stevanovic and O. Kostic. 2008. The potential of *Festuca rubra* and *Calamagrostis epigejos* for the revegetation of fly ash deposits. Sci. Total Environ. 407: 338–347.

Mogil'nitskii, G.M., R.T. Sagatelyan, T.N. Kutishcheva, S.V. Zhukova, S.I. Kerimov and T.B. Parfenova. 1987. Disruption of the protective properties of the polyvinyl chloride coating under the effect of microorganisms. Prot. Met. (Engl. Transl) 23: 173–175.

Mondal, S. and D. Palit. 2019. Effective role of microorganism in waste management and environmental sustainability. *In*: M.K. Jhariya, A. Banerjee, R.S. Meena, D.K. Yadav (eds), Sustainable Agriculture, Forest and Environmental Management. Springer Nature, Singapore, pp. 485–515.

Mortberg, M. and H.Y. Neujahr. 1985. Uptake of phenol by *Trichosporon cutaneum*. J. Bacteriol. 161: 615–619.

Mucha, K., E. Kwapisz, U. Kucharska and A. Okruszeki. 2010. Mechanism of aniline degradation by yeast strain *Candida methanosorbosa* BP-6. Pol. J. Microbiol. 59(4): 311–315.

Mustafa, Y.A., H.M. Abdul-Hameed and Z.A. Razak. 2015. Biodegradation of 2,4-dichlorophenoxyacetic acid contaminated soil in a roller slurry bioreactor. Clean-Soil Air Water. 43: 1115–1266.

Navarro-Cano, J.A., M. Verdú and M. Goberna. 2018. Trait-based selection of nurse plants to restore ecosystem functions in mine tailings. J. Appl. Ecol. 55: 1195–1206.

Nowack, B., S. Rainer and R.H. Brett. 2006. Critical assessment of chelant-enhanced metal phytoextraction. Environ. Sci. Technol. 40: 5225e5232.

Pandey, V.C. 2013. Suitability of *Ricinus communis* L. cultivation for phytoremediation of fly ash disposal sites. Ecol. Eng. 57: 336–341.

Pandey, V.C. and N. Singh. 2015. Aromatic plants versus arsenic hazards in soils. J. Geochem. Explor. 157: 7780–7784.

Pandey, V.C., D.N. Pandey and N. Singh. 2015. Sustainable phytoremediation based on naturally colonizing and economically valuable plants. J Cleaner Product. 86: 37–39.

Pandey, V.C., A. Rai and J. Korstad. 2019. Aromatic crops in phytoremediation. *In*: V.C. Pandey, K. Bauddh (eds), Phytomanagement of Polluted Sites. Elsevier, pp. 255–275.

Panesar, A.S., A. Kumar and Kalpana. 2019. Phytoremediation: An ecofriendly tool for *In-Situ* remediation of contaminated soil. J. Pharm. Phytochem. 8(1S): 311–316.

Pathak, H., J. Ladha, P. Aggarwal, S. Peng, S.D. Singh, Y. Singh, et al. 2003. Trends of climatic potential and on-farm yields of rice and wheat in the Indo-Gangetic Plains. Field Crops Res. 80(3): 223–234.

Patlolla, A.K. and P.B. Tchounwou. 2005. Cytogenetic evaluation of arsenic trioxide toxicity in Sprague–Dawley rats. Mutation Res/Genet Toxicol. Environ. Mutagenesis. 587(1–2): 126–133.

Philp, J.C. and R.M. Atlas. 2005. Bioremediation of contaminated soils and aquifers. *In*: R.M. Atlas and J.C. Philp (eds), Bioremediation: Applied Microbial Solutions for Real-World Environmental Cleanup. American Society for Microbiology (ASM) Press, Washington, pp. 139–236.

Poschenrieder, C., G. Benet, C. Isabel and O.J. Barcel. 2008. A glance into aluminum toxicity and resistance in plants. Sci. Total Environ. 400: 356e368.

Praveen, A., C. Pandey, N. Marwa and D.P. Singh. 2019. Rhizoremediation of polluted sites: Harnessing plant-microbe interactions. *In*: A.C. Pandey and K. Bauddh (eds), Phytomanagement of Polluted Sites. Elsevier Science, Amsterdam, pp. 389–407.

Priyanka, N. and T. Archana. 2011. Biodegradability of polythene and plastic by the help of microorganism: A way for brighter future. J. Environ. Anal. Toxicol. 1: 111–115.

Olivares, R.A., R. Carrillo-González, M.D.C.A. González-Chávez and R.M. Soto Hernández. 2013. Potential of castor bean (*Ricinus communis* L.) for phytoremediation of mine tailings and oil production. J. Environ. Manag. 114: 316–323.

Rajan C.S. 2005. Nanotechnology in ground water remediation. Int. J. Environ. Sci. Dev. 2(3): 182–187.

Rajkumar, M., S. Sandhya, M.N. Prasad and H. Freitas. 2012. Perspectives of plant-associated microbes in heavy metal phytoremediation. Biotechnol. Adv. 30: 1562–1574.

Rhind, S.M. 2009. Anthropogenic pollutants: A threat to ecosystem sustainability? Philos. Trans. R. Soc. Lond. B. Biol. Sci. 364(1534): 3391–3401.

Rostami, S. and A. Azhdarpoor. 2019. The application of plant growth regulators to improve phytoremediation of contaminated soils: A review. Chemosphere 220: 818–827.

Rowell, D.L. 1994. Soil Science: Methods and Applications. Longman Group UK Ltd., London.

Roychowdhury, R., M. Roy, S. Zaman and A. Mitra. 2019. Bioremediation potential of microbes towards heavy metal contamination. Int. J. Res. Anal. Rev. 6(1): 1088–1094.

Rummel, C.D., A. Jahnke, E. Gorokhova, D. Kühnel and M. Schmitt-Jansen. 2017. Impacts of biofilm formation on the fate and potential effects of microplastic in the aquatic environment. Environ. Sci. Technol. Lett. 4: 258–267.

Safiyanu, I., A. Abdulwahid Isah and U.S. Abubakar. 2015. Review on comparative study on bioremediation for oil spills using microbes. Res. J. Pharma. Biol. Chem. Sci. 6(6): 783–90.

Saharan, B.S. and V. Nehra. 2011. Plant growth promoting rhizobacteria: A critical review. Life Sci. Med. Res. 21(1): 30.

Salt, D.E., M. Blaylock, P.B.A.N. Kumar, V. Dushenkov, B.D. Ensley, I. Chet, et al. 1995. Phytoremediation: A novel strategy for the removal of toxic metals from the environment using plants. Biotechnol. 13: 468e475.

Sandermann, H. 1994. Higher plant metabolism of xenobiotics: the "green liver" concept. Pharmacogenet. 4(5): 225–241.

Sekara, A., M. Poniedzialek, J. Ciura and E. Jedrszczyk. 2005. Zinc and copper accumulation and distribution in the tissues of nine crops: Implications for phytoremediation. Pol. J. Environ. Stud. 14: 829e835.

Seo, J.S., Y.S. Keum and Q.X. Li. 2009. Bacterial degradation of aromatic compounds. Int. J Environ. Res. Public Health. 6: 278–309.

Seung-Kyu, K. 2019. Trophic transfer of organochlorine through food-chain in coastal marine ecosystem, Environ. Eng. Res. 25(1): 43–51.

Sherchan, D.P. and K.B. Karki. 2005. Plant nutrient management for improving crop productivity in Nepal. In: Improving Plant Nutrient Management for Better Farmer Livelihood, Food Security and Environmental Sustainability, Procs. Regional Workshop, Beijing, China 12–16: 41–57.

Siddiqui Manzer, H., Mohammed H. Al-Whaibi and Mohammed O. Basalah. 2011. Interactive effect of calcium and gibberellin on nickel tolerance in relation to antioxidant systems in *Triticum aestivum* L. Protoplasma. 248(3): 503–11.

Silva-Castro G.A., I. Uad, A. Rodríguez-Calvo, J. González-López and C. Calvo. 2015. Response of autochthonous microbiota of diesel polluted soils to land-farming treatments. Environ. Res. 137: 49–58.

Sims, G.K. and R.G. Kanissery. 2019. Anaerobic biodegradation of pesticides. *In*: P.K. Arora (ed.), Microbial Metabolism of Xenobiotic Compounds. Springer Nature Singapore, pp. 33–54.

Singh, P., R. Jain and N. Srivastava. 2017. Current and emerging trends in bioremediation of petrochemical waste: A review. Crit. Rev. Environ. Sci. Technol. 14: 203–232.

Singh, R. 2017. Biodegradation of xenobiotics—A way for environmental detoxification. Int. J. Dev. Res. 7(7): 14082–14087.

Singh, S., M. Zacharias, S. Kalpana and S. Mishra. 2012. Heavy metals accumulation and distribution pattern in different vegetable crops. J. Environ. Chem. Ecotoxicol. 4: 75e81. https://doi.org/10.5897/JECE11.076.

Sivaperumal, P., K. Kamala and R. Rajaram. 2017. Bioremediation of industrial waste through enzyme producing marine microorganisms. Adv. Food Nutr. Res. 80: 165–179.

Smith, E., P. Thavamani, K. Ramadass, R. Naidu, P. Srivastava and M. Megharaj. 2015. Remediation trials for hydrocarbon-contaminated soils in arid environments: evaluation of bioslurry and biopiling techniques. Int Biodeterior. Biodegrad. 101: 56–65.

Solanki, P., M. Narayan and A.K. Rabha. 2018. Assessment of cadmium scavenging potential of *Canna indica* L. Bull. Environ. Contam. Toxicol. 101: 446–451.

Speight, J.G. 2014. The Chemistry and Technology of Petroleum, 5th Ed. CRC Press, Taylor & Francis Group, Boca Raton, FL.

Speight, J.G. 2017a. Sources and Types of Inorganic Pollutants. Environmental Inorganic Chemistry for Engineers 231–282. doi:10.1016/b978-0-12-849891-0.00005-9.

Speight, J.G. 2017b. Environmental Organic Chemistry for Engineers. Butterworth Heinemann, Elsevier, Cambridge.

Srivastava, N.K., L.C. Ram and R.E. Masto. 2014. Reclamation of overburden and lowland in coal mining area with fly ash and selective plantation: a sustainable ecological approach. Ecol. Eng. 71: 479–489.

Srivastava, S. and M. Kumar. 2019. Biodegradation of polycyclic aromatic hydrocarbons (PAHs): A sustainable approach. *In*: S. Shah, V. Venkatramanan and R. Prasad (eds), Sustainable Green Technologies for Environmental Management. Springer, Singapore, pp. 121–131.

Stephenson, C. and C.R. Black. 2014. One step forward, two steps back: The evolution of phytoremediation into commercial technologies. Bioscience Horizons: The Int. J. Student Res. 7: hzu009, https://doi.org/10.1093/biohorizons/hzu009

Surajit, D. and H.R. Dash. 2014. Microbial bioremediation: A potential tool for restoration of contaminated areas. *In*: S. Das (ed.), Microbial Biodegradation and Bioremediation. Elsevier, pp 1–21.

Tassi. E., P. Joel, P. Gianniantonio and B.F. Meri. 2008. The effects of exogenous plant growth regulators in the phytoextraction of heavy metals. Chemosphere 71: 66e73.

Tripathi, S., V. Kumar, P. Srivastava, R. Singh, D. Sanayaima, A. Kumar, et al. 2020. Phytoremediation of organic pollutants: Current status and future directions. *In*: P. Singh, A. Kumar, A. Borthakur (eds), Abatement of Environmental Pollutants: Trends Strategies. Elsevier, pp. 81–105.

Tuna, A.L., C. Kaya, M. Dikilitas and D. Higgs. 2008. The combined effects of gibberellic acid and salinity on some antioxidant enzyme activities, plant growth parameters and nutritional status in maize plants. Environ. Exp. Bot. 62: 1e9.

Ungureanu, G., S. Santos, R. Boaventura and C. Botelho. 2015. Biosorption of antimony by brown algae *S. muticum* and *A. nodosum*. Environ. Eng. Manag. J. 14: 455–463.

USEPA-SAB. 2001. Monitored natural attenuation: USEPA research program-An EPA science advisory board review. EPA-SAB-EEC-01-004.

Vallini, G., S. Di Gregorio and S. Lampis. 2005. Rhizosphere induced selenium precipitation for possible applications in phytoremediation of se polluted effluents. Z. Naturforsch. C. J. Biosci. 60(3–4): 349–356.

Velma, V., S.S. Vutukuru and P.B. Tchounwou. 2009. Ecotoxicology of hexavalent chromium in freshwater fish: A critical review. Rev. Environ. Health. 24(2): 129–145.

Vargas-García, M.C., M.J. López, F. Suárez-Estrella and J. Moreno. 2012. Compost as a source of microbial isolates for the bioremediation of heavy metals: in vitro selection. Sci. Total. Environ. 431: 62–67.

Verma, J.P. and D.K. Jaiswal. 2016. Book review: Advances in biodegradation and bioremediation of industrial waste. Front Microbiol. 6: 1–2. doi: 10.3389/fmicb.2015.01555.

Walker, J.D., R.R. Colwell and L. Petrakis. 1975. Degradation of petroleum by an alga, *Prototheca zopfii*. J. Applied Microbial. 30(1): 79–81.

Wang, J.L. and C. Chen, 2006. Biosorption of heavy metals by Saccharomyces cerevisiae: A review. Biotechnol. Adv. 24: 427–451.

Wang. S. and X. Shi. 2001. Molecular mechanisms of metal toxicity and carcinogenesis. Mol. Cell Biochem. 222: 3–9.

Whelan, M.J., F. Coulon, G. Hince, J. Rayner, R. McWatters, T. Spedding, et al. 2015. Fate and transport of petroleum hydrocarbons in engineered biopiles in polar regions. Chemosphere 131: 232–240.

White, O., J.A Eisen, J.F. Heidelberg, E.K. Hickey, J.D. Peterson, R.J. Dodson, et al. 1999. Genome sequence of the radio resistant bacterium *Deinococcus radiodurans* R1. Science 286(5444): 1571–1577

Wu, S., K. Cheung, Y. Luo and M. Wong. 2006. Effects of inoculation of plant growth promoting rhizobacteria on metal uptake by *Brassica juncea*. Environ. Pollut. 140: 124–135.

Ye, S., G. Zeng, H. Wu, C. Zhang, J. Liang, J. Dai, et al. 2017. Co-occurrence and interactions of pollutants, and their impacts on soil remediation—A review. Crit. Rev. Environ. Sci. Technol. 47: 1528–1553.

Yedjou, C.G. and P.B. Tchounwou. 2007. In-vitro cytotoxic and genotoxic effects of arsenic trioxide on human leukemia (HL-60) cells using the MTT and alkaline single cell gel electrophoresis (Comet) assays. Mol. Cell Biochem. 301(1–2): 123–30.

Yoon, J., X. Cao, Q. Zhou and L.Q. Ma. 2006. Accumulation of Pb, Cu, and Zn in native plants growing on a contaminated Florida site. Sci. Total Environ. 368(2–3): 456–464.

Zhou, Q.X. and Y.F. Song. 2004. Remediation of Contaminated Soils: Principles and Methods. Beijing, Science Press.

CHAPTER 20

Phytoremediation of Contaminated Agro-ecosystem through Plants and Fungi

Shazia Iram

Department of Environmental Sciences, Fatima Jinnah Women University,
The Mall, 46000, Rawalpindi, Pakistan.

1. INTRODUCTION

Among other developing countries, Pakistan also falls in the list of those countries that have disorganized wastewater disposal system like wastewater is disposed of near the vicinity of rivers, canal, agricultural drains and rivers. Surface and subsurface horizons are mainly affected by the heavy metal contamination (Hussain et al. 2006). So, this contamination is a serious menace to human beings as well as ecosystem as it directly effects the food chain as shown in Fig. 1.

Another dilemma also associated with metals is that they have a long persistent time as they do not degrade easily because of their ability to make complexes with organic matter, iron oxides, phosphates, carbonates and clay minerals. So, to overcome this problem, there are various techniques to tackle this out but negative externality associated with those remediation techniques is that they are very high priced. But among all of them, phytoremediation is widely accepted. According to this technique, to eradicate heavy metals from soil, plants mainly hyper-accumulators are used. The effects of fungi, native plants and chelating agents for metal remediation is environmental friendly, cost effective and is also effective for phytomining (Akhtar et al. 2014).

---

*Corresponding author: iram.shazia@gmail.com

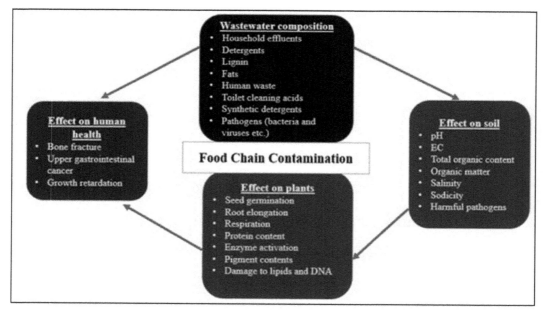

Figure 1   Food chain contamination by heavy metals.

2.   SOIL, PLANTS AND UNTREATED WASTEWATER CONTAMINATED BY HEAVY METALS

Irrigation of agricultural soil by wastewater is helpful as it assists to improve the soil fertility and alleviates water shortage, but the disadvantage is that extensive use of this water contaminates the soils and in turn plants.

This dilemma is also associated with Pakistan that the soils of some regions (Lahore, Kasur, Gujranwala and Multan) are irrigated by the wastewater coming from municipality and industries and the toxicity level reaches to its epic. The concentrations varies because of the source variation and extent of metals. Numerous factors are associated with the amassing of metals like climatic conditions, plant species and reduction and oxidation states, etc.

2.1   Irrigation of Agricultural Land by Wastewater

Eighty percent of wastewater is used for agricultural purposes that is crucial for the people living in the suburbs since the wastewater contains almost all of the nutrients required by plants. Excess irrigation using wastewater leads to accumulation of heavy metals in soils, which is up taken up by plants and metals enters into the food chain, which leads to health hazards (Zafar et al. 2007).

Cadmium, a toxic, non-essential plant metal, is exceedingly phytotoxic. It is involved in inhibiting DNA mediated conversion in microorganisms, meddlesome positive interaction among microbes, plants and distressing enzymatic activities. Amongst industries, fabric industry is a major factor to uplift metal concentration in water and printing of fabrics releases toxic metals, mainly, Cd, Zn, Ni, Cu and Cr.

2.2   Soil Contamination with Heavy Metals

Irrigation of land with wastewater is a 400 years old practice but persists till today. Worldwide, twenty millions of arable land is being watered by wastewater. According to Asian and African statistics, 50% of vegetables is being watered by wastewater. The metropolitan cities of Pakistan, mainly, Gujranwala, Lahore, Multan and Kasur have elevated heavy metals' concentration due to extensive use of wastewater for longer time period (Mahmood 2010).

According to a study (Farid et al. 2015) conducted in Faisalabad (Pakistan), Ni, Cd and Pb level increased from the expected level. One other reason of soil contamination with heavy metals is pesticide as cadmium concentration in soil increased because the farmers used pesticides. Bio-solids are the best contributor to heavy metal contamination because they have a positive relation with metal amassing. Chromium, another heavy metal, also increased in the soil and the basic factors behind its accumulation are the characteristics of adsorption of soil, clay and organic matter content and iron content. Heavy metal from residential, industrial and agricultural fields are the prime factor behind the impurity of soil as shown in Fig. 2 (Iram et al. 2012).

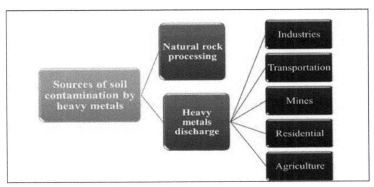

Figure 2   Sources of soil contamination by heavy metals.

2.3 Crops/Vegetables' Contamination by Heavy Metals

As we know, soil watered by contaminated water results in heavy metals that deteriorates the soil health and soil environment, and then when they enter into food chain, they disrupt the human body systems. Higher concentration of metals when taken up by the plant damages the plant growth by degenerating the important cell organelles and disordering the membranes. Heavy metals act as mutagenic element and affect the imperative functions of physiology like effect on respiratory process, photosynthesis, formation of proteins and metabolism of carbohydrate.

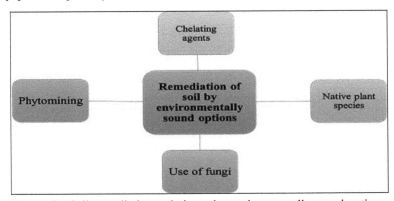

Figure 3   Soil remediation techniques by environmentally sound options.

Soil requires a healthy biota, which mainly includes microbes that help to maintain the soil's positive health, but with the higher concentration of heavy metals the microbes are badly affected and are not able to accept that change of burdened amount of heavy metals. Extensive amount of research is required in the field of scrutinizing the magnitude of plant contamination. According to Azizollahi et al. (2019) cadmium accumulation was positively related to summer season due to its higher temperature. So, now due to global warming the temperature of earth is increasing at a greater pace, that's why there is more probability of food chain contamination by these noxious, long lived heavy metals.

Some of the best and most opted methods for heavy metal remediation that will improve the contaminated soil status include the use of native species of plants, fungi and the finest use of chelating agents (Akhtar et al. 2014).

3. FUNGAL DIVERSITY IN SOILS CONTAMINATED WITH METALS

Fungal variety arising from agricultural soils and waste water analysis accommodates monitoring and evaluating changes of fungal diversity because of environmental pollution. Different factors which affect fungal diversity are nutrient quality, soil texture and bio-chemical pollution of the soils. Soil samples irrigated with wastewater showed metal accumulations higher than permissible limits, which might affect soil characteristics and fungal diversity.

If heavy metals' accumulation continued to increase with wastewater for irrigation, it would pose a risk to microbial health and, likewise, it will affect the crop production and eventually ecosystem of the agricultural soil. Therefore, it is essential to use treated waste water for irrigation purposes. A prolonged accumulation of heavy metals causes significant variation of microbial diversity, population and their activity.

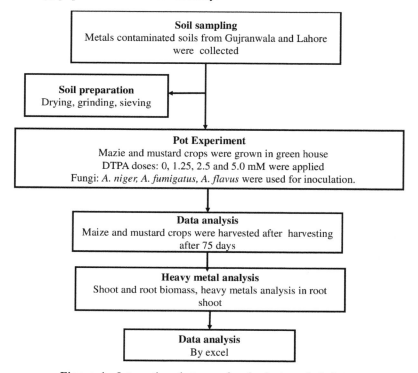

Figure 4   Interactions between fungi, plants and chelates.

Changes in the diversity of soil fungi cannot be understood entirely because of heavy metals' stress in the soil. It is recognized fact that metals cannot be degraded chemically. More research is needed to know the phenomenon of these fungal diversity stress to increase the management of soil or better production of crops. Bioremediation/phytoremediation techniques are essential to restore the soils contaminated with heavy metals.

Heavy metals' contamination causes the environmental stress and this might be owing to heavy metals pollution. The soil samples from polluted sites were more affected by wastewater which was used for irrigation due to high metal concentrations and later affected the population density of fungi. The difference between the sampling sites in microbial population and diversity is closely related with the degree of heavy metal pollution. Normally, greater the soil and water heavy metal pollution, lesser will be the microbial diversity.

It is possible to isolate metal resistant fungi from the soil which was treated with wastewater. Many fungi belong to genera *Penicillium, Aspergillus, Alternaria, Fusarium, Geotrichum, Monilla Rhizopus* and *Trichoderma*. Isolation procedure is given in Fig. 4. Bioremediation is an efficient technique due to high efficiency, low cost and eco-friendly nature. Several developments have been made in metal-microbe interaction and their use for metal accumulation/detoxification. Fungi are capable of absorbing, leaching and transforming heavy metals due to their resistance to heavy metals.

4. EFFECT OF CHELATES AND TOLERANT FUNGAL STRAINS FOR METAL EXTRACTION AND SOLUBILIZATION

Due to persistent properties of the metals, the remediation of heavy metals contaminated soil is difficult, to which current technologies present costly and, sometime, disturbing solutions. Phytoextraction is a developing technology which is environmental friendly and very economic for heavy metals contaminated soils. There are a lot of factors for phytoextraction but a critical requirement for the uptake of metals by plants is bioavailability of metals to the plants' root. This can be done by addition of chelates. The chelating agent forms complexes with the free metal ions in the soil solution. They do the further dissolution of metals in the sorbed phase, hence equilibrium is attained between the free metal ions and the insoluble phases.

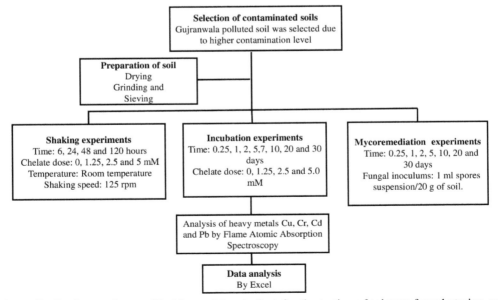

Figure 5 Batch experiments (Shaking and Incubation) for the testing of tolerant fungal strains, and their solubilizing efficiency for metal extraction.

The complete procedure is given in Fig. 5. Soil is heterogeneous combination of mineral and organic components. The moisture content, pH and redox potential of soils vary according to the environment. Due to many complex biological and chemical interactions, heavy metals go in soil. The reactions are precipitation-dissolution, oxidation-reduction, and volatilization. Surface solution phase complexes take place in soil. The chelating agents should overcome these reactions so extraction of target metal can be done. They have greater stability and solubility in aqueous phase than other reactions with metals.

5. METALS' SOLUBILIZATION BY EDTA

The levels of solubilized Cu, Pb, Cd and Cr increased with application of EDTA to the contaminated soil. The addition of EDTA to contaminated soils increases the soluble or exchangeable fraction of heavy metals in the soil solution making them available for uptake by plants. Significant amounts of the soluble fraction of Cu, Pb, Cd and Cr could be observed in the soil solution for several hours following the application of EDTA. This shows that the solubilizing effect of EDTA did not decrease with time, which could be due to its high environmental persistence. The major factors which affect remediation process are the physiochemical characterization of soil, chelating agent and heavy metals' solubilization. Adoption of appropriate chelating agent is the first step in chelate improved phytoremediation. EDTA is efficient for the Pb solubilization. Pb solubilization increased with EDTA concentration. Pb solubilization also increases lower pH, as lower pH increases leaching of heavy metals.

The extraction of Pb by EDTA and NTA is also dependent on soil containing 70% silt and clay. EDTA removed 68% of Pb and took less time, whereas NTA removed 19% Pb and took longer time as compared to EDTA. Soil comprises more than 70% sand and can solubilize 85% Pb, probably because soil having more clay will bind to the metal, but sand in soil will not have adsorptive capacity like clay and therefore more Pb would be chelated. The addition of chelating agents depends upon sorption and solubilization of metals with soil. Another factor is the relation between clay content of soil and sorption of metals. Clay content can be manipulated for achieving chelation.

In a study conducted by Akhtar and Iram (2014), solubilization of heavy metals (Pb, Cu, Cr and Cd) with chelating agents (NTA, DTPA and EDTA) was assessed. 1.25, 2.5 and 5.0 mM/ kg of soil was the rate of application of chelating agents. Experiments for incubation were done for 30 days.

EDTA (chelating agents) enhanced the metals' uptake, desorbed metals from the soil solid phase and produced complexes of soluble water. Metal has the ability to divide in two phases: irreversible phase and reversible phase. Metals will extract first from the reversible phase to irreversible phase in the presence of chelates. Chelating agent has the ability to free the metal from bound phase. They can unlock metals more efficiently from the irreversible phase.

The solubility of Cd increased with the application of DTPA and concentration of metal related to the concentration of chelator (Akhtar and Iram 2014). Strong effect of chelator produced after 30 days application of DTPA and solubility was observed. At 5.0 mM DTPA per kg of soil, the solubility of Cr, Pb, Cu and Cd was greater as compared to control (Fig. 6).

The heavy metals solubilization was enhanced by application of NTA, which increased the availability of heavy metals for plants. In the above mentioned study, after application of NTA chelate a major concentration was detected in soil solution. It was because of NTA persistence in environment. For remediation of soils polluted with metals chelates concentration, incubation period and shaking time are major factors for developing effective model. Thus, a direct conclusion about effect of heavy metals on solubility cannot be drawn when less amount of complexing agents are used, like in conventional agricultural practices where these are applied to recover nutrient deficiency.

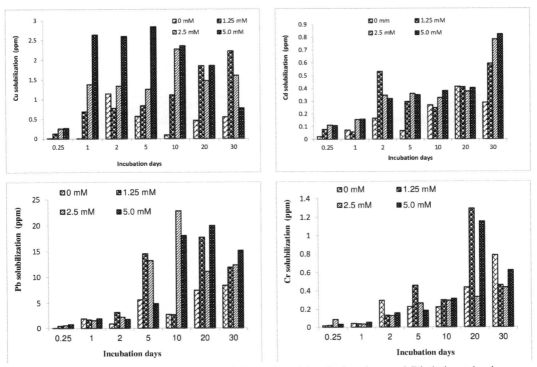

Figure 6   Solubilized Cu, Pb, Cd, and Cr at altered incubation days and Diethylenetriamine Pentaacetic Acid (DTPA) concentrations.

6.   HEAVY METALS' SOLUBILIZATION BY FUNGI

Microscopic organisms such as fungi play a vital role in decontamination of heavy metal polluted soil due to their extensive range of transmuting features. Some major properties include the resistivity or tolerance, solubility and virulent behavior. Fungi are the most dominant entity of the soil environment. They are able to convert the insoluble phase metals to soluble phase. However, several studies have examined the soluble actions of fungi related to heavy metals. In terms of resistance, fungi show great opposition to metal pollution, which may be caused by substantial industries. Other changing circumstances of environment such as low pH are important for the metabolic properties of microbes and microbial community.

Some physiological characteristics of fungal communities enable them to solubilize a variety of heavy metals. For example, mycelium surface of fungi and extracellular metabolites contrived from fungi make it metal soluble. During organic matter decomposition, fungi discharge metals and increase their accessibility by percolating metals from inorganic material.

According to Akhtar and Iram (2014), *Aspergillus niger* was solubilized Cu (17 mg kg^{-1}), Cd (34 mg kg^{-1}) and Cr (150 mg kg^{-1}), while *Aspergillus* and *Penicillium* fungi are the most dynamic species for leaching. Chelating agents are another main solubilizer for removal of heavy metals. One of the major source of chelating agents are fungi such as *Aspergillus nigerr*. This fungi showed probable actions in producing citric, malic, oxalic and tartaric acids and fungal producing organic acids act as chelating agents. These agents are also called biological chelates. As compared to chemical chelates, the biological chelates have low or zero cost. Biological chelates remove Pb (30%), Cu (56%), Cd (100%) and Zn (19%). Table 1 shows the bioconcentration factors (Cd, Pb, Cu and Cr) of mustard and maize crop in Lahore and Gujranwala contaminated soils amended with *Aspergillus* species and Diethylenetriamine Pentaacetic Acid (DTPA).

Table 1 Bio-concentration factors of Copper (Cu), Cadmium (Cd), Chromium (Cr) and Lead (Pb)
of mustard and maize crops in Lahore and Gujranwala unclean soils
amended with *Aspergillus* species and DTPA

BCF Bio-concentration factor									
		Cd		Pb		Cu		Cr	
Locations	*Treatments*	*Maize*	*Mustard*	*Maize*	*Mustard*	*Maize*	*Mustard*	*Maize*	*Mustard*
Gujranwala	Control	0.30	11.90	0.99	0.32	0.38	0.41	2.08	2.27
	Aspergillus niger	0.45	6.38	1.36	0.41	0.37	0.55	2.70	14.72
	Aspergillus flavus	0.57	4.52	1.88	0.17	0.50	0.42	3.43	50.18
	Aspergillus fumigatus	1.59	24.84	1.30	0.29	0.42	0.63	9.51	73.97
	DTPA 1.25mM	1.60	23.68	1.82	0.52	0.25	1.14	7.09	50.13
	DTPA 2.5 mM	1.16	33.60	1.93	0.54	0.23	0.91	61.50	90.53
	DTPA 5.0 mM	1.52	31.91	1.94	0.35	1.02	0.53	11.07	44.44
Lahore	Control	0.54	49.65	3.5	3.25	0.24	110.98	9.48	37.17
	Aspergillus niger	1.32	49.91	2.67	4.65	0.67	121.98	19.78	107.22
	Aspergillus flavus	0.55	141.96	6.58	11.05	0.65	484.18	19.70	46.68
	Aspergillus fumigatus	0.69	159.84	3.92	5.35	0.15	15.66	9.33	36.03
	DTPA 1.25 mM	0.59	40.86	22.63	11.03	0.24	10.79	33.44	22.81
	DTPA 2.5 mM	2.46	88.38	7.74	9.85	0.13	54.12	26.81	20.64
	DTPA 5.0 mM	0.91	249.86	10.58	12.38	0.19	75.97	31.80	90.48

7. BIOLOGICALLY, CHEMICALLY AND NATURALLY BOOSTED PHYTOEXTRACTION PROSPECTIVE OF NATIVE PLANTS

The study indicated that local plants can accumulate the toxic heavy metals and also uptake them from Gujranwala and Lahore contaminated soil. This particular capability was patterned and the route of identification is given in Fig. 7. Seeds of Bajra (*Pennisetum americanum*), Barley (*Hordeum vulgare*), Wheat (*Triticum aestivum*), Sunflower (*Helianthus annuus*), Soybean (*Glycine max*), Mustard (*Brassica campesstris)* and Maize (*Zea mays*) are grown in vitro. The accumulation and uptake of heavy metals' ability vary widely in different studied plants. The decreased dry and fresh weight, shoots and root lengths of maize plant were observed in current study associated with the Cu, Cd, Pb and Cr treatment at higher concentrations (Fig. 8). Entirely decreased shoot, root, dry biomasses and seed germination was perceived by Gupta (2011) when doses treatment was applied such as 50, 100, 150 and 200 mg L^{-1} of Cd and Cu. The most essential metal for plant growth is copper by regulating usual digestion of plants. Nouri et al. (2009) described that Cu is necessary to plant development; however, this metal also showed accumulation in leaves and shoots structures such as copper level beyond the 2 mg kg^{-1}. Beyond this level, copper metal was intensely toxic to living cells. Sometimes, it also act as a growth inhibiting factor and major cause of death, especially to plants.

Owing to the higher concentration treatment of Cd, Cr, Cu and Pb, the decreased shoot length, root length, dry weight and fresh weight of mustard plant was noticed in this study. There are some environmental factors which increase the heavy metals' accumulation and dissemination in plants. Such dynamics include bioavailability and chemical, pH, cation exchange capacity (CEC), temperature, redox, plant species, dissolved oxygen and root secretion. Roots are acutely affected due to Cu, which leads to the reduction in roots' growth and lessens the seedling plasma membrane permeability. The study determined that the accumulation rate of different metals in plants also vary. For example, Cd and Pb accumulated in higher concentration in maize roots as paralleled to Cu and Cr. Due to the high mobility rate of Cd, it can be easily absorbed by root surfaces of plants and make its way to move towards higher plant parts such as wood tissues and leaves.

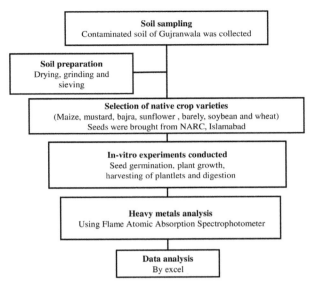

Figure 7 In vitro experiments to explore the natural and chemically enhanced phytoextraction potential of native plant species.

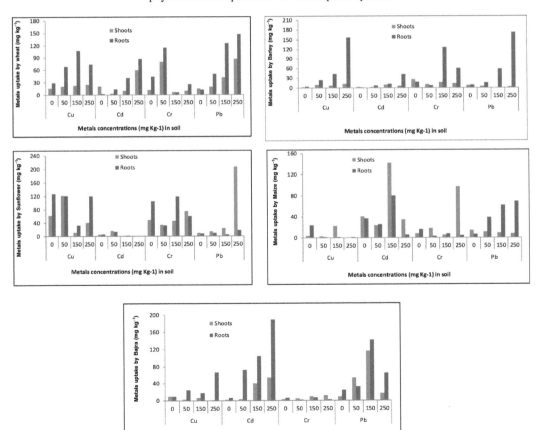

Figure 8 Accumulation of selected heavy metals in shoot and root of wheat, barley, maize, sunflower and bajra.

Copper caused the antagonistic effects on germination of wheat seed as it relaxed the seed growth due to which it took about 6 days for a tiny radicle to appear in the seeds. A significant difference in barley growth was observed in different concentrations of Cd, Cr, Cu and Pb. The increased shoot length and root length of control crop was observed in application of Cr, Cu and Pb. But two specific heavy metals, Cu and Pb, showed an inhibitory effect on barley as decreased seed germination rate as well as slow roots and shoots growth. For instance, in shoots of sunflower plant, the accumulation and uptake process for Pb at varying concentrations was high. This indicated greater potential to stock Pb in sunflower roots.

The current study concluded that the most effective phytoremediation plants among observed plants were mustard and maize. The best plant used for phytoremediation of Cd polluted soils was canola and for Pb and Cd polluted soils' remediation the finest plant was maize. The diverse changes in the plants' development parameters were determined due to increased concentrations of chromium, cadmium, copper and lead. Among all metals, the most affected metal was cadmium. It had a more inhibitory effect on seed germination, seedling development, elongation of shoot and root among all the selected local plants (Akhtar and Iram 2017).

8. RELATION BETWEEN PLANT GROWTH, CONTAMINANTS AND SOIL

Mustard and maize plants were grown in Lahore and Gujranwala's metal contaminated soils followed by the amendment of different doses of DTPA and inoculums of three specified fungal species including *Aspergillus niger, Aspergillus fumigates* and *Aspergillus flavus* grown in contaminated soil of Gujranwala. Both the plant species showed highest biomass growth of shoot in case of fungal amendment as compared to the chemical amendment with DTPA at 1.25, 2.5 and 5.0 mMkg^{-1} soil.

Aspergillus niger, Aspergillus flavus and *Aspergillus fumigatus* exhibited excellent results and crop growth was maximum. The other potential shown was the highest phytoremediation of heavy metals' (Cd, Cr, Pb and Cu) contaminated soils. Contaminated soils of Lahore and Gujranwala treated with fungal inoculum amplified shoot biomass of mustard and maize crops more efficiently as compared to the chemical modification (DTPA).

Among the three fungal strains, *Aspergillus fumigatus* has the maximum potential to enhance the shoot growth. Thus, it was concluded that *Aspergillus fumigatus* could be very beneficial for phytoremediation of polluted calcareous soils. Literatures reported that the shoot, grain and root harvest of crops treated with microbial inoculums and chemicals was satisfactory as compared to control samples. In contaminated soil, microbial inoculums not just prolonged the nutrients uptake by the plants, specifically NPK, but in addition to these they improved the soil properties, organic matter substance and nitrogen content in the soil. Pot experiments conducted for mustard and maize crops in green house demonstrated that soils treated with *Aspergillus fumigatus* and DTPA applied at the rate of 1.25 mM kg^{-1} improved root growth, which can increase phytoextraction potential of toxic metals from contaminated soils with satisfactory efficacy. Experiment showed that addition of DTPA and fungal inoculums to the two soils tempted Cu, Pb, Cr and Cd solubilization in the soil, which eventually stirred phytoaccumulation and enabled phytoextraction.

One of the prominent environmental threat disclosed during the study was the possibility of metals leaching down the soil profile and impacting the groundwater quality. Few studies reported that risk can be reduced by the utilization of synthetic chelators.

Mycorrhizal species play a vital role for excellent plant growth, particularly in soils contaminated with heavy metals as they make most of the mineral supplements such as NPK available for plant usage. In addition, they have the ability to improve soil texture that counterattacks the harms of wind and water erosion. Fungal inoculums including *Aspergillus niger, Aspergillus flavus* and *Aspergillus fumigatus* isolated from the polluted sites were mulled over in comparison with DTPA that showed compacted potential for progressive maize crop growth and phytoremediation of soils contaminated with Cd, Cr, Pb and Cu.

Results of the study also depicted that shoot growth of mustard and maize crops extended more professionally because of expansion of fungal strain, in comparison with chemical treatments with DTPA (1.25, 2.5 and 5.0 mM kg^{-1} of soil) and to the control soil. The increase in shoot height could be associated with activation of phosphorus present in soil chemistry because of phosphorus solubilizing fungal isolates. Study disclosed that *Aspergillus fumigatus* has the highest potential for the improvement of the shoot biomass. Thus, it is concluded that *Aspergillus fumigatus* is useful for phytoremediation of contaminated soils as fungi has the potential to boost the biomass production and heavy metal removal.

9. CONCLUSION

The study revealed that by adding the finest amount of chelate, i.e. DTPA, it boosts the uptake of heavy metals (Cr, Cu, Pb and Cd) and endorses the phytoextraction. For the soils of Lahore and Gujranwala, maize is the best opted native plant species to remove all four heavy metals with greater proficiency. Bio-concentration factor (BCF) is a term referring to the uptake of metals by hyperaccumulator plant tissue and their subsequent removal. So, in this case maize crop has high BCF for specific metals (Cd, Cr and Cu) as compared to mustard crop which has high BCF for Pb. By using the chelates, it will assist in the better uptake and translocation of metals from soil to native plant species, thus refining the phytoextraction technology.

Another term, phytoextraction rate (PR), is used to describe the process of plant accumulation proficiency of metals and subsequent agricultural practices to heighten plant growth. So, mustard plant showed higher PR of Cu and Cd, while in the case of maize crop PR was higher for Pb and Cr. By comparing the RF with EF (extraction factor), we can know about the harmful impacts caused by unwanted leaching of metal multiplexes to groundwater. According to the study by Akhtar and Iram (2017) EF value for Cu, Cd and Pb were greater for mustard crop and most of the Cr was extracted by maize crop. A study showed that mustard plant has enough ability to remove Pb from plants.

Phytoremediation is an affordable and effective technique used to remove inactive heavy metals. Maize extract Cu, Cd, Pb and Cr efficiently from roots larger than shoots. Due to high biomass and accumulation capability of maize and mustard, the uptake ability to remove metal is highly considerable. From results, it can be concluded that maize is very suitable for phytoextraction of contaminated soils. Phytoremediation is a suitable technique for removing heavy metals and metal complexes from contaminated soils.

By interpreting all the results, it can be concluded that phytoremediation is economically feasible than physical and chemical techniques. Through phytoremediation, it is possible to use plant biomass for bioenergy, biofuel, and biogas recovery of metals.

▇ References

Akhtar, S., S. Iram, M.H. Hassan, V. Suther and I. Ahmad. 2014. Heavy metal concentration in peri-urban soils and crops under untreated wastewater. Int. J. Sci. Eng. Res. 5(9): 523–535.

Akhtar, S. and S. Iram. 2014. Effect of chelating agents on heavy metal extraction from contaminated soils. Int. J. Sci. Eng. Res. 5(9): 536–546.

Akhtar, S. and S. Iram. 2017. *In-vitro* assessment of heavy metal removal from contaminated agricultural soil by native plant species. Pakistan J. Anal. Environ. Chem. 18(2): 120–128.

Azizollahi, Z., S.M. Ghaderian and A.K. Ghotbi-Ravandi. 2019. Cadmium accumulation and its effects on physiological and biochemical characters of summer savory (*Satureja hortensis* L.). Int. J. Phytoremediat. 21(2): 1241–1253.

Farid, G., N. Sarwar, U. Saifullah, A. Ahmad, A. Ghafoor and M. Rehman. 2015. Heavy metals (Cd, Ni and Pb) contamination of soils, plants and waters in Madina town of Faisalabad metropolitan and preparation of Gis based maps. Adv. Crop Sci. Tech. 4: 199–205.

Gupta, D. 2011. Toxicity of copper and cadmium on germination and seedling growth of maize (*Zea mays* L.) seeds. Indian J. Sci. Res. 2(3): 67–72.

Hussain, I., A. Ghafoor, S. Ahmad, M.A. Aziz and H.R. Ahmad. 2006. Effect of lime on retention, transport of lead and chromium, and their fractionation in soil applied in irrigation. *In*: Abstracts of the 11th Congress of Soil Science, 28–31 March 2006, NARC, Islamabad. Islamabad: Soil Science Society of Pakistan.

Iram, S., K. Parveen, J. Usman, K. Nasir, N. Akhtar, S. Arouj and I. Ahmad. 2012. Heavy metal tolerance of filamentous fungal strains isolated from soil irrigated with industrial wastewater. Biologija. 58(3): 107–116.

Mahmood, T. 2010. Phytoextraction of heavy metals: The process and scope for remediation of contaminated soils. Soil Environ. 29(2): 91–109.

Nouri, J., N. Khorasani, B. Lorestani, M. Karami, A.H. Hassani and N. Yousefi. 2009. Accumulation of heavy metals in soil and uptake by plant species with phytoremediation potential. Environ. Earth Sci. 59(2): 315–323.

Zafar, S., F. Aqil and I. Ahmad. 2007. Metal tolerance and biosorption potential of filamentous fungi isolated from metal contaminated agricultural soil. Bioresour. Technol. 98(13): 2557–2561.

CHAPTER 21

Potential of Aquatic Macrophytes in Phytoremediation of Heavy Metals: A Case Study from the Lake Sevan Basin, Armenia

Lilit Vardanyan[1]* and Jaysankar De[2]

[1]Soil and Water Sciences Department, Institute of Food and Agricultural Sciences, University of Florida, Gainesville, Florida, 32611, USA.
[2]Department of Food Science and Human Nutrition, Institute of Food and Agricultural Sciences, University of Florida, Gainesville, Florida, 32611, USA.

1. INTRODUCTION

Contamination of soil, groundwater, sediments, surface water, and air with heavy metals is a major threat to the environment. Aquatic systems often act as final receptacles to these metals, of which the concentration in interstitial waters might increase several thousand times beyond their initial concentrations by effluents from wastes (Bastian and Hammer 1993). Heavy metals are especially toxic due to their ability to bind with proteins and prevent DNA replication (Kar et al. 1992, Nies 1999, Fernandez-Pol 2002, Gür and Topdemir 2008). Because they are stable and cannot be degraded or destroyed, they tend to accumulate in soils, water, and sediments (Ramaiah and De 2003, De et al. 2007, 2009). The principal anthropogenic sources of heavy metals are industrial point sources (mines, foundries, and smelters, etc.) and diffuse sources (combustion by-products, traffic, etc.). Pollutants enter aquatic systems via numerous pathways, including effluent discharge, industrial, urban, and agricultural run-off, as well as airborne deposition. Many technologies have been used to reduce aquatic pollution, but they are generally costly, labor-intensive and generate secondary waste. Phytoremediation is an alternative approach to circumvent this shortcoming of other technologies (Flathman and Lanza 1998, Rock et al. 2000, Kamal et al. 2004).

*Corresponding author: lilitvardanyan@ufl.edu

The main goal of phytoremediation is the removal of metals from contaminated soils and waters using natural or induced metal tolerance/accumulation capacities of some plant species or populations originating mainly from contaminated areas (Baker and Brooks 1989, Salt et al. 1998, McCutcheon and Schnoor 2003). Phytoremediation by using certain species could offer a low cost and low technology alternative to currently available clean up technologies (Baker et al. 1991). Some heavy metals (e.g. Cu, Fe Zn, Mn, Ni, Mo, Se, etc.) are essential for normal plant growth, although both essential and non-essential metals (e.g. Hg, Cd, Pb, As) can result in growth inhibition and toxicity symptoms (Poschenrieder et al. 2006). Most of the metals induce production of free radicals and reactive oxygen species (ROS) leading to oxidative stress (Van Assche et al. 1990, Meharg 1994, Gratão et al. 2005). Phytoremediation is thus a biological process in which living plants are used to remove, accumulate, degrade, or contain environmental contaminants. This passive remediation technique is based on the natural ability of vegetation to utilize nutrients, which are transported by capillary action from the soil and groundwater through a plant's root system. With advances in biological, chemical, and engineering technologies, phytoremediation has the potential to serve as a sustained, ecologically sound method to remediate contaminated soil and groundwater (Brown and Hall 1990, Susarla et al. 2002, Vardanyan et al. 2008). There are several mechanisms helping this process, such as phytostabilization, phytotransformation, phytostimulation, phytovolatilization, phytoextraction, and rhizofiltration (Prasad and Freitas 2003, Rai 2009).

Several studies have demonstrated that aquatic macrophytes can perform as biological filters and they carry out a purifying function by accumulating dissolved metals and toxins in their tissue (Mejare and Bulow 2001, Prasad and Freitas 2006, Vardanyan and Ingole 2006). This ability makes them capable to be used for the phytoremediation process (Welsh and Denny 1980, Say et al. 1981, Heisey and Damman 1982, Bishop and DeWaters 1986, Brix and Schierup 1989, Gardea-Torresday et al. 2005). Macrophytes actively take up metals from the sediments through their roots and translocate them to the shoots, which are available for grazing by aquatic animals, including fish. These may also be available for epiphytic phytoplankton and herbivorous and detritivorous invertebrates (Gibbons et al. 1994, Ravera 2001, Cardwell et al. 2002), representing a major route of bioaccumulation of heavy metals in the aquatic food chain. It is, therefore, of interest to assess the levels of heavy metals in macrophytes due to their importance in ecological processes.

The immobile nature of macrophytes makes them a particularly effective bioindicator of heavy metal pollution, as they reflect the actual environmental contamination prevailing at that site. From an ecological perspective, the apparent lack of translocation of metals from roots to shoots means that the likelihood of bioaccumulation along trophic levels is reduced (Matagi et al. 1998). This is because the higher concentrations of heavy metals are found in plant roots and organs that are unavailable for ingestion and heavy metals are thus tied up, or stored, in areas that are unlikely to be transferred into other areas of the ecosystem and its biota. Studies on the total ecosystem effects involving not only macrophytes but also sediment and other biota are thus necessary to provide a complete picture of the effects of heavy metals on aquatic ecosystems. Significant differences obtained for the heavy metal concentrations between macrophyte species suggest that the interactions and behaviors of heavy metals with macrophytes are different for each species (Prasad and Frietas 2003, Vardanyan and Ingole 2006). The factors affecting distribution and abundance of submerged aquatic vegetation and the effect of this vegetation on water quality are poorly understood. Also, the problem with the selectivity of accumulation of elements by the macrophytes is yet to be studied systematically (Monni et al. 2000, Vardanyan and Ingole 2006).

The objective of this study was to investigate the concentrations of different heavy metals in thirteen of the most abundant aquatic macrophytes occurring in the catchment zone of the Lake Sevan, Armenia. Lake Sevan is the major source of water for drinking and irrigation in

Armenia. This study focused on the fact that shore macrophyte vegetation of the Lake Sevan had decreased immensely due to the direct loss of littoral area by lowering the lake level, the consequential increase of shore erosion and the unstable growing conditions due to the water level fluctuations.

2. MATERIALS AND METHODS

2.1 Sampling Site

Lake Sevan is situated in the northeastern part of Armenia. It is the largest lake in the Caucasus Region and one of the largest freshwater high mountain lakes of Eurasia (Fig. 1).

Figure 1 Location of the Lake Sevan (Armenia).

Before the increased artificial outflow which began in 1933, the surface of the Lake Sevan was at an altitude of 1916.20 m above the mean sea level with a surface area of 1,416 km^2 and volume of 58.5 km^3 (Babayan et al. 2005). The decrease in water-level (about 20 m) influenced an array of hydrological and ecological conditions. The watershed is known to contain more than 1,500 species of flower and seed-producing and more than 250 species of spore-producing plants such as mosses and lichens. In addition, many endemic (local varieties specific to the Sevan Basin) and relict (representatives of old disappearing flora) species can be found in the watershed (Barseghyan 1990). Many of these endemic and endangered plants have highly restricted areas of coverage and are sensitive to changes in environmental conditions.

Prior to the 1960s, surface water in Armenia, and especially in the Sevan Basin, was of remarkably high quality as per the international standards. Groundwater resources were well protected from pollution resulting from agricultural or industrial practices. Spring water usually was of good quality and could be used for drinking without further treatment. However, the situation changed drastically with the emergence of rapid industrialization. The discharge of industrial pollutants, domestic sewerage and agricultural run-off into the lake increased the organic loading. Pollution inflicted from sewage, industry, and agriculture sharply increased since 1960s. Decomposition of organic matter decreased the oxygen concentrations of the water body. In Lake Sevan in the 1970s, oxygen saturation in the bottom water of the profundal zone during the stratification period was close to analytical zero (Babayan et al. 2005). Deterioration of oxygen conditions may seriously contaminate the water, endangering the plant and animals living therein. Dumping of garbage has been a major problem for urban areas, especially for Sevan, Gavar, and Martuni areas.

Extensive use of water resources associated with increased pollution of Lake Sevan influenced the ecosystem of the lake, from physical conditions to primary production and fish community. Livestock overgrazing on the Lake Lichk area resulted in degradation of vegetation and serious deterioration of the waterfowl habitats. As a result of the intensive exploitation of the lake over the years, its ecological system has been disturbed, with falling water level and resultant swamping bringing about qualitative changes, and the state of its native fish life, the most sensitive index of the health of the lake, has changed. Besides hydrological factors, anthropogenic impact, recreational load, settlements growth, industrial and agricultural development affected the lake water quality and quantitative and qualitative development of the hydrobionts. The largest artificial forest of the country is situated along the shoreline of the lake. Diversity of aquatic associations (plankton, benthos and ichthyofauna) is qualitatively poor with only a few dominant species, which simplified studies on ecological relationships such as food web, etc. About 1,600 species of vascular plants (50% of Armenia's flora) have been registered from the lake basin. Of them, 48 species are in the Red Data Book of Armenia (1990). The dominant communities of the Sevan basin are mountain steppe, sub-alpine and alpine vegetation (Barseghyan 1990). A total of 28 rivers empty into Lake Sevan but only four (Gavaraget, Argichi, Makenis and Masrik) of them flow throughout the year. They all play an important role in maintaining the flow of water to and from the lake.

2.2 Water and Plant Samples

Samples were collected from four sites (Fig. 2) along the Lake shore, where the respective rivers meet the lake: site A (Gavaraget), site B (Argichi), site C (Makenis) and site D (Masrik) on June 20, 2008.

All samples were stored in clean polyethylene bottles under freezing temperature until analysis. Measurements of temperature, dissolved oxygen (DO), and pH were performed on the field.

Ten representatives of similar sizes from each of the sampled thirteen macrophyte species (Table 1) were collected from all sites.

The plants were thoroughly washed at the sampling site with a jet stream of tap water until the surfaces appeared to be clean. The bulk water was removed by vigorous shaking of the plants and then by air-drying for 3 hours. Plants were packed into paper bags and transported to the laboratory. The samples were washed several times with deionized water, air-dried, and then placed in a drying oven at 70°C for 6 hours. Microscopic inspection of plant surfaces did not show any visible deposits of solid matter on it.

A composite sample from all the ten representative plants for each of the 13 plant species was powdered by mechanically grinding them together by using a stainless-steel grinder. The materials were then powdered by homogenization. Before analysis, aliquots of this powder were again dried overnight at 70°C.

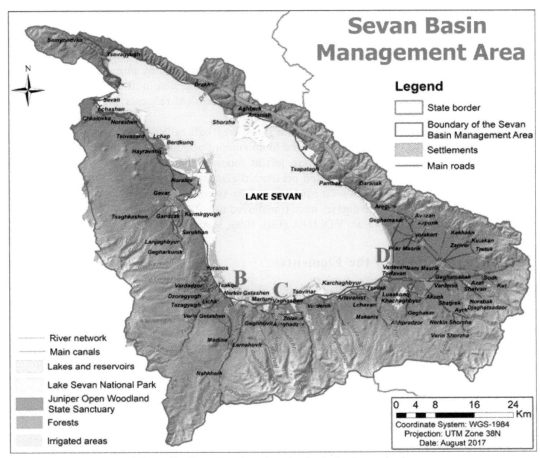

Figure 2 Sampling locations in the Lake Sevan basin.
Source: Ministry of Nature Protection of Armenia (1998)

Table 1 Macrophyte samples from Sevan Lake (Armenia)

No*	Name	Family	Sampling site
1	*Potamogeton natans* L.	Potamogetonaceae	C
2	*Veronica anagallis-aquatica* L.	Scrophulariaceae	A
3	*Lemna minor* L.	Lemnaceae	B
4	*Veronica beccabunga* L.	Scrophulariaceae	D
5	*Ranunculus rionii* (Lagger.)	Ranunculaceae	B
6	*Myriophyllum spicatum* L.	Haloargaceae	B
7	*Alisma plantago-aquatica* L.	Alismataceae	D
8	*Potamogeton pectinatus* L.	Potamogetonaceae	C
9	*Schoenoplectus lacustris* subsp. *tabernaemontani*	Cuperaceae	C
10	*Glyceria plicata* Fries	Gramineae	A
11	*Potamogeton crispus* L.	Potamogetonaceae	A
12	*Potamogeton perfoliatus* L.	Potamogetonaceae	C
13	*Juncus bufonius* L.	Juncaceae	B

*These numbers are used in Figure 3.

2.3   Digestion of the Samples

The acidified and filtered water samples were analyzed directly, whereas plant materials were subjected to an acid mineralization procedure prior to ICP-MS (Elan 6000, Perkin Elmer-Sciex, Canada) measurements. Glass and plasticwares were decontaminated by immersing them in 10% (v/v) Extran R solution (MERCK) for 2 days, followed by immersion in diluted HNO_3 (10% v/v) for 3 days and finally rinsing with Milli-Q water. All chemical reagents used in this process were of at least analytical grade.

The dried plant samples were milled in a metal-free ball mill (ZrO_2). Five ml HNO_3 and 1 ml H_2O_2 (high analytical grade) were added to an amount of 500 mg sample in a PTFE-vessel. The samples in tightly closed vessels were left at room temperature overnight, and then the vessels were put into a microwave-heated pressure digestion system (MLS 1200, MLS GmbH, Germany). They were digested for 20 minutes at 800 W and were left to cool down to room temperature. Following that the digests were transferred to a 25 ml volumetric flask and filled up to the mark with ultra-pure water (VDLUFA (Ed), 1996, Methodenbuch Bd. VII, Methode 2.1.1).

2.4   Concentrations of the Elements

The concentration of metals in water, plant, and sediment samples were analyzed by Inductively Coupled Plasma-Optical Emission Spectrometry (ICP-OES; EN ISO 11885; for Ni, Zn), ICP-MS (DIN 38406-29; for Cd, Co and Cu) and by Cold Vapor Atomic Emission Spectrometry (CV-AAS; DIN EN 1483; for Hg). The ICP-MS was equipped with a Scott type Rhyton spray chamber and a quadrupole mass filter. Rh (10 ppb) served as an internal standard. The samples were diluted as appropriate by 2% HNO_3-solution and measured against external calibration curves using freshly prepared standards. Hg was measured by flow-injection hydride generation Atomic Absorption Spectrometry (AAS). The samples were mixed with a sodium-borohydride solution in an automated system. Elementary Hg formed due to the reduction was removed from the solution by a stream of Ar and transferred to a quartz-cell, using the method described in EN ISO 11885 (1997). All the analyses were carried out in triplicate and the average results are reported.

3.   RESULTS

3.1 Water Samples

General water parameters (i.e., heavy metals excluded) showed little differences between the four selected sites. The average temperature was $15 \pm 0.7°C$ and the pH ranged between 7.25–7.95. Details of other parameters are shown in Table 2.

Concentrations of heavy metals such as Cd, Co, Cu, Ni, Pb, Tl, Zn and Hg in water were estimated from the investigated rivers (Table 3). Concentrations of Co (0.5 ppb), Cd (0.5 ppb), Tl (0.1 ppb) and Hg (<0.3 ppb) in water were in the same range for the all sites. For Cu, the highest concentration (2.4 ppb) was recorded in Gavaraget and the lowest (1.3 ppb) was in Makenis. For Ni, the highest concentration was in Masrik (3.5 ppb) and the lowest in Gavaraget (1.4 ppb). For Pb, the highest (1.3 ppb) concentration was in Gavaraget and at the rest of the sites the value was always <0.5 ppb. Zinc was the highest (34.6 ppb) in Gavaraget and the lowest (5 ppb) in Argichi, Masrik and Makenis. It was obvious that the water in the Gavaraget was most polluted in terms of load of heavy metals (Table 3), which could be attributed to the discharge of untreated sewage from the city of Gavaraget into the river.

Table 2 Water parameters of the rivers

Index/river	Gavaraget	Argichi	Makenis	Masrik
pH	7.25	7.6	7.8	7.95
HCO_3^- (g/m³)	158.5	98.6	92	171.3
Cl^- (g/m³)	26.4	12.5	13.3	17.4
Ca^{2+} (g/m³)	23.7	20.8	18.6	30.4
Mg^{2+} (g/m³)	18.4	9.6	9.3	18.4
SO_4^{2-} (g/m³)	28.2	21.5	22.3	26.3
Total Ions (g/m³)	259	165	157	267
NH_4^+ (g/m³)	0.29	0.15	0.05	0.14
NO_2^- (g/m³)	0.008	0.005	0.002	0.009
NO_3^- (g/m³)	2.86	1.86	1.12	2.63
N_{min} (g/m³)	3.16	2.02	1.18	2.8
PO_4^{3-} (g/m³)	0.26	0.11	0.08	0.14
BOD_5 (gO₂/m³)	3.83	3.51	2.2	3.65

Table 3 Occurrence of heavy metals in the rivers

Rivers / Heavy metals*	Gavaraget	Argichi	Makenis	Masrik
Cd	<0.5	<0.5	<0.5	<0.5
Co	<0.5	<0.5	<0.5	<0.5
Cr	3.4	2.3	3.4	2.9
Cu	2.4	1.9	1.3	1.8
Ni	1.4	1.5	1.7	3.5
Pb	1.3	<0.5	<0.5	<0.5
Tl	<0.1	<0.1	<0.1	<0.1
Zn	34.6	5.0	5.0	5.0
Hg	<0.3	<0.3	<0.3	<0.3

*Concentration in ppb

3.2 Plants

In general, there was no specific pattern in the accumulation of heavy metals by the 13 different species of aquatic macrophytes (Figs. 3A–D). Mercury (<0.01 mg/kg) was below detection limit in all the plants. The maximum Co (8.94mg/kg) was recorded in *Lemna minor* L., whereas Cu (19.80 mg/kg) and Mo (2.98 mg/kg) were the highest in *Batrachium rionii* (Lagger). Cd was the highest (0.60 mg/kg) in *Veronica anagallis-aquatica* L.

The highest concentrations of Pb (5.25 mg/kg) and Zn (129 mg/kg) were in *Myriophyllum spicatum* L. Tl was the highest (0.14 mg/kg) in *L. minor,* whereas Ni was the maximum (15.8 mg/kg) in *Potamogeton perfoliatus* L. (Figs. 3A–D). The minimum accumulation of Co (0.41 mg/kg) was recorded in *V. beccabunga*. The lowest accumulation of Cd (0.025 mg/kg) and Mo (0.65 mg/kg) was recorded in *Schoenoplectus lacustris* subsp. *tabernaemontani*. *M. spicatum* showed the highest accumulation of total heavy metals, whereas *P. perfoliatus*, *Ranunculus rionii* (Lagger.), and *V. anagallis-aquatica* also showed high potential of heavy metal accumulation. *Schoenoplectus lacustris* subsp. *tabernaemontani* and *Juncus bufonius* L. showed minimal accumulation of heavy metals. Our results indicated that the macrophytes accumulated toxic heavy metals like Cd, Co or Pb in concentrations several hundred times than that of the water bodies from which they were collected, where these metals were hardly detectable.

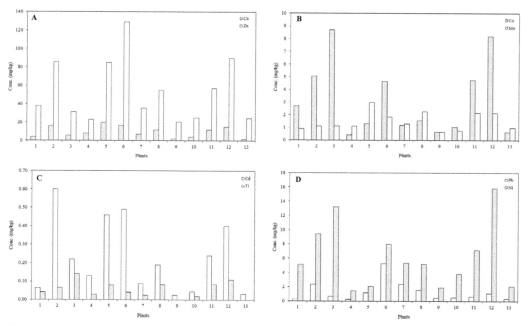

Figure 3 (A–D)   Concentrations of heavy metals in 13 species of plant samples collected from the Lake Sevan basin rivers (see Table 1 for the list of plant species). **A:** Concentrations of Cu and Zn, B: Concentrations of Co and Mo, C: Concentrations of Cd and Ti, and D: Concentrations of Pb and Ni.

4.   DISCUSSION

Factors such as light intensity, oxygen concentration and temperature are known to influence the uptake of minerals (Devlin 1967). Moreover, the energy derived from photosynthesis and the oxygen released can improve conditions for the active absorption of elements. However, interactions between metals are often complex, and they are dependent on the metal concentration and pH of the growth medium (Balsberg-Pohlsson 1989). The effects of trace elements in an aquatic ecosystem can be assessed by changes in the community structure, physiological activity and ultrastructural components of macrophytes (Chester and Stoner 1974, Bohn 1975, Gunterspergen et al. 1989, Blaylock and Huang 2000, Kumar et al. 2006). However, comparison of metal content in macrophytes is often difficult because of differences in sampling time (age of plants) and presence of pollution sources. Moreover, the metal data cannot be extrapolated from one species to another or even within the same species, largely due to different accumulation rates across different plant species. In eutrophic lakes, such as Lake Sevan, high local concentrations of metals often occur as a result of the strong reducing (low oxygen) environment coupled with industrial and municipal discharges (Vardanyan 2002). In addition, the Sevan basin rivers have inflow and outflow of freshwater that may reduce the rate of metal accumulation in aquatic macrophytes.

Results from this study showed that the macrophytes concentrate heavy metals from the surrounding water and thus act in purifying the water. However, an increase in the concentration of these microelements in water is toxic for hydrobionts. One of the major properties of heavy metals is their ability to interact with several organic compounds that produce relatively strong complex ionic compounds such as cyanide, radonide and thiosulfates by combining with ions (Babayan 1988). Many of them produce toxic salts that occur in very low concentrations.

Therefore, methods with high determination accuracies, i.e. where determination error ratio is minimal, are required for their determination.

Compared with other heavy metals, copper has one of the most toxic effects on the growth and development of plants. A 1 mg/l (ppm) concentration of copper may cause a withering of a plant (Balsberg-Pohlsson 1989). Copper affects the oxidizing system of a cell, while it is an essential element in several enzyme systems. Copper is known to reduce photosynthesis rates and respiration of aquatic moss, *Fontinalis antipyretica* (Vazguez et al. 2000). Although the microelements Zn, Mo and Cu play a leading part in photosynthesis activity of plants and contribute in transfer of assimilators from leaves to generative organs and roots in plants, even a slight increase in the concentration of these elements in nutrient medium may inhibit the growth and development of plants (Balsberg-Pohlsson 1989, Guilizzoni 1991). The microelements Co, Zn and Cu increase the drought-resistance of plants. Several microelements directly participate in the build-up of enzymes that act as catalysts in protein, carbohydrate, and other cellular compounds exchange processes (Vangronsveld and Clijsters 1994).

Heavy metals generally penetrate from aquatic medium into humans through water–plant–human or water–plant–animal–human biological chains (Smirnova 1984). Therefore, finding solution to the problem of toxin tolerance in an aquatic medium is essential for an ecosystem and its components. There are reports with a variety of data detailing the effects of heavy metals on water and water plant structure and their properties, their enzymatic activity and nutrition pattern (Wittman 1979, Pascoe et al. 1994, Orth et al. 2001, Cardwell et al. 2002). A negative correlation was revealed between heavy metal content in growth medium and plants' submerged organs and green biomass (Zayed et al. 1998). However, these peculiarities are not always well-defined and greatly depend on the plant species and compound forms of certain metals (Clemens et al. 2002). The effect of microelements on metabolism process is closely related with composition of the nutrient medium where the plants develop. For example, our results indicate that plants of one species taken from different riverbank sites show different rates of heavy metal content (Vardanyan and Ingole 2006). In many cases, the concentrations reported in the literature (Schlacher-Hoenlinger and Schlacher 1998,Weiss et al. 2006, Gupta et al. 2008, Mishra and Tripathi 2008) are supported by the data presented in this paper.

High concentrations of heavy metal salts in storage reservoirs may depend on both geochemical peculiarities of the area and water inflow. If, in the former case, aquatic organisms have evolved some properties enabling it to adapt to the medium, then in the latter case aquatic organisms are exposed to injurious effects because of water pollution by heavy metals from industrial waste. Water plants, by accumulating heavy metals in their tissues, play an important role in their transport in aquatic systems. Experiments showed that plants accumulate 2.5 kg Zn from 1 ha water surface (Fritioff and Greger 2006). Submerged vegetation accumulates heavy metals (Cu and Cr) 4-9 times more than riverside vegetation does. The accumulation ratio of magnesium and copper in respect to/against their content in the medium is highest (about 10 times) in moss *(Fontinalis).*

In this study, the Hg concentration was not high in the plant samples, which suggested that the concentration of Hg in the water and soil at the study sites was low. Submerged water plants, particularly hornwort *(Ceratophyllum),* accumulate mercury, particularly inorganic mercury compounds, and they are so rapid in doing so that a balance with the solution is established after two hours (Vardanyan and Ingole 2006). Riverside vegetation also take up and accumulate mercury very rapidly, especially by the submerged portion of their stem. Reeds can take up mercury from the soil and bottom deposits and then release mercury vapor into the atmosphere via evaporation from the leaves, especially under optimal lighting and high transpiration factors (James et al. 1980).

Study on microelement accumulating properties of water plants revealed that they accumulate by order of Mn, Ni, Cu, Mo, V, Sr, Ba, Fe, Al (Greger 1999). Chemical analysis of higher plants taken from industrial sewage polluted storage reservoirs indicate that accumulation of heavy metals by aquatic plants reduces concentration of these microelements in the water (Samecka et al. 2001). Our results substantiate these studies.

This study was targeted to gain a better understanding of the importance of aquatic plants in heavy metal accumulation and detoxification mechanisms that may lead to elaboration of new pollution control and prevention facilities aimed to reserve the lake's ecosystem. Many of the plant species investigated in the study have been recognized as medicinal and edible ones.

5. CONCLUSION

Taking in consideration the role of macrophytes in the Sevan basin and the fact that above mentioned issues have not been properly addressed in the region, we proceeded to study the heavy metal accumulation peculiarities in macrophyte. Such discussions and investigations have become especially urgent over the last few years keeping metal pollution of the environment (especially aquatic regime) in view. Macrophytes are at the priority level in the list of cleaning and detoxification methods. Important functions of the lake ecosystem include the direct retention of nutrients and toxic substances during the growth period as well as by forming biotic structures for biofilms degrading organic and toxic substances, for epiphytic algae and macro invertebrates, and for providing measures for protection and redevelopment of juvenile fishes.

From the present observations, it is concluded that there is a uniform pattern of heavy metal variation in the macrophytes of Lake Sevan basin. In general, values of some metals like zinc and lead were high in almost all the specimens. Since the investigations on the heavy metals accumulation in macrophytes of Sevan Lake were carried out after a long time, the results presented here could be very useful for environmental monitoring and checking the health of the water body. The data presented here is indispensable information for studies of related nature. The aquatic macrophytes were found to be the potential source for accumulation of heavy metals from water and wetlands. Therefore, such studies should become an integral part of the sustainable development of the ecosystems and pollution assessment program. Among various future tasks related to habitat technology, priority should be given to the restoration and creation of coastal nurseries such as macrophyte beds. There is an urgent need to study more of those specific macrophytes which are "responsible" for cleaning the water body from toxic heavy metals. It would be advantageous to understand the similarities and differences between accumulation pathways across different macrophytes for effective deployment in phytoremediation, as well as to identify new macrophytes capable of such activities. Most importantly, this study is significant, especially as the water from this habitat is used for drinking and irrigation.

Acknowledgments

This study was carried out as a part of the SEMIS international project, funded by the Volkswagen Foundation, Germany.

References

Babayan, A., S. Hakobyan, K. Jenderedjian, S. Muradyan and M. Voskanov. 2005. Lake Sevan: Experience and Lessons Learned Brief. ILEC, Managing Lakes and their Basins for Sustainable Use: A report for Lake Basin Managers and Stakeholders. International Lake Environment Committee Foundation: Kusatsu, Japan. 347–362.

Babayan, G.G. 1988. Dynamics of some heavy metals and types of their migration in lake sevan. PhD Thesis. Yerevan.

Baker, A.J.M. and R.R. Brooks. 1989. Terrestrial higher plants which accumulate metallic elements—A review of their distribution, ecology and phytochemistry. Biorecovery 1: 81–126.

Baker, A.J.M., R.D. Reeves and S.P. McGrath. 1991. *In situ* decontamination of heavy metal polluted soils using crops of metal-accumulating plants—A feasibility study. *In*: R.E. Hinchee and R.F. Olfenbuttel (eds), *In Situ* Bioreclamation, Applications and Investigations for Hydrocarbon and Contaminated Site Remediation. Battelle Memorial Institute, Columbus, OH. Butterworth-Heinemann, Boston, MA, pp. 600–605.

Balsberg-Påhlsson, A.M. 1989. Toxicity of heavy metals (Zn, Cu, Cd, Pb) to vascular plants. A literature review. Water Air Soil Pollut. 47: 287–319.

Barseghyan, A.M. 1990. Marsh Aquatic Vegetation of Armenian SSR. Publ. of NAS RA, Armenia.

Bastian, R.K. and D. Hammer. 1993. The use of constructed wetlands for wastewater treatment and recycling. *In*: G.A. Moshiri (ed.), Constructed Wetlands for Water Quality Improvement. Lewis Publishers, CRC Press, Boca Raton, Fl, p. 59.

Bishop, P. and J. DeWaters. 1986. Heavy metal removal by aquatic macrophytes in a temperate climate aquatic treatment system. Proceedings: International Conference on Innovative Biological Treatment of Toxic Wastewaters, Arlington, VA, 100–117.

Blaylock, M.J. and J.W. Huang. 2000. Phytoextraction of metals. *In*: I. Raskin and B.D. Ensley (eds), Phytoremediation of Toxic Metals: Using Plants to Clean Up the Environment, John Wiley and Sons, New York, pp. 53–70.

Bohn, A. 1975. Arsenic in marine organisms from west greenland. Mar. Pollut. Bull. 6: 87–89.

Brix, H. and H.H. Schierup. 1989. The use of aquatic macrophytes in water pollution control. Ambio 18: 100–107.

Brown, M.T. and I.R. Hall. 1990. Ecophysiology of metal uptake by tolerant plants. *In*: A.J. Shaw (ed.), Heavy Metal Tolerance in Plants: Evolutionary Aspects. CRC Press, Boca Raton, FL, pp. 95–104.

Cardwell, A.J., D.W. Hawker and M. Greenway. 2002. Metal accumulation in aquatic macrophytes from southeast Queensland, Australia. Chemosphere 48: 653–663.

Chester, R. and J.H. Stoner. 1974. The distribution of zinc, nickel, manganese, cadmium, copper and iron in some surface waters from the world ocean. Marine Chem. 2: 17–32.

Clemens, S., M.G. Palmgren and U. Kramer. 2002. A long way ahead: understanding and engineering plant metal accumulation. Trends Plant Sci. 7: 309–315.

De, J., K. Fukami, K. Iwasaki and K. Okamura. 2009. Distribution of heavy metals at the Uranouchi Inlet, Kochi prefecture, Japan. Fisheries Science. 75: 413–423.

Devlin, R.M. 1967. Plant Physiology. Reinhold, New York, 564–580.

EN ISO 11885. 1997. Detennination of 33 elements by inductively coupled plasma atomic emission spectroscopy.

Fernandez-Pol, Jose A. US Patent 6441009 – Agent and method of preventing and treating heavy metal exposure and toxicity. US Patent Issued on August 27, 2002.

Flathman, P.E. and G.R. Lanza. 1998. Phytoremediation: current views on emergent green technology. J. Soil Contam. 7: 415–432.

Fritioff, A. and M. Greger. 2006. Uptake and distribution of Zn, Cu, CD, and Pb in an aquatic plant *Potamogeton natans*. Chemosphere 63: 220–227.

Gardea-Torresday, J.L., J.R. Peralta-Videa, G. De La Rosa and J.G. Parsons. 2005. Phytoremediation of heavy metals and study of the metal coordination by X-ray absorption spectroscopy. Coord. Chem. Rev. 249: 1797–1810.

Gibbons, M.V., H.L. Gibbons, Jr., and M.D. Sytsma. 1994. A Citizen's Manual for Developing Integrated Aquatic Vegetation Management Plans. Washington State Department of Ecology, Olympia, WA.

Gratão, P.L., M.N.V. Prasad, P.F. Cardoso, P.J. Lea and R.A. Azevedo. 2005. Phytoremediation: Green technology for the cleanup of toxic metals in the environment. Brazilian J. Plant Physiol. 17: 53–64.

Greger, M. 1999. Metal availability and bioconcentration in plants. *In*: M.N.V. Prasad and J. Hagemeyer (eds), Heavy Metal Stress in Plants—From Molecules to Ecosystems, Berlin Heidelberg, Germany, Springer Verlag, pp. 1–27.

Guilizzoni, P. 1991. The role of heavy metals and toxic materials in the physiological ecology of submerged macrophytes. Aquatic Bot. 41: 87–109.

Gunterspergen, G.R., F. Stearns and J.A. Kadlec. 1989. Wetland vegetation. *In*: D.A. Hammer (ed), Constructed Wetlands for Wastewater Treatment: Municipal Industrial and Agricultural. Lewis Publishers, Chelsea, Michigan, pp. 73–88.

Gupta, S., S. Nayek, R.N. Saha and S. Satpati. 2008. Assessment of heavy metal accumulation in macrophyte, agricultural soil, and crop plants adjacent to discharge zone of sponge iron factory. Environ. Geol. 55: 731–739.

Gür, N. and A. Topdemir. 2008. Effects of some heavy metals on in vitro pollen germination and tube growth of apricot (*Armenica vulgaris* Lam.) and Cherry (*Cerasus* avium L.). World Appl. Sci. J. 4: 195–198.

Heisey, R.M. and A.W.H. Damman. 1982. Copper and lead uptake by aquatic macrophytes in eastern Connecticut, U.S.A. Aquat. Bot. 14: 213–229.

James, R., P. Wells, B. Kaufman and J.D. Jones. 1980. Heavy metal contents in some macrophytes from saginaw bay (Lake Huron, U.S.A). Aquat. Bot. 9: 185–193.

Kamal, M., A.E. Ghaly and C.R. Mahmoud. 2004. Phytoaccumulation of metals by aquatic plants. Environ. Pollu. 29: 1029–1039.

Kar, R.N., B.N. Sahoo and L.B. Sukla. 1992. Removal of heavy metal from mine water using sulphate reducing bacteria. Pollution Res. 11: 1–13.

Kumar J.I.N., H. Soni and R.N. Kumar. 2006. Biomonitoring of selected freshwater macrophytes to assess lake trace element contamination: A case study of Nal Sarovar Bird Sanctuary, Gujarat, India. J. Limnology, 65: 9–16.

Matagi, S.V., D. Swai and R. Mugabe. 1998. A review of heavy metal removal mechanisms in wetlands. African J. Trop. Hydrobiol. Fisheries 8: 23–35.

McCutcheon, S.C. and L.L. Schnoor (eds). 2003. Phytoremediation: Transformation and Control of Contaminants. Wiley-Interscience, Inc. Hoboken, New Jersey.

Meharg, A.A. 1994. Integrated tolerance mechanisms—constitutive and adaptive plant—responses to elevated metal concentrations in the environment. Plant Cell Environ. 17: 989–993.

Mejare, M. and L. Bulow. 2001. Metal-binding proteins and peptides in bioremediation and phytoremediation of heavy metals. Trends Biotechnol. 19: 67–73.

Mishra, V.K. and B.D. Tripathi. 2008. Concurrent removal and accumulation of heavy metals by the three aquatic macrophytes. Bioresource Technol. 99: 7091–7097.

Monni, S., M. Salemaa and N. Millar. 2000. The tolerance of *Empetrum nigrum* to copper and nickel. Environ. Pollu. 109: 221–229.

Nies, D.H. 1999. Microbial heavy-metal resistance. Appl. Microbiol. Biotechnol. 51: 730–50.

Orth, R.J., D.J. Wilcox, L.S. Nagey, J.R. Whiting and J.R. Fishman. 2001. 2000 Distribution of Submerged Aquatic Vegetation in the Chesapeake Bay and Tributaries and the Coastal Bays. VIMS Special Scientific Report Number 142. Final report to U.S. EPA, Chesapeake Bay Program, Annapolis, MD. Grant No. CB993777-02-0. http://www.vims.edu/bio/sav/sav00

Pascoe, G.A., R.J. Blanchet and G. Linder. 1994. Bioavailability of metals and arsenic to small mammals at a mining waste-contaminated wetland. Environ. Contam. Toxicol. 27: 44–50.

Poschenrieder, C., R. Tolrà and J. Barceló. 2006. Can metals defend plants against biotic stress? Trends Plant Sci. 11: 288–295.

Prasad, M.N.V. and H. Freitas. 2003. Metal hyperaccumulation in plants—biodiversity prospecting for phytoremediation technology. Electron. J. Biotechn. 6: 285–321.

Prasad, M.N.V. and H. Freitas. 2006. Metal tolerant plants—Biodiversity prospecting for promoting phytoremediation technologies. *In*: M.N.V Prasad, K.S. Sajwan and R. Naidu (eds), Trace Elements in the Environment: Biogeochemistry, Biotechnology and Bioremediation. CRC Press, Florida, USA (Taylor and Francis) Chap. 25, pp. 483–506.

Rai, P.K. 2009. Heavy metal phytoremediation from aquatic ecosystems with special reference to macrophytes. Critical Rev. Environ. Sci. Technol. 39: 697–753.

Ramaiah, N. and J. De. 2003. Unusual rise in mercury resistant bacteria in coastal environs. Microbial. Ecol. 45: 444–454.

Ravera, O. 2001. Monitoring of the aquatic environment by species accumulator of pollutants: A review. Scientific and legal aspects of biological monitoring in freshwater J. Limnol. 60: 63–78.

Rock, S., B. Pivetz, K. Madalinski, N Adams and T. Wilson. 2000. Introduction to phytoremediation. U.S. Environmental Protection Agency, Washington, D.C., EPA/600/R-99/107 (NTIS PB2000-106690).

Salt, D.E., R.D. Smith and I. Raskin. 1998. Phytoremediation. Ann. Rev. Plant Physiol. Plant Mol. Biol. 49: 643–668.

Samecka, A., A. Cymerman and J. Kempers. 2001. Concentrations of heavy metals and plant nutrients in water, sediments and aquatic macrophytes of anthropogenic lakes (former open cut brown coal minces) differing in stage of acidification. Sci. Total Environ. 281: 87–98.

Say, P.J., J.P.C. Harding and B.A. Whitton. 1981. Aquatic mosses as monitors of heavy metal contamination in the river etherow, Great Britian. Environ. Pollu. (Series B) 2: 295–307.

Schlacher-Hoenlinger, M.A. and T.A. Schlacher. 1998. Differential accumulation patterns of heavy metals among the dominant macrophytes of a mediterranean seagrass meadow. Chemosphere 37: 1511–1519.

Smirnova, N.N. 1984. The role of higher aquatic plants in heavy metal migration in water reservoirs. Limnology of mountainous water reservoirs. Yerevan 288–289.

Susarla, S., V.F. Medina and S.C. McCutcheon. 2002. Phytoremediation: An ecological solution to organic chemical contamination. Ecol. Eng. 18: 647–658.

The Red Book of Armenian SSR: Plants of Armenian SSR, 1990. Institute of Botany. Hayastan, pp. 284.

Van Assche, F., J. Vangronsveld and H. Clijsters. 1990. Physiological aspects of metal toxicity in plants. *In*: J. Barcelo (ed.), Environmental Contamination. CEP Consultants, Edinburgh, pp. 246–250.

Vangronsveld, J. and H. Clijsters. 1994. Toxic effects of metals. *In*: M.E. Farago (ed), Plants and the Chemical Elements: Biochemistry, Uptake, Tolerance and Toxicity. VCH, New York, pp. 149–177.

Vardanyan, L.G. 2002. Heavy metals concentration in some dominant aquatic macrophytes from Lake Sevan. XXI century: Ecological science in Armenia. Materials of the III Republican Youth Scientific Conference, Yerevan 117–121.

Vardanyan L.G. and B. Ingole. 2006. Studies on heavy metal accumulation in aquatic macrophytes from sevan (Armenia) and carambolim (India) lake systems. Environ. Int. 32: 208–218.

Vardanyan L., K. Schmieder, H. Sayadyan, T. Heege, J. Heblinski, T. Agyemang, et al. 2008. Heavy metal accumulation by certain aquatic macrophytes from Lake Sevan (Armenia). *In*: M. Sengupta and R. Dalwani (eds), Proceedings of TAAL-2007: The 12th World Lake Conference. Jaipur, India. pp. 1028-1038. 2008.

Vazquez, M.D., J.A. Fernandes, J. Lopez and A. Carballeira. 2000. Effects of water acidity and metals concentration on accumulation and within-plant distribution of metals in the aquatic bryophyte fontinalis antipyretica. Water Air Soil Pollu. 120: 1–19.

Weiss, J., M. Hondzo, D. Biesboer and M. Semmens. 2006. Laboratory study of heavy metal phytoremediation by three wetland macrophytes, Int. J. Phytoremed. 8: 245–259.

Welsh, R.P.H. and P. Denny. 1980. The uptake of lead and copper by submerged aquatic macrophytes in two english lakes. J. Ecol. 68: 443–455.

Wittman, G. 1979. Toxic metals. *In*: U. Forstner and G.T.W. Wittman (eds), Metal Pollution in the Aquatic Environment. Springer-Verlag, Berlin, pp. 3–70.

Zayed, A., S. Gowthaman and N. Terry. 1998. Phytoaccumulation of trace elements by wetland plants: I. Duckweed. J. Environ. Quality. 27: 715–721.

CHAPTER 22

Microbial Surfactants:
Current Perspectives and Role in Bioremediation

Pooja Shivanand[1]*,
Nur Bazilah Afifah Binti Matussin[1] and Lee Hoon Lim[2]

[1]Environmental and Life Sciences, Faculty of Science, Universiti Brunei Darussalam, Jalan
Tungku Link, Gadong BE1410, Negara Brunei Darussalam
[2]Chemical Sciences, Faculty of Science, Universiti Brunei Darussalam, Jalan Tungku Link,
Gadong BE1410, Negara Brunei Darussalam.

1. INTRODUCTION

Surfactants are molecules consisting of both hydrophilic and hydrophobic moieties, which enable them to reduce interfacial tension between the molecules themselves and their interface (Ron and Rosenberg 2001, Chandran and Das 2010, Fracchia et al. 2012, Bhardwaj 2013). These surface-active agents produced from microorganisms and plants possess biological properties such as antimicrobial, antiviral, and anticancer activities and find suitable applications in pharmaceutical, cosmetic and textile industries, oil recovery, wastewater treatment and bioremediation (Randhawa and Rahman 2014, Bouassida et al. 2018).

A wide variety of biosurfactants produced by various microorganisms have been reported, among which surfactin, a bacterial cyclic lipopeptide, was first reported and named in 1968 (Arima et al. 1968). In the late 1960s, a biosurfactant study attracted the attention of researchers due to its properties as a hydrocarbon dissolution agent and its associated applications, which could reduce dependence on chemical surfactants such as carboxylates, sulphonates and sulphate esters (De et al. 2008). These biomolecules have multi-functionality and low toxicity compared to chemical surfactants (Anna et al. 2002, Rufino et al. 2013). Rapid growth in industry and

*Corresponding author: pooja.shivanand@ubd.edu.bn; poojashivanand@outlook.com; Tel: +673 246 0923;
ORCID ID: 0000-0002-5740-6234

transportation sectors in the past decades have led to prevailing environmental hazards like petroleum products. It is imperative to understand the mechanism of degradation of these hydrocarbons in order to devise effective remediation tools. Biosurfactants from microorganisms are excreted extracellularly, either totally or partially, assisting in the adhesion of oil droplets and subsequent degradation of hydrocarbons, thereby offering an eco-friendly alternative to chemical remediation technology. Hence, biosurfactant production by various microorganisms has been studied extensively over the past few years (Parthipan et al. 2017, Keikha 2018).

2. CLASSIFICATION OF BIOSURFACTANTS

Biosurfactants are classified according to their composition and source of production (Rikalović et al. 2015). The structure of a biosurfactant includes a hydrophilic domain consisting of amino acids or peptides, mono-, di- or polysaccharides and a hydrophobic domain consisting of unsaturated or saturated fatty acids (Chen et al. 2015). Based on their chemical structure, biosurfactants are broadly classified into five classes: glycolipids, phospholipids, lipopeptides, polymeric and particulate biosurfactants (Randhawa and Rahman 2014, Chen et al. 2015).

2.1 Glycolipids

As one of the most common biosurfactants, glycolipids comprise long chain fatty acids or hydroxyl fatty acids in combination with mono- or disaccharide moieties (Fracchia et al. 2012, Mnif and Ghribi 2016). These compounds are further divided into several groups depending on the nature of their carbohydrate moieties, such as rhamnolipids, trehalose lipids, sophorolipids, cellobiose lipids, and mannosylerythritol lipids (Fracchia et al. 2012, Mnif and Ghribi 2016). Table 1 summarizes the general structures and producer microorganisms of different biosurfactants (Kulakovskaya et al. 2009, Perfumo et al. 2010, Fracchia et al. 2012, Hoffmann et al. 2012, Mongkolthanaruk 2012, Morita et al. 2015).

Rhamnolipids are composed of either one or two rhamnose sugar molecules linked to β-hydroxy fatty acids (Uzoigwe et al. 2015, Mnif and Ghribi 2016). They are exclusively produced by *Pseudomonas* sp. (Peter and Singh 2014, Li et al. 2016). These rhamnolipids enhance the cell-surface hydrophobicity of *P. aeruginosa,* which aids in bacterial growth and increases the rate of bioremediation. They are also easily biodegradable and considered to be nontoxic (Mohan et al. 2006, Mnif and Ghribi 2016).

Trehalose lipids are biosurfactants commonly found in *Rhodococcus* and other *actinomycetes*. They are composed of two glucose units connected by an α,α-1,1-glycosidic linkage. These lipids are mostly produced when hydrocarbons are used as a source of nutrients for microbes to grow on. The lipids help overcome the low solubility of hydrocarbons (Franzetti et al. 2010) and have potential research applications as antitumour agents (Fracchia et al. 2012, Duarte et al. 2014).

Sophorolipids are produced by different strains of *Torulopsis* and consist of two glucose units connected by a (1,2)-β linkage and a 1-hydroxyoleic acid glycone (Hoffmann et al. 2012). Sophorolipids have the ability to increase surfactant-organic interactions, which can be utilized in the flushing of hydrocarbons from soil (Kang et al. 2010). Cellobiose lipids excreted by *Candida, Geotrichum, Kurtzmanomyces* and *Pseudozyma* were studied for their ability to dissolve organic compounds for consumption by microbes (Kulakovskaya et al. 2009). These lipids consist of two glucose units linked by a 1,4'-β-glycoside bond and a fatty acid aglycone (Kulakovskaya and Kulakovskaya 2014). They serve as natural biocontrol agents against fungal spoilage (Kulakovskaya et al. 2009, Morita et al. 2011). Mannosylerythritol lipids are commonly excreted by yeasts and consist of 4-O-β-D-mannopyranosylerythritol or 1-O-β-D-mannopyranosylerythritol linked to fatty acids. These lipids have antioxidant and healing agent properties, which make them suitable candidates for pharmaceutical use (Morita et al. 2015, Yu et al. 2015).

Table 1 General structures of different types of glycolipids and lipopeptides with their producers (Kulakovskaya et al. 2009, Perfumo et al. 2010, Fracchia et al. 2012, Hoffmann et al. 2012, Mongkolthanaruk 2012, Morita et al. 2015)

Type of glycolipid	General structure	Producer
Rhamnolipid		*Pseudomonas* sp.
Trehalose lipid		*Actinomycete* sp. *Rhodococcus* sp.
Sophorolipid		*Torulopsis bombicola,* *Candida* sp., *Trichosporon asahii*
Cellobiose lipid		*Cryptococcus humicola,* *Saccharomyces cerevisiae*
Mannosylerythritol lipid		Basidiomycetous yeast (*Pseudozyma*)
Type of lipopeptide		
Surfactin		*Bacillus subtilis*
Iturin		*Bacillus subtilis, Bacillus methylotrophicus*
Fengycin		*Bacillus subtilis*

2.2 Lipopeptides

Lipopeptides are classified as cyclic and linear based on their chemical structure. Cyclic lipopeptides contain a cyclic component formed with a carboxyl group in the C-terminus bonded to the amino group of the peptide chain or a hydroxyl group of the fatty acid chain. The best known cyclic lipopeptides are surfactin, iturins, fengycins, lichenysins, viscosins, amphisin and putisolvins (Vandana and Singh 2018). Linear lipopeptides contain a linear chain of amino acids and fatty acids bonded to the α-amino group or other hydroxyl groups. Lipopeptides are mainly produced by *Bacillus* sp. (such as *B. subtilis*, *B. licheniformis* and *B. polymyxa*), *Pseudomonas*, *Streptomyces*, *Aspergillus*, *Serratia* and *Actinoplanes* species (Biniarz et al. 2017).

Surfactin is composed of a peptide chain formed by seven α-amino acids bonded to a hydroxyl fatty acid by a lactone bond to form a cyclic lipopeptide. The typical sequence of amino acids in the peptide ring is L-Glu^1-L-Leu^2-D-Leu^3-L-Val^4-L-Asp^5-D-Leu^6-L-Leu^7 (Dhiman et al. 2016). The cyclic lipopeptide surfactin is mainly produced by *Bacillus* sp. (Vandana and Singh 2018). It acts as an inhibitor of fibrin clot formation and has antibacterial and antitumour properties (Walia 2015).

Iturin is a cyclic lipo-heptapeptide that contains a β-amino fatty acid in its extended chain. The molecular mass of iturin is ~1.1 kDa, consisting of 7 amino acids linked from C14–C17. The lipopeptide profile and bacterial hydrophobicity vary with strain, with iturin A being the only lipopeptide produced by all *B. subtilis* strains (Dhiman et al. 2016). Iturin was found to be a strong antifungal agent with constrained antibacterial activity against *Micrococcus* and *Sarcina* strains (Walia 2015).

Fengycin is a deca-peptide with an internal lactone ring in the peptidic moiety and a β-hydroxy fatty acid chain. Fengycins are classified into fengycin A and B. These are also known as plipastatin (Walia 2015, Dhiman et al. 2016). Fengycin is produced by *Bacillus subtilis,* and plipastatin is produced from *Bacillus cereus*. Fengycins are found to have strong antifungal activities (Walia 2015).

Lichenysin is produced by *Bacillus licheniformis* and acts synergistically, exhibiting excellent stability towards temperature, pH and salt. Furthermore, this molecule has a similar structure and physiochemical properties as that of surfactin (Walia 2015, Dhiman et al. 2016).

3. HIGH-MOLECULAR-WEIGHT BIOSURFACTANTS

Biosurfactants with high molecular weights are generally grouped together as polymeric biosurfactants. They are produced by a number of bacteria and include lipoproteins, proteins, polysaccharides, lipopolysaccharides or complexes containing several of these structural types. Polymeric biosurfactants include alasan, liposan, lipomann, emulsan and other polysaccharide-protein complexes (Siñeriz et al. 2001). The most commonly studied biopolysaccharides are emulsan and alasan. Emulsan isolated from *Acinetobacter calcoaceticus RAG-1 ATCC 31012* was reported to be composed of hydrophobic fatty acid chains and an anionic polysaccharide backbone. It was found that alasan isolated from *Acinetobacter radioresistens* has a molecular weight of 1000 kDa (Walzer et al. 2006, Perfumo et al. 2010). Emulsan is a valuable emulsifying agent for hydrocarbons in water, even at a concentration as low as 0.01 to 0.1%. Moreover, emulsan and biodispersants produced by *Acinetobacter calcoaceticus* comprise a heteropolysaccharide backbone where fatty acids are covalently linked. Liposan is an extracellular water-soluble emulsifier commonly produced by *Candida* sp. and *Yarrowia* sp. and is composed of carbohydrate and protein complexes (Shekhar et al. 2015).

4. PROPERTIES OF BIOSURFACTANTS

4.1 Critical Micelle Concentration

Critical micelle concentration (CMC) is the amount of surfactant molecule in the bulk stage beyond which aggregates of surface-active agents are formed (Anna et al. 2002). Amphiphilic molecules can disperse on the surface of the aqueous phase so that the polar moiety interacts with the aqueous phase (Shekhar et al. 2015). The CMCs for microbial surfactants generally range from 5–100 mg/l. A low concentration of surfactants is sufficient to reduce the surface tension because these surface-active agents have low CMC values. At low concentrations, biosurfactants reside on the surface of water. As the surface becomes crowded, excess biosurfactant molecules assemble as micelles (Roy 2017). Biosurfactants with lower CMCs can attain higher emulsification potential, offering industrial and biotechnological applications (Sharma et al. 2016).

4.2 Increase in Surface Area of Hydrophobic Water-insoluble Substrates

At the phase boundary, biosurfactants reduce surface tension, which allows microorganisms to grow on water-immiscible substrates, thus making substrates more readily available for uptake and metabolism (Suryanti et al. 2009). The number of fatty acid chains can significantly affect the hydrophilic-lipophilic balance (HLB) of biosurfactants, and it is likely that cells interact with different substrates by modulating the overall properties of their surface to shift from hydrophilic to hydrophobic and vice versa (Banat et al. 2000, Perfumo et al. 2010).

4.3 Increase in Bioavailability of Hydrophobic Water-insoluble Substrates

Microbial growth on bound substrates is improved by biosurfactants by desorbing them from surfaces or by increasing their apparent water solubility. Low-molecular-weight biosurfactants that have low CMCs increase the apparent solubility of hydrocarbons by incorporating them into the hydrophobic cavities of micelles (Ron and Rosenberg 2001).

4.4 Emulsifier Production

The growth of microbes on hydrocarbons induces emulsifier production. Microorganisms that cannot produce emulsifiers grow poorly on hydrocarbons. The emulsifier (lipid-carbohydrate-lipid) complex is associated with the bacterial cell wall during the log phase of growth and generally shows its extracellular emulsification activity in the stationary phase (Muller-Hurtig et al. 1993, Beal and Betts 2000).

Biosurfactants form stable emulsions with a lifespan of several months to years. They can also stabilize or destabilize emulsions. High molecular mass biosurfactants, such as liposan, have the ability to emulsify edible oil but do not reduce surface tension, whereas sophorolipids produced by *T. bombicola* have been shown to reduce surface tension but are not good emulsifiers (De et al. 2008, Sharma et al. 2016).

4.5 Reduction of Surface Tension

Biosurfactants can reduce the surface tension of water from 72 to 35 mN/m and lower the interfacial tension of water against n-hexadecane from 40 to 1 mN/m. It has been reported that the CMC value of biosurfactants is 10–40 times lower than that of synthetic surfactants, which makes biosurfactants more efficient and effective (De et al. 2008). Surfactin produced by *Bacillus subtilis* can lower the surface tension of water to 25 mN/m and the interfacial tension of water against *n*-hexadecane to less than 1 mN/m (Liu et al. 2015).

4.6 Toxicity

Although biosurfactants are thought to be environmentally friendly, some experiments have shown that under certain circumstances, microbial surfactants can exhibit certain toxicity. For example, lipopeptides synthesized by *Bacillus* subtilis ATCC 6633 were reported to cause rupture of erythrocytes (Gudina et al. 2013). However, biosurfactants do not pose harmful effects to organs, nor do they interfere with blood coagulation in normal clotting time (De et al. 2008).

In a study by Sharma et al. (2016) on biosurfactant produced by *Enterococcus. faecium* MRTL9, it was found that 6.25 mg ml^{-1} of biosurfactant resulted in 90% viability of mouse fibroblast cells. Furthermore, *L. jensenii* and *L. rhamnosus* producing biosurfactant were observed to have low toxicity on eukaryotic cells, and biosurfactant concentrations ranging from 25 to 100 mg/ml revealed no toxicity (Sharma et al. 2016).

Different concentrations of biosurfactants synthesized by *E. faecium* MRTL9 ranging from 1.25 to 5 mg/ml were tested on different plants, such as *Brassica nigra* and *Triticum aestivum*. It was found that vital growth parameters such as root elongation, vigour index and germination index were excellent when treated with biosurfactant compared to sodium dodecyl sulphate. Moreover, the vital growth parameter increased with increasing concentration of biosurfactant (Saharan et al. 2012, Sharma et al. 2016).

5. APPLICATION OF BIOSURFACTANT

Biosurfactants are notable for their unique properties, which provide various functionalities. Other than having the reported antimicrobial and anticancer properties important in medical research, biosurfactants can also be used as major immunomodulatory molecules and adhesive agents and are researched for use in vaccines and gene therapy (Kumar et al. 2006, Donio et al. 2013). These compounds find applications in the bioremediation of oil spills and microbial-assisted oil recovery.

5.1 In Medicine

5.1.1 Anticancer Activity

From experiments conducted by Ramalingam et al., biosurfactant was able to induce apoptosis in HeLa cells, as it can lead to the induction of the intrinsic apoptosis pathway through G2/M phase cell cycle arrest by inducing condensed nuclei, apoptotic bodies and permeable mitochondrial membranes. Glycolipids have no toxicity against normal human cells (Kim et al. 2007, Ramalingam et al. 2016). Surfactin was found to decrease the viability of both T47D and MDA-MB-231 breast cancer cell lines, and it also caused cell cycle arrest at the G1 phase, which led to the inhibition of cell proliferation (Cao et al. 2009, Duarte et al. 2014).

5.1.2 Antimicrobial Activity

Naturally produced glycolipids and lipopeptides are found as a mixture of congeners, with increased antimicrobial potency. The crude surfactin and rhamnolipid extracts have antimicrobial activity against a broad spectrum of opportunistic and pathogenic microorganisms, including antibiotic-resistant *Staphylococcus aureus* and *Escherichia coli* strains and the pathogenic yeast *Candida albicans* (Ndlovu et al. 2017).

Research conducted by Mani et al. (2016) indicates possible use of biosurfactants in various therapeutic and biomedical applications, as the isolated biosurfactant from marine source *S. saprophyticus* revealed broad physicochemical stabilities and possessed excellent antimicrobial activities (Khopade et al. 2012, Mani et al. 2016). Biosurfactant produced by *Bacillus subtilis* has also shown antimicrobial activity against microorganisms with multidrug-resistant profiles, including *Staphylococcus aureus*, *Staphylococcus xylosus*, *Enterococcus faecalis* and *Klebsiella pneumonia* (Furtado et al. 2005, Ghribi et al. 2012).

5.1.3   Anti-adhesive Properties

Biosurfactants extracted from *Lactobacilli* have been found to have anti-adhesive activities against several pathogenic microorganisms (Pascual et al. 2008, Zakaria 2013). The adsorption of biosurfactants to a substratum surface modifies the surface hydrophobicity, interfering with microbial adhesion and desorption processes. In that sense, the release of biosurfactants by probiotic bacteria *in vivo* can be considered a defence mechanism against other colonizing strains in the urogenital and gastrointestinal tracts (Rufino et al. 2011). The effects of biosurfactants on decreased microbial adhesion and detachment from different surfaces can be utilized in fields of medicine; for example, their antitumour activities, surface activity and anti-adhesive properties can be suitable for preventing microbial colonization of implants or urethral catheters (Janek et al. 2012, Duarte et al. 2014).

5.1.4   Antiviral Properties

The antiviral potential of biosurfactants involves sorption to the viral lipid envelope and capsid breakdown, leading to the creation of ion channels (Kim et al. 1998, Rodrigues et al. 2006). Surfactin is reported to have an effect against human immunodeficiency virus 1 (HIV-1). The antiviral action was suggested to be due to physicochemical interactions between the membrane-active surfactant and the virus lipid membrane, which causes permeability changes and at higher concentrations leads to disintegration of the mycoplasma membrane (Sen 2010).

5.1.5   Drug Delivery Agents

Research carried out by Nakanishi et al. (2009) reported that nano-vectors with a biosurfactant are useful tools for delivery of foreign DNA and oligonucleotides into mammalian cells in gene transfection, drug delivery and gene therapy due to their high transfection efficiency and low toxicity (Nakanishi et al. 2009).

5.1.6   Immunomodulatory Action

It has been reported that sophorolipids decrease sepsis-related mortality at 36 h *in vivo* in a rat model of septic peritonitis by modulation of nitric oxide, adhesion molecules and cytokine production. IgE production *in vitro* in U266 cells was also decreased, possibly by affecting plasma cell activity (Kiama 2015). Sophorolipids downregulate important genes involved in IgE pathobiology by decreasing IgE production in U266 cells. This finding demonstrates the utility of sophorolipid as an anti-inflammatory agent and a novel potential therapy in diseases of altered IgE regulation. Further studies are needed to screen more biosurfactants with immunomodulatory action.

6.   IN AGRICULTURE

6.1   Fertilizer

Biosurfactants can be used for plant pathogen elimination and for increasing the bioavailability of nutrients for beneficial plant-associated microbes. Many rhizosphere and plant-associated microbes produce biosurfactants that play a vital role in motility, signalling and biofilm formation, indicating that biosurfactants govern plant-microbe interactions (Ron and Rosenberg 2001, Berti et al. 2007, Sachdev and Cameotra 2013)

6.2 Pesticides

Rhamnolipid works as a pesticide by disrupting cell membranes; the targeted fungal pest zoospores are especially vulnerable because they lack the protective cell wall present in the fungal pest's life stages. It was also found that biosurfactant produced by *Pseudomonas fluorescens* MFS03 can be used for the clean-up of insecticide residues in spinach varieties *Amaranthus tricolor* (Perfumo et al. 2010, Govindammal and Rengasamy 2013). Moreover, *Pseudomonas stutzeri* MCs AS01 was shown to biodegrade the pesticide chlorpyrifos (CPS) (Kavitha et al. 2016).

7. IN INDUSTRIES

7.1 Detergent in Laundry

Biosurfactants are amphiphilic in nature and tend to partition preferentially at the interface between different phases, such as air/water, water/stain and stain/fabric. The low surface tension of water makes it easier to lift oil, which is the basis for cleaning dirt and grease off dirty dishes, clothes and other surfaces by forming emulsions (Mishra et al. 2009, Bouassida et al. 2018). Research conducted by Ebrahimipour et al. (2017) reported that the combination of lipase and biosurfactant produced by the bacterium *O. intermedium* strain MZV101 was effective in detergent application (Cherif et al. 2011, Ebrahimipour et al. 2017). However, it is suggested that biosurfactants have high sensitivity to water hardness and can cause fabric yellowing (Tai and Nardello-Rataj 2001).

7.2 Cosmetics and Healthcare Products

The cosmetic and healthcare industries use biosurfactants in a variety of products, including insect repellents, antacids, acne pads, contact lens solutions, hair colour and care products, deodorants, nail care products, lipstick, eye shadow, mascara, toothpaste, denture cleaners, baby products, foot care products, antiseptics and depilatory products (Gudina et al. 2013). Rhamnolipids have been patented to make liposomes and emulsions important in the cosmetic industry. These lipids are known to have advantages over synthetic surfactants, such as low irritancy and compatibility with skin (Vecino et al. 2017).

7.3 Food Additives

The particular combination of characteristics such as emulsification, anti-adhesion and antimicrobial activities by biosurfactants suggests their application as multipurpose ingredients or additives (Ranasalva et al. 2014). Biosurfactants are used in controlling agglomeration of fat globules, stabilization of aerated systems, improvement of textures and shelf life of starch-containing products and enhanced consistency and texture of fat-based products. Biosurfactants contribute to controlling consistency, retarding staling and solubilizing flavour oils in bakery and ice cream formulations. Rhamnolipids are reportedly used to improve dough stability, texture, and volume and in the conservation of bakery products. L-Rhamnose has considerable potential as a precursor for flavouring agents. It is used industrially as a precursor of high-quality flavour components such as furaneol (Rikalović et al. 2015).

8. BIOSURFACTANTS AS TOOLS OF BIOREMEDIATION

8.1 Remediation of Metal-contaminated Soil

The presence of heavy metals in both urban and agricultural run-offs resulting from activities such as mining, manufacturing and the use of synthetic products contaminates the environment. Microbial cells may chelate metals from solution; however, there is little information concerning the use of biosurfactants to chelate metals (De et al. 2008, Shekhar et al. 2015). It was found that surfactants can remove metals from surfaces in a number of ways. First, non-ionic metals can form complexes with biosurfactants, enhancing surface removal by Le Chatelier's principle. Furthermore, the use of anionic surfactants in contact with metals can lead to the desorption of metals from surfaces. In a study, soil and sediments contaminated with Zn^{2+}, Cu^{2+}, and Cd^{2+} were treated using surfactin from *Bacillus subtilis* in contaminated materials, where heavy metals associated with carbonates, oxides and organic fractions could be removed using surfactin with the aid of NaOH (Vandana and Singh 2018).

8.2 Hydrocarbon Degradation

The hydrophobic nature of hydrocarbons is a substantial limiting factor in biodegradation due to their low solubility (Matvyeyeva et al. 2014). Bacterial surfaces are usually hydrophilic, which allows effective interactions with water-soluble compounds. However, the interaction between hydrophilic bacterial surfaces and hydrophobic hydrocarbons is difficult (Aparna et al. 2011, Silva et al. 2014). Biosurfactants are able to modify and translate the hydrophobic hydrocarbons to a form that is acceptable by the hydrophilic bacterial cell surface (Ilori et al. 2005, Das and Chandran 2011). These biosurfactants enhance the emulsification of hydrocarbons by reducing surface tension and forming micelles, thereby increasing the surface area and bioavailability of hydrocarbons for the bacteria to encapsulate and degrade (Shahaby et al. 2015). Biosurfactants have shown an enhanced rate of degradation of up to 90%, as listed in Table 2 (Obayori et al. 2009, Chandran and Das 2010, Reddy et al. 2010, Das and Chandran 2011, Saravana and Amruta 2013, Kaur et al. 2016, Parthipan et al. 2017, Patowary et al. 2017).

Table 2 Biodegradation of hydrocarbons by biosurfactants

Producer microorganism	Type of hydrocarbon	Percentage degradation (%)	References
Bacillus subtilis A1	Alkane	97	(Parthipan et al. 2017)
Brevibacterium sp. PDM-3	Phenanthrene	93.92	(Reddy et al. 2010)
Pseudomonas sp.	Tetradecane	98	(Saravana and Amruta 2013)
	Dodecane	98	
	Naphthalene	96	
	Naphthalene 1,6,7-Trimethyl	96	
Pseudomonas sp. LP1	Crude oil	92.34	(Obayori et al. 2009)
	Diesel oil	95.29	
P. aeruginosa and *R. erythropolis* (consortia)	Hydrocarbon	90	(Das and Chandran 2011)
Pseudomonas sp. GBS.5	Tetradecane	80	(Kaur et al. 2016)
	Hexadecane	92	
	Heptadecane	96	
	Docosane	97	
P. aeruginosa PG1	Crude oil	81.8	(Patowary et al. 2017)
Trichosporon asahii	Diesel oil	95.01	(Chandran and Das 2010)

8.3   Microbial Enhanced Oil Recovery (MEOR)

MEOR is a tertiary recovery process that involves microorganisms or their metabolic end products to recover residual oil entrapped in mature oil fields. The low permeability of some reservoirs and high viscosity of oil lead to poor oil recovery. Additionally, high capillary forces retaining the oil in the reservoir rock can lead to high interfacial tension between water and oil. According to a study, approximately 30% of oil can be recovered using current enhanced oil recovery processes, such as water flooding, pressurization and chemical surfactants, which are energy and cost intensive. As a result, biosurfactants are widely studied for recovering oil cost effectively because biosurfactants can function in a wide range of temperatures, unlike chemical surfactants (Banat 1995). The two different processes in MEOR include *in situ* and *ex situ* processes. *In situ* is where indigenous microorganisms producing biosurfactants are stimulated and injected into the reservoir. On the other hand, *ex situ* is when laboratory-produced biosurfactants are injected into the reservoir (El-Sheshtawy et al. 2015).

In a study conducted by Pereira et al. (2013), 1 g/l of biosurfactant produced by *Bacillus subtilis* could recover between 19 and 22% of oil, whereas using the same concentration of chemical surfactants could only recover 9 to 12% of oil (Pereira et al. 2013). Biosurfactants produced by *Fusarium* sp. were found to recover 48% of oil (Qazi et al. 2013). Previous studies have recorded 20–60% oil recovery from different species, such as *Alcaligenes* sp., *Bacillus* sp., *Fusarium* sp. and *Pseudomonas* sp. (Jain and Sharma 2007).

9.   OTHER APPLICATIONS

9.1   Biosurfactant-assisted Greener Synthesis of Nanoparticles

Surfactant-mediated synthesis of nanoparticles is evolving as a green technology and is considered to be a potential method for the size-controlled synthesis of nanoparticles. Biosurfactants can readily form a variety of liquid crystals in aqueous solutions, which is useful in high-performance nanomaterial production (Sena et al. 2018). Rhamnolipid has been reported to be used to stabilize nanozirconia particles. The advantage of using biosurfactants to assist in the production of nanoparticles is that the processes are efficient and cost effective (Hatzikioseyian 2010, Gudina et al. 2013). In a study by Kiran et al. (2016), silver nanoparticles stabilized by glycolipids produced by *B. casei* were effective and advantageous over chemical surfactants.

9.2   Biosurfactant-mediated Disruption of Biofilms

Sponge-associated marine bacteria are emerging as a potential source of anti-biofilm compounds. Biosurfactant produced by the sponge-associated marine Acinetobacter was gauged for the control of pathogenic biofilms (Kiran et al. 2016). In one study, marine *Brevibacterium casei* MSA19-synthesized glycolipids acted as a potential substance against biofilms produced by *Vibrio* sp., *E. coli* and *Pseudomonas* sp. Glycolipid extracted from *Lysinibacillus fusiformis* S9 showed inhibition of biofilm formation by pathogenic bacteria such as *E. coli* and *Streptococcus* mutants, where biosurfactants were not bactericidal but effective in biofilm inhibition (Gudina et al. 2013, Shekhar et al. 2015).

10.   FUTURE PROSPECTS

As these molecules have emerged as potential agents of social utility in many industrial and environmental processes as well as in biomedical and therapeutic applications, it is essential to devise large-scale and cost-effective methods of production. Newly isolated and studied organisms giving higher yields can bring further innovations in the production process. Moreover, new sources of potential producers are inevitable for the development of diverse production processes using cheaper substrates such as agro-wastes and/or raw materials. Further investigations are required to address safety issues for the use of biosurfactants in medical research because only limited reports are available. Despite the many laboratory-based successes in biosurfactants, production at an industrial scale remains a challenging issue because the composition of the final product is affected by the availability of nutrients, suggesting further trials on large-scale production.

Over the years, biosurfactants have become broadly pertinent in various industries as alternatives to synthetic surfactants. This review provided an overview of properties and applications of microbial surfactants. Due to the widespread use of biosurfactants in the pharmaceutical, cosmetic, petroleum, textile and food industries, there has been a continuous increase in their demand worldwide. Biosurfactants have high biodegradability and high toxicity.

Acknowledgments

The authors gratefully acknowledge the facilities and support provided by Universiti Brunei Darussalam.

References

Anna, L.M.S., G.V. Sebastian, E.P. Menezes, T.L.M. Alves, A.S. Santos, N. Pereira, et al. 2002. Production of biosurfactants from *Pseudomonas aeruginosa* PA 1 isolated in oil environments. Brazillian J. Chem. Eng. 19: 159–166.

Aparna, A., H. Smitha and G. Srinikethan. 2011. Effect of addition of biosurfactant produced by *Pseudomonas sps.* on biodegradation of crude oil. *In*: 2nd International Conference on Environmental Science and Technology. IACSIT Press, Singapore, pp. 71–75.

Arima, K., A. Kakinuma and G. Tamura. 1968. Surfactin, a crystalline peptidelipid surfactant produced by *Bacillus subtilis*: Isolation, characterization and its inhibition of fibrin clot formation. Biochem. Biophys. Res. Commun. 31: 488–494. doi: 10.1016/0006-291X(68)90503-2.

Banat, I.M. 1995. Biosurfactants production and possible uses in microbial enhanced oil recovery and oil pollution remediation: a review. Bioresour. Technol. 51: 1–12.

Banat, I.M., R.S. Makkar and S.S. Cameotra. 2000. Potential commercial applications of microbial surfactants. Appl. Microbiol. Biotechnol. 53: 495–508.

Beal, R. and W.B. Betts. 2000. Role of rhamnolipid biosurfactants in the uptake and mineralization of hexadecane in *Pseudomonas aeruginosa*. J. Appl. Microbiol. 89: 158–168.

Berti, A.D., N.J. Greve, Q.H. Christensen and M.G. Thomas. 2007. Identification of a biosynthetic gene cluster and the six associated lipopeptides involved in swarming motility of *Pseudomonas syringae* pv. tomato DC3000. J. Bacteriol. 189: 6312–6323.

Bhardwaj, G. 2013. Biosurfactants from fungi: A review. J. Petrol Environ. Biotechnol. 4: 1–6.

Biniarz, P., M. Lukaszewicz and T. Janek. 2017. Screening concepts, characterization and structural analysis of microbial-derived bioactive lipopeptides: A review. Crit. Rev. Biotechnol. 37: 393–410.

Bouassida, M., N. Fourati, I. Ghazala, S. Ellouze-Chaabouni and D. Ghribi. 2018. Potential application of *Bacillus subtilis* SPB1 biosurfactants in laundry detergent formulations: Compatibility study with detergent ingredients and washing performance. Eng. Life Sci. 18: 70–77.

Cao, X., A.H. Wang, R.Z. Jiao, C.L. Wang, D.Z. Mao, L. Yan, et al. 2009. Surfactin induces apoptosis and G(2)/M arrest in human breast cancer MCF-7 cells through cell cycle factor regulation. Cell Biochem. Biophys. 55: 163–171.

Chandran, P. and N. Das. 2010. Biosurfactant production and diesel oil degradation by yeast species *Trichosporon Asahii* isolated from petroleum hydrocarbon contaminated soil. Int. J. Eng. Sci. Tech. 2: 6942–6953.

Chen, W.C., R.S. Juang and Y.H. Wei. 2015. Applications of a lipopeptide biosurfactant, surfactin, produced by microorganisms. Biochem. Eng. J. 103: 158–169.

Cherif, S., S. Mnif, F. Hadrich, S. Abdelkafi and S. Sayadi. 2011. A newly high alkaline lipase: an ideal choice for application in detergent formulations. Lipids Health Dis. 10:221. doi: 10.1186/1476-511x-10-221

Das, N. and P. Chandran. 2011. Microbial degradation of petroleum hydrocarbon contaminants: An overview. Biotechnol. Res. Int. 2011: 941810. doi: 10.4061/2011/941810

De, S., S. Malik, A. Ghosh, R. Saha and B. Saha. 2008. A review on natural surfactants. RSC Adv. 5: 65757. doi: 10.1039/C5RA11101C

Dhiman, R., K. Meena, A. Sharma and S. Kanwar. 2016. Biosurfactants and their screening methods. Res. J. Recent. Sci. 5: 1–6.

Donio, M.B.S., F.A. Ronica, V.T. Viji, S. Velmurugan, J.S.C.A. Jenifer, M. Michaelbabu, P. Dhar and T. Citarasu. 2013. *Halomonas* sp. BS4, a biosurfactant producing halophilic bacterium isolated from solar salt works in India and their biomedical importance. Springer Plus 2: 149. doi: 10.1186/2193-1801-2-149

Duarte, C., E.J. Gudina, C.F. Lima and L.R. Rodrigues. 2014. Effects of biosurfactants on the viability and proliferation of human breast cancer cells. AMB Express 4: 40. doi: 10.1186/s13568-014-0040-0

Ebrahimipour, G., H. Sadeghi and M. Zarinviarsagh. 2017. Statistical methodologies for the optimization of lipase and biosurfactant by *Ochrobactrum intermedium* strain MZV101 in an identical medium for detergent applications. Molecules 22: 1460. doi: 10.3390/molecules22091460

El-Sheshtawy, H.S., I. Aiad, M.E. Osman, A.A. Abo-Elnasr, A.S. Kobisy. 2015. Production of biosurfactant from *Bacillus licheniformis* for microbial enhanced oil recovery and inhibition the growth of sulfate reducing bacteria. Egypt. J. Pet. 24: 155–162.

Fracchia, L., M. Cavallo, M.G. Martinotti and I.M. Banat. 2012. Biosurfactants and bioemulsifiers biomedical and related applications-present status and future potentials. *In*: D.N. Ghista (ed.), Biomedical Science, Engineering and Technology. IntechOpen, Rijeka, pp. 325–370.

Franzetti, A., I. Gandolfi, G. Bestetti, T.J.P. Smyth and I.M. Banat. 2010. Production and applications of trehalose lipid biosurfactants. Eur. J. Lipid Sci. Technol. 112: 617–627.

Furtado, G.H., S.T. Martins, A.P. Coutinho, G.M. Soares, S.B. Wey and E.A. Medeiros. 2005. Incidence of vancomycin-resistant *Enterococcus* at a university hospital in Brazil. Rev. Saude. Publica. 39: 41–46.

Ghribi, D., L. Abdelkefi-Mesrati, I. Mnif, R. Kammoun, I. Ayadi, I. Saadaoui, et al. 2012. Investigation of antimicrobial activity and statistical optimization of *Bacillus subtilis* SPB1 biosurfactant production in solid-state fermentation. J. Biomed. Biotechnol. 2012: 373682. doi: 10.1155/2012/373682

Govindammal, M. and P. Rengasamy. 2013. Biosurfactant as a pesticide cleaning agent in leafy vegetables produced by *Pseudomonas fluorescens* isolated from mangrove ecosystem. Gold. Res. Thoughts 2: 1–7.

Gudina, E.J., V. Rangarajan, R. Sen and L.R. Rodrigues. 2013. Potential therapeutic applications of biosurfactants. Trends Pharmacol. Sci. 34: 667–675.

Hatzikioseyian, A. 2010. Principles of bioremediation processes. *In*: G. Plaza (ed.), Trends bioremediation phytoremediation. Research Signpost, Trivandrum, India, pp. 23–54.

Hoffmann, N., J. Pietruszka and C. Söffing. 2012. From sophorose lipids to natural product synthesis. Adv. Synth. Catal. 354: 959–963.

Ilori, M.O., C.J. Amobi and A.C. Odocha. 2005. Factors affecting biosurfactant production by oil degrading *Aeromonas* spp. isolated from a tropical environment. Chemosphere 61: 985–992.

Jain, M. and G. Sharma. 2007. Microbial enhanced oil recovery using biosurfactant produced by *Alcaligenes faecalis*. Iran J. Biotechnol. 7: 216–223.

Janek, T., M. Lukaszewicz, A. Krasowska. 2012. Antiadhesive activity of the biosurfactant pseudofactin II secreted by the Arctic bacterium *Pseudomonas fluorescens* BD5. BMC Microbiol. 12: 24.

Kang, S.W., Y.B. Kim, J.D. Shin and E.K. Kim. 2010. Enhanced biodegradation of hydrocarbons in soil by microbial biosurfactant, sophorolipid. Appl. Biochem. Biotechnol. 160: 780–790.

Kaur, H., S. Khan, S. Gupta, N. Gupta. 2016. Diesel oil degradation using biosurfactant produced by *Pseudomonas* sp. J. Adv. Biol. Biotechnol. 8: 1–9.

Kavitha, D., M. Sureshkumar and B. Senthilkumar. 2016. Screening of pesticide degrading and biosurfactant producing bacteria from chlorpyrifos contaminated soil. Int. J. Pharma. Bio. Sci. 7: 525–532.

Keikha, M. 2018. *Williamsia* spp. are emerging opportunistic bacteria. New Microbes New Infect. 21: 88–89.

Khopade, A., B. Ren, X.Y. Liu, K. Mahadik, L. Zhang and C. Kokare. 2012. Production and characterization of biosurfactant from marine *Streptomyces* species B3. J. Colloid Interface Sci. 367: 311–318.

Kiama, C. 2015. Isolation and characterization of hydrocarbon biodegrading fungi from oil contaminated soils in Thika, Kenya. Master of Science Thesis. Jomo Kenyatta University of Agriculture and Technology, Nairobi, Kenya.

Kim, K., S.Y. Jung, D.K. Lee, J.K. Jung, J.K. Park, D.K. Kim and C.H. Lee. 1998. Suppression of inflammatory responses by surfactin, a selective inhibitor of platelet cytosolic phospholipase A2. Biochem. Pharmacol. 55: 975–985.

Kim, S.Y., J.Y. Kim, S.H. Kim, H.J. Bae, H. Yi, S.H. Yoon, et al. 2007. Surfactin from *Bacillus subtilis* displays anti-proliferative effect via apoptosis induction, cell cycle arrest and survival signaling suppression. FEBS Lett. 581: 865–871.

Kiran, G.S., A.S. Ninawe, A.N. Lipton, V. Pandian and J. Selvin. 2016. Rhamnolipid biosurfactants: evolutionary implications, applications and future prospects from untapped marine resource. Crit. Rev. Biotechnol. 36: 399–415.

Kulakovskaya, E. and T. Kulakovskaya. 2014. Structure and occurrence of yeast extracellular glycolipids. *In*: E. Kulakovskaya and T. Kulakovskaya (eds), Extracellular Glycolipids of Yeasts. Elsevier, Waltham, MA, USA, pp. 1–13.

Kulakovskaya, T., A. Shashkov, E. Kulakovskaya, W. Golubev, A. Zinin, Y. Tsvetkov, et al. 2009. Extra-cellular cellobiose lipid from yeast and their analogues: structures and fungicidal activities. J. Oleo. Sci. 58: 133–140.

Kumar, M., V. Leon, A.S. Materano and O.A. Ilzins. 2006. Enhancement of oil degradation by co-culture of hydrocarbon degrading and biosurfactant producing bacteria. Pol. J. Microbiol. 55: 139–146.

Li, J.L., X. Sun, L. Chen and L.D. Guo. 2016. Community structure of endophytic fungi of four mangrove species in Southern China. Mycology 7: 180–190.

Liu, J.F., S.M. Mbadinga, S.Z. Yang, J.D. Gu and B.Z. Mu. 2015. Chemical structure, property and potential applications of biosurfactants produced by *Bacillus subtilis* in petroleum recovery and spill mitigation. Int. J. Mol. Sci. 16: 4814–4837.

Mani, P., G. Dineshkumar, T. Jayaseelan, K. Deepalakshmi, C.G. Kumar and S.S. Balan. 2016. Anti-microbial activities of a promising glycolipid biosurfactant from a novel marine *Staphylococcus saprophyticus* SBPS 15. 3 Biotech. 6: 163.

Matvyeyeva, O.L., O. Vasylchenko and O. Aliieva. 2014. Microbial biosurfactants role in oil products biodegradation. Int. J. Environ. Bioremediat. Biodegrad. 2: 69–74.

Mishra, M., P. Muthuprasanna, K.S. Prabha, P.S. Rani, I.A. Satish, I.S. Chandrian, et al. 2009. Basics and potential applications of surfactants—A review. Int. J. Pharm. Tech. Res. 1: 1354–1365.

Mnif, I. and D. Ghribi. 2016. Glycolipid biosurfactants: main properties and potential applications in agriculture and food industry. J. Sci. Food Agric. 96: 4310–4320.

Mohan, P.K., G. Nakhla and E.K. Yanful. 2006. Biokinetics of biodegradation of surfactants under aerobic, anoxic and anaerobic conditions. Water Res. 40: 533–540.

Mongkolthanaruk, W. 2012. Classification of *Bacillus* beneficial substances related to plants, humans and animals. J. Microbiol. Biotechnol. 22: 1597–1604.

Morita, T., T. Fukuoka, T. Imura and D. Kitamoto. 2015. Mannosylerythritol lipids: Production and applications. J. Oleo. Sci. 64: 133–141.

Morita, T., Y. Ishibashi, T. Fukuoka, T. Imura, H. Sakai, M. Abe, et al. 2011. Production of glycolipid biosurfactants, cellobiose lipids, by *Cryptococcus humicola* JCM 1461 and their interfacial properties. Biosci. Biotechnol. Biochem. 75: 1597–1599.

Muller-Hurtig, R., F. Wagner, R. Blaszczyk and N. Kosaric. 1993. Biosurfactants for environmental control. *In*: N. Kosaric and F. Sukan (eds), Biosurfactants: Production, Properties, Applications. CRC Press, New York, USA, pp. 447–469.

Nakanishi, M., Y. Inoh, D. Kitamoto and T. Furuno. 2009. Nano vectors with a biosurfactant for gene transfection and drug delivery. J. Drug Deliv. Sci. Tech. 19: 165–169.

Ndlovu, T., M. Rautenbach, J.A. Vosloo, S. Khan and W. Khan. 2017. Characterisation and antimicrobial activity of biosurfactant extracts produced by *Bacillus amyloliquefaciens* and *Pseudomonas aeruginosa* isolated from a wastewater treatment plant. AMB Express. 7: 108–113.

Obayori, O.S., M.O. Ilori, S.A. Adebusoye, G.O. Oyetibo, A.E. Omotayo and O.O. Amund. 2009. Degradation of hydrocarbons and biosurfactant production by *Pseudomonas* sp. strain LP1. World J. Microbiol. 25: 1615–1623.

Parthipan, P., E. Preetham, L.L. Machuca, P.K. Rahman, K. Murugan and A. Rajasekar. 2017. Biosurfactant and degradative enzymes mediated crude oil degradation by bacterium *Bacillus subtilis* A1. Front. Microbiol. 8: 193. doi: 10.3389/fmicb.2017.00193

Pascual, L.M., M.B. Daniele, F. Ruiz, W. Giordano, C. Pajaro and L. Barberis. 2008. *Lactobacillus rhamnosus* L60, a potential probiotic isolated from the human vagina. J. Gen. Appl. Microbiol. 54: 141–148.

Patoway, K., R. Patoway, M.C. Kalita and S. Deka. 2017. Characterization of biosurfactant produced during degradation of hydrocarbons using crude oil as sole source of carbon. Front. Microbiol. 8: 279. doi: 10.3389/fmicb.2017.00279

Pereira, J.F.B., E.J. Gudiña, R. Costa, R. Vitorino, J.A. Teixeira, J.A.P. Coutinho, et al. 2013. Optimization and characterization of biosurfactant production by *Bacillus subtilis* isolates towards microbial enhanced oil recovery applications. Fuel. 111: 259–268.

Perfumo, A., T.J.P. Smyth, R. Marchant and I.M. Banat. 2010. Production and roles of biosurfactants and bioemulsifiers in accessing hydrophobic substrates. *In*: K. Timmis (ed.), Handbook of hydrocarbon and Lipid Microbiology. Springer Berlin, Berlin, Heidelberg, pp. 1502–1512.

Peter, J. and D.P. Singh. 2014. Characterization of emulsification activity of partially purified rhamnolipids from *Pseudomonas fluorescens*. Int. J. Inno. Sci. Res. 3: 88–100.

Qazi, M., M. Subhan and N. Fatima. 2013. Role of biosurfactant produced by *Fusarium* sp. BS-8 in enhanced oil recovery (EOR) through sand pack column. Int. J. Biosci. Biochem. 3: 598–604.

Ramalingam, V., K. Varunkumar, V. Ravikumar and R. Rajaram. 2016. Development of glycolipid biosurfactant for inducing apoptosis in HeLa cells. RSC Adv. 6: 64087–64096.

Ranasalva, N., R. Sunil and G. Poovarasan. 2014. Importance of biosurfactant in food industry. IOSR J. Agri. Veter. Sci. 7: 6–9.

Randhawa, K.K.S. and P.K. Rahman. 2014. Rhamnolipid biosurfactants-past, present, and future scenario of global market. Front. Microbiol. 5: 454.

Reddy, M.S., B. Naresh, T. Leela, M. Prashanthi, N. Madhusudhan, G. Dhanasri, et al. 2010. Biodegradation of phenanthrene with biosurfactant production by a new strain of *Brevibacillus* sp. Bioresour. Technol. 101: 7980–7983.

Rikalović, M.G., M.M. Vrvić and I.M. Karadžić. 2015. Rhamnolipid biosurfactant from *Pseudomonas aeruginosa*–from discovery to application in contemporary technology. J. Serb. Chem. Soc. 80: 279.

Rodrigues, L., I.M. Banat, J. Teixeira and R. Oliveira. 2006. Biosurfactants: Potential applications in medicine. J. Antimicrob. Chemother. 57: 609–618.

Ron, E.Z. and E. Rosenberg. 2001. Natural roles of biosurfactants. Environ. Microbiol. 3: 229–236.

Roy, A. 2017. Review on the biosurfactants: properties, types and its applications. J. Fundam. Renewable Energy Appl. 8: 248. doi:10.4172/20904541.1000248.

Rufino, R., J. Luna, L. De Sarubbo, L.R.M. Rodrigues, J.A.C. Teixeira and de G.M. Campos-Takaki. 2013. Antimicrobial and anti-adhesive potential of a biosurfactants produced by *Candida* species. *In*: A. Andrade (ed.), Practical Applications in Biomedical Engineering. InTechOpen, pp. 245–257.

Rufino, R.D., J.M. Luna, L.A. Sarubbo, L.R. Rodrigues, J.A. Teixeira and G.M. Campos-Takaki. 2011. Antimicrobial and anti-adhesive potential of a biosurfactant Rufisan produced by *Candida lipolytica* UCP 0988. Colloids Surf. B Biointerfaces 84: 1–5.

Sachdev, D.P. and S.S. Cameotra. 2013. Biosurfactants in agriculture. Appl. Microbiol. Biotechnol. 97: 1005–1016.

Saharan, B., R. Sahu and D. Sharma. 2012. A review on biosurfactants: fermentation, current developments and perspectives. Genet. Eng. Biotechnol. J. 2011: 1–14.

Saravana, P. and S. Amruta. 2013. Analysis of biodegradation pathway of crude oil by *Pseudomonas* sp. isolated from marine water sample. Arch. Appl. Sci. Res. 5: 165–171.

Sen, R. 2010. Biosurfactants, Vol 672. Springer, New York, NY.

Sena, H.H., M.A. Sanches, D.F.S. Rocha, W.O.P.F. Segundo, E.S. Souza and J.V.B. Souza. 2018. Production of biosurfactants by soil fungi isolated from the Amazon forest. Int. J. Microbiol. 2018: 8. doi: 10.1155/2018/5684261

Shahaby, A., A. Alharthi and A. Tarras. 2015. Bioremediation of petroleum oil by potential biosurfactant-producing bacteria using gravimetric assay. Int. J. Curr. Microbiol. Appl. Sci. 4: 390–403.

Sharma, D., B.S. Saharan and S. Kapil. 2016. Biosurfactants of probiotic lactic acid bacteria. *In*: D. Sharma, B.S. Saharan and S. Kapil (eds), Biosurfactants of Lactic Acid Bacteria. Springer International Publishing, Cham, pp. 17–29.

Shekhar, S., A. Sundaramanickam and T. Balasubramanian. 2015. Biosurfactant producing microbes and their potential applications: A review. Crit. Rev. Environ. Sci. Technol. 45: 1522–1554.

Silva, E.J., N.M.R. Silva, R.D. Rufino, J.M. Luna, R.O. Silva and L.A. Sarubbo. 2014. Characterization of a biosurfactant produced by *Pseudomonas cepacia* CCT6659 in the presence of industrial wastes and its application in the biodegradation of hydrophobic compounds in soil. Colloids Surf B Biointerfaces. 117: 36–41.

Siñeriz, F., R. Hommel, C. Leipzig and G. Kleber. 2001. Production of biosurfactants. Encyclopedia of Life Support Systems 5: 386–392.

Suryanti, V., S. Hastuti, W.T. Dwi and M.D. Ika. 2009. Biosurfactants production by *Pseudomonas aeruginosa* using soybean oil as substrate. Indo. J. Chem. 9: 107–112.

Tai, L.H.T. and V. Nardello-Rataj. 2001. The main surfactants used in detergents and personal care products. OCL-OL CORP. 8(2): 141–144.

Uzoigwe, C., J.G. Burgess, C.J. Ennis and P.K.S.M. Rahman. 2015. Bioemulsifiers are not biosurfactants and require different screening approaches. Front. Microbiol. 6: 245–251.

Vandana, P. and D. Singh. 2018. Review on biosurfactant production and its application. Int. J. Curr. Microbiol. App. Sci. 7: 4228–4241.

Vecino, X., J.M. Cruz, A.B. Moldes and L.R. Rodrigues. 2017. Biosurfactants in cosmetic formulations: Trends and challenges. Crit. Rev. Biotechnol. 37: 911–923.

Walia, N.K. 2015. Lipopeptides: Biosynthesis and applications. J. Microbial. Biochem. Technol. 07: 103–107.

Walzer, G., E. Rosenberg and E.Z. Ron. 2006. The *Acinetobacter* outer membrane protein A (OmpA) is a secreted emulsifier. Environ. Microbiol. 8: 1026–1032.

Yu, M., Z. Liu, G. Zeng, H. Zhong, Y. Liu, Y. Jiang, et al. 2015. Characteristics of mannosylerythritol lipids and their environmental potential. Carbohydr. Res. 407: 63–72.

Zakaria, G.E. 2013. Antimicrobial and anti-adhesive properties of biosurfactant produced by *Lactobacilli* isolates, biofilm formation and aggregation ability. J. Gen. Appl. Microbiol. 59: 425–436.

Index

Chlorpyrifos (OP Pesticide)   189
Composting   304
Contaminants   53, 210, 372
   —environment   31, 41, 83, 254
   —soil   358
   —water   359
Contamination   232, 270
Corn steep liquor   316
Cosmetic   427
Cost effective biotreatment   318
Critical Micelle Concentration (CMC)   424
Cyanobacteria   217

D

Decolourization   193
Decomposition   273
Degradation   29
Denitrification   336, 337
Design–Build–Test–Learn (DBTL)   74
Detergent   427
Detoxification   166, 280
Drinking water   232
Dyes   288

E

Ecotoxicity   29
EDTA (Chelating Agents)   400
Elytrigia repens   33, 35, 36, 37
Emulsification   166, 428
Emulsifiers   6, 424
Endomycorrhiza   202
Environment   181, 397
Environmental
   —contaminants   408
   —pollution   357
   —segments   357
Enzymes   107, 317
Exopolysaccharides (EPS)   91
Experimental strategies   14
Explosive material   300
Explosives   277, 305

F

Fabrication   145
Fengycin   423
Filamentous fungi   212
Fungi   3, 399
Fungicides   344
Fusarium   259

G

Genetic engineering   254
Gliocladium   259
Glycolipids   421
Green technology   429
Groundwater   231
Groundwater pollution   231

H

Hazardous chemicals   270
Hazardous substances   28, 34
Healthcare   427
Health problems   232
Heavy metals   28, 29, 30, 31, 32, 33, 34, 35,
    36, 37, 38, 40, 128, 182, 184, 203, 232,
    269, 339, 340, 360, 373, 395, 407, 428
Herbicides   275, 343
Holobiont   213
Hydrocarbon   103, 275, 299
Hydrocarbon solubility   105
Hydrophilic   165
Hydrophilic tail   165
Hydrophobic   165
Hydrophobic head   165
Hydrophobic pollutants   1
Hydroxylatio   303
Hyperaccumulator   213, 395

I

Immobilization   107
Industrialization   360
Insecticide   347
Iturin   423

L

Landfarming   304
Landfills   233
Lichenysin   423
Lignocellulosic wastes   324
Lipopeptides   423
Low technology   408

M

Magnetic oxides   115
Magnetite   107
Magnetosome nanoparticles   149
Metabolomics   73
Metagenomics   72
Metal binding   41

Milton Keynes UK
Ingram Content Group UK Ltd.
UKHW050449071024
449327UK00014B/297